The Earth's mantle plays a crucial role in a variety of geologic processes and provides researchers important insights into the development of our planet. Interdisciplinary in scope, this book is a comprehensive overview of the composition, structure, and evolution of the mantle layer.

Written by internationally recognized scientists from the Research School of Earth Sciences at the Australian National University, and dedicated to the memory of A. E. ("Ted") Ringwood, this book draws on perspectives from cosmochemistry, isotope geochemistry, fluid dynamics and petrology, seismology and geodynamics, and mineral and rock physics.

The book begins with a discussion of the accretion and differentiation of the Earth, including the cosmochemical initial conditions, scenarios for core segregation, constraints on the age of the Earth, the dating of core formation, and the subsequent differentiation and outgassing responsible for both the continental crust and the atmosphere. It also reviews the evolution of the Earth, emphasizing the 'plate' and 'plume' modes of mantle convection. Finally, the book describes experimental constraints on magma genesis and the structure and physical properties of the modern mantle.

Striking a balance between matters of consensus and continuing controversy, *The Earth's Mantle* will provide researchers and graduate students with an authoritative review of this important part of our planet.

The Earth's Mantle

Composition, Structure, and Evolution

Edited by

Ian Jackson

Australian National University

CAMBRIDGE
UNIVERSITY PRESS

PUBLISHED BY THE PRESS SYNDICATE OF THE UNIVERSITY OF CAMBRIDGE
The Pitt Building, Trumpington Street, Cambridge, United Kingdom

CAMBRIDGE UNIVERSITY PRESS
The Edinburgh Building, Cambridge CB2 2RU, UK   http://www.cup.cam.ac.uk
40 West 20th Street, New York, NY 10011-4211, USA   http://www.cup.org
10 Stamford Road, Oakleigh, Melbourne 3166, Australia
Ruiz de Alarcón 13, 28014 Madrid, Spain

First published 1998
First paperback edition 2000

Printed in the United States of America

Typeset in Times Roman

A catalog record for this book is available from the British Library

Library of Congress Cataloging in Publication data
The Earth's mantle : composition, structure, and evolution / edited by
Ian Jackson.
p.   cm.
Includes bibliographical references and index.
1. Earth–Mantle I. Jackson, Ian, 1950—
QE509.E234   1998
551.1'16–dc21   97-1195

ISBN 0 521 56344 5 hardback
ISBN 0 521 78566 9 paperback

Contents

III   Structure and Mechanical Behaviour of the Modern Mantle

Chapter Outlines

Chapter 3   *Primordial Solar Noble-Gas Component in the Earth: Consequences for the Origin and Evolution of the Earth and its Atmosphere*

Part II   Dynamics and Evolution of the Earth's Mantle

Chapter 4   *Understanding Mantle Dynamics through Mathematical Models and Laboratory Experiments*

Contributors

Victoria C. Bennett
Research School of Earth Sciences
Australian National University
Canberra ACT, 0200
Australia

I. H. Campbell
Research School of Earth Sciences
Australian National University
Canberra ACT, 0200
Australia

Geoffrey F. Davies
Research School of Earth Sciences
Australian National University
Canberra ACT, 0200
Australia

Martyn R. Drury
Department of Geology
Geodynamics Research Institute
Utrecht University
P. O. Box 80.021
Utrecht 3508TA
The Netherlands

Trevor J. Falloon
Geology Department
University of Tasmania
GPO Box 252C
Hobart, Tasmania 7001
Australia

John D. Fitz Gerald
Research School of Earth Sciences
Australian National University
Canberra ACT, 0200
Australia

David H. Green
Research School of Earth Sciences
Australian National University
Canberra ACT, 0200
Australia

R. W. Griffiths
Research School of Earth Sciences
Australian National University
Canberra ACT, 0200
Australia

Masahiko Honda
Research School of Earth Sciences
Australian National University
Canberra ACT, 0200
Australia

Ian Jackson
Research School of Earth Sciences
Australian National University
Canberra ACT, 0200
Australia

Paul Johnston
Research School of Earth Sciences
Australian National University
Canberra ACT, 0200
Australia

B. L. N. Kennett
Research School of Earth Sciences
Australian National University
Canberra ACT, 0200
Australia

Kurt Lambeck
Research School of Earth Sciences
Australian National University
Canberra ACT, 0200
Australia

Malcolm T. McCulloch
Research School of Earth Sciences
Australian National University
Canberra ACT, 0200
Australia

Ian McDougall
Research School of Earth Sciences
Australian National University
Canberra ACT, 0200
Australia

Hugh St. C. O'Neill
Research School of Earth Sciences
Australian National University
Canberra ACT, 0200
Australia

Herbert Palme
Mineralogisch-Petrographisches
    Institut
Universität zu Köln
Zulpicher Strasse 49b
50674 Köln
Germany

Sally M. Rigden
Department of Geological Sciences
Queen's University
Kingston, Ontario
Canada K7L 3N6

J. S. Turner
Research School of Earth Sciences
Australian National University
Canberra ACT, 0200
Australia

R. D. van der Hilst
Department of Earth, Atmospheric &
    Planetary Sciences
Massachusetts Institute of Technology
Cambridge, MA 02139-4307
USA

Dedication

Questions concerning the origin and evolution of the Earth have inevitably been of special significance to our species and accordingly have occupied the minds of many of our most influential thinkers. During the past four decades, phenomenal progress has been made towards answers to these fundamental questions, and the late Professor A. E. ("Ted") Ringwood was consistently at the forefront of this research. He participated in, and capitalized upon, the progressive development during this period of equipment now capable of reproducing in the laboratory the extreme conditions of pressure and temperature that prevail within the Earth's interior. He boldly exploited the opportunities offered by these technological developments to explore key aspects of the chemical behaviour of geological materials. Perhaps foremost amongst his achievements was the demonstration of the occurrence and importance of pressure-induced phase transformations. He combined such findings with insights emerging from increasingly detailed seismological probing of the internal structure of the Earth, as well as with the perspective offered by exploration of the solar system, in imaginative and compelling new models to describe the chemical composition, internal structure, origin, and evolution of our planet and its Moon. Preparation of a volume on *The Earth's Mantle: Structure, Composition, and Evolution*, written by his colleagues at the Australian National University (ANU), therefore seemed a fitting tribute to this brilliant earth scientist who will be long remembered for his many seminal contributions in this field of intellectual endeavour.

Ted Ringwood was born in Melbourne, Australia, in 1930 and was educated at the Geelong Grammar School and the University of Melbourne, graduating with his Ph.D. in 1956. He was amongst the first generation of distinguished Australian scientists to be educated to Ph.D. level within Australia. After a postdoctoral fellowship with Francis Birch at Harvard University, Ted joined the Department of Geophysics at the ANU as a Senior Research Fellow in 1959. His rapidly growing stature was recognized through his appointment in 1963 as Personal Professor, and in 1967 as Professor of Geochemistry, a position he filled with distinction until his premature death in 1993.

During the late 1960s and early 1970s, with the support of the late Professor J. C. Jaeger, then head of the Department of Geophysics and Geochemistry of the Research School of Physical Sciences at ANU, Ted Ringwood argued the ultimately successful case for the formation of a new Research School of Earth Sciences (RSES). The new school's mandate was to expand into carefully selected new areas such as geophysical fluid dynamics, ore genesis, and environmental geochemistry, all of which are now integral parts of the school's research activity. Ringwood also recognized and promoted opportunities for the enhancement of existing research activities, notably in mineral/rock physics and noble-gas geochemistry. Jaeger and Ringwood were responsible for the inspired appointment in 1973 of Professor A. L. Hales as foundation director of RSES. Ringwood himself later served a term (1978–83) as director of the school.

There was a close and instructive symbiosis between Ringwood's research into the chemical composition, origin, and evolution of the Earth and Moon and his applied-research interests. The same crystal-chemical principles and experimental methods that elucidate the high-pressure behaviour of silicate materials in the Earth's deep interior provided the basis for the Synroc strategy for safe immobilization of high-level nuclear wastes in a durable ceramic wasteform. He also patented procedures for the fabrication of new cutting-tool materials in the form of diamond- and boron-nitride-based composites.

The impact of Ringwood's research, as reported in more than 300 papers, two books, and several patents, has been recognized through frequent citation and many awards, including Fellowship of the Australian Academy of Science (1966), the American Geophysical Union (AGU) (1969), and the Royal Society of London (1972). He was a recipient of the Bowie Medal of the AGU (1974), the Day Medal of the Geological Society of America (1974), the Holmes Medal of the European Union of Geosciences (1985), the Wollaston Medal of the Geological Society of London (1988), the Feltrinelli Award of the Italian National Academy (1991), and the Hess Medal of the AGU (1993), amongst others.

Above all, Ted Ringwood will deservedly be remembered as a bold, original, and lateral thinker and an excellent communicator. He was a powerful, often irresistible, advocate for the causes to which he was committed, and a feisty debater. He was supportive of and loyal to those who worked closely with him, impatient with mediocrity, and intensely proud of his country and the life-style that it offers. It is with a real sense of loss that his colleagues in the Research School of Earth Sciences at the Australian National University dedicate this volume to his memory.

A. E. ("Ted") Ringwood (1930–93)

Preface

It has been our goal in assembling this volume to produce an overview of the composition, structure, and evolution of the Earth's mantle that will be authoritative, up-to-date, and forward-looking, yet thoroughly readable. It is our hope that it will prove useful to all those interested in the Earth's mantle and its workings, from beginning graduate students to experienced researchers. The volume consists of 11 chapters contributed by the staff of the Research School of Earth Sciences at the Australian National University and their collaborators, arranged into three parts, as follows:

Accretion and Differentiation of the Earth
Dynamics and Evolution of the Earth's Mantle
Structure and Mechanical Behaviour of the Modern Mantle

Recent progress towards consensus on many of the major issues surrounding the composition, structure, and evolution of the Earth's mantle makes this volume particularly timely. This Preface is intended to provide the reader a brief connected account of the topics addressed in this volume, not necessarily in their order of appearance, and an indication of the general philosophy adopted in assembling the material.

It is now widely accepted that the planet Earth was accreted from a hierarchy of planetesimals that formed in our solar nebula over a range of radial distances from the Sun. Through studies of chondritic meteorites, which display uniform relative abundances of refractory elements and systematic depletions of volatile species, the bulk composition of the silicate Earth can be constrained, albeit within significant residual uncertainties, especially in the important Mg/Si/Al ratios. The abundances of the siderophile elements in the Earth's mantle differ markedly from those that would be expected on the basis of the metal–silicate distribution coefficients measured at low pressure. Thermodynamic equilibrium between the core and the mantle in a homogeneously accreted Earth would therefore be excluded, unless the distribution coefficients should prove to have a particular pressure dependence. It is suggested in this volume that it is more likely that the Earth was formed by heterogeneous accretion. In this scenario, core formation began with the sequestering into the core of almost the entire siderophile-element inventory of the early, volumetrically dominant, highly reduced, strongly devolatilized component of the proto-Earth. This was followed by incorporation into the core of siderophile elements derived from a subsidiary, more oxidized, volatile-rich component. Following the completion of core formation, at about 4.5 billion years ago (4.5×10^9 years ago, or 4.5 Ga), a late-stage 'veneer' of chondritic material was added to the Earth's mantle.

Isotopic constraints indicate that another 200 million years probably elapsed before it became possible to preserve a continental crust enriched in light rare-earth elements (LREEs). It seems that part of the Earth's mantle was at least as LREE-depleted in the early Archaean (before 3.8 Ga) as are the source regions for modern mid-ocean-ridge basalts (MORBs) and that the continental crust probably has grown progressively, albeit episodically, through geological time, with recycling of continental crust back into the mantle playing a subsidiary role. Studies of the crust/mantle distribution of incompatible trace elements and of the Earth's inventory of radiogenic argon (^{40}Ar) suggest that about half of the Earth's mantle has been stripped of its incompatible and volatile elements. Studies of noble-gas isotopes (particularly Ne, Ar, and ^{129}Xe) indicate that much of the degassing to form the atmosphere must have occurred within the first few hundred million years of the Earth's history. A striking contrast between the correlated He and Ne isotopic ratios in MORB and ocean-island-basalt (OIB) source regions provides strong evidence of continuing outgassing from the OIB source region of a primordial (solar) noble-gas component.

Convection within the Earth's present-day mantle is also becoming increasingly well understood as the superposition of two main modes. The dominant plate-scale flow is driven mainly by the gravitational instability of the cold, stiff upper thermal boundary layer or lithosphere. Plumes represent the second, subsidiary mode of mantle convection, arising from instabilities in a bottom-heated lower thermal boundary layer. The dynamics of both the plate-scale and plume-related flows are becoming increasingly accessible to study through a combination of laboratory experiments and numerical modelling, although complete three-dimensional calculations, with realistic Rayleigh numbers and appropriate temperature- and depth-dependent rheologies that incorporate the complicating effects of phase transformations, melting, and chemical differentiation, remain to be performed.

Considerations of the relative magnitudes of mid-ocean-ridge topography and hotspot-swell topography suggest that plumes originate from the core–mantle boundary (CMB), rather than from another thermal boundary layer at the base of an upper mantle strongly heated from below. It has been argued that plume-head diameters comparable to those of continental flood-basalt provinces provide additional evidence of a CMB origin for mantle plumes. However, recent modelling indicates that the highest temperatures may be strongly localized in the near-axial region of the uppermost layer of the plume head. Under these circumstances, an explanation of flood-basalt eruptions in terms of laterally extensive partial melting of plume heads seems to require a major-element chemistry for plumes that is substantially enriched relative to pyrolite – consistent with previous inferences from enriched trace-element signatures. Alternatively, a mechanism would be required that would allow ascent of plume-head material to shallower levels, especially within the continental lithosphere.

Increasingly detailed knowledge of the relevant phase equilibria and of the elastic properties of mantle minerals suggests that phase transformations in an isochemical (pyrolite) model mantle provide an adequate explanation for the seismologically well constrained radial structure, within the residual uncertainties in the temperature and pressure dependence of elastic (especially shear) moduli. Strong compositional layering, with its implication of an additional pair of thermal boundary layers in the mid-mantle, separating convection above and below, not only is not required but also would be difficult to reconcile with the radial velocity models. The superimposed large-scale lateral variability in wave speeds, as revealed by seismic tomography, is most pronounced in the outer few hundred kilometres of the mantle, where it is closely correlated with surface tectonics and probably is primarily of thermal origin. On smaller scales, compositional heterogeneity and anisotropy probably are more important. Tomographic studies suggest that most subducting lithospheric plates penetrate into the lower mantle, but not without significant distortion and deflection within the transition zone, plausibly explained by interactions with the viscosity structure of the mantle and by the effects of chemical layering within the down-going slab and the influence of trench migration.

The likelihood that the Clapeyron slopes for the phase transformations in the normative pyroxene-garnet component of the mantle are comparable in magnitude, but of opposite sign, to those for the relatively well understood olivine $\rightarrow$ wadsleyite and ringwoodite $\rightarrow$ perovskite $+$ magnesiowüstite transformations suggests that transformational buoyancy probably does not strongly perturb the dominantly thermal convection of the mantle. Chemical buoyancy associated with the basaltic layer of subducting oceanic lithosphere and any such material ascending in plumes might be at least equally as important as transformational buoyancy; its effect would be to resist the descent of slabs, but to promote the ascent of plumes.

According to recent analyses of sea-level changes consequent upon deglaciation, the effective viscosity of the mantle increases about 30-fold with increasing depth, from a minimum value of about 3×10^{20} Pa $\cdot$ s in the upper mantle to 10^{22} Pa $\cdot$ s in the deep mantle. The rheology of the dominant upper-mantle mineral olivine is becoming increasingly well understood through detailed laboratory experimentation. Of particular interest is the possibility of a transition from dislocation creep to diffusion creep in lithospheric shear zones and deep in the upper mantle, and there is evidence for significant weakening produced by small amounts of melt or water. However, the weakening effects of small degrees of partial melting might actually be more than offset by the hardening effects induced by strong partitioning of dissolved water-related defect species from olivine into the melt. There are also preliminary indications of the relative strengths of the key high-pressure minerals of the transition zone and lower mantle, and there have been some intriguing insights into transformation mechanisms and their consequences for the origin of deep-focus earthquakes. Microphysical rheological models for the deep mantle, based on extrapolation of experimentally determined flow laws for analogue materials, are most readily reconciled with those derived from analyses of glacial-rebound phenomena within a whole-mantle convection scenario in which there are no mid-mantle thermal boundary layers.

All of these inferences are consistent with a model of whole-mantle convection in which old cold slabs descend well into the lower mantle, though at speeds substantially less than those for the upper mantle, because of the higher viscosity. In this scenario, plumes rise from the thermal boundary layer at the base of the mantle, with narrow conduits or tails feeding large mushroom-shaped heads that grow by entrainment of surrounding mantle. The small relative motions of 'hotspots' require that their sources be located within a region of relatively high viscosity, most plausibly the very deep mantle. The inferred substantial increase in viscosity with depth in the Earth's mantle results in much longer circulation times, and less effective mixing, for material transported along streamlines penetrating deep into the lower mantle. Long residence times and ineffective mixing are necessary to explain the survival (on timescales of $\sim$2 billion years) of chemical and isotopic heterogeneity, including the observation that the degassing of primordial He and Ne and of radiogenic ^{40}Ar is far from complete.

This unifying view of the composition, structure, and dynamics of the modern mantle is a persistent theme throughout much of this volume. However, there remain substantial areas of uncertainty and controversy, which are also addressed. Some of the areas of residual uncertainty have already been mentioned in the preceding summary. One that is central to the admissibility or exclusion of the model of whole-mantle convection is the efficiency of mixing. The noble-gas data mentioned earlier and the evidence for the extraction into the continental crust of only about 50% of the Earth's budget of incompatible elements require the presence of a reservoir (or regions having a significant volume) that is essentially inaccessible to the processes of melt extraction and outgassing on timescales comparable to the age of the Earth. Strictly layered convection, in which the circulation patterns of the upper mantle and lower mantle would be separated by a pair of thermal boundary layers impermeable to matter, but conductive of heat, obviously would provide an effective mechanism for the chemical and isotopic isolation of the lower mantle. The upper mantle, with its substantially lower average viscosity and its direct participation in the processes of mid-ocean-ridge volcanism, would be both more homogeneous and much more depleted. However, strict layering is difficult to reconcile with the wide range of geophysical evidence summarized earlier. Nevertheless, it remains to be demonstrated conclusively that whole-mantle convection in a mantle with depth-dependent viscosity is compatible with the survival of chemically and isotopically distinct materials for timescales on the order of 2 billion years.

Another issue of central importance for an understanding of the evolution of the Earth is the possibility that even if whole-mantle convection is currently operative, it may not always have been so. There may have been episodic layering of the mantle in the past, as well as involvement of the cool upper boundary layer of the early Earth in processes distinctly different from those of modern plate tectonics. The proportion of the basaltic layer within the differentiated lithosphere, as well as the equilibrium and kinetic controls on its successive transformations to denser assemblages, may be very influential in deciding the fate of oceanic lithosphere. These factors might be important in determining whether all or part of the lithosphere can sink into the mantle, how and where any such subduction is initiated, and whether or not subduction can be sustained through the transition zone and into the lower mantle.

Also presented here are the very different implications of two alternative views of magma genesis in mid-ocean-ridge, ocean-island, and flood-basalt settings. The advocates of one model consider that the parental magmas for MORBs are olivine tholeiites derived at relatively low pressure from passive upwelling and melt pooling along a $\sim 1,300°C$ adiabat. The common presence of picrites in ocean-island settings could then be interpreted as evidence for the substantially higher potential temperatures expected of deep-seated plumes. The alternative view, that MORBs are generally derived from picritic precursors, is also argued in some detail in this volume. Under these circumstances, the potential temperature for the typical upper mantle would be much higher ($\sim 1,450°C$), indistinguishable from that for hotspot magmatism. Hotspots would be attributed not to higher temperatures, but rather to melting that was fluxed by locally higher concentrations of volatiles, derived ultimately from old subducted lithosphere deep in the upper mantle.

The mixture of emerging consensus and continuing vigorous debate, represented by the contributions that follow, seems appropriate for a volume in honour of Ted Ringwood. Moreover, this blend faithfully represents a living science in which the prevailing hypotheses are subject to continual testing against new observations, followed by revision or replacement as appropriate.

Finally, I thank the following individuals outside the authorship group for their thoughtful reviews of the various chapters: Bill Compston, Steve Eggins, Oli Gudmundsson, Anton Hales, Sue Kesson, Mervyn Paterson, Malcolm Sambridge, John Stone, Geoff Taylor, Sharon Webb, and Greg Yaxley. The editorial assistance of Kay Provins, and the enthusiastic involvement of Catherine Flack from CUP, are greatly appreciated.

Ian Jackson
Canberra

Part One

Accretion and Differentiation of the Earth

1

Composition of the Silicate Earth: Implications for Accretion and Core Formation

HUGH ST. C. O'NEILL and HERBERT PALME

Our knowledge of the constitution and composition of the Earth's mantle has advanced enormously during the last 30 years.... As a result of these developments many new and important boundary conditions for the origin of the Earth have emerged. I do not believe that the significance of these boundary conditions, mainly of a geochemical nature, [has] been adequately recognised in many recent discussions of the origins of terrestrial planets in general and of the Earth in particular.

— A. E. Ringwood (1979)

1.1. Introduction

The formation of our solar system followed the collapse and fragmentation of a dense interstellar molecular cloud. As interstellar matter always has some angular momentum, the development of a central star by direct infall was not possible, and instead a rotating disk resulted. Material within the disk lost angular momentum through viscous dissipation or other processes, leading ultimately to the growth of a central star, our Sun. Only a tiny fraction of the mass of the solar system ($\sim$0.1%) was left behind in the disk, eventually to form the planets and asteroids.

The duration of the initial collapse phase was short, less than 1 million years. After this phase, the remnants of the accretion disk may have persisted for as long as 10 million years before the planets were assembled. This history derives from astronomical observations and is consistent with isotopic evidence from meteorites (Podosek and Cassen, 1994). The mixture of gas and grains that made up the proto-solar accretion disk is known as the *solar nebula*. The solar nebula plays a key role in our understanding of the chemical compositions of the Earth, the planets, the asteroids, and other constituents of the solar system. Many of the important and distinctive features of the Earth's composition that are responsible for its structure and evolution were inherited from the processes that operated during the condensation of the solar nebula into solid matter.

The growth of solid bodies in the solar nebula began with the aggregation of tiny dust grains that either formed by condensation or were of interstellar origin. The small, micrometer-size dust particles suspended in the nebula tended to settle to the midplane. Grains then grew by coagulation due to different settling rates and because of turbulent motion in the nebula. This produced centimetre-size bodies that in turn grew to metre- and kilometre-size blocks by collisional coagulation, produced by gas drag on particles of different sizes (Weidenschilling, 1988). Once the diameters of the planetesimals had reached 1–10 km, gravitational forces determined further growth. Computer simulations show that fast runaway accretion will produce planetary 'embryos', bodies of some 10^{23} kg (i.e., about 2% of the mass of the Earth), within about 10^5 years (Wetherill, 1990, 1994). Three major conclusions derived from this model are pertinent to the growth of the Earth and other planets: (1) Planetesimals appear to grow from small particles through a hierarchy of increasingly larger bodies – an evolving size distribution. Planetesimals (or planets) do not grow by accreting dust onto a single nucleus. (2) Bodies up to kilometre size are composed of material derived from 'local feeding zones' and thus should retain chemical signatures that are characteristic of the heliocentric distances at which they formed. (3) At some point there was clearing of the nebula through intense activity of the early Sun (e.g., T-Tauri phase), leading to removal of gas and fine dust. Metre-size and kilometre-size bodies were not affected. This is potentially an effective mechanism for fractionation of volatile elements. However, it is not known how far accretion had proceeded when the clearing of the nebula took place.

The next step, formation of the inner planets by accumulation of embryos, took millions of years to complete. At this stage, some radial mixing of embryos formed at different heliocentric distances would be expected, so that material formed far out in the asteroid belt may have made significant contributions to the growing Earth. Such components may have been very different in volatile-element content and oxidation state compared with the material in the 'indigenous' embryos at 1 AU (Wetherill, 1994). Clearly, this may have had important consequences not only for the bulk composition of the Earth but also for any differentiation that may have accompanied accretion, particularly core formation. This model for planetary accretion is depicted in Figure 1.1.

In addition to samples from the Earth and the Moon, a wide variety of other solar-system material has been delivered to the Earth as meteorites and is accessible for laboratory study. Some rare types of meteorites (the 'SNC' meteorites, discussed later) are believed to be samples from Mars, and a few have come from the Moon. Most meteorites, however, are thought to be derived from asteroids (Wood and Morfill, 1988). Asteroids are small planetary bodies, or planetesimals, many of which are concentrated in the gap between Mars and Jupiter at 2–3.2 AU, which interrupts the regular spacing of the planets as described by the Titius–Bode law. They are remnants from the era during which the planets of the inner solar system, including the Earth, formed. Accretion of a planet in the position of the present

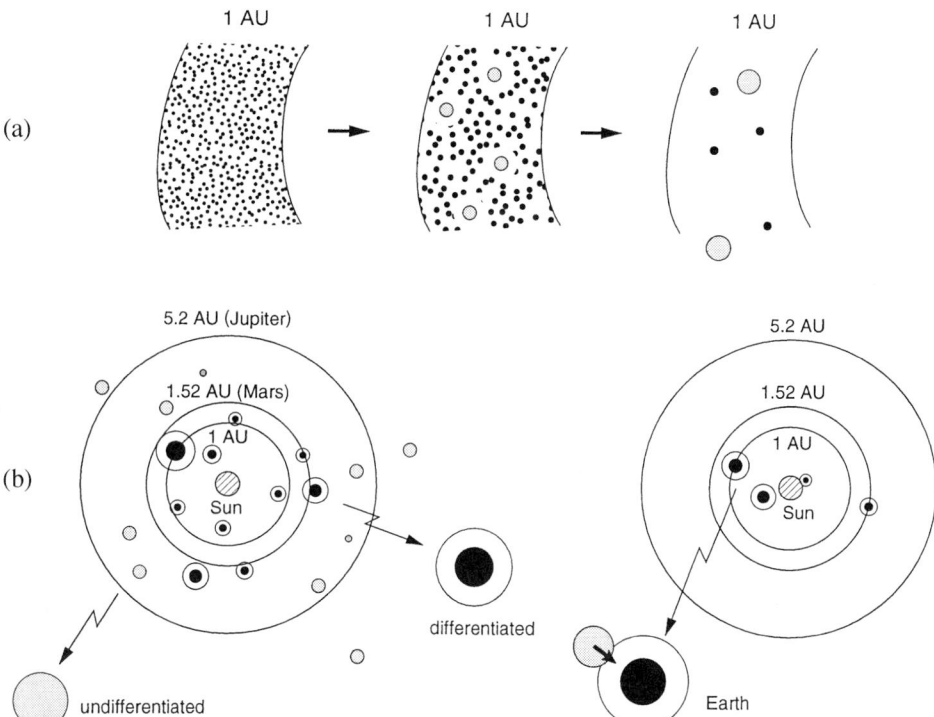

Figure 1.1. Formation of the Earth involved two major steps: (a) Within the local feeding zone of the Earth, at 1 AU (astronomical unit, the distance from the Earth to the Sun), bodies the size of the Moon or Mars ('planetary embryos') were formed within 10^5 years by collisions among kilometre-size planetesimals. (b) In the second step, the terrestrial planets grew by collisions among these embryos. This step took much longer (millions of years) and may have involved bodies formed at distances ranging from 0.5 to more than 3 AU, leading to substantial mixing of materials originating at different heliocentric distances (Wetherill, 1994). Whether or not a planetesimal or planetary embryo was differentiated (i.e., had been heated sufficiently to undergo partial melting and core formation) probably was a function of the size of the body, the timing of the accretion process, and its bulk composition. All of those factors may have depended on heliocentric distance. Thus, both differentiated and undifferentiated planetesimals and planetary embryos may have contributed to the growing Earth, depending on the heliocentric distances at which they formed. Bodies originating from the asteroid belt beyond 2.5 AU may have added undifferentiated, oxidized, and volatile-rich components late in the accretion process, as required by heterogeneous-accretion models.

asteroid belt probably was hindered by the early growth of Jupiter, so that accretion did not progress past the planetesimal stage (i.e., bodies of $\sim 10^{21}$ kg). The spectral properties of the asteroid Vesta suggest that it could have been the parent body for the 'HED' suite (howardites, eucrites, diogenites) of meteorites (Binzel and Xu, 1993).

Analyses of all these different kinds of solar-system materials demonstrate that the solar nebula was rather homogeneous in isotopic composition. Exceptions to that rule are provided by rare exotic grains of pre-solar origin found in the most primitive of the chondritic meteorites (e.g., Anders and Zinner, 1993), some isotopic

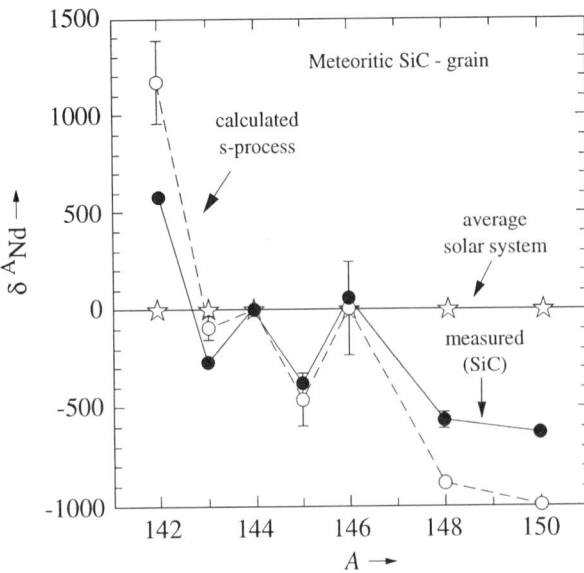

Figure 1.2. The isotopic composition of solar-system Nd, as found in terrestrial and lunar rocks and all bulk samples of meteorites, is uniform to better than one part in 10^4 (after allowing for variable ^{143}Nd from the decay of ^{147}Sm). This contrasts dramatically with the isotopic composition of Nd in a pre-solar SiC grain. The deviation of its Nd isotopic composition from the terrestrial ratio is given in parts per million and all ratios are normalized to ^{144}Nd. Full symbols are measured values (see Richter, Ott, and Begemann, 1992). Error bars are, in most cases, smaller than symbol sizes. The calculated s-process production ratios are also indicated. The important implication is that if the solar nebula had not been well mixed, bodies originating from different parts of it would have consisted of materials with different isotopic compositions for Nd, and, by analogy, for most other elements. From Palme and Beer (1993), with permission.

anomalies in Ca- and Al-rich inclusions in carbonaceous chondrites (e.g., Lee, 1988), and certain enigmatic variations in oxygen isotopes (Clayton, 1993) and noble-gas isotopic compositions (McDougall and Honda, Chapter 3, this volume). In view of the huge isotopic anomalies found in pre-solar grains, it is of great significance that the isotopic compositions of well-studied elements such as Sr, Nd, Hf, and Os (and, by inference, of most other elements) are, within 0.01%, identical in all bulk samples of solar-system material that are available for analysis.

An example of an isotopic anomaly is given in Figure 1.2, in which the isotopic composition of Nd in a pre-solar SiC grain is compared with the isotopic composition of solar-system Nd (i.e., from terrestrial, lunar, and bulk samples of the many different types of meteorites, which differ by less than one part in 10^4). The Nd isotopic composition of the SiC grain is derived from a specific nucleosynthetic source, the s-process, as indicated by comparison with calculated s-process yields (Palme and Beer, 1993). This grain was formed in the interior of a star and was ejected into the interstellar medium, and it finally found its way into a meteorite, more or less undisturbed. By contrast, solar-system Nd must have been derived by mixing of Nd from more than one nucleosynthetic process and thus has a distinctive isotopic composition. This is probably true of many elements. Thus, the rare interstellar grains, with their huge isotopic anomalies (not, of course, confined to Nd), are truly 'exceptions that prove the rule', the rule being that the materials now in the inner solar system share a common origin, consistent with derivation from a homogenized nebula that was efficiently mixed on the usual scale at which meteorites are sampled for conventional isotopic analysis.

Interestingly, current research is increasingly revealing isotopic inhomogeneity on a finer scale, and it seems that in some primitive meteorites different nucleo-synthetic sources are still discernible in individual mineral grains. For example, the Cr isotopic composition of the Cr-bearing phases of carbonaceous chondrites is highly variable, although the average isotopic compositions of the bulk meteorites are identical with those of terrestrial rocks and other solar-system samples (Rotaru, Birck, and Allègre, 1992). Nonetheless, on all except this smallest of sampling of scales, the dominant picture remains one of essential uniformity in isotopic constitution.

The major deduction from the hypothesis of a well-mixed nebula is that all solar-system material, including that forming the Earth, was derived from a nebula material that had a uniform, characteristic *chemical* and isotopic composition. This is the solar composition as observed in the solar photosphere.

Variations in the oxygen isotopic compositions of solar-system materials (Clayton, 1993) might appear to contradict this hypothesis of a homogeneous neb-ula. However, the pattern of these variations has not been satisfactorily explained, and it now seems likely that the mass-independent isotopic fractionation observed in meteoritic oxygen may have resulted from physical processes within the solar nebula (e.g., condensation) (Thiemens, 1988). This is supported by recent findings of such effects in stratospheric and mesospheric CO_2 (Thiemens et al., 1995).

Although the solar nebula was essentially homogeneous in composition, the com-positional diversity of the bodies in the inner solar system (especially as sampled by meteorites) indicates that a substantial amount of chemical differentiation must have taken place either during its condensation or subsequently during accumula-tion of the condensed matter into planetesimals and planetary embryos. The material that now forms asteroids and the material that formed the precursors to the larger planets can reasonably be supposed to have shared a common early history, so that meteorites derived from asteroids should document the type and extent of the frac-tionation processes that occurred during these early stages of planet-building and affected the Earth-forming precursor materials. Identifying the fundamental chem-ical fractionation processes responsible for the surprisingly large compositional diversity of meteorites provides the framework not only for constraining composi-tional models of the Earth but also for appreciating the significance of the Earth's composition in chronicling the history of its accretion and early differentiation.

Classification of

Meteorites

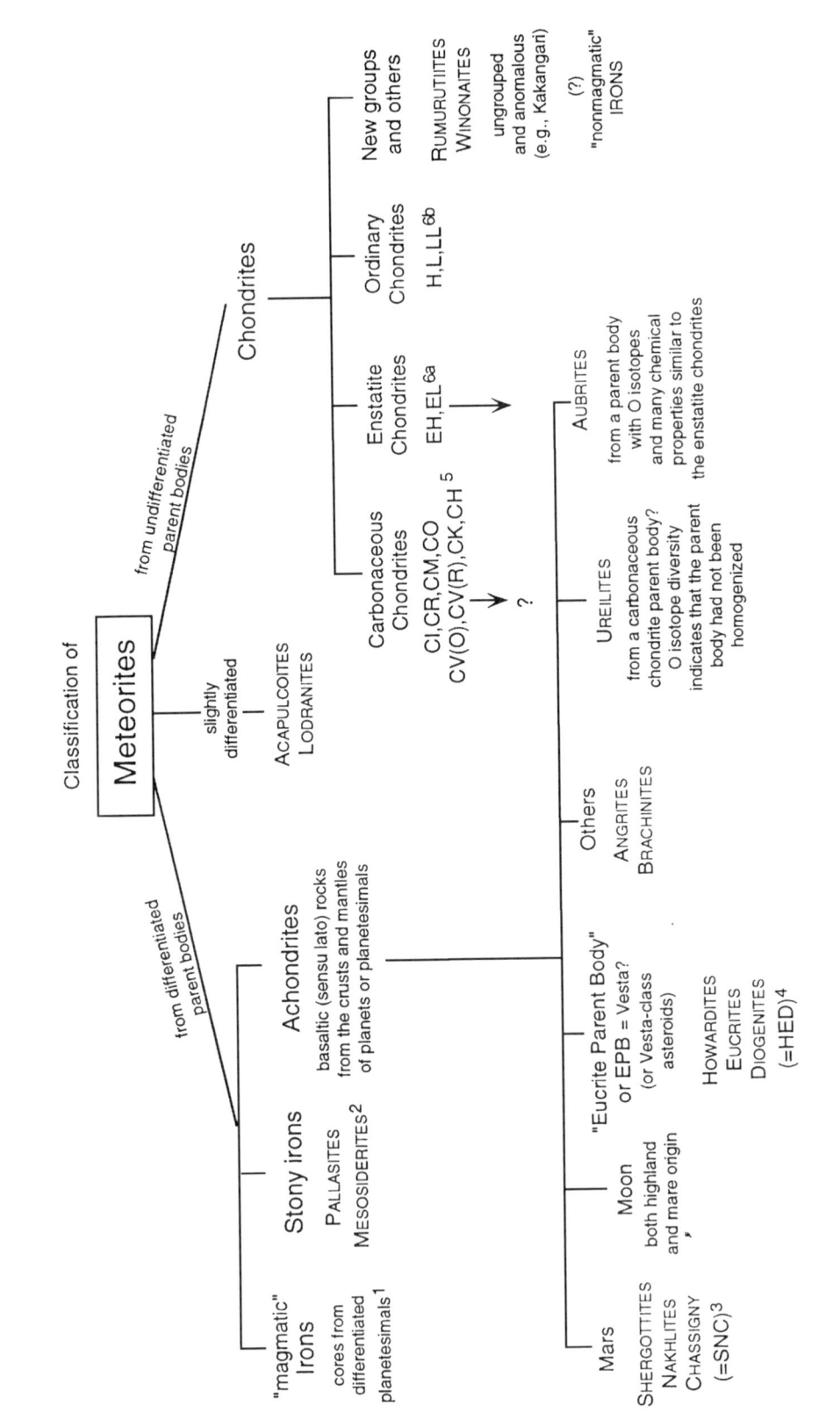

from differentiated parent bodies

slightly differentiated

from undifferentiated parent bodies

"magmatic" Irons
cores from differentiated planetesimals[1]

Stony irons
PALLASITES
MESOSIDERITES[2]

Mars
SHERGOTTITES
NAKHLITES
CHASSIGNY
(=SNC)[3]

Moon
both highland and mare origin

"Eucrite Parent Body" or EPB = Vesta? (or Vesta-class asteroids)
HOWARDITES
EUCRITES
DIOGENITES
(=HED)[4]

Achondrites
basaltic (sensu lato) rocks from the crusts and mantles of planets or planetesimals

Others
ANGRITES
BRACHINITES

ACAPULCOITES
LODRANITES

UREILITES
from a carbonaceous chondrite parent body? O isotope diversity indicates that the parent body had not been homogenized

AUBRITES
from a parent body with O isotopes and many chemical properties similar to the enstatite chondrites

Chondrites

Carbonaceous Chondrites
CI,CR,CM,CO
CV(O),CV(R),CK,CH [5]
?

Enstatite Chondrites
EH,EL [6a]

Ordinary Chondrites
H,L,LL [6b]

New groups and others
RUMURUTIITES
WINONAITES
ungrouped and anomalous (e.g., Kakangari)
(?)
"nonmagmatic" IRONS

1.2. The Meteorite Record
1.2.1. Classification of Meteorites and the Diversity in Their Chemical Compositions

Most meteorites formed around 4.5 billion years ago, as revealed by the standard radiogenic-isotope geochronometers. Only for a comparatively short fraction of the time since then (i.e., on the order of a few million years) do they show evidence of having been exposed to cosmic rays. For most of their existence, therefore, the fragments of rocks that we now find as meteorites were buried in the interiors of their parent planetesimals. That there were variations in the chemical compositions and thermal histories of these parent bodies is evidensed by the many different types of meteorites. A classification scheme for meteorites is depicted in Figure 1.3.

The fundamental distinction among meteorites is between those derived from parent bodies that had undergone igneous differentiation and metal segregation, leading to core formation, and those derived from essentially undifferentiated parent bodies. The latter are particularly useful, because they preserve directly the chemical compositions of their parent bodies.

The meteorites from differentiated bodies include the products of partial melting (e.g., the basaltic meteorites, such as the eucrites), residues from partial melting (lodranites and ureilites), the metallic cores of planetesimals (the 'magmatic' irons), samples from the core–mantle boundaries of the planetesimals (the pallasite stony

Figure 1.3. The diversity of meteorites.

[1]There are approximately 12 major groups of irons, plus anomalous or ungrouped examples. Originally, four major groups of irons were recognized on chemical grounds (I, II, III, and IV). Further work led to subdivision (e.g., IVA, IVB). Some of these subdivisions have subsequently been recombined (e.g., IAB) and recombined again (IAB + IIICD). However, the fundamental division of the irons is into 'magmatic' types (which show chemical evidence for differentiation by fractional crystallization in the core of an asteroid-size parent body) and 'nonmagmatic' types (at least some of which may be close to being compositionally primitive condensates from the solar nebula, and therefore of chondritic affinity).

[2]Mesosiderites are breccias involving a mixture of iron meteorites with material of possible HED origin. Main-group pallasites may come from the core–mantle boundary of the EPB (O isotopes). Pallasites of the Eagle Station type come from a body with a different O isotope signature.

[3]Shergottites are (presumably) Martian basalts; nakhlites and Chassigny are cumulate rocks of hypabyssal origin.

[4]Eucrites are basalts; diogenites are hypabyssal cumulates. Most eucrites and diogenites are breccias. Howardites are polymict breccias of eucrites and diogenites.

[5]The second letter usually stands for the name of a representative member of the class (e.g., I for Ivuna); see Table 1.1. The H in CH is an exception.

[6a]H is for high Fe, and L is for low Fe.

[6b]H is for high Fe, L is for low Fe, and LL is for (very) low total Fe and low metallic Fe. Currently, chondrite groups are always referred to by their letter(s), except anomalous meteorites and the newly defined groups (rumurutiites, winonaites). Rumurutiites are now R chondrites.

irons), and so forth. The meteoritic record probably samples a continuum in the differentiation of planetesimals, ranging from the completely undifferentiated, through those that had just begun to melt and show the traces of incipient differentiation (e.g., acapulcoites), to the completely differentiated, of which the eucrite parent body (EPB) is thought to be an example.

Meteorites from undifferentiated bodies are termed *chondrites*. Meteorite nomenclature is encumbered with some interesting but potentially confusing historical baggage, of which the term "chondrite" is a good example. The term comes from *chondrules*, the name given to near-spherical objects typically of sub-millimetre-to-millimetre dimensions that are interpreted to be agglomerations of nebula dust that were melted by rapid heating and then were rapidly cooled. That much about the origin of chondrules was discovered by Sorby (1877), using a petrologic microscope and thin-section process that he himself had invented. Although, some 120 years later, the cause of the melting still is not generally agreed upon, it evidently was a common process, at least in the inner solar system, as chondrules are major components of many chondritic meteorites. For example, chondrules and chondrule fragments account for 65–75%, by volume, of ordinary chondrites with primitive (unmetamorphosed) textures, much of the remainder being *matrix*. In many chondrites, however, chondrules are difficult to recognize, because subsequent recrystallization accompanying thermal metamorphism in the meteorite parent body has obscured their shapes. Even more extensive metamorphism has led to complete erasure of chondrules in some groups of chondritic meteorites. The CI carbonaceous chondrites do not contain chondrules – they are composed mostly (>95%) of matrix. Thus the presence of chondrules plays no part in the current definition of a chondritic meteorite.

Meteoriticists recognize three major groups of chondritic meteorites, the ordinary chondrites (OCs), the enstatite chondrites (ECs), and the carbonaceous chondrites (CCs), plus a number of minor groups and a few anomalous individuals. These major groups are further subdivided on the basis of their chemistry and textures. A brief overview of chondritic-meteorite classification is given in Table 1.1, and some of the key chemical characteristics of the chondrite groups are listed in Table 1.2. In keeping with other irregular aspects of meteorite nomenclature, most of the carbonaceous-chondrite subgroups are not particularly rich in carbon. An indication of the relative abundances of the different types of meteorites, as sampled here on the Earth, is provided by the statistics in Table 1.3.

Chemically, the most primitive chondrites are thought to be the CI carbonaceous chondrites, which are the richest in volatile elements. The chemical composition of this group of meteorites is accorded special significance, as it appears to be the same as that in the solar photosphere (Figure 1.4) for all elements except the most volatile (the 'ice-forming elements' H, C, O, and N, and the rare gases) and the light elements actively being consumed by nuclear reactions in the interior of the Sun (Li, Be, and B). Because the present mass of the solar system is almost completely contained in the Sun, the solar photosphere, being representative of the Sun, also

Table 1.1. *Comparative properties of chondritic meteorites*

Classification	Chemistry[a]	Oxidation state	Texture	Mineralogy[b]	Metamorphic grade[c]
Carbonaceous chondrites (CCs)					
CI (C1) Ivuna (Orgueil)	Solar abundances of elements heavier than oxygen, except rare gases	Highly oxidized, water, no metal, but rare FeO-poor olivine grains	No chondrules; >95% matrix; very fine grained, brecciated	Mainly hydrous silicates, with magnetite, dolomite, pyr, sulfates, plus a few isolated ol and px grains	Aqueous alteration, type 1
CM (C2) Mighei (Murray) (Murchison)	Enriched in refractories (1.1 × CI); depleted in volatiles (0.5 × CI)	Oxidized, with water, but forsteritic aggregates with metal inclusions	~10% chondrules, 70% matrix; inclusions, mineral and lithic fragments, accretionary rims around chondrules and inclusions, very refractory CAIs	Phyllosilicate matrix, abundant isolated ol grains, low in FeO; several % carbonates, magnetite, pyr, pntld; a little kamacite	Aqueous alteration, type 2
CO (C3O) Ornans	Refractory elements as in CM; depleted in volatiles (0.2 × CI)	Less oxidized than CM, only traces of water	Small (0.2–0.5 mm) chondrules (40%), fine-grained anhydrous matrix (35%), high-temperature components, superrefractory inclusions	Progressive change in ol composition, with metamorphism from fa 12 to fa 34; kamacite, taenite, tro	Comprises a metamorphic sequence, reflected in FeO of olivine, type 3
CV (C3V) Vigarano (Allende)	Enriched in refractories (1.3 × CI); lower in Fe, Na, K than CO; depleted in volatiles (0.23 × CI)	Two subgroups: oxidized CV(O), Allende; reduced CV(R), Vigarano	Chondrules larger than in CO; 40% chondrules, 35% anhydrous matrix, 5–10% CAIs (up to cm size)	Wide variety of ol, fa 0.3 to fa 45; CV(O): magnetite, Ni-rich sulfate; CV(R): metal, Ni-poor sulfide	Type 3
CR Renazzo	Refractory elements as in CI; depleted in volatiles (0.24 × CI)	Reduced, with metal, but phyllosilicates, mostly low-FeO olivine	Chondrules (40–60%) of mm-size in matrix, fine-grained dark inclusions and abundant CAIs	ol (fa 1–4), px (fs 1–4); much metal (4–75% Ni); CAIs contain grossite	Alteration of mtx, type 3

(cont.)

Table 1.1. (cont.)

Classification	Chemistry[a]	Oxidation state	Texture	Mineralogy[b]	Metamorphic grade[c]
CK Karoonda	Refractory elements (1.21 × CI); volatiles between CO and CV	Oxidized, with NiO in olivine	Recrystallized (metamorphic sequence); 10–15% chondrules (Karoonda); matrix: ol (25 μm)	ol (fa ∼ 30), opx, cpx, plag, magnetite, pntld	Types 4–6
CH ALH 85085	CI RLEs; high total Fe (1.4 × CI) and other nonvolatile metals; very low in volatiles (0.1 × CI)	Reduced, with metal, low-FeO olivine	Very small constituents (fragments, chond-rules, clasts, CAIs); little matrix, few chondrules (∼10%)	ol (fa 1–3), px, metal, tro, pntld; CAIs with melilite and grossite	Type 3
Ordinary chondrites (OCs)					
H Tieschitz (H3)	H, L, LL: refractories (0.8 × CI); Mg/Si (0.9 × CI); (1 × CI); other volatiles Na, Mn, Au 1–0.1 × CI; highly volatile elements depend on petrologic type	fa 16–20, FeNi, $Fe^{2+}/\sum Fe$ ∼ 0.38	H, L, LL: texture depends on metamorphic grade; ∼90% of the silicate in unequilibrated types is composed of chondrules and chondrule fine-grained matrix; refractory fragments; the rest is inclusions	Abundances in wt%: ol (36.2), opx (24.5), cpx (4), fspar (10), apat (0.5), chr (0.6), ilm (0.2), tro (6.1), NiFe (7.5)	Types 3–6; rare type 7, e.g., Shaw (incipient partial melting)
L Bruderheim (L6)	L: depleted in Fe and siderophiles (0.75 × CI)	fa 23–26, FeNi, $Fe^{2+}/\sum Fe$ ∼ 0.66	rare Unequilibrated → equilibrated: coarsening of matrix; chondrules increasingly difficult to see; transformation of glass to	Abundances in wt%: ol (47), opx (22.7), cpx (4.6), (10.7), apat (0.6), chr (0.6), ilm (0.2), tro (6.1), NiFe (7.5)	
LL Chainpur (LL3)	LL: depleted in Fe and siderophiles (0.6 × CI)	fa 27–32, FeNi, $Fe^{2+}/\sum Fe$ ∼ 0.88	plag and of monoclinic low-Ca px to opx		

Enstatite chondrites (ECs)

EH Abee (EH4)	Refractories (0.8 × CI); Mg/Si (0.7 × CI); high in volatiles (0.7 × CI)	Very reduced; fs <0.2, several % Si in metal	Texture depends on metamorphic grade; unequilibrated → equilibrated	opx (60–80%), metal (13–28%), troilite (5–17%), oldhamite, alabandite, daubreelite	Types 3–5
EL Hvittis (EL6)	Refractories (0.7 × CI); Mg/Si (0.84 × CI); low in volatiles				Classically type 6 only; new finds types 3 and 5
Others					
Acapulcoites Acapulco	Refractories similar to OCs; higher in volatiles than OCs; chemical composition more variable than in OCs (especially S, Cr, LREE)	fa 10, FeNi	Equilibrated texture, ca. 1,000°C; no chondrule (except relicts in Monument Draw)	Acapulco abundances in wt%: ol (17.7), opx (53.4), cpx (1), fspar (9.4), apat (1.0), chr (1.0), tro (8), NiFe (11.0)	More equilibrated than type-6 OCs (type 7?)
Rumurutiites Rumuruti	Depleted in refractories (0.9×CI, Mg norm.); depleted in volatiles (0.3×CI); high S	fa 37–45; <1% metal; very oxidized, 0.25 wt% NiO in olivine	Most members are breccias, clasts of various petrologic types, rare chondrules	Rumuruti abundances in wt%: ol (70), opx (0.5), cpx (5.2), fspar (14.5), pyr (4.4), pntld (3.3), phosphate (0.4), chr (0.4)	Some are breccias (3–6); others (3.6–3.9)
Winonaites Winona	Not a very coherent group, similar to IAB inclusions (except siderophiles); high in volatiles	Winona: fa 5, trace FeNi; Pontlyfni: fa 0.5	Equilibrated, similar to IAB silicate inclusions (800–1,000°C)	ol, opx, cpx, fspar, apat, chr, tro, metal, graphite	More equilibrated than type-6 OCs

[a] Depletion/enrichment factors relative to CI carbonaceous chondrites are normalized to Mg.

[b] Mineral abbreviations: ol, olivine (fa refers to mole fraction Fe_2SiO_4); px, pyroxene (fs refers to mole fraction $FeSiO_3$); opx, orthopyroxene; cpx, clinopyroxene; fspar, feldspar; apat, apatite; chr, chromite; ilm, ilmenite; tro, troilite; pntld, pentlandite; plag, plagioclase; pyr, pyrrhotite; grossite is $CaAl_2O_4$.

[c] In the classification of the chondrites, "petrologic type" is equivalent to metamorphic grade:

Type 3 chondrites are the most primitive and are thought to be nearly unaltered aggregates of nebular material.

Type 3 → type 1: increasing degrees of aqueous alteration, no temperature significance; seen in carbonaceous chondrites only.

Type 3 → type 6: increasing degrees of thermal metamorphism (up to 1000°C). Type 7: beginning of melting.

Sources: Data form Dodd (1981), Wasson and Kallemeyn (1988), Kallemeyn et al. (1991), Palme and Beer (1993), Bischoff et al. (1993a,b), Schulze et al. (1994), Zipfel et al. (1995).

Table 1.2. Element abundances in chondritic meteorites

Body	Refractory elements: Ca/Mg (by weight)	Major components		Moderately volatile elements			Oxygen isotopes	
		Si/Mg (by weight)	Fe/Mg (by weight)	Mn/Mg (by weight)	Se (ppm)	Zn (ppm)	$\delta^{18}O$ (‰)	$\delta^{17}O$ (‰)
Sun	0.099	1.08	1.91	0.015				
Carbonaceous chondrites								
CI	0.103	1.09	1.90	0.020	21.3	323	~16.4	~8.8
CM	0.109	1.10	1.79	0.015	12.7	185	4.5 to 13	−1 to 5
CO	0.109	1.10	1.71	0.011	7.6	100	−2.5 to −0.5	−6 to −4
CV	0.131	1.08	1.62	0.010	8.3	116	−4 to −5	−8 to −1
CR	0.098	1.10	1.79	0.013	7.4	106	~2	~−1
CK	0.116	1.06	1.59	0.010	6.9	99	~−1	~−4
CH	0.112		3.30	0.0089	3.9	36		
Rumurutiites	0.091	1.29	1.85	0.017	14.3	150	5.28	4.97
Ordinary chondrites								
LL	0.085	1.24	1.21	0.017	9.9	46	4.9	3.9
L	0.088	1.24	1.44	0.017	9.0	50	4.6	3.0
H	0.089	1.21	1.96	0.017	7.7	47	4.1	2.9
Acapulcoites	0.072	1.11	1.47	0.018	10.1	202	2.5–4	0.2–0.9
Enstatite chondrites								
EH	0.080	1.58	2.74	0.021	25.5	250	5.6	3.0
EL	0.072	1.32	1.56	0.012	13.5	17	5.3	2.7

Sources: Data from Figure 1.5 and Clayton (1993).

Table 1.3. *Frequencies of meteorite types (%)*

Body	World falls ($n = 835$)	Antarctic finds ($n = 1,100$)	Sahara finds ($n = 316$)
Chondritic meteorites			
Ordinary chondrites (OC)	80.0	87.9	94.3
Carbonaceous chondrites (CC)	4.6	2.7	2.5
Enstatite chondrites (EC)	1.6	0.7	1.0
Rumurutiites	0.001	0.003	0.003
Acapulcoites	0.001	0.006	0
Winonaites	0	0.004	0
Achondrites			
Achondrites	7.9	5.8	0.3
Stony iron and iron meteorites			
Mesosiderites	1.1	0.7	1.0
Irons	4.8	2.2	0.6

Source: Data from Bischoff and Geiger (1995).

Table 1.4. *CI (type 1) carbonaceous chondrites – all known samples*

Name	Locality	Date of fall	Amount recovered[a]
Alais	Gard, France	15/3/1806	>6kg[b]
Orgueil	Tarn-et-Garonne, France	14/5/1864	>10 kg
Tonk	Rajasthan, India	22/1/1911	7.7 g
Ivuna	Tanzania	16/12/1938	704.5 g
Revelstoke	British Columbia, Canada	31/3/1965	~1 g

[a] By contrast, about 2 tons of Allende (CV3) probably were recovered after the 8/2/1969 fall.
[b] 'Very little preserved', probably about 100 g extant.
Source: Data from Graham, Bevan, and Hutchison (1985).

represents the average composition of the solar nebula. The abundances of some elements have not been measured in the solar photosphere, and for many other elements the precision of the chemical analyses of meteorites is much superior to the precision of photospheric measurements (e.g., Anders and Grevesse, 1989); consequently, it is the meteorite abundance measurements that are assumed to give the best figures for the composition of the solar system. Considering, then, the importance of CI chondrites, it is well to emphasize how rare these meteorites are. A complete list of the five known falls is given in Table 1.4 (CI chondrites may be too fragile, both chemically and physically, to survive long enough at the

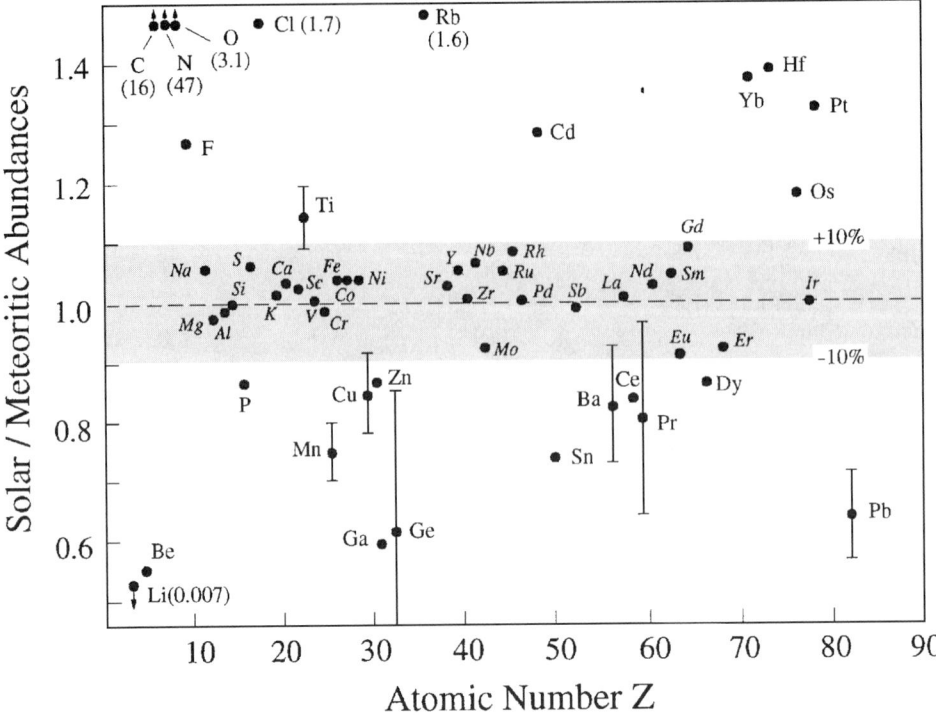

Figure 1.4. Comparison between the abundances of the elements measured in CI carbonaceous chondrites and in the solar photosphere. The most volatile of the elements (H, C, N, and O and the noble gases) are not fully condensed in CI chondrites, and the light elements Li and Be (and B) are consumed by nuclear reactions in the interior of the Sun. Error bars are shown only when they do not overlap with the solar/meteoritic = 1 line. The agreement between the solar-photosphere measurements and CI measurements has steadily improved with time; in particular, the measurement of the photosphere abundance of Fe is now in full agreement with the meteorite data. Only for two elements (Mn and Pb) do there remain significant unexplained discrepancies. To make the comparison, solar-photosphere abundances are normalized to 10^{12} atoms of H, and the meteoritic abundances to 10^6 atoms of Si. The conversion factor from the solar-abundance scale to the meteoritic scale is 35.5, estimated by comparing the abundances on the two scales of the best-determined elements. (From Palme and Beer, 1995, with permission.)

Earth's surface to become 'finds'). For obvious reasons, most chemical data are from Orgueil, supplemented by some from Ivuna and Alais.

The chondrites are also classified according to 'petrologic type', which corresponds to metamorphic grade. The meteorites with the most primitive textures, which have suffered only minimal alteration in their parent bodies, are assigned a 'petrologic type' of 3. Petrologic types 4, 5, and 6 describe the progressive textural and mineral-compositional effects of increasing thermal metamorphism. Type-6 chondrites are well-equilibrated metamorphic rocks, the thermal histories of which are suited to investigation by the standard techniques of mineral geothermometry (although the effects of slow cooling need to be taken into account): Their peak metamorphic temperatures are estimated to have been about 950°C. A few L and LL ordinary chondrites evidently were heated to slightly higher temperatures and show evidence of incipient partial melting; these are sometimes assigned to petrologic type 7. Acapulcoites might be considered as extreme examples of type-7 chondrites, or as transitional to achondrites. Type-6 chondrites show only ghostly reminders of the original chondrules in their textures. The abundances of the most volatile elements (e.g., noble gases, Bi, In, etc.) correlate with petrologic type, and these elements are presumed to have been driven off during metamorphism – to where is a mystery.

Petrologic types 2 and 1 describe increasing aqueous alteration (in that order), which is approximately correlated with increasing water content. CIs are all type 1 (consistent with their high water contents); thus, whereas CIs may be chemically the most primitive chondrites, they are not considered so either texturally or mineralogically.

Because the moderately rare petrologic type-3 chondrites best preserve in their textures and their mineralogy the nearest thing to a direct, if incomplete, record of how the solar nebula actually condensed, they are of particular interest. Consequently, they are further divided into subtypes 3.0 (the most primitive) through 3.9. The study of these least-altered textural types reveals that chondrites are principally composed of the following materials:

1. Refractory inclusions (e.g., CAIs, for calcium-aluminium-rich inclusions), which approximate in composition and mineralogy the first material to condense from the solar nebula prior to the condensation of the major Mg silicates and Fe-Ni metal phases
2. Chondrules and obvious chondrule fragments, compositionally and texturally somewhat diverse.
3. Mineral fragments (e.g., of olivine), most of which probably came from the disaggregation of chondrules, but some of which may have formed as individual grains in the solar nebula
4. Metal, whose origin probably was similar to that of the mineral fragments
5. Matrix, composed of very fine grained assemblages, the constituent minerals of which are characteristic of the chondrite group
6. Fragments from other chondrites

The various proportions of these materials in the different classes of chondrites are indicated in Table 1.1. An important observation is that the bulk compositions of chondrites, particularly carbonaceous chondrites, are surprisingly insensitive to the proportions of chondrules to matrix in the sample. This reflects an essential chemical similarity between chondrules and matrix, indicating that both derived from the same batch of material, and thus implying a local process for the formation of the chondrules.

A further complication is that many meteorites, both chondrites and achondrites, are breccias, composed of fragments of different rocks cemented together. Some of these are simply mixtures of fragments of the same lithology (monomict breccias); others are composed of fragments or clasts of different lithologies from the same parent body (polymict breccias, e.g., the howardites from the HED association), and these types are reasonably interpreted as having originated from the regoliths of their parent bodies. Lunar meteorites, for example, clearly had a regolith origin. Other breccias are composed of different petrologic types of the same group (genomict breccias). Primitive breccias are (mostly) type-3 chondrites, composed of clasts often from different compositional groups. An extreme example is Kaidun, in which host material of CR composition contains fragments of a CI or CM carbonaceous chondrite, both groups of enstatite chondrites (EH and EL), and material similar to R chondrites (Sears, 1996). The question raised by the existence of such breccias is how much of the material in any given chondrite has been recycled from other chondrites.

The lack of differentiation in the parent planetesimals of the chondrites presumably was related to their sizes and/or the distance from the Sun at which they formed. At a given heliocentric distance, the smaller a planetesimal, the less likely should it be for it to heat up sufficiently to melt and differentiate; this applies to any of the likely sources of heat, such as the radioactive decay of short-lived isotopes (e.g., ^{26}Al, ^{60}Fe) or electrical or solar heating. The efficiency of the latter two processes decreases with increasing distance from the Sun. Longer accretion times at greater heliocentric distances would also have reduced the extent of heating by ^{26}Al and other short-lived isotopes (Grimm and McSween, 1993). Gravitational heating probably was not important for asteroid-size bodies. Large impacts at that stage of the accretion process would be expected to have removed more material than they added (Sonnett and Reynolds, 1979). Therefore, the chondrites, coming from small parent bodies, probably are composed of material derived only from their local feeding zone in the solar nebula. Later evolution of the orbits of small planetesimals to larger eccentricities because of interactions with Jupiter led to collisions and some radial mixing of meteorite parent bodies formed in different environments. To some extent this may be documented in the meteorite breccias containing material presumed to be from more than one parent body.

Meteorites in general are rare objects. The total number of all types, either observed to fall or found, was only about 2,500 prior to the mid-1970s, of which some 1,000 were falls (Table 1.3). Since then, the number of specimens available to science has more than doubled, firstly through the discovery of a treasure trove of meteorites on the aprons of the Antarctic ice sheet, and secondly because of systematic collection of meteorites from other desert areas (e.g., the Sahara and Nullarbor Plain). The huge increase in the number of specimens available for study has considerably extended the known compositional range of chondrites, resulting in the recognition of several new groups. The value of meteorite groups as indicators of solar-system chemical processes does not increase in proportion to the frequency with which their members arrive and are recovered on Earth; this probably has more to do with stochastic processes out in the asteroid belt. For example, the most abundant meteorites, the ordinary chondrites (Table 1.3), are very poorly represented in the asteroid belt, as judged by the mapping of asteroid compositions by reflectance spectroscopy. Rather, the rarer types and even the unique chondrites should all be viewed as having the potential to increase our understanding of the processes responsible for the variations in the chemistry of nebular materials and thus, by implication, the materials from which the Earth and other terrestrial planets accreted.

The abundances of several elements with representative cosmochemical properties are shown for the different types of chondritic meteorites in Figure 1.5. Before elaborating further on the significance of these abundance patterns, it is necessary to discuss a cosmochemical classification scheme for the elements.

1.2.2. The Cosmochemical Classification of the Elements

The geochemical classification of elements is based on their behaviour during igneous differentiation processes, such as partial melting or fractional crystallization, and thus essentially considers the element's affinity for silicate melt versus crystalline silicate (occasionally also oxide) phases. Ionic size and charge are the most important parameters in determining this partitioning behaviour. The tendency of an element to be partitioned into the melt phase is known as its *incompatibility*.

In contrast, the cosmochemical behaviour of an element largely depends on two other properties: its volatility in a gas of solar composition, and its affinity for metallic versus silicate or oxide phases. This latter property is of great importance in two quite different contexts: in the solar-nebula environment, and later in the core-forming process in differentiated bodies. Thus the cosmochemical classification of the elements must describe gas/solid or gas/liquid equilibria, on one hand, and metal/silicate or sulphide/silicate equilibria, on the other. Consideration of the latter type of partitioning classifies the elements into those that are lithophile (which preferentially partition into silicates or oxides), siderophile (partition into Fe-rich metal), and chalcophile (partition into sulphide). The distinction between siderophile and chalcophile elements is of minor importance and becomes redundant above the temperature of the Fe-FeS eutectic, where sulphide and metal melt to one S-containing metallic liquid (at least in C-poor systems).

A convenient measure of the volatility of an element in the nebular environment is its condensation temperature. These temperatures are calculated by assuming thermodynamic equilibrium between solid condensates and a gas of solar composition (whose main constituent is H_2). Detailed descriptions of the methodology have been given by Grossman and Larimer (1974) and Saxena and Ericsson (1986). The results of such calculations are conveniently reported as '50% condensation temperatures', defined as the temperature at which 50% of the solar abundance of the element is condensed (most elements are calculated to condense over a fairly narrow temperature interval). Major elements condense as minerals (e.g., Mg and Si as Mg_2SiO_4, olivine), whereas trace elements condense in solid solution with major phases (e.g., Au in FeNi metal, Mn in olivine). A gas of solar composition is very reducing, such that Fe condenses as metal, and Mg silicates are initially FeO-free. Only at rather low temperatures is solid Fe metal calculated to react with oxygen and

Refractory Lithophile Elements

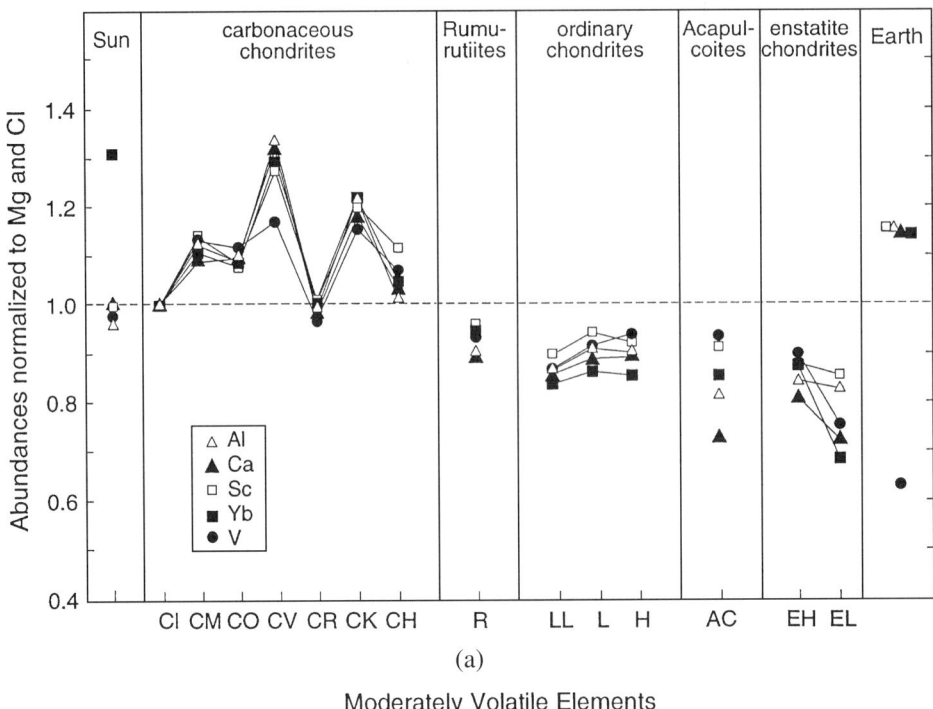

(a)

Moderately Volatile Elements

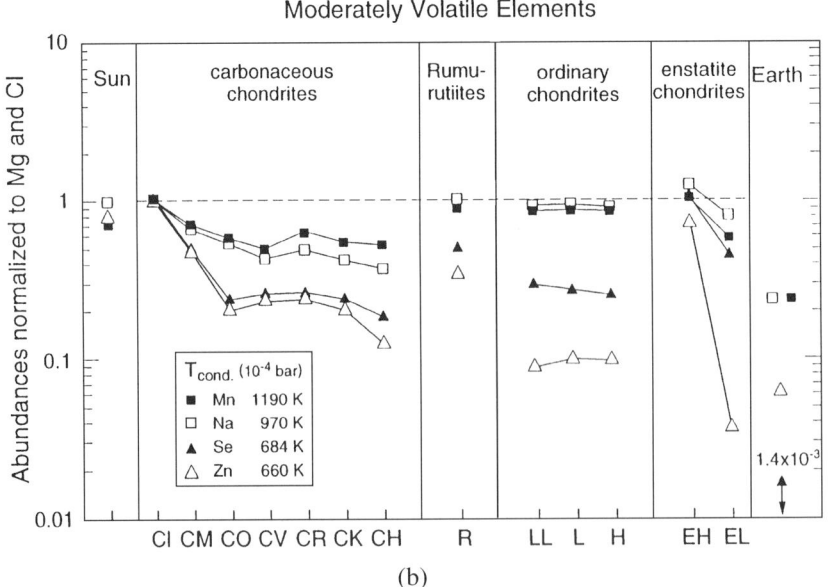

(b)

Figure 1.5. Element/Mg weight ratios of characteristic (a) refractory lithophile elements and (b) moderately volatile elements in chondritic (i.e., undifferentiated) meteorites. The groups are arranged in the order of decreasing O/Mg ratios, reflecting increasingly more reducing conditions. Data from Wasson and Kallemyn (1988), Kallemyn et al. (1991), Palme and Beer (1993), Bischoff et al. (1993b), Schulze et al. (1994), McDonough and Sun (1995), Zipfel et al. (1995). Solar abundances are taken from Anders and Grevesse (1989). The best match between solar and meteoritic abundances is with CI chondrites, as can be seen both from the abundances of the refractory elements (Ca, Al, Sc, Yb, and V) and those of the moderately volatile elements (Mn, Na, Zn, and Se). Enstatite chondrites, although high in volatile elements (especially the EH group), are depleted of refractory elements. The apparent disagreement between the solar abundances and the CI abundances (and those for

Mg silicates to form FeO-containing silicates. According to calculations by Saxena and Ericsson (1983), the fayalite content of olivine will reach 20 mol% only at 425 K, given equilibrium with gas and solid (incidentally, this suggests that the FeO content of chondrites is a result of the reheating that caused the chondrule formation). Calculated condensation temperatures for some elements are given in Table 1.5.

This seemingly quantitative volatility scale based on calculated condensation temperatures must be used with great care. The pitfalls are of two sorts: Firstly, the calculations undoubtedly contain inaccuracies and errors, because of inaccuracies and errors in the available thermochemical data. Secondly, volatility-related processes may not have occurred under the strict conditions of thermodynamic equilibrium with gas of canonical solar composition, which the calculation assumes.

The likelihood of errors in the available thermochemical data can be appreciated by considering the nature of the equilibrium between gas and solid. For each element, one needs to know (1) the identities of all the important gas species in the solar nebula containing not only the element of interest but also the other elements that occur in these species (as a function of temperature and pressure), (2) thermochemical data for these species, (3) into which phases the element condenses and, for trace elements, the nature of the substitution, and (4) the chemical potentials of the appropriate components in the condensed phases (e.g., for trace elements, standard-state thermochemical data plus activity coefficients at infinite dilution). Much of this information, particularly for items 3 and 4, is simply unavailable. In addition, the calculation depends not only on the accuracy with which the abundance of the element in question is known but also on the abundances of all the other elements that occur both in its main nebular gaseous species and in the phases into which it condenses. Prudence enjoins that a certain amount of caution be attached to the results of condensation calculations.

The other complication is that volatility-related fractionations may not have occurred under the simplified conditions that are assumed for the canonical condensation calculation. For many lithophile trace elements that condense below $\sim$1,000 K, diffusion rates into the pre-existing silicates usually must have been too slow for equilibrium to have been maintained, and a fractional-condensation process may better describe the physical reality. Volatility strongly depends on the oxygen fugacity, and for most elements there are large differences in volatilities between reducing and oxidizing conditions (Wulf, Palme, and Jochum, 1995). This is important because there is abundant evidence for volatility-related fractionation processes at oxygen fugacities higher than that defined by the canonical solar H_2/H_2O ratio. Higher oxygen fugacities could have been produced during reheating of nebular

Figure 1.5 (*cont.*). other chondrites) for Yb and Mn is thought to be due to inaccurate determinations of solar abundances (see Figure 1.4). Na and Mn are depleted to the same extent in all chondrite groups. The Na and Mn abundances are low in carbonaceous chondrites, but relatively high in ordinary chondrites. Zn shows an opposite tendency, being lower in ordinary chondrites than in carbonaceous chondrites. Se (hence also S, with which it correlates in meteorites) is conspicuously high in enstatite chondrites.

Table 1.5. *Condensation temperatures for some elements (50%*
condensation temperatures; pressure 10^{-4} *bar)*

Element	Condensation temperature (K)	Major host phase
Refractory components		
Re, Os, W	1,800	Refractory metal alloy
Zr	1,750	ZrO_2
Al	1,680	Corundum
Ti	1,590	Perovskite
Lu, Th	1,590	Perovskite
Ca	1,520	Melilite
Ir, Ru, Mo	1,600	Refractory metal alloy
Yb	1,420	Perovskite
Major components		
Mg	1,340	Forsterite, enstatite
Si	1,311	Forsterite, enstatite
Ni	1,354	FeNi metal
Fe	1,336	FeNi metal
Pd	1,334	FeNi metal
Cr[a]	~1,300	Forsterite, enstatite, FeNi metal
Moderately volatile elements		
P	1,267	Fe_3P
Li	1,225	Enstatite
Au	1,225	FeNi metal
Mn[b]	1,190	Forsterite
Na[b]	970	Anorthite
K[b]	1000	Anorthite
Ga	997	FeNi metal
B[c]	964	Anorthite
Se	684	FeS
S	648	FeS
Highly volatile elements		
Pb	496[d]	FeS
Bi	451[d]	FeS
Tl	428[d]	FeS

[a]Calculated from data in Li et al. (1995).
[b]Empirical observations of abundances in meteorites suggest that the order of condensation temperatures should be Mn = Na > K.
[c]Lauretta and Lodders (1997).
[d]Condensation temperature at 10^{-5} bar.
Source: Adapted from Wasson (1985).

material if the material did not maintain equilibrium with the H-rich nebular gas, either because that gas had dispersed or because of other physical or kinetic impediments.

Condensation temperatures for most elements depend on the total pressure of the system, which may have varied both spatially and temporally in the nebula. A few elements, such as S, have pressure-independent condensation temperatures. The ratios of S to other moderately volatile trace elements, such as Zn, are thus pressure-dependent.

Despite these uncertainties, most calculated condensation temperatures probably are sufficiently accurate for qualitative comparisons, as shown by those major fractionation trends in meteorites that clearly correlate with calculated condensation temperatures. On this basis, several groups of cosmochemical elements can be defined:

Refractory elements have condensation temperatures above those of the major phases in meteorites (which are the Mg silicates and FeNi metal). The principal constituents of the refractory component are Al, Ca, and Ti oxides and silicates, such as perovskite ($CaTiO_3$) and gehlenite ($Ca_2Al_2SiO_7$). Large numbers of trace elements are expected to condense together with the refractory phases. The most refractory elements (Hf, Zr, Sc, and Y), however, condense as separate phases ahead of Ca and Al oxides. Condensation temperatures for the rare-earth elements (REEs), Nb, Ta, U, Th, Sr, Ba, and V, are calculated by assuming a solid solution with perovskite and melilite. Refractory metals, such as Re, Os, Ir, Ru, Pt, Rh, Mo, and W, will condense as alloys, as reviewed by Palme and Boynton (1993). Fractionations among refractory elements are observed in certain inclusions in carbonaceous chondrites. Some meteorites contain 'ultra-refractory' inclusions, with high concentrations of only the most refractory elements (Zr, Hf, Lu, etc.) (e.g., Palme et al., 1982). More commonly, refractory inclusions in carbonaceous chondrites have patterns that show strong excesses of refractory elements condensed from a fractionated gas, which itself had suffered an earlier loss of an ultra-refractory component (Boynton, 1975).

Apparently those high-temperature fractionations were local phenomena in the solar nebula. They occurred only on the scale of the inclusions and are not observed in the bulk meteorites. All chondritic meteorites have identical refractory-element ratios (e.g., Wasson and Kallemeyn, 1988). From the perspective of this study, this observation is the most important finding to come from the study of chondrite chemistry. The rule of constant ratios of refractory lithophile elements (RLEs) can be shown to hold true to better than ±5%; several apparent anomalies discussed in the past have now been shown to have been due to analytical error. It is therefore a central postulate in reconstructing planetary bulk compositions that refractory elements always occur in the same relative abundances as in meteorites. Thus the abundances of Al, Ca, and Ti are linked to the abundances of REEs, Sc, Hf, U,

Th, and so forth. The refractory component makes up about 5% of the condensable matter of the solar nebula.

The *common elements* Mg, Si, Cr, Fe, Ni, and Co have similar condensation temperatures, although they condense in different phases: Mg and Si form forsterite (Mg_2SiO_4) and enstatite ($MgSiO_3$), whereas Fe, Ni, and Co condense as metal alloys. Mg silicates and FeNi metal together are volumetrically the most important components, accounting for at least 90% of chondritic meteorites. It is cosmochemically important that forsterite and metallic iron have very similar condensation temperatures. This is probably the reason for the limited fractionation of metal from silicate in objects formed in the solar nebula, as discussed later. Condensation of Mg silicates and metallic iron changed the opacity and therefore the heat balance of the nebular disk considerably. More radiation was absorbed and converted to heat, leading to evaporation. This thermostat may have been responsible for a delay in cooling, thus allowing fractionation of a refractory component and perhaps also such limited olivine–metal separation as is observed (Tscharnuter and Boss, 1993).

Cr was previously considered to condense into Fe-Ni metal (e.g., Wasson, 1985). However, calculations using newly available thermodynamic data (Li, O'Neill, and Seifert, 1995) predict that it should mostly condense into olivine and pyroxene (as $Cr_2^{2+}SiO_4$ and $Cr^{2+}SiO_3$ components). This lithophile behaviour under solar-nebula conditions resulted in a restricted range of Cr/Mg ratios in all chondrites. The relatively low abundance of Cr in the Fe-metal-rich CH group demonstrates empirically that Cr does not correlate with Fe and was indeed mainly lithophile in nebula processes.

Moderately volatile elements are elements with condensation temperatures lower than those of the Mg silicates and Fe-Ni alloy, down to and including that of FeS (troilite). This component contains a variety of elements of different geochemical character: lithophile elements (Li, Mn, P, Rb, K, Na, F, and Zn), siderophile elements (Au, As, Ag, Ga, Ge, Sb, Sn, Te), and chalcophile elements (S, Se, and Cu).

Highly volatile elements comprise an even smaller component of meteorites. These elements condense after S, and the include Cl, Br, I, Cs, In, Tl, Bi, and Pb. These elements are so volatile that their abundances in ordinary chondrites depend on the degree of subsequent metamorphism in their parent bodies. Thus, type-3 ordinary chondrites (little or no metamorphic reheating) have higher concentrations than do type-6 ordinary chondrites (extensive reheating that resulted in complete metamorphic recrystallization).

Ice-forming elements condense at temperatures below 300 K. These elements are H, C, N, and the rare gases. Only tiny fractions of the solar abundances of these elements are contained in even the most volatile-rich meteorites (CI), and therefore meteorites cannot provide a useful guide to their detailed cosmochemical behaviour.

Oxygen has a unique status in the volatility scheme. Some O is among the first material to condense with the refractory elements (e.g., the O in Al_2O_3), and more condenses with Mg and Si. For most purposes, this O can be considered to remain bound to the lithophile element with which it condenses. Next, more O condenses by oxidation of Fe metal, forming the FeO component of silicates. This oxidation begins near the condensation temperature of S (depending on nebular pressure). Further oxidation, with falling temperature, yields Fe_3O_4 (magnetite). However, the amount of O condensing with either the lithophile elements or Fe plus related metals is only ~15% of the total amount in the nebula. The remaining 85% is predicted to condense as H_2O in the ice-forming regime. The amount of O that condenses by oxidation of Fe and in the ices represents the most important chemical variable in planet-building, because it will determine the overall oxidation state of the planet and the amount of unoxidized metal that is available to form the core.

1.3. Cosmochemical Components and the Composition of Chondritic Meteorites
1.3.1. Refractory Elements

A good example of the different ways in which elements may sometimes behave cosmochemically, as compared with their more familiar geochemical behaviour, is provided by the REEs. The compatibility of a REE (i.e., its mineral/melt partition coefficient) is a smooth function of ionic radius and thus of atomic number. This leads to the familiar smooth REE fractionation patterns when the REE abundances are normalized to chondritic values. The cosmochemical volatility of the REEs, however, is not a smooth function of ionic radius. For example, the volatilities of the two heaviest REEs, Yb and Lu, are greatly different, with Yb being one of the most volatile of the REEs, and Lu the most refractory. The difference in volatilities is related to the different gaseous species that are stable under the reducing conditions of the solar nebula: The principal gaseous species for Yb is the monatomic gas Yb(g), and those for Lu are the oxide species LuO(g) and LuO_2(g). Fractionations of refractory elements based on differences in volatilities are observed in Ca- and Al-rich inclusions in carbonaceous chondrites, as reviewed by Palme and Boynton (1993). Under oxidizing conditions, Yb also forms gaseous oxide species, and the difference in volatility largely disappears (Boynton, 1975).

The distinctive pattern of volatility-related fractionations in the REEs and other RLEs is not observed in bulk chondrite meteorites. The relative abundances of refractory elements are, as pointed out previously, the same in all chondrite groups, with constant ratios of Ca/Al, Ti/Al, Ti/Sc, and Ti/REE, and inter-element REE ratios (such as Yb/Lu, etc.) within analytical uncertainty. This is strictly true only for lithophile refractory elements and siderophile refractory elements separately. The ratio of these two refractory components (e.g., as expressed by Ir/Al) is not constant. For example, the Al/Sc ratio in CI chondrites is 1,466, and in H chondrites 1,446 (i.e., the same within analytical uncertainty), whereas the Al/Ir ratio in H chondrites, at 1.44×10^5, is significantly below the CI ratio of 1.88×10^5 (CI-chondrite data from Palme and Beer, 1993; chondrite data from Kallemeyn et al., 1986). This divergence is due to another fractionation process that separates metal from silicate, as discussed later.

Possible exceptions to the rule of constant RLE ratios are the EL enstatite chondrites. It has been suggested that this group may show small deficiencies in some RLE ratios, such as Ca/Al and La/Sm, as compared with the CI ratios (Kallemeyn and Wasson, 1986). This, however, is not supported by a more recent study of EL chondrites from Antarctica (Zhang, Benoit, and Sears, 1995). In any case Ca and La (along with some other usually lithophile elements) exhibit partly chalcophile behaviour in these meteorites, which equilibrated under extreme (for meteorites) conditions of low oxygen fugacity (fO_2) and high sulphur fugacity (fS_2). Thus, even if future work substantiates the anomalous RLE ratios in EL chondrites, the case could be made that the anomalies are due to the non-lithophile behaviour of the affected elements under exceptional circumstances and therefore provide another example of an exception that proves the rule.

1.3.2. Mg and Si

Although some Mg and Si condense with the refractory elements in phases such as melilite, the canonical nebular calculation shows that $\sim$95% of these elements condense as olivine (Mg_2SiO_4) and pyroxene ($MgSiO_3$) at somewhat lower temperatures. Mg and Si can thus be considered to be slightly volatile, relative to the refractories. Chondritic meteorites show differences within a factor of 2 in their abundances of these elements (i.e., Al/Si in CV chondrites is twice as high as in enstatite chondrites). When compared with the solar or CI composition, carbonaceous chondrites are usually enriched in the refractory component relative to Mg and Si, whereas the ordinary chondrites (and related groups) and enstatite chondrites are depleted. Because it is customary for cosmochemists to normalize element abundances relative to either Mg or Si (to allow for different volatile-element abundances or terrestrial alteration, for example), it is usual to express the depletion or enrichment of Mg and Si relative to the refractory elements in the inverse manner, that is, as an enrichment or depletion of the refractory component relative to Mg and Si. This does not, however, carry any implication as to which component was added or subtracted.

Mg and Si are not completely correlated in meteorites, and this has important implications for evaluating the significance of the composition of the Earth. The carbonaceous chondrites show a well-defined trend of essentially constant Mg/Si with increasing Al/Si (Figure 1.6). Larimer and Wasson (1988) believe that this trend is due to addition of a high-temperature refractory component, specifically CAIs (Ca- and Al-rich inclusions), which are indeed more abundant in the high-Al/Si types (i.e., CV and CK). CAIs are rather varied in composition, but have a mean containing about 10% Mg, 15% Si, and 18% Al by weight. Consequently, to

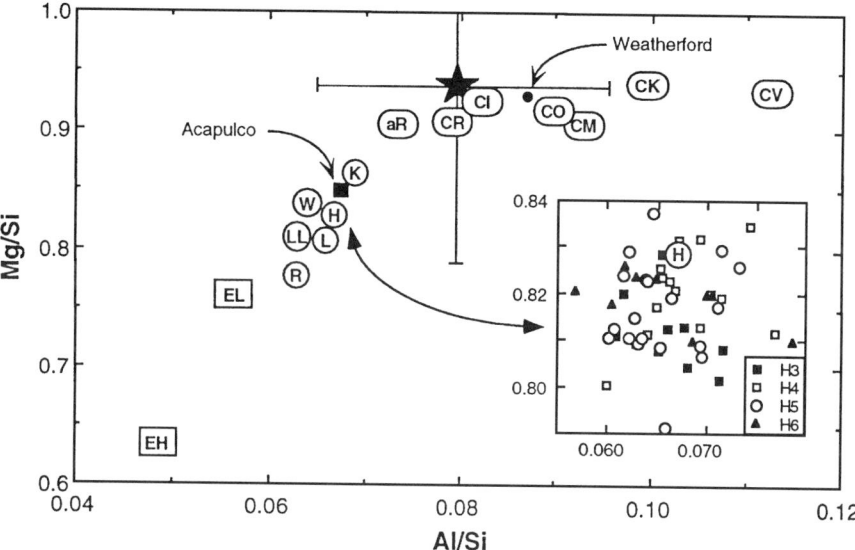

Figure 1.6. Mg/Si vs. Al/Si ratios in chondrites. Al is chosen as representative of the refractory lithophile elements (RLEs). Abbreviations as in Table 1.1 and as follows: W, Winona; K, Kakangari (a unique type-3 chondrite); aR, al-Rais (anomalous CR). Data mainly from Wasson and Kallemeyn (1988) and Jarosewich (1990), supplemented by data from Graham, Easton, and Hutchison (1977), Weisberg et al. (1993), Kallemeyn et al. (1994), and Palme, Weckwerth, and Wolf (1996). The solar-photosphere value (star) with error bars is from Anders and Grevesse (1989). This value is in good agreement with the CI value, but is not as accurately known. To illustrate within-group scatter, analyses of individual meteorites of the H group OCs from Jarosewich (1990) have been plotted in the inset. At least two trends are evident: (1) the carbonaceous-chondrite trend of increasing Al/Si at nearly constant Mg/Si, and (2) a trend originating from the CI and CR chondrites, running through the field of ordinary chondrites and rumurutiites to the two enstatite-chondrite groups. An alternative interpretation is that the ordinary chondrites form a single narrowly defined field in Mg-Si-Al space that is not genetically related to the ECs, and thus the trend is but coincidence.

derive the CV Al/Si ratio from the CI composition would require only the addition of 2% CAI material. A possible argument against this is that CAIs (or calculated compositions of high-temperature nebular condensates) have low Mg/Si ratios, such that a 2% addition of CAI material to the CI composition should slightly lower the Mg/Si ratio, whereas, if anything, the mean composition of CV chondrites appears to have a slightly elevated Mg/Si ratio (Figure 1.6).

The ordinary chondrites and enstatite chondrites present even more interesting problems. The trend is less well defined than that for the carbonaceous chondrites and is largely delineated by the two enstatite-chondrite groups (EH and EL), although the new data for the R group (rumurutiites) do tend to support the existence of the trend. An alternative reading of the data is that ordinary chondrites (plus rumurutiites and winonaites) form a single field in Mg/Si-versus-Al/Si space, implying a process that results in a quantized reduction of the Mg/Si ratio, with the Mg-Si-RLE pattern in the enstatite chondrites being produced by another process.

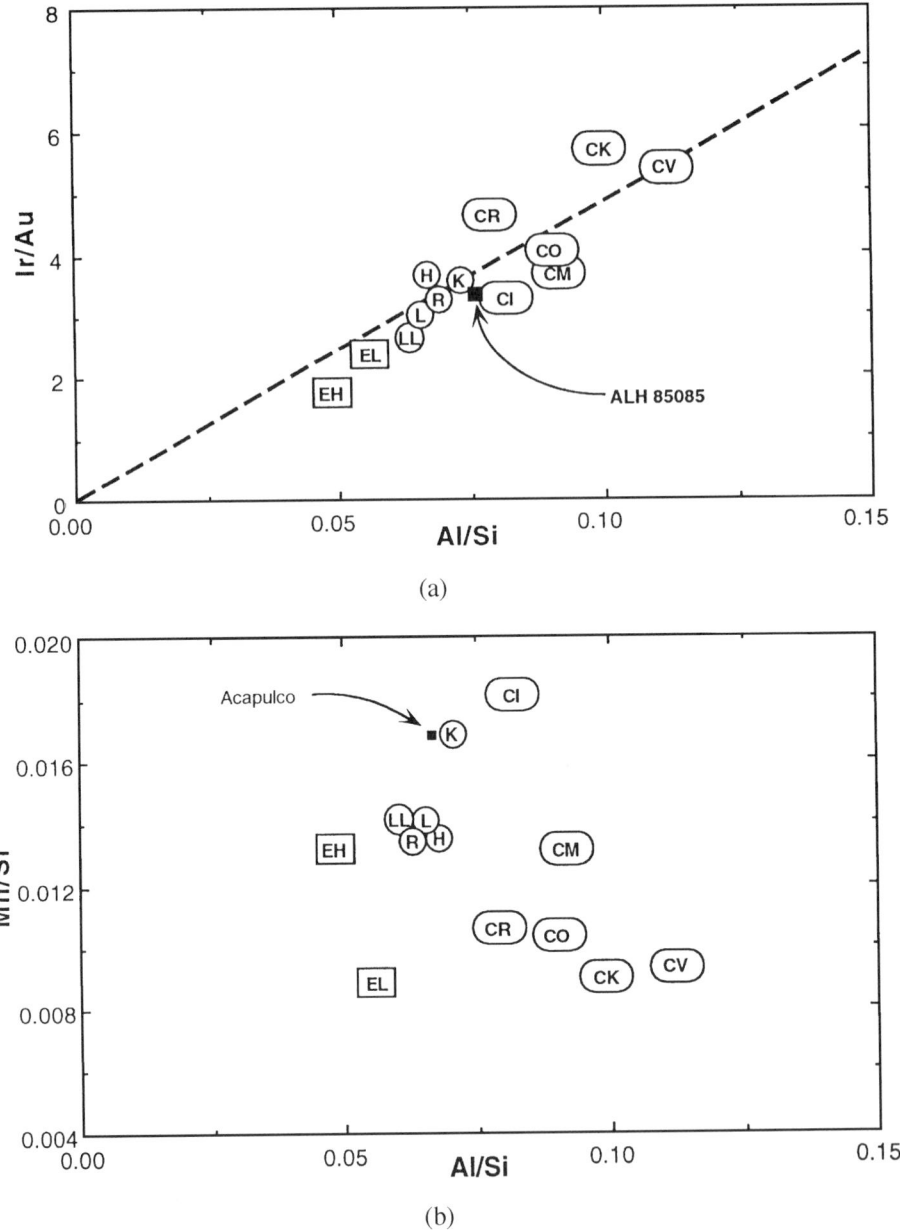

Figure 1.7. The plot of Ir/Au versus Al/Si in chondrites (a) shows good correlation, with Ir/Au = 48 (±2) Al/Si; $R^2 = 0.73$ (regression constrained to pass through the origin). Some of the scatter may be due to the observed scatter in the analytical data for Au within chondrite classes. Ir and Al are refractory elements, whereas Au and Si are both cosmochemically slightly volatile, with similar calculated condensation temperatures under solar-nebula conditions (Table 1.5) (Wasson, 1985). Neither the Ir-Al pair nor the Au-Si pair share any other properties, and they are held in different phases and even components in chondrites. Ir and Au are highly siderophile elements, present in trace amounts, whereas Al and Si are lithophile and are major elements. The correlation extends over both enrichments and depletions of the slightly volatile elements (Au and Si) with respect to their refractory partners (Ir and Al). This correlation shows that the process that caused both the depletion and enrichment of Si with respect RLEs in chondrites affected Au similarly and thus was

In any event, if an ordinary-chondrite–enstatite-chondrite trend exists, it is different from that exhibited by the carbonaceous chondrites and must have resulted from a different cause. Larimer and Wasson (1988) suggest that the trend is best explained by *subtraction* of material containing two components: a refractory CAI-like component (as for the carbonaceous-chondrite trend), and olivine (Mg_2SiO_4). A specific example of such nebular material would be amoeboid olivine inclusions or aggregates (AOIs or AOAs), which occur primarily in carbonaceous chondrites, but have also been described from Semarkona, an extremely primitive LL3.0 ordinary chondrite (Grossman and Wasson, 1983). The whole question of how and why this happens, together with the puzzle of why the two components should remain correlated at the same ratio in all groups on this trend (not to mention what happens to the large amount of subtracted material), poses one of the most important unsolved problems in our understanding of the formation of the inner solar system.

That the variation in the RLE/Si or RLE/Mg ratio in chondrites is primarily due to the volatility of Si and Mg, and not to some silicate fractionation process, is proved by the remarkable correlation of Ir/Au with Al/Si (Figure 1.7a). Nevertheless, the variation in RLE/Si or RLE/Mg is surprisingly poorly correlated with the depletion in moderately volatile elements (Figure 1.7b), even if attention is confined to only the carbonaceous chondrites. It seems that there were different volatility-related processes in the solar nebula.

1.3.3. Oxygen Content and Oxidation State

The oxidation state of equilibrated meteorites is reflected in the average Mg# [molar Mg/(Mg + Fe)] of olivine (or pyroxene in olivine-free types). It is, however, important to realize that unmetamorphosed chondritic meteorites are not equilibrium assemblages, but comprise mixtures of oxidized and reduced materials, and olivine of variable Mg#. Thus they do not conform to a well-defined oxidation state (this has not prevented attempts to measure an obscure quantity known as the 'intrinsic oxygen fugacity' in such meteorites). Nevertheless, a putative oxidation state can still be defined for such meteorites based on their net oxygen content. Thus, an unequilibrated meteorite that contained sufficient water to oxidize all metallic Fe to FeO, should it ever be equilibrated, would be thought of as being 'oxidized'.

There is a significant variation in the degree of oxidation in chondrite meteorites that is not in any simple way related to their other properties. The exception is the inverse correlation between Fe (and other siderophile elements) and oxidation state in the three 'classic' ordinary-chondrite groups (H, L, and LL). Originally, this

Figure 1.7 (*cont.*). entirely related to volatility. The fractionation is independent of metal/non-metal ratios, as especially well illustrated by ALH 85085, a CH chondrite (Al/Si in ALH 85085 estimated from data of Scott, 1988). (b) There is, however, almost no correlation between the Si enrichment/depletion and the depletion of the moderately volatile elements, here represented by Mn. This is true even if a correlation is looked for within the carbonaceous-chondrite clan.

trend was illustrated by plotting reduced Fe (i.e., metallic Fe and Fe in sulphide) versus oxidized Fe in silicate (Urey and Craig, 1953, fig. 1); however, it is better illustrated by the negative correlation seen when Ni/Si or Ir/Si (representing total metal content) is plotted against the fayalite content of the olivine (in equilibrated types), as, for example, illustrated by Wasson (1972, fig. 1.8a,b). The significance of this trend is unclear; the new group of R chondrites, which are similar to ordinary chondrites in many other respects (Palme et al., 1996), do not follow this trend. They have high siderophile contents (similar to the H group), but also high oxidation states (an even higher fayalite component in olivine than the LL group).

Carbonaceous chondrites, in particular the CIs, are generally oxidized and contain water and magnetite (Fe_3O_4, with substantial Fe^{3+}). It is important to note that the presence of Fe-rich metal is not thermodynamically compatible with the presence of water even at low temperatures in small planetary bodies. Standard free-energy data show that the equilibrium constant for the reaction

$$\underset{\text{metal}}{Fe} + \underset{\text{water}}{4H_2O} = \underset{\text{magnetite}}{Fe_3O_4} + \underset{\text{gas}}{4H_2} \tag{1}$$

requires $P_{H_2} = 635$ bar (in SI, 63.5 MPa) at 300 K. Thus the chondrites that contain metallic Fe (ordinary chondrites and enstatite chondrites) usually are almost entirely water-free. In the carbonaceous chondrites (the overwhelming majority of which are unequilibrated) there is a negative correlation between Fe-rich metal and hydrous minerals such as the matrix phyllosilicates. Metal is absent from the most water-rich carbonaceous chondrites (CI and CM), and phyllosilicates are rare in the more metal-rich CV types. The exception to this general trend is the highly unequilibrated CR group, whose members contain both abundant metal and hydrous phases, but even in these cases metal is sparse in the parts of the material that have suffered aqueous alteration (Kallemeyn, Rubin, and Wasson, 1994). Sulphide is, however, present in intimate association with hydrous phases in the carbonaceous-chondrite matrix. The anti-correlation between water and metallic Fe is an important constraint on Earth-forming models, as it seems unlikely that the same material could simultaneously have delivered Fe metal and water to the growing Earth (especially if that material was already formed into thermally metamorphosed planetesimals).

The most reduced meteorites are the enstatite chondrites, which contain remarkably little oxidized iron (less than 0.1% FeO in pyroxenes in equilibrated types) (Wasson et al., 1994) and contain metal with up to several percent Si. Many of the normally lithophile elements in these meteorites are present largely as reduced species and thus behave at least partly as chalcophile elements (e.g., Ti and the REEs).

1.3.4. Nebular Metal–Silicate Fractionation

There are substantial variations in the Fe/Mg ratios in undifferentiated meteorites (Figure 1.8). Not only do L and LL chondrites record a significant depletion of

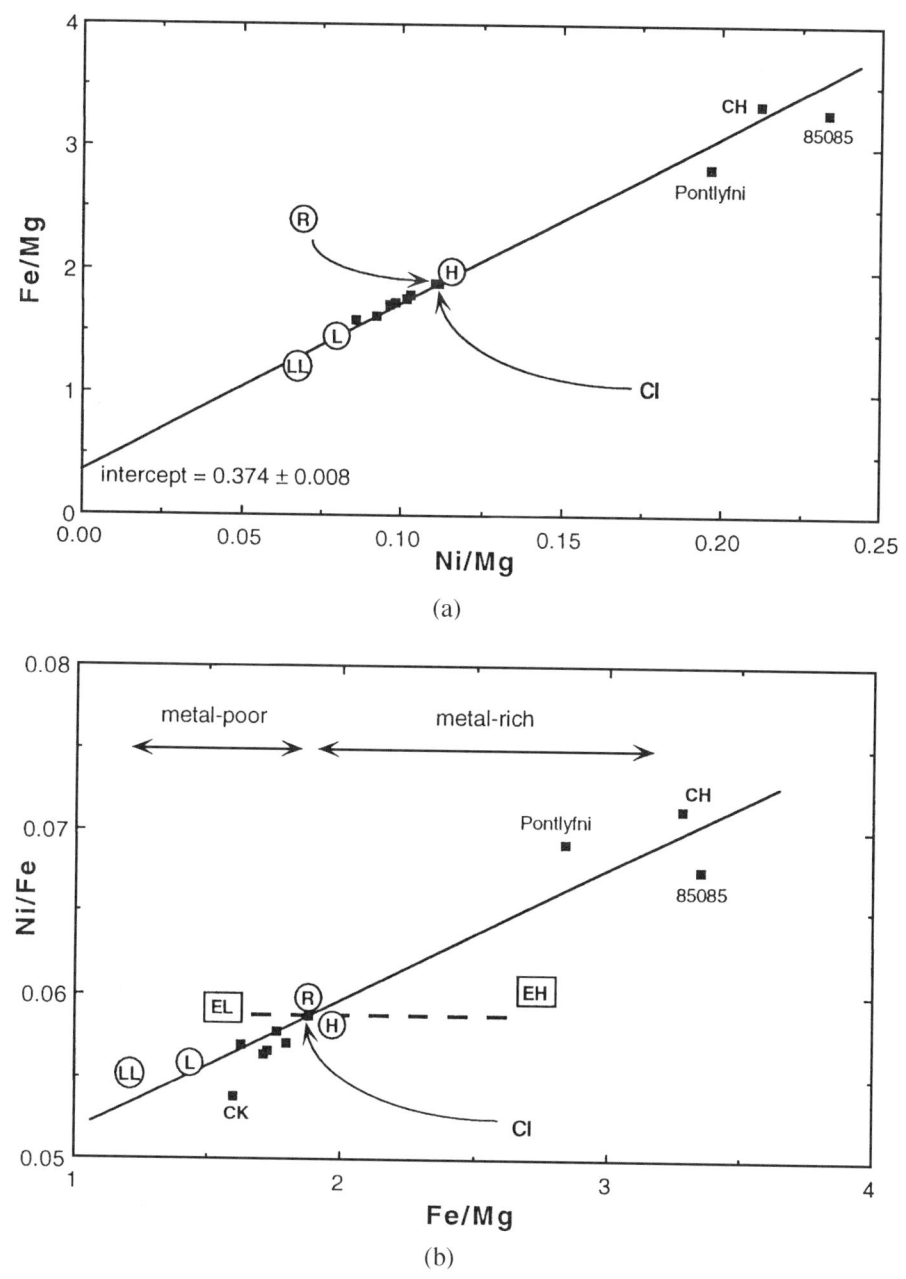

Figure 1.8. Metal/silicate fractionation in chondrites. Compared with the CI composition, chondrites can be metal-rich (the new CH group of carbonaceous chondrites, the H group of ordinary chondrites, and Pontlyfni, a unique Fe-rich "winonaite") as well as metal-poor (all the rest). CH refers to the mean of Acfer 182, 207, and 214, from Bischoff et al. (1993a); AH 85085 is plotted separately. (a) Fe/Mg vs. Ni/Mg, mean values for groups apart from the enstatite chondrites (not plotted). Fe refers to total Fe. The value of the intercept can be interpreted as the Fe^{2+}/Mg ratio of a parental silicate component at the time of metal–silicate fractionation, and it corresponds to an Mg number of 0.860 ± 0.003. (b) Fractionation of silicate containing Fe^{2+}/Mg = 0.374 from metal changes the bulk Ni/Fe in the chondrite. Unlike all other groups, the enstatite-chondrite groups (EH and EL) are related to the CI composition by fractionation at zero Fe^{2+}/Mg. Note that all chondrites can be related to the CI composition, confirming further that CI chondrites do indeed record the solar-nebula abundances.

Fe relative to the CI ratio, but so do CV and, to a lesser extent, CO chondrites. In a recent study, Wolf et al. (1996) found Fe/Mg ratios as low as 1.15 in bulk samples of Efremovka, a CV meteorite (the CI ratio is 1.90). With the exception of the H-group ordinary chondrites and the EH enstatite chondrites, the 'classic' ordinary- and carbonaceous-chondrite groups are deficient in Fe/Mg relative to the CS. However, a recently discovered group of meteorites, the CH carbonaceous chondrites, have Fe/Mg ratios much greater than those of CIs, as do a number of other 'anomalous' or ungrouped meteorites or parts of meteorites with chondritic affinities. The CH chondrites have Fe/Mg ratios as high as 3.3 (Bischoff et al., 1993b).

Other siderophile elements vary in parallel with Fe in these meteorites, indicating a metal/silicate fractionation process. A remarkable observation made by Larimer and Anders (1970) is that all the ordinary- and carbonaceous-chondrite groups follow a single trend in the Fe/Mg-versus-Ni/Mg diagram, within compositional uncertainty. Assuming an uncertainty in Fe/Mg of 1%, and in Ni/Mg of 4%, all the data in Figure 1.8a (10 chondrite groups, plus AH 85085, Pontlyfni, and the 'CH' value) can be fitted to a straight line:

$$\frac{Fe}{Mg} = \left(\frac{Fe}{Mg}\right)_0 + \frac{(Fe/Mg)_{CI} - (Fe/Mg)_0}{(Ni/Mg)_{CI}} \frac{Ni}{Mg} \qquad (2)$$

where $(Fe/Mg)_0$ is the intercept at $Ni/Mg = 0$ and is the only parameter to be determined. This equation constrains the data to pass through the CI composition exactly. The regression gives a value for $(Fe/Mg)_0$ of 0.374 ± 0.008, with a χ^2 of 1.30. The low value for χ^2 indicates that all data are well fitted by this one-parameter equation, within the assumed uncertainties. If we assume that all Ni was in the metal phase at the time of metal fractionation, then this intercept gives the Fe^{2+}/Mg ratio in the silicate at that time. The observed ratio at the intercept corresponds to a silicate molar $100 \, Mg/(Mg + Fe^{2+})$ (i.e., Mg#) of 86.0 ± 0.3. [Note that in these chondrites, $(Fe^{2+}/Mg)_{olivine} \cong (Fe^{2+}/Mg)_{bulk}$, because olivine is the dominant ferromagnesian phase, and Fe^{2+} and Mg are fairly evenly partitioned between olivine and the other ferromagnesian phases such as low-Ca pyroxenes.] As a result of this process, the Ni/Fe ratio of chondrites increases with increasing Fe/Mg (Figure 1.8b). The slope of the trend line gives the Fe/Ni ratio of the metal at the time of extraction (Fe/Ni = 13.6).

The discovery that the new metal-rich CH meteorites, as well as anomalous (un-grouped) Fe-rich chondrites, fall on the same trend line lends this correlation extra significance. The trend implies that there was an early metal/silicate fractionation that occurred when all chondrites (apart from the enstatite chondrites) had the same bulk Fe^{2+}/Mg ratio (i.e., oxidation state). This ratio differs from the various ratios found in the meteorites at present. For example, the equilibrated LL chondrites now have olivine Mg#'s of 70 ± 3. These LL chondrites cannot have been derived from material with the CI bulk Fe/Mg ratio by metal/silicate fractionation involving silicate with such a low Mg#. Conversely, winonaites such as Pontlyfni have olivine with a very high Mg# (Pontlyfni has Mg# 99), but they cannot have been derived from CI (solar) material by metal/silicate fractionation with such a high Mg#. The inference is that the currently observed oxidation states of the ordinary chondrites (and perhaps also the carbonaceous chondrites) were acquired by further condensation of O some time after the metal/silicate fractionation. If true, this would seem to imply that the negative correlation between oxidation state and siderophile-element content (i.e., the Urey-Craig relationship discussed earlier), which is observed in the three 'classic' ordinary-chondrite groups (H, L, and LL), is a coincidence and is of no cosmochemical significance.

The two enstatite-chondrite groups (EH and EL) can be connected through the CI composition by a different trend that extrapolates to Fe/Mg = 0.0 at Ni/Mg = 0 (Larimer and Anders, 1970) (Figure 1.8b), implying metal/silicate fractionation under very reducing conditions, with essentially no Fe^{2+} in silicate. This is in line with their current highly reduced oxidation states. Thus the reduced nature of the enstatite-chondrite groups is a primary feature, whereas the reduction of winonaites such as Pontlyfni must have been due to a fundamentally different, secondary process.

The nebular metal/silicate fractionation process as evinced by the variable Fe/Mg ratios does not correlate with the abundances of S or chalcophile elements in the ordinary chondrites, indicating that this latter component condensed afterwards (Larimer and Anders, 1970). An observation of some significance for attempting to constrain the bulk Earth composition or other planetary compositions is that the metal/silicate fractionation also is not correlated with the Mg-Si-RLE fractionations, but appears to have operated as an independent process subsequent to those fractionations. There is thus no useful constraint on the amount of metal in the Earth's core that can be deduced from cosmochemical principles.

1.3.5. Volatile-Element Fractionations

Both the moderately volatile elements and the highly volatile elements are depleted, relative to CI chondrites, in every other meteorite [i.e, in all the other chondrite groups, as well as in meteorites from differentiated parent bodies such as the HED group (eucrites, etc.), Martian meteorites, and every group of iron meteorites]. There are no known examples of solar-system material with concentrations of volatile elements higher than the CI concentrations. This indicates that the depletion was not a simple local redistribution (i.e., vaporization–recondensation) and also was fundamentally different from the siderophile-element fractionation discussed in the preceding section, because the latter resulted in both metal-rich and metal-poor materials. Low volatile-element abundances are characteristic of the inner solar system, and the Earth should be no exception (e.g., Palme, Larimer, and Lipschutz, 1988).

Broadly speaking, the volatile elements show good correlations of increasing depletion with decreasing condensation temperatures in all chondrites. This is

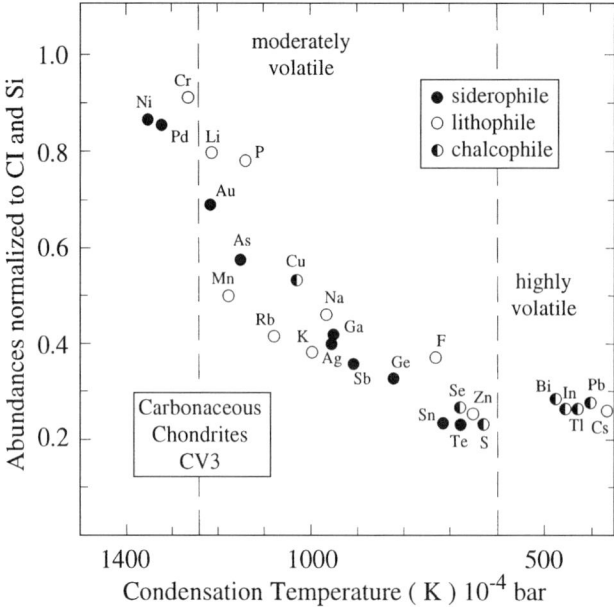

Figure 1.9. The abundances of moderately volatile elements in bulk CV chondrites decrease more or less continuously with decreasing condensation temperature, independent of the geochemical character of the elements. This strongly suggests that cosmochemical volatility alone was responsible for the depletion sequence. Such patterns may reflect progressive loss of volatiles during nebular cooling and, for the CV pattern shown here, can be quantitatively modelled (Cassen, 1996). For sources of data, see Palme et al. (1988).

illustrated for the CV carbonaceous-chondrite group in Figure 1.9. It is remarkable that the monotonic decrease in the CV/CI ratios is independent of the geochemical behaviour of the elements. Siderophile elements (Au, As, Ge, etc.), lithophile elements (Na, Mn, K, F, etc.), and the chalcophile elements Se and Zn all appear to follow a single sequence. In detail, however, the patterns of depletion are different in the different chondrite groups. This can be seen by examining the relative depletions of four representative moderately volatile elements, Mn, Na, Zn, and Se (which can be taken to stand in for S), in Figure 1.5. The abundances of the two elements with the lower condensation temperatures, Zn and Se, are more variable than those of Mn and Na, elements with higher condensation temperatures. In fact, the depletion of Na is similar to that of Mn in all chondrite groups, an observation that is of some significance for the history of the Earth, as will be emphasized later. Compared with the carbonaceous chondrites, ordinary and R chondrites show much lower depletions of Mn and Na, but greater depletions of Zn. Enstatite chondrites are different again. The obvious conclusion is that the volatile-element depletions occurred under different conditions and may have involved different processes in different parts of the nebula. The order in which the meteorite classes in Figure 1.5 are plotted corresponds roughly to their order of decreasing oxygen content, as measured by their O/Si ratios (i.e., carbonaceous chondrites < ordinary chondrites < enstatite chondrites). To some extent the depletion of Zn follows this trend, which accords with increasing Zn volatility with increasing degrees of reduction (Wulf et al., 1995). The EH enstatite chondrites appear to disrupt this trend, but EH chondrites have nearly the complete inventory of moderately volatile elements.

The kind of depletion pattern shown in Figure 1.9 evidently is not due to simple closed-system condensation/volatilization processes, which would result in much more severe fractionations among the elements (e.g., consider equilibrium condensation: virtually all Na and Mn should have condensed well before any Zn or Se condensed). The pattern is best explained by progressive removal of the nebular gas during cooling, so that the fractions of condensable elements gradually decreased with decreasing condensation temperatures (Wai and Wasson, 1977). Recently, calculations based on simplified nebular models have been able to reproduce quantitatively the depletion patterns for the CO and CV carbonaceous chondrites (Cassen, 1996). Insofar as there is as yet no quantitative explanation for the different patterns of depletion shown by the CM, ordinary, and enstatite chondrites, the reasons for these differences are not well understood.

1.3.6. Ice-forming Elements

The chondrites richest in H, C, and N are the CIs. However, even in these volatile-rich meteorites the amounts of these elements are (together with those of O and the noble gases) but small fractions of their solar-system abundances. The CI C/Mg ratio is 0.34, as compared with the solar ratio of 4.6; $(N/Mg)_{CI}$ is 0.03, versus the solar 1.6; and $(O/Mg)_{CI}$ is ~5, versus the solar 14.6 (Palme and Beer, 1993). The original abundance of H in CI chondrites is controversial, because some terrestrial water may have been absorbed into these samples; for example, according to Kaplan (1971), epsomite, $MgSO_4 \cdot 7H_2O$, an important constituent of CI chondrites, could originally have been present as $MgSO_4 \cdot H_2O$. Thus the H content could have been considerably less than the $\sim2.0\%$ currently observed, perhaps only 0.5% (Kaplan, 1971). A figure of 2.0% H would imply $(H/Mg)_{CI} = 0.2$, versus the solar 10^3. Clearly, only a paltry fraction of the solar H has condensed in CI chondrites. The extent to which the noble gases are depleted in chondrites is highly variable, but always extreme, relative to the solar composition. For all of these elements, there seems no good reason to attach any particular significance to the exact amounts which happen to be present and preserved in chondrites, beyond the observation that the order of depletion is $H \gg N > C > O$.

1.3.7. Oxygen Isotopes

The variations in the oxygen isotopic composition of solar-system materials (Clayton, 1993) provide a useful tool for examining the possibility that different types of meteorites may have come from a single parent body. Because O has three stable isotopes, it is possible to allow for the effects of mass fractionation. (The significance of the isotopic variability of C and N in solar-system materials is more difficult to evaluate, as they each have only two stable isotopes.) Mass fractionation should change O isotopes along a line with a constant slope of 0.47 in plots of $\delta^{17}O$ versus $\delta^{18}O$. Samples that do not plot on the same mass-fractionation line usually are assumed not to have come from the same parent body, although there are some exceptions: non-mass-dependent inhomogeneities in O isotopes are preserved in texturally primitive CV chondrites, such as Allende, and some parent bodies may be breccias that escaped homogenization (this may be the case for the ureilites) (Clayton and Mayeda, 1996). Generally, though, the simple negative inference appears reasonably robust. The logical inference which might be derived from the opposite kind of relationship, i.e., in which O isotopes from different meteorites do plot on a common mass-fractionation line, calls for more caution. A similarity in oxygen isotopes need not point to the same parent body; for example, it may imply only an origin in the same part of the solar nebula. Even that, however, is uncertain if the variation in oxygen isotopes among solar-system materials is due to physical processes rather than original inhomogeneity in the nebula.

This particular problem has a major consequence regarding the origin of the Earth. The Earth and Moon have identical O isotopes, which perhaps is not surprising, as many authorities believe that the Moon derived from the same region of the solar nebula as the Earth (i.e., at ~ 1 AU), if not from the Earth directly. More striking, however, is the observation that the Earth and the Moon also share an identity in oxygen isotopic composition with the two enstatite-chondrite groups (EH and EL) and the aubrites (achondrites, which have other chemical similarities to the enstatite chondrites, consistent with a close genetic association). Does this imply a genetic relationship among the Earth, the Moon, and the enstatite chondrites? In nearly all other regards, apart from the stable-isotope similarities, the enstatite chondrites are the furthest removed from either the Earth or the Moon of any chondrite group, as discussed later. Here we take the view that because the cause of the oxygen isotopic variations among solar-system materials is not understood, it is not appropriate to assign any genetic consequence to oxygen isotopic similarities. For a discussion of the ramifications of adopting the opposite view (i.e., that the Earth was made from enstatite chondrites), see Javoy (1995).

A basic conclusion from the study of the distribution of oxygen isotopes in solar-system materials is that none of the terrestrial planets could have been assembled by mixing together known varieties of meteorites (e.g., Clayton and Mayeda, 1996). Meteorites are not the 'building blocks of the planets'; rather, as we argue here, they record the types of chemical fractionation processes that operated in the early solar system, which we presume also to have affected the planetesimals, which were the true building blocks of the planets. The variations in the oxygen isotopic compositions of chondritic materials cannot be correlated with any of these chemical fractionation processes (e.g., volatile-element depletion, oxidation state, etc.). Rather, oxygen isotopic composition seems to be another independent variable superimposed on all the other compositional variables.

1.3.8. Cosmochemical Constraints on the Composition of the Bulk Earth: A Summary

From the foregoing discussion of the meteorite evidence we deduce that the following general principles constrain the bulk chemical composition of the Earth:

1. Except for the most volatile, 'ice-forming' elements (H, C, N, and O and the rare gases), the Earth has the same isotopic composition of the elements as the rest of the solar-system materials known to us, including the chondritic (i.e., undifferentiated) meteorites. Therefore, all of these materials share a common origin in a well-mixed solar nebula. It follows that the fundamental chemical fractionation processes discernible in the chondrites probably also affected the materials that formed the Earth.

2. The RLEs are not fractionated from each other in the bulk material of undifferentiated meteorites. Only differentiation within a planetary body (i.e., partial melting, etc.) appears to have fractionated these elements. The RLEs are therefore postulated to be present in the bulk Earth in strictly chondritic proportions, relative to each other. The main corroborating evidence for this is the smoothness of the chondrite-normalized REE patterns in terrestrial rocks (e.g., compared with the spiky patterns seen in some CAIs), which shows that the material that formed the Earth did not suffer any volatility-related fractionation at temperatures above the condensation temperatures of the major components (Fe-Ni metal and the Mg silicates). Identification of which elements compose the RLE group can be made empirically from the meteorite data, and it accords very well with that deduced from calculation of the condensation sequence of the elements from the solar nebula.

3. Mg and Si are not RLEs. There were at least two processes that fractionated Mg and Si from the RLEs in the chondrite meteorites: the one responsible for the carbonaceous-chondrite trend seen in the Mg/Si-versus-Al/Si diagram (Figure 1.6), and the other for the ordinary-chondrite–enstatite-chondrite trend (if it be a trend). Possibly, different processes may have been responsible for the ordinary-chondrite Mg-Si-RLE compositional field and the enstatite-chondrite compositions, in which case the minimum number of cosmochemical processes fractionating Mg and Si from the RLEs would be three. The important point is that there was no unique process responsible for setting Mg-Si-RLE abundances in solar-system material. Therefore, the Mg-Si-RLE ratios in the protoplanetary material from which the Earth accreted might have been affected by some combination of the carbonaceous-chondrite process and the ordinary-chondrite–enstatite-chondrite process(es). Consequently, we argue that the meteoritic record cannot offer a precise constraint on the bulk Earth's Mg/RLE and Si/RLE ratios.

4. The volatility of elements under cosmochemical conditions can be determined empirically from meteorite abundances. The moderately volatile and highly

volatile elements are depleted relative to Mg and Si in all chondrite groups, apart from the CIs, and these elements can be expected to be depleted in the bulk Earth.

5. Siderophile elements can be fractionated from RLEs, Mg, and Si by a process of metal/silicate fractionation that is not volatility-related. Both metal-rich and metal-poor (relative to CI) chondrites are known. Therefore, the bulk Mg/Fe ratio in the Earth cannot be exactly constrained a priori, and the amount of oxidized Fe in the silicate portion of the Earth cannot be constrained by the relative sizes of the core and the mantle.

6. The major fractionation processes identified earlier tend not to be correlated with each other, but seem largely to have acted independently. We should therefore not be surprised if the materials that formed the Earth, even though we expect them to have been subjected to the same fractionation processes, do not match exactly the composition of any particular chondrite in all respects.

7. The abundances of the ice-forming elements (H, C, N, and O) and the noble gases are poorly constrained in meteorites. Importantly, there appears to us to be no reason to connect current abundances in meteorites with those that might be expected in planet-forming material. A further complication is that free H_2 and He are light enough to be continually lost from the Earth's atmosphere to space (the mean residence time of ^{4}He in the atmosphere is only 10^6 years). Massive loss of H from Venus is indicated by the very high D/H ratio in the atmosphere of that planet, two orders or magnitude higher than the terrestrial ratio (Donahue and Hodges, 1992). The D/H ratio on Earth is similar to that in hydrated minerals in meteorites (Geiss and Reeves, 1981) and is within the range inferred for comets (Eberhardt et al., 1987). Taken at face value, this indicates that the Earth's present H inventory has not experienced fractionation by progressive loss of H of the kind occurring on Venus, although this does not exclude loss of H and other volatiles by some mass-independent process (e.g., catastrophic loss of an early atmosphere after a massive impact). The amount of H in planetary embryos would have been subject to a constraint that is potentially of great importance for Earth-building models, namely, that hydrous phases are not compatible with Fe-rich metal in small equilibrated bodies.

1.4. Composition of the Bulk Silicate Earth
1.4.1. Compositions of the Bulk Earth and Bulk Silicate Earth

The fundamental division in the Earth is between its outer silicate portion, consisting of the mantle plus the crust, which, following Hart and Zindler (1986), we call the Bulk Silicate Earth (BSE), and its Fe-rich metal core, which is 32% of the Earth by weight. We assume that the core is chemically isolated from the BSE and has been for most of the Earth's history. The justification for this view is the absence of any noticeable change in the siderophile-element contents of mantle-derived rocks with time, over the entire geological record (i.e., beginning nearly 4.0 billion years ago) [e.g., Ni and Co (Delano and Stone, 1985), Mo (Newsom et al., 1986), Sn (Jochum, Hofmann, and Seufert, 1993)]. This view is also consistent with the current understanding of the evolution of Pb and Os isotopes in the Earth (e.g., McCulloch and Bennett, Chapter 2, this volume).

The isolation of the core from the mantle means that the chemistry of the core is not amenable to direct investigation, except by imposing some broad-scale constraints from geophysical observations such as density and seismic velocities. Rather, the chemistry of the core must be inferred from a global mass balance between the composition of the whole Earth, deduced from cosmochemical principles, and the composition of the BSE. The principle is simple: If the abundance of an element in the bulk Earth can be deduced from cosmochemical constraints, then the difference between that deduced value and its abundance in the BSE will give the amount of the element in the core. It is therefore necessary first to determine the composition of the BSE.

Extensive lists of BSE abundances of the elements have been provided: Ringwood and Kesson (1977), Ringwood (1979), Jagoutz et al. (1979), Wänke, Dreibus, and Jagoutz (1984), Sun (1982), and McDonough and Sun (1995). The last is comprehensive, apart from H, O, and the noble gases. Hart and Zindler (1986) and Allègre et al. (1995) have considered the BSE major elements in detail. For most elements, agreement among all these estimates is good (Table 1.7). Our aim here is to review the methodology by which a BSE composition can be estimated. This is necessary for any appreciation of its accuracy, reliability, and robustness.

1.4.2. The Nature of the Problem: Differentiation of the BSE and the Major Geochemical Reservoirs

The BSE is composed of a number of geochemical reservoirs: the continental crust; the sub-continental lithospheric mantle; the oceanic crust; the sub-oceanic lithosphere; the depleted, well-stirred upper mantle, which is the source region for mid-ocean-ridge basalt (MORB); the enriched, heterogeneous, source regions for ocean-island basalt (OIB), which may include subducted oceanic crust; and, possibly, primitive mantle that has never been differentiated. For a few extremely incompatible elements, such as hydrogen, the noble gases, and Cl, Br, and I, the oceans and atmosphere also need to be included. The choice of what to define as a geochemical reservoir is somewhat arbitrary. For practical purposes, we choose to divide the BSE into reservoirs that, though they may not be chemically homogeneous (e.g., the continental crust), nevertheless can be treated as having each a reasonably well defined mean composition. The assumptions we make about the relationships among these reservoirs, insofar as they pertain to chemistry, are illustrated in Figure 1.10.

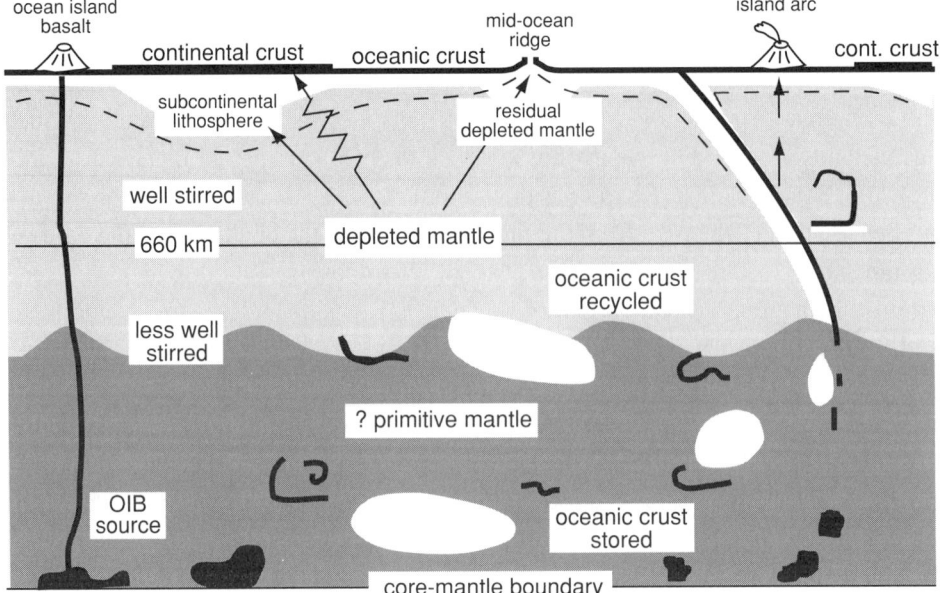

Figure 1.10. Geochemical reservoirs in the mantle. Two geophysically defined boundaries are thought to have geochemical significance: (1) that between the lithosphere (the rigid thermal boundary layer at the surface) and the well-stirred upper mantle and (2) the 660-km seismic discontinuity, which separates the upper mantle from the stiffer, less well stirred lower mantle and may also act as a partial barrier, inhibiting the transfer of material between the lower mantle and the upper mantle. The upper mantle beneath the lithosphere is a chemically homogeneous reservoir, the source of MOR basalt. It has been depleted, relative to the primitive-mantle, by extraction of the continental crust. A primitive-mantle reservoir, shown as being preserved in the lower mantle, may not exist, because of mixing with the depleted mantle (i.e., primitive mantle + depleted mantle → less depleted mantle). This does not affect the mass balance.

For each element M there is a mass balance summing the mean concentrations $c_M^{\text{res } i}$ in each reservoir i, multiplied by the mass fraction of the reservoir in the BSE, $m^{\text{res } i}$:

$$c_M^{\text{BSE}} = \sum_i c_M^{\text{res } i} \cdot m^{\text{res } i} = c_M^{\text{cc}} \cdot m^{\text{cc}} + c_M^{\text{dm}} \cdot m^{\text{dm}} + \cdots + c_M^{\text{BSE}} \cdot (1 - m^{\text{cc}} - m^{\text{dm}} \cdots)$$

(3)

where the last term is the contribution from the primitive (i.e., undifferentiated) mantle, which has BSE composition by definition. The superscript 'cc' stands for continental crust, and 'dm' for depleted mantle. Obviously, if we knew the masses of all the important reservoirs and were able to evaluate their mean chemical compositions, the solution would be straightforward. In practice, we often know neither, and it becomes a question of how best to proceed under the limitations of this imperfect knowledge.

The masses of some reservoirs, namely, the oceans, the atmosphere, and the continental crust, are known independently of geochemical data. Here we shall use a value of 0.0054 for the fraction m^{cc}, from Taylor and McLennan (1995). The subcontinental lithosphere can be constrained approximately as occupying no more than 200 km of depth beneath the continents, giving $m^{scl} \leq 0.04$.

The masses of the other mantle reservoirs cannot be known directly, but must be inferred from geochemical data. The most important reservoir is the depleted mantle (the MORB source). The key concept in mantle geochemistry is that the depleted mantle is in its depleted state primarily by virtue of having had the continental crust extracted from it, probably as the net result of several processes. If we ignore all other reservoirs (which are quantitatively less important), equation (3) reduces to the approximate relationship

$$c_M^{BSE} = \frac{c_M^{dm} m^{dm} + c_M^{cc} m^{cc}}{m^{dm} + m^{cc}} \tag{4}$$

For pairs of RLEs with an isotopic parent/daughter relationship (Sm-Nd, Lu-Hf, and the U-Th-Pb system), this mass balance yields a value for m^{dm} (mass fraction of depleted mantle) of about 0.5 (e.g., McCulloch and Bennett, Chapter 2, this volume). Similar conclusions can be reached from data on the outgassing of the noble gases, as discussed later. The uncertainty in the result stems from the uncertainties in the geochemical data and from the nature of the approximation embodied in equation (4) (i.e., ignoring other possible reservoirs such as the OIB source), but it probably is about ± 0.15. The depleted mantle is thus large enough to fill the entire volume of the upper mantle (0.29 of the total mantle) and also, probably, a significant portion of the lower mantle.

Because the ratio m^{cc}/m^{dm} is approximately 0.01, equation (4) can be further simplified to

$$c_M^{BSE} \approx c_M^{dm} \left[1 + (m^{cc}/m^{dm}) D_M^{cc/dm} \right] \tag{5}$$

where $D_M^{cc/dm} = c_M^{cc}/c_M^{dm}$ (i.e., the empirically observed partition coefficient between crust and mantle). If $D_M^{cc/dm} \ll m^{dm}/m^{cc}$, that is, if $D_M^{cc/dm} \ll \sim 100$, as expected for major elements and compatible and moderately incompatible trace elements, the second term on the left-hand side of equation (5) can be ignored, and the BSE concentration is effectively the same as that in the depleted-mantle. This is important, as it is the depleted mantle that is directly accessible to us as peridotite samples. It is also apparent that for these elements, precise knowledge of either their crustal abundances or the mass of the depleted-mantle reservoir is not an issue.

An important assumption behind the simplification of equations (4) and (5) is that the isotopically distinctive OIB source reservoirs can be neglected (i.e., the sum of their mass fractions is small compared with their compositional distinctiveness relative to the depleted mantle). Again, this assumption is likely to be best met for the major elements and the compatible and moderately incompatible trace elements. For highly incompatible elements, however, the lack of knowledge of the size of the OIB source reservoirs must be taken into account.

The depleted mantle is the source from which the MORB magmas that form the oceanic crust are extracted. The extraction of basalt from the depleted-mantle reservoir depletes it further, on the more local scale of the oceanic crust; this results in material we shall call 'residual depleted mantle'. Upper-mantle material that has not yet suffered such further depletion we shall call 'primitive depleted mantle' (PDM). This is essentially the same as the 'primitive upper mantle' of Hart and Zindler (1986). The oceanic crust is chemically complementary to its associated residual depleted mantle, so that these two reservoirs must, when taken together, sum to the original PDM from which they formed. Much of the oceanic crust and its associated residual depleted mantle probably is returned fairly directly to the well-stirred depleted mantle, whence it came, by subduction (i.e., the classic plate-tectonic cycle):

$$\text{depleted mantle} \underset{\text{seafloor spreading}}{\longrightarrow} \text{(oceanic crust}$$

$$\text{+ residual depleted mantle)} \underset{\text{subduction}}{\longrightarrow} \text{depleted mantle}$$

However, some of the most highly incompatible elements are selectively extracted and returned to the near-surface environment by island-arc magmatism. This may be an important part of the process forming the continental crust. Apart from this, the oceanic crust and the residual depleted mantle need not be thought of as forming distinct reservoirs, unless they become separated after subduction. If separated, former oceanic crust may be preserved from remixing if it is stored in the viscous, less well stirred lower mantle. Oceanic crust stored in the lower mantle for a couple of billion years may comprise part of the OIB source regions (Hofmann and White, 1982; Campbell, Chapter 6, this volume). In any event, OIBs sample mantle reservoirs with diverse long-lived isotopic heterogeneities.

The lithosphere is the thermal boundary layer at the Earth's surface. Rheologically it is stiff, and mantle material in the lithosphere is chemically isolated from the well-stirred (asthenospheric) upper mantle. Whereas sub-oceanic lithosphere is eventually subducted on a timescale of $\sim$100 million years, material in the lithosphere beneath the continents is effectively trapped. Geochemically, it can be thought of as fossilized 'residual depleted mantle' (ex-asthenosphere). A complication is that sub-continental lithospheric mantle often has been affected by one or more later episodes of metasomatism that have added a small fraction of a component highly enriched in incompatible elements to the original composition of the residual depleted mantle (Frey and Green, 1974).

The lower mantle is less accessible to chemical investigation than the upper mantle, but it is by no means terra incognita, geochemically speaking. Whereas MOR basalts sample the top of the mantle, plume-related basalts (flood basalts and OIBs) are thought to sample the thermal boundary layer at its bottom. It is possible that the portion of the depleted mantle that is in the lower mantle, shown in the cartoon of Figure 1.10 as physically distinct from the primitive mantle, has to some extent been mixed in with the primitive mantle, which therefore may no longer exist. Geochemically this can be viewed as a mixing of the depleted MORB-source mantle with the primitive mantle to form a less-depleted lower-mantle reservoir. Whether or not that has occurred does not matter for present purposes, because it would not affect the mass balance described by equations (3)–(5). What is important is that the mantle is not fundamentally layered in its major-element composition (e.g., by some process of whole-Earth differentiation such as crystal settling from an early magma ocean). The absence of geophysical evidence for such a complication is discussed by Davies (Chapter 5, this volume) and by Jackson and Rigden (Chapter 9, this volume). The geochemical evidence against it is best appreciated after deriving the BSE composition, and will be briefly discussed later.

1.4.3. The Empirical Constraints on Geochemical-Reservoir Compositions

The continental crust obviously is chemically heterogeneous, and there is thought to be a fundamental difference between the relatively silica-rich upper crust and the more mafic lower crust. Immature sediments derived from the erosion of wide areas, such as shales, greywackes, and loess, can provide reasonable estimates for upper crustal compositions, at least for elements not readily dissolved during weathering (Taylor and McLennan, 1995). Lower crustal compositions are less well constrained. Recent estimates of mean crustal abundances have been provided by Taylor and McLennan (1995) and Rudnick and Fountain (1995).

The chemistry of the various mantle reservoirs can be examined in two ways: by studying mantle-derived partial melts (i.e., basalts and related rocks) and, more directly, by looking at those actual mantle rocks and rock fragments that are available at the Earth's surface. It is worth listing such mantle samples to emphasize the diversity available:

1. Mantle sections in ophiolites: sub-oceanic lithosphere, probably mostly of back-arc-basin affinity.
2. Other massive peridotites, variously known as alpine, orogenic, or simply high-temperature peridotites: some have chemistry appropriate for MORB-source mantle.
3. Abyssal peridotites, dredged from the ocean floor: the residue from extraction of the oceanic crust.
4. Spinel-peridotite (rarely garnet-peridotite) xenoliths from alkali basalts (*sensu lato*): fragments of lithosphere from depths as great as 80 km. Most examples described in the literature are from the sub-continental lithosphere, but suites from several ocean-island settings (e.g., Hawaii, Tahiti, Samoa) are indistinguishable from sub-continental suites.
5. Garnet-peridotite xenoliths from kimberlites and lamproites: these fragments sample deeper levels in the sub-continental lithosphere.

6. Minerals and rare polymineralic assemblages found as inclusions in diamond: small, possibly unrepresentative samples, but occasionally from great depth, even including extremely rare examples from the lower mantle.

Peridotites have the physical properties (e.g., density and seismic wave speeds) that are required of mantle material to satisfy geophysical constraints. Chemically, mantle peridotite can also be shown to meet the requirements to be the parent material for basalt (e.g., Green and Falloon, Chapter 7, this volume). Another obvious reason for believing that these rocks have come from the mantle is that the chemical compositions of their constituent minerals show that they equilibrated at mantle pressures (P) and temperatures (T). The exceptions are some massive peridotites that were re-equilibrated in the crust at lower P–T conditions. The study of mineral-phase equilibria as a function of pressure and temperature is a well-developed field, commonly known as 'geothermometry and geobarometry'. For upper-mantle peridotites, the established techniques of geothermometry and geobarometry have yielded estimates for their pressures and temperatures of equilibration to within ± 0.3 GPa (i.e., a depth interval of ± 10 km) and $\pm 30°$C, and sometimes those techniques can reveal some of the P–T history of the peridotite's journey towards the Earth's surface, when this is recorded in mineral zoning.

Each of the different kinds of mantle samples adds a different perspective to the mantle story. Ophiolites and massive peridotites are particularly useful, because they allow an appreciation of the spatial distribution and field relations of the samples that are selected for geochemical analysis. By contrast, mantle xenoliths transported to the surface in mantle-derived magmas are small (not more than a few kilograms) and necessarily sample the mantle blindly, but they have considerably extended the range of tectonic settings that can be sampled. Because they were transported rapidly from the mantle to the Earth's surface, xenoliths retain in their mineralogy a pristine signature of the P–T environment from which they were plucked. Xenoliths from recent eruptions can be expected to be fresh, little altered by low-temperature processes. Conversely, massive peridotites typically will have been serpentinized during their journey to the surface; however, because they cooled while still at relatively high pressure, they may better preserve the mantle's sulphide component.

Generally, mantle xenoliths are brought to the surface by two types of magmatic activity. Magmas that are broadly of the alkali-basalt family often contain xenoliths that equilibrated in the mantle's spinel-lherzolite facies, at depths ranging from 40 to 60 km. Rare samples contain garnet (e.g., Ionov et al., 1993), indicating that they equilibrated at slightly higher pressures. However, it is clear that this type of volcanism does not exhume xenoliths from depths greater than about 80 km, and therefore it samples only the lithospheric mantle. The much less common kimberlitic/lamproitic type of magmatism brings to the surface xenoliths that equilibrated at pressures up to about 7 GPa (corresponding to a depth of about 220 km) and temperatures up to 1,400°C (e.g., Danchin, 1979). However, these xenoliths all appear to have suffered considerable metasomatism (e.g., as manifested in high Ti/Al ratios). Kimberlitic xenoliths also include samples of ancient sub-continental lithosphere, some of which appears anomalously silica-rich and Fe-poor compared with the vast majority of residual depleted mantle (Boyd, 1989), probably attributable to having suffered a different style of melt extraction. Being highly depleted (once the effects of metasomatism are accounted for), these anomalous compositions do not require a different primitive-mantle composition.

The vast majority of mantle peridotites, from all the different provenances listed earlier, belong to a single compositional clan and can plausibly be interpreted as all having originated from the 'residual depleted mantle', differing from one another mainly by having suffered various degrees of melt extraction (model calculations indicate typically up to 30%) (e.g., Nickel and Green, 1984; Frey, Suen, and Stockman, 1985). The most direct proof of such an origin is that this kind of mantle peridotite occurs in the mantle sections of ophiolites and is stratigraphically situated in the appropriate place for residual depleted mantle. Most xenoliths and many massive peridotites have spent considerable time fossilized in the lithosphere and bear the marks of cryptic metasomatism (i.e., of the type that causes enrichment in the light rare-earth elements, LREEs); comparison with unmetasomatized examples shows that this usually did not affect either their major-element chemistry or the abundances of compatible and moderately incompatible trace elements (e.g., Ni or the heavy rare-earth elements, HREEs). The chemistry of these mantle peridotites is characterized by MgO contents in the range 35–46 wt%, SiO_2 at 43–46 wt%, remarkably constant abundances of FeO (8 ± 1 wt%), Cr_2O_3 (0.4 ± 0.1 wt%), Co (100 ± 10 ppm), and the highly compatible, highly siderophile trace element Ir (4 ± 1 ppb), and high Ni contents ($2,200 \pm 500$ ppm). Bulk-rock Mg#'s are ≥ 89. The Cr, Ni, and Ir contents are diagnostic. Compositions beyond these limits usually can be explained by metasomatism, by infiltrating basaltic or perhaps carbonatitic melts (lithospheric samples) or by the characteristic Fe-Ti metasomatism of high-temperature 'fertile' peridotites (Nixon et al., 1981). Other mantle-derived samples are interpreted as cumulates or other fractionation products within the mantle, as frozen partial melts, or possibly as ancient recycled but unrehomogenized basalt from the oceanic crust (the eclogite xenoliths in kimberlites, thought to comprise <1% of the sub-continental lithosphere) (Schulze, 1989). A list of detailed geochemical studies (since 1978) of reasonably undepleted suites of mantle peridotites is given in Table 1.6. Data from earlier studies were summarized by Maaløe and Aoki (1977).

Mantle-derived partial melts may sample reservoirs not accessed directly by xenoliths or massive peridotites, such as the OIB source reservoir. Another advantage of examining partial melts is that the processes of melting and melt aggregation ensure that any local inhomogeneities in the mantle are smoothed out. This is likely to be important for the more highly incompatible trace elements, which may be mobile on the scale at which peridotites can be sampled. Finally, some types of

Table 1.6. A database of mantle peridotite suites

Abbreviation for suite	Reference	Locality/country	No. of samples
Xenoliths from alkali basalts: spinel-facies lherzolites, except some garnet-facies samples from PaliA and Vitim			
Hung2	Embey-Isztin et al. (1988)	Transdanubian Volcanic Region, Hungary	16
Hung1	Downes et al. (1992)	Transdanubian Volcanic Region, Hungary	14
SCarl[a]	Frey and Prinz (1978)	San Carlos, Arizona	6
HessD	Hartmann and Wedepohl (1990)	Hessian Depression, northwest Germany	28
KilbH	Roden et al. (1988)	Kilbourne Hole, New Mexico	15
SEChi	Qi et al. (1995)	Southeastern China	8
Sardi	Dupuy et al. (1987)	Sardinia	11
Mongo	Press et al. (1986), Stosch and Seck (1980)	Tariat Depression, Mongolia	12
PaliA[b]	Stern et al. (1986)	Patagonia, Chile	7
Vitim	Ionov et al. (1993)	Vitim Volcanic Field, Baikal region, Russia	19
SEAus	Nickel and Green (1984)	Victoria, Australia	11
Massive peridotites: mostly spinel facies, some plagioclase, rare garnet (Ronda)			
EPyre	Bodinier et al. (1988)	Eastern Pyrenees, France	31
Dinar	Lugovic et al. (1991)	Central Dinaric Ophiolite Belt, Yugoslavia	29
Zar (w, d)[c]	Kurat et al. (1993), Bonatti et al. (1986)	Zabargad Island, Red Sea, Egypt	17,14
Lanzo	Bodinier (1988)	Lanzo, western Alps, Italy	17
Horom	Frey et al. (1991)	Horoman, Hokkaido, Japan	7
Ronda	Frey et al. (1985)	Southern Spain	15

Baldi	Hartmann and Wedepohl (1993)	Baldissero, Ivrea Complex, northern Italy	14
Balmu	Hartmann and Wedepohl (1993)	Balmuccia, Ivrea Complex, northern Italy	16
ExLig	Rampone et al. (1995)	External Ligurides, northern Appenines, Italy	20

Xenoliths from kimberlites: garnet facies, both high- and low-temperature

PreSA[d]	Danchin (1979)	Premier mine, Transvaal, South Africa	31

[a] One datum from Jagoutz et al. (1979).

[b] These appear to be the most primitive known samples (unfortunately, no trace-element data).

[c] Divided into hydrous (minor amphibole, 'w') and anhydrous ('d') suites.

[d] Some metasomatism (high K and Ti), no trace-element data, but many undepleted samples, with near-chondritic Ca/Al ratios.

Note: Suites were selected from the literature on the following criteria: (a) samples unmetasomatized, or with only moderate metasomatism (low K, no phlogopite, minor amphibole). (b) at least 6 samples, with a high proportion having low degrees of depletion (less than 42% MgO); (c) some trace-element data for compatible and moderately incompatible elements (except PaliA); (d) publication since 1977. For earlier data, see Maaløe and Aoki (1977). Samples with MgO > 46 wt% (dunites), with CaO/Al$_2$O$_3$ > 1 (generally clinopyroxene-rich or wehrlites), or with anomalously high FeO^t (usually identified as of cumulate or metasomatic origin) were excluded.

basalts, particularly the MORBs, must, over time, be derived from such huge volumes of the mantle that we can be certain that a quantitatively significant fraction of the total mantle is involved in their genesis. MORB plays a key part in the strategy for estimating the mantle composition. Its relative uniformity implies a well-mixed source (as well as a fundamental process). The current worldwide production rate for MORB is estimated to be 20 km^3 per annum, and MORB is the result of approximately a 10% partial melting of its peridotitic source (e.g., Hofmann, 1988). Thus, over the course of geological time (4.5×10^9 years), and at the current production rates, creation of the oceanic crust should have processed 3×10^{24} kg of mantle peridotite, equivalent to 75% of the entire mantle (mass of 4×10^{24} kg). This estimate is a minimum, because the rate of production of oceanic crust may have been much greater in the past. (Note that this calculation does not imply that >75% of the mantle is depleted, as we have not considered recycling.)

The main conclusion of this section is that although our knowledge of the chemistry of the Earth's mantle is incomplete, we should have enough pieces of the jigsaw in place to get an accurate impression of the whole picture. In the following section we outline in detail how this can be done.

1.4.4. A Procedure for Estimating the BSE Composition

We shall assume that the RLEs are present in the BSE in their chondritic relative abundances. This means that the BSE is characterized by a single RLE/Mg ratio, chondrite-normalized: (RLE/Mg)$_N$. The justification is that the chondrite-normalized abundances of the RLEs in terrestrial samples do not show the patterns of depletion that would be expected either for volatility-related fractionation (as discussed earlier) or for metal–silicate (or sulphide–silicate) fractionation that would be evident if some of these elements had been partitioned into the core. An exception that proves this particular rule is V, which is usually classified as a RLE in meteorites, but which clearly has siderophile tendencies in the Earth, and will here be treated as a siderophile element. Apart from V, the least lithophile of the RLEs (as judged from their silicate/sulphide distribution in the enstatite chondrites and from thermodynamic calculations) include Ti, Ca, and U, and the most staunchly lithophile include Al, Sc, and Th. Any tendency towards siderophile or chalcophile behaviour among the RLEs would have produced sub-chondritic ratios for Ca/Al, Ti/Al, Ca/Sc, U/Th, and so forth, for which there is no evidence in the available mantle samples.

With the assumption of constant (RLE/Mg)$_N$ for the BSE, it remains to determine the magnitude of (RLE/Mg)$_N$ and the abundances of Mg and Si and of all the volatile and siderophile/chalcophile elements, which, judging from the evidence in chondrites, will be expected to be depleted in the BSE, on account of their volatility or their being partitioned into the core or both. For this exercise it is convenient to divide the elements into five groups:

1. Major elements, conveniently considered as oxide components: CaO, Al$_2$O$_3$, MgO, SiO$_2$, and FeO^t (the latter being total iron in all its oxidation states expressed as FeO). Because Ca and Al are RLEs, the abundances of CaO and Al$_2$O$_3$ will give, with MgO, the all-important value of (RLE/Mg)$_N$.
2. Compatible and moderately incompatible volatile and siderophile elements (except those concentrated in the mantle's sulphide phase): Li, Na, V, Cr, Mn, Ni, Co, Zn, Ga, and Ge.
3. Chalcogenides and highly chalcophile elements, concentrated in mantle sulphide, such as S, Se, Cu, and the platinum-group elements.
4. Midrange incompatible elements with incompatibilities not exceeding those of the most incompatible of the RLEs [which are, in order of increasing compatibility, Th, U, Ba, Nb (or Ta), and La]: these include P, K, Mo, and W.
5. Very incompatible elements (more incompatible than Th, etc.), such as Cl, Br, and the noble gases.

1.4.4.1. Group 1: Major Elements

Accurate abundances of the five major elements in mantle reservoirs cannot be inferred from the study of partial melts (e.g., basalts), as explained later. Instead, the major elements must be assessed from actual mantle samples, the different kinds of which were discussed in the preceding section. We shall assume that such samples are from the residual depleted mantle.

The concentrations of major elements (and the compatible and moderately incompatible elements, as discussed in the next section) in peridotite suites of different provenances (Table 1.6) follow simple chemical trends (e.g., linear variation with the abundance of MgO: a Bowen diagram, which is a variation on the Harker diagram, which uses SiO$_2$ as the abscissa). Some examples for three elements (Fe as FeO^t, Ni, and V) are given in Figure 1.11. These trends are interpreted as being primarily due to melt extraction (e.g., Nickel and Green, 1984; Frey et al., 1985; Press et al., 1986). There is, however, a further complication.

The least depleted samples should be those that retain the greatest amounts of the components preferentially extracted into basalt (i.e., CaO, Al$_2$O$_3$, and TiO$_2$) and the least amounts of MgO. In theory, it should be possible to account for the effects of melt extraction quantitatively by invoking the fundamental assumption of constant RLE ratios in the primitive mantle and by noting also that the abundances of the moderately incompatible RLEs (e.g., Ca, Al, Ti, Sc, and the HREEs) in the PDM should be unaffected by the extraction of the continental crust. Then, on a plot of the chondrite-normalized element abundance (RLE/Mg)$_N$ versus [MgO], the trends for all of the RLEs should intersect at a single point, corresponding to the primitive-mantle composition (Figure 1.12a). For nearly all known suites, whether massive peridotites or xenoliths, the required convergence does not occur at reasonable MgO concentrations (Figure 1.12b). Many suites, in fact, have conspicuously higher than chondritic Ca/Al ratios at all values of MgO (Palme and Nickel, 1985).

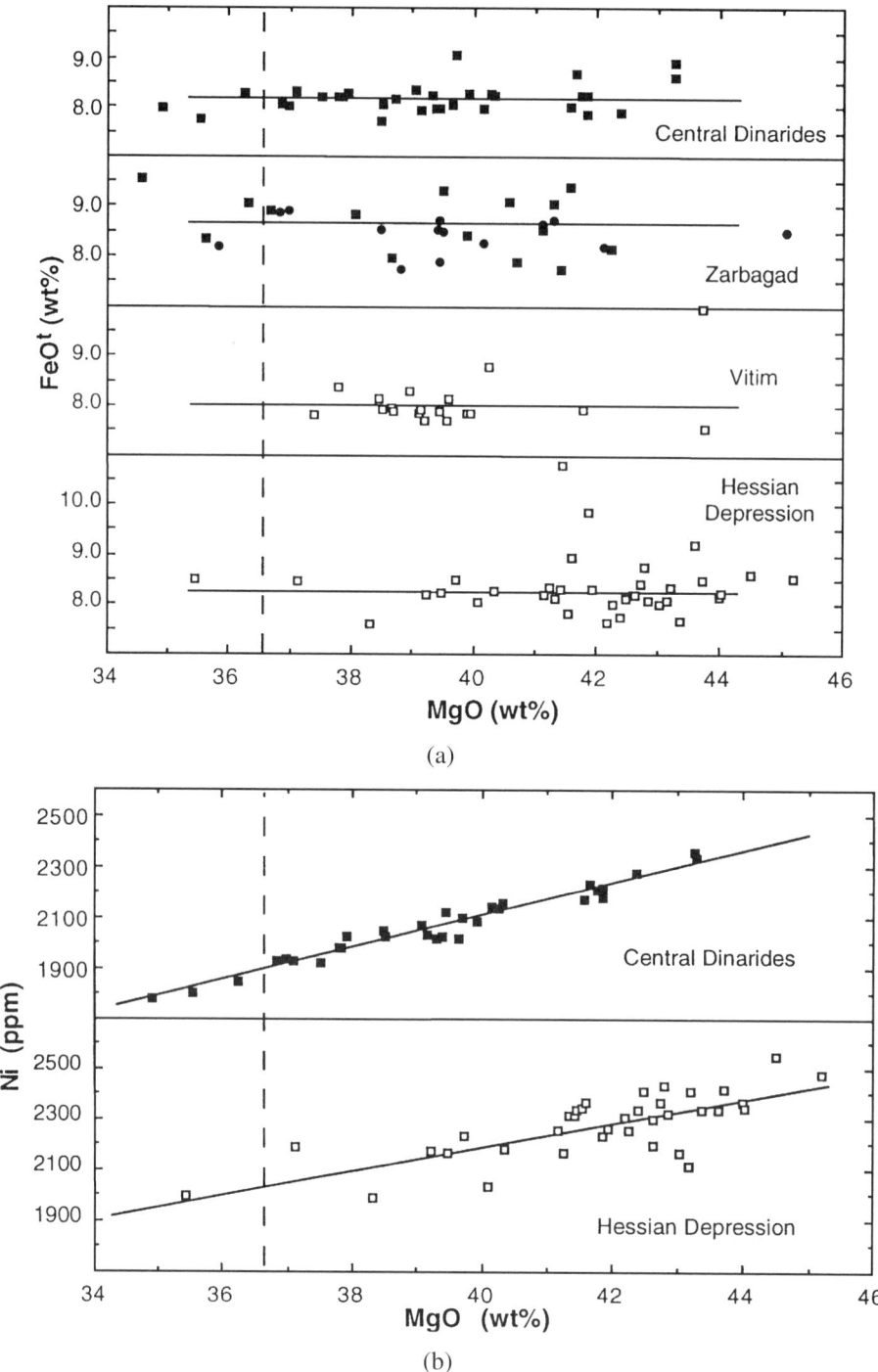

Figure 1.11. Examples of Bowen diagrams for (a) FeO^t, (b) Ni, and (c) V in suites of mantle samples (massive peridotites shown as solid symbols, and xenoliths from alkali basalts as open symbols). Compatible and moderately incompatible elements produce linear trends on such diagrams, within the scatter of the data. A compatible element has a positive slope (e.g., Ni); an incompatible element has a negative slope (e.g., V). The FeO^t arrays have effectively zero slope. The trends are mainly due to the extraction of partial melt, but may also reflect modal heterogeneity (modal banding) in the peridotites. The abundances of the elements in the primitive depleted mantle (PDM) are assumed to be given by the

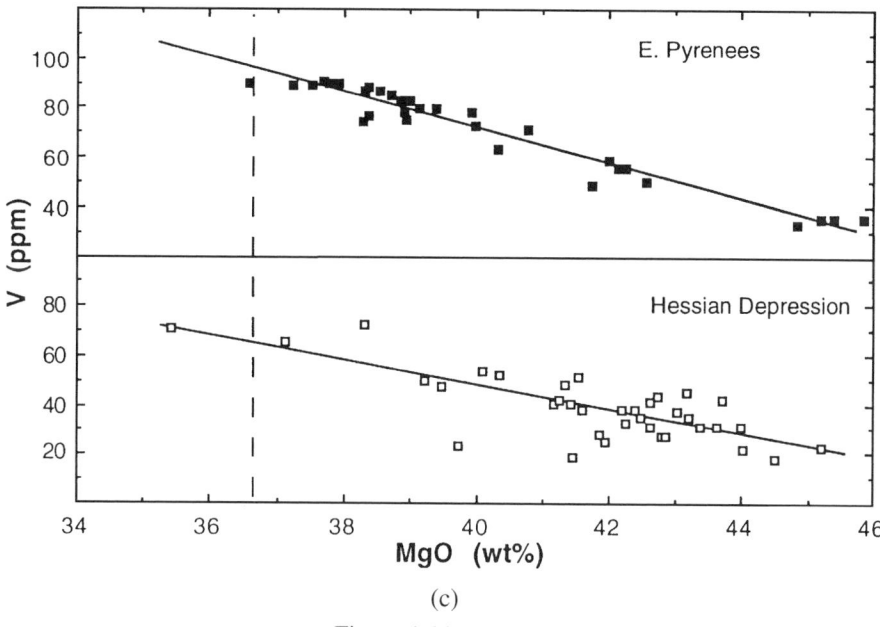

(c)

Figure 1.11c (*cont.*).

This apparent failure of the model not only prevents identification of the primitive-mantle MgO value but also raises doubts about the assumption of a constant RLE ratio for the PDM (i.e., in this context, the upper mantle). Geochemists have either accepted this variation at face value, attempting to explain it by invoking upper/lower-mantle differentiation due to majorite fractionation (Palme and Nickel, 1985), or argued that it is an artifact of the sampling process. Hart and Zindler (1986) suggested that the peridotites selected by geochemists for analysis are biased towards specimens that are modally enriched in clinopyroxene, and they corrected their dataset accordingly. McDonough and Sun (1995) favoured a variation in modal spinel. The latter explanation would imply a correlation between Ca/Al and Cr content, which is not observed.

In our view, the problem originates because peridotites are metamorphically textured rocks that exhibit pronounced gneissose banding on the scale of centimeters to meters. This banding is a result of modal heterogeneity in the distribution of all the phases in the peridotite, with, for example, olivine-rich and olivine-poor layers, often aligned along a pervasive tectonic fabric. The origin of the layering is obscure, but one suggestion is that it may be due to metamorphic segregation by pressure solution (Dick and Sinton, 1979). Other possible explanations are listed in the review by Spray (1989). Note that this modal banding is a phenomenon

Figure 1.11 (*cont.*). intercepts of the trends with the PDM value of 36.6 wt% MgO (indicated by vertical dashed lines). Samples plotting at MgO values lower than 36.6 wt% are thought to be enriched in pyroxene, because of either metasomatism or modal heterogeneity; nevertheless, such samples usually plot close to the lines as defined by the samples with MgO > 36.6 wt%. Data sources are given in Table 1.6.

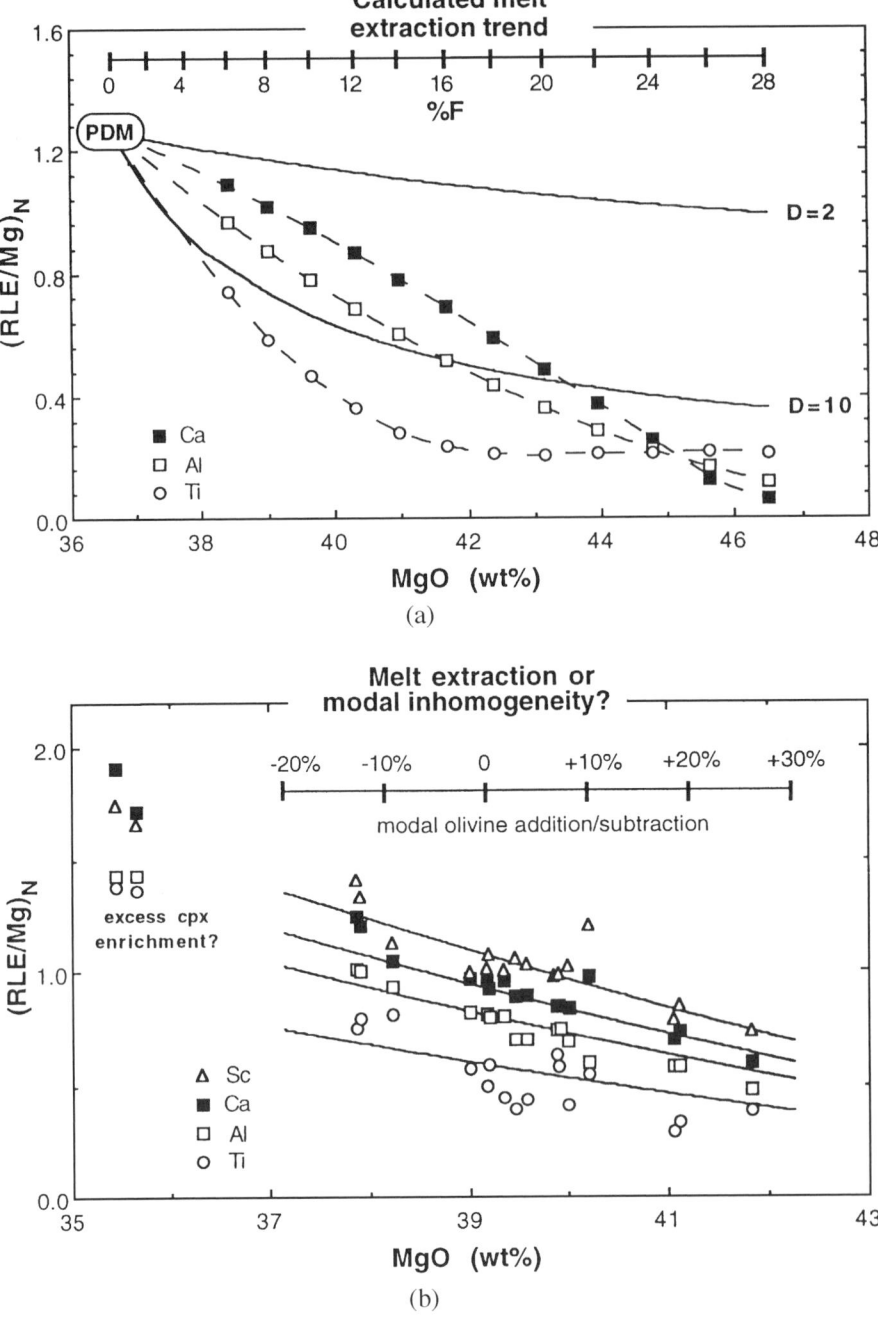

Figure 1.12. (a) Calculated melt-extraction trends illustrating how the abundances of Ca, Al, Ti, and two moderately incompatible trace elements (with $D_M^{\text{melt/source}} = 2$ and 10) might be expected to decrease with increasing fraction of extracted melt (F), starting from an original primitive depleted mantle (PDM) composition, at which all chondrite-normalized RLE ratios [i.e., $(\text{RLE/Mg})_N$] have the same value. The trends for Ca, Al, and Ti were calculated using the compositions of accumulated fractional melts from the parameterization by Niu and Batiza (1991) of various experimental studies, for initial melting at 1.8 GPa. (b) An actual example of a putative 'melt-extraction trend' in the Balmuccia massive peridotite, plotted on a diagram of $(\text{RLE/Mg})_N$ versus MgO. Data are from Hartmann and Wedepohl

entirely separate from that seen in the layers of pyroxenite and garnet-pyroxenite of the aluminous augite suite, caused by diking of melts into the peridotite, nor does it include hypothetical layers of remixed crustal material of the 'marble-cake-mantle' variety (Allègre and Turcotte, 1986). It does include, however, the almost ubiquitous clinopyroxene- and orthopyroxene-rich layers of Cr-diopside websterite and bronzitite, which clearly do not have the compositions of partial melts and which are chemically and isotopically close to being in equilibrium with the adjacent peridotite. The banding in massive peridotites is easily appreciated when these rocks are seen in the field. It sometimes may not be so evident in xenoliths (but see Irving, 1980), but because xenolith suites show major element chemical trends identical with those for massive peridotites, it seems fair to assume that they generally are also derived from a source with the same kind of modal heterogeneity. The conclusion is that in most mantle peridotites there is perhaps no such thing as a homogeneous, representative sample. Modal heterogeneity superimposed on melt extraction is a possible mechanism by which the super-chondritic Ca/Al ratios in many upper-mantle samples might be explained (Figure 1.13).

An additional problem is that even though the trends in individual peridotite suites may seem linear to quite high MgO values, modelling of likely melt-extraction trends from experimental melting studies predicts some curvature (Figure 1.12a). This curvature is too slight to be resolved among the natural scatter in the peridotite analytical data, but it is sufficient to confound the hope of being able to extrapolate back to the primitive depleted-mantle composition using simple linear trends of wt% element or oxide versus wt% MgO relations. The modelling in Figure 1.12a shows that only peridotites that have experienced less than about 10% melt extraction (MgO $\leq$ 39 wt%) fall in the approximately linear region. Such peridotites are fairly rare, and most trends have been established by data with MgO > 39 wt%.

In light of these complications, we suggest the following alternative algorithm to extract the composition of the primitive depleted mantle from the peridotite data. Firstly, we solve for the abundances of the five major elements by invoking 5 constraints (i.e., 5 equations):

Figure 1.12 (*cont.*). (1993) and are for samples all collected from one small locality (the quarry at Balmuccia); despite the fact that they all came from a small volume of mantle, the range in MgO contents would, if the trend were entirely due to the extraction of partial melt, imply a range in F (melt fraction) of about 15%. The trends for the individual RLEs (Ca, Al, Ti, and Sc) cannot be extrapolated back to a single point of intersection, as required for a PDM composition. The relationship $(Sc/Mg)_N > (Ca/Mg)_N > (Al/Mg)_N > (Ti/Mg)_N$ at all MgO contents is typical for most peridotite suites and is in the order expected for moderate degrees of melt extraction (F < 25%). The curves drawn through the data are produced by simple addition or subtraction of olivine of constant composition (Mg# 0.90, containing no RLEs) to one arbitrarily chosen sample (BAM-4). This represents the simplest conceivable form of modal heterogeneity and appears to explain the data at least as well as a conventional melt-extraction trend.

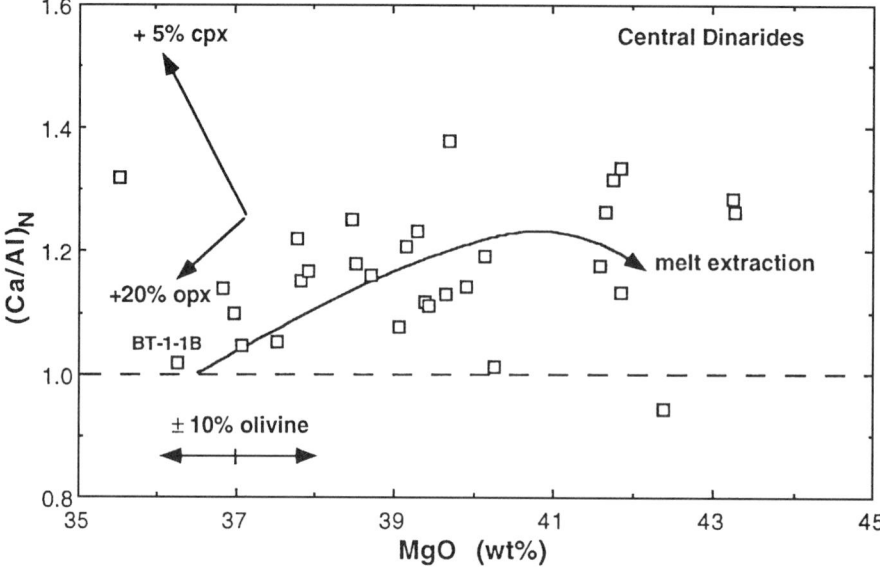

Figure 1.13. Chemical variations in mantle peridotites may be due to a combination of melt extraction and modal heterogeneity; data from Lugovic et al. (1991) for the central Dinaric ophiolite belt. These samples are from a wide area, in contrast to the data in Figure 1.12b, and therefore perhaps more plausibly represent variable degrees of depletion by melt extraction. The effect of melt extraction on the Ca/Al ratio is modelled from Niu and Batiza (1991). The possible effects of modal heterogeneity are illustrated by compositional vectors for the addition of 5% clinopyroxene and 20% orthopyroxene (assuming a typical pyroxene composition for fertile peridotite equilibrated at 900°C in the spinel-lherzolite facies) and −10% to +10% olivine addition or subtraction. Addition or subtraction of any phase (olivine, clinopyroxene and orthopyroxene, spinel, and sulphide) should be considered a possibility. The whole array can be explained by modal heterogeneity of about ±10% in olivine and a few percent in clinopyroxene, superimposed on the melt-extraction trend. Sample BT-1-1B is a fair match in major elements for the PDM.

1. $CaO + Al_2O_3 + MgO + SiO_2 + FeO^t = 98.54$ wt%. The uncertainty is probably only ±0.04%. The extremely good robustness of this constraint derives from the fact that the abundances of minor components which constitute most (1.36%) of the remaining 1.46% (i.e., Na_2O, TiO_2, Cr_2O_3, MnO, and NiO, plus the excess O in Fe_2O_3) vary little among reasonably fertile peridotites, and all other components (H_2O, CO_2, S, K_2O, and the entire remaining trace-element inventory) sum to <0.1%.

2. In massive peridotites and alkali-basalt-derived lherzolite xenoliths, FeO^t is constant, irrespective of MgO content (e.g., data in Figure 1.11). This is expected, as FeO^t should hardly be affected by melt extraction at low degrees of partial melting ($D_{Fe}^{liq/sol} \cong 1$), and only slightly affected by modal segregation. The mean value is 8.07% FeO^t, with a standard error of the mean of 0.06 (Figure 1.14); also see the earlier data summarized by Maaløe and Aoki (1977) and McDonough (1990).

3. Molar $100Mg/(Mg + Fe^t)$ (i.e., the bulk-rock Mg#) ought not to be affected by

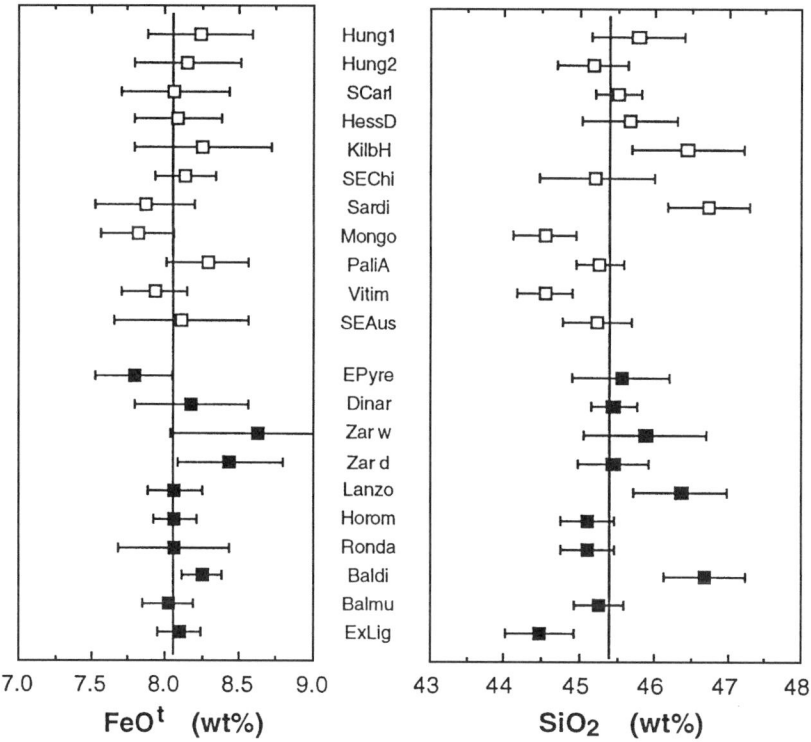

Figure 1.14. PDM abundances of FeO^t (mean values, independent of MgO content) and SiO$_2$ (values at 36.6 wt% MgO, calculated from regression of SiO$_2$-vs.-MgO data for each peridotite suite). Data from Table 1.6.

modal segregation, even if this is followed by subsequent re-equilibration of the segregated rocks at different P and T, because mineral Mg#'s are nearly the same in olivine and pyroxenes (as shown by experimental phase-equilibrium studies and empirical studies of coexisting phases in peridotites). Hence $(Mg\#)_{ol} = (Mg\#)_{bulk}$, to a very good approximation, in spinel lherzolites. Extraction of partial melt causes an increase in Mg# (as MgO increases, FeO^t remains constant), so that the most primitive samples should have the lowest Mg#'s. However, metasomatism by infiltration of melt may lower the Mg#, and that effect must be filtered out. Histograms of olivine compositions in unmetasomatized spinel lherzolites from a wide variety of environments show that fairly primitive mantle samples have Mg#'s in the range 88.8 to about 89.6. Because the uncertainty with which olivine Mg#'s are determined in individual samples typically is about 0.2, we interpret this distribution as implying an Mg# for the PDM of 89.0, which value we adopt. With 8.07% FeO^t, this gives MgO $= 36.64$ wt%.

4. Although not as invariant as FeO^t, SiO$_2$ should also be fairly insensitive to modal segregation processes. The SiO$_2$ concentration from SiO$_2$-versus-MgO trends at 36.6% MgO is 45.4 ± 0.3 wt% (Figure 1.14). The uncertainty allows for a $\pm 1\%$ uncertainty in MgO.

5. Ca and Al are RLEs, and therefore the CaO/Al_2O_3 ratio is constrained at the chondritic value of 0.792. With the other constraints, this gives 4.6% Al_2O_3 and 3.7% CaO, and thus the $(RLE/Mg)_N$ ratio is 1.28.

Table 1.7 compares the major-element compositions of the PDM and (with a slight correction for the continental crust) of the BSE, obtained in this way, with recent literature estimates. An advantage of the method presented here is that the algorithm can be reduced to a simple linear equation:

$$[Al_2O_3](1 + 0.792) = 98.54 - [MgO] - [SiO_2] - [FeO^t] \qquad (6)$$

where

$$[MgO] = (0.5612Mg\#)[FeO^t]/(1 - Mg\#) \qquad (7)$$

This makes the effects of the uncertainties transparent and easily evaluated. The main weakness of the method is that the final results for [CaO], $[Al_2O_3]$, and the $(RLE/Mg)_N$ ratio are particularly sensitive to the chosen value of Mg#. However, the values for [CaO] and $[Al_2O_3]$ look very reasonable when compared with actual analyses of least-depleted samples, and we suggest that propagation of errors, with uncertainties in $[SiO_2]$ of ± 0.3 wt%, in $[FeO^t]$ of ± 0.06 wt%, and in Mg# of ± 0.001 (corresponding to an uncertainty in [MgO] of ± 0.4 wt%), leads to uncertainties in $[Al_2O_3]$ of 0.3 wt% and in $(RLE/Mg)_N$ of $\pm 10\%$.

A similar approach to estimating the major-element composition of the mantle has been used by Hart and Zindler (1986) and by Allègre et al. (1995), but in each case using a much smaller database and invoking a different set of five constraints. Those two factors account for the slightly different results (Table 1.7).

The garnet peridotites brought to the surface in kimberlites and lamproites have come from deeper in the mantle than the xenoliths hosted in alkali basalts, and those from the best-known province, the Kaapvaal craton in southern Africa, show somewhat different chemical trends (e.g., Boyd, 1989). Most of these xenoliths are very depleted, and the differences in chemistry are plausibly attributable to different styles of melt extraction, rather than to different PDM compositions. This is illustrated for the major elements in Figure 1.15, for a suite of samples from the Premier kimberlite pipe in South Africa, which includes a number of samples rather less depleted than the norm for xenoliths from that kind of provenance.

It is important to understand the reason that the abundances of the major elements in planetary mantles are not accurately constrained by the composition of their derived partial melts. Basically, FeO and MgO behave sufficiently similarly during mantle melting processes for them to be counted, for many purposes, as a single thermodynamic component in the chemical system encompassing typical

Table 1.7. *Major-element composition of the BSE and comparison with previous estimates*

Element	Continental crust[a]	Continental crust[b]	This study Depleted mantle	This study Primitive mantle	Ringwood (1975)[c]	Ringwood (1979)	Sun (1982)[d]	Jagoutz et al. (1979)[d]	Wänke et al. (1984)	Palme & Nickel (1985)	Hart & Zindler (1986)	McDonough & Sun (1995)
MgO	4.4	5.3	36.64	36.33	38.1	38.1	38.0	38.3	36.8	35.5	37.8	37.8
Al_2O_3	15.8	15.9	4.62	4.73	4.6	3.3	4.3	4.0	4.1	4.8	4.1	4.4
SiO_2	59.1	57.3	45.44	45.56	45.1	45.1		45.1	45.6	46.2	46.0	45.0
CaO	6.4	7.4	3.71	3.75	3.1	3.1	3.5	3.5	3.5	4.4	3.2	3.5
FeO^t	6.6	9.1	8.07	8.17	7.9	8.0	8.4	7.8	7.5	7.7	7.5	8.1
Total	92.3	95.0	98.6	98.54	98.8	97.6		98.7	97.5	98.6	98.6	98.8
$(RLE/Mg)_N$				1.29 (11)	1.02–1.19	1.02	~1.15	1.03–1.14	1.1–1.4	1.3–1.5	1.06	1.17
Mg#			89.0	88.8	89.6	89.5	89.0	89.7	89.7	89.1	89.9	89.3

[a]Rudnick and Fountain (1995).
[b]Taylor and McLennan (1995).
[c]Column 8 of Table 5.2 in Ringwood (1975); alumina is stated as being 'probably too high'.
[d]Partial-melting model.

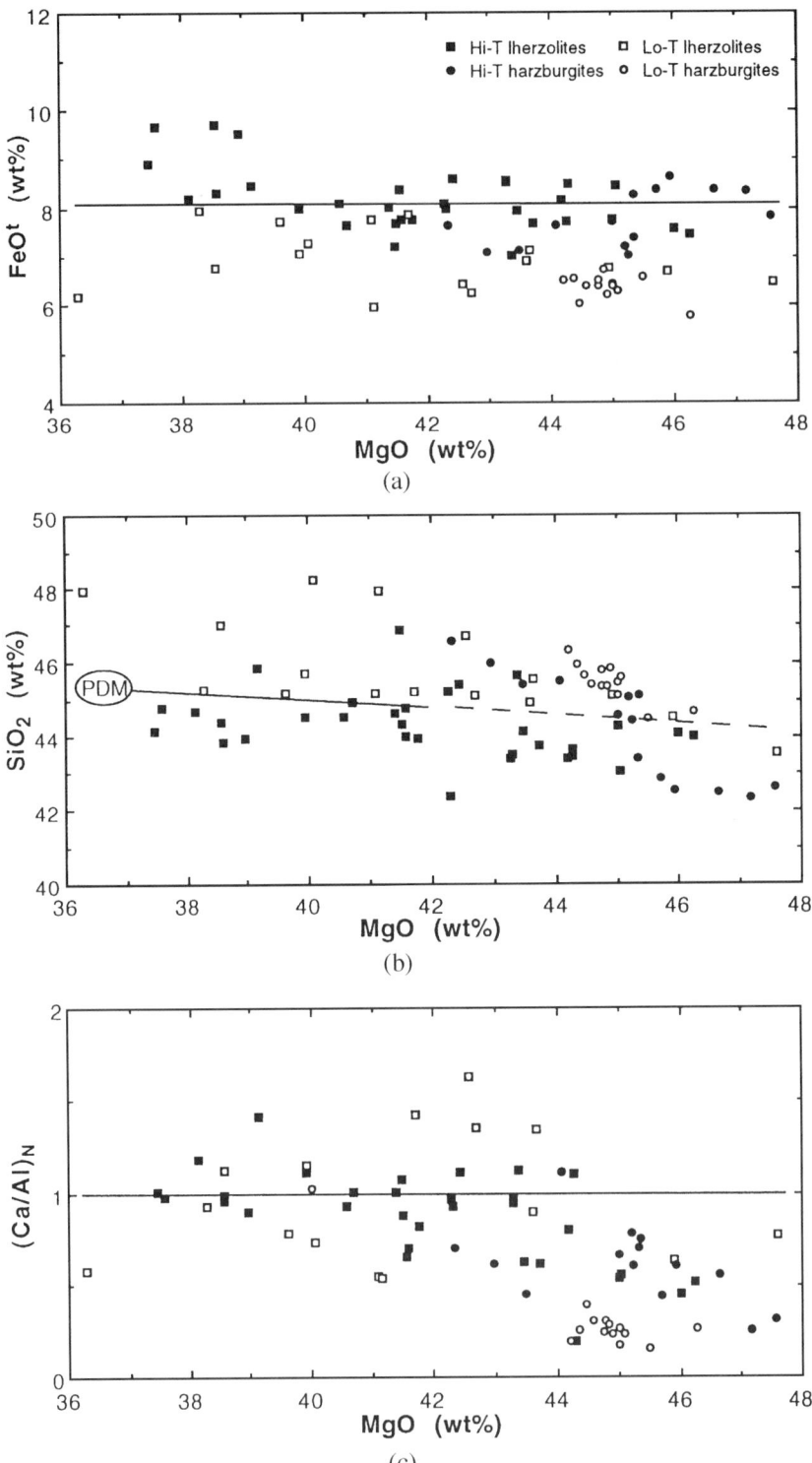

Figure 1.15. Major-element variations in peridotite xenoliths from kimberlites. All samples from the Premier mine, Transvaal, South Africa. Data from Danchin (1979). The samples are divided into "Lo-T" lherzolites and harzburgites, which equilibrated at <1,200°C

planetary mantle compositions. For any cosmochemically reasonable major-element composition, there are then effectively four components in the system CaO-*MF*O-Al_2O_3-SiO_2 (where MF = Mg + Fe^{2+}) that will crystallize in the upper-mantle basalt source region as an assemblage of four solid phases: olivine, orthopyroxene, clinopyroxene, and an aluminous phase, which, depending on pressure, will be plagioclase, spinel, or garnet (Green and Falloon, this volume). The beginning of melting will add a fifth phase to the system, which, by the phase rule, will then be isobarically invariant – that is, for a given pressure, the composition of the melt will be completely fixed (will not depend on Ca/Al and MF/Si ratios, etc.) and will remain so until one of the four solid phases is melted out. In practice, that does not happen until large degrees of melting are reached (probably >20%). Thus, in the model system, only very extensive melts will have compositions that will reflect the major-element composition of the source. Although in more complex natural systems the addition of minor components (principally Na_2O, Cr_2O_3, and TiO_2) removes the invariance of the melting reaction, the principle still stands. For example, the ratio (*MF*)/Si is buffered by olivine and orthopyroxene, and neither the Mg/Si ratio nor (RLE/Mg)$_N$ is determinable from an examination of basalts. Ca/Al ratios even in primary magmas do not directly reflect the Ca/Al ratio of their source (except in the case of high-degree melts, such as some komatiites), but rather record the influences of minor elements (e.g., Cr/Al or Na/Ca ratio), as well as physical factors such as the depths at which melting begins and ceases. Likewise, although the Mg/Fe ratio in a basalt does potentially reflect that in its source, the practicalities of the matter are that the relationship is used the other way around – because the Mg/Fe ratio of the mantle is known from mantle samples, the Mg/Fe ratios in basalts are used to identify whether or not the magma is primary (i.e., unaffected by fractional crystallization). One upshot of all this is that the major-element compositions of other planetary bodies, such as the Moon, Mars, Venus, and even the eucrite parent body, for which we do not have direct mantle samples, remain very poorly known. This is manifested in the long-running and still unresolved controversy over the Moon's major-element composition (e.g., O'Neill, 1991a).

Figure 1.15 (*cont.*). (calculated from mineral geothermometry) and are from the ancient lithospheric craton, and "Hi-T" lherzolites and harzburgites, which equilibrated at >1,200°C and may be asthenospheric. These samples have generally suffered considerable metasomatism, evinced by high Ti and K values. (a) At all MgO values the Hi-T suite has FeO^t contents virtually identical with the PDM value, as estimated from the massive peridotites and alkali-basalt xenoliths (8.07 wt%, indicated by a horizontal line), except for some anomalous Fe-rich 'fertile' samples. The cratonic Lo-T suite has lower FeO^t values at high MgO contents, but possibly trending towards the PDM value; this might indicate a different style of depletion. (b) SiO_2 contents are similar to the trend defined by the massive peridotites and alkali-basalt xenoliths. (c) The (Ca/Al)$_N$ ratio for these xenoliths shows evidence of a depletion trend different from that generally shown by the massive peridotites and alkali-basalt xenoliths (cf. Figures 1.12 and 1.13).

1.4.4.2. Group 2: Compatible and Moderately Incompatible Trace Elements

If we define 'compatible' as describing an element with mean empirical $D_M^{cc/dm} < 1$, there are only a few: the major element Mg and the trace elements Ni, Cr, and Co. (For reasons to be explained later, we choose to use a different category for those chalcophile elements that empirically appear to be compatible.) There are no compatible RLEs (which is unfortunate, as a compatible RLE would be very useful for estimating the BSE composition). Fe and Mn have $D_M^{cc/dm} \cong 1$.

The slightly incompatible trace elements that have $D_M^{cc/dm}$ values in the range 1–10 include the RLEs Ti and Sc and the HREEs and the non-RLEs Li, Na, V, Co, Zn, Ga, and Ge. As for the major elements, the abundances of these elements in upper-mantle peridotites plot as linear trends on Bowen diagrams (element abundance versus [MgO]). Their abundances in the PDM were estimated using the following method.

Firstly, we treated each suite of peridotites (both xenoliths and massive peridotites) individually. For each suite (data sources in Table 1.6), the data for each element E were fitted by linear regression to an equation of the form

$$[E (\text{or } EO_{x/2})] = A + B[MgO] \tag{8}$$

This equation was then used to extrapolate the data to the PDM value of 36.6 wt% MgO; for examples, see Figure 1.11. The uncertainty in the derived value was taken to be a combination of the standard error of the estimate from the regression and the value of B from the regression (i.e., the effect that the slope of the correlation has, assuming an uncertainty of ± 1 wt% MgO), except for FeO^t, Cr, and Zn, which used only the standard error of the estimate from the regression (the slope was assumed to be effectively zero). For each element, the mean, standard deviation, and χ^2 value of the fit were then calculated from the individual values for each suite, weighted according to the calculated uncertainties. It should be noted that this method does not take into account the effects of the modal banding in peridotites discussed earlier. However, for the elements considered here, the effects of modal banding are relatively unimportant and are largely subsumed in the scatter of the data. Elements concentrated in one phase (e.g., Na and perhaps V in clinopyroxene) will be the most strongly affected. A feeling for the likely magnitude of the modal-banding effect is given by the observation that the Ca/Al ratio, when extrapolated to 36.6% MgO, is typically ∼10% above the chondritic ratio (e.g., Palme and Nickel, 1985).

The reason for considering suites of peridotites one by one, rather then lumping all data together into a global correlation, is that different suites may have been depleted by melt extraction under different conditions of P–T or volatile-element content (i.e., under different tectonic regimes), or may have been subject to different degrees of modal banding. That potentially could result in different slopes on the Bowen diagrams, even though the starting composition (i.e., the presumed PDM) was the same. The results for individual suites are summarized in Figure 1.16, and the means and standard errors of the entire database are given in Table 1.8.

Table 1.8. *PDE and BSE abundances of FeO^t, SiO$_2$, and compatible or mildly incompatible volatile and siderophile minor and trace elements*

Element or oxide	Number of suites	Number of samples	Mean PDM^a	Standard error	χ^2	BSE^b	Depletion^c factor
FeO^t	21	326	8.09 wt%	0.05 wt%	1.19		
FeO^{td}	21	326	8.10 wt%	0.05 wt%	0.40	8.19 wt%	0.15
SiO$_2$	21	326	45.33 wt%	0.10 wt%	1.63	45.47 wt%	
Li^e	5	70	2.1 ppm	($\sim$0.4) ppm	—	2.2 ppm	0.62
Na	20	313	2,380 ppm	80 ppm	2.1	2,590 ppm	0.23
V	13	208	85 ppm	2.4 ppm	2.3	86 ppm	0.69
Cr	21	305	2,540 ppm	40 ppm	0.90	2,520 ppm	0.42
Mn^f	6		1,050 ppm	$\sim$50 ppm		1,050 ppm	0.24
Ni	17	252	1,875 ppm	22 ppm	0.96	1,860 ppm	0.077
Co	15	214	102.4 ppm	1 ppm	0.65	102 ppm	0.090
Zn	13	196	53.3 ppm	0.9 ppm	1.61	53.5 ppm	0.076
Ga	7	81	4.1 ppm	0.2 ppm	1.42	4.4 ppm	0.18
Other compatible and/or chalcophile elements							
Cu^g	9	151	12–44 ppm			18 ppm	0.066
Ge^h			1.1 ppb	$\sim$0.1 ppb		1.2 ppb	0.016
Cd^h			40 ppb	$\sim$20 ppb		$\sim$20 ppb	0.03
Ag^h			$\sim$8 ppb	$\sim$4 ppb		$\sim$8 ppb	0.02
Ir		92^i	3.9 ppb	1 ppb		3.9 ppb	0.004
Other HSE^j							0.004

^{a}PDM = primitive depleted mantle (i.e., average MORB source mantle).

^{b}BSE composition obtained from PDM by correcting for continental crust, using equation (4) in the text.

^{c}Normalized to Mg and the value in CI chondrites, i.e., (element/Mg)$_{BSE}$/(element/Mg)$_{CI}$.

^{d}Overall mean of the simple mean value of FeO^t in each suite, i.e., not extrapolated to 36.6 wt% MgO. This is identical with the value obtained by extrapolating each suite to 36.6% MgO, also listed.

^{e}See also Ryan and Langmuir (1987). Many peridotite data are reported to ±0.5 ppm (i.e., ±25%), so analytical errors probably predominate.

^{f}Mn is usually reported, along with the major elements, as MnO, to the nearest 0.01 wt%. Reported values are extremely constant, independant of MgO content, at 0.12–0.14% (i.e., 920 to 1,085 ppm Mn). This way of reporting the Mn data makes it probable that rounding errors are larger than the true dispersion of Mn in mantle samples. The value chosen is from rare studies that listed their Mn data to greater precision (including a high proportion by INAA).

^{g}Mean value of 18 ppm from O'Neill (1991b); see discussion in text.

^{h}From BVSP (1981) and McDonough and Sun (1995). Data for Ge in peridotites are rare, but the Ge/Si ratio in the BSE is very constant (data from Malvin and Drake, 1987). Data for Cd and Ag are very sparse; the estimate for Ag is little more than a guess.

^{i}Data listed by O'Neill et al. (1995).

^{j}The other highly siderophile elements are presumed to be present in chondritic ratios to Ir (O'Neill et al., 1995). Their BSE abundances (in ppb) are therefore as follows: Ru, 6.0; Rh, 1.1; Pd, 4.7; Os, 4.2; Pt, 8.4; Re, 0.3; Au, 1.22.

Note: Data sources for peridotite suites listed in Table 1.6.

The almost-trivial correction for continental-crust extraction can be made using equation (4) and the mean crustal abundances of these elements, as given by Taylor and McLennan (1995) and Rudnick and Fountain (1995).

The abundances of the slightly incompatible trace elements in the sources of mantle-derived partial melts can also be estimated from their observed abundances in the melts, either by assuming a particular melting model (e.g., Sun, 1982) or by using a technique to be described later for the more incompatible elements. This approach is inherently less precise than the direct one of looking at peridotites

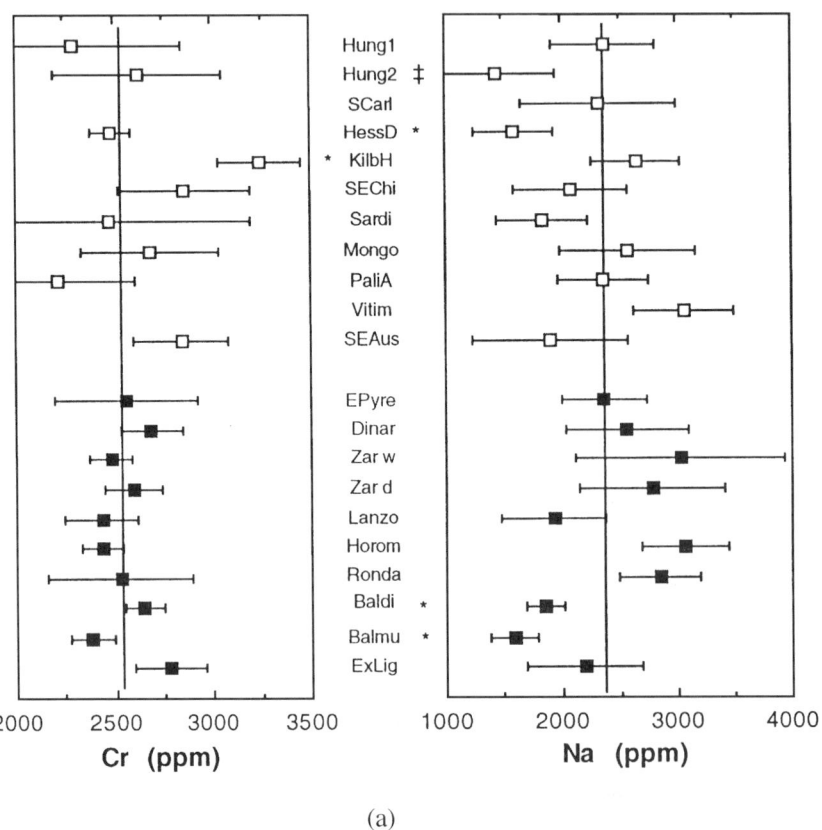

(a)

Figure 1.16. Some compatible and moderately incompatible elements in the PDM: calculated values at 36.6 wt% MgO for individual peridotite suites. Data sources are listed in Table 1.6. *These suites were given half weights in determining the grand mean.‡ These suites were excluded. (a) Cr and Na. Cr is nearly constant; only the KilbH suite appears anomalous. A sample of the same suite analyzed by Jagoutz et al. (1979) gave 2,460 ppm (KH1), in agreement with the results from other suites. Thus the anomaly may be analytical. Na shows considerably more scatter. Data for three suites analyzed in the same laboratory (HessD, Balmu, and Baldi) are all low, but older analyses in the literature for Balmu indicate this may be real. (b) Ni and Co are essentially constant. (c) V and Zn. V shows some scatter. Unlike the situation for Na, there is evidence that this may be an artifact: There is a statistically very significant correlation between the calculated value of V at 36.6% MgO in each suite and the slope V/MgO. This kind of effect is just what is to be expected if the V/MgO correlation is sensitive to the modal-inhomogeneity problem. Zn is mostly constant, but is probably affected by metasomatism in some suites.

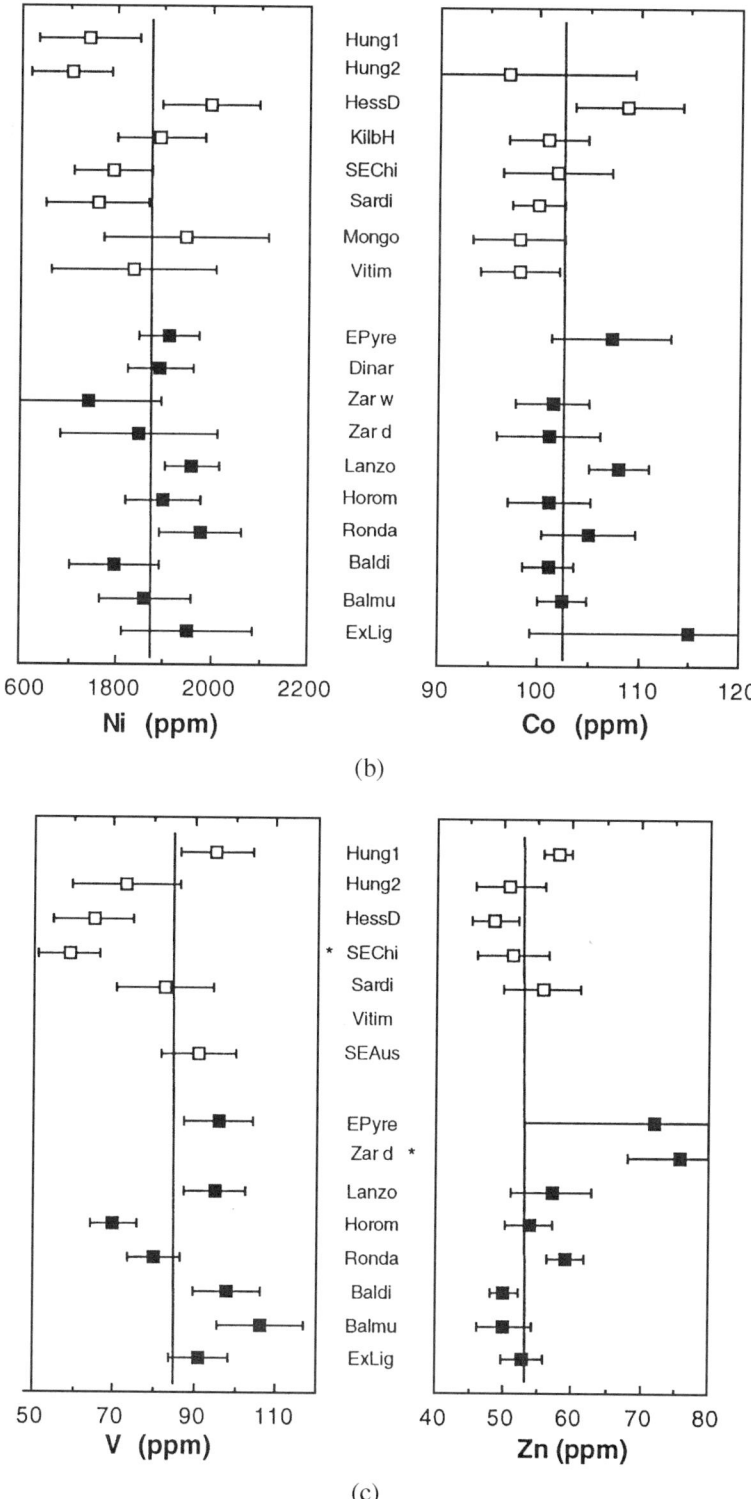

Figure 1.16 (*cont.*).

themselves, and it involves more assumptions. It should be most accurate when applied to those komatiites whose chondritic Ca/Al and Al/Ti ratios suggest that they were derived during such extensive melting that the only residual phase was olivine. The source abundances of compatible elements such as Ni and Cr are difficult to estimate accurately, as their concentrations in melts are heavily affected by low-pressure fractional crystallization of olivine and spinel. Undoubtedly though, the value of the partial-melt approach is that it brings more of the mantle into the picture. Because abundances inferred from melts are indistinguishable, within their probable uncertainties, from those derived from peridotites (Sun, 1982; McDonough and Sun, 1995), they add considerably to the robustness of the latter.

1.4.4.3. Group 3: Chalcogenides and Chalcophile Elements

This group needs special consideration because of the special properties of the mantle sulphide phase. The S in sub-solidus mantle is entirely concentrated in the sulphide phase. Because it is self-evidently a major structural component of that phase, S behaves as a major element, and its abundance in a partial melt reflects the S saturation level of that melt, not the S abundance of the source. The most probable value for the S content of the upper mantle is ∼200 ppm (O'Neill, 1991b). Se and Te substitute as trace elements for S. The available data are consistent with the chondritic Se/S ratio.

The other chalcophile elements are defined, for present purposes, as those whose mantle complements are substantially contained in the mantle sulphide phase. They include the six platinum-group elements (PGEs) Ru, Rh, Pd, Os, Ir, and Pt, plus Re and Au (these eight elements are conveniently considered together as a group, the highly siderophile elements, HSEs), probably Cu and Ag, and, with varying degrees of likelihood, elements such as As, Sb, Pb, and so forth. The uncertainty in classifying these latter elements is due to our ignorance of their distributions among the phases of a peridotite.

The apparent compatibility of these chalcophile elements is primarily due to their retention in the mantle sulphide phase during partial melting, until the latter is melted out (which usually occurs in the Earth only for partial melts of the very highest degree, e.g., komatiites, or for fluxed remelting of a previously depleted source, e.g., boninites). Establishing their abundances will require accurate knowledge of the behaviour of sulphide in mantle processes, which knowledge is presently inadequate. In particular, the possible mobility of sulphide in orogenic peridotites (Lorand, Keays, and Bodinier, 1993; Pattou, Lorand, and Gros, 1996) and, for xenoliths, vaporization of sulphide during eruption (O'Neill et al., 1995) can be anticipated to cause difficulties. Nevertheless, the simple empirical observation is that the abundance of Ir (the most often analyzed PGE) varies in mantle samples with a standard deviation of only 30% (O'Neill et al., 1995). The evolution of Os isotopes in the BSE is consistent with the chondritic Re/Os ratio to within ±10% (e.g., Meisel, Walker, and Morgan, 1996). We thus assume that the BSE has chondritic relative abundances of the PGEs, Re, and Au, but depleted to 0.8% of the absolute chondritic values.

The abundance of Cu in the depleted mantle raises a particular problem. Unlike other moderately compatible elements (Group 2), there appear to be differences between the Cu abundances in massive peridotites and most, but not all, of the xenolith suites from alkali basalts. The Cu-versus-MgO correlations in massive peridotites consistently extrapolate to PDM values around 30 ppm, whereas those for the xenoliths generally extrapolate to <20 ppm, albeit with much scatter. This scatter is markedly greater than that observed for the correlations between Group 2 compatible and moderately incompatible elements versus [MgO]. The value 30 ppm is relatively high when chondrite-normalized [$(Cu/Mg)_N = 0.11$], and its acceptance would imply Cu/Ni and Cu/Co ratios greater than chondritic, which would be difficult to explain. However, the Cu abundances in massive peridotites are correlated with S and may have been affected by the S mobility postulated by Lorand (1991). In xenoliths, Cu is not correlated with S, and its abundance in the xenoliths and also its abundances inferred from correlations in basalts and komatiites point to a substantially lower depleted mantle (and hence BSE) abundance of about 15 to 20 ppm (O'Neill, 1991b). We provisionally adopt the median of these latter values (18 ppm).

1.4.4.4. Group 4: Midrange Incompatible Trace Elements

This group comprises elements that are more incompatible than those of Group 2, but still have incompatibilities less than those of the most incompatible RLEs. These are, in order of increasing incompatibility, La, U, Ba, and Th, with Nb and Ta having incompatibilities approximately equivalent to that of U during MORB and OIB genesis, but being more compatible during island-arc-basalt (IAB) genesis and the processes that formed the continental crust (Hofmann et al., 1986). Of the volatile and siderophile elements, this group includes K, Rb, and probably Cs; P, Mo, Sn, and W; plus some elements that might be better considered in Group 2 or Group 3 if more were known about their properties, such as As, Sb, Cd, In, Bi, and possibly Pb.

Because these elements are so incompatible, we can expect that their extraction into the continental crust and their complementary depletion in the mantle have been too severe for us to derive primitive-mantle abundances directly from their current abundances as observed in mantle lherzolites. For the same reason, their enrichments in the OIB source reservoirs may be too high to justify ignoring the OIB sources in the approximation of equation (5). In addition, a further consequence of their incompatibility is the ease with which their abundances in mantle samples can be perturbed by metasomatism and late-stage alteration, plus, in the case of xenoliths, infiltration of the host magma. This is shown by the wide scatter of the data for the more incompatible RLEs (particularly U) in mantle samples (e.g., Jochum et al., 1989).

The abundances of these elements are best estimated from their abundances in mantle-derived partial melts. This is done using the analogue-element ratio technique (Hofmann and White, 1983). The argument is simple, but powerful: For any type of differentiation process (e.g., partial melting in the mantle), the concentration of an incompatible element E in the melt is given by some equation of the type

$$c_E^{\text{melt}} = c_E^{\text{source}} \times f\left(F, D_E^{\text{melt/source}}\right) \tag{9}$$

where F is the degree of melting, and $D_E^{\text{melt/source}}$ is the modally weighted melt/source distribution coefficient, which may itself be a function of F. Because we are dealing here with incompatible elements, their relative abundances will not be affected by modest amounts of late-stage fractional crystallization. Thus, if for two elements the ratio $c_{E1}^{\text{melt}}/c_{E2}^{\text{melt}}$ remains constant in basalts, while the absolute values of c_{E1}^{melt} and c_{E2}^{melt} vary by several orders of magnitude, then the most likely explanation is that $D_{E1}^{\text{melt/source}} = D_{E2}^{\text{melt/source}}$ over the range of F appropriate for the processes under consideration and that $c_{E1}^{\text{source}}/c_{E2}^{\text{source}} = c_{E1}^{\text{melt}}/c_{E2}^{\text{melt}}$.

The argument can be extended to other differentiation processes, particularly those responsible for the formation of the continental crust. We can then set up a mass balance for element $E1$ in all the quantitatively important geochemical reservoirs in the BSE to which we have access:

$$c_{E1}^{\text{BSE}} = c_{E2}^{\text{Re s.1}}\left(\frac{c_{E1}^{\text{Re s.1}}}{c_{E2}^{\text{Re s.1}}}\right)m^{\text{Re s.1}} + c_{E2}^{\text{Re s.2}}\left(\frac{c_{E1}^{\text{Re s.2}}}{c_{E2}^{\text{Re s.2}}}\right)m^{\text{Re s.2}} + \cdots$$

$$= \sum_i c_{E2}^{\text{Re s.}i}\left(\frac{c_{E1}^{\text{Re s.}i}}{c_{E2}^{\text{Re s.}i}}\right)m^{\text{Re s.}i} \tag{10}$$

If we find a constant ratio c_{E1}/c_{E2} in all reservoirs, equation (10) reduces to

$$c_{E1}^{\text{BSE}} = \left(\frac{c_{E1}}{c_{E2}}\right)\sum_i c_{E2}^{\text{Re s.}i}m^{\text{Re s.}i} \tag{11}$$

The BSE mass balance for $E2$ is

$$c_{E2}^{\text{BSE}} = \sum_i c_{E2}^{\text{Re s.}i}m^{\text{Re s.}i} \tag{12}$$

Substitution of (12) into (11) gives

$$c_{E1}^{\text{BSE}} = c_{E2}^{\text{BSE}}\left(\frac{c_{E1}}{c_{E2}}\right) \tag{13}$$

Therefore, if a constant ratio between two elements $E1$ and $E2$ can be identified in all reservoirs, and if $E2$ is a RLE, whose abundance in the BSE is assumed to be known, the BSE concentration of $E1$ can be determined. An advantage of the method is that basalts sample a large volume of mantle, providing a smoothing effect.

For the method to work accurately, the two elements being ratioed must have similar mean partition coefficients not only during partial melting in the mantle but also during all the other important processes of the Earth's differentiation, particularly those responsible for the formation of the continental crust. A cautionary example is provided by the systematics of the Nb/U ratio. Hofmann et al. (1986) showed that Nb/U is constant over several orders of magnitude of Nb and U in both MORBs and OIBs, at about 47 ± 10; from the foregoing argument, that constancy implies that Nb and U have similar incompatibilities during mantle melting and that this ratio is also the ratio in the sources for both MORBs and OIBs. However, because Nb and U are both RLEs, we know that their ratio in the BSE should be 30, as found in chondritic meteorites. It turns out that the high ratio in the MORB- and OIB-source mantle is compensated for by a low ratio in the continental crust of about 10 (also in IABs, the 'smoking gun' for their role in the origin of the continents), indicating that Nb and U did not behave congruently during the processes responsible for crust formation. It is worth noting, however, that mistaking the MORB–OIB ratio of 47 as primitive would have produced an error in the calculated BSE content of Nb or U of only 50%, which is, within the context of the present exercise, not totally unacceptable.

The volatile and siderophile elements whose mantle abundances have been estimated by the ratio method are listed in Table 1.9, together with their normalizing RLEs. The important heat-producing element K poses a slight problem, as its incompatibility falls between those of U and La. Thus the K/U ratio is quite constant in MORB at 1.27×10^4 (Jochum et al. 1983), but is about 1.0×10^4 in the continental crust (Taylor and McLennan, 1995; Rudnick and Fountain, 1995); the K/La ratio is equally constant at about 330 in MORB (e.g., Schilling et al., 1980) and in OIB and continental flood basalt (CFB), but rather higher in the continental crust. An incompatibility for K between those for U and La is also seen in lunar rocks (O'Neill, 1991a, fig. 3). Hence for K we adopt the mean of the estimates from the K/U and K/La ratios. The K/U ratio is very useful in planetary geochemistry, as it can be determined by remote sensing (γ-ray spectroscopy) and provides a ready measure of the depletion of the bulk planet in the moderately volatile elements.

A constant ratio preserved between two elements during all the pertinent large-scale differentiation processes which have occurred in the evolution of the BSE would seem to imply that the elements must share similar chemical properties. For some element pairs, the similarity is expected from their known basic chemistry; for example, the constant Rb/Ba ratio comes as no surprise, because of the general congruence in the crystal-chemical properties of Rb and Ba (e.g., similar ionic radii, electronegativities, polarizabilities). For other element pairs on the list, however, the match is purely empirical and cannot be justified by crystal chemistry. For example, P is traditionally ratioed to Nd (McDonough, 1985), with which it shares no obvious chemical similarities at all. Indeed, in spinel lherzolites, P is evenly distributed among olivine, orthopyroxene, and clinopyroxene, and because olivine is modally dominant, most of the P is actually in the olivine. Nd, by contrast, is almost entirely concentrated into clinopyroxene. Consequently, it would seem

Table 1.9. *Abundances of incompatible volatile and siderophile elements in the BSE*

Element (E)	Normalizing RLE	Ratio E/RLE	Abundance BSE	Depletion Factor[a]	References
Elements with constant normalizing RLE ratios in crust and mantle					
B	(K)	$1.0 \pm 0.3 \times 10^{-3}$	0.26 ppm	0.17	Chaussidon and Jambon (1994)
K[b]	U	1.3×10^4	290 ppm	0.23	Jochum et al. (1983)
K	La	330	230 ppm	0.18	This study
Rb[c]	Ba	0.09 ± 0.02	0.605 ppm	0.12	Hofmann and White (1983)
P	Nd	65 ± 10	90 ppm	0.036	McDonough (1985); Langmuir et al. (1992)
Mo	Ce	0.027 ± 0.012	50 ppb	0.024	Sims et al. (1990)
Mo	Nb[d]	0.05 ± 0.015	35 ppb	0.017	Fitton (1995)
W	Th	0.19 ± 0.03	16 ppb	0.077	Newsom et al. (1996)
In	Y	0.003 ± 0.001	14 ppb	0.078	Yi et al. (1995)
Sn	Sm	0.32 ± 0.06	144 ppb	0.038	Jochum et al. (1993)
Elements with differing analogue RLE ratios in crust and mantle					
Cs	Ba	1.1×10^{-3e} 3.6×10^{-3f}	18 ppb	0.042	McDonough et al. (1992)
As	Ce	9.6×10^{-3e} 0.08^f	50 ppb	0.012	Sims et al. (1990)
Sb	Ce	1.2×10^{-3e} 7×10^{-3f}	5 ppb	0.016	Sims et al. (1990)
Pb	Ce	0.040^e 0.11^f	110 ppb	0.019	Hofmann et al. (1986)
Pb[g]	(U)		120 ppb	0.021	McCulloch & Bennett (Chapter 2, this volume)

[a] Normalized to Mg and the value in CI chondrites, i.e., (element/Mg)$_{BSE}$/(element/Mg)$_{CI}$.
[b] K is intermediate in incompatibility between U and La (see text). We adopt a mean value of 260 ppm and a depletion factor of 0.21.
[c] Independently constrained by the Rb/Sr-isotope system, with Rb/Sr $= 0.029 \pm 0.002$.
[d] Mo/Nb appears more nearly constant than Mo/Ce, but more data are needed.
[e] Mantle.
[f] Crust.
[g] Abundance far better constrained from BSE μ, $(^{238}U/^{204}Pb) = 8.5 \pm 0.5$.

unlikely that $D_P^{\text{melt/source}}$ could remain the same as $D_{Nd}^{\text{melt/source}}$ over a wide range of degrees of melting, F, as the ratio method requires, or over different P–T regimes of melting. We surmise that such complications are subsumed in the scatter of the data. Given our present state of knowledge, we can only point to the empirical evidence and say that the method works better than it should. For other elements (e.g., As, Sb, Mo, and W), almost nothing is known about their distribution among the upper-mantle minerals, including how much may be held in the mantle sulphide phase. This is important, as the behaviour of the sulphide phase is decoupled from that of the silicates during melting, so that a chalcophile element held in sulphide will show changes in mean $D_M^{\text{melt/source}}$ as a function of F that will be quite different from those for any RLE. Our current state of ignorance does not inspire confidence, and clearly, more work is needed on element distributions among the major mantle mineral phases.

Satisfactory analogue RLEs have not been found for four elements: Cs, As, Sb, and Pb. Although for each of these elements a constant analogue-element ratio can be identified for the mantle sources of MORBs and OIBs (e.g., Pb/Ce) (Hofmann et al., 1986), the ratios for those analogues appear different in the continental crust, and also usually in IABs. The possible reasons are still matters of debate, but may be due to (1) different behaviours of the element and its analogue during continental growth and arc magmatism (e.g., different solubilities in an aqueous fluid phase transporting one element preferentially from subducting oceanic crust) or (2) the role of the mantle sulphide. For these elements, an estimate of the BSE value can still be obtained from equation (2) if estimates of the appropriate values for crustal abundances and depleted-mantle abundances are made. This has been done for As and Sb by Sims, Newsom, and Gladney (1990, appendix). Estimates that have to be obtained in this way are obviously far less robust.

It is observed that the values of an analogue-element ratio measured in a large number of samples generally follow a log-normal distribution. The optimum estimate of the ratio's value is thus its geometric mean (i.e., from the mean of the logarithms of the ratios), and the uncertainty attached to this value is the geometric standard deviation (from the standard deviation of the logarithms of the ratios). Because these standard deviations always greatly exceed analytical uncertainty, it is apparent either that these ratios are not being sampled from a single uniform source (i.e., the scatter in the data reflects real scatter in the source rocks) or that different processes are in fact altering the ratio to some extent, or both. Accordingly, we believe that the standard error of the mean ($\sigma / \sqrt{N}$) is not the appropriate measure of the uncertainty; that is, the uncertainty in an analogue-element ratio should not be considered as tending to zero as N tends to infinity. Mean ratios and what we consider to be more reliable estimates of their uncertainties are given in Table 1.9.

A historical note: The method of analogue-element ratios appears to have been first used with samples returned from the Moon (e.g., Wänke et al., 1973), as a means of comparing the Moon's chemistry with that of the BSE. An approach like this, based on basalts accessible at a planet's surface, is at present a necessity for planetary matter, because mantle samples are not available from any planetary body except from the Earth. Fortunately, the method is particularly suited to extraterrestrial applications, because the Moon, the meteorite parent bodies, and probably the other terrestrial planets all appear to have had simpler tectonic histories than the Earth; most importantly, these bodies have no equivalent of the Earth's highly evolved continental crust. Because they have not suffered the same plenitude of differentiation processes, these other planetary bodies are more likely to have maintained a constant analogue-element ratio in each reservoir in their silicate portions (their mantle–crust systems). For example, the closest analogue to the Earth's crust in terms of enrichment in the incompatible elements is the KREEP component on the Moon (the acronym is for enrichment in K, REEs, and P). KREEP is probably residual from early fractional crystallization of a large part of the Moon (Palme and Wänke, 1975), according to the lunar magma-ocean model. The sequence of enrichment of the incompatible elements is generally similar to that in the Earth's crust, although there are some differences that must reflect different processes on Earth, plausibly due either to the higher pressures achievable in the Earth, or to the influence of water. Indeed, as shown by Hofmann et al. (1986), analogue element ratios, especially those involving two RLEs (e.g., Nb/U), can be useful tools for unravelling the Earth's more complex differentiation history.

1.4.4.5. Group 5: Very Incompatible Elements

Obviously, elements that are more incompatible than the most incompatible RLE (probably Th) cannot be normalized to a suitable RLE. Such elements include Cl, Br, I, the rare gases, and N. Their extreme incompatibility means that they have become so highly concentrated in the continental crust, oceans, or atmosphere that their BSE abundance largely depends on their abundance in these reservoirs. Other incompatible elements that cannot be ratioed satisfactorily to an RLE because of their distinctive geochemical properties are H, C, and O. There is no alternative but to estimate the abundances of all these elements using the approximate relationship of equation (4) (in which the oceans and atmosphere are counted as continental crust). For example, the amount of H_2O in the hydrosphere is 1.7×10^{21} kg (oceans, pore water in sediments, and ice). If this water has come from degassing of half the mantle (i.e., $m^{dm} = 0.5$), then it alone contributes 850 ppm to the BSE inventory. Michael (1988) argued that the incompatibility of H_2O during mantle melting processes was intermediate between those for Ce and Nd, and he suggested that $H_2O/(Ce + Nd) = 95 \pm 30$, from which we obtain a PDM value of 250 ppm H_2O. Allowing another 0.5 wt% for the mean crustal content of chemically bound water gives a BSE abundance of 1,160 ppm H_2O. Because the inventory of H_2O in the exosphere is quite well constrained, the uncertainty of this estimate is about $\pm 20\%$. Other highly incompatible elements are determined similarly, if not as accurately (Table 1.10).

There are three elements that need special consideration. C can occur in the mantle as a carbonate component under oxidized condition, or as graphite or diamond or possibly methane under reducing conditions, and it exhibits entirely different geochemical behaviours accordingly (e.g., diamond or graphite is thought to be insoluble in silicate melt, and thus presumably is infinitely 'compatible'). The current estimate (150 ppm), based on the work of Zhang and Zindler (1993), assumes that such 'compatible' C is negligible in the residual depleted mantle.

Apart from the rare gases, the most incompatible element is iodine. However, unlike the rare gases, I is not concentrated in a readily measurable reservoir such as the atmosphere; rather, the main exospheric reservoir seems to be the organic material in marine sediments. Within this reservoir, the I has been further concentrated at

Table 1.10. *Abundances of highly incompatible volatile elements in the BSE*

Element	BSE	Depletion factor	References
H	130 ppm[a]	2.8×10^{-3}–0.01[b]	See text
C	150 ppm	2×10^{-3}	Zhang and Zindler (1993)
N	3 ppm	4×10^{-4}	Zhang and Zindler (1993)
F	20 ppm	0.15	Jagoutz et al. (1979); Schilling et al. (1980)
S	200 ppm	1.6×10^{-3}	O'Neill (1991b)
Cl	30 ppm	0.020	Jambon et al. (1995)
Br	80 ppb	0.010	Jambon et al. (1995)
I	7 ppb	7×10^{-3}	See text

[a]Corresponding to 1,160 ppm H_2O.
[b]Depending on whether the H_2O content of CI chondrites is 18 wt%, as analyzed, or 5 wt%, because of terrestrial hydration (Kaplan, 1971).

the sediment/seawater interface by early diagenetic alteration (e.g., Price, Calvert, and Jones, 1970). To estimate the mean amount of I in the whole reservoir, we adopt the I/(organic C) ratio in the main source of organic carbon in marine sediments, namely, plankton: These little organisms have I/(organic C) $= 10^{-3}$. The total organic-C content of all sediments is 1.2×10^{16} kg (Hunt, 1972), giving 1.2×10^{19} kg of I in this reservoir, which contributes 6 ppb towards the integrated BSE abundance. Other crustal sources plus the residual I in the PDM (Déruelle, Dreibus, and Jambon, 1992) bring this total up to 7 ppb.

The oxygen content of the BSE is the sum of the oxygen attached to the major and trace elements in their familiar oxidation states (i.e., MgO, SiO_2, Cr_2O_3, NiO, H_2O, CO_2, etc.), with all Fe counted as FeO. This gives a total of 44.35 wt%. We then subtract the small amount of oxygen that is needed to convert FeO to FeS for a mantle abundance of 200 ppm S (i.e., 0.01%) and add the extra oxygen required for the PDM's observed Fe_2O_3 content of 0.3 wt% (Canil et al., 1994). The total is now 44.37 wt%. Fe_2O_3 behaves as a moderately incompatible component, so ignoring its crustal abundance should not alter matters significantly.

Looked at in this way, it can be seen that the 'oxidation state' (i.e., what would be the chemical potential of oxygen at specified P and T) of the BSE is very sensitive to tiny changes in the oxygen content. A more useful measure of oxidation state than oxygen content is provided by the $Fe^{3+}/\sum Fe$ ratio, which is 0.033 (Canil et al., 1994).

1.4.5. Mantle Compositions from 'Primitive' Mantle Samples

An alternative approach to estimating BSE abundances, pioneered by Jagoutz et al. (1979), is to search for samples of the mantle that are chemically 'primitive' (i.e., specimens that appear to have escaped differentiation). Such samples would be recognizable because they would have a single, constant, chondrite-normalized abundance for all of the RLEs, including a flat chondrite-normalized REE pattern. No such philosopher's stone has ever been found, although certain specimens appear to come quite close. The best-known example is a xenolith from San Carlos, Arizona, SC1 (Jagoutz et al., 1979; Wänke et al., 1984; Jochum et al., 1989). Another is Ib/8, from Dreiser Weiher, Germany (Stosch and Seck, 1980). These samples and their like are all spinel-lherzolite xenoliths transported from the sub-continental lithosphere in alkali basalts. The problem with this approach is in determining whether such samples are truly primitive or were once depleted and then metasomatically re-enriched (as were many of their companions from the same localities), the enrichment being by chance just sufficient to bring the REE pattern back to approximate flatness. That this is likely may be deduced from the following observations:

1. Most such samples have somewhat non-chondritic Ca-Al-Ti ratios (e.g., the Ca/Al of SC1 is 1.17 times the CI ratio).
2. In detail, none of the REE patterns is truly flat. SC1 is slightly hump-shaped, with $(Sm/Yb)_N = 1.18$ (Jochum et al., 1989).
3. Radiogenic isotopes (particularly Nd) show that these samples have not had primitive parent–daughter ratios for the entire span of the Earth's history (e.g., $\varepsilon_{Nd} \neq 0$) (Jagoutz, Carlson, and Lugmair, 1980).
4. Th is always depleted [e.g., $(Th/Nd)_N \ll 1$]. Although it is the most incompatible RLE, Th appears to be less affected than most by contamination/alteration effects.

Nevertheless, if the constancy of ratios is maintained in the LREE-enriching metasomatism, this 'quasi-primitive-xenolith' approach still should give approximately correct values for the BSE abundances of volatile/siderophile elements with incompatibilities similar to those covered by the REEs. Most element abundances for the BSE listed by Jagoutz et al. (1979) and Wänke et al. (1984) are indeed similar to those of McDonough and Sun (1995) and those given here. The objection to this approach is that it uses only a tiny fraction of the available information.

1.4.6. Evidence against Gross Compositional Layering in the Mantle

The relevant alternative view to the 'dominant paradigm' for the chemical structure of the mantle pictured in Figure 1.10, and assumed here, is that there is a substantial difference in major-element chemistry, particularly Mg/Si and Mg/Fe ratios, between the upper mantle above the 660-km seismic discontinuity and the lower mantle below that discontinuity. This kind of layering may have occurred either as a direct result of inhomogeneous accretion without subsequent mixing (which has not, to our knowledge, been seriously suggested in recent times) or by some process akin to crystal fractionation from an early magma ocean (which has been suggested). The latter process would also imply gross layering of trace elements and would invalidate the conclusions drawn here concerning BSE volatile- and siderophile-element abundances. For example, the nearly chondritic Ni/Co ratio observed in the upper mantle would be a fortuitous consequence of olivine flotation into the upper mantle (Murthy, 1991).

Three kinds of evidence have been put forward in support of the idea of a lower mantle with a composition different from that of the upper mantle. The first is the apparent lack of a match between the seismic and other geophysical properties observed for the lower mantle and the laboratory-measured properties of lower-mantle minerals ($MgSiO_3$-rich perovskite and magnesiowüstite) for an assemblage with the upper-mantle composition (meaning, effectively, with the upper mantle's Mg/Si and Mg/Fe ratios). This topic is discussed by Jackson and Rigden in Chapter 9 of this volume, and they conclude that there is no such mismatch.

The second line of evidence is that crystal fractionation from an early magma ocean would inevitably have led to layering. The fluid-dynamic reasons why this was not inevitable (and in fact was unlikely) are discussed by Tonks and Melosh (1990) and Solomatov and Stevenson (1993). In fact, these studies show that the evident lack of layering in the mantle (as discussed later) can no longer be used as an argument against the existence of an early magma ocean. The presence of such an ocean seems unavoidable if the accretion of the Earth did involve giant impacts (Tonks and Melosh, 1993).

The third kind of evidence is that the upper-mantle composition violates the cosmochemical constraints on BSE compositions that have been obtained from the meteorite record. Here we argue that there are, in fact, no such constraints (for Mg/Si ratios), or else that the apparent deviation from the constraints (e.g., the tendency towards a non-chondritic Ca/Al ratio noted by Palme and Nickel, 1985) is best accounted for by means other than gross compositional layering.

The geochemical evidence against gross compositional layering of the mantle seems to us conclusive. This evidence is the observation that the upper-mantle composition, as deduced from the depleted-mantle/continental-crust model, conforms, within uncertainty, to the cosmochemical requirement of having strictly chondritic ratios for the RLEs. These ratios would not have survived large-scale differentiation by fractional crystallization. The most precise line of argument comes from the observed secular evolution of ε_{Nd} and ε_{Hf} towards their present values in the depleted-mantle reservoir, which required chondritic Sm/Nd and Lu/Hf ratios to within about $\pm 10\%$ (the ε_{Nd} evidence is discussed by McCulloch and Bennett, Chapter 2, this volume). Experimental data (e.g. Kato, Ringwood, and Irifune, 1988) show that Sm/Nd and Lu/Hf would not have remained unfractionated if there had been any significant $MgSiO_3$-perovskite or majoritic garnet fractionation. The near-chondritic Re/Os ratio in the upper mantle (presumably from a late veneer added after core formation, as discussed later) argues against differentiation of the mantle by olivine flotation, given that Os, but not Re, appears to be compatible in olivine (O'Neill et al., 1995).

Much of the intellectual driving force for postulating major compositional differences between the upper mantle and the lower mantle arose out of misconceptions regarding the nature of the constraints imposed by cosmochemical arguments and the study of chondritic meteorites. A particularly egregious misconception was that the bulk Earth should possess exactly the solar (or CI) Mg/RLE and Si/RLE ratios. That seems inherently unlikely, given that CI material probably is quite rare amongst material in the inner solar system and that the bulk Earth obviously is not CI-like in its abundances of the moderately volatile elements, the highly volatile elements, or the ice-forming elements, nor probably with respect to its siderophile/lithophile-element ratio. The study of meteorite cosmochemistry attests to the variety of chemical-fractionation processes operating in the early solar nebula. Because of this variety, cosmochemical arguments can only set limits on planetary compositions, and these limits are much broader than desirable for characterizing our planet. Thus an estimation of the composition of the Earth's mantle needs to be derived as empirically as possible from the study of Earth materials, as in the approach used here. Once the BSE composition is estimated, however, its significance can be illuminated by reference to the cosmochemical observations. This is the goal of the next section.

1.5. The Composition of the BSE in Context: Implications for Accretion
1.5.1. Significance of the Mg-Si-RLE Ratio of the BSE

The estimated Bulk Silicate Earth composition is compared with the trends shown by the chondrite meteorites on the Mg/Si-versus-Al/Si diagram in Figure 1.17. The BSE composition plots above the carbonaceous-chondrite trend. One possible interpretation is that the BSE ratios are the same as those of the bulk Earth (i.e., implying negligible Si in the core). Such ratios can be envisaged as being achieved by a cosmochemical fractionation process that combined the process responsible for the carbonaceous-chondrite trend with the reciprocal of the process responsible for the ordinary-chondrite–enstatite-chondrite trend; that is, if the ordinary-chondrite–enstatite-chondrite trend was caused by subtraction of material from the solar (CI) composition, then the Earth may have incorporated some of that material.

Alternatively, it is possible that the core contains some Si (experimental studies indicate entirely negligible solubility of Mg and Al into Fe-rich metal at reasonable core-forming conditions of temperature, pressure, and oxygen fugacity) (e.g, O'Neill, Canil, and Rubie, 1997). The diagram shows that 5% Si in the core would bring the composition of the bulk Earth onto the carbonaceous-chondrite trend, in the vicinity of the CV composition. We shall see later, however, that the pattern of depletion of volatile elements in the Earth differs somewhat from that in the CV chondrites, so perhaps not too much should be made of this observation.

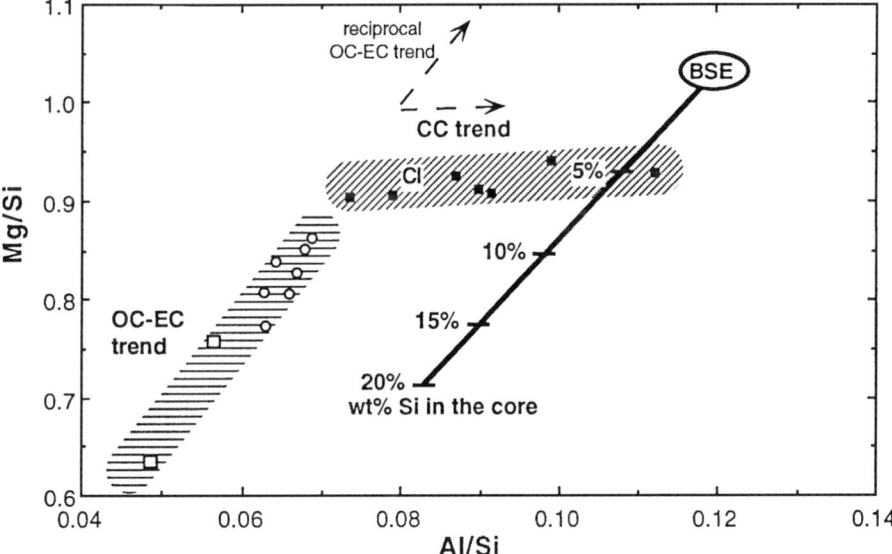

Figure 1.17. Mg-Si-RLE ratios in the bulk Earth compared with those in chondrites. The shaded regions show the ordinary-chondrite/enstatite-chondrite trend (OC-EC) and the carbonaceous-chondrite trend (CC), as shown in detail in Figure 1.6. The effect of some Si in the core is also shown. With a core content of 5 wt% Si, the bulk Earth would fall on the carbonaceous-chondrite trend, close to CV chondrites. Note, however, that the BSE is more depleted than the CVs in moderately volatile elements, and the bulk Earth must have a higher metal/silicate ratio (i.e., the Earth is not a CV chondrite). Alternatively, the BSE composition could be produced by a combination of the process that caused the carbonaceous-chondrite trend and the reciprocal of the process responsible for the ordinary-chondrite–enstatite-chondrite compositions (shown as vectors).

1.5.2. Depletion of Moderately and Highly Volatile Lithophile Elements

The moderately and highly volatile lithophile elements are the alkali metals (Li, Na, K, Rb, and Cs), the halides (F, Cl, Br, and I), B, Mn (that this element has negligible siderophile tendencies in the Earth will be argued in the next section), and possibly Zn and Cd. For the less volatile of these elements there is excellent correlation of their depletions in the BSE with their calculated condensation temperatures (Figure 1.18a). Moreover, this trend is approximately parallel with that found in the chondrites (e.g., the CV trend, shown in Figure 1.18b). For the more volatile elements that condense below about 900 K (as calculated at 10^{-4} bar) there are some intriguing differences. The heavier halides (Cl, Br, and I), and also N and perhaps C, show markedly enhanced depletions, whereas Zn and In are considerably less depleted (Zn and In are also potentially siderophile or chalcophile, so some of the bulk-Earth Zn and In might perhaps be in the core; consequently, their BSE abundances record the maximum possible influence of volatility on their depletion). These two groups of elements differ in two ways that might have produced their differences in volatility-related depletions. The members of the group consisting of the heavy halides and N and C are highly incompatible and therefore may have

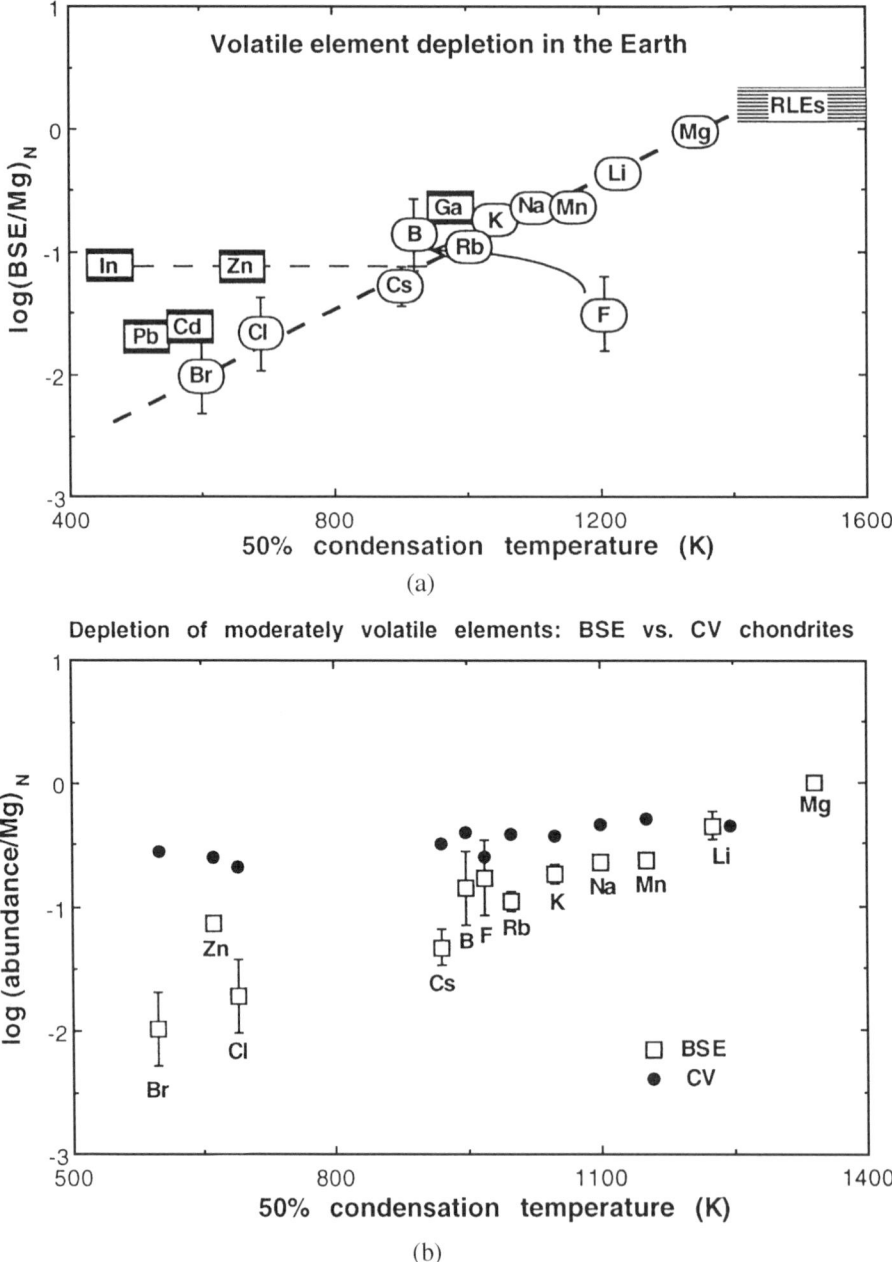

Figure 1.18. Depletion of volatile lithophile elements in the BSE. (a) The depletion follows
a smooth pattern for calculated condensation temperatures down to 900 K. At lower temperatures there is more scatter: The heavier halides show additional depletion. Elements
in square boxes are potentially chalcophile; therefore, their depletions might have been
attributed, a priori, to partitioning into the core, as well as volatility. It is very significant
that there is no indication for that in the BSE abundances of Ga, Zn, and In. The abundance of Zn, in particular, is very accurately known. Cd is less chalcophile than Zn, but
more depleted. (b) The pattern above 900 K parallels that of the volatile-element-depleted
carbonaceous-chondrite groups (e.g., CV). Below 900 K, the BSE values for the Zn-In
group follow the chondrite trend (i.e., it is the more depleted halides that are anomalous).

been concentrated early into the accreting Earth's exosphere, from which they may have preferentially been lost during subsequent giant impacts. Zn and, to a lesser extent, In are, by comparison, relatively compatible and could not have been lost preferentially from the Earth in this way. Alternatively, the halide grouplet consists of elements whose volatilities, relative to other volatile elements, are predicted to increase with increasing fO_2, whereas the volatilities of Zn and In decrease with increasing fO_2. The fO_2 of the accreting Earth would have been much higher than that of the solar nebula, under which the CV volatility-related depletion trend was established. Thus the difference in behaviour of the two groups may reflect volatile loss under more oxidizing conditions. We tentatively favour the latter explanation, because we shall argue below that all these volatile elements were, in any case, added late in the accretion process. Siderophile elements whose volatilities increase with increasing fO_2 also appear rather more depleted than their fellows (e.g., P, As).

Elements such as Pb and Cd, which can volatilize either as monatomic species or as oxide species, show intermediate depletions, although their depletions relative to Zn and In may also have been due to their chalcophile tendencies, leading to greater partitioning into the core. In the case of Pb, the distinction is important, as the separation of radiogenic Pb from the RLE parents U and Th is conventionally taken to record 'the age of the Earth' (e.g., McCulloch and Bennett, Chapter 2, this volume). If Pb was lost to the core, then such an age is actually the age of core formation. On the other hand, if Pb was lost by reason of its volatility, such an age may reflect the timing of the final stages of accretion.

It is the pattern of depletion of the Zn-In grouplet, rather than that of the halides, that, when considered in conjunction with the pattern established by the moderately volatile elements (those with calculated condensation temperatures greater than 900 K at 10^{-4} bar), best mimics the pattern of depletion in chondrites (Figure 1.18b). As previously remarked, this trend rather distinctively flattens out in the temperature range 800–400 K (Figures 1.9 and 1.18b). We interpret the similarity in the shapes of these trends to imply the following: (1) The depletion of Zn and In in the BSE was by volatility alone, and despite the potential for these two elements to behave as chalcophiles, their BSE abundances have not been significantly affected by partitioning into the core. (2) The halides were depleted further by some process that did not affect Zn and In; for example, by volatile loss at an fO_2 higher than that of the solar nebula. It should be noted that the BSE is more depleted in the moderately and highly volatile elements than is any chondrite group (the most depleted groups being the CV and CK carbonaceous chondrites).

We showed earlier that the fractionation of Mg and Si from the RLEs is related to volatility (Figure 1.7a), but perhaps surprisingly is not correlated with depletion of the moderately volatile elements (Mn, Na, etc.) as illustrated in (Figure 1.7b). That the BSE should be depleted by both volatility-related processes may make intuitive sense, but we should not be tempted into trying to correlate the two processes, in light of the lack of correlation in other solar-system materials. In other words, the extent of the depletion of the moderately volatile elements cannot be used to determine the total amount of Si in the bulk Earth, hence the core.

An important observation recently reported by Humayun and Clayton (1995a,b) is that K shows remarkable isotopic homogeneity in various solar-system bodies, including the Earth (and the Moon). Therefore, the depletion of K in the Earth was not by a mass-dependent fractionation process such as Raleigh fractionation. This is in agreement with the depletion trends of Figure 1.18, in which there is no sign that the degree of depletion increases with the mass of the volatile element (e.g., the Li-Na-K-Rb-Cs depletion, or Cl vs. Br). Catastrophic loss of an early-formed exosphere would be one process that would have been mass-independent.

1.5.3. Mn/Na Systematics

Whereas Na is widely recognized to be a moderately volatile lithophile element, the status of Mn regarding its depletion in the Earth's mantle (volatile, siderophile, or both) is still disputed. Figure 1.19 shows that in nearly all of the groups of chondritic meteorites, the ratio Mn/Na remains constant to within ±30%, even

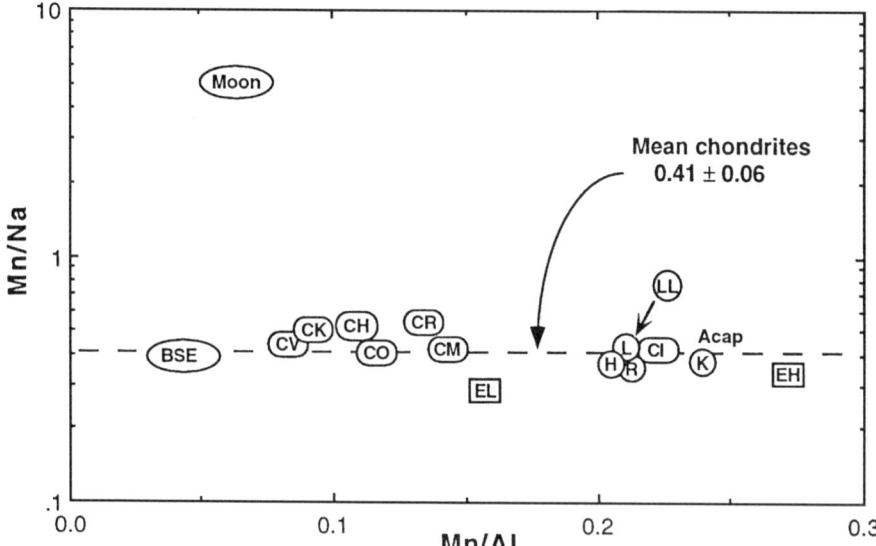

Figure 1.19. Mn/Na vs. Mn/Al in chondrites and in the BSE (symbols as in Figure 1.6). Acap is Acapulco. The Mn/Na ratio is constant in all chondrites, irrespective of the degree of volatile-element depletion, represented by the Mn/Al ratio. This suggests that Mn and Na had similar volatilities in the solar nebula. The BSE has the same Mn/Na ratio, whereas the Moon (also Mars and the eucrite parent body) has a much higher Mn/Na ratio, indicating secondary loss of Na at conditions more oxidizing than those in the nebula. Chondrite data are mainly from Wasson and Kallemeyn (1988), supplemented by data from Graham et al. (1977), Kallemeyn et al. (1994), and Palme et al. (1996). The slightly lower values for the enstatite chondrites (EL and EH) are supported by the recent data of Zhang et al. (1995) and may be due to greater volatility of Mn, compared with Na, at conditions more reducing than those of the canonical solar nebula.

while the abundance of Mn or Na relative to the RLEs (represented by Mn/Al in the diagram) decreases threefold (the enstatite chondrites may have slightly lower Mn/Na ratios, as discussed later).

It can be concluded that Mn and Na show similar volatilities under cosmochemical conditions. This is approximately consistent with the results from the canonical condensation calculation, insofar as the calculation predicts that both Mn and Na are moderately volatile elements with 50% condensation temperatures falling within the less volatile part of this group. In detail, however, the difference in the calculated 50% condensation temperatures (1,190 vs. 970 K, Table 1.5) would seem to imply a somewhat greater difference in volatilities than that which is observed. This may be due to inadequacies in the condensation calculation. It is worth noting, nonetheless, that the condensation calculation does predict that Mn condenses as a lithophile element, and not as a siderophile with the Fe-Ni-Co group. This prediction has been confirmed by the abundances of Mn in the recently discovered group of metal-rich CH chondrites.

The BSE Mn/Na ratio plots exactly on the chondrite trend (Figure 1.19). This may be coincidence (with several processes that fractionate Mn from Na in different directions summing to this particular ratio), but the simplest (Occamist) interpretation is that both Mn and Na are depleted in the Earth for the same reason that they are in chondrites, namely, volatility. There is thus no reason to believe that the depletion of Mn in the BSE is due to its having been partitioned into the core.

The common Mn/Na ratio in the chondrites and in the Earth has further significance. The main gaseous species involved in cosmochemical condensation/fractionation processes are monatomic Mn(g) and Na(g). Mn occurs in condensed phases as the MnO component (i.e., divalent Mn^{2+}, such as Mn_2SiO_4 in olivine), whereas Na is monovalent and occurs as the Na_2O component (e.g., as $NaAlSi_3O_8$ in plagioclase). Consequently, their condensation/volatilization behaviour can be described by the reactions:

$$MnO(c) = Mn(g) + \frac{1}{2}O_2(g) \tag{14}$$

$$\frac{1}{2}Na_2O(c) = Na(g) + \frac{1}{4}O_2(g) \tag{15}$$

and therefore the ratio of their volatilities [e.g., expressed as $(p_{Na}/a_{NaO_{\frac{1}{2}}})/(p_{Mn}/a_{MnO})$] is proportional to $(fO_2)^{1/4}$. This implies that Mn and bulk Na can have similar volatilities only at one particular fO_2 (temperature and composition of solids assumed constant). Na will become relatively more volatile than Mn at higher fO_2 and vice versa at lower fO_2.

On the basis of the evidence of the constant Na/Mn ratio in the chondrites, it appears that the Mn and Na abundances were set while fO_2 was being controlled by the H-C-O ratios of the solar nebula. [There is some evidence that the enstatite chondrites formed under slightly more reducing conditions in the nebula; hence the

slightly lower Mn/Na ratios in both the EH and EL groups. More-reducing conditions are also implied by the distinctive enstatite chondrite nebula metal/silicate
fractionation trends for enstatite chondrites discussed in Section 1.3.4. Small differences in oxidation states of this kind are not necessarily due to compositional
differences in the nebula, but may be a result of different nebular pressures, hence
condensation temperatures.] That the Earth's mantle has the same Mn/Na ratio as
the carbonaceous chondrites and ordinary chondrites implies that the terrestrial
abundances of Mn and Na were also set under solar-nebular conditions, not during
any subsequent heating event, such as during accretion. The reason one can make
these statements with confidence is that fO_2 in the solar nebula was buffered by
the nebular H-C-O ratios, because of the dominance of the nebular composition
by these elements. After dispersion of the solar nebula, the fO_2 for a condensed
assemblage would have been internally controlled. In planetary material, this internal buffering presumably was at the fO_2 of the metal/silicate equilibrium [i.e.,
$FeO_{(in\ olivine/orthopyroxene)} = Fe + \frac{1}{2}O_2$], but could have been at more oxidizing conditions if Fe metal was not present. Note that the Mn/Na ratio in the BSE is that
of the ordinary chondrites and carbonaceous chondrites, not that of the enstatite
chondrites.

The Moon provides an example of a body with an emphatically non-chondritic
Mn/Na ratio, with Na being more depleted than Mn by an order of magnitude.
This observation implies that the loss of volatile elements in the Moon, unlike that
in the Earth, did not occur under nebular conditions, but during some later event,
under much more oxidized conditions (e.g., a giant impact). A quantitative explanation has been provided elsewhere (O'Neill, 1991a). Somewhat unexpectedly, Mars
also appears to have super-chondritic Mn/Na (Wänke and Dreibus, 1988), as does
the parent body of the HED achondrites. As for the Moon, this indicates loss of
moderately volatile elements under post-nebular conditions. The chondritic (hence
'primitive') Mn/Na ratio in the BSE is thus unusual for differentiated bodies!

In Figure 1.20 we compare the contrasting behaviours of Cr and Mn in the BSE.
Cr does not appear to have been depleted in the BSE by volatility. The variation
in Cr/Mg ratios in chondrites is small, as expected from the essentially lithophile
behaviour of Cr under solar-nebula conditions and its co-condensation into Mg
silicates. The Cr/Mg ratio of the BSE is so much lower that it is reasonable to
ascribe the loss of Cr to an internal terrestrial process (i.e., partitioning of Cr into
the core). A similar plot for Mn demonstrates not only the variability of the Mn/Mg
ratio in chondrites but also that the pattern of that variation trends towards the BSE
ratio.

1.5.4. Radioactive Heating and the Earth's Heat Budget

The BSE abundances of K, Th, and U, the main heat-producing elements in the
Earth, imply that radiogenic heat is currently being produced at a rate of 21.3×10^{12}

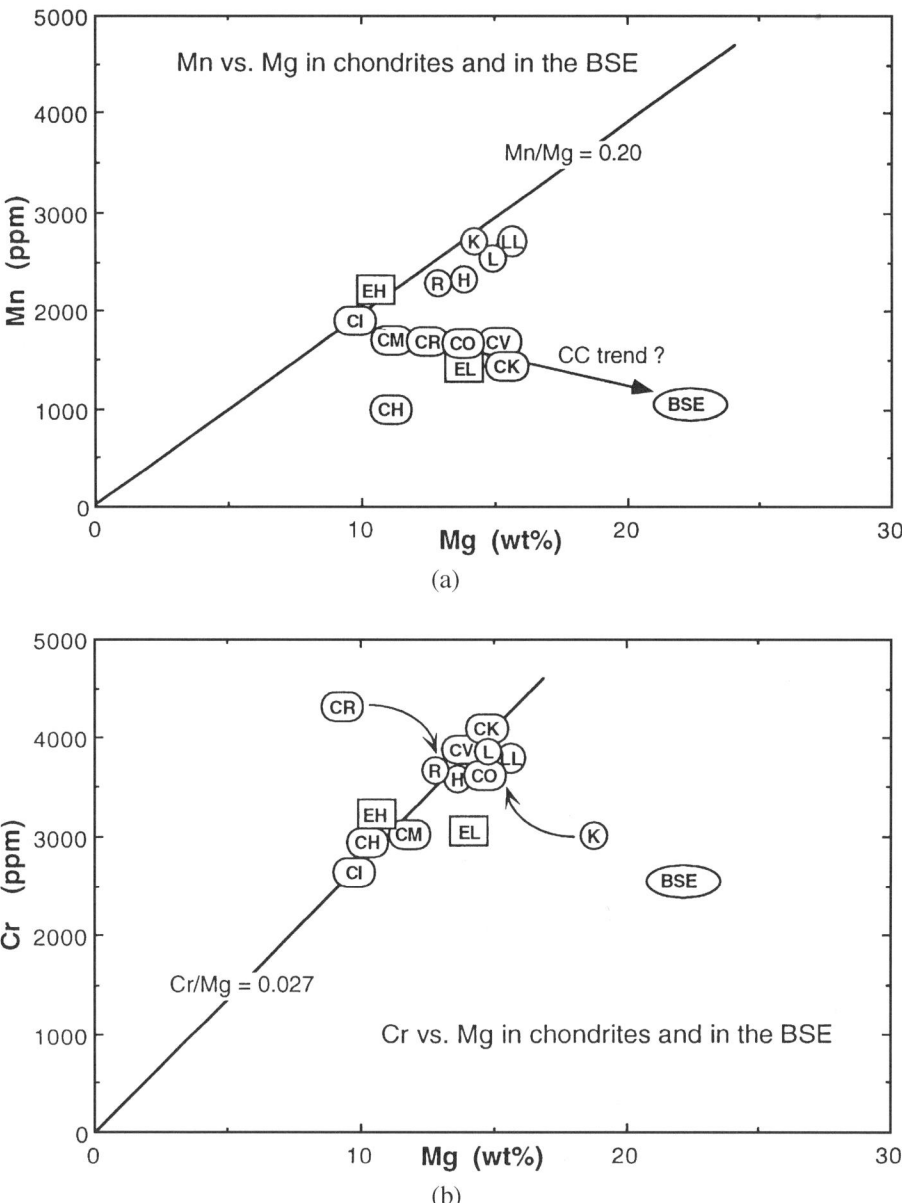

Figure 1.20. (a) Relationships of Mn and Mg in chondritic meteorites and in the BSE. The Mn/Mg ratio is poorly defined in chondrites, but a trend is discernible in the carbonaceous-chondrite data. Mn/Mg in the BSE falls on the extension of that trend, consistent with depletion in the Earth by volatility. There is no reason to assume that any Mn has been lost to the Earth's core; in fact, the trend shown here [and the similarity of the Mn/Na ratios in the BSE and in chondritic meteorites (Figure 1.19)] strongly suggests that Mn did not have significant siderophile tendencies during the core-forming process in the Earth. Meteorite data from Wasson and Kallemeyn (1988) and other sources listed in Figure 1.6. (b) Analogous plot of Cr vs. Mg. The Cr/Mg ratio in chondritic meteorites is constant, indicating both similar condensation temperatures and lithophile behaviour by Cr. Only the volumetrically unimportant group of EL chondrites has a somewhat lower ratio (0.022). The BSE is clearly depleted in Cr, with Cr/Mg ratios 50% lower than those for CI chondrites. In contrast to the Mn/Mg ratio, this depletion does not fall on the extension of any meteorite trend and can be ascribed to loss of Cr to the Earth's core.

W in the BSE, as calculated from the data of Stacey (1992, Table 6.2), corrected for our BSE values of U, Th, and K. Additional minor amounts of heat come from the core (2.8×10^{12} W) (Stacey, 1992) and from gravitational energy released by chemical differentiation and thermal contraction (2.7×10^{12} W) (Stacey, 1992), giving a total heat production of 27×10^{12} W. The heat lost from the Earth, as estimated from a global analysis of heat-flow measurements, is 44.2×10^{12} W (Pollack, Hurter, and Johnson, 1993). The deficit between heat production and heat loss must be made up by cooling of the BSE; the inferred rate is 12 K per 100 million years.

Radiogenic heat production was greater in the past; heat production would have been in balance with the present-day rate of heat loss $\sim$2.5 billion years ago. It is therefore reasonable to expect that the Earth must have been able to lose heat more efficiently prior to this (i.e., in the Archaean), perhaps implying a different style of global tectonism.

1.5.5. K in the Core?

It is particularly important to note that the depletion of K in the BSE lies squarely on the lithophile-element volatility trend of Figure 1.18. The possibility that some K may have been partitioned into the Earth's core was raised by Murthy and Hall (1970), on the basis of the observation that the alkali metals show considerable chalcophile tendencies in the enstatite chondrites (e.g., some K occurs in the sulphide mineral djerfisherite, and Na in caswellsilverite). The presence of the radioactive element K in the core would have profound implications for the Earth's heat budget. Currently, the decay of ^{40}K accounts for 18% of radioactive heat production in the BSE. Because K in the BSE is depleted relative to Mg and Si by a factor of 5, the presence of the missing K in the core would double the radioactive heat production in the bulk Earth. Thus the Earth, rather than cooling, would be heating up. Moreover, because ^{40}K has a relatively short half-life compared with the age of the Earth, the amount of heat from the radioactive decay of any ^{40}K in the core would have been much greater in the past, dominating mantle heat production. A greater degree of bottom-heating of the mantle from radioactive heat production in the core might be expected to have led to different global tectonic styles (e.g., greater plume activity from the core–mantle boundary).

The empirical arguments against K in the core therefore need to be emphasized with some force. They are as follows:

1. The depletion of K in the BSE conforms exactly with the patterns of depletion of other moderately volatile *lithophile* elements, including other alkali metals, in chondrites (Figure 1.18). The chondrite patterns are known to have been established by volatility alone. Although the metal/silicate partitioning behaviour of K relative to Na, for example, has not been studied in detail, it seems unlikely that

their partition coefficients would have exactly the right magnitude to mimic the chondritic Na/K ratio by chance. Note also in Figure 1.18 that the depletion trend for the moderately volatile elements in the BSE includes B and F, which are not known to become chalcophile or siderophile under any reasonable conditions. The depletion of F is particularly significant: Because F substitutes for oxygen in silicates (i.e., as an anion), F should become less siderophile with decreasing oxygen fugacity, rather than more siderophile. Compare the metal/silicate partitioning reactions:

$$\underset{\substack{\text{component} \\ \text{in silicate}}}{K_2O} = \underset{\text{in metal}}{2K} + \frac{1}{2}O_2 \tag{16}$$

$$\underset{\substack{\text{component} \\ \text{in silicate}}}{MgF_2} + O_2 = \underset{\text{in metal}}{2F} + \underset{\substack{\text{component} \\ \text{in silicate}}}{MgO} \tag{17}$$

Analogous reactions can be written involving sulphides rather than metals.

2. Lodders and Fegley (1995) have suggested that the process that may have been responsible for the removal of alkali-bearing sulphide from enstatite chondrites in order to form the aubrites (enstatite achondrites) may also have been the mechanism for removal of alkalis to the Earth's core. By analogy with the enstatite-chondrite/aubrite relationship, these authors concluded that 30% of the initial Na and 74% of K were removed to the core. This would produce a change in the Na/K ratio of the residue, leading to a non-chondritic ratio in the mantle of the Earth, which is not observed. Moreover, under the unusual conditions of high fS_2 and low fO_2 of the enstatite chondrites, where K does indeed become partially chalcophile, so do a number of RLEs, including Ca, REEs, and, importantly, Ti, but not Al or Sc. Thus, if conditions in the Earth were such that K was lost to the core, it seems reasonable to infer that Ca, REEs, and Ti would likewise have been lost, resulting in sub-chondritic Ca/Al, Ti/Al, or Yb/Sc ratios in the BSE composition, for which there is no evidence.

3. The amount of heat lost by the core over geological time has been calculated to be 3.8×10^{29} J, exclusive of any hypothetical contribution from radioactive elements in the core (Stacey, 1992, p. 337). This implies a heat flux from core to mantle of 2.8×10^{12} W. Davies (1988) and Sleep (1990) have independently estimated the heat flux from plumes at 2.5×10^{12} W. The plume flux is reasonably supposed to be the mechansim by which heat is removed from the thermal boundary layer at the core–mantle boundary. Thus the non-radiogenic heat flux out of the core is accounted for well. There is no sign of additional radiogenic heat from the core.

1.5.6. K and the Size of the Depleted-Mantle Reservoir

The amount of ^{40}Ar produced from ^{40}K is

$$^{40}\text{Ar} = [\lambda_{\text{Ar}}/(\lambda_{\text{Ca}} + \lambda_{\text{Ar}})]^{40}\text{K}(e^{\lambda t} - 1) \tag{18}$$

For 260 ppm K in the BSE (Table 1.9), this gives 1.43×10^{17} kg of ^{40}Ar produced over the age of the Earth (4.5×10^9 years). The mass of the atmosphere is 5.1×10^{18} kg, of which 1.3% by mass is Ar. This is nearly all ^{40}Ar, implying that it is not primeval, but has been derived from ^{40}K decay (McDougall and Honda, Chapter 3, this volume). Thus the amount of ^{40}Ar in the atmosphere corresponds to 0.46 of the total amount produced in the BSE over geological time. Apart from the ^{40}Ar in the atmosphere, there is a small fraction of the total ^{40}Ar that has been produced since differentiation that is still locked up in the continental crust and in the depleted mantle. The amount of this ^{40}Ar will depend on the mean age of degassing, but is unlikely to be greater than about 0.05 of the total produced. Allowing for this, the fraction of the depleted mantle in the BSE (i.e., m_{dm}) is about 0.50.

1.5.7. The Metal/Silicate Ratio in the Bulk Earth

Compared with the CI composition, metal is fractionated from silicate in most chondrite meteorites, as recorded in their varying Fe/Mg ratios. Reasonably narrow limits on the extent of metal/silicate fractionation in the Earth can be obtained from a simple mass balance:

$$\left(\frac{\text{Fe}}{\text{Mg}}\right)_{\text{bulk Earth}} = \frac{0.68c_{\text{Fe}}^{\text{BSE}} + 0.32c_{\text{Fe}}^{\text{core}}}{0.68c_{\text{Mg}}^{\text{BSE}}} \tag{19}$$

where

$$c_{\text{Fe}}^{\text{core}} \cong 1 - c_{\text{Ni}}^{\text{core}} - \text{LC} = 1 - 0.06 - 0.1\ (\pm 0.1) \tag{20}$$

and LC represents the putative light component in the core, generally thought to be about 10% by weight (e.g., Poirier, 1994), but conservatively assessed an uncertainty of $\pm 10\%$ (as discussed later). This gives a value for (Fe/Mg)$_{\text{bulk Earth}}$ of 2.1 ± 0.2, tending to higher values than the CI ratio (1.9), but well within the range of chondrites in general (Figure 1.8). Although in many respects the composition of the bulk Earth seems to resemble most closely that of the CV chondrites, the resemblance does not extend to the metal/silicate ratio: To match the Fe/Mg value of the CV chondrites (1.6) would require the light component to be 32 wt% of the core, an implausibly high amount.

1.5.8. The Pattern of Siderophile-Element Depletion in the BSE

The cosmochemically siderophile elements are expected to be depleted in the BSE by virtue of having been partitioned into the core. Before discussing this in detail,

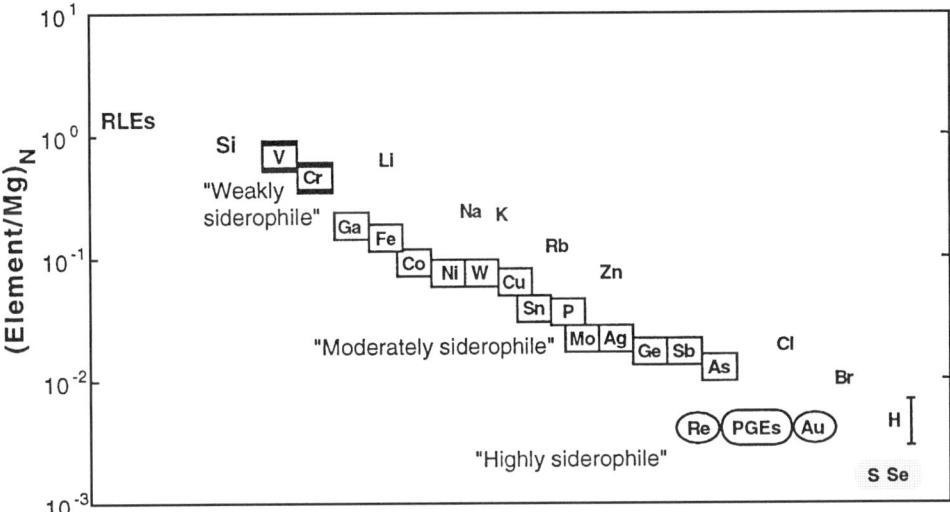

Figure 1.21. Patterns of siderophile-element depletion in the BSE. For comparison, some lithophile elements are also plotted. Note that this is not a graph; the horizontal axis is not defined, and thus the order in which the elements are plotted is arbitrary, chosen to display the perceived pattern. The pattern shows four major features: (1) V and Cr, which are not siderophiles in carbonaceous chondrites or ordinary chondrites, are depleted relative to Mg, by approximately a factor of 2. That is much more than would be predicted from their known metal/silicate partition coefficients, which typically are <1 (normalized to $D_{Fe}^{met/sil} = 14$). These we call 'weakly siderophile elements'. (2) The 'moderately siderophile elements' (Ga to As) have typical low-pressure $D_M^{met/sil}$ values in the range $10–10^4$ (at $D_{Fe}^{met/sil} = 14$). However, they are mostly depleted by less than that, and the range of their depletions is compressed to within an order of magnitude. Some of these elements are cosmochemically volatile, although any volatility-related effect in the pattern is certainly too subtle to be immediately obvious. (3) The 'highly siderophile elements' have $D_M^{met/sil} > 10^5$ (some probably $\gg 10^5$), but are only depleted by a factor of 250, all (within uncertainty) by similar amounts. The usual explanation is that these elements were added in a late veneer after loss of metal to the core had ceased. (4) S and Se are even more depleted than the highly siderophile elements.

it is useful first to obtain some feeling for the pattern of this depletion. The pattern is illustrated in Figure 1.21. There are three points to be noted:

1. Two elements, V and Cr, that are not usually siderophile in chondrites (except in the enstatite chondrites) appear to have been depleted by core formation in the Earth. Their BSE abundances are very well determined.

2. Most moderately siderophile elements fall in a fairly narrow range of depletions, between 0.09 and 0.015 of their Mg-normalized CI abundances. Fe and Ga are slightly less depleted. Such a narrow range of depletions is somewhat surprising, given that the low-pressure metal/silicate distribution coefficients for this group of elements differ by as much as 1,000-fold. Furthermore, for many of these elements (e.g., Sn, Ge, As, Sb), volatility may also have been a contributing factor in their BSE depletions. In detail, however, elements that are highly

volatile cosmochemically, such as Sn, appear hardly, if at all, more depleted than refractory elements of similar siderophile/chalcophile tendencies, such as Mo.

3. The eight highly siderophile elements (HSEs, consisting of Au, Re, and the six platinum-group elements, PGEs) are somewhat more depleted than the preceding group. All seem to have suffered similar amounts of depletion (to 0.35% of the CI values, to Mg-normalized) (O'Neill et al., 1995). They are, however, less depleted than S and Se.

1.5.9. The HSEs, Sulphur, and the Late-Veneer Hypothesis

This last feature of the foregoing pattern leads to the late-veneer hypothesis (Kimura, Lewis, and Anders, 1974). The most robust evidence for this hypothesis is the chondritic or near-chondritic Re/Os ratio in the BSE, as deduced from study of Os isotopes in the mantle (e.g., Meisel et al., 1996). However, all eight HSEs appear to be present in the BSE in chondritic relative abundances. All the HSEs have (by definition) very high metal/silicate or metal/sulphide partition coefficients, in excess of 10^4, but there is no reason to suppose that all eight HSEs would share the same value of $D_{HSE}^{met/sil}$. In fact, it is likely that the values of $D_{HSE}^{met/sil}$ for the eight HSEs at any given $P–T$ conditions and fO_2 or fS_2 would cover several orders of magnitude (O'Neill et al., 1995). Results from recent experiments on the partitioning of Ir and Pd at atmospheric pressure have demonstrated a difference between $D_{Ir}^{met/sil}$ and $D_{Pd}^{met/sil}$ of five orders of magnitude (Borisov, Palme, and Spettel, 1994; Borisov and Palme, 1995). Therefore, the modest levels of depletion of the HSEs relative to their extreme siderophile tendencies, as well as their chondritic (or nearly chondritic) relative abundances, argue that they were added to the Earth's mantle after core formation had completely ceased, as a late veneer.

An important point to note is that the extent of depletion of S in the mantle is even greater than those of the HSEs (O'Neill et al., 1995). Assuming 200 ppm S in the BSE, the S/Ir ratio in the BSE would be 0.6×10^5, which would be noticeably less than the CI ratio of 1.14×10^5 (but would be quite similar to that in the somewhat volatile-depleted CM chondrites, 0.55×10^5, or that in the EL group of enstatite chondrites). Two other observations are pertinent. Firstly, the S abundances in chondrites show that S is cosmochemically only moderately volatile (Table 1.5), not highly volatile. For instance, its depletion in the ordinary chondrites is not correlated with metamorphic grade, whereas the depletions of the more volatile chalcophile elements, such as Tl, Pb, Bi, and In, are so correlated. Despite this, S is far more depleted (by nearly an order of magnitude) in the BSE than these elements. S is also much more depleted than volatile lithophile elements such as Cl and Br. This argues that there is more to the extreme depletion of S than just its volatility. Secondly, although S is depleted relative to Ir and the other HSEs, the Se/S ratio in the BSE appears to be chondritic (Morgan, 1986), indicating that the current amounts of S and Se in the mantle are not residual from sulphide extraction or sulphide-containing liquid-metal extraction, which would fractionate Se from S. This conclusion would be further supported if future study of Te in the BSE should confirm that the Te/S ratio is indeed chondritic.

The main objection to the late-veneer hypothesis is that there is no evidence for it on the Moon (Drake, 1987). That objection would be removed if, as seems possible, core formation in the Moon occurred subsequently (O'Neill, 1991a).

McDonough (1995) has suggested that the late-veneer component, rather than being a meteoritic influx, might be some original core material reintroduced back into the mantle, and reoxidized, early in the Earth's history. For most purposes, the consequences of this alternative would be similar to the consequences of the original scenario. However, it seems less attractive to us for the following reason: Mass balance indicates that the core should have an Fe/Ir ratio some 30% less than chondritic (i.e., 2.7×10^5). If that were divided by the S/Ir ratio of 0.6×10^5, which we inferred for the late-veneer component, it would imply an Fe/S ratio in the core of ~ 4.5, or about 17 wt% S (allowing for 5% Ni in the core). That appears to be too large an amount of S to be compatible with the core's density: If S were the only element contributing to the light component in the core, the amount required would be $\sim 10\%$, according to a recent estimate (Poirier, 1994). Moreover, such a large amount of S in the core would also be difficult to reconcile with the Earth's general volatile-element depletion, as deduced, for example, from studies of moderately volatile elements such as Na and Zn to be discussed later. Hence, we believe that the original meteoritic late-veneer hypothesis is still to be preferred.

1.5.10. Nature of the Late Veneer

An assumption among some commentators is that the late veneer was an oxidizing component responsible for adding the Earth's most volatile components, such as H_2O. This is unlikely. Only CI chondrites have sufficiently high H_2O/Ir ratios to have been able to add all the BSE's 1,160 ppm H_2O, and then only if the upper limit of 18 wt% H_2O in CIs is accepted (see the discussion in Section 1.3.6). However, the BSE Ir/S ratio clearly shows that the late veneer was volatile-depleted. A similar conclusion can be reached by considering the BSE Ir/N ratio (1.3×10^{-3}), which is considerably greater than those for the volatile-rich CI and CM carbonaceous chondrites ($\sim 0.3 \times 10^{-3}$) (Wasson and Kallemeyn, 1988). The currently observed Ir/N ratio in the BSE is very much a lower limit for that in the late veneer, as an unknown fraction of the N (which is not a siderophile or chalcophile element) in the BSE probably was present before addition of the late veneer.

Further insight into the nature of the late veneeer may perhaps come from detailed examination of the Os isotopic composition of the BSE. Recently, Meisel et al. (1996) have shown that different chondrite groups may have slightly different $^{187}Os/^{188}Os$ ratios, reflecting small differences in initial Re/Os ratios (note that only isotopic analysis of parent–daughter pairs is precise enough to resolve the small differences being discussed here). Ordinary chondrites and enstatite chondrites have present-day $^{187}Os/^{188}Os$ values of 0.1287 ± 0.0009 (1 s.d.), whereas two carbonaceous chondrites have the value 0.1263. The former value is closer to the value that Meisel and associates deduced for the primitive depleted mantle, namely, 0.1292 ± 0.0007. Hence Meisel et al. (1996) argued that the material of the late veneer may show greater similarities to the ordinary chondrites or the ensatite chondrites than to the carbonaceous chondrites. However, the data must be interpreted with considerable caution. Even higher $^{187}Os/^{188}Os$ ratios are characteristic of OIB sources (typically ~0.135) (Widom and Shirey, 1996), so that, when discussing the fine detail, it is by no means clear exactly what is the mean $^{187}Os/^{188}Os$ ratio of the BSE (this is not to suggest that it is not approximately chondritic). Re is considerably less siderophile than Os or any of the other HSEs (O'Neill et al., 1995). Hence it is possible that a small fraction of the Re now in the BSE may predate the late veneer. The combination of some residual Re plus that added in the late veneer would result in a slightly super-chondritic Re/Os ratio in the BSE.

The reality of a late veneer seems likely, regardless of how the preceding 99.5% of the Earth may have accreted. Hence the abundances of the HSEs in the BSE are best treated independently of the more general problems posed by the patterns of depletion of the other siderophile elements. These problems will be addressed in the next section.

1.6. Core Formation: The Geochemical Evidence
1.6.1. The Physical Setting

Core formation in the Earth may be anticipated to have been intimately associated with accretion. Accordingly, the chemical relationship between the core and the BSE, as expressed in the pattern of chondrite-normalized siderophile element abundances in the BSE (Figure 1.21), potentially holds information on some of the details of accretion. For example, consider these two issues: (1) Does the BSE siderophile-element pattern reflect metal/silicate equilibrium at high pressures and temperatures, which can be obtained only in an Earth-size planet? An alternative possibility, at least conceptually, is that the pattern preserves the record of metal/silicate equilibrium established at low pressures and moderate temperatures in the asteroid-size planetesimals, from which the Earth is thought to have accreted. (2) To what extent was the Earth accreted from materials of essentially uniform composition (i.e., indigenous materials condensed from the solar nebula at 1 AU)? Is there any evidence for input of materials of different compositions, formed at greater heliocentric distances?

Core formation is essentially the process of segregation of the metallic component of chondritic material from the silicate component, followed by settling of the denser metal to the centre of the planetary body. The existence of 'magmatic' iron-meteorites demonstrates empirically that core formation can occur in asteroid-size bodies. [Most iron-meteorite groups have trace element patterns that reflect partial melting or fractional crystallization inside a parent body (e.g., Kelly and Larimer, 1977; Scott, 1979). These are the 'magmatic irons'. Other origins are proposed for the 'nonmagmatic irons' (e.g., impact melting for the IAB–IIICD groups) (Choi, Ouyang, and Wasson, 1995). The IVB group may have originated as a high-temperature metal condensate from the solar nebula (Kelly and Larimer, 1977).] The HED achondrites also show that core formation occurred on the eucrite parent body.

The critical requirement for core formation in asteroid-size bodies seems to have been that the body should have been heated to a sufficient temperature to allow essentially complete melting of the metal phase. The origin of the magmatic irons by crystallization from an initially molten state is demonstrated by their chemical-fractionation trends (Scott, 1979), and the requirement for molten metal is also expected from theoretical considerations, which show that separation of solid metal from solid silicate is too sluggish a process to have led to core formation over reasonable times (Stevenson, 1990).

The heat sufficient to melt asteroid-size bodies usually is ascribed to the decay of short-lived radioisotopes, such as ^{26}Al and ^{60}Fe, although there are several less well established possibilities, such as heating by solar radiation from a superluminous sun or by electrical induction from an intense solar wind (Sonnett and Reynolds, 1979). There is a clear expectation that the effects of all of these postulated forms of heating should diminish with increasing heliocentric distance; for radiogenic heating by ^{26}Al and other short-lived isotopes, that is because the expected accretion times are longer at greater heliocentric distances (Grimm and McSween, 1993). All of these proposed heat sources would be confined to the initial million years or so of solar-system formation (e.g., the half-life of ^{26}Al is 7.2×10^5 years).

In what is currently thought to be physically the most probable model for accretion of a planet the size of the Earth, accretion would proceed through an evolving size distribution of planetesimals (Figure 1.1). The existence of cores from asteroid-size bodies, presumably formed at >2.5 AU, points towards the likelihood of core formation in similarly sized planetesimals at 1 AU, because of the greater heating at the lesser heliocentric distance. Metal/silicate segregation in such planetesimals would have occurred at low pressures and probably at temperatures not too much above the low-pressure silicate solidus (say ~1,400°C), and those conditions would be reflected in the patterns of depletion of the siderophile elements in the residual silicate.

Subsequently, planetesimals would have collided and formed larger bodies. Planetesimals that were not melted and differentiated by the early sources of heat may nevertheless have undergone core formation at that stage if enough energy was available from impacts. Calculations indicate that once a body reaches the 'planetary-embryo' size of ~10^{23} kg (~2% of the mass of the Earth), impacts of planetesimals will be energetic enough to trigger substantial melting in the regions of the impacts

(Tonks and Melosh, 1992). Gravitational energy released by sinking of metal blobs will then produce more heat, triggering further melting.

If metal cores formed by early heating already exist, they can either merge (e.g., Benz, Cameron, and Melosh, 1989) or disperse into an emulsion in the molten silicate of a transient magma ocean (Stevenson, 1990). In the merging-cores scenario, the metal cannot further equilibrate with the silicate, which will thus preserve the signature of their low-pressure metal/silicate equilibrium. The dispersal of the metal of previously formed cores would allow chemical re-equilibration in the larger bodies. This introduces a number of other possibilities, which will now be briefly considered. The fluid dynamics of metal segregation in the Earth have been more fully discussed by Stevenson (1990) and Karato and Murthy (1997).

At the low pressures pertaining in asteroid-size bodies, the temperature needed to melt Fe-Ni-rich metal probably (depending on the metal's S content) was high enough that much of the silicate would also have been molten. The near-coincidence in the melting temperatures of Fe-Ni metal and chondritic silicate compositions at low pressures ensures that metal segregation and core formation in small bodies will necessarily be processes involving both molten metal and at least partially molten silicate. Whether or not the silicate has to be molten for there to be efficient metal/silicate segregation in a body the size of the Earth is not known for certain. The effect of pressure on the melting points of silicates is much greater than for metals (e.g., Boehler, 1996). Thus the temperature of the silicate solidus in the interior of a large planet will be far above that of Fe-Ni-rich metal. (This enables the outer core of the Earth to be molten, while the thermal boundary layer at the base of the mantle is sub-solidus.) Can molten metal segregate from sub-solidus silicate? The dihedral angle θ_D of liquid Fe in an olivine matrix is $\gg 60°$, the threshold value for wetting behaviour and the formation of an interconnected melt topology. This implies that extensive melting would indeed be necessary for efficient segregation of metal in the upper mantle (Stevenson, 1990). It is possible, though, that θ_D becomes $< 60°$ at very high pressures (i.e., in the lower mantle, where olivine is no longer stable, and $MgSiO_3$-rich perovskite is the main silicate phase), allowing efficient segregation, but there is as yet no experimental proof of this. The textures recovered from high-pressure experiments have been mostly the results of short run times and large temperature gradients and in no way represent textural equilibrium.

It is therefore most probable, but not certain, that the conditions necessary for core formation in the Earth would have been met in an early magma ocean, of the kind that is believed to have formed after the giant impacts expected during the latter stages in the Earth's accretion. The steep increase in solidus temperature with pressure in the Earth's mantle means that extensive melting of the lower mantle would have been accompanied by very high temperatures at the surface of the magma ocean, and thus very rapid loss of heat by radiation. A magma ocean extending deep into the lower mantle of an Earth-size planet probably would have been a transitory phenomenon, if it occurred at all.

Any metal present in the magma ocean would have been in the form of immiscible liquid drops, the hydrodynamically stable size of which is ~1 cm (Stevenson, 1990). According to the discussion by Stevenson (1990), pools formed by the rain-out of these metal drops as the magma ocean cooled would have accumulated at the interface between the magma ocean and the underlying, mostly solid lower layer. Eventually the pools would have sunk as large diapirs. The extent to which chemical equilibrium between metal and silicate can be achieved during such processes will be determined by the usual interplay involving the length scale of the process, time, and diffusion rates. Diffusion coefficients are typically on the order of 10^{-4} cm^2/s in liquid Fe, 10^{-8} cm^2/s in liquid silicates, and $>10^{-11}$ cm^2/s in solid silicates. The ~1-cm-size droplets in the magma ocean would have equilibrated with surrounding molten silicate by diffusion over times on the order of hours. However, large masses of metal would have been unable to re-equilibrate with silicate in the time taken to sink to the core. In this scenario, metal/silicate equilibrium should record the $P–T$ conditions of the interface at which the pools accumulate. Such $P–T$ conditions ($P > 20$ GPa?) are likely to be very much greater than those attained in asteroid-size parent bodies.

In stark contrast, metal/silicate equilibration at pressures approaching that of the core–mantle boundary would be achieved if the metal were to percolate through the deep mantle, as may be possible if metal can wet solid silicate at those depths.

There is a further complication: The stochastic nature of the accretion process in its final stages makes it likely that there were several giant impacts, with resulting magma oceans, and therefore several potential episodes of core formation as the Earth grew. Because the pressure at the centre of a growing planet increases as its radius (R) squared, whereas the mass increases as R^3, the mean pressure at which metal/silicate equilibrium could have occurred may have been relatively high (e.g., Newsom and Slane, 1993).

The likelihood is that accretion and core formation was a complex business, making it inappropriate to seek a single, precise set of physical conditions at which metal/silicate equilibrium can be said to have 'happened'. Nevertheless, we can distinguish two possible scenarios. In the first, the pattern of the BSE siderophiles is set by essentially a single kind of process, operating on a single kind of accreting material. We call this type of scenario 'homogeneous accretion'. Conceptually feasible examples of conditions under which metal/silicate equilibrium might be established during homogenous accretion are as follows:

1. at low pressures in planetesimals or small planetary embryos accreting from indigenous material condensed at 1 AU, without subsequent re-equilibration as accretion proceeds through the size-distribution hierarchy
2. at the interface between the bottom of a magma ocean and a mostly unmelted lower mantle (perhaps several episodes of this)
3. at the core–mantle boundary in a growing Earth

The important property of this kind of scenario for the chemistry of the process is that it seems reasonable to seek a single set of physical conditions at which metal/silicate equilibrium would have occurred, *on average*. More complex scenarios, involving mixing of several components, some of which might have undergone metal/silicate segregation in different ways, we call 'heterogeneous accretion'. For these, we would not expect to find a signature of metal/silicate equilibrium corresponding to any single $P–T$ condition. Heterogeneous-accretion scenarios are essentially mixing models.

An application of Occam's razor favours the conceptually simpler models of the homogeneous-accretion class. Hence, heterogeneous-accretion models should be accepted only by default, that is, if none of the homogeneous-accretion scenarios are able to account for the BSE siderophile-element abundances. It is thus necessary first to consider the possible homogeneous-accretion models in detail. The strategy is as follows:

1. We calculate the abundances of siderophile elements in the core by mass balance, subtracting the observed abundances in the BSE (Section 1.4) from the abundances expected on the basis of cosmochemical constraints for the bulk Earth (Section 1.3). This gives quantitative estimates for the mean metal/silicate distribution coefficients that are required to explain the siderophile-element pattern in the BSE.
2. We then compare these required values of metal/silicate distribution coefficients with actual values (experimentally measured or calculated from thermochemical data), as functions of the unknowns of pressure, temperature, and composition of the metal. (Here, the composition of the metal is considered an unknown because of the unknown nature of the 'light component' in the core. The nature and amount of the light component will affect the relative oxygen fugacity at which metal/silicate equilibrium will occur.) Finding a match would reveal the mean conditions (P, T, etc.) under which metal–silicate segregation occurred. We show that no such match can be found within the $P–T$ ranges thus far covered by measurements and reliable calculations, and we argue that such a match is, on the basis of the currently available evidence, somewhat unlikely.
3. We then show that the siderophile element pattern in the BSE can be produced by a plausible heterogeneous accretion model.

1.6.2. The Composition of the Earth's Core Derived from Mass Balance Considerations

The simple principle is that the amount of an element in the Earth's core is given by the difference between the amount expected in the bulk Earth, as deduced from cosmochemical constraints, and its observed abundance in the BSE (as given in Tables 1.7–1.10). Quantitatively, the mass of a siderophile element M in the BSE is $0.68c_M^{BSE} M_{Earth}$, and the amount in the core is $0.32c_M^{core} M_{Earth}$, because the BSE and the core comprise 68% and 32% of the mass of the Earth, respectively. The amount of Mg in the bulk Earth (core plus mantle) is $0.68c_{Mg}^{BSE} M_{Earth}$, because we assume no Mg in the core. This latter assumption is supported by experimental evidence (e.g., Ringwood and Hibberson, 1991; O'Neill et al., in press). We then assume that the ratio M/Mg is the same in the bulk Earth as in CI chondrites, but corrected for volatility and for metal enrichment, so that

$$\frac{0.68c_M^{BSE} + 0.32c_M^{core}}{0.68c_{Mg}^{BSE}} = v_M s_M \frac{c_M^{CI}}{c_{Mg}^{CI}} \tag{21}$$

The factor v_M is the correction for terrestrial volatility; it is 1.29 for refractory elements (Table 1.7), and $\sim$1 for the common siderophile elements Fe, Ni, and Co, and also for Cr (Figure 1.20b). For the moderately volatile siderophile elements, reasonable estimates for v_M can be inferred from the empirical correlation for moderately volatile lithophile elements in Figure 1.18; evidently, v_M should be between about 0.2 and 1. The correction factor for metal enrichment, s_M, is likely to vary between $\sim$1.0 for the volatile siderophile/chalcophile elements and 1.1 ± 0.1 for the common and refractory siderophiles (deduced from the bulk-Earth Fe/Mg ratio; see Section 1.5.7). The concentration of M in the Earth's core is then

$$c_M^{core} = 2.125 \left[v_M s_M c_{Mg}^{BSE} \left(c_M^{CI}/c_{Mg}^{CI} \right) - c_M^{BSE} \right] \tag{22}$$

and therefore the mean metal/silicate distribution coefficient required for core–mantle equilibrium is given by

$$\left(D_M^{met/sil} \right)_{required} = c_M^{core}/c_M^{BSE} = 2.125 \left[v_M s_M \frac{c_M^{CI}/c_{Mg}^{CI}}{c_M^{BSE}/c_{Mg}^{BSE}} - 1 \right] \tag{23}$$

where

$$\frac{c_M^{CI}/c_{Mg}^{CI}}{c_M^{BSE}/c_{Mg}^{BSE}}$$

is the inverse of the BSE depletion factor (e.g., as given in Tables 1.8 and 1.9). Propagation of uncertainties through this equation suggests that values of $(D_M^{met/sil})_{required}$ probably will be constrained to between $\pm 20\%$ and $\pm 60\%$ for most siderophile elements, depending mainly on volatility.

In the foregoing, the nature of the 'silicate' was not specified. Because the fluid dynamics of metal/silicate separation probably require some melting of the silicate matrix, it is convenient to normalize the partitioning relations to silicate melt. The concentration of a siderophile element M in the silicate melt is then given by

$$c_M^{silmelt} = c_M^{BSE} \left[D_M^{sol/melt}(1 - F) + F \right] \tag{24}$$

where F is the degree of melting, and $D_M^{sol/melt}$ is the mean Nernst partition coefficient during partial melting (i.e., a measure of the incompatibility of the element during mantle melting). The required metal/silicate distribution coefficient is then

$$\left(D_M^{met/sil}\right)_{required} = 2.125\left[v_M s_M \frac{c_M^{CI}/c_{Mg}^{CI}}{c_M^{BSE}/c_{Mg}^{BSE}} - 1\right] \Big/ \left[D_M^{sol/melt}(1 - F) + F\right]$$

(25)

Note that this formalism is usable even if no melt is actually present (i.e., as $F \to 0$).

The next task is to find matches between the values for $(D_M^{met/sil})_{required}$ and the observed values for $D_M^{met/sil}$ as functions of the physical variables of pressure, temperature, and other relevant parameters, such as composition of the metal and relative oxygen fugacity, for as many elements M as possible. That will demand a thorough appreciation of how the values for $D_M^{met/sil}$ depend on the physical variables, which can be achieved only from an understanding of the thermochemical basis for metal–silicate partitioning equilibria.

1.6.3. Metal/Silicate Distribution Coefficients

Metal/silicate distribution coefficients $(D_M^{met/sil})$ do not behave like Nernst solid/liquid trace-element partition coefficients of the kind familiar from igneous petrology, because their values depend strongly on oxygen fugacity. The oxygen fugacity of a metal/silicate system is, in turn, a strong function of temperature, pressure, and the compositions of the metal and silicate phases. For reasonable core-forming models, the range of appropriate fO_2 values can cover several orders of magnitude. The importance of fO_2 is easily appreciated if the partitioning equilibrium is written out explicitly as a mass-balanced chemical reaction:

$$\underset{\text{silicate}}{MO_{x/2}} = \underset{\text{metal}}{M} + \frac{x}{4} O_2$$

(26)

where x is the effective valence state of M in its oxide component in the silicate (assumed here, for convenience, to be single phase, like a silicate melt; the principles are easily extended to polyphase systems, e.g., by calculating the concentration of M in one of the phases). Therefore, at equilibrium,

$$RT \ln K_{26} = RT \ln\left[\frac{a_M^{met}(fO_2)^{x/4}}{a_{MO_{x/2}}^{sil}}\right] = -\Delta G^\circ(26)$$

(27)

where a_M^{met} is the activity of M in the metal, and $a_{MO_{x/2}}^{sil}$ is the activity of the component $MO_{x/2}$ in the silicate melt. For both phases, activities are directly proportional to mole fractions X at low dilution (Henry's law), so that

$$D_M^{met/sil} = k_M \frac{X_M^{met}}{X_{MO_{x/2}}^{sil}} = (fO_2)^{-x/4} k_M (\gamma_{MO_{x/2}}^{sil}/\gamma_M^{met})\exp\{-\Delta G^\circ(26)/RT\}$$ (28)

where k_M is a constant used to convert mole fractions to element concentrations, and γ_M^{met} and $\gamma_{MO_{x/2}}^{sil}$ are the activity coefficients for M and $MO_{x/2}$ in metal and silicate melt, respectively. At constant $T–P$ and composition of the silicate, the

fO_2 of the system can vary by several orders of magnitude, depending on whether the metal is nearly pure Fe-Ni alloy or contains a light component such as S or O. Values for $D_M^{met/sil}$ measured at one P–T condition cannot be extrapolated to other P–T conditions without making some assumption as to how fO_2 changes as a function of P and T. Such changes can also be of several orders of magnitude.

A simple expedient to circumvent the fO_2 problem is to use the partitioning behaviour of one element, usually, for convenience, Fe, to define fO_2:

$$\underset{\text{silicate}}{\text{FeO}} = \underset{\text{metal}}{\text{Fe}} + \frac{1}{2} O_2 \tag{29}$$

(e.g., O'Neill, 1992). Subtraction of reaction (29) from (26) to eliminate O_2 gives the two-element distribution reaction

$$\underset{\text{metal}}{\frac{x}{2}\text{Fe}} + \underset{\text{silicate}}{MO_{x/2}} = \underset{\text{silicate}}{\frac{x}{2}\text{FeO}} + \underset{\text{metal}}{M} \tag{30}$$

which is independent of fO_2. The two-element distribution coefficient $KD_{M-\text{Fe}}^{met/ox}$ is defined as

$$KD_{M-\text{Fe}}^{met/sil} = \left(X_M^{met}/X_{MO_{x/2}}^{sil}\right)/\left(X_{\text{Fe}}^{met}/X_{\text{FeO}}^{sil}\right)^{x/2}$$

$$= k_M/(k_{\text{Fe}})^{x/2} D_M^{met/sil}/\left(D_{\text{Fe}}^{met/sil}\right)^{x/2} \tag{31}$$

Note that $k_M/(k_{\text{Fe}}) = (A_M/M_{MO_{x/2}})/(A_{\text{Fe}}/M_{\text{FeO}})$, where A stands for atomic weight and M for molecular weight. This ratio is close to unity for Ni and Co. The two-element distribution coefficient is related to the thermodynamic equilibrium constant, $K_{M-\text{Fe}}^{met/ox}$, by the introduction of activity coefficients, such that

$$K_{M-\text{Fe}}^{met/sil} = KD_{M-\text{Fe}}^{met/sil}\left(\gamma_M^{met}/\gamma_{MO_{x/2}}^{sil}\right)/\left(\gamma_{\text{Fe}}^{met}/\gamma_{\text{FeO}}^{sil}\right)^{x/2} \tag{32}$$

where

$$RT \ln\left(K_{M-\text{Fe}}^{met/sil}\right) = -\Delta G^\circ = -\Delta H^\circ + T\Delta S^\circ - \int \Delta V^\circ dP \tag{33}$$

Hence at constant temperature and pressure, $KD_{M-\text{Fe}}^{met/sil}$ depends only on the values of the activity coefficients, which are functions of the major-element compositions in the metal and silicate phases. For Fe-rich metal, $\gamma_{\text{Fe}}^{met} \approx 1$, by definition. Also, conveniently, $\gamma_{\text{FeO}}^{sil}$ is not far from unity in silicate melts and varies little with temperature or melt composition (e.g., $\gamma_{\text{FeO}}^{sil} \approx 1.7 \pm 0.2$) (Holzheid, Palme, and Chakraborty, 1997). Thus, provided that changes in the composition of the metal are assumed to be small,

$$D_M^{met/sil} \propto \left(D_{\text{Fe}}^{met/sil}\right)^{x/2} \tag{34}$$

This relationship can be used to extrapolate values of $D_M^{met/sil}$ measured at any concentration of FeO in the silicate to values appropriate for modelling the equilibrium between the metal now in the Earth's core and in the Earth's mantle (e.g.,

$D_{\text{Fe}}^{\text{met/sil}} \approx 14$). A reasonably small amount of a 'light component' in the metal (e.g., S, Si, or O) will not affect two-element partitioning relationships significantly unless its influences on the activity coefficients γ_M^{met} and $\gamma_{\text{Fe}}^{\text{met}}$ are very different. For many elements there is sufficient experimental evidence for all three of these likely light-component elements to suggest that such effects are indeed small, at the likely concentrations of the light component.

There is a key point to be made here concerning our attempts to understand the chemical factors that may underlie the BSE siderophile-element pattern. It is not the absolute way in which $D_M^{\text{met/sil}}$ varies with pressure, temperature, fO_2, or composition of melt or metal that is important, but rather the way in which $D_M^{\text{met/sil}}$ varies relative to $D_{\text{Fe}}^{\text{met/sil}}$. This is because the appropriate value of $D_{\text{Fe}}^{\text{met/sil}}$ is always constrained by the BSE/core mass balance.

A further convenience in using Fe as the normalizing element is that its partition coefficient between silicate melt and residue is near unity, so that its BSE/core partitioning behaviour is not sensitive to the degree of partial melting at which metal separation occurs.

1.6.4. *Sources of Metal/Silicate Partitioning Data*

There is a wealth of data on the partitioning of many of the siderophile elements between silicate melt and Fe-rich metal. The field of extractive metallurgy (particularly, for obvious reasons, that concerned with steelmaking) is an important source of data for temperatures <2,000 K at low pressures (e.g., Ward, 1962; von Bogdandy and Engell, 1971; Alcock, 1976). The metallurgical literature is replete with data for the more common elements, such as Ni, Cr, Mn, and P (important in steel-making), as well as data on the influence of potential light-component elements such as S, Si, and (in small concentrations) O. From such sources, metal/silicate distribution coefficients and their dependences on temperature and oxygen fugacity can often be obtained directly. High-quality information is also available for many of the (metallurgically) less common metals, although some thermodynamic manipulation to obtain values for $D_M^{\text{met/sil}}$ in appropriate form is often required. The metallurgical literature also discusses other aspects of the physical chemistry of metal/silicate separation processes, such as mechanisms of segregation and reaction kinetics.

Over the past 15 years, that information has been supplemented by many experimental studies specifically undertaken with the goal of furthering our understanding of core formation (e.g., Jones and Drake, 1986). Such studies have examined the metal/silicate partitioning behaviour of a number of elements (e.g., W) poorly covered in the metallurgical literature (Ertel et al., 1996), often under conditions rather different from those encountered in the typical blast furnace (e.g., for different fO_2 values, as well as the effects of S in the metallic liquid). Such studies have also refined the accuracy of the metallurgical data and increased our knowledge of how partition coefficients depend on details such as melt composition (e.g., Holzheid and Palme, 1996). A more recent development, made possible by technical advances in experimental petrology, is the extension of such studies to very high pressures (e.g., to 25 GPa) (Hillgren, Drake, and Rubie, 1994; O'Neill et al., in press).

Meteorites can also yield abundant empirical data on metal/silicate partitioning relations. Partitioning between solid metal and silicate phases can be inferred directly from equilibrated chondrites; historically such studies have provided semi-quantitative estimates of the siderophile and/or chalcophile tendencies for the entire periodic table. The study of magmatic-iron meteorites can rather usefully reveal the partitioning behaviours of siderophile elements between solid and liquid metal phases.

Ideally, a model explaining core formation should be able to account for the BSE abundances of *all* the siderophile elements. In practice, however, the problem can be obscured by 'noise' in the empirical observations or can be overwhelmed by the kind of superfluous complexity that bedevils any model that seeks to describe events so far removed from common physical experience. It is therefore worth emphasizing three groups of siderophile elements which are of special importance because of the constraints imposed by their BSE abundances on any model. These groups are (1) Fe, Ni and Co, (2) Cr and V, and (3) Mo and W.

1.6.4.1. Fe-Ni-Co

1. Ni and Co are almost identical to Fe in cosmochemical volatility (e.g., Grossman, 1972), so they can both be assumed to be present in the bulk Earth in strictly chondritic proportions relative to Fe.
2. Ni and Co, like Fe, occur only in the divalent (2+) state in silicates and oxides under relevant conditions. Their metal/silicate partitioning behaviour will therefore not change relative to that of Fe as a function of fO_2.
3. Ni and Co, like Fe, are compatible in the solid silicates and oxides of the mantle, which means that their chemical potentials, and hence bulk metal/silicate ($\pm$ oxide) distribution coefficients, will not be much affected by the extent of partial melting in the silicate at the time of metal separation (unlike highly incompatible elements) (e.g., O'Neill, 1991a, p. 1147). It is also their compatible behaviour that enables their BSE abundances to be so very accurately known.
4. The Fe, Ni, and Co metal/silicate distribution coefficients are well known (except at very high pressures), not only from studies published in the geological literature but also from a wealth of metallurgical investigations. Moreover, their metal/silicate distribution coefficients can also be calculated with useful accuracy from standard thermodynamic data.
5. The influences of potential minor components (S, Si, and O) in the metal phase on the partitioning behaviour of Fe, Ni, and Co are well known, and the effects of such components are similar for all three elements.

1.6.4.2. Cr and V

The BSE abundances of both elements are well known. Cr and V usually are not siderophile under cosmochemical conditions (except in the enstatite chondrites), so it seems reasonable to suppose that their apparently siderophile behaviour in the Earth holds particularly interesting clues.

1.6.4.3. Mo, W, and the Mo/W Ratio

Mo and W are the only moderately siderophile elements (i.e., whose abundances are not entirely controlled by the late veneer) that are also cosmochemically refractory. Therefore, as in the cases of Fe, Co, and Ni, their BSE abundances should not have been depleted by any volatility-related processes and should not require any volatility correction. Unlike Fe, Co, and Ni, both Mo and W behave as highly incompatible elements during partial melting. However, whereas the absolute values of their abundances in the BSE should be very sensitive to the degree of silicate melting at which metal/silicate segregation took place, the ratio of their abundances should remain unaffected by this variable [equation (25)]. Nor should the ratio of their abundances have been affected by large-scale mantle differentiation (if it is insisted that this cannot anyway be ruled out) (Section 1.4.6). There are now also ample metal/silicate partitioning data for both elements, at least at low pressure (e.g., Ertel et al., 1996). Whereas very high concentrations of S in the 'metal' phase (i.e., when Fe/S ratios are near unity) have large effects on both Mo and W partitioning relative to Fe, the effects are similar for the two elements and therefore have a relatively minor influence on Mo/W ratios (Lodders and Palme, 1991).

1.6.5. Homogeneous-Accretion Models
1.6.5.1. Low Pressures and Moderate Temperatures (<2,000 K)

This scenario is readily tested with the abundant metal/silicate partitioning data available. The result of such a test, assuming a temperature of $1,600°C$ (and therefore essentially complete melting of the BSE at low pressures) and no S (or other light element) in the metal phase, is illustrated in Figure 1.22. It is remarkable that the BSE abundance of only one siderophile element (W) is in agreement with its known partitioning behaviour. The moderately siderophile and highly siderophile elements are overabundant in the BSE, but V and Cr, the two weakly siderophile elements, are more depleted than expected. Making an assumption that other conditions prevail (e.g., a S-rich metal phase) does not significantly improve matters.

The conclusion that the upper mantle is grossly out of chemical equilibrium with the metal that formed the Earth's core was first emphasized by Ringwood in his classic paper on the chemical evolution of terrestrial planets (Ringwood, 1966), and it has become known as the siderophile-element anomaly.

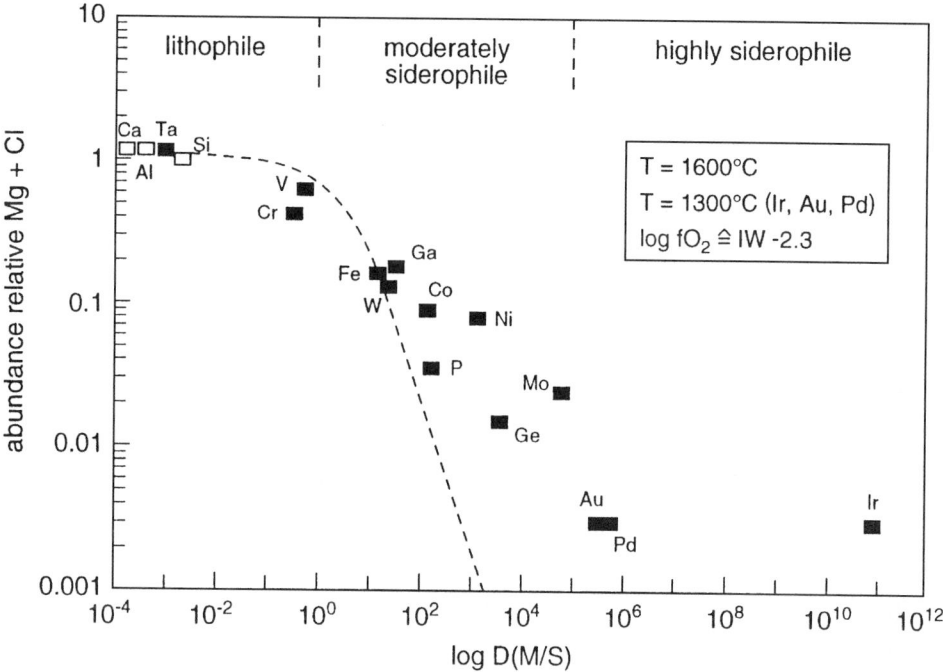

Figure 1.22. Test of the hypothesis that the pattern of siderophile elements in the BSE can be explained by core formation at low pressures and temperatures in the approximate range 1,300–1,600°C. The dashed curve indicates the values of the metal/silicate partition coefficients [i.e., $(D_M^{met/sil})_{required}$] that would be required to produce a given depletion of a siderophile element through equilibrium core formation. The symbols show the experimentally measured values of $D_M^{met/sil}$, normalized to $D_{Fe}^{met/sil} = 14$ and plotted at the observed BSE abundances (not corrected for volatility). If the hypothesis were correct, these symbols would plot on the dashed curve. Including a volatility correction would move the symbols upwards and thus worsen the agreement for *all* the moderately volatile siderophile elements. The hypothesis fails the test miserably. Experimentally measured values of $D_M^{met/sil}$ compiled by A. Holzheid et al. (in press) from the literature.

1.6.5.2. Very High Temperatures (>2,000 K)

Murthy (1991) reinvigorated the debate about the significance of the siderophile-element anomaly by suggesting that many of its features can be explained by simple equilibrium core formation, but at much higher temperatures than previously presumed – for example, at present-day core–mantle boundary temperatures (~2,700 K). This idea can be tested if the various values of $D_M^{met/sil}$ can be extrapolated to higher temperature, which can be done as follows: Combination of equations (31)–(34) gives

$$D_M^{met/sil} = (k_{Fe})^{x/2}/k_M (D_{Fe}^{met/sil})^{x/2} [(\gamma_{Fe}^{met}/\gamma_{FeO}^{sil})^{x/2}/(\gamma_M^{met}/\gamma_{MO_{x/2}}^{sil})]$$

$$\times \exp\left\{ \left(-\Delta H° + T\Delta S° - \int \Delta V° dP \right) \Big/ RT \right\} \qquad (35)$$

For a constant value of $D_{Fe}^{met/sil}$, this equation can be written for two temperatures T_1 and T_2. If we assume that the activity coefficients are independent of temperature (or change in similar ways for M and Fe), then dividing one equation by the other and eliminating the constant terms will give

$$(D_M^{met/sil})_{T_2} = (D_M^{met/sil})_{T_1} \exp\left\{-\Delta H_{M-Fe}^\circ \left(\frac{1}{T_1} - \frac{1}{T_2}\right)\right\} \qquad (36)$$

The first point to note is that the extrapolation proceeds as the inverse of temperature. On an inverse temperature scale, the difference between 1,373 K and 1,873 K (over which range experimental measurements are available) is equivalent to an extrapolation from 1,873 K to 3,000 K (i.e., higher than the present-day temperature at the base of the mantle). Hence, extrapolation to temperatures of $\sim$3,000 K should be reasonably accurate, provided the nature of the melt or metal does not change in unexpected ways that affect M and Fe differently (there has been no evidence for any such marked changes in experiments thus far, which have now reached 2,800 K) (e.g., O'Neill et al., in press). Secondly, within the assumptions used in deriving equation (36), whether $D_M^{met/sil}$ increases or decreases with temperature (at constant $D_{Fe}^{met/sil}$) will depend on the sign of the standard-state enthalpy of the M–Fe exchange reaction, ΔH_{M-Fe}°. Generally, one expects that if M is more siderophile than Fe, then ΔH_{M-Fe}° will be negative, causing $D_M^{met/sil}$ to decrease with increasing temperature (relative to $D_{Fe}^{met/sil}$), whereas if M is less siderophile than Fe, increasing temperature should cause $D_M^{met/sil}$ to increase; that is, all $D_M^{met/sil}$ should converge to unity as T goes to infinity. As pointed out by Murthy (1991), this is in the direction required to explain the main features of the BSE siderophile-element anomaly (namely, that V and Cr are more depleted than predicted, but the moderately and highly siderophile elements are less depleted) (Figure 1.22). However, the thermodynamic equation used by Murthy (1991) was incorrect (not merely oversimplified), and the magnitude of the effect with a thermodynamically valid extrapolation [equation (36)] is not sufficient to explain the siderophile-element anomaly at reasonable temperatures (e.g., <5,000 K). For the key elements, Ni and Co, this has been demonstrated experimentally by Seifert, O'Neill, and Brey (1988) for metal/olivine equilibria, and by Holzheid and Palme (1996) for metal/silicate-melt equilibria.

There are also exceptions to this expected convergence. Some elements are siderophile not by virtue of their standard-state properties (i.e., the sign of ΔH_{M-Fe}°) but because they have high values of $\gamma_{MO_{x/2}}^{sil}$. An excellent example is provided by Mo and W (Ertel et al., 1996): For both W and Mo, ΔH_{M-Fe}° is close to zero, and the more siderophile nature of Mo versus W is due to a much larger value of $\gamma_{MoO_2}^{sil\ melt}$ versus $\gamma_{WO_2}^{sil\ melt}$. The experimental evidence shows that the ratio $\gamma_{MoO_2}^{sil\ melt}/\gamma_{WO_2}^{sil\ melt}$ does not change significantly with temperature, and hence the ratio $D_{Mo}^{melt/sil}/D_W^{melt/sil}$ remains high ($\sim$10^3) at all T–fO$_2$ conditions (Ertel et al., 1996) and shows no convergence towards the values required for the BSE

$[(D_{Mo}^{melt/sil} / D_W^{melt/sil})_{required} = 3 \pm 1]$ (Table 1.9). High temperature alone cannot explain the siderophile-element anomaly.

1.6.5.3. High Pressures

The equivalent of equation (36) in pressure is

$$(D_M^{met/sil})_{P_2} = (D_M^{met/sil})_{P_1} \exp\left\{ -\int_{P_1}^{P_2} \frac{\Delta V_{M-Fe}^{\circ} \, dP}{RT} \right\} \qquad (37)$$

Unlike the case for ΔH_{M-Fe}°, there is no reason to expect the sign of ΔV_{M-Fe}° to be either positive or negative according to whether M is more or less siderophile than Fe, so that increasing pressure, unlike increasing temperature, should not exert a systematic tendency on the way the values of $D_M^{met/sil}$ change relative to $D_{Fe}^{met/sil}$. Yet to explain the siderophile-element anomaly (as displayed in Figures 1.21 and 1.22), ΔV_{M-Fe}° would have to be negative for the siderophile elements that are less siderophile than Fe (V and Cr), but positive for all siderophile elements that are more siderophile than Fe. This seems inherently unlikely.

In lieu of data on partial molar volumes for siderophile-element components in silicate melts, some feeling for the likely direction of the effect that pressure will have on the siderophile tendencies of an element, relative to Fe, can be obtained by considering standard-state molar-volume data for idealized exchange reactions of the type

$$\frac{x}{2}Fe + MO_{x/2} = \frac{x}{2}FeO + M \qquad (38)$$
$$\text{metal} \quad \text{oxide} \quad \text{oxide} \quad \text{metal}$$

where each constituent is in its standard state. The required molar-volume data are readily available (e.g., Robie, Hemingway, and Fisher, 1979). This admittedly naive exercise predicts that the oxide components NiO, Ga_2O_3, and SnO_2 may become increasingly lithophile with increasing pressure (in the required direction for Ni and Sn, but wrong direction for Ga), but that MoO_2, WO_2, GeO_2, Cu_2O, As_2O_3, and Sb_2O_3 will become more siderophile (wrong direction for all). The effects of pressure therefore do not appear promising as a means of explaining the siderophile-element anomaly.

Having said that, it is worth pointing out that the effects of very high pressures (>20 GPa) on Fe-Ni-Co equilibria between Fe-rich metal and an essentially near-solidus BSE composition do appear to account reasonably well for the BSE abundances of Ni and Co (O'Neill et al., in press); however, the same experimental study showed that the BSE abundance of Cr cannot be similarly explained. It remains to be seen what happens to other critical siderophile elements (e.g., V, W, and Mo) at such pressures.

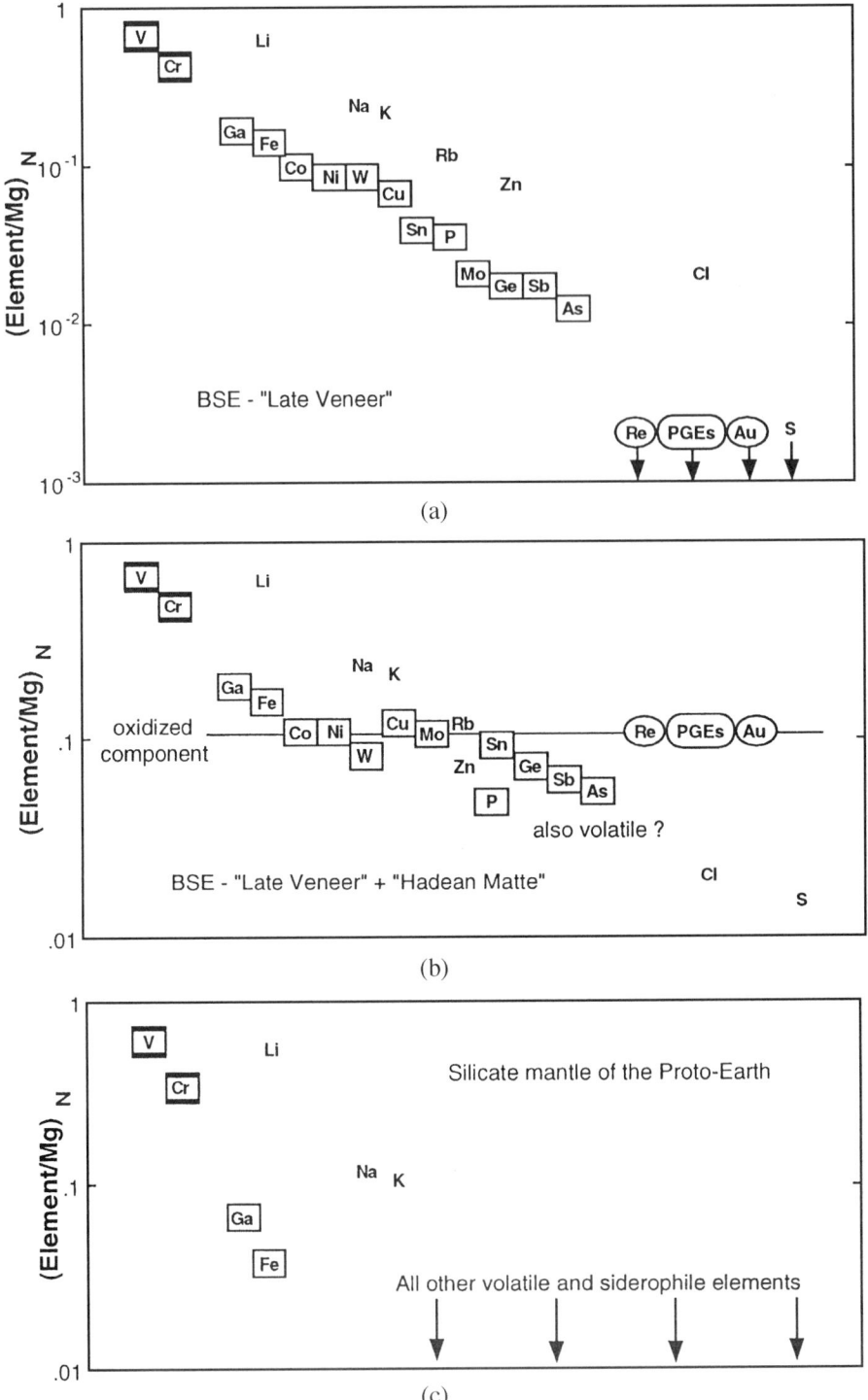

Figure 1.23. A heterogeneous-accretion model for the composition of the BSE, based on O'Neill (1991b). Starting with the observed BSE pattern of the siderophile and volatile elements (Figure 1.21), the proposed composition of the silicate portion of the accreting Earth is shown in reverse chronological order. (a) Firstly, the contribution from the late

1.6.6. Heterogeneous Accretion

Most of the recently proposed heterogeneous-accretion models are somewhat similar to one another, in that they explain the BSE siderophile-element abundances by adding a late, oxidized component (not to be confused with the 'late veneer', in the terminology used here) to an earlier reduced component (e.g., Jagoutz et al., 1979; Newsom and Sims, 1992). The description given here follows the quantitative description of O'Neill (1991b).

The accretion of the Earth is proposed to have occurred in three steps:

1. Accretion of the first 85–90% of the Earth (the 'proto-Earth') from planetesimals and/or planetary embryos condensed from the indigenous material in the solar nebula at $\sim$1 AU. This material was poor in volatiles and was quite reduced. Core formation under reducing conditions stripped the proto-Earth's mantle of *all* moderately siderophile and highly siderophile elements, nearly all Fe, and also partitioned some V and Cr into the core.
2. Exogenous material (10–15% of the Earth) was then added, perhaps as a single planetary embryo, which would therefore have involved a giant impact. This material was volatile-rich and oxidizing, and is envisaged as having originated at much greater heliocentric distances than the indigenous component. The S in that material separated to the core as sulphide, in a second stage of core formation, which was quantitatively minor but established the pattern for the moderately siderophile elements in the BSE.
3. A late-veneer component (probably reducing) that accounted for 0.6% ($\pm$0.1%) of the Earth was added, as discussed in Section 1.5.9. This sets the abundances of the highly siderophile elements and S.

To discuss this model in more detail, it is convenient to start with the present-day composition of the BSE and then subtract or add, as appropriate, the various components in reverse chronological order. Thus, the last component to be added was the late veneer (Sections 1.5.9 and 1.5.10), and we subtract it first. The amount of the late veneer is small ($\sim$0.6%), and subtracting it from the present-day BSE

Figure 1.23 (*cont.*). veneer is subtracted. (b) Secondly, 0.2% of a sulphide phase whose composition is in equilibrium with the composition derived in part (a) is added back in. This sulphide is the 'Hadean matte', presumed separated to the core in a last episode of core formation. Late-stage loss of sulphide is necessary to account for the extreme depletion of S in the BSE. Adding back the Hadean matte yields the $\sim$10% of the 'oxidized component'. This component may have been accreted to the Earth as an undifferentiated planetary embryo, or as planetesimals, composed of material accreted at >2.5 AU. Some elements (Cl, Zn, P, As, S) are depleted below the $\sim$10% level, perhaps by volatilization under oxidizing conditions. (c) Subtracting the 'oxidized component' yields the composition of the proto-Earth's mantle. This may largely be composed of 'indigenous' material, condensed from the solar nebula at $\sim$1 AU. This composition will be in equilibrium with core-forming Fe-rich metal at very reducing conditions, accounting for the BSE depletions in Cr and V.

composition will have little effect on other elements, apart from the highly sidero-phile elements, and S, Se, and Te. This is shown in Figure 1.23a.

We have argued that all the S in the present-day BSE was added in the late veneer (Section 1.5.9); consequently, the BSE prior to the arrival of the late veneer must have been completely stripped of sulphide, by loss to the core. That sulphide has been named the 'Hadean matte' (O'Neill, 1991b). The most parsimonious assumption is that the Hadean matte was originally in chemical equilibrium with the mantle, but the mean pressure and temperature at which it separated from the mantle have not yet been constrained.

Because the late veneer had a reduced oxidation state (suggested by its low S/Ir ratio, gross depletion in N, etc.; see Section 1.5.10), the BSE at the time of separation of the Hadean matte would have been in a slightly more oxidized state than is the present-day mantle.

Loss of the Hadean matte to the core was responsible for the depletion trend of the moderately siderophile elements (Figure 1.21). The hypothesis is that adding the Hadean-matte composition back into the BSE should bring the moderately and highly siderophile elements back to chondritic abundances, relative to one another (Figure 1.23b). The amount of Hadean matte needed to do this is rather small: 0.2 wt% of the present-day mantle, using the matte composition adopted by O'Neill (1991b). Thus, quantitatively the Hadean matte is a very minor component of the core and does not have much effect on the light-component problem (see Section 1.8).

The chondritic relative abundances of the moderately and highly siderophile elements shown in Figure 1.23b suggest that they were added by a chondritic, volatile-rich, and highly oxidized component. Unlike the late veneer, this compo-nent must have been fully oxidized (i.e., all Fe, Ni, and other moderately siderophile elements in the silicate portion). This is a firm part of the model that is necessi-tated by the fact that in the present-day mantle the Fe, Ni, and so forth, are in the oxidized state and constitute too large a part of the Earth to have been oxidized after accretion (O'Neill, 1991b). This constraint also rules out "inefficient-core-formation" models, in which the Ni and the other moderately siderophile elements in the BSE composition would be seen as having originated in metal left stranded in the mantle, as would have happened if the metal-segregation process had been less than perfectly efficient.

The volatile-rich oxidized component, first specifically identified by Jagoutz et al. (1979), is the 'component B' of Wänke et al. (1984) and was identified with the 'giant impactor' associated with the formation of the Moon by O'Neill (1991a,b), in order to account for the siderophile-element abundances in the Moon. However, we shall here refer to this component by the value-neutral name of 'oxi-dized component'.

Some of the more volatile lithophile elements, particularly Cl and Br, are notice-ably more depleted, relative to CI values, than are the other elements in the oxidized component (Figure 1.23b). This indicates either that the oxidized component was itself depleted in some moderately volatile elements or that some fractions of the moderately volatile elements were lost during the process by which the oxidized component was added to the proto-Earth. As already remarked, the pattern of depletion of these moderately volatile elements is not similar to that observed in chondrites. Elements such as P, Cl, and Br, which become more volatile under oxidizing conditions, appear anomalously depleted. Other highly volatile elements, such as Zn and In, which become less volatile with increasing oxygen fugacities, appear undepleted. This is consistent with the idea that some volatile elements were lost during the process that added the oxidized component, rather than such depletion being an intrinsic characteristic of this component (Section 1.5.2).

Subtraction of the oxidized component gives the composition of the proto-Earth's mantle, shown in Figure 1.23c. This composition is refractory, containing <0.1 of the chondrite-normalized abundances of Na and K, and no elements more volatile. It is calculated to contain only $\sim$2 wt% FeO and will thus be in equilibrium with Fe-rich metal at very low fO$_2$. This is consistent with the proto-Earth's mantle being completely stripped of all Ni, Co, and the other moderately volatile elements during core formation. Also, at these reducing conditions and an assumed temperature of 1,600° C, O'Neill (1991b) calculated values for the Cr and V metal/silicate partition coefficients ($D_{Cr}^{melt/sil}$ and $D_{V}^{melt/sil}$) of $\sim$1.7, which give good agreement with the values required to explain the depletions of Cr and V in the BSE. The effect of pressure on the partitioning behaviour of Cr and V at such low fO$_2$ values has not yet been investigated. More detailed investigation of Cr and V partitioning might yield clues as to how the main episode(s) of core formation occurred.

Why 2% FeO in the proto-Earth's mantle? A reasonable explanation is that some oxidation of Fe metal was produced by the partitioning of Si into the metal phase, according to the reaction

$$\underset{\text{silicate}}{SiO_2} + \underset{\text{metal}}{2Fe} = \underset{\text{metal}}{Si} + \underset{\text{silicate}}{2FeO} \tag{39}$$

which proceeds to the right hand side with increasing temperature (e.g., Ward, 1962). The mass balance of this equilibrium implies that the 2% FeO in the proto-Earth's mantle corresponds to 0.8% Si in the proto-Earth's core. Taking account of the oxygen similarly released by the partitioning of Cr into the core reduces this further. Such a small amount of Si in the core would not significantly affect the bulk-Earth Mg-Si-RLE relations and would be far below the 5% required to account for the BSE deviation in Mg/Si ratio from the carbonaceous-chondrite Mg-Si-RLE trend (Figure 1.17).

1.7. The Timing of Core Formation

Recent data on the W isotopic compositions of terrestrial rocks and meteorites have demonstrated the usefulness of the ^{182}W–^{182}Hf system for dating core formation (Harper et al., 1991; Lee and Halliday, 1995; Harper and Jacobsen, 1996). ^{182}Hf decays with a half-life of 9×10^6 years to ^{182}W. Because Hf is an RLE and W is a refractory but moderately siderophile element, the system should not be disturbed by volatility-related processes. However, metal/silicate separation (i.e., core formation) in the presence of live ^{182}Hf would have greatly influenced the ^{182}W/^{183}W ratios in the resulting metal and silicate reservoirs. Early data on the W isotopic composition in some iron meteorites have revealed a 3–5‰ depletion of ^{182}W relative to terrestrial W (Harper et al., 1991). This was interpreted by Harper and Jacobsen (1996) as reflecting rapid core formation in the Earth, because early removal of W into the core would have increased the Hf/W ratio in the Earth's mantle, producing the observed excess of ^{182}W relative to the iron meteorites. In that interpretation, it was assumed that the ^{182}W/^{184}W ratio in the iron meteorites was that of the present-day solar system. However, Lee and Halliday (1995) recently determined, for the first time, the precise W isotopic composition in chondritic meteorites. Surprisingly, they found that the chondritic ^{182}W/^{184}W ratio was identical with that of the Earth's mantle. That suggested an interpretation opposite to the Harper and Jacobsen model: In the Earth, the separation of W to the core was delayed until the decay of ^{182}Hf to ^{182}W was effectively complete. Core formation in the Earth must have been a relatively late event. To be consistent with the W isotopic data, a single-step core-formation model would require about 50–60 million years to have elapsed between the formation of the iron meteorites and the formation of the Earth's core (see McCulloch and Bennett, Chapter 2, this volume). This is graphically displayed in Figure 1.24.

The data place extremely tight constraints on core-formation models. Obviously, the data immediately show that the BSE could not have inherited its W abundance from core formation in early forming, asteroid-size planetesimals (this is ironic, as W is the only siderophile whose abundance agrees well with that predicted from low-pressure metal/silicate distribution coefficients, Figure 1.22). The cores from such planetesimals could not have merged, but must indeed have dispersed during subsequent collisions. In fact, any isolation of W from the material comprising the BSE within 50 million years after the iron-meteorite core-forming event would have produced an observable anomaly in the isotopic composition of the W remaining in the BSE. Thus the data are not consistent with core formation in the homogeneous-accretion scenario, in which the influx of new material to the accreting Earth is accompanied by synchronous loss of metal into the core.

The constraints imposed by the W isotopic data can be more easily satisfied by heterogeneous-accretion models of the type discussed in the preceding section. Accretion of the Earth through impacts of planetary embryos would have been a comparatively slow process lasting millions of years (Section 1.1). Addition of oxidized, undifferentiated material more than 50 million years after the time of iron-meteorite formation would have produced no anomaly in the W isotopic

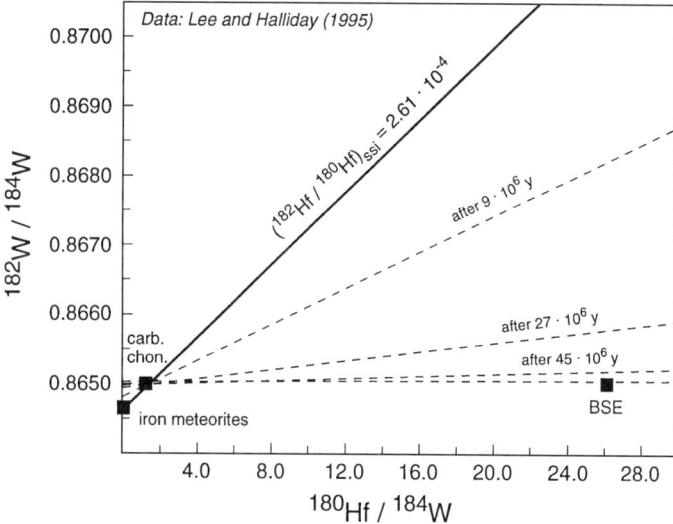

Figure 1.24. The timing of core-formation in the Earth: the evidence from the Hf–W isotopic system. ^{182}Hf decays to ^{182}W, with a half-life of 9×10^6 years. Because W and Hf are both refractory elements, the W/Hf ratio in most types of carbonaceous chondrites is near the solar ratio. (Some variation of W/Hf in chondrites is to be expected from cosmochemical metal/silicate fractionation; see Figure 1.8. Thus the metal-rich CH chondrites have a significantly higher W/Hf ratio than other carbonaceous chondrites.) Core formation in the iron-meteorite parent bodies occurred early ($< \sim 9 \times 10^6$ years), and therefore the W in these meteorites is deficient in ^{182}W. With the assumption of no radiogenic ^{182}W in the iron meteorites, the difference in the ^{182}W/^{184}W ratio between the iron meteorites and the carbonaceous chondrites constrains the solar-system initial ^{182}Hf/^{180}Hf ratio to be $(2.61 \pm 0.13) \times 10^{-4}$. Radiogenic ^{182}W in the iron meteorites would imply even higher $(^{182}$Hf/^{180}Hf$)_{ssi}$. A problem that has not yet been resolved is that this $(^{182}$Hf/^{180}Hf$)_{ssi}$ is at least an order of magnitude greater than model predictions (Harper and Jacobsen, 1996). Despite the abundance of W in the BSE being greatly depleted relative to Hf by core formation, the isotopic composition of the remaining W in the BSE is identical with that in undifferentiated chondrites. The present-day values of ^{182}W/^{184}W expected in the silicate portion of a planetary body after single-stage metal-separation events, occurring at 9, 27, and 45 million years, respectively, are plotted as functions of the resulting ^{180}Hf/^{184}W ratio (dashed lines). The chondritic ^{182}W/^{184}W ratio in the BSE implies that the depletion of W must have occurred after the complete decay of ^{182}Hf (i.e., after about six half-lives). The chondritic isotopic composition of W in the BSE is consistent with a heterogeneous-accretion model, in which the W presently found in the BSE is added subsequent to core formation, from an undifferentiated 'oxidizing component', after $\sim 50 \times 10^6$ years.

composition provided that two conditions are met: (1) if the W now in the mantle was indeed derived from the late component and (2) if prior to the arrival of the late component, all W, including all the radiogenically produced W, had been lost to the core.

Evidence for the timing of the accretion of the Earth can also be gleaned from U-Th-Pb isotopic systematics, as discussed by McCulloch and Bennett (Chapter 2, this volume) and by Galer and Goldstein (1996). The average ^{238}U/^{204}Pb ratio (μ) of the BSE is about 65 times larger than the CI ratio, mainly, we believe, because of incomplete condensation of Pb in pre-terrestrial matter and volatile loss during the accretion process, but perhaps also because of some partitioning of Pb into the core. This view differs slightly from that of Galer and Goldstein (1996), who did not consider the possibility of volatile loss during accretion and therefore ascribed a larger share of Pb depletion in the BSE to partitioning into the core. Whatever its cause, the Pb loss displaced the Pb isotopic ratios for rocks of upper-mantle origin from the 4.556×10^9-year geochron (i.e., the curve expected for the evolution of Pb isotopes beginning at the time of formation of the first solids in the solar system at 4.556×10^9 years ago). The magnitude of the displacement can be used to determine when the Pb loss occurred. Both McCulloch and Bennett (Chapter 2, this volume) and Galer and Goldstein (1996) calculated a time interval of 80 ± 40 million years, entirely consistent with the W isotopic data.

1.8. The Enigma of the 'Light Component' in the Core
1.8.1. The Problem from a Geochemical Perspective

Geophysical observations regarding the density of the Earth's core, when compared with laboratory measurements of the equation-of-state of iron metal extrapolated to core pressures and putative core temperatures, indicate that the core is less dense than would be expected for a binary liquid Fe-Ni alloy (e.g., Birch, 1964; Ahrens, 1982). The core is therefore thought to contain a significant amount of a light element, or combination of light elements. A recent review (Poirier, 1994) puts the core's density deficit at around $10 \pm 2\%$ compared with liquid Fe-Ni alloy. The candidate elements that are sufficiently abundant in the solar composition to have an effect on core density, and which are known to dissolve into Fe-rich liquid metal in sufficient quantities at $P–T$ conditions obtainable in the Earth, are H, C, N, O, Si, and S. All of these (except N) have had their proponents at various times (Poirier, 1994). Poirier estimated that a 10% density deficit would require 11 wt% S, or 8.2 wt% O, or 18 wt% Si dissolved in the core. We shall adopt these figures for the purposes of discussion, although the complexity of the procedures by which they are deduced may permit some doubts as to their robustness.

The problem from a geochemical perspective is that no single one of the aforementioned candidates can satisfy the geochemical constraints on the light element's identity. The most important of these constraints is the observation that the bulk Earth is heavily depleted in the cosmochemically volatile elements, as deduced from the BSE abundances of the volatile *lithophile* elements (Figure 1.18). This constraint can be immediately applied to the proposition that S forms the main part of the light component (e.g., Murthy and Hall, 1970; Ahrens, 1979). It has long been recognized that the S abundance in the bulk Earth probably is insufficient to account for the density deficit (e.g., Ringwood, 1979). The argument can be put very simply. Sulphur is certainly cosmochemically more volatile than Na and K (e.g., Figure 1.5b, in which Se can be taken as standing in for S). Sulphur should therefore be more depleted than Na or K in the bulk Earth. Even in the enstatite-chondrite meteorites, which formed under conditions of unusually high fS_2 (and which thus exhibit many geochemical features, such as the chalcophile behaviour of Ca, Ti, and REEs, that are not seen in the chemistry of the Earth's mantle), S is more depleted than Na and K. If the depletion of S in the bulk Earth were the same as that of Na and K, then the amount of S in the core would be limited to ~6 wt%, or about half the amount required for the density deficit. In fact, the depletion pattern for lithophile elements in the BSE is closest to that shown by the carbonaceous chondrites (Figure 1.18), implying that the volatility-related depletion factor for S in the bulk Earth can more realistically be modelled using the BSE abundance of Zn (Dreibus and Palme, 1996), an approach which yields an upper limit for S in the core of about 1.7 wt%. For the Earth's core to contain the 11 wt% or so of S that the density argument seems to require, about half of the S in the solar composition would need to have condensed and accreted to the bulk Earth. No mechanism by which that could have been done, while at the same time accounting for the observed depletions of Na, K, Zn, and the other volatile lithophile elements in the BSE (Figure 1.18), has yet been proposed.

The volatility argument also rules out H, C, and N, which are cosmochemically much more volatile than S. It has not been possible to construct a cosmochemical scenario in which the required amounts of these elements could be condensed without also postulating that S would be nearly fully condensed. Of course, if S were nearly fully condensed, there would be too much S in the core to be compatible with its density.

The volatility objection to the choice of S led Ringwood (1979) to propose O as the light element. Although O is only slightly soluble in liquid Fe near its melting point at a pressure of 1 atm, Ringwood (1979, 1984) demonstrated experimentally that sufficient O can become soluble at higher (superliquidus) temperatures and high pressures, at least in the simple system Fe-O, to make the candidacy of O a finite possibility.

However, what might be termed the *available* O (i.e., the O that is not inextricably bound up in non-siderophile oxide components such as Al_2O_3, CaO, SiO_2, and MgO) is cosmochemically more volatile than S. The first of this available O to condense from the solar nebula was the O that oxidized Fe in the previously condensed Fe-Ni alloy, forming, for example, the Fe_2SiO_4 component in olivine. Nebular-condensation calculations (Grossman and Larimer, 1974; Saxena and Ericsson, 1986) show that this process becomes significant only at temperatures below the S condensation temperature. (Still more available O condenses at even lower temperatures as H_2O or CO_2.) Secondly, recent experimental studies have shown that the amount of oxygen that can dissolve in liquid Fe in equilibrium with oxide with an Mg# similar to that in the Earth's mantle does not increase with pressure.

Thus the amounts of oxygen that can dissolve in Fe-rich metal at realistic temperatures are limited to about 1%, at least at pressures to 25 GPa (O'Neill et al., in press).

That leaves Si. One superficially attractive feature of having Si in the core is that Si is depleted in the Earth's mantle as compared with the solar (or CI) abundances (Figure 1.17). This has led some authors to propose that the 'missing' Si is in the core (e.g., MacDonald and Knopoff, 1958). This argument does not withstand quantitative scrutiny. The amount of Si in the core needed to reconcile the BSE Mg-Si-RLE ratios with the carbonaceous-chondrite trend is only 5 wt% (Figure 1.17), which is well below the 18 wt% required (according to Poirier, 1994) to produce the 10% density deficit. Si is not a very effective agent for reducing the density of liquid Fe. With 18 wt% Si in the core, the bulk-Earth composition would plot in a region of Mg/Si–Si/Al space that is not occupied by any other known solar-system material (Figure 1.17). This may not be a conclusive argument against having ∼18 wt% Si in the core – but it certainly is not an argument for it.

Perhaps the most important objection to Si as the main light element in the core is that to get anywhere near this 18% into Fe metal would imply unrealistically low oxygen fugacity, at which virtually no FeO component would be left in coexisting silicate phases. Empirically, the lack of oxidized Fe components in silicates coexisting with Si-containing metal is seen in the equilibrated enstatite chondrites, which in any case have at most only a few percent Si in their Fe metal, in equilibrium with enstatite with Mg# > 99 (i.e., <1% of the $FeSiO_3$ component) (Zhang et al., 1995). Although the solubility of Si in Fe-rich metal in equilibrium with silicate of fixed Mg# increases with temperature, thermodynamic analysis (e.g., O'Neill, 1991b, p. 1166) indicates that to reach 18 wt% Si in metal, in equilibrium with mantle olivine with Mg# 98, would require a temperature >3,500°C at atmospheric pressure. The effect of pressure has been experimentally demonstrated to be minimal (O'Neill et al., in press). There is, however, another problem in supposing that Si constitutes the core's light component: the mass balance for oxygen.

This constraint arises because Si condensed from the solar-nebula into silicates (e.g., Grossman and Larimer, 1974), that is, as an SiO_2 component, as, for example, in Mg_2SiO_4 (olivine). The amount of Si entering Fe-rich metal under all reasonable solar-nebula $T–fO_2$ conditions would have been inconsequential, mainly because of the solubility of Si in Fe-rich metal being a sensitive function of temperature: Any significant solubility would require temperatures too high for any plausible solar-nebula pressure. Thus, we must assume that Earth-forming material initially condensed with nearly all its Si in an oxidized state, in silicates. If initially all Si exists as silicate and all Fe is in the metallic state, the dissolution of Si into Fe-rich metal can be described by the reaction

$$SiO_2 + 2Fe = Si + 2FeO \qquad (40)$$

silicate   metal   metal   silicate

The reduction of sufficient SiO_2 component to achieve 18 wt% Si in Fe-rich metal would liberate enough oxygen, given the relative masses of the core and mantle in the Earth ($0.32M_E$ and $0.68M_E$), to produce a mantle concentration of FeO of 43 wt%. That is five times the currently observed abundance of FeO in the Earth's mantle. Of course, buildup of FeO in the silicate would bring the reaction to a halt long before that level was reached anyway (Le Chatelier's principle).

The important conclusion is that there is, from a geochemical view, no single candidate that can adequately account for the mystery light element in the core. The objection to Si is that its partitioning into metal would evolve a superfluity of O, which is ironic when it is remembered that the fundamental objection to O was that there would not have been enough available. It therefore seems logical to examine the possibility that a combination of Si and O might provide a way around this dilemma.

1.8.2. A Combination of Si and O as the Light Component?

The geochemical objections would seem to be answered if Si and O could simultaneously become soluble in Fe-rich metal, in sufficient amounts. For example, consider what could happen if Si and O in the ratio given by the stoichiometry of SiO_2 could be dissolved in Fe-rich metal. If the 10% density deficit could be satisfied by a linear combination between 8.2 wt% of O alone and 18 wt% of Si alone [i.e., wt% O = $8.2 - (8.2/18)$ wt% Si], dissolution of stoichiometric SiO_2 would involve 5.9 wt% O and 5.0 wt% Si. The amount of SiO_2 required to be lost from the mantle would be 5 wt%, which is about 11% of the present-day amount. As pointed out earlier, such an amount is easily accommodated within the rather loose constraints that cosmochemical arguments can impose on the bulk Earth's Si content. Actually, the ratio of Si to O dissolved in liquid Fe can, in principle, have any value, and the dissolved component can be referred to as Si_xO. A value of x slightly greater than 0.5 is the most appropriate for avoiding conflict with the oxygen mass balance constraint. The possibility of solving the light-component problem in this way was raised by O'Neill (1991b).

Unfortunately, the evidence from steel-making practice does not favour simultaneous dissolution of Si and O in liquid Fe. Indeed, Si has been widely used as a deoxygenating agent in steel-making, and consequently the physical chemistry of the equilibrium between Si and O in liquid Fe is well understood, at least at atmospheric pressure and at temperatures up to 2,000 K (e.g., Ward, 1962). The equilibrium between silica and the silicon and oxygen in molten iron can be represented as

$$[Si]_{1wt\%} + 2[O]_{1wt\%} = SiO_2 \tag{41}$$

where the square brackets refer to Si and O dissolved in liquid Fe. The standard state

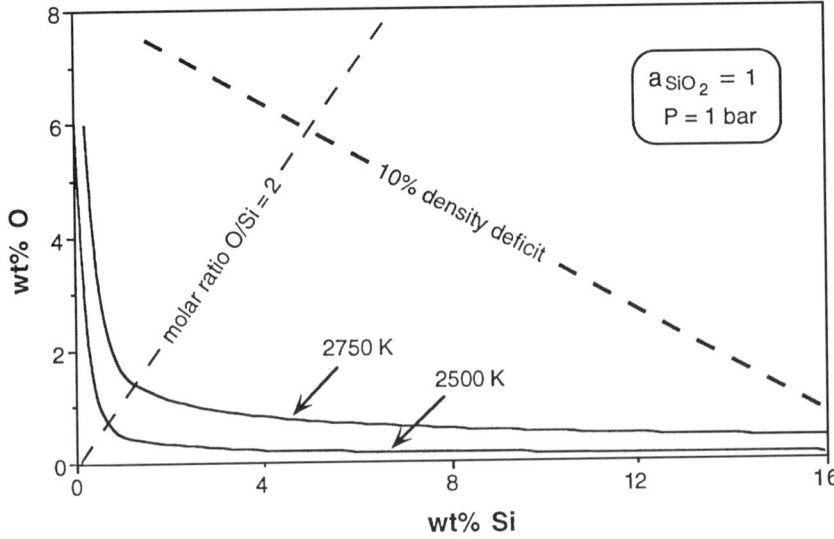

Figure 1.25. The mutual dependence of the solubilities of Si and O in liquid Fe at low pressures, from steel-making literature [equations (40)–(42)]. The solubilities are drawn for $a_{SiO_2} \approx 1$. Lower activities of silica (as in the BSE) result in lower solubilities. The solubilities of Si and O are largely mutually exclusive, which is why Si is used as a deoxygenating agent in steel-making. Also shown is the combination of Si and O solubilities needed to account for a 10% density deficit in the Earth's outer core, according to Poirier (1994).

for the latter is defined to be 1 wt% dissolved in the Fe. The equilibrium constant is

$$K_{Si-O} = a_{SiO_2}/[c_{Si}][c_O]^2 \qquad (42)$$

which, from experimental data (Ward, 1962) on silica-saturated slags ($a_{SiO_2} \approx 1$), has the value of

$$\log K_{Si-O} = 29,150/T - 11.01 \qquad (43)$$

The amounts of Si and O in molten Fe at silica saturation predicted by this equation, extrapolated slightly in temperature from the original measurements (which extend to 2,000 K), are presented in Figure 1.25, together with the amounts proposed by Poirier (1994) as being needed to satisfy the core's density deficit. The nature of the equilibrium constant K_{Si-O} obviously is such that the solubilities of Si and O are largely mutually exclusive (which is, of course, the reason silicon is effective as a deoxygenating agent), and it seems that significant solubilities of Si and O at the Si/O ratios needed to satisfy the oxygen mass balance could not be achieved at atmospheric pressure, except by extrapolating the trends shown on this diagram to excessively high temperatures (at which some RLEs, such as Ti, may be expected to partition into the metal).

The question, then, is whether or not their mutual exclusivity persists at higher pressures. A series of experiments to measure the combined solubilities of Si and O in liquid Fe, in equilibrium with model mantle silicates, has recently been undertaken (O'Neill et al., in press). The results show that the mutual exclusivity between

Si and O seen at low pressures persists to at least 25 GPa, with no indication that still higher pressures would reverse the relationship.

There is thus, at present, no neat solution to the light-component problem. Perhaps the search for a neat solution is misconceived: As Stevenson (1981) has remarked, 'There is no reason to believe that the core is a particularly clean system'; a stew of light elements, each in an amount at the upper limit given by the various cosmochemical arguments discussed earlier (e.g., $\sim$1% S and O, $\leq$5% Si, plus some C), would add up to about half of that required (Poirier, 1994) to account for the presumed 10% density deficit. That raises the question of how robust the estimate of the density deficit really is. In other words, how well known is the equation-of-state for liquid Fe-Ni alloys at core pressures?

1.9. Summary: A Model for the Earth's Accretion and Core Formation

The abundances of the siderophile elements in the BSE clearly are not those expected on the basis of metal/silicate equilibria at low pressure, and indeed they are considerably different from the abundances found in the silicate portions of small differentiated planetesimals like the parent body of the eucrite basaltic achondrites (e.g., Dreibus and Wänke, 1980). According to Dreibus and Wänke (1980), the silicate eucrite parent body (EPB) is calculated to be undepleted in Cr (relative to CI, Mg-normalized), but depleted by a factor of $\sim$500 in Ni (i.e., the silicate EPB has $\sim$40 ppm Ni, versus 1,860 ppm in the BSE), and it has a Ni/Co ratio of 3, versus 18 in the BSE and 21 in chondrites. The pattern of depletion of these elements in the EPB is consistent with their low-pressure, moderate-temperature metal/silicate partitioning relations. It is the Earth that appears anomalous.

The Earth is, of course, a much larger body than the EPB. Size can affect the chemistry of core formation in two ways. Firstly, in larger bodies, metal/silicate segregation may occur at very high pressures, under which conditions (it can be speculated) the siderophile properties of the elements (i.e., their metal/silicate distribution coefficients) will differ significantly from their familiar low-pressure behaviours. This venerable proposition is increasingly being put to the test experimentally. So far, the experimental data suggest that high pressures ($>$25 GPa) may explain the pattern of depletion of the important siderophile elements Ni and Co relative to Fe, but not the depletions of Cr, W, and Mo.

Secondly, the large size of the Earth makes it reasonable to suppose that the Earth incorporated material that had originally condensed at great heliocentric distances, in addition to the 'indigenous' material at 1 AU (Wetherill, 1994). We have shown that such a heterogeneous-accretion model, in which $\sim$15% of 'exogenous' oxidized material would be added at a late stage of accretion to the 'indigenous' refractory, reduced material, could explain the pattern of the siderophile-element abundances in the BSE. This explanation, in particular, accounts well for the robust constraints set by the BSE's Fe-Ni-Co systematics, the W/Mo ratio, and its depletions in the 'weakly siderophile' elements Cr and V, whose abundances are otherwise difficult to explain. However, heterogeneous accretion should not be viewed as an exclusive alternative to high-pressure metal/silicate equilibria as the mechanism for establishing the BSE pattern of the siderophiles; perhaps both factors were involved. An optimum solution to the general problem of core formation in the Earth awaits better constraints on where and how metal/silicate segregation is likely to have taken place (e.g., Stevenson, 1990).

The composition of the postulated refractory component that constituted the proto-Earth (i.e., the first ∼85% of the Earth to accrete, as summarized in Figure 1.23c) was not like the composition of any chondrite meteorite. It was much more depleted in volatile elements, being free of all elements more volatile than K. It consisted of the refractory elements (lithophile and presumably also siderophile) plus the common elements (Mg, Si, Cr, and the Fe-Ni-Co group), with about 50% condensation of Li, the least volatile of the moderately volatile lithophile elements, but less than 10% condensation of Mn, Na, and K, the next elements in the condensation sequence. We propose that this was the indigenous material at 1 AU. Virtually all Fe was initially present in this material as metal. Although unlike any chondrite meteorites, the composition and oxidation state of this high-temperature material is essentially that expected theoretically from equilibrium condensation of the solar nebula (allowing for the usual metal/silicate and Mg/Si fractionations), if condensation were halted at the temperature region in which Na and K have just begun to condense. The composition of the postulated high-temperature component is thus simpler to account for than that of actual chondrites.

Theoretical modelling suggests that the indigenous material at 1 AU accreted through an evolving size distribution of bodies. It is reasonable to expect that those bodies had undergone differentiation and possessed metallic cores by the time they had reached the size of planetesimals (10^{21} kg), if not the size of planetary embryos (10^{23} kg). Certainly core formation had taken place in the planetesimal-size asteroids that were the parent bodies of the magmatic-iron meteorites. The evidence from the ^{107}Pd–^{107}Ag (Chen and Wasserburg, 1996), Re–Os (Shen, Papanastassiou, and Wasserburg, 1996), and ^{182}Hf–^{182}W isotopic systems confirms that core formation in those bodies happened early, within about ∼10 million years after nucleosynthesis. In the Earth, however, the evidence from the W isotopes shows that at least 50 million years would elapse before core formation was complete. In the time scale of the planetary-accretion process, that is thought to be well after the planetary embryos had coalesced to form the proto-Earth. The W evidence therefore requires that any ^{182}W produced in the proto-Earth's mantle was efficiently lost to the core. This might have been facilitated if accretion of planetary embryos was accompanied by rehomogenization of the embryos' cores and silicate mantles within the proto-Earth on a scale fine enough to allow chemical re-equilibration. Such a scenario can also provide the high temperatures needed to account for the exchange of Si and Fe between metal and silicate [equation (39)] and for the Cr and V depletions. At that stage, the silicate portion of the proto-Earth would have contained almost no Ni, Co, nor most of the other moderately siderophile elements, either because they had been lost to the core or because of their volatility.

The 'oxidized component' supplying these latter elements was then added late, after more than 50×10^6 years, judging from the evidence of the W and Pb isotopes. In addition to being oxidized, that component would have contained, in contrast to the postulated indigenous material at 1 AU, a nearly full complement of the moderately volatile and highly volatile elements, and presumably it would also have been relatively rich in the ice-forming elements, including water. Such oxidized, volatile-rich material, plausibly formed at greater heliocentric distance than 1 AU (e.g., in the asteroid belt), would have been delivered to the vicinity of 1 AU by stochastic processes. There is no reason to suppose that such volatile-rich material did not contain S in abundances within the range found in chondritic materials. S is a crucial element for the heterogeneous-accretion model: Most of the added S would have segregated into a sulphide phase and been lost to the core in the Hadean matte, thereby accounting for the pattern of depletion of the moderately siderophile elements and the greater depletions of the very highly siderophile elements that we see today (Figures 1.21 and 1.23). This second-stage loss of material to the core would have occurred under oxidizing conditions, which may have been important for the physical process of sulphide segregation, by facilitating percolation through a decrease in the dihedral angle for sulphide/silicate mixtures (O'Neill, 1991b). This kind of second-stage loss of S would be a natural consequence of the heterogeneous-accretion model, and some process of that nature is surely required to account for the extraordinary degree of depletion of S (and Se) in the BSE.

There is a suggestion in the pattern of depletion of the more volatile elements (Figure 1.18) that the addition of the oxidized component was accompanied by a certain amount of volatile loss under oxidizing conditions, which depleted the BSE of elements like Cl, Br, and N (and possibly P) relative to Zn, In, and H_2O. That volatilization was evidently a mass-independent process.

The oxidized component added the present-day BSE inventory of W, which consequently has a chondritic $^{182}W/^{184}W$ ratio, as observed. W is not sufficiently siderophile to have been affected by the removal of the Hadean matte, but Pb may be; alternatively, Pb may have been lost by volatilization.

The final stage in the accretion of the Earth was the arrival of the late veneer. The sub-chondritic S/Ir ratio of the late veneer is not consistent with CI-like abundances of volatile elements, and the depletion of volatiles implied by that ratio (and also perhaps by the low CI-normalized abundance of N in the BSE) argues against the late veneer being responsible for the H_2O or C content of the BSE. The sub-chondritic S/Ir ratio is more consistent with fairly reduced, rather than oxidized, material, so that the effect of the late veneer would have been to lower the oxidation state of the mantle, compared with the oxidation state at which the Hadean matte segregated. This may be important in preventing the present-day mantle's sulphide phase from segregating to the core.

1.10. Acknowledgements

We thank Ian Jackson for his enormous editorial patience. Detailed reviews and comments by Ian Jackson, Sue Kesson, Ian Campbell, and Geoff Davies are gratefully acknowledged. Astrid Holzheid kindly supplied Figure 1.22.

References

Ahrens, T. J. 1979. Equations of state of iron sulfide and constraints on the sulfur content of the Earth. *J. Geophys. Res.* 84:985–98.

Ahrens, T. J. 1982. Constraints on core composition from shock-wave data. *Phil. Trans. R. Soc. London* A306:37–47.

Alcock, C. B. 1976. *Principles of Pyrometallurgy*. London: Academic Press.

Allègre, C. J., Poirier, J.-P., Humler, E., and Hofmann, A. W. 1995. The chemical composition of the Earth. *Earth Planet. Sci. Lett.* 134:515–26.

Allègre, C. J., and Turcotte, D. L. 1986. Implications of a two-component marble-cake mantle. *Nature* 323:123–7.

Anders, E., and Grevesse, N. 1989. Abundances of the elements: meteoritic and solar. *Geochim. Cosmochim. Acta* 53:197–214.

Anders, E., and Zinner, E. 1993. Interstellar grains in primitive meteorites: diamond, silicon carbide, and graphite. *Meteoritics* 28:490–514.

Basaltic Volcanism Study Project 1981 Permagon Press Inc., New York. 1286 pp.

Benz, W., Cameron, A. G. W., and Melosh, H. J. 1989. The origin of the Moon and the single-impact hypothesis. II. *Icarus* 81:113–31.

Binzel, R. P., and Xu, S. 1993. Chips off of asteroid 4 Vesta: evidence for the parent body of basaltic achondrite meteorites. *Science* 260:186–91.

Birch, F. 1964. Density and composition of mantle and core. *J. Geophys. Res.* 69:4377–88.

Bischoff, A., and Geiger, T. 1995. Meteorites from the Sahara: find locations, shock classification, degree of weathering and pairing. *Meteoritics* 30:113–22.

Bischoff, A., Palme, H., Ash, R. D., Clayton, R. N., Schultz, L., Herpers, U., Stoeffler, D., Grady, M. M., Pillinger, C. T., Spettel, B., Weber, H., Grund, T., Endress, M., and Weber, D. 1993a. Paired Renazzo-type (CR) carbonaceous chondrites from the Sahara. *Geochim. Cosmochim. Acta* 57:1587–603.

Bischoff, A., Palme, H., Schultz, L., Weber, D., Weber, H. W., and Spettel, B. 1993b. Acfer 182 and paired samples, an iron-rich carbonaceous chondrite: similarities with ALH85085 and its relationship to CR chondrites. *Geochim. Cosmochim. Acta* 57:2631–48.

Bodinier, J. L. 1988. Geochemistry and petrogenesis of the Lanzo peridotite body, western Alps. *Tectonophysics* 149:67–88.

Bodinier, J. L., Dupuy, C., and Dostal, J. 1988. Geochemistry and petrogenesis of eastern Pyrenean peridotites. *Geochim. Cosmochim. Acta* 52:2893–907.

Boehler, R. 1996. Experimental constraints on melting conditions relevant to core formation. *Geochim. Cosmochim. Acta* 60:1109–12.

Bonatti, E., Ottonello, G., and Hamlyn, P. R. 1986. Peridotites from the island of Zabargad (St. John), Red Sea: petrology and geochemistry. *J. Geophys. Res.* 91:599–631.

Borisov, A., and Palme, H. 1995. The solubility of Ir in silicate melts: new data from experiments with $Ir_{10}Pt_{90}$ alloys. *Geochim. Cosmochim. Acta* 59:481–5.

Borisov, A., Palme, H., and Spettel, B. 1994. Solubility of palladium in silicate melts: implications for core formation in the Earth. *Geochim. Cosmochim. Acta* 58:705–16.

Boyd, F. R. 1989. Compositional distinction between oceanic and cratonic lithosphere. *Earth Planet. Sci. Lett.* 96:15–26.

Boynton, W. V. 1975. Fractionation in the solar nebula: condensation of yttrium and the rare earth elements. *Geochim. Cosmochim. Acta.* 39:569–84.

BVSP. 1981. *Basaltic Volcanism Study Project*. Elmsford, NY: Pergamon Press.

Canil, D., O'Neill, H. St. C., Pearson, D. G., Rudnick, R. L., McDonough, W. F., and Carswell, D. A. 1994. Ferric iron in peridotites and mantle oxidation states. *Earth Planet. Sci. Lett.* 123:205–20.

Cassen, P. 1996. Models for the fractionation of moderately volatile elements in the solar nebula. *Meteoritics* 31:793–806.

Chaussidon, M., and Jambon, A. 1994. Boron content and isotopic composition of oceanic basalts: geochemical and cosmochemical implications. *Earth Planet. Sci. Lett.* 121:277–91.

Chen, J. H., and Wasserburg, G. J. 1996. Live [107]Pd in the early solar system and implications for planetary evolution. In: *Earth Processes: Reading the Isotopic Code*, ed. A. Basu and S. Hart, pp. 1–20. AGU Monograph 95. Washington, DC: American Geophysical Union.

Choi, B.-G., Ouyang, X., and Wasson, J. T. 1995. Classification and origin of IAB and IIICD iron meteorites. *Geochim. Cosmochim. Acta* 59:593–612.

Clayton, R. N. 1993. Oxygen isotopes in meteorites. *Annu. Rev. Planet. Sci.* 21:115–49.

Clayton, R. N., and Mayeda, T. K. 1996. Oxygen isotope composition of achondrites. *Geochim. Cosmochim. Acta* 60:1999–2017.

Danchin, R. V. 1979. Mineral and bulk chemistry of garnet lherzolite and garnet harzburgite xenoliths from the Premier Mine, South Africa. *Proc. 2nd Int. Kimberlite Conf.* 2:104–26.

Davies, G. F. 1988. Ocean bathymetry and mantle convection. 1. Large-scale flow and hotspots. *J. Geophys. Res.* 93:10467–80.

Delano, J. W., and Stone, K. 1985. Siderophile elements in the Earth's upper mantle: secular variations and possible causes for their overabundance. *Lunar Planet. Sci.* 16:181–2.

Déruelle, B., Dreibus, G., and Jambon, A. 1992. Iodine abundances in oceanic basalts: implications for Earth dynamics. *Earth Planet. Sci. Lett.* 108:217–27.

Dick, H. J. B., and Sinton, J. M. 1979. Compositional layering in alpine peridotites: evidence for pressure solution creep in the mantle. *J. Geol.* 87:403–16.

Dodd, R. T. 1981. *Meteorites*. Cambridge University Press.

Donahue, T. M., and Hodges, R. R., Jr. 1992. Past and present water budget of Venus. *J. Geophys. Res.* 97:6083–91.

Downes, H., Embey-Isztin, A., and Thirlwall, M. F. 1992. Petrology and geochemistry of spinel peridotite xenoliths from the western Pannonian Basin (Hungary): evidence for an association between enrichment and texture in the upper mantle. *Contrib. Min. Pet.* 109:340–54.

Drake, M. J. 1987. Siderophile elements in planetary mantles and the origin of the Moon. *J. Geophys. Res.* 92:377–86.

Dreibus, G., and Palme, H. 1996. Cosmochemical constraints on the sulfur content in the Earth's core. *Geochim. Cosmochim. Acta* 60:1125–30.

Dreibus, G., and Wänke, H. 1980. The bulk composition of the eucrite parent asteroid and its bearing on planetary evolution. *Z. Naturforsch.* 35:204–16.

Dupuy, C., Dostal, J., and Bodinier, J. L. 1987. Geochemistry of spinel peridotite inclusions in basalts from Sardinia. *Min. Mag.* 51:561–8.

Eberhardt, P., Dolder, U., Schulte, W., Krankowsky, D., Lämmerzahl, P., Hoffman, J. H., Hodges, R. R., Berthelier, J. J., and Illiano, J. M. 1987. The D/H ratio in water from comet P/Halley. *Astron. Astrophys.* 187:435–7.

Embey-Isztin, A., Scharbert, H. G., Dietrich, H., and Poultidis, H. 1988. Petrology and geochemistry of peridotite xenoliths in alkali basalts from the Transdanubian Volcanic Region, West Hungary. *J. Petrol.* 30:79–105.

Ertel, W., O'Neill, H. St. C., Dingwell, D. B., and Spettel, B. 1996. Solubility of tungsten in a haplobasaltic melt as a function of temperature and oxygen fugacity. *Geochim. Cosmochim. Acta* 60:1171–80.

Fitton, J. G. 1995. Coupled molybdenum and niobium depletion in continental basalts. *Earth Planet. Sci. Lett.* 136:715–21.

Frey, F. A., and Green, D. H. 1974. The mineralogy, geochemistry and origin of lherzolite inclusions in Victorian basanites. *Geochim. Cosmochim. Acta* 38:1023–59.

Frey, F. A., and Prinz, M. 1978. Ultramafic inclusions from San Carlos, Arizona: petrologic and geochemical data bearing on their petrogenesis. *Earth Planet. Sci. Lett.* 38:129–76.

Frey, F. A., Shimizu, N., Leinbach, A., Obata, M., and Takazawa, E. 1991. Compositional variations within the lower layered zone of the Horoman Peridotite, Hokkaido, Japan: constraints on models for melt–solid segregation. In: *Journal of Petrology special issue: Orogenic Lherzolites and Mantle Processes*, ed. M. A. Menzies, C. Dupuy, and A. Nicolas, pp. 221–7.

Frey, F. A., Suen, C. J., and Stockman, H. W. 1985. The Ronda high temperature peridotite: geochemistry and petrogenesis. *Geochim. Cosmochim. Acta* 49:2469–91.

Galer, S. J. G., and Goldstein, S. L. 1996. Influence of accretion on lead in the Earth. In: *Earth Processes: Reading the Isotopic Code*, ed. A. Basu and S. Hart, pp. 75–98. AGU Monograph 95. Washington, DC: American Geophysical Union.

Geiss, J., and Reeves, H. 1981. Deuterium in the solar system. *Astron. Astrophys.* 93:189–99.

Graham, A. L., Bevan, A. W. R., and Hutchison, R. 1985. *Catalogue of Meteorites*, 4th ed. London: British Museum (Natural History).

Graham, A. L., Easton, A. J., and Hutchison, R. 1977. Forsterite chondrites; the meteorites Kakangari, Mount Morris (Wisconsin), Pontlyfni, and Winona. *Min. Mag.* 41:201–10.

Grimm, R. E., and McSween, H. Y. 1993. Heliocentric zoning of the asteroidal belt by aluminium-26 heating. *Science* 259:653–5.

Grossman, J. N., and Wasson, J. T. 1983. Refractory precursor components of Semarkona chondrules and the fractionation of refractory elements among chondrites. *Geochim. Cosmochim. Acta* 47:759–71.

Grossman, L. 1972. Condensation in the primitive solar nebula. *Geochim. Cosmochim. Acta* 36:597–619.

Grossman, L., and Larimer, J. W. 1974. Early chemical history of the solar system. *Rev. Geophys. Space Sci.* 12:71–101.

Harper, C. L., and Jacobsen, S. B. 1996. Evidence for [182]Hf in the early solar system and constraints on the timescale of terrestrial accretion and core formation. *Geochim. Cosmochim. Acta* 60:1131–53.

Harper, C. L., Völkening, J., Heumann, K. G., Shih, C.-Y., and Wiesmann, H. 1991. New cosmochronometric constraints on terrestrial accretion, core formation, the astrophysical site of the r-process, and the origin of the solar system. *Lunar Planet. Sci.* 23:489–90.

Hart, S. R., and Zindler, A. 1986. In search of a bulk-earth composition. *Chem. Geol.* 57:247–67.

Hartmann, G., and Wedepohl, K. H. 1990. Metasomatically altered peridotite xenoliths from the Hessian Depression (northwest Germany). *Geochim. Cosmochim. Acta* 54:71–86.

Hartmann, G., and Wedepohl, K. H. 1993. The composition of peridotite tectonites from the Ivrea Complex, northern Italy: residues from melt extraction. *Geochim. Cosmochim. Acta* 57:1761–82.

Hillgren, V. J., Drake, M. J., and Rubie, D. C. 1994. High-pressure and high-temperature experiments on core–mantle segregation in the accreting earth. *Science* 264:1442–5.

Hofmann, A. W. 1988. Chemical differentiation of the Earth: the relationship between mantle, continental crust, and oceanic crust. *Earth Planet. Sci. Lett.* 90:297–314.

Hofmann, A. W., Jochum, K. P., Seufert, M., and White, W. M. 1986. Nb and Pb in oceanic basalts: new constraints on mantle evolution. *Earth Planet. Sci. Lett.* 79:33–45.

Hofmann, A. W., and White, W. M. 1982. Mantle plumes from ancient oceanic crust. *Earth Planet. Sci. Lett.* 57:421–36.

Hofmann, A. W., and White, W. M. 1983. Ba, Rb and Cs in the Earth's mantle. *Z. Naturforsch.* 38:256–66.

Holzheid, A., and Palme, H. 1996. The influence of FeO on the solubilities of cobalt and nickel in silicate melts. *Geochim. Cosmochim. Acta* 60:1181–93.

Holzheid, A., Palme, H., and Chakraborty, S. 1997. The activities of NiO, CoO and FeO in silicate melts. *Chem. Geol.* 139:21–38.

Humayun, M., and Clayton, R. N. 1995a. Precise determination of the isotopic composition of potassium: application to terrestrial rocks and lunar soils. *Geochim. Cosmochim. Acta* 59:2115–30.

Humayun, M., and Clayton, R. N. 1995b. Potassium isotope cosmochemistry: genetic implications of volatile element depletion. *Geochim. Cosmochim. Acta* 59:2131–48.

Hunt, J. M. 1972. Distribution of carbon in crust of earth. *Bull. Am. Assoc. Pet. Geol.* 56:2273–7.

Ionov, D. A., Ashchepkov, I. V., Stosch, H.-G., Witt-Eickschen, G., and Seck, H. A. 1993. Garnet peridotite xenoliths from the Vitim Volcanic Field, Baikal region: the nature of the garnet-spinel peridotite transition zone in the continental mantle. *J. Petrol.* 34:1141–75.

Irving, A. J. 1980. Petrology and geochemistry of composite ultramafic xenoliths in alkalic basalts and implications for magmatic processes within the mantle. *Am. J. Sci.* 280A:389–426.

Jagoutz, E., Carlson, R. W., and Lugmair, G. W. 1980. Equilibrated Nd–unequilibrated Sr isotopes in mantle xenoliths. *Nature* 286:708–10.

Jagoutz, E., Palme, H., Baddenhausen, H., Blum, K., Cendales, M., Dreibus, G., Spettel, B., Lorenz, V., and Wänke, H. 1979. The abundances of major, minor and trace elements in the earth's mantle as derived from primitive ultramafic nodules. In: *Proceedings of the 10th Lunar and Planetary Science Conference*, pp. 2031–50. Elmsford, NY: Pergamon Press.

Jambon, A., Déruelle, B., Dreibus, G., and Pineau, F. 1995. Chlorine and bromine abundance in MORB: the contrasting behaviour of the Mid-Atlantic Ridge and East Pacific Rise and implications for chlorine geodynamic cycle. *Chem. Geol.* 126:101–17.

Jarosewich, E. 1990. Chemical analyses of meteorites: a compilation of stony and iron meteorite analyses. *Meteoritical Soc.* 25:323–37.

Javoy, M. 1995. The integral enstatite chondrite model of the earth. *Geophys. Res. Lett.* 22:2219–22.

Jochum, K. P., Hofmann, A. W., Ito, E., Seufert, H. M., and White, W. M. 1983. K, U and Th in mid-ocean ridge basalt glasses and heat production, K/U and K/Rb in the mantle. *Nature* 306:431–6.

Jochum, K. P., Hofmann, A. W., and Seufert, H. M. 1993. Tin in mantle-derived rocks: constraints on Earth evolution. *Geochim. Cosmochim. Acta* 57:3585–95.

Jochum, K. P., McDonough, W. F., Palme, H., and Spettel, B. 1989. Compositional constraints on the continental lithospheric mantle from trace elements in spinel peridotite xenoliths. *Nature* 340:548–50.

Jones, J. H., and Drake, M. J. 1986. Geochemical constraints on core formation in the Earth. *Nature* 322:221–8.

Kallemeyn, G. W., Rubin, A. E., Wang, D., and Wasson, J. T. 1989. Ordinary chondrites:

bulk composition, classification, lithophile-element fractionation, and composition–petrographic type relationship. *Geochim. Cosmochim. Acta* 53:2747–67.

Kallemeyn, G. W., Rubin, A. E., and Wasson, J. T. 1991. The compositional classification of chondrites: V. The Karoonda (CK) group of carbonaceous chondrites. *Geochim. Cosmochim. Acta* 55:881–92.

Kallemeyn, G. W., Rubin, A. E., and Wasson, J. T. 1994. The compositional classification of chondrites: VI. The CR carbonaceous chondrite group. *Geochim. Cosmochim. Acta* 58:2873–88.

Kallemeyn, G. W., and Wasson, J. T. 1986. Compositions of enstatite (EH3, EH4,5 and EL6) chondrites: implications regarding their formation. *Geochim. Cosmochim. Acta* 50:2153–64.

Kaplan, I. R. 1971. Hydrogen (1). In: *Handbook of Elemental Abundances in Meteorites*, ed. B. Mason. London: Gordon & Breach.

Karato, S., and Murthy, V. R. 1997. Core formation and chemical equilibrium in the Earth. I. Physical considerations. *Phys. Earth Planet. Int.* 100:61–79.

Kato, T., Ringwood, A. E., and Irifune, T. 1988. Experimental determination of element partitioning between silicate perovskites, garnets and liquids: constraints on early differentiation of the mantle. *Earth Planet. Sci. Lett.* 89:123–45.

Kelly, W. R., and Larimer, J. W. 1977. Chemical fractionations in meteorites. VIII. Iron meteorites and the cosmochemical history of the metal phase. *Geochim. Cosmochim. Acta* 41:93–111.

Kimura, K., Lewis, R. S., and Anders, E. 1974. Distribution of gold and rhenium between nickel-iron and silicate melts: implications for the abundances of the siderophile elements in the Earth and Moon. *Geochim. Cosmochim. Acta* 38:683–701.

Kurat, G., Palme, H., Embey-Isztin, A., Touret, J., Ntaflos, T., Spettel, B., Brandstätter, F., Palme, C., Dreibus, G., and Prinz, M. 1993. Petrology and geochemistry of peridotites and associated vein rocks of Zabargad Island, Red Sea, Egypt. *Mineral. Petrol.* 48:309–41.

Langmuir, C. H., Klein, E. M., and Plank, T. 1992. Petrological systematics of mid-ocean ridge basalts: constraints on melt generation beneath ocean ridges. In: *Mantle Flow and Melt Generation at Mid-Ocean Ridges*, ed. J. Phipps Morgan, D. K. Blackman, and J. M. Stinton, pp. 183–280. Washington, DC: American Geophysical Union.

Larimer, J. W., and Anders, E. 1970. Chemical fractionations in meteorites. III. Major element fractionations in chondrites. *Geochim. Cosmochim. Acta* 34:367–87.

Larimer, J. W., and Wasson, J. T. 1988. Siderophile-element fractionation. In: *Meteorites and the Early Solar System*, ed. J. F. Kerridge and M. S. Matthews, pp. 416–35. Tucson: University of Arizona Press.

Lauretta, D. S., and Lodders, K. 1997. The cosmochemical behavior of beryllium and boron. *Earth Planet. Sci. Lett.* 146:315–27.

Lee, D.-C., and Halliday, A. N. 1995. Hafnium-tungsten chronometry and the timing of terrestrial core formation. *Nature* 378:771–4.

Lee, T. 1988. Implications of isotopic anomalies for nucleosynthesis. In: *Meteorites and the Early Solar System*, ed. J. F. Kerridge and M. S. Matthews, pp. 1063–89. Tucson: University of Arizona Press.

Lee, T., and Larimer, J. W. 1979. The condensation and fractionation of refractory lithophile elements. *Icarus* 40:446–54.

Li, J.-P., O'Neill, H. St. C., and Seifert, F. 1995. Subsolidus phase relations in the system $MgO-SiO_2-Cr-O$ in equilibrium with metallic Cr, and their significance for the petrochemistry of chromium. *J. Petrol.* 36:107–32.

Lodders, K., and Fegley, B., Jr. 1995. The origin and evolution of the terrestrial alkali element budget. In: *Volatiles in the Early Solar System*, ed. K. A. Farley, pp. 99–105. AIP Conference Proceedings 341. Woodbury, NY: American Institute of Physics.

Lodders, K., and Palme, H. 1991. On the chalcophile character of molybdenum: determination of sulfide/silicate partition coefficients of Mo and W. *Earth Planet. Sci. Lett.* 103:311–24.

Lorand, J. P. 1991. Sulphide petrology and sulphur geochemistry of orogenic lherzolites: a comparative study between Pyrenean bodies (France) and the Lanzo Massif (Italy). In: *Orogenic Lherzolites and Mantle Processes*, ed. M. A. Menzies. *J. Petrol.* 32:77–95.

Lorand, J. P., Keays, R. R., and Bodinier, J. L. 1993. Copper and noble metal enrichments across the lithosphere–asthenosphere boundary of mantle diapirs: evidence from the Lanzo lherzolite massif. *J. Petrol.* 34:1111–40.

Lugovic, B., Altherr, R., Raczek, I., Hofmann, A. W., and Majer, V. 1991. Geochemistry of peridotites and mafic igneous rocks from the Central Dinaric Ophiolite Belt, Yugoslavia. *Contrib. Min. Pet.* 106:201–16.

Maaløe, S., and Aoki, K.-I. 1977. The major element composition of the upper mantle estimated from the composition of lherzolites. *Contrib. Min. Pet.* 63:161–73.

MacDonald, G. J. F., and Knopoff, L. 1958. On the chemical composition of the outer core. *Geophys. J. Royal Astron. Soc.* 1:284–97.

McDonough, W. F., McCulloch, M. T., and Sun, S.S. 1985. Isotopic and geochemical systematics in Tertiary–Recent basalts from southeastern Australia and implications for the evolution of the sub-continental lithosphere. *Geochim. Cosmochim. Acta* 49:2051–67.

McDonough, W. F. 1990. Constraints on the composition of the continental lithospheric mantle. *Earth Planet. Sci. Lett.* 101:1–18.

McDonough, W. F. 1995. An explanation for the abundance enigma of the highly siderophile elements in the Earth's mantle. *Lunar Planet. Sci.* 26:927–8.

McDonough, W. F., and Sun, S.-S. 1995. The composition of the Earth. *Chem. Geol.* 120:223–53.

McDonough, W. F., Sun, S.-S., Ringwood, A. E., Jagoutz, E., and Hofmann, A. W. 1992. Potassium, rubidium, and cesium in the Earth and Moon and the evolution of the mantle of the Earth. *Geochim. Cosmochim. Acta* 56:1001–12.

Malvin, D. J., and Drake, M. J. 1987. Experimental determination of crystal/melt partitioning of Ga and Ge in the system forsterite-anorthite-diopside. *Geochim. Cosmochim. Acta* 51:2117–28.

Mason, B. 1962–3. The carbonaceous chondrites. *Space Sci. Rev.* 1:621–46.

Meisel, T., Walker, R. J., and Morgan, J. W. 1996. The osmium isotopic content of the Earth's primitive upper mantle. *Nature* 383:517–20.

Michael, P. J. 1988. The concentration, behaviour and storage of H_2O in the suboceanic upper mantle: implications for mantle metasomatism. *Geochim. Cosmochim. Acta* 52:555–66.

Morgan, W. J. 1986. Ultramafic rocks: clues to the Earth's late accretionary history. *J. Geophys. Res.* 91:12375–87.

Murthy, V. R. 1991. Early differentiation of the Earth and the problem of mantle siderophile elements: a new approach. *Science* 253:303–6.

Murthy, V. R., and Hall, H. T. 1970. The chemical composition of the core: possibility of sulfur in the core. *Phys. Earth Planet. Int.* 2:276–82.

Newsom, H. E., and Sims, K. W. W. 1992. Chemical fractionation in the continental crust: clues from As, Sb, W, Mo, and Pb in lower crustal xenoliths. *Abstracts, V. M. Goldschmidt Conference*, p. A-75.

Newsom, H. E., Sims, K. W. W., Noll, P. D., Jr., Jaeger, W. L., Maehr, S. A., and Beserra, T. B. 1996. The depletion of tungsten in the bulk silicate earth: constraints on core formation. *Geochim. Cosmochim. Acta* 60:1155–69.

Newsom, H. E., and Slane, F. A. 1993. Pressure–temperature regimes and core formation in the accreting Earth. *Geophys. Monograph 76, IUGG* 16:129–33.

Newsom, H. E., White, W. M., Jochum, K. P., and Hofmann, A. W. 1986. Siderophile and chalcophile element abundances in oceanic basalts, Pb isotope evolution and growth of the Earth's core. *Earth Planet. Sci. Lett.* 80:299–313.

Nickel, K. G., and Green, D. H. 1984. The nature of the upper-most mantle beneath Victoria, Australia, as deduced from ultramafic xenoliths. In: *Kimberlites*, ed. J. Kornprobst, pp. 161–78. Proceedings of the Third International Kimberlite Conference. Amsterdam: Elsevier.

Niu, Y., and Batiza, R. 1991. An empirical method for calculating melt compositions produced beneath mid-ocean ridges: application for axis and off-axis (seamounts) melting. *J. Geophys. Res.* 96:21753–77.

Nixon, P. H., Rogers, N. W., Gibson, I. L., and Grey, A. 1981. Depleted and fertile mantle xenoliths from southern African kimberlites. *Earth Planet. Sci. Lett.* 9:285–309.

O'Neill, H. St. C. 1991a. The origin of the Moon and the early history of the Earth – a chemical model. Part 1: The Moon. *Geochim. Cosmochim. Acta* 55:1135–57.

O'Neill, H. St. C. 1991b. The origin of the Moon and the early history of the Earth – a chemical model. Part 2: The Earth. *Geochim. Cosmochim. Acta* 55:1159–72.

O'Neill, H. St. C. 1992. Siderophile elements and the Earth's formation. *Science* 257:1282–5.

O'Neill, H. St. C., Canil, D., and Rubie, D. C. In press. Metal-oxide equilibria to 2500°C and 25 GPa: implications for core formation and the light component in the Earth's core. *J. Geophys. Res.*

O'Neill, H. St. C., Dingwell, D. B., Borisov, A., Spettel, B., and Palme, H. 1995. Experimental petrochemistry of some highly siderophile elements at high temperatures, and some implications for core formation and the mantle's early history. *Chem. Geol.* 120:255–73.

Palme, H., and Beer, H. 1993. Abundances of the elements in the solar system. In: *Astronomy and Astrophysics*. Vol. 3, Subvol. a: *Instruments; Methods; Solar System*, ed. H. H. Voigt, pp. 196–221. Berlin: Springer-Verlag.

Palme, H., and Boynton, W. V. 1993. Meteoritic constraints on conditions in the solar nebula. In: *Protostars and Planets* III, ed. E. H. Levy and J. I. Lunine, pp. 979–1004. Tucson: University of Arizona Press.

Palme, H., Larimer, J. W., and Lipschutz, M. E. 1988. Moderately volatile elements. In: *Meteorites and the Early Solar System*, ed. J. F. Kerridge and M. S. Matthews, pp. 436–61. Tucson: University of Arizona Press.

Palme, H., and Nickel, K. G. 1985. Ca/Al ratio and composition of the Earth's upper mantle. *Geochim. Cosmochim. Acta* 49:2123–32.

Palme, H., Schultz, L., Spettel, B., Weber, H. W., Wänke, M.-L., Christophe, M., and Lorin, J. C. 1981. The Acupulco meteorite: chemistry, mineralogy and irradiation effects. *Geochim. Cosmochim. Acta* 45:727–52.

Palme, H., and Wänke, H. 1975. A unified trace-element model for the evolution of the lunar crust and mantle. *Proc. Lunar Sci. Conf.* 6:1179–202.

Palme, H., Weckwerth, G., and Wolf, D. 1996. The composition of a new R-chondrite and the classification of chondritic meteorites. *Lunar Planet. Sci.* 27:991–2.

Palme, H., and Wlotzka, F. 1976. A metal particle from a Ca,Al-rich inclusion from the meteorite Allende, and the condensation of refractory siderophile elements. *Earth Planet. Sci. Lett.* 33:45–60.

Palme, H., Wlotzka, F., Nagel, K., and El Goresy, A. 1982. An ultra-refractory inclusion from the Ornans carbonaceous chondrite. *Earth Planet. Sci. Lett.* 61:1–12.

Pattou, L., Lorand, J. P., and Gros, M. 1996. Non-chondritic platinum-group element ratios in the Earth's mantle. *Nature* 379:712–15.

Podosek, F. A., and Cassen, P. 1994. Theoretical, observational, and isotopic estimates of the lifetime of the solar nebula. *Meteoritics* 29:6–25.

Poirier, J.-P. 1994. Light elements in the Earth's outer core: a critical review. *Phys. Earth Planet. Int.* 85:319–37.

Pollack, H. N., Hurter, S. J., and Johnson, J. R. 1993. Heat flow from the Earth's interior:

analysis of the global data set. *Rev. Geophys.* 31:267–80.

Press, S., Witt, G., Seck, H. A., Eonov, D., and Kovalenko, V. I. 1986. Spinel peridotite xenoliths from the Tariat Depression, Mongolia. I: Major element chemistry and mineralogy of a primitive mantle xenolith suite. *Geochim. Cosmochim. Acta* 50:2587–99.

Price, N. B., Calvert, S. E., and Jones, P. G. W. 1970. The distribution of iodine and bromine in the sediments of the southwestern Barents Sea. *J. Marine Res.* 28:22–34.

Qi, Q., Taylor, L. A., and Zhou, X. 1995. Petrology and geochemistry of mantle peridotite xenoliths from SE China. *J. Petrol.* 36:55–79.

Rampone, E., Hofmann, A. W., Piccardo, G. B., Vannucci, R., Bottazzi, P., and Ottolini, L. 1995. Petrology, mineral and isotope geochemistry of the external Liguride peridotites (northern Apennines, Italy). *J. Petrol.* 36:81–105.

Richter, S., Ott, U., and Begemann, F. 1992. S-process isotope anomalies: neodymium, samarium and a bit more strontium. *Lunar Planet. Sci.* 23:1147–8.

Ringwood, A. E. 1966. The chemical evolution of terrestrial planets. *Geochim. Cosmochim. Acta* 30:41–104.

Ringwood, A. E. 1975. *Composition and Petrology of the Earth's Mantle.* New York: McGraw-Hill.

Ringwood, A. E. 1979. *Origin of the Earth and Moon.* Berlin: Springer-Verlag.

Ringwood, A. E. 1984. The Earth's core: its composition, formation and bearing upon the origin of the Earth. *Proc. R. Soc. Lond.* A395:1–46.

Ringwood, A. E. 1990. Earliest history of the Earth-Moon system. In: *Origin of the Earth*, ed. H. E. Newsom and J. H. Jones, pp. 101–34. Oxford University Press.

Ringwood, A. E., and Hibberson, W. 1991. Solubilities of mantle oxides in molten iron at high pressures and temperatures: implications for the composition and formation of Earth's core. *Earth Planet. Sci. Lett.* 102:235–51.

Ringwood, A. E., and Kesson, S. E. 1977. Basaltic magmatism and the bulk composition of the Moon. II. Siderophile and volatile elements in Moon, Earth and chondrites: implications for lunar origin. *The Moon* 16:425–64.

Robie, R. A., Hemingway, B. S., and Fisher, J. R. 1979. *Thermodynamic Properties of Minerals and Related Substances at 298.15 K and 1 Bar (10^5 Pascals) Pressure and at Higher Temperatures.* Geological Survey Bulletin 1452. Washington, DC: U.S. Government Printing Office.

Roden, M. F., Irving, A. J., and Murthy, V. R. 1988. Isotopic and trace element composition of the upper mantle beneath a young continental rift: results from Kilbourne Hole, New Mexico. *Geochim. Cosmochim. Acta* 52:461–73.

Rotaru, M., Birck, J.-L., and Allègre, C. J. 1992. Clues to early solar system history from chromium isotopes in carbonaceous chondrites. *Nature* 358:465–70.

Rudnick, R. L., and Fountain, D. M. 1995. Nature and composition of the continental crust: a lower crustal perspective. *Rev. Geophys.* 33:267–309.

Ryan, J. G., and Langmuir, C. H. 1987. The systematics of lithium abundances in young volcanic rocks. *Geochim. Cosmochim. Acta* 51:1727–41.

Saxena, S. K., and Ericsson, G. 1983. Low- to medium-temperature phase equilibrium in a gas of solar composition. *Earth Planet. Sci. Lett.* 65:7–16.

Saxena, S. K., and Ericsson, G. 1986. Chemistry of the formation of the terrestrial planets. In: *Chemistry and Physics of Terrestrial Planets*, vol. 6, ed. S. K. Saxena, pp. 30–105. Berlin: Springer-Verlag.

Schilling, J.-G., Bergeron, M. B., and Evans, R. 1980. Halogens in the mantle beneath the North Atlantic. *Phil. Trans. R. Soc. Lond.* A297:147–78.

Schulze, D. J. 1989. Constraints on the abundance of eclogite in the upper mantle. *J. Geophys. Res.* 94:4205–12.

Schulze, H., Bischoff, A., Palme, H., Spettel, B., Dreibus, G., and Otto, J. 1994. Mineralogy and chemistry of Rumuruti: the first meteorite fall of the new R-chondrite group. *Meteorites* 29:275–86.

Scott, E. R. D. 1979. Origin of iron meteorites. In: *Asteroids*, ed. T. Gehrels, pp. 892–925. Tucson: University of Arizona Press.

Scott, E. R. D. 1988. A new kind of primitive chondrite, Allan Hills 85085. *Earth Planet. Sci. Lett.* 91:1–18.

Sears, D. 1996. Is Kaidun really the Rosetta stone? *Meteor. Planet. Sci.* 31:543–4.

Seifert, S., O'Neill, H. St. C., and Brey, G. 1988. The partitioning of Fe, Ni and Co between olivine, metal, and basaltic liquid: an experimental and thermodynamic investigation, with application to the composition of the lunar core. *Geochim. Cosmochim. Acta* 52:603–16.

Shen, J. J., Papanastassiou, D. A., and Wasserburg, G. J. 1996. Precise Re-Os determinations and systematics of iron meteorites. *Geochim. Cosmochim. Acta* 60:2887–900.

Sims, K. W. W., Newsom, H. E., and Gladney, E. S. 1990. Chemical fractionation during formation of the Earth's core and continental crust: clues from As, Sb, W, and Mo. In: *Origin of the Earth*, ed. H. E. Newsom and J. H. Jones, pp. 291–317. Oxford University Press.

Sleep, N. H. 1990. Hotspots and mantle plumes: some phenomenology. *J. Geophys. Res.* 95:6715–36.

Solomatov, V. S., and Stevenson, D. J. 1993. Suspension in convecting layers and style of differentiation of a terrestrial magma ocean. *J. Geophys. Res.* 98:5375–90.

Sonnett, C. P., and Reynolds, R. T. 1979. Primordial heating in asteroids. In: *Asteroids*, ed. T. Gehrels, pp. 822–48. Tucson: University of Arizona Press.

Sorby, H. C. 1877. On the structure and origin of meteorites. *Nature*, April 5, 1877:495–8.

Spray, J. G. 1989. Upper mantle segregation processes: evidence from alpine-type peridotites. In: *Magmatism in the Ocean Basins*, ed. A. D. Saunders and M. J. Norry, pp. 29–39. Oxford: Blackwell Scientific.

Stacey, F. D. 1992. *Physics of the Earth*. Brisbane: Brookfield Press.

Stern, C. R., Saul, S., Skewes, M. A., and Futa, K. 1986. Garnet peridotite xenoliths from the Pali-Aike alkali basalts of southernmost South America. In: *Kimberlites and Related Rocks*, ed. J. Ross, pp. 735–44. Proceedings of the Fourth International Kimberlite Conference. Oxford: Blackwell Scientific.

Stevenson, D. J. 1981. Models of the Earth's core. *Science* 214:611–19.

Stevenson, D. J. 1990. Fluid dynamics of core formation. In: *Origin of the Earth*, ed. H. E. Newsom and J. H. Jones, pp. 229–30. Oxford University Press.

Stosch, H.-G., Lugmair, G. W., and Kovalenko, V. I. 1986. Spinel peridotite xenoliths from the Tariat Depression, Mongolia. II: Geochemistry and Nd and Sr isotopic composition and their implications for the evolution of the subcontinental lithosphere. *Geochim. Cosmochim. Acta* 50:2601–14.

Stosch, H.-G., and Seck, H. A. 1980. Geochemistry and mineralogy of two spinel peridotite suites from Dreiser Weiher, West Germany. *Geochim. Cosmochim. Acta* 44:457–70.

Sun, S.-S. 1982. Chemical composition and origin of the earth's primitive mantle. *Geochim. Cosmochim. Acta* 46:179–92.

Taylor, S. R., and McLennan, S. M. 1995. The geochemical evolution of the continental crust. *Rev. Geophys.* 33:241–65.

Thiemens, M. H. 1988. Heterogeneity in the nebula: evidence from stable isotopes. In: *Meteorites and the Early Solar System*, ed. J. F. Kerridge and M. S. Matthews, pp. 899–923. Tucson: University of Arizona Press.

Thiemens, M. H., Jackson, T., Zipf, E. C., Erdman, P. W., and van Egmond, C. 1995. Carbon dioxide and oxygen isotope anomalies in the mesosphere and stratosphere. *Science* 270:969–72.

Tonks, W. B., and Melosh, H. J. 1990. The physics of crystal settling and suspension in a turbulent magma ocean. In: *Origin of the Earth*, ed. H. E. Newsom and J. H. Jones, pp. 151–74. Oxford University Press.

Tonks, W. B., and Melosh, H. J. 1992. Core formation by giant impacts. *Icarus* 100:326–46.

Tonks, W. B., and Melosh, H. J. 1993. Magma ocean formation due to giant impacts. *J. Geophys. Res.* 98:5319–33.

Tscharnuter, W. M., and Boss, A. P. 1993. Formation of the protosolar nebula. In: *Protostars and Planets III*, ed. E. H. Levy and J. I. Lunine, pp. 921–38. Tucson: University of Arizona Press.

Urey, H. C., and Craig, H. 1953. The composition of the stone meteorites and the origin of the meteorites. *Geochim. Cosmochim. Acta* 4:36–82.

von Bogdandy, L., and Engell, H.-J. 1971. *The Reduction of Iron Ores*. Berlin: Springer-Verlag.

Wai, C. M., and Wasson, J. T. 1977. Nebular condensation of moderately volatile elements and their abundances in ordinary chondrites. *Earth Planet. Sci. Lett.* 36:1–13.

Wänke, H., Baddenhausen, H., Dreibus, G., Jagoutz, E., Kruse, H., Palme, H., Spettel, B., and Teschke, F. 1973. Multielement analyses of Apollo 15, 16, and 17 samples and the bulk composition of the moon. *Proc. 4th Lunar Sci. Conf.*, pp. 1461–81. Elmsford, NY: Pergamon Press.

Wänke, H., and Dreibus, G. 1988. Chemical composition and accretional history of terrestrial planets. *Phil. Trans, R. Soc. Lond.* A235:545–57.

Wänke, H., Dreibus, G., and Jagoutz, E. 1984. Mantle chemistry and accretion history of the Earth. In: *Archean Geochemistry*, ed. A. Kröner, pp. 1–24. Berlin: Springer-Verlag.

Ward, R. G. 1962. *An Introduction to the Physical Chemistry of Iron and Steel Making*. London: Edward Arnold.

Wasson, J. T. 1972. Formation of ordinary chondrites. *Rev. Geophys. Space Phys.* 10:711–59.

Wasson, J. T. 1985. *Meteorites: Their Record of Early Solar-System History*. San Francisco: W. H. Freeman.

Wasson, J. T., and Kallemeyn, G. W. 1988. Compositions of chondrites. *Phil. Trans. R. Soc. London* A325:535–44.

Wasson, J. T., Kallemeyn, G. W., and Rubin, A. E. 1994. Equilibration temperatures of EL chondrites: a major downward revision in the ferrosilite contents of enstatite. *Meteoritics* 29:658–62.

Weidenschilling, S. J. 1988. Formation processes and time scales for meteorite parent bodies. In: *Meteorites and the Early Solar System*, ed. J. F. Kerridge and M. S. Matthews. pp. 348–71. Tucson: University of Arizona Press.

Weisberg, M. K., Prinz, M., Clayton, R. N., and Mayeda, T. K. 1993. The CR (Renazzo-type) carbonaceous chondrite group and its implications. *Geochim. Cosmochim. Acta* 57:1567–86.

Wetherill, G. W. 1990. Formation of the Earth. *Annu. Rev. Earth Planet. Sci.* 18:205–56.

Wetherill, G. W. 1994. Provenance of the terrestrial planets. *Geochim. Cosmochim. Acta* 58:4513–20.

Widom, E., and Shirey, S. B. 1996. Os isotope systematics in the Azores: implications for mantle plume sources. *Earth Planet. Sci. Lett.* 142:451–65.

Wolf, D., Spettel, B., Nazarov, M., El Goresy, A., and Palme, H. 1996. Incomplete sampling of metal in the CV-chondrite Efremovka. *Meteor. Planet. Sci.* 31:A153–4.

Wood, J. A., and Morfill, G. E. 1988. A review of solar nebula models. In: *Meteorites and the Early Solar System*, ed. J. F. Kerridge and M. S. Matthews, pp. 329–47. Tucson: University of Arizona Press.

Wulf, A. V., Palme, H., and Jochum, K. P. 1995. Fractionation of volatile elements in the early solar system: evidence from heating experiments on primitive meteorites. *Planet. Space Sci.* 43:451–86.

Yi, W., Halliday, A. N., Lee, D.-C., and Christensen, J. N. 1995. Indium and tin in basalts, sulfides, and the mantle. *Geochim. Cosmochim. Acta* 59:5081–90.

Zhang, Y., Benoit, P. H., and Sears, D. W. G. 1995. The classification and complex thermal history of the enstatite chondrites. *J. Geophys. Res.* 100:9417–38.

Zhang, Y., and Zindler, A. 1993. Distribution and evolution of carbon and nitrogen in Earth. *Earth Planet. Sci. Lett.* 117:331–45.

Zipfel, J., Palme, H., Kennedy, A. K., and Hutcheon, I. D. 1995. Chemical composition and the origin of the Acapulco meteorite. *Geochim. Cosmochim. Acta* 59:3607–27.

2

Early Differentiation of the Earth: An Isotopic Perspective

MALCOLM T. McCULLOCH and VICTORIA C. BENNETT

2.1. Introduction

During the past 4.5 billion years the Earth has undergone a complex process of differentiation that has resulted in the formation of a metallic core, a magnesium-rich silicate mantle, and a siliceous continental crust. One of the major challenges for the Earth sciences is not only to document the present-day chemical and physical states of these distinctive regions, but also to unravel the complex series of events associated with their origin and long-term evolution. Isotope geochemistry can provide important constraints on many of these events, as the formation and subsequent differentiation of the Earth were accompanied by chemical fractionation of the parent–daughter elements that comprise many of the naturally occurring radioactive-decay systems. The isotopic compositions of these daughter elements will therefore yield constraints on the timing and magnitude of fractionation events and thus provide insights into the processes responsible for the differentiation of the Earth. In the first part of this chapter we shall show how isotopic systematics, particularly the isotopic compositions of Pb and Sr preserved in ancient terrestrial samples, can be used to constrain the timescales for the Earth's accretion and the formation of its core. The second part of this chapter discusses the constraints on the formation and evolution of the continental crust and upper mantle that are provided by the Sm-Nd isotopic systematics of early Archaean rocks.

Probably the most fundamental question associated with the formation of the Earth is its age. Did the Earth form essentially contemporaneously with its meteorite parent bodies at 4.56 Ga, or was a substantial time interval of ≥ 100 million years required for planetary accretion? The difficulty in establishing a precise constraint on the Earth's age arises from the lack of a geological record for the first 500 million years of Earth history. The only known remnants from that period are some 4.27-Ga detrital zircons preserved in an Archaean sediment from the Jack Hills region of Western Australia (Compston and Pidgeon, 1986), whereas the oldest known rocks are rare 4-Ga gneisses (Bowring, Williams, and Compston, 1989) preserved in isolated outcrops in northern Canada. From about 3.7–3.8 Ga, the geological record becomes more abundant, as rocks of this age are preserved on several continents (e.g., Nutman et al., 1993, 1996). The absence of intact terrestrial materials formed during the earliest period of the Earth's history means that our knowledge of that period is heavily dependent upon the information that can be deduced from the isotopic compositions of ancient samples. The parent isotopes of the commonly used radioactive systems (e.g., U, Rb, and Sm) have half-lives of $\geq 10^9$ years, and hence the isotopic compositions preserved in their daughter elements (Pb, Sr, and Nd) in the oldest rocks record the first 500–1,000 million years of parent/daughter fractionation. Some events that occurred early in the Earth's history, such as core formation, left permanent imprints on the elemental concentrations of modern rocks as well, such as the abundances of the platinum-group elements (Pt, Pd, Os, Ir, Ru, Rh) (O'Neill and Palme, Chapter 1, this volume). It is unclear, however, which chemical signatures are attributable to early Earth processes and which reflect subsequent mantle differentiation.

Knowledge of the composition and evolution of the continental crust is important, as it reflects the complex and dynamic history of continental growth and depleted-mantle evolution. On the basis of the present-day chemical compositions of the continental crust and the upper portion of the Earth's mantle, as sampled from mid-ocean-ridge basalts (MORBs), it is now well established that the bulk continental crust is enriched, by up to several orders of magnitude relative to the upper mantle, in elements that are strongly partitioned into a melt phase (Taylor and McLennan, 1985). These elements are referred to as highly *incompatible* elements, in contrast to the compatible elements, which are retained in the crystalline phases of the mantle during melting events. Furthermore, it is also apparent that to the first order the continental crust and upper mantle can be regarded as broadly complementary reservoirs, with the extraction of incompatible elements into an enriched continental crust being responsible for the depletion observed in the upper mantle (Figure 2.1). What is not apparent from their present-day chemical compositions is the long-term history of the continental crust and complementary depleted mantle. Did the continental crust and the depleted mantle both increase in mass with time, or was the continental crust largely formed early in the Earth's history, perhaps as a product of early Earth differentiation? If the continental mass has increased in size, was that growth continuous or episodic? What is the relationship between crustal growth and large-scale mantle convection? These questions will be considered in the second part of this chapter.

2.2. Timescales for the Earth's Accretion and Core Formation

The Earth is believed to have formed by the accretion of a hierarchy of smaller bodies and planetesimals, with melting and differentiation resulting in chemical segregation of the core and silicate portions of the Earth. Precise ages, such as a 4.565 ± 0.004-billion-year Pb-Pb age (Chen and Tilton, 1976), have been determined for a number of chondritic meteorites, a class of meteorites considered to be among

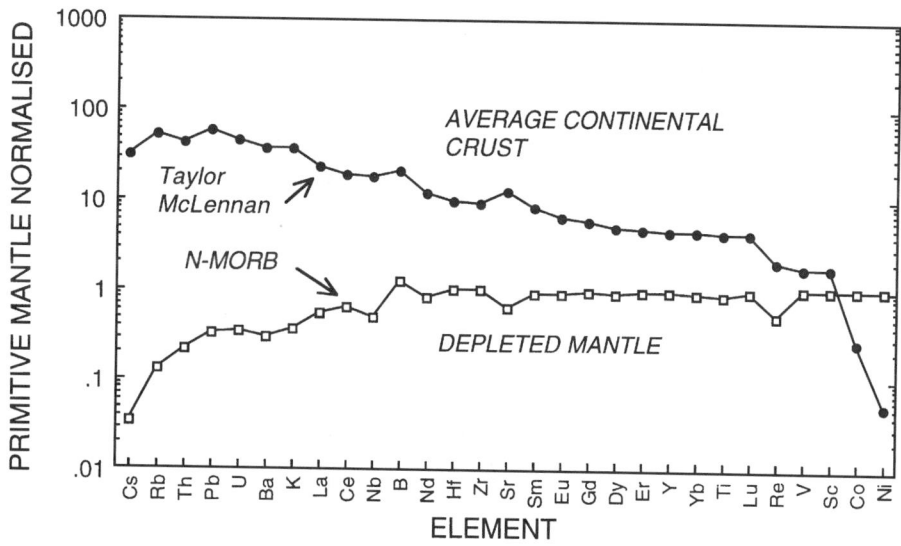

Figure 2.1. Trace-element abundances in the continental crust and the depleted upper mantle, the latter inferred from MORB chemistry. The highly incompatible trace elements are on the left side of the diagram and are enriched in the continental crust by factors of 30 to 70 compared with primitive-mantle abundances, whereas normal mid-ocean-ridge basalts (N-MORBs) are depleted of these elements. This well-established dichotomy (e.g., Taylor and McLennan, 1985) between the abundances of incompatible elements in the continental crust compared with the upper mantle is the basis for assuming that the differentiation of the continental crust and the depletion of the mantle are complementary.

the most primitive objects in the solar system. Numerous studies, as reviewed by Dalrymple (1991), have indicated that most meteorites were formed in the narrow time interval between 4.54 Ga and 4.57 Ga. Thus, by analogy with primitive meteorites, the earliest time at which the accretion of the Earth could have commenced was about 4.56 Ga. However, questions remain: How long did the accretionary process take? Was there a substantial time lag between accretion and early differentiation of the Earth? Conclusive answers to these questions remain elusive, but several lines of evidence that are considered here appear to indicate that the accretion of the Earth and its segregation into metallic core and silicate mantle were not completed until at least 70 million years after the formation of chondrites.

2.2.1. Pb Isotopic Constraints on the Timing of
Core Formation and Accretion

Early measurements of Pb isotopic compositions in terrestrial ore minerals by Nier and colleagues (Nier, 1938; Nier, Thompson, and Murphey, 1941) provided the motivation for the development of Pb isotopic systematics for the Earth. Gerling (1942) and, subsequently, Holmes (1946) and Houtermann (1946), using Nier's analyses, described a general model for Pb isotopic evolution in the Earth that became known as the Holmes-Houtermann Pb isotopic model. The models of Gerling, Holmes, and Houtermann, in particular that of a single-stage evolution for the Earth, were more fully developed by Russell and Farquhar (1960). These models have continued to be refined by subsequent workers as new analytical techniques, more data, and a better understanding of the complexities of the behavior of Pb in the Earth have become available; for a complete historical account, the reader is referred to Dalrymple (1991).

The U-Pb isotopic system has the unique advantage, relative to other radiogenic isotopes, of being a paired system, with the parent isotopes ^{238}U and ^{235}U decaying to the stable daughter isotopes ^{206}Pb and ^{207}Pb, with half-lives of 4.47×10^9 years and 0.704×10^9 years, respectively. A single-stage model of Pb isotopic evolution, from the time of formation of the Earth (of age T_{Earth}) to some later time represented by minerals of age t that have behaved as closed systems with respect to U-Pb isotopic evolution, yields, for the 'initial' ratios (i.e., at the time the mineral formed),

$$^{206}\text{Pb}/^{204}\text{Pb}(t) = {}^{206}\text{Pb}/^{204}\text{Pb}(T_{\text{Earth}}) + {}^{238}\text{U}/^{204}\text{Pb}(e^{\lambda T_{\text{Earth}}} - e^{\lambda t}) \qquad (1)$$

and

$$^{207}\text{Pb}/^{204}\text{Pb}(t) = {}^{207}\text{Pb}/^{204}\text{Pb}(T_{\text{Earth}}) + {}^{235}\text{U}/^{204}\text{Pb}(e^{\lambda' T_{\text{Earth}}} - e^{\lambda' t}) \qquad (2)$$

where $^{207}\text{Pb}/^{204}\text{Pb}(T_{\text{Earth}})$ and $^{206}\text{Pb}/^{204}\text{Pb}(T_{\text{Earth}})$ describe the Earth's initial Pb composition at its time of formation, and ^{238}U and ^{235}U are the present-day abundances. The decay constants for ^{238}U and ^{235}U are $\lambda = 0.155125$ Æ$^{-1}$ and $\lambda' = 0.98485$ Æ$^{-1}$, respectively (Æ$^{-1} = 10^{-9}$ yr^{-1}). By combining equations (1) and (2), the U/Pb ratio can be eliminated, and the age of the Earth derived from the following parametric relationship:

$$\frac{^{207}\text{Pb}/^{204}\text{Pb}(t) - {}^{207}\text{Pb}/^{204}\text{Pb}(T_{\text{Earth}})}{^{206}\text{Pb}/^{204}\text{Pb}(t) - {}^{206}\text{Pb}/^{204}\text{Pb}(T_{\text{Earth}})} = \frac{^{235}\text{U}(e^{\lambda' T_{\text{Earth}}} - e^{\lambda' t})}{^{238}\text{U}(e^{\lambda T_{\text{Earth}}} - e^{\lambda t})} \qquad (3)$$

where $^{235}\text{U}/^{238}\text{U} = 1/137.88$. Although it was apparent to Gerling, Holmes, and Houtermann in the 1940s that the age of the Earth could be calculated from these types of equations, the limitation on these early efforts was the number of unknowns that remained to be accurately determined. In particular, estimates of the Earth's primordial Pb isotopic composition were lacking. Even so, Gerling, Holmes, and Houtermann all made credible estimates of T_{Earth}, suggesting ages greater than or equal to 3.0 billion years. A significant advance was the Pb isotopic study by Patterson et al. (1953) of the mineral troilite (FeS) from the Canyon Diablo meteorite that formed Meteor Crater in Arizona. These data revealed the most primitive Pb isotopic compositions yet measured and allowed the first direct estimates for primordial Pb in the solar system. Patterson (1956) advanced that work a step further by measuring the Pb isotopic compositions of several meteorites and combining these data with an estimate of average modern terrestrial Pb obtained from marine sediments. He argued that they were part of a single isochron and indicated an age for formation of the Earth of 4.55 ± 0.07 billion years. Subsequent revision of the U-decay constants has reduced this age to 4.48 ± 0.07 billion years. The work of Patterson clearly demonstrated the antiquity of both the Earth and meteorites.

It is now recognized that the Earth's crust and upper mantle experienced a complex series of events that caused significant fractionation of U from Pb; the younger a rock is, the more complex is its source history. Thus estimates of the Earth's age based, for example, on the Pb isotopic systematics in modern basalts (Allègre et al., 1995) should be regarded with caution. The most reliable constraints for the Earth's age are given by the Pb isotopic compositions preserved in the oldest terrestrial rocks (Figure 2.2), as the source regions for such samples generally will have had the least complicated histories. The Pb isotopic composition registers the

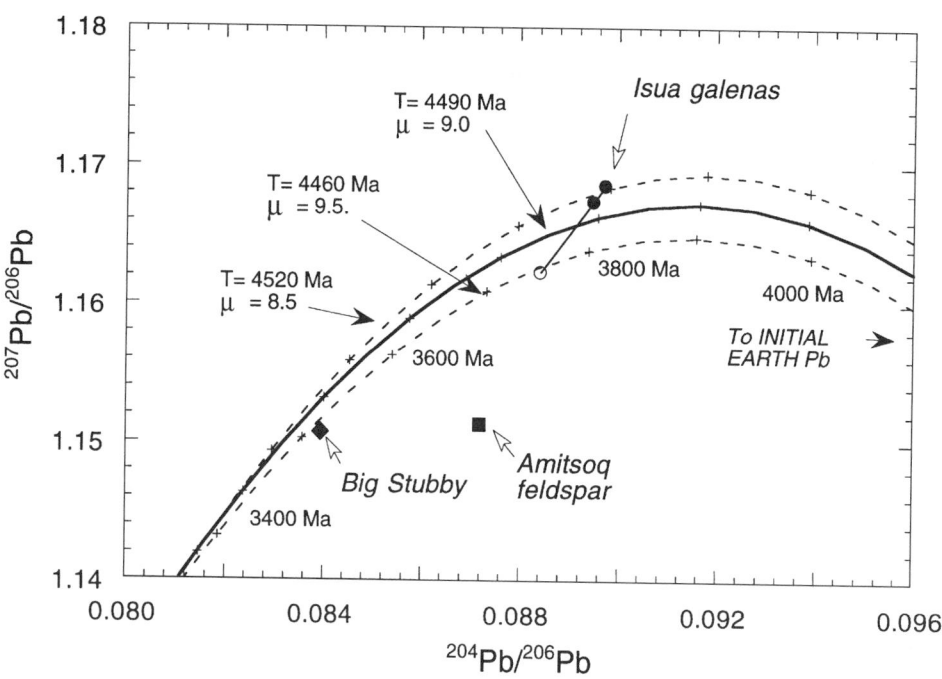

Figure 2.2. Plot of primitive terrestrial ^{207}Pb/^{206}Pb versus ^{204}Pb/^{206}Pb compositions for galenas from Isua, Southwest Greenland (Appel et al., 1978; Richards and Appel, 1987), and Big Stubby, Pilbara (Pidgeon, 1978; Richards, 1986), and feldspar from the Amîtsoq gneisses (Gancarz and Wasserburg, 1977). The curves show Pb isotopic evolution for $\mu = 8.5, 9.0,$ and 9.5. The best fit to both the Isua and Big Stubby data is given by an age $T_{Earth} = 4.49 \pm 0.03$ Ga and $\mu = 9.0$. The initial Earth Pb is calculated at 4.49 Ga by allowing for evolution from a 4.56-Ga Canyon Diablo composition (Tatsumoto et al., 1973) with $\mu = 1$. The straight line connects Isua galenas (Richards and Appel, 1987) having the most primitive (solid symbols) and least primitive (open symbol) Pb isotopic compositions. The possibility that even more primitive Pb isotopic compositions were present prior to formation of this linear Isua Pb array cannot be discounted, although that would imply $\mu < 8.5$ for the early Earth.

mean age of U/Pb fractionation, and it is therefore important to understand the processes that may have fractionated Pb from U during the formation of the Earth. The daughter element Pb has a strongly chalcophile ('sulfur-loving') nature and as a result was extensively partitioned into the Earth's core, whereas the parent element U was excluded. That contributed to the silicate portion of the Earth having a substantially higher (50 to 100 times higher) U/Pb ratio than primitive-solar-system materials such as carbonaceous chondrites. Therefore, if the major terrestrial U/Pb fractionation took place only during core formation, as was first argued by Oversby and Ringwood (1971), then the U-Pb isotopic system should provide constraints on the mean age of core formation. It is likely, however, that there had also been some U/Pb fractionation during the accretion of the Earth, as Pb is volatile compared with U. The relative contributions of volatile loss versus core formation to the apparent 50–100-fold depletion of Pb in the Earth relative to chondrites are still uncertain, but comparisons with other elements having volatilities similar to that of Pb (O'Neill and Palme, Chapter 1, this volume) indicate that volatile loss of Pb during accretion may have contributed an amount of U/Pb fractionation similar to that caused by core formation. Thus the Pb isotopic composition of the early Earth probably registers a mean age for the combined processes of accretion and core formation.

The least ambiguous record of the Pb isotopic composition of the early Earth is provided by the most primitive, oldest, terrestrial Pb compositions, which are measured in ancient Pb-rich and U-poor minerals such as galenas and feldspars. In these mineral phases the measured Pb isotopic composition generally closely approximates the initial Pb composition, as there has been only minimal production of in situ radiogenic Pb from post-crystallization U decay. Here we solve equation (3) and determine T_{Earth} using the most primitive and reliable initial terrestrial Pb isotopic compositions with precisely known age t. Those that meet our criterion for 'primitiveness', that is, with $^{206}\text{Pb}/^{204}\text{Pb} < 12$, are shown in Figure 2.2. The Pb isotopic compositions of galenas from Isua (Richards and Appel, 1987) are shown in solid symbols and have Pb model ages of 4.51 billion (10^9) years ($\mu_2 = 8.72$) and 4.49 billion years ($\mu_2 = 9.11$) (μ will be defined later). In addition to the need for reliable initial Pb compositions, an equally important consideration in the application of equation (3) is the uncertainty regarding the crystallization age t of the early Archaean samples. Precise ion-probe U-Pb dating of zircons in units associated with the galenas that yield the primitive Pb isotopic compositions has recently been undertaken at Isua, Southwest Greenland (Nutman et al., 1996), and in the Pilbara Block, Western Australia, for the Duffer dacite (McNaughton, Compston, and Barley, 1993), the host of the Big Stubby galena. The improved chronology has been particularly helpful in the reinterpretation of initial Pb compositions from the Isua supra-crustal belt (Appel, Moorbath, and Taylor, 1978; Richards and Appel, 1987), where it has now been shown (Nutman et al., 1996) that this complex includes at least two volcanic sequences with ages of 3.708 ± 0.003 and ≥3.79 billion years. The older sequence is most probably age-equivalent to a felsic unit in the area that yielded a precise date of 3.807 ± 0.002 Ga (Compston et al., 1986), although the sequence may be somewhat older if the felsic unit is interpreted as an intrusive gneiss sheet (Rosing et al., 1996). Nutman et al. (1996) have shown that the Isua galenas yielding the most primitive initial Pb compositions are from a fault closely associated with the older mafic and ultramafic rocks in the western part of the belt that probably belong to the ~3.807-Ga package. The Isua galenas measured by Richards and Appel (1987) form a linear array (Figure 2.2), and thus the possibility that even more-primitive Pb compositions may be present cannot be excluded. The appropriate time t for the T_{Earth} calculation is 3.807 Ga, and obviously some caution is still required in interpreting the Pb-isotopic compositions from Isua galenas.

The most primitive terrestrial initial Pb compositions are plotted as $^{207}Pb/^{206}Pb$ versus $^{204}Pb/^{206}Pb$, which emphasizes the more rapid evolution of ^{207}Pb compared with ^{206}Pb in the early Archaean (Gancarz and Wasserburg, 1977). It is apparent from Figure 2.2 that plausible ranges for $^{238}U/^{204}Pb$ (the present-day value of this ratio is conventionally referred to as μ) of 8.5–9.5 will be required to be consistent with the evolution of the Isua and Big Stubby Pb compositions, as well as satisfy the constraints imposed by younger conformable Pb compositions (Stacey and Kramers, 1975). Given this range of μ values, the Pb-Pb age of the Earth, T_{Earth}, is estimated to be between 4.52 and 4.46 billion years. A secondary effect that must be taken into account in applying equation (3) is that this age is clearly less than the 4.56-billion-year age of chondritic meteorites, and thus the values for the initial Pb isotopic composition of the Earth, indicated by comparison with the reference material, Canyon Diablo meteorite (Tatsumoto, Knight, and Allègre, 1973), must be adjusted to take into account the Pb isotopic evolution from the time of meteorite formation (at 4.56 Ga) to the time of U/Pb fractionation in the Earth (at ~4.50 Ga). The average μ for carbonaceous chondrites is low (e.g., Chen and Tilton, 1976) and is based on the correlation of Pb/U versus K/U; a μ_1 value of ~0.7 has been inferred by Allègre et al. (1995) for the proto-Earth prior to core formation. Assuming that Pb is slightly more volatile than K (O'Neill and Palme, Chapter 1, this volume) implies $\mu_1 \cong 1$. With this minor correction to the value used for the Earth's initial Pb composition, the best estimate of the Earth's age, using equation (3), is $T_{Earth} = 4.49 \pm 0.03$ billion years. This is in good agreement with a young age for the Earth that was independently determined by Galer and Goldstein (1996) using similar constraints, and it is interpreted as being a mean age for accretion and core formation. Our estimate is based on the initial Pb compositions for the Isua supra-crustal belt, as well as Big Stubby, and is consistent with $\mu_2 \approx 9$. It is noted that the initial Pb composition derived from a 3.6-Ga Amîtsoq gneiss feldspar, a sample previously used by Gancarz and Wasserburg (1977) to determine the Pb evolution of the early Earth, is not compatible with this data set (Figure 2.2). This might be due to metamorphic disturbance (Gancarz and Wasserburg, 1977) or attributable to the fact that the sample is a porphyritic granite consisting of material with a complex prehistory.

2.2.2. Rb-Sr Isotopic Constraints on Volatile–Refractory-Element Fractionation

Volatile elements were lost during the accretion of the Earth, resulting in significant terrestrial fractionation between volatile and refractory elements relative to the solar-system and primitive-meteorite compositions. That resulted in preferential depletion of the volatile element Rb relative to the more refractory element Sr, and thus the bulk silicate Earth was characterized by an $^{87}Rb/^{86}Sr$ ratio lower than that of its precursor bodies. In contrast to the behaviour of the U-Pb system, it is highly unlikely that either Rb or Sr was secreted in the Earth's core. Fractionation of Rb/Sr during planetary accretion is recorded by the difference between the Earth's initial $^{87}Sr/^{86}Sr$ isotopic ratio, referred to as BEBI ('bulk-Earth best initial'), and the initial compositions in other planetary bodies. Based on the same reasoning as outlined for Pb isotopes, the most rigorous constraints on the Earth's initial Sr composition are those derived from the most primitive terrestrial Sr isotopic compositions. With that in mind, McCulloch (1994) reported a *measured* $^{87}Sr/^{86}Sr$ initial ratio of 0.70050 ± 0.00001 for a 3.46-Ga barite from the North Pole Dome in the Pilbara Block of Western Australia. Determination of accurate, reliable initial Sr isotopic compositions in ancient samples is extremely difficult owing to potential mobility and exchange of Sr, as well as the common problem of multiple episodes of secondary Rb addition, leading to erroneous corrections for in situ ^{87}Sr decay. Thus, unlike the situation for the U-Pb and Sm-Nd systems, very few reliable initial Sr data exist for Precambrian rocks. Analysis of barites, however, largely avoids these problems, as barites have very high Sr concentrations and no Rb, so that uncertainties arising from secondary disturbances in their Rb/Sr ratios are minimized. A complication with using barite is that this mineral is thought to have precipitated during seafloor hydrothermal alteration and therefore does not directly sample the Sr isotopic composition of the mantle. Modern seawater Sr isotopic compositions are much more radiogenic than mantle values, because of the river input of radiogenic Sr derived from ancient continental crust, which is characterized by high $^{87}Sr/^{86}Sr$ values. At 3.5 Ga, the time at which the barite precipitated, the continental crust was not much older, on average, than the barite, and therefore the continental contribution to seawater was not highly radiogenic. For that reason, during the early Archaean there probably was only a minimal difference in Sr isotopic composition between the upper mantle and seawater. The only other datum for the early Archaean is a less precise initial Sr ratio determined from a 3.46-Ga basaltic komatiite from the Onverwacht Group, South Africa (Jahn and Shih, 1974), which yields almost the same initial Sr isotopic composition for the early Archaean of 0.70045 ± 0.00005. The similarity, within errors, between the seawater-derived barite and the komatiite argues that this is a reasonable estimate for the Archaean mantle, and these data currently provide the best estimate of the Sr isotopic composition for the early Archaean Earth.

In order to obtain an estimate of the Earth's initial Sr, it is necessary to correct for the early Archaean evolution of the Sr isotopic composition from 4.49 Ga (the time of the Earth's accretion) to 3.46 Ga (the time of deposition of the Pilbara barite). This requires an estimate of the $(^{87}Rb/^{86}Sr)_{Earth}$ ratio for the early Archaean mantle, which is constrained by the limits of 0.085 for the bulk-Earth mantle versus 0.04 for the average depleted mantle. Thus, assuming a partially ($\sim$30%) depleted early Archaean mantle, we get a value for $(^{87}Rb/^{86}Sr)_{Earth}$ of 0.07 ± 0.01 (McCulloch, 1994) for the pre-3.4-Ga depleted mantle, consistent with other observations, such as positive ε_{Nd} values (as discussed later). Based on this conservative estimate, a relatively well defined $I_{Sr(BEBI)}$ value for the Earth of 0.69940 ± 0.00010 is calculated. This value of $I_{Sr(BEBI)}$ is significantly higher than the initial Sr composition of achondrites, defined by data from the meteorite Angra dos Reis (referred to as ADOR), which has $^{87}Sr/^{86}Sr = 0.69893 \pm 0.00002$ (Wasserburg et al., 1977; Lugmair and Galer, 1992). Achondrites are differentiated, Rb-poor meteorites and therefore provide a good upper limit for the initial Sr ratio in the primitive solar system. The difference between BEBI and ADOR values is interpreted as representing a time period of high Rb/Sr evolution prior to the lowering of the terrestrial Rb/Sr ratio by accretionary fractionation. Moreover, the BEBI value is significantly greater than the initial Sr ratio for the Moon (referred to as LUNI, and equal to 0.69900 ± 0.00002) (Nyquist et al., 1973), and thus it allows chronological constraints to be placed on the first 100 million years of evolution of the Earth–Moon system. If we consider the ADOR value as the zero reference point on a timeline for the accretion of planetary bodies, then the difference in initial Sr compositions between ADOR and BEBI can be translated into a mean time interval for Earth accretion relative to ADOR:

$$I_{Sr(BEBI)} - I_{Sr(ADOR)} = \lambda \Delta T (^{87}Rb/^{86}Sr)_p \qquad (4)$$

where ΔT is the formation interval, $(^{87}Rb/^{86}Sr)_p$ is the parent/daughter ratio in the planetary precursor bodies, and $\lambda = 1.42 \times 10^{-11}$ yr^{-1}. By far the most significant uncertainty in applying the ΔT–I_{Sr} method to planetary accretion is in estimating the $(^{87}Rb/^{86}Sr)_p$ ratio for the terrestrial precursors. In this case, the use of chondritic-meteorite compositions for the Rb/Sr ratio cannot be readily justified, because of accumulating evidence that the volatile/refractory ratio in the asteroid belt, the source area for chondritic meteorites, may not be representative of the solar nebula from which the terrestrial precursors accumulated (Taylor and Norman, 1990; O'Neill and Palme, Chapter 1, this volume). The ratio of volatile/involatile elements is expected to have increased radially outwards from the Sun following the T-Tauri stage, therefore implying a lower $(^{87}Rb/^{86}Sr)_p$ for proto-Earth materials as compared with chondrites. Unfortunately, the volatile/involatile ratio for the terrestrial planetesimals cannot be directly ascertained, as those bodies presumably were swept up and accreted into the Earth–Moon system.

An upper limit on the extent of volatile/refractory-element fractionation of 10-fold is given by comparison of the bulk-Earth ($\sim$0.085) and chondritic ($\sim$0.8) ^{87}Rb/^{86}Sr ratios. An intermediate estimate for $(^{87}$Rb/^{86}Sr$)_p$ of $\sim$0.3 is assumed here, as this implies depletion of Rb/Sr in the terrestrial region of the solar nebula by a factor of 2–3 compared with chondrites. For $I_{Sr(BEBI)} = 0.69940$, this corresponds to a ΔT of $\sim$100 million years relative to ADOR (Figure 2.3). A substantially greater depletion in the volatile content of the terrestrial planetesimals is unlikely, as that would yield $\Delta T > 100$ million years, implying a date for accretion of the Earth significantly more recent than the $\sim$4.49-Ga Pb-Pb date obtained previously. Thus, although the Rb-Sr isotopic systematics do not give definitive constraints for the Earth's age, they are nevertheless consistent with a 100 ± 50-million-year mean

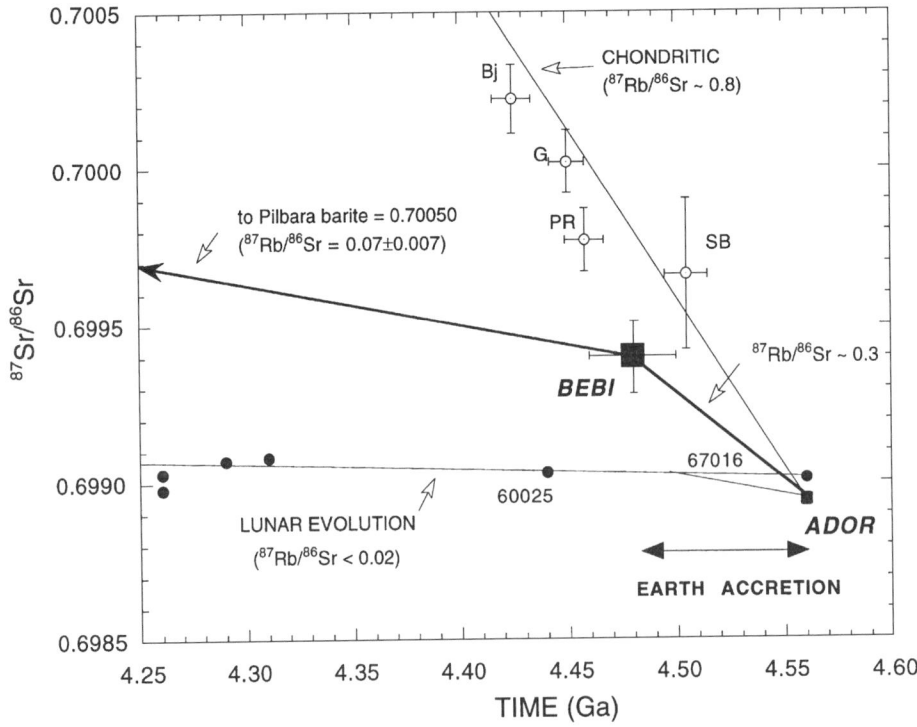

Figure 2.3. Plot of Sr composition versus time for chondritic meteorites, the Moon, and the Earth. The BEBI Sr composition was determined from analysis of ancient barites (McCulloch, 1994). Values for the Moon (Alibert et al., 1994) and the isotopically primitive but chemically differentiated achondrite meteorite ADOR (Lugmair and Galer, 1992; McCulloch, 1994) are shown as solid symbols; data for ordinary chondrites are shown as open symbols; Bj, Bjurbole; G, Guarena; PR, Peace River; SB, Soko Banja (Brannon et al., 1987). Estimates of the mean formation interval for the Earth using the Rb-Sr system range from $\sim$80 to 120 million years and assume an ^{87}Rb/^{86}Sr ratio of 0.4–0.3, based on $\sim$50% loss of volatiles during accretion. (Adapted from McCulloch, 1994.)

accretion interval for the Earth, with the larger errors mainly reflecting uncertainties (by about a factor of 2) in the volatile/refractory-element fractionation in the solar nebula.

2.2.3. Constraints from Short-Lived 'Extinct' Isotopic Systems: ^{182}Hf-^{182}W and ^{146}Sm-^{142}Nd

In the preceding section we considered constraints derived from the long-lived isotopic systems in which the parent isotopes (^{238}U, ^{235}U, and ^{87}Rb) have half-lives similar to or much greater than the age of the Earth and thus are still 'alive'. There are, however, some short-lived isotopic systems, such as ^{182}Hf-^{182}W and ^{146}Sm-^{142}Nd, that have half-lives of 10^7–10^8 years, and thus the parent isotopes (e.g., ^{182}Hf and ^{146}Sm) are no longer present. These 'extinct' systems can, in principle, provide high-resolution constraints on the first 100 million years of Earth history. The temporal resolution (ΔT) possible for an isotopic system is dependent on the analytical precision with which isotopic differences can be measured, the parent/daughter ratio in a given system, and the decay constant that governs how rapidly isotopic differences will form. Thus, if the parent isotope was present in sufficient abundance and there was subsequent fractionation of parent–daughter elements, then theoretically the short-half-life systems can be used to resolve small time differences.

2.2.3.1. ^{182}Hf-^{182}W Systematics

Recent, and still controversial, evidence that a long time interval was required for formation of the Earth's core has come from investigations of variations in the isotopic abundance of the daughter element ^{182}W. At the time of solar-system formation, the short-half-life (9 million years) parent isotope ^{182}Hf was present. As ^{182}W formed from the now-extinct ^{182}Hf, any variations in ^{182}W ratios must therefore reflect fractionation events that occurred while ^{182}Hf was alive, that is, very early in solar-system history. Measurements of iron meteorites, terrestrial rocks, and chondritic meteorites (Harper and Jacobsen, 1992; Lee and Halliday, 1995) have demonstrated that the Earth has a ^{182}W/^{184}W composition identical with that for chondrites, but different from that for iron meteorites. As W is moderately siderophile and Hf is lithophile, large fractionation of the Hf/W ratio resulted from planetary core formation. If that occurred within the first $\sim$60 million years of solar-system evolution, while the parent ^{182}Hf was still alive, then significant deviations in the ^{182}W/^{184}W isotopic composition should be detectable. That phenomenon is observed in iron meteorites, which represent the very rapidly (a few million years) formed cores of small planetary bodies. The Earth, however, does not have a ^{182}W anomaly relative to chondrites. The most straightforward interpretation of the absence of terrestrial ^{182}W anomalies is that the Earth's core and mantle segregated after ^{182}Hf became extinct, that is, at some time after $\sim$4.50 Ga (see

O'Neill and Palme, Chapter 1, this volume). Thus this constraint is consistent with the approximately 4.50-billion-year age for the Earth, derived from the Pb-Pb and Rb-Sr isotopic systematics already outlined.

2.2.3.2. ^{146}Sm-^{142}Nd Systematics

The Sm-Nd isotopic system is unique among the commonly used isotopic systems in having both extinct and live parent–daughter pairs: the commonly used, long-lived ^{147}Sm-^{143}Nd system (discussed later), with a half-life of 106×10^9 years, and ^{146}Sm-^{142}Nd, with a half-life of 103×10^6 years. Owing to its short half-life, ^{146}Sm effectively became extinct; that is, for present-day measurement techniques, the ^{142}Nd/^{144}Nd composition became invariant at ~4.30 Ga. Variations in ε_{142} (differences in ^{142}Nd/^{144}Nd ratios are given as ε_{142} values, which are the deviations, in parts per 10^4, of ^{142}Nd/^{144}Nd ratios relative to the present-day ^{142}Nd/^{144}Nd abundance) must therefore represent the effects of differentiation processes operative prior to 4.30 Ga, that is, during the first 250 million years of the Earth's history. Furthermore, both elements are refractory and lithophile, making them immune to volatility effects and fractionation during core formation. Thus the ^{146}Sm-^{142}Nd isotopic system has the potential to provide constraints on the magnitude and duration of early chemical differentiation within the silicate Earth (e.g., due to the presence of an early magma ocean). Positive or negative ε_{142} values are possible if rare-earth-element (REE) fractionated reservoirs were formed within the first 100–200 million years of Earth history and survived intact as closed systems until they could be sampled later via magmatic processes, such as the formation of the oldest terrestrial rocks (3.96–3.70 Ga). This scenario is shown in Figure 2.4, where ε_{142} values are plotted versus the time of reservoir fractionation and isolation. In these calculations it is assumed that the solar-system value of ^{146}Sm/^{144}Sm $= 0.007$ at 4.56 Ga is the same as that indicated from meteorite studies (e.g., Prinzhofer, Papanastassiou, and Wasserburg, 1992). The magnitude of any ε_{142} anomaly is directly dependent on the degree of fractionation f, where $f = $ (Sm/Nd)/(Sm/Nd)$_{CHUR} - 1$, as well as on the time T when the reservoir fractionated and became isolated. Following Harper and Jacobsen (1992), the ε_{142} effects can be calculated from the following relationship:

$$\varepsilon_{142}(t) = f Q_{142} \left[{}^{146}Sm/{}^{144}Sm \right] \left[e^{-\lambda_{146}(T_0 - T)} - e^{-\lambda_{146}(T_0 - t)} \right] \qquad (5)$$

where Q_{142} is a constant (354×10^{-9} yr^{-1}), $T_0 = 4.56$ billion years, T is the age of mantle differentiation, and t is the crystallization age of the rock.

The lunar basalts sampled by the *Apollo 17* mission (Nyquist et al., 1995) show small ε_{142} effects of about 25 ppm (Figure 2.4), together with very positive ε_{143} values of more than +7 (Figure 2.5). This is consistent with crystallization of the lunar magma ocean at about 4.3 Ga, with f values of 0.25–0.4, that is, with extreme Sm/Nd fractionation (Figure 2.4). The data for the ancient terrestrial samples analyzed by McCulloch and Bennett (1993) are broadly consistent with the lunar

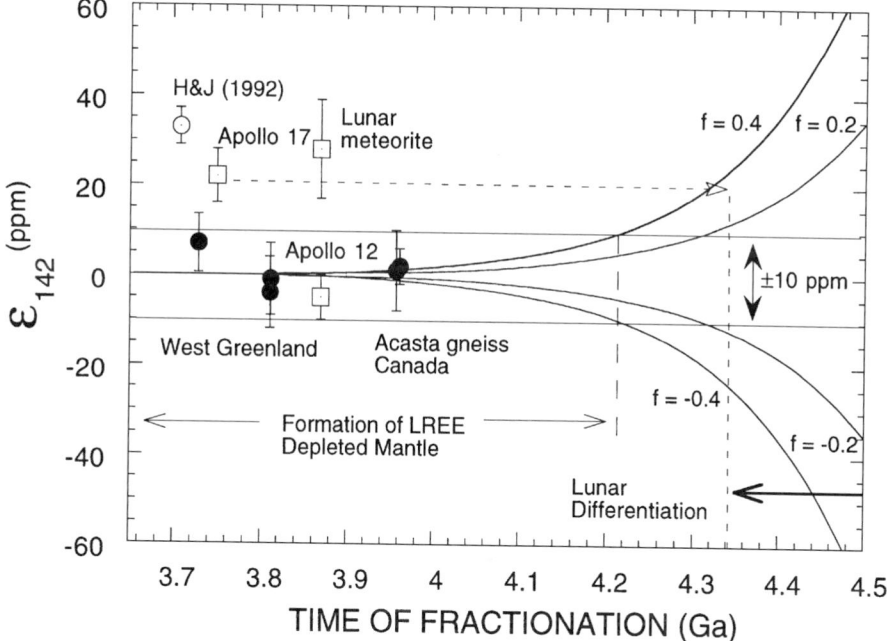

Figure 2.4. Variation of ε_{142} with time, calculated using an initial $^{146}Sm/^{144}Sm = 0.007$ (Prinzhofer et al., 1992). If a mantle depleted of light rare-earth elements (LREEs) ($f > 0.2$) formed in the first 200–300 million years of Earth history, it would be expected to record positive ε_{142} values greater than the analytical resolution of ± 10 ppm. Despite a number of investigations (e.g., McCulloch and Bennett, 1993) of the oldest preserved terrestrial samples (solid symbols), only a single positive ε_{142} value of 33 ± 4 ppm has thus far been reported (Harper and Jacobsen, 1992). The general absence of widespread ε_{142} excesses therefore suggests that LREE-fractionated terrestrial reservoirs were not preserved until after $\sim$4.3 Ga, following the main decay of ^{146}Sm. Lunar data are from Nyquist et al. (1995).

findings of Nyquist et al. (1995), in having ε_{143} values that are $\sim$50% of those for the Moon and no measurable ε_{142} effects (i.e., ε_{142} effects less than 10 ppm). These observations are consistent with the average f value for the terrestrial mantle being lower than for the lunar mantle. Harper and Jacobsen (1992) have, however, reported an ε_{142} of 33 ppm for a terrestrial sample with $\varepsilon_{143}(3.8\,Ga) = +3.5$, the latter ε_{143} value being significantly smaller than that found in *Apollo 17* basalts. This raises the problem of decoupling the two Sm-Nd decay schemes, as in any simple model positive ε_{142} anomalies should be accompanied by large ε_{143} values. Despite the efforts of a number of laboratories (Goldstein and Galer, 1992; McCulloch and Bennett, 1993; Regelous and Collerson, 1995; Sharma et al., 1996), including additional analyses by Harper and Jacobsen, no other early Archaean samples with unambiguous ^{142}Nd variations have been identified. Although there are 'strong hints of ^{142}Nd excess' (Sharma et al., 1996), at this time the Harper and Jacobsen finding appears to be inconsistent with other measurements for the Earth and Moon and may not be representative of the terrestrial early Archaean mantle.

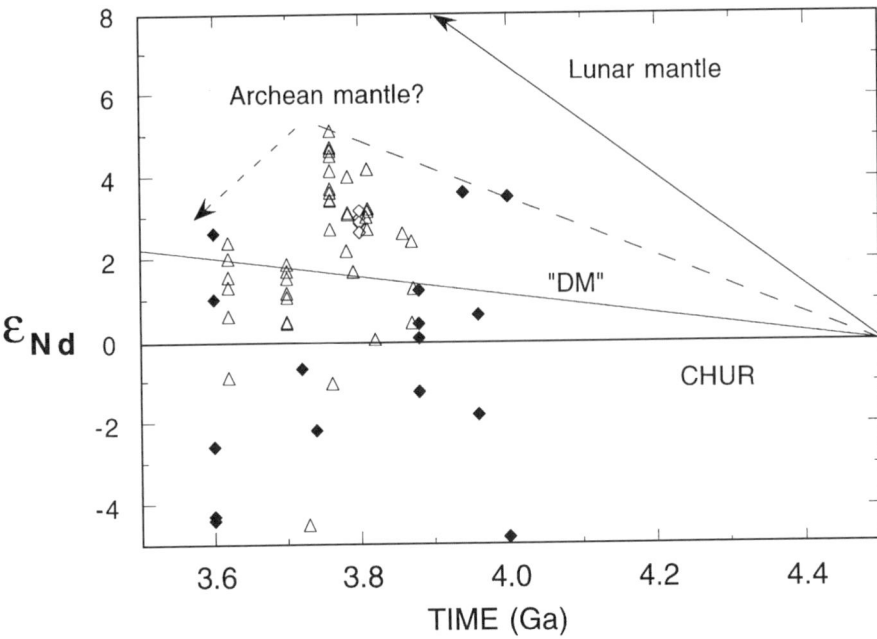

Figure 2.5. Diagram of ε_{Nd} (the deviation, in parts per 10^4, of the ^{143}Nd/^{144}Nd ratio relative to a chondritic source at a given time) versus time for early Archaean rocks, including Greenland samples (open triangles) (Amîtsoq gneisses, Akilia metagabbros, and Isua supracrustals), Acasta gneisses (solid diamonds) from Canada, and residual mantle samples from Labrador (open diamonds). The solid lines indicate lunar evolution, chondritic evolution (CHUR), and the trajectory to the present-day upper mantle assuming linear evolution (DM); the dashed line indicates possible Archaean mantle evolution, with rapid depletion being buffered at ∼3.75 Ga. The ^{147}Sm/^{144}Nd ratio for a reservoir can be calculated from the relationship $\varepsilon_{Nd} = Q_{Nd}\{[(^{147}\text{Sm}/^{144}\text{Nd})_{dm}/(^{147}\text{Sm}/^{144}\text{Nd})_{pm})] - 1\}T$, where Q_{Nd} is a constant (25.13), T is in 10^9 years, and $(^{147}\text{Sm}/^{144}\text{Nd})_{pm} = 0.1967$ (i.e., CHUR, the chondritic value). For the present-day depleted mantle (MORB) having $\varepsilon_{Nd} = +12$, and for $T = 4.3 \times 10^9$ years (see Figure 2.4), we get $(^{147}\text{Sm}/^{144}\text{Nd})_{dm} = 0.22$, indicating LREE depletion caused by extraction of the continental crust from the mantle. A similar or even greater ^{147}Sm/^{144}Nd ratio is indicated by the early Archaean ε_{Nd} value, implying the formation of a strongly LREE-depleted mantle early in the Earth's history. (Adapted from Bennett et al., 1993.)

The absence of clearly resolvable terrestrial ε_{142} effects, together with the presence of strongly positive ε_{143} values (as discussed in the following section), provides a firm upper limit on the age of terrestrial mantle differentiation. These constraints require that highly fractionated, LREE-depleted reservoirs were formed and preserved only after ∼4.3 Ga, by which time decay of ^{146}Sm was essentially complete. In the interval from ∼4.5 Ga to 4.3 Ga the Earth probably was well mixed and constantly rehomogenized by increased tectonic activity related to higher heat flow, possibly due to the effects of giant impacts. The Nd isotopic results are therefore compatible with a young age for the Earth, in that they do not require an extremely ancient (pre-4.49-Ga) event. The general absence of ^{142}Nd anomalies and the consequent lack of correlated ^{142}Nd-versus-^{143}Nd effects also argue against early differentiation processes associated with the Earth's formation being the primary causes of depletion of the early Archaean upper mantle.

2.3. Nd Isotopic Constraints on Crust–Mantle Evolution

Rocks preserved in the stable cratonic portions of the continental crust represent an archive, albeit incomplete, of the composition of the Earth's mantle and crust over the past ~4 billion years. The radiogenic initial ^{143}Nd isotopic compositions preserved in these rocks, as compared with bulk-Earth estimates, provide some of the strongest direct evidence of the long-term history of the chemical depletion of the mantle. Because the processes that shaped the evolution of the Earth's continental crust and mantle were so intimately related (Figure 2.1), tracking the Nd isotopic composition of the mantle through time, particularly for the Archaean (prior to 2.6 Ga), should help to unravel the complex interplay among changes in the continental mass, recycling rates, and the changing mass of the depleting mantle reservoir. The utility of Nd isotopic records produced by in situ decay of ^{147}Sm to ^{143}Nd stems from several factors. Primarily, the REEs Sm and Nd are both refractory and lithophile and therefore did not participate in core formation, nor were they fractionated during the Earth's accretion. Therefore the bulk silicate portion of the Earth (the mantle plus crust) is thought to have chondritic relative abundances of Sm and Nd as well as the other REEs. Thus, in contrast to other isotopic systems, Sm-Nd measurements in chondrites (Jacobsen and Wasserburg, 1984) provide us a well-defined baseline for comparison. Sm and Nd were, however, fractionated during the magmatic differentiation processes between the crust and mantle (Figure 2.1). Extraction of LREE-enriched continental crust (low Sm/Nd) resulted in a complementary LREE depletion of the upper mantle (high Sm/Nd). This fractionation leads to the evolution of radiogenic ^{143}Nd/^{144}Nd compositions ($\varepsilon_{Nd} = +10$ to $+12$) in the upper mantle and correspondingly unradiogenic or low ^{143}Nd/^{144}Nd ratios in felsic crustal rocks, as compared with the bulk Earth.

Another important trait of the Sm-Nd system is the relative immunity of these two elements, compared with other parent–daughter pairs, to secondary alteration after crust formation. This increases the feasibility of being able to make accurate corrections for in situ decay, allowing the determination of accurate initial isotopic compositions. We shall review the Nd isotopic database from ancient rocks and then discuss these data in terms of relative volumes of depleted and enriched reservoirs and constraints on continental growth rates and processes.

2.3.1. Evidence for Depleted Mantle in the Early Archaean

Beginning with the first studies of Archaean samples (Hamilton, Evensen, and O'Nions, 1979; McCulloch and Compston, 1981), Nd isotopic investigations of ancient rocks established and have continued to reveal the depleted nature of the mantle in the early Earth. Even the very oldest (ca. 4-Ga) crustal remnants are characterized by positive ε_{Nd} values (Bowring and Housh, 1995). In fact, almost all pre-3.7-Ga gneisses have positive initial values, and some have values as positive as $\varepsilon_{Nd} = +4$. These include members of the Acasta gneiss complex in Canada (Bowring, Williams, and Compston, 1989), the Uivak gneisses of Labrador (Collerson et al., 1991), and the Itsaq gneiss complex (containing the Amîtsoq gneisses, the Isua supra-crustal belt, and the Akilia association of supracrustal rocks) of Greenland (Jacobsen and Dymek, 1988; Bennett, Nutman, and McCulloch, 1993). However, the reliability of the very large positive ε_{Nd} values for the isotopic compositions in early Archaean rocks has been questioned. Determination of accurate initial isotopic compositions in Archaean terranes presents numerous difficulties, requiring precise independent determination of the age of each sample in conjunction with an understanding of its geological context. Early Archaean rocks are invariably preserved in complex granite-gneiss terranes that have undergone a prolonged history of metamorphism and tectonism. Therefore the question arises whether these rocks behaved as closed systems preserving the initial isotopic ratios or have been perturbed by subsequent, perhaps subtle, alteration. Alteration is not an uncommon problem. For example, in the course of Rb-Sr isotopic studies, disturbance of Archaean initial Sr ratios generally is recognized as being pervasive, thus explaining the paucity of reliable initial Sr compositions for that period (Section 2.2.2). For the Sm-Nd system, the geochemical similarities and rather immobile nature of these elements in most geologic environments have led to the expectation that they usually will have been immune to secondary disturbance. This has been demonstrated for many Precambrian gneiss terranes, but there are exceptions, as shown, for example, by McCulloch and Black (1984) in high-grade metamorphic rocks from Enderby Land in Antarctica, where resetting of the Sm-Nd system during younger granulite-grade metamorphism has been documented.

It is impossible a priori to know if the Sm-Nd systematics of ancient gneisses have been disturbed; however, various criteria can be used to evaluate the integrity of datasets. Bennett et al. (1993) reported sets of internally consistent but relatively high initial Nd ratios corresponding to an ε_{Nd} up to +4 for a suite of pre-3.76-Ga rocks from the Itsaq complex of Southwest Greenland. This area of Greenland contains the most extensive exposures of well-preserved, locally homogeneous ancient gneisses. The crystallization ages for all of the samples were determined by high-precision U-Pb analysis of individual zircons using the SHRIMP (sensitive, high-resolution ion microprobe) developed by W. Compston and colleagues at the Australian National University. Only those samples with simple concordant zircon populations yielding well-defined crystallization ages, with least evidence for either metamorphic overprinting or Pb mobility, were used in the Nd isotopic studies. Although unmodified zircon U-Pb characteristics do not prove that the whole-rock Sm-Nd system also has been undisturbed, it is evidence that large amounts of fluids have not interacted with the samples. In some cases, zircons from 3.8 Ga or earlier with high U concentrations (~1,000 ppm), and therefore abundant radiation damage, showed no evidence of Pb mobility (Nutman et al., 1996), implying minimal disturbance of the rocks.

As shown in Figure 2.5, ε_{Nd} values greater than +3 are observed only in the oldest (pre-3.76-Ga) gneisses from several localities. The Nd isotopic compositions from the large literature dataset indicate that the rapid Nd evolution rate was subsequently moderated, such that for the next billion years, between 3.7 Ga and about 2.6 Ga, the Archaean mantle was characterized by similar or even slightly lower ε_{Nd} compositions than in the early Archaean. We note here that, at least for Greenland, the younger gneisses (3.7 Ga or later) with the more moderate ε_{Nd} values are in close proximity to and have identical post-crystallization histories as the older gneisses (predating 3.7 Ga) with the more extreme ε_{Nd} values, making it difficult to argue that one set of samples is disturbed whilst the other is not.

Recent Hf isotopic data highlight the complexities of working in early Archaean terranes. In some, but not all, Proterozoic and Phanerozoic rocks the extent of Lu/Hf fractionation during crust formation was approximately twice that for Sm/Nd, resulting in the ε_{Hf} values often being about twice the ε_{Nd} values. Thus, it was suggested that a test of the reliability of the Nd data would be to determine initial Hf isotopic compositions in zircons from the same samples. Initial Hf results (Vervoort et al., 1996) for two samples of Itsaq gneiss zircons from the study of Bennett et al., (1993) did not show the expected $\varepsilon_{Hf} = 2\varepsilon_{Nd}$ relationship. This was interpreted by Vervoort et al., (1996) as indicating that the Hf system had remained closed, while the Sm-Nd system had been disturbed. The correlation between ε_{Hf} and ε_{Nd} values is, however, not straightforward, with MORBs, for example, having a wide range of Hf compositions (Salters and Hart, 1991) but relatively constant Nd compositions, equivalent to $\varepsilon_{Hf} = (1.3-2.5)\varepsilon_{Nd}$. That indicates a much more complex relationship between Hf and Nd isotopic compositions than had generally been assumed. Given the limited number of samples measured thus far and our lack of knowledge of the relative behaviour of Hf and Nd during Archaean crust formation and hence fractionation processes and of the behaviour of Hf in Archaean zircons, it is premature to conclude that the highly positive ε_{Nd} values resulted from disturbance of the Nd system. These new Hf findings do, however, highlight the need for continuing detailed studies of early Archaean rocks.

Although there is ongoing controversy as to whether or not ε_{Nd} values were as positive as +3 to +4 in the early Earth, it is quite clear, considering the Hf and Nd isotopic evidence, that ε_{Nd} values of at least +2 at 3.8 Ga were prevalent in the early Archaean upper mantle. The presence of ε_{Nd} values of +2 or greater in pre-3.7-Ga rocks is still extremely important, as they require a major early differentiation of at least portions of the Earth's mantle – within the first 200–400 million years of Earth history. As shown in Figure 2.5, to have generated $\varepsilon_{Nd} = +2$ at 3.8 Ga would have required a minimum mean $^{147}Sm/^{144}Nd$ ratio in the upper mantle of 0.22. A value of $\varepsilon_{Nd} = +4$ at 3.8 Ga would have required a minimum of $^{147}Sm/^{144}Nd \geq 0.26$.

For comparison, present-day MORB-source mantle, after extraction of the whole of the continental crust, is characterized by $^{147}Sm/^{144}Nd \cong 0.22$. Thus, from the Nd isotopic constraints it is apparent that during the early Archaean, parts of the Earth's upper mantle were at least as depleted of some lithophile elements as at present, and during some periods may have been even more depleted.

2.3.2. Mantle Depletion and Growth of the Continental Crust

The evolutionary histories of the Earth's mantle and continental crust are intimately related, and thus well-constrained models for the evolution of both the crust and the mantle are necessary for a proper understanding of either system. As already argued, the present-day chemical compositions of the continental crust and depleted mantle are broadly complementary. Therefore the concentration of a trace element C is given by the simple mass-balance relationship

$$M_{pm}C_{pm} = M_{cc}C_{cc} + M_{dm}C_{dm} \tag{6}$$

$$M_{pm} = M_{cc} + M_{dm} \tag{7}$$

where the subscripts cc, dm, and pm indicate continental crust, depleted mantle, and primitive mantle, respectively. M_{dm} and M_{cc} are the masses of the depleted mantle and continental crust, respectively, and M_{pm} refers to only that portion of the 'primitive' mantle affected by crustal extraction. These simple mass-balance relationships assume that the continental crust is the major enriched incompatible-element reservoir. The source regions for mantle plumes constitute another possible enriched reservoir, but as they are composed of mainly peridotite or recycled basalt, they are not considered to provide a significant reservoir for incompatible elements relative to the continental crust (cf. Hofmann et al., 1986). Therefore, after rearrangement of equation (6), the mass of the depleted mantle relative to the continental crust is given by

$$M_{dm}/M_{cc} = (C_{cc}/C_{pm} - 1)/(1 - C_{dm}/C_{pm}) \tag{8}$$

For the purpose of estimating the mass of the depleted mantle it is preferable to use the abundances of trace elements that are both highly incompatible and refractory, so that the effects of volatile loss during accretion will be insignificant, whilst having large enrichments in the continental crust. Using the element Th as a typical example, we have $(C_{cc}/C_{pm})_{Th} = 55$ and $(C_{dm}/C_{pm})_{Th} = 0.2$ (McCulloch and Bennett, 1994), and hence, from equation (8), $M_{dm}/M_{cc} = 70$. For a present-day value of $M_{cc} = 0.006\,M_{mantle}$, this gives $M_{dm} = 0.41\,M_{mantle}$; that is, the chemically depleted portion of the mantle comprises about 40% of the total mass of the mantle. A limitation of this approach is that it requires knowledge of the primitive-mantle concentrations, which are themselves model-dependent (e.g., Sun and McDonough, 1989). A more robust approach is to apply the mass-balance approach to elemental ratios ($R = C^i/C^j$), such that

$$M_{dm}/M_{cc} = (C^j_{cc}R_{cc} - C^j_{pm}R_{dm})/(C^j_{pm}R_{pm} - C^j_{dm}R_{dm}) \qquad (9)$$

For $M_{pm} \approx M_{dm}$ (i.e., $M_{dm} \gg M_{cc}$) we have

$$M_{dm}/M_{cc} = (C^j_{cc}/C^j_{dm})(R_{cc} - R_{dm})/(R_{pm} - R_{dm}) \qquad (10)$$

An important case is that where $i = $ Sm and $j = $ Nd, as the long-term average Sm/Nd ratio can be independently constrained from Nd isotopic compositions. For the depleted mantle, we have, from Figure 2.5, $(^{147}Sm/^{144}Nd)_{dm} \approx 0.22$. From studies of the REE abundances in the crust (Taylor and McLennan, 1985), $(^{147}Sm/^{144}Nd)_{cc} \approx 0.12$, and for Nd the ratio $C_{cc}/C_{pm} \approx 26$. Using these parameters, equation (10) gives $M_{dm}/M_{cc} = 85$, and hence $M_{dm} = 0.5 M_{mantle}$. Thus, from the geochemical constraints, using either equation (8) or (10) and allowing for uncertainties in individual geochemical parameters, the present-day depleted mantle is constrained to be $45 \pm 10\%$ of the mass of the total mantle. This is in good agreement with an estimate by Hofmann et al. (1986), using similar constraints from other elemental ratios, and with an estimate by O'Neill and Palme (Chapter 1, this volume) of $\sim 40\%$ depleted mantle, based on the amount of ^{40}Ar in the atmosphere produced by decay of ^{40}K in the mantle (see McDougall and Honda, Chapter 3, this volume). We emphasize that these geochemical constraints do not specify the spatial distribution of the depleted mantle. However, as the trace-element concentrations for the depleted mantle are derived from modern MORBs, which are presumed to be the melting products of the upper mantle, it is generally assumed that the depleted mantle resides mainly within the upper half of the mantle at depths $\lesssim 1,000$ km. The mantle above the 660-km discontinuity constitutes only about 30% of the total mass of the mantle, indicating that a significant portion of the lower mantle is also likely to have been geochemically depleted as a result of the formation of the continental crust. These arguments therefore indicate that the 660-km discontinuity does not currently represent the major geochemical boundary with respect to the distribution of depleted mantle.

An important feature of equation (10) is that the parameter R_{dm} or $(^{147}Sm/^{144}Nd)_{dm}$ can be determined at various times during the history of the Earth based on the evolution of the Nd isotopic compositions in the depleted mantle. For the early Archaean mantle, assuming $\varepsilon_{Nd} \approx +4$, we have an upper limit for the time-averaged $(^{147}Sm/^{144}Nd)_{dm}$ of ~ 0.26 (Figure 2.5), indicating that for some periods $M_{dm}/M_{cc} \approx 30$ (i.e., the ratio M_{dm}/M_{cc} was much smaller than it is today). A direct corollary from this simple analysis is that although the growth of the continental crust and of the depleted part of the mantle were closely linked, their relative masses may differ by a factor of about 3. This would have important implications for mantle dynamics. For example, some quantitative models of mantle dynamics (e.g., Tackley et al., 1993) suggest that the early Archaean mantle may have had a greater propensity for layering (see Davies, Chapter 5, this volume), providing a mechanism for

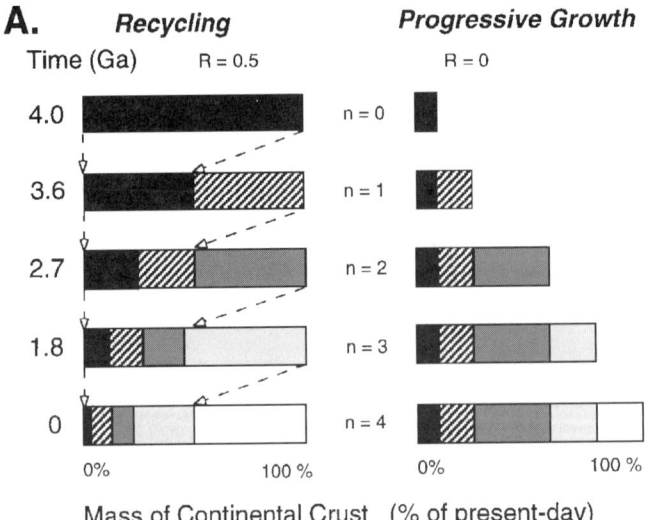

Mass of Continental Crust   (% of present-day)

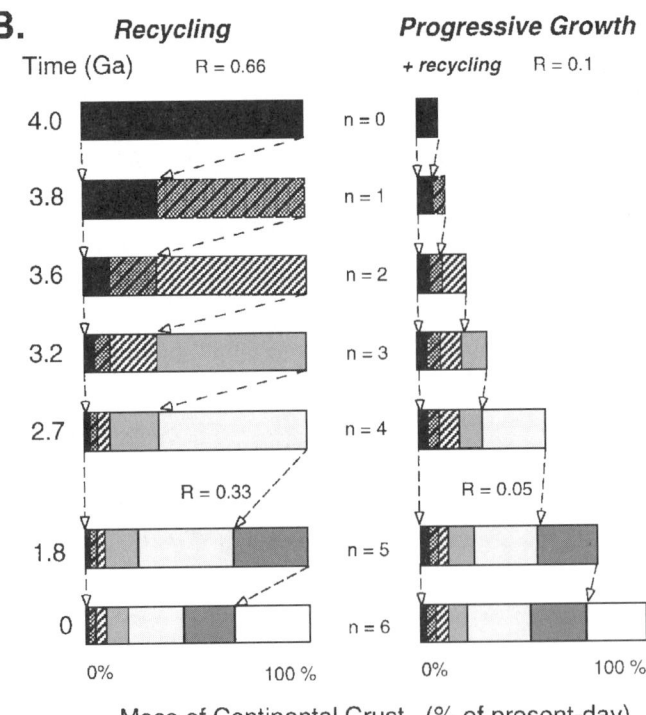

Mass of Continental Crust   (% of present-day)

Figure 2.6. (A) Schematic diagram comparing recycling models and progressive-growth models for the continental crust. In the steady-state recycling model, the total mass of continental crust remains approximately constant, with losses due to recycling being matched by new additions of crust from the mantle. In this example, the recycling rate $R = 0.5$, and thus half of the pre-existing crust is recycled during each crust-formation–recycling episode n. For the progressive-growth model, the mass of the continental crust increases episodically because of new crustal additions. In the latter case, the present-day age-distribution pattern is representative of the overall growth rate. (B) Optimized version of the recycling model, with additional growth–recycling steps n and with $R = 0.66$ in the Archaean, decreasing to $R = 0.33$ in the post-Archaean. Although the mass of crust of a given age decreases rapidly, isolating portions of the mantle and extracting crust from smaller mantle volumes. Clearly, inconsistencies in the Nd-Hf isotopic databases for the depleted mantle will have to be resolved before this line of argument can be pursued further. The following section describes the constraints on the rate of growth of the continental crust.

2.3.3. Progressive Growth versus Recycling of the Continental Crust

The present-day age distribution of the continental crust provides a boundary condition on crustal growth. However, because of the possibility for both growth and destruction of the continents by recycling back into the mantle, the time-dependent history of the growth of the continental crust is not uniquely determined, and that has resulted in two extreme viewpoints regarding development of the continental crust. The first view is that the mass of the continental crust has increased progressively through time and that the return flux of bulk continental crust back into the mantle has not been significant (Figure 2.6A). This type of model has been considered by many authors (e.g., Taylor and McLennan, 1985), with recent proponents being McCulloch and Bennett (1994). We note that this type of model does not exclude recycling of juvenile island-arc crust at subduction zones nor, of course, recycling of basaltic oceanic crust. This is because young, primitive island-arc crust and oceanic crust have Nd isotopic compositions similar to that of the depleted mantle and, with rapid recycling and mixing rates, would not have significantly affected its long-term isotopic evolution.

The alternative view (Armstrong, 1981, 1990; Bowring and Housh, 1995) is that a continental crust of approximately present-day size formed very early in the Earth's history (prior to 4 Ga) and since then has remained constant in mass. We know from geochronological studies that the continents are composed of rocks of many different ages, and new crust can be seen forming today in island arcs, so it is clear that there have been ongoing additions to the continents. In the no-growth or steady-state recycling model, the new additions to the continental crust must therefore be matched by subtraction of an equivalent mass of continental crust that is recycled (i.e., returned to the mantle). That steady-state balance between new additions to the continental crust and losses via recycling is an attractive feature of the no-growth model. The possibility of significant portions of continental crust undergoing recycling into the mantle has also become popular with the recognition of the important role that subduction of oceanic crust plays in the Earth's evolution (e.g., McCulloch, 1994).

Figure 2.6. (*cont.*) at 3.8 Ga the pre-4-Ga crust still constituted about one-third of the total crust. The almost complete absence of pre-4-Ga zircons in the earliest Archaean sediments from 3.7 Ga or earlier is therefore the prime argument against large-scale crustal recycling. The preferred model of dominantly progressive growth of the continental crust, being modulated with, at most, modest amounts of recycling, is shown, with $R = 0.1$–0.05.

Isotopic evolution curves alone cannot discriminate between these two very different scenarios of crustal growth, for the variables of continental mass and depleted-mantle mass are interrelated and do not allow a unique solution. Production of large amounts of continental crust early in the Earth's history would have resulted in a complementary depletion in the early Archaean mantle that subsequently could have been buffered by recycling of enriched continental crust back into the mantle. This is mathematically equivalent to progressively growing the continental crust and the complementary depleted mantle through time without recycling. However, the distinction between these two different modes for the production and preservation of continental crust is important, not only for an understanding of the evolution of continental crust but also for an understanding of how the style of mantle convection may have changed through time.

If the mass of the continental crust has increased through time by a factor of at least 5–10, as proposed, for example, by McCulloch and Bennett (1994), then the mass fraction of depleted mantle will be expected to have increased by a commensurate amount. That implies major changes in the style of mantle convection and in the degree of isolation of the upper- and lower-mantle reservoirs through time. The critical issue is which of these two competing models provides the best description of the growth and evolution of the continental crust. Indirect evidence for recycling of continental crust into the Earth's mantle has been found in a few ocean-island basalts (e.g., Woodhead and McCulloch, 1989), but these studies indicated that only limited amounts (<2%) of continentally derived sediments were recycled. Strong evidence against the recycling model comes from the lack of old (pre-4.0-Ga) zircons preserved in ancient clastic sediments. If a mass of pre-4.0-Ga continental crust equal in size to the present-day continental crust had formed and later had been systematically eroded and recycled into the mantle, then some evidence of its existence should still remain in the ancient sedimentary record. This concept is illustrated in Figure 2.6A. Here we assume that the present-day age structure of the continental crust is strongly episodic, as shown, for example, by McCulloch and Wasserburg (1978), with periods of new crustal growth occurring predominantly at times of $\sim$4.0 Ga, $\sim$3.6 Ga, $\sim$2.7 Ga, and $\sim$1.8 Ga; see also Nelson and DePaolo (1985) and Taylor and McLennan (1985). It is also assumed that the formation of new crust and the recycling of pre-existing crust are interrelated processes. For this purpose, the total mass of continental crust $M_{cc}(t)$ can be considered as a mixture of two components: $m_{cc}(t)$, the mass of new crust formed at time t, and $m_{cc}\langle\tau\rangle$, the mass of pre-existing crust with a mean age τ. If we define the recycling factor as

$$R = 1 - [m_{cc}\langle\tau\rangle_{n+1}/m_{cc}\langle\tau\rangle_n] \tag{11}$$

then $R=0$ represents the no-recycling case, and $R=1$ represents total recycling at each stage n. From Figure 2.6A it can be seen that for the steady-state recycling case, the present-day fraction of crust of age τ is given by $(1-R)^n$, where n is the integer number of crustal-growth–recycling episodes. For example, for $R=0.5$ and

Table 2.1. *Models for growth and recycling of crust*

Stage (n)	Time (Ga)	Progressive growth — Fraction	Progressive growth — Sum	Progressive growth + recycling[a] — Fraction added	Progressive growth + recycling[a] — (>4 Ga)	Progressive growth + recycling[a] — Sum	Recycling — Fraction	Recycling — (>4 Ga)	Recycling — Sum
					$R = 0.10$			$R = 0.66$	
$n = 0$	>4.0	0.08	0.08	0.08	$(0.08)^b$	0.08	1	(1)	1.0
$n = 1$	3.8	0.04	0.12	0.05	(0.072)	0.12	0.66	(0.33)	1.0
$n = 2$	3.6	0.10	0.22	0.11	(0.065)	0.22	0.66	(0.11)	1.0
$n = 3$	3.2	0.09	0.31	0.11	(0.058)	0.31	0.66	(0.037)	1.0
$n = 4$	2.6	0.25	0.56	0.28	(0.052)	0.56	0.66	(0.013)	1.0
					$R = 0.05$			$R = 0.33$	
$n = 5$	1.8	0.16	0.72	0.19	(0.05)	0.72	0.33	(0.008)	1.0
$n = 6$	0	0.28	1.0	0.32	(0.047)	1.0	0.33	(0.006)	1.0

[a]The preferred model.
[b]Numbers in parentheses are the remaining mass fractions of pre-4-Ga crust.

$n = 4$, the pre-4.0-Ga crust would constitute about one-sixteenth of the present-day fraction of continental mass. This is approximately the same fraction as proposed by McCulloch and Bennett (1994) in their progressive-growth model. However, a major dilemma with the recycling model becomes apparent in considering the crustal age distribution in the early Archaean. At 3.6 Ga, the pre-4.0-Ga crustal component is shown in Figure 2.6A as comprising about 50% of the total crust; at 2.7 Ga, the pre-4.0-Ga crust still makes up about 25% of the total crust. If we make the reasonable assumption that recycling of Archaean continental crust occurred by erosion and transportation of sediments into subduction zones, then clearly some evidence of large proportions of pre-4.0-Ga crust would be expected to be preserved in the Archaean sedimentary record. Among the most persistent forms of evidence of older crustal materials are detrital zircon populations preserved in sediments. However, despite intensive searches using both U-Pb (e.g., Nutman et al., 1993, 1996) and Lu-Hf measurements (Stevenson and Patchett, 1990) of detrital zircons in early Archaean sediments, no evidence for significant amounts of pre-4.0-Ga crust has been found.

An optimized version of the recycling model is shown in Figure 2.6B and Table 2.1; it includes several more episodes of crustal growth (4.0 Ga, 3.8 Ga, 3.6 Ga, 3.2 Ga, 2.7 Ga, and 1.8 Ga) and hence recycling ($n = 6$). The estimate for the time interval between episodes has been reduced from 400 million years to 200 million years for the earliest Archaean. This model features what we believe to be a realistic upper limit for the number (n) of growth–recycling stages, based on the observed periodicity in crustal ages. We have also assumed that the recycling rate was greater in the Archaean, with $R = 0.66$ prior to 2.7 Ga, compared with 0.33 for the post-Archaean, which has the effect of increasing the amount of late-Archaean 2.7-Ga crust preserved. This combination of optimized parameters for the crustal recycling model now produces a present-day age-distribution pattern very similar to that proposed earlier (McCulloch and Bennett, 1994) based on the distribution of Nd model ages. In particular, 2.7-Ga crust now makes up a larger fraction of the present-day crustal distribution, consistent with observations (e.g., McCulloch and Wasserburg, 1978). However, the dilemma of preservation of un-realistically large proportions of pre-4.0-Ga continental crust, although somewhat ameliorated, remains a major obstacle for crustal recycling models. For example, for the formation at 3.8–3.9 Ga of the oldest identified clastic sediments, the re-cycling model shown in Figure 2.6B predicts that about 33% of the crust should have consisted of pre-4.0-Ga material. However, despite intensive ion-probe inves-tigations of detrital zircon populations in the Greenland metasediments, no zircons predating 3.9 Ga have been identified (Nutman et al., 1996). Similarly, at 3.6 Ga, the recycling model predicts that pre-4.0-Ga crust should have constituted $\sim$11% of the total crust, a still-significant fraction. Ion-probe investigations of single zir-con grains from detrital populations in Archaean sediments from several continents (e.g., Maas and McCulloch, 1991; Mueller, Wooden, and Nutman, 1992; Nutman et al., 1993, 1996) have identified pre-4.0-Ga crust from only one sediment locality in Western Australia that has a deposition age of $\sim$3.0 billion years. In this case, the proportion of pre-4.0-Ga zircons is constrained to be $\ll$1% of the total sediment. The difficulty with the persistence of large volumes of pre-4-Ga crust also cannot be overcome by further increasing the value of R or n for the early Archaean. The values used here of $R = 0.66$ and $n = 6$ are considered to be realistic upper limits, as it is necessary to preserve sufficient early Archaean crust to have generated the degree of mantle depletion required by Sm-Nd isotopic systematics. For the limits of $R = 1$ and $n \rightarrow \infty$, continental crust would have been recycled instantaneously and would not have left the isotopic signature of depletion in the mantle.

It is therefore concluded that the progressive-growth model provides the best description thus far for the growth of continental crustal through time. The question nevertheless remains as to a realistic upper limit for the amount of crustal recycling that can occur during orogenic episodes. Constraints on maximum values for R are subjective, but considerations of the upper limits for the sediment component in ocean-island basalts and of the volume of sediment entering modern subduction zones suggest that $R < 0.1$. Thus, for 10% recycling at each stage, reducing to 5% in the post-Archaean, the factor $(1 - R)^n$ still will have a significant effect on the present-day proportions of pre-4-Ga crust (Figure 2.6B). For $n = 6$, pre-4.0-Ga crust would be reduced by 40% of its initial abundance as a result of repeated recycling episodes, despite the low values of R. In the progressive-growth model of McCulloch and Bennett (1994), pre-4.0-Ga crust was assumed to account for 8% of the total crust in the preferred model (Table 2.1) (McCulloch and Bennett, 1994). That limit was constrained mainly by the need to produce early Archaean ε_{Nd} values greater than +3, whilst having the amount of pre-4-Ga crust small enough so that it would not dominate the early Archaean sedimentary record. If we assume, however, that progressive growth of the continental crust was accompanied by limited mantle recycling, then the initial fraction of pre-4-Ga crust ($\sim$8%) would, via repeated recycling, have been reduced to a present-day proportion of $\sim$4.5% for $R = 0.1$–0.05 (Figure 2.6B). This lower proportion is more compatible with the extremely rare present-day observations of pre-4-Ga crust. An important outcome of this first-order modelling is that by about 3.4 Ga both the recycling and crustal-growth models predict similarly small quantities ($<$6%) of pre-4-Ga crust remaining. Thus the lack of evidence for abundant pre-4-Ga crust in earliest Archaean sediments remains the single most compelling piece of evidence against large-scale crustal recycling.

2.4. Summary

From Pb and Sr isotopic systematics, together with the lack of anomalies observed in the ^{182}W that was produced from short-lived ($T_{1/2} = 9$ million years) ^{182}Hf, it is concluded that the Earth formed over a period of at least 70–100 million years. Evidence from ^{142}Nd, formed from ^{146}Sm ($T_{1/2} = 103$ million years), indicates that a LREE-depleted mantle reservoir was not preserved until about 4.3 Ga, further supporting a young age for the Earth. The interpretation of this formation interval (ΔT) is not straightforward. Essentially, the U-Pb system records the integrated history of accretion and core formation, whilst Rb-Sr records accretionary volatile/refractory fractionation, with neither system discriminating between various rates of core formation and accretion. Thus the maximum ΔT of 100 million years inferred for the Earth does not necessarily imply that the Earth accreted continuously throughout that interval. For example, the Earth may have formed essentially catastrophically in a period of less than 10^7 years, but starting about 70 million years after the differentiation of the most primitive solar-system bodies (i.e., at $\sim$4.49 Ga). Alternatively, if accretion was essentially continuous throughout that period, then the Rb/Sr ratio used to calculate ΔT would represent a time-averaged ratio for both the planetesimals and the proto-Earth during that period. We emphasize that the isotopic data for the early Earth recorded a complex series of events. For example, scenarios are possible in which the Earth began to form at 4.56 Ga, and then, as a result of a later impact with a Mars-size body, which led to the formation of the Moon, the Earth's isotopic character may have been modified by partial core/mantle re-equilibration and/or mixing with the impactor (O'Neill and Palme, Chapter 1, this volume). In summary, the U-Pb, Rb-Sr, and Sm-Nd isotopic constraints are all consistent with a relatively young mean age for formation of the Earth of $\sim$4.50 Ga. Core formation probably occurred either contemporaneously with accretion of the Earth or shortly thereafter, possibly by $\sim$4.49 $\pm$ 0.03 Ga, but is not strictly defined by the Pb-Pb age, because of volatile loss of Pb during accretion. These constraints, summarized in Figure 2.7, are also consistent with the lack of clear

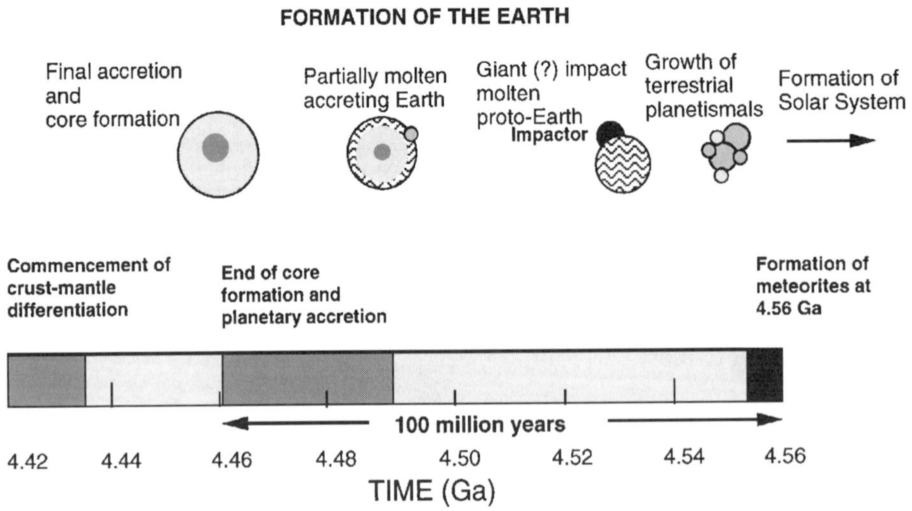

FORMATION OF THE EARTH

Figure 2.7. Summary of the constraints for the timing of formation of meteorites, lunar rocks, and the Earth.

terrestrial isotopic effects due to decay of short-lived systems such as ^{146}Sm and ^{182}Hf.

It is clear that the most profound long-term chemical effect that was recorded by the upper mantle was the formation of the continental crust. Although making up less than 0.6% of the mass of the mantle, the continental crust is nevertheless a major reservoir, containing up to 50% of the Earth's highly incompatible lithophile elements. The formation of the Earth's continental crust is thus the single most important outcome of mantle differentiation. It has been shown that the evolution of the mantle and that of the continental crust were closely interrelated, although that relationship is likely to have changed with time. It is argued here that the most viable model is of predominantly progressive but episodic growth of the continental crust that may have been modulated by the recycling of limited amounts of continental crust back into the mantle. Based on this premise, progressive growth of the chemically depleted upper mantle can be related to the growth of the continental crust using mass-balance constraints, such that $M_{dm}/M_{cc} = 80$. Using this ratio, together with the growth rate for the continental crust (Table 2.1), the Nd isotopic evolution of the depleted mantle can be calculated. This is shown in Figure 2.8, where the volume of depleted mantle is assumed to have increased from 10% to 20% at 3.6 Ga, from 20% to 30% at 2.7 Ga, and from 30% to the present-day volume of ~45% at 1.8 Ga. Although only three discrete episodes of mantle growth are shown here, a more complicated history is possible (e.g., Figure 2.6B), as discussed, for example, by Davies (1995, and Chapter 5, this volume).

If it is assumed that the mantle is stratified, with the chemically depleted portion residing in the uppermost mantle, then the approximate depth to which the

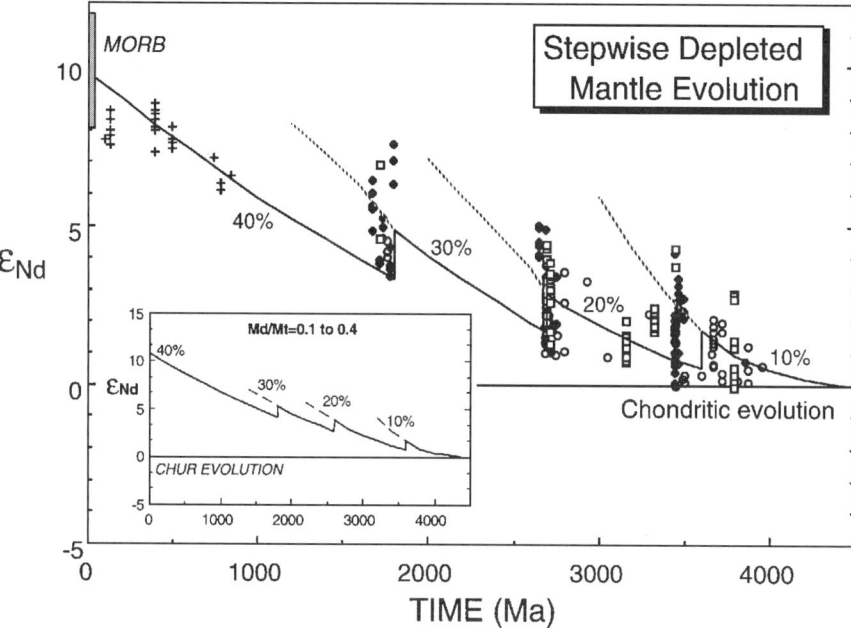

Figure 2.8. Diagram showing the initial Nd isotopic compositions, expressed as ε_{Nd} values, for selected Precambrian gabbros and basalts (solid symbols) and granitoids (open symbols) and Phanerozoic ophiolites (crosses). Also shown is the Nd isotopic evolution calculated for stepwise increases (10%, 20%, 30%, and 40%) in the mass of depleted mantle, inferred to occur at the same time as major crustal-growth episodes (i.e., at 3.6 Ga, 2.7 Ga, and 1.8 Ga). The dashed lines show Nd isotopic evolution trajectories for a mantle composition that was not buffered (in this model, by additions of undepleted mantle material).The inset shows how stepwise growth of depleted mantle can approximate a smoothly depleting mantle. (Adapted from McCulloch and Bennett, 1994.)

depleted mantle extends can be calculated as a function of time. This is illustrated in Figure 2.9, with the depth to the base of the depleted mantle increasing from ~220 km to 400 km at 3.6 Ga, from 400 km to 660 km at 2.7 Ga, and from 660 km to ~800–1,000 km after 1.8 Ga. The relatively shallow depth of less than 220 km at which the depleted mantle is inferred to have resided for the first billion years of Earth history (i.e., until 3.6 Ga), however, raises a number of questions. Early Archaean basaltic komatiites with chemically depleted isotopic signatures (i.e., positive ε_{Nd} values) are, from high-pressure experiments, inferred to have been generated from depths of ~400 km. By ~3.4 Ga, the date of the Barberton and Pilbara komatiites, the depleted mantle is assumed to have extended to a depth of ~400 km, which may account for the paucity of such higher-temperature, isotopically depleted magmas prior to that time. It is also possible that the distribution of the very early Archaean depleted mantle was laterally heterogeneous, in which case localized pockets might have extended to greater depths. Clearly, our knowledge of the temporal distribution of the major reservoirs in the Earth's mantle and of the relationship between heat transfer from the lower mantle and partial melting in the

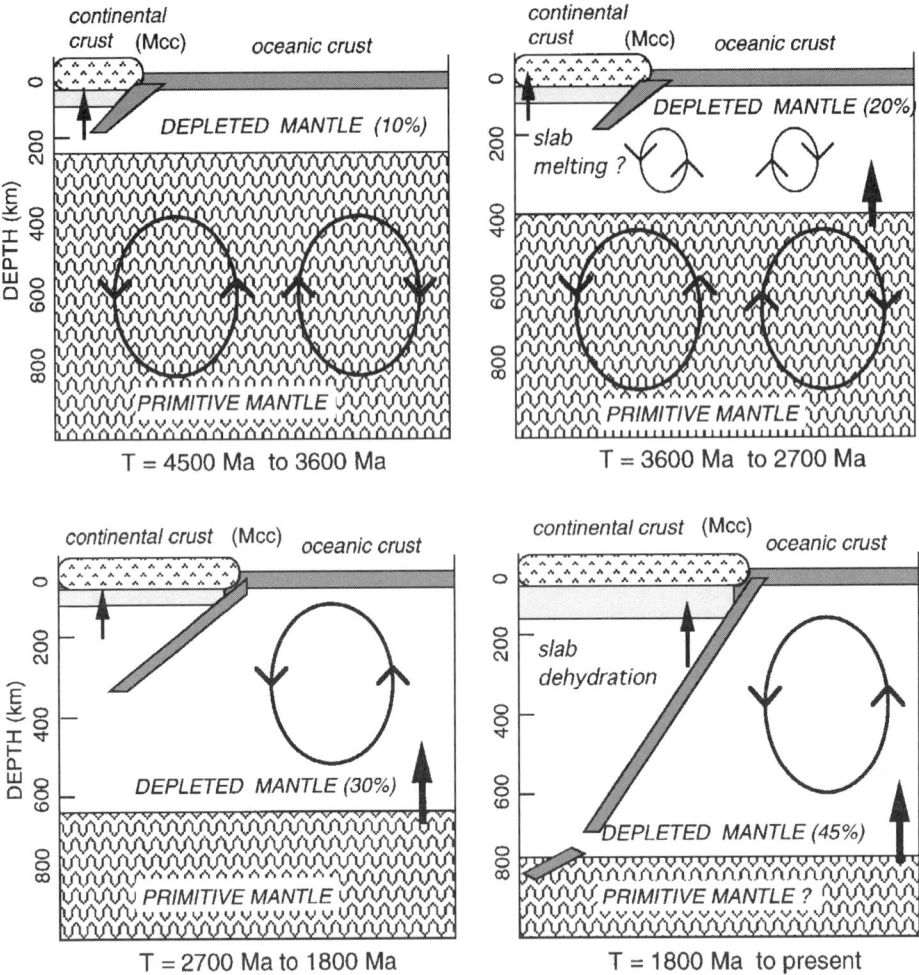

Figure 2.9. Illustration of changing convection regimes in the mantle, based on the assumption that the Earth's continental crust and depleted mantle have both increased in mass with time. The relationship $M_{dm}/M_{cc} \approx 80$ implies that the depleted mantle now makes up $\sim 45 \pm 10\%$ of the mass of the total mantle – more than the fraction of the total mantle that is situated above the 660-km discontinuity. Periods of episodic crustal growth may have been accompanied by additions of more primitive lower mantle into the upper mantle (large arrows). That would have resulted in an increasing mass of depleted upper mantle through time, which would have been buffered to an approximately constant composition. (Adapted from McCulloch and Bennett, 1993.)

upper mantle is still very limited. Further work will be required to better understand both the chemical and physical constraints.

Acknowledgements

This manuscript has benefitted from the incisive and constructive editorial comments provided by the editor of this volume, Dr. I. Jackson. We thank A. Nutman for his generosity in sharing his expertise in Greenland Archaean geology.

References

Alibert, C., Norman, M., and McCulloch, M. T. 1994. An ancient Sm-Nd age for a ferroan noritic anorthosite clast from lunar breccia 67016. *Geochim. Cosmochim. Acta* 58:2921–6.

Allègre, C. J., Manhes, G., and Gopel, C. 1995. The age of the Earth. *Geochim. Cosmochim. Acta* 59:1445–56.

Allègre, C. J., Poirier, J. P., Humler, E., and Hofmann, A. W. 1995. The chemical composition of the Earth. *Earth Planet. Sci. Lett.* 134:515–26.

Appel, P. W. U., Moorbath, S., and Taylor, P. N. 1978. Least radiogenic terrestrial lead from Isua, West Greenland. *Nature* 272:524–6.

Armstrong, R. L. 1981. Radiogenic isotopes: the case for recycling on a near-steady-state no-continental-growth Earth. *Phil. Trans. R. Soc. London* A301:443–72.

Armstrong, R. L. 1990. The persistent myth of crustal growth. *Australian J. Earth Sci., Geol. Soc. Aust.* 38:613–40.

Bennett, V. C., Nutman, A. P., and McCulloch, M. T. 1993. Nd isotopic evidence for transient, highly depleted mantle reservoirs in the early history of the Earth. *Earth Planet. Sci. Lett.* 119:299–317.

Bowring, S. A., and Housh, T. 1995. The Earth's early evolution. *Science* 269:1535–40.

Bowring, S. A., Williams, I. S., and Compston, W. 1989. 3.96 gneisses from the Slave Province, NW Territories, Canada. *Geology* 17:971–5.

Brannon, J. C., Podosek, F. A., and Lugmair, G. W. 1988. Initial $^{87}Sr/^{86}Sr$ and the Sm-Nd chronology of chondritic meteorites. *Proc. Lunar Planet. Sci. Conf.* 18:555–64.

Chen, J. H., and Tilton, G. R. 1976. Isotopic lead investigations on the Allende carbonaceous chondrite. *Geochim. Cosmochim. Acta* 40:635–43.

Collerson, K. D., Campbell, L. M., Weaver, B. L., and Palacz, Z. A. 1991. Evidence for extreme fractionation in early Archaean ultramafic rocks from northern Labrador. *Nature* 349:209–14.

Compston, W., Kinny, P. D., Williams, I. S., and Foster, J. J. 1986. The age and lead loss behaviour of zircons from the Isua supracrustal belt as determined by ion microprobe. *Earth Planet. Sci. Lett.* 80:71–81.

Compston, W., and Pidgeon, R. T. 1986. Jack Hills, evidence of more very old detrital zircons in Western Australia. *Nature* 321:766–9.

Dalrymple, G. B. 1991. *The Age of the Earth.* Stanford University Press.

Davies, G. F. 1995. Punctuated tectonic evolution of the Earth. *Earth Planet. Sci. Lett.* 136:363–79.

Galer, S. J. G., and Goldstein, S. L. 1996. Influence of accretion on lead in the Earth. In: *Earth Processes: Reading the Isotopic Code*, eds. A. Basu and S. Hart, pp. 75–98. AGU monograph 95. Washington, DC: American Geophysical Union.

Gancarz, A. J., and Wasserburg, G. J. 1977. Initial Pb of the Amîtsoq gneiss, West Greenland, and implications for the age of the Earth. *Geochim. Cosmochim. Acta* 41:1283–301.

Gerling, E. K. 1942. Age of the Earth according to radioactivity data. *Comptes Rendus (Doklady) de l'Académie des Sciences de l'URSS* 34:259–61.

Goldstein, S. L., and Galer, S. J. G. 1992. On the trail of early mantle differentiation: $^{142}Nd/^{144}Nd$ ratios of early Archaean rocks. *EOS, Trans. AGU* 73:323.

Hamilton, P. J., Evensen, N. M., and O'Nions, R. K. 1979. Sm-Nd systematics of Lewisian gneisses: implications for the origin of granulites. *Nature* 277:25–8.

Harper, C. L., and Jacobsen, S. B. 1992. Evidence from coupled ^{147}Sm-^{143}Nd and ^{146}Sm-^{142}Nd systematics for very early (4.5 Gyr) differentiation of the Earth's mantle. *Nature* 360:728–32.

Hofmann, A. W. 1988. Chemical differentiation of the Earth: the relationship between mantle, continental crust and oceanic crust. *Earth Planet. Sci. Lett.* 90:297–314.

Hofmann, A. W., Jochum, K. P., Seufert, M., and White, W. M. 1986. Nb and Pb in oceanic basalts: new constraints on mantle evolution. *Earth Planet. Sci. Lett.* 79:33–45.

Holmes, A. 1946. An estimate of the age of the Earth. *Nature* 159:127–8.

Houtermann, F. G. 1946. The isotopic abundances in natural lead and the age of uranium. *Naturwissenschaften* 33:185–6.

Jacobsen, S. B., and Dymek, R. F. 1988. Nd and Sr isotope systematics of clastic matasediments from Isua, West Greenland: identification of pre-3.8 Ga differentiated crustal components. *J. Geophys. Res.* 93:338–54.

Jacobsen, S. B., and Wasserburg, G. J. 1984. Sm-Nd isotopic evolution of chondrites and achondrites. II. *Earth Planet. Sci. Lett.* 67:137–50.

Jahn, B., and Shih, C. 1974. On the age of the Onverwacht Group, Swaziland Sequence, South Africa. *Geochim. Cosmochim. Acta* 38:873–85.

Lee, D.-C., and Halliday, A. N. 1995. Hafnium-tungsten chronometry and the timing of terrestrial core formation. *Nature* 378:771–4.

Lugmair, G. W., and Galer, S. J. G. 1992. Age and isotopic relationships among the angrites Lewis Cliff 86010 and Angra dos Reis. *Geochim. Cosmochim. Acta* 56:1673–94.

Maas, R., and McCulloch, M. T. 1991. The provenance of Archaean clastic metasediments in the Narryer Gneiss Complex, Western Australia; trace element geochemistry, Nd isotopes and U-Pb ages for detrital zircons. *Geochim. Cosmochim. Acta* 55:1913–32.

McCulloch, M. T. 1994. Primitive $^{87}Sr/^{86}Sr$ from an Archaean barite and conjecture on the Earth's age and origin. *Earth Planet. Sci. Lett.* 126:1–13.

McCulloch M. T., and Bennett V. C. 1993. Evolution of the early Earth: constraints from ^{142}Nd-^{143}Nd isotopic systematics. *Lithos* 30:237–55.

McCulloch, M. T., and Bennett, V. C. 1994. Progressive growth of the Earth's continental crust and depleted mantle: geochemical constraints. *Geochim. Cosmochim. Acta* 58:4717–38.

McCulloch M. T., and Black, L. 1984. Sm-Nd isotopic systematics of Enderby Land granulites and evidence for the redistribution of Sm and Nd during metamorphism. *Earth Planet. Sci. Lett.* 71:46–58.

McCulloch, M. T., and Compston, W. C. 1981. Sm-Nd age of Kambalda and Kanowna greenstones and heterogeneity in the Archaean mantle. *Nature* 294:322–7.

McCulloch M. T., and Wasserburg, G. J. 1978, Sm-Nd and Rb-Sr chronology of continental crust formation. *Science* 200:1003–11.

Machado, N., Brooks, C., and Hart, S. R. 1986. Determination of initial $^{87}Sr/^{86}Sr$ and $^{143}Nd/^{144}Nd$ in primary minerals from mafic and ultramafic rocks: experimental procedure and implications for isotopic characteristics of the Archaean mantle under the Abitibi greenstone belt, Canada. *Geochim. Cosmochim. Acta* 50:2335–8.

McNaughton, N., Compston, W., and Barley, M. E. 1993. Constraints on the age of the Warrawoona Group, eastern Pilbara Block, Western Australia. *Precambrian Research* 60:69–98.

Mueller, P. A., Wooden, J. L., and Nutman, A. P. 1992. Provenance of zircons from the Wyoming Craton. *Geology* 20:327–30.

Nelson B. T., and DePaolo, D. J. 1985. Rapid production of continental crust 1.7–1.9 b.y. ago: Nd and Sr isotopic evidence from the basement of the North America and isotopic evolution of the Proterozoic mantle. *Geo. Soc. Am. Bull.* 96:746–54.

Nier, A. O. 1938. Variations in the relative abundances of the isotopes of common lead from various sources. *J. Am. Chem. Soc.* 60:1571–6.

Nier, A. O., Thompson, R. W., and Murphey, B. F. 1941. The isotopic composition of lead and the measurement of geologic time. III. *Phys. Rev.* 60:112–16.

Nutman, A. P., Bennett, V. C., Friend, C. R. L., and Rosing, M. T. In press. ≈3710 and ≥3790 Ma volcanic sequences in the Isua (Greenland) supracrustal belt; structural and isotopic implications. *Chem. Geol.*

Nutman, A. P., Friend, C. R. L., Kinny, P. D., and McGregor, V. R. 1993. Anatomy of an early Archaean gneiss complex: 3900 to 3600 Ma crustal evolution in southern West Greenland. *Geology* 21:415–18.

Nutman, A. P., McGregor, V. R., Friend, C. R. L., Bennett, V. C., and Kinny, P. D. 1996. The Itsaq gneiss complex of southern West Greenland; the world's most extensive record of early crustal evolution (3,900–3,600 Ma). *Precambrian Research* 78:1–39.

Nyquist, L. E., Hubbard, N. J., Gast, P. J., Bansal, B. M., Weismann, H., and Jahn, B. 1973. Rb-Sr systematics for chemically defined Apollo 15 and 16 materials. *Proc. Lunar Sci. Conf.* 4:1823–46.

Nyquist, L. E., Weismann, H., Bansal, B., Shih, C.-Y., Kieth, J. E., and Harper, C. L. 1995. ^{146}Sm-^{142}Nd formation interval for the lunar mantle. *Geochim. Cosmochim. Acta* 59:2817–37.

Oversby, V. M., and Ringwood, A. E. 1971. Time of formation of the Earth's core. *Nature* 234:463–6.

Patterson, C. C. 1956. Age of meteorites and the Earth. *Geochim. Cosmochim. Acta* 10:230–7.

Patterson, C. C., Brown, H., Tilton, G. R., and Ingham, M. G. 1953. Concentration of uranium and lead and the isotopic composition of lead in meteoritic material. *Phys. Rev.* 92:1234–5.

Pidgeon, R. T. 1978. Big Stubby and the early history of the Earth. In: *Short Papers of the 4th ICOG*, ed. R. E. Zartman. U.S. Geological Survey Open File Report 78–701, p. 476.

Prinzhofer, A., Papanastassiou, D. A., and Wasserburg, G. J. 1992. Samarium-neodymium evolution of meteorites. *Geochim. Cosmochim. Acta* 56:797–815.

Regelous, M., and Collerson, K. D. 1995. ^{143}Nd-^{142}Nd systematics of early Archaean rocks from northern Labrador and implications for crust-mantle evolution of the North Atlantic Craton. *EOS, Trans. AGU* 76:687.

Richards, J. R. 1986. Lead isotopic signatures: further examination of comparisons between South Africa and Western Australia. *Trans. Geol. Soc. South Africa* 89:285–304.

Richards, J. R., and Appel, P. W. U. 1987. Age of the "least radiogenic" galenas at Isua, West Greenland. *Chem. Geol.* 66:181–91.

Ringwood, A. E. 1986. Terrestrial origin of the Moon. *Nature* 322:323–8.

Rosing, M. T., Rose, N. M., Bridgwater, D. T., and Thomsen, H. S. 1996. A reappraisal of the >3.7 Ga Isua (Greenland) supracrustal sequence. *Geology* 24:43–6.

Russell, R. D., and Farquhar, R. M. 1960. *Lead Isotopes in Geology*. New York: Interscience.

Rutherford, E. 1929. Origin of actinium and the age of the Earth. *Nature* 123:313–14.

Salters, V. J. M., and Hart, S. R. 1991. The mantle sources of ocean ridges, islands and arcs: the Hf isotope connection. *Earth Planet. Sci. Lett.* 104:364–80.

Sharma, M., Papanastassiou, D. A., Wasserburg, G. J., and Dymek, R. F. 1996. The issue of the terrestrial record of ^{146}Sm. *Geochim. Cosmochim. Acta* 60:2037–47.

Stacey, J. S., and Kramers, J. D. 1975. Approximation of terrestrial lead isotope evolution by a two-stage model. *Earth Planet. Sci. Lett.* 26:207–21.

Stevenson, D. J. 1987. Origin of the Moon: the collision hypothesis. *Annu. Rev. Earth Planet. Sci.* 15:271–315.

Stevenson, R. K., and Patchett, J. P. 1990. Implications for the evolution of continental crust from Hf isotope systematics of Archean detrital zircons. *Geochim. Cosmochim. Acta* 54:1683–97.

Sun, S. S., and McDonough, W. F. 1989. Chemical and isotopic systematics of oceanic basalts: implications for mantle composition and processes. In: *Magmatism in the Ocean Basins*, ed. A. D. Saunders and M. J. Norry, pp. 313–45. London: Geological Society.

Tackley, P. J., Stevenson, D. J., Glatzmaier, G. A., and Schubert, G. 1993. Effects of an endothermic phase transition at 670 km depth in a spherical model of convection in the earth's mantle. *Nature* 361:699–704.

Tatsumoto, M., Knight, R. J., and Allègre, C. J. 1973. Time differences in the formation of meteorites as determined from the ratio of lead-207 to lead-206. *Science* 180:1278–83.

Taylor, S. R., and McLennan, S. M. 1985. *The Continental Crust: Its Composition and Evolution.* Oxford: Blackwell.

Taylor, S. R., and Norman, M. D. 1990. Accretion of differentiated planetesimals to the Earth. In: *Origin of the Earth*, ed. H. E. Newsom and J. H. Jones, pp. 29–43. Oxford University Press.

Vervoort, J. D., Patchett, P. J., Gehrels, G. E., and Nutman, A. P. 1996. Constraints on early Earth differentiation from hafnium and neodymium isotopes. *Nature* 379:624–7.

Wasserburg, G. J., Tera, F., Papanastassiou, D. A., and Huneke, J. C. 1977. Isotopic and chemical investigations on Angra dos Reis. *Earth Planet. Sci. Lett.* 35:294–316.

Woodhead, J. D., and McCulloch, M. T. 1989. Ancient seafloor signals in the Pitcairn Island lavas and evidence for large amplitude, small length scale mantle heterogeneities. *Earth Planet. Sci. Lett.* 94:257–73.

3

Primordial Solar Noble-Gas Component in the Earth: Consequences for the Origin and Evolution of the Earth and Its Atmosphere

IAN McDOUGALL and MASAHIKO HONDA

3.1. Introduction

The abundances and isotopic compositions of the noble gases helium, neon, argon, krypton, and xenon trapped in mantle-derived samples provide important constraints on hypotheses concerned with the origin and evolution of the Earth's atmosphere, crust, mantle, and core. In particular, identification of the noble-gas composition of the primordial Earth is critically important for an understanding of how and when the Earth acquired its volatiles and how its atmosphere evolved. Analyses of samples derived from the mantle have been particularly helpful over the past decade or so, not only for the purpose of determining the Earth's primordial components and its outgassing history but also in relation to the identification and characterization of mantle reservoirs.

In this chapter we review the evidence concerning the primordial noble-gas components in the Earth, principally from studies of mantle-derived samples, but also drawing on information provided by noble-gas studies of meteorites, lunar samples, and the Sun. In recent years, recognition of a remarkable correlation between helium-isotope and neon-isotope systematics in mantle-derived samples has provided strong evidence for a primordial solar component within the Earth. We shall review that evidence and subsequently explore the consequences, especially in regard to the composition and abundances of the heavier noble gases in the Earth, in an attempt to improve our understanding of the origin and evolution of our planet and its atmosphere.

We shall provide background information concerning the noble gases and the known noble-gas components in the solar system prior to discussing the question of primordial components in the Earth. This will lead in to a consideration of constraints provided by the noble gases and other isotopic data on models for the origin and evolution of the Earth and what we can conclude from such data about the mantle, together with some comments on possible future research directions. For more comprehensive recent reviews of noble-gas geochemistry, the reader is referred to works by Kurz (1991), Matsuda (1994), and Ozima (1994), and the book by Ozima and Podosek (1983) remains a valuable source of basic information.

A chapter of this kind is particularly appropriate in a volume in honour and memory of Ted Ringwood, because of his great interest in and appreciation of the value of noble-gas studies in constraining the history of the solar system. Indeed, Ted was a strong advocate of the idea that the Research School of Earth Sciences at the Australian National University should direct its research into noble-gas geochemistry; his support in the planning and implementation of that new venture in the mid-1980s was unwavering. Ted Ringwood shared the excitement attendant upon our discoveries of a strong link between the helium and neon isotopic compositions in mantle-derived samples and the implications flowing from those findings.

3.2. Background Information

The noble gases and their stable isotopes, together with their abundances in air, are listed in Table 3.1. With the notable exception of argon, the noble gases are present in air in the concentration range of parts per million or less. We exclude the heaviest noble gas, radon, from this discussion because its isotopes are radioactive, with short half-lives (<4 days), and thus have only transient existences as parts of the decay chains of uranium and thorium.

The noble gases being considered – helium, neon, argon, krypton, and xenon – span a very wide mass range (Table 3.1). Each element has isotopes covering a wide proportional mass range, so that in principle the noble gases can be sensitive tracers of physical processes. Except for helium, each noble gas has three or more stable isotopes, often permitting variations in isotopic composition to be examined in terms of physical processes such as mass fractionation and/or in terms of nuclear processes.

Each of the noble gases has two or more isotopes, with at least one isotope primordial, and at least one whose abundance is augmented with time owing to radioactive decay, fission, or nucleogenic reactions. Thus, the isotope ratios of particular interest include ^{3}He/^{4}He, ^{20}Ne/^{22}Ne, ^{21}Ne/^{22}Ne, ^{40}Ar/^{36}Ar, ^{129}Xe/^{130}Xe, and ^{136}Xe/^{130}Xe. The isotopes ^{3}He, ^{20}Ne, ^{22}Ne, ^{36}Ar, ^{38}Ar, and ^{130}Xe are considered to be primordial, having been produced during nucleosynthesis and incorporated within the Earth when it formed; the abundances of these isotopes have not been significantly augmented subsequently by processes such as spallation or fission. In contrast, the isotope ^{4}He is continually produced by decay of uranium and thorium, ^{21}Ne is incremented by nucleogenic processes, ^{40}Ar is derived from decay of ^{40}K, ^{129}Xe was augmented from decay of the now-extinct ^{129}I (half-life of 17 million years), and several xenon isotopes (^{131}Xe, ^{132}Xe, ^{134}Xe, and ^{136}Xe) increase because of spontaneous fission of ^{244}Pu (extinct, half-life of 82 million years) and ^{238}U. Thus, in principle, temporal information is also available from noble-gas studies. It should be noted that some of the isotopes of krypton also have a fissiogenic component.

Table 3.1. *Abundances of noble gases and their isotopes in air*

Element	Volume fraction in dry air	Percentage abundance (atomic)
Helium	5.24×10^{-6}	
^{3}He		0.00014
^{4}He		~100
Neon	1.82×10^{-5}	
^{20}Ne		90.50
^{21}Ne		0.268
^{22}Ne		9.23
Argon	9.34×10^{-3}	
^{36}Ar		0.336
^{38}Ar		0.063
^{40}Ar		99.60
Krypton	1.14×10^{-6}	
^{78}Kr		0.347
^{80}Kr		2.257
^{82}Kr		11.523
^{83}Kr		11.477
^{84}Kr		57.00
^{86}Kr		17.398
Xenon	8.7×10^{-8}	
^{124}Xe		0.095
^{126}Xe		0.089
^{128}Xe		1.919
^{129}Xe		26.44
^{130}Xe		4.07
^{131}Xe		21.22
^{132}Xe		26.89
^{134}Xe		10.43
^{136}Xe		8.86

Source: Adapted from Ozima and Podosek (1983).

Studies of noble gases can provide unique information because of the volatile nature of these elements. The noble-gas isotopic compositions in the Earth can be greatly affected by degassing processes, as the noble gases are readily outgassed, whereas the parent elements of those noble-gas isotopes generated over time are relatively involatile. Thus, the history of the Earth's degassing and the related evolution of its atmosphere and mantle have become focal points of noble-gas geochemistry. However, a number of critical questions remain largely unanswered. In particular, identification of the noble-gas elemental and isotopic compositions of the primordial Earth, the timing and rate of outgassing of the Earth's primordial volatiles, the extent of outgassing, and which portions of the Earth's interior have been outgassed are questions yet to be fully addressed. These questions are to some degree interdependent.

3.3. Noble Gases in the Solar System

Before we discuss the question of the primordial noble gases within the Earth, it is necessary to briefly summarize information concerning the noble-gas components found in the solar system. Two main compositions are generally recognized, designated as solar and planetary.

The term 'solar' is used to refer to those noble-gas compositions observed in samples from the lunar surface, recording a history of solar-wind implantation. This component is presumed to closely represent the noble gases in the Sun, the main noble-gas reservoir in the solar system. There are good reasons for believing that the bulk noble-gas composition of the solar nebula, from which the solar system formed, had essentially the same isotopic and elemental noble-gas compositions as the Sun has today (Wieler et al., 1983). This notion is further supported by the recent observations from the *Galileo* probe that the atmosphere of Jupiter, believed to have formed from the solar nebula, has a solar-like noble-gas composition (Niemann et al., 1996). The solar component can be subdivided into a solar-wind (SW) component and a component associated with solar energetic particles (SEP) emitted from the Sun, with energies higher than those of the solar wind. However, we shall use the SW component as representative of the solar component in the following discussion, because the SEP component is thought to be quite minor (less than a few percent) (Wieler, Baur, and Signer, 1986). Estimated solar abundances (gram/gram) of representative isotopes for each of the noble gases are shown in Figure 3.1a. As might be expected from the cosmic abundances of the elements, decreasing abundances are found with increasing mass, a pattern that is especially marked for krypton and xenon.

The term 'planetary' is used to refer to the noble-gas compositions observed in primitive meteorites, such as carbonaceous chondrites. This term arose from recognition of similar patterns of abundance for the noble-gas elements in carbonaceous chondrites and in the Earth's atmosphere (Signer and Suess, 1963), as shown in Figure 3.1b, in which the abundance patterns are presented relative to the primordial abundance of argon (^{36}Ar), normalized to the solar abundances. The term has often been taken to imply that meteoritic (planetary) noble gases were the precursors of the noble gases in our terrestrial atmosphere and possibly other planetary atmospheres. That supposition will be examined in this chapter.

The absolute abundances of the noble gases in carbonaceous chondrites and in the atmospheres of Earth, Venus, and Mars display highly variable depletions with respect to solar abundances (Figure 3.1a). These depletions no doubt are directly related to the processes by which proto-planetary material acquired volatiles from the solar nebula (e.g., O'Neill and Palme, Chapter 1, this volume), as well as the

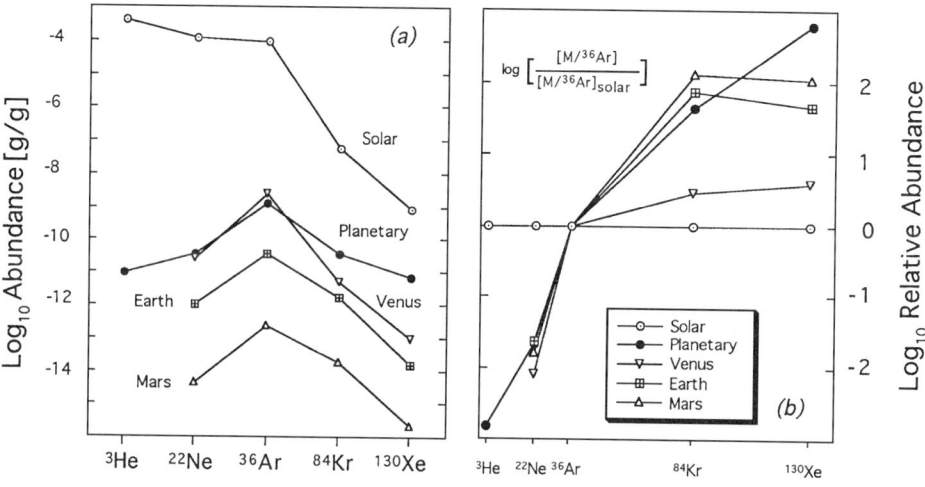

Figure 3.1. Patterns of abundance of noble-gas elements in the atmospheres of the terrestrial planets (Earth, Venus, Mars) and in the solar and planetary components, as compiled by Pepin (1991) using primordial isotopes. Because helium escapes from the planetary atmospheres, helium data for the planetary atmospheres are not shown in the figures. (a) Absolute abundances of noble gases in solar-system matter ('Solar'), in volatile-rich CI chondrites ('Planetary'), and in terrestrial-planet atmospheres. (b) Relative abundances $[M/^{36}Ar]/[M/^{36}Ar]_{solar}$, normalized to ^{36}Ar, with respect to the solar relative abundance, where M is ^3He, ^{22}Ne, ^{84}Kr, or ^{130}Xe.

processes involved in the formation of the atmospheres of the terrestrial planets. From Figure 3.1b it is clear that the planetary (meteoritic) component and the atmospheres of Earth and Mars have quite similar relative abundance patterns for neon, argon, and krypton. The heavier noble gases in carbonaceous chondrites and in the Earth and Mars atmospheres are systematically enriched relative to the lighter noble gases, with similar degrees of fractionation with respect to the solar component (Figure 3.1b), despite the very large absolute and relative depletions of the noble gases compared with solar abundances (Figure 3.1a). Xenon in the Earth and Mars atmospheres is slightly depleted relative to the planetary (meteoritic) pattern. The depletion of xenon in the Earth's atmosphere (the so-called missing xenon) and in the Mars atmosphere could perhaps be explained if it were found that some xenon remains trapped in the cores of the planets (Stevenson, 1985). However, experimental findings on noble-gas solubilities at high pressures (Matsuda et al., 1993) do not appear to support this hypothesis.

The solar and planetary (meteoritic) components also have distinct isotopic characteristics compared with the Earth's atmosphere. Helium, neon, and argon isotopic compositions of these components are listed in Table 3.2. Solar and planetary $^3He/^4He$ ratios are both more than two orders of magnitude greater than the $^3He/^4He$ ratio observed in the present-day atmosphere. However, the $^3He/^4He$ in the Earth's atmosphere has little fundamental significance, because ^3He and ^4He are lost from the top of the atmosphere at different rates. Torgersen (1989) has

Table 3.2. *Isotopic compositions of He, Ne, and Ar in the solar system*

Reservoir	$^3He/^4He$ ($\times 10^{-6}$)	$R/R_A{}^a$	$^{20}Ne/^{22}Ne$	$^{21}Ne/^{22}Ne$	$^{38}Ar/^{36}Ar$	$^{40}Ar/^{36}Ar$
Solar[b]	457	326	13.8	0.0328	0.1825	$\sim 3 \times 10^{-4d}$
Planetary (meteoritic)[c]	143	102	8.2	0.024	0.188	$\sim 3 \times 10^{-4d}$
Earth atmosphere[c]	1.4	1	9.8	0.0290	0.1880	295.5

[a] $R/R_A = (^3He/^4He)_{observed}/(^3He/^4He)_{atmosphere}$; $(^3He/^4He)_{atmosphere} = 1.4 \times 10^{-6}$.
[b] Isotopic compositions of the solar wind (Geiss et al., 1972; Benkert et al., 1993).
[c] From Ozima and Podosek (1983) and references therein.
[d] $^{40}Ar/^{36}Ar$ ratio observed in ureilites (Göbel et al., 1978).

estimated residence times for ^3He and ^4He in the atmosphere of 0.4–0.8 and 0.9–1.8 million years, respectively. However, it is clear that atmospheric helium is a mixture of mantle helium, outgassed mainly from mid-ocean ridges, and crustal helium, enriched in radiogenic ^4He.

The planetary (meteoritic), atmospheric, and solar neon components also have quite different isotopic compositions. Several distinct neon-isotope compositions have been identified in carbonaceous chondrites, including Ne-A ($^{20}Ne/^{22}Ne = 8.2$), Ne-Q ($^{20}Ne/^{22}Ne = 10.7$), and Ne-E ($^{20}Ne/^{22}Ne \approx 0$) (Wieler et al., 1992; Wieler, 1994). Some of these components are considered to be of exotic origin, having been synthesized in galactic nucleosynthetic environments and transported into the primitive solar system in pre-solar carbonaceous dust grains (e.g., Anders and Zinner, 1993; Wieler, 1994). There is continuing debate regarding the origin of these neon components, but in this chapter we shall follow the conventional view that Ne-A is the most significant neon composition in primitive chondrites (Pepin, 1967). The atmospheric component in Figure 3.2 lies between the solar and planetary (meteoritic) components: the $^{20}Ne/^{22}Ne$ ratio in the Earth's atmosphere ($= 9.8$) is 29% lower than the solar ratio ($= 13.8$) and 20% higher than the planetary (meteoritic) ratio ($= 8.2$). Thus, if solar or planetary (meteoritic) components were involved in formation of the primordial Earth, the large differences in their isotopic compositions should allow them to be identified through measurements of neon isotopic compositions in samples derived from the mantle, as will be discussed later.

The isotopic compositions of the essentially primordial argon and krypton components in the Earth's atmosphere differ little from those found in the solar and planetary components. Thus, the ratio of ^{38}Ar to ^{36}Ar in the Earth's atmosphere ($= 0.188$) is only about 3% higher than the solar ratio ($= 0.1825$) – noting that both isotopes are primordial, not augmented by radioactive decay like ^{40}Ar. The krypton isotopic data from lunar samples show relatively large scatter. Nevertheless, solar

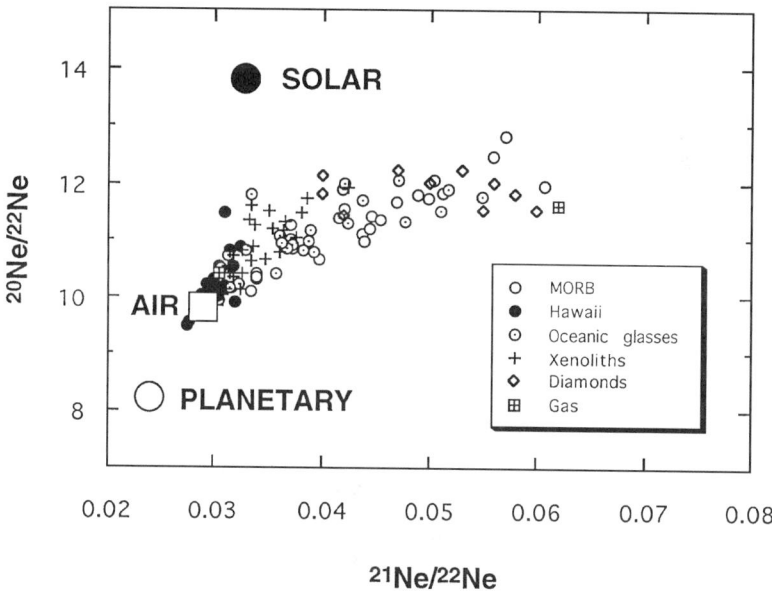

Figure 3.2. Three-isotope plot for neon. Atmospheric ('AIR'), solar, and planetary (meteoritic) values are shown, as well as neon isotopic compositions observed in mantle-derived samples. Data are from mid-ocean-ridge basalt (MORB), Hawaiian basaltic glasses from Loihi and Kilauea, basaltic oceanic glasses from Rocard seamount in the central Pacific, Shona ridge in the South Atlantic, and the Lau Basin, xenolith samples from the islands of Réunion and Samoa, diamonds, and gas samples from the Yellowstone hotspot and CO_2 well gases. Data with neon isotopic compositions that deviate from atmospheric values by more than two standard deviations are shown in the figure. See text for data sources.

krypton appears to exhibit a mass-dependent fractionation toward light-isotope enrichment with respect to the Earth's atmospheric krypton (∼1% per mass unit). On the other hand, planetary (meteoritic) krypton is fractionated such that the heavy isotopes are increased relative to the Earth's atmospheric krypton composition by about 0.3% per mass unit.

In contrast, as shown in Figure 3.3, large variations in xenon isotopic ratios are found. In Figure 3.3, atmospheric and planetary (meteoritic) components are plotted as per-mil differences relative to solar xenon; $\delta_{130} = 1,000 \, (R/R_s - 1)$, where $R = {}^{i}Xe/{}^{130}Xe$ for atmospheric or planetary xenon, and R_s is the same ratio for solar xenon, assumed to be similar in composition to the primordial solar-nebula component (Pepin, 1991). The planetary xenon isotopic composition is based on the bulk xenon composition observed in CI carbonaceous chondrites. The relative enrichment in heavier xenon isotopes in planetary xenon, as compared with solar xenon, is believed to reflect exotic noble-gas components derived from pre-solar grains formed in stellar atmospheres (e.g., Anders and Zinner, 1993). The isotopic composition of xenon in the Earth's atmosphere exhibits a marked mass-dependent fractionation (except for ${}^{129}Xe$) toward heavy-isotope enrichment with respect to solar xenon, averaging about 3.5% per mass unit. The apparent excess of ${}^{129}Xe$

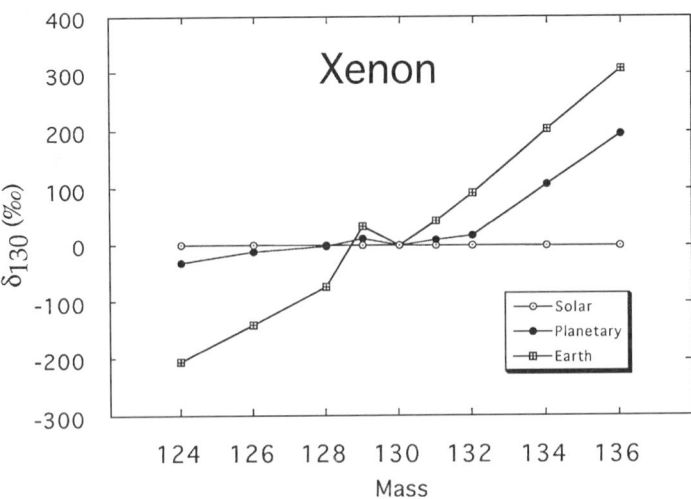

Figure 3.3. Isotopic compositions for planetary, solar, and Earth-atmospheric xenon, shown as per-mil variations of observed isotopic ratios, normalized to ^{130}Xe, in atmospheric and planetary components compared with the corresponding solar ratios. Data compiled by Pepin (1991).

above the level predicted by interpolation of the fractionation pattern is believed to have resulted from decay of the now-extinct radionuclide ^{129}I (half-life of 17 million years), which either was originally present in the Earth or was derived from material accreted at a relatively late stage of Earth formation.

It is important to note that the heavy noble gases (neon, argon, krypton, and xenon) in the Earth's atmosphere appear to be isotopically fractionated with respect to the solar component, with relative enrichment in the heavy isotopes. However, note that the largest relative isotope fractionation is observed for neon, and the next largest fractionation is found for xenon. Argon and krypton show only slight isotope fractionation. In addition, the differences in noble-gas compositions between the atmospheric and the planetary (meteoritic) components are not very systematic. Indeed, the differences among the atmospheric, solar, and planetary noble-gas compositions make it very difficult to understand how the Earth's atmospheric noble-gas compositions could have been derived from either the solar or planetary (meteoritic) compositions by simple processes.

In summary, except for helium, the relative abundances of the noble gases in the Earth's atmosphere are rather similar to the planetary pattern (Figure 3.1). The isotopic composition of atmospheric neon lies essentially between the compositions for the planetary and solar components (Figure 3.2), whereas atmospheric xenon is rather more strongly mass-fractionated than either solar or planetary xenon (Figure 3.3). In contrast, the isotopic compositions of argon (excluding ^{40}Ar, as it is augmented by decay of ^{40}K) and of krypton in the atmosphere and in the Earth as a whole seem to differ little from the solar and planetary compositions.

3.4. Noble Gases in the Mantle

We shall now briefly review the current state of knowledge regarding the evolution of the Earth and its atmosphere from the perspective of noble-gas geochemistry.

3.4.1. Helium

The abundance of helium in the atmosphere is very low (5.2 ppm), owing to its short residence time in the atmosphere. Thus, in analysis of terrestrial samples, contamination by atmospheric helium rarely poses a problem. Consequently, the helium isotopic compositions measured in samples recently derived from the mantle are in most cases considered to be closely representative of their mantle sources.

Helium-isotope measurements in basaltic glass samples dredged from the oceanic basins show at least two distinctive signatures, as summarized in Figure 3.4. The mid-ocean-ridge basalts (MORBs), thought to be derived from the depleted-mantle source, show remarkably uniform $^3He/^4He$ ratios of $1.2 \pm 0.2 \times 10^{-5}$, or $R/R_A = 8.5 \pm 1.4$ (cf. Kurz, 1991), where $R/R_A = (^3He/^4He)_{observed}/(^3He/^4He)_{atmosphere}$, and $(^3He/^4He)_{atmosphere} = 1.4 \times 10^{-6}$. In contrast, the $^3He/^4He$ ratios of intraplate plume-related ocean-island basalts (OIBs) are much more variable, ranging as high as 4.6×10^{-5} (or $R/R_A = 33$) for the Loihi seamount in Hawaii (Kurz et al., 1982, 1983; Rison and Craig, 1983; Honda et al., 1993b), which is located over a 'hotspot', an upwelling plume of material from deep within the Earth. Some OIBs, such as those from Tristan da Cunha and Gough Island (Kurz, Jenkins, and Hart, 1982) and St. Helena (Graham et al., 1992), exhibit $^3He/^4He$ ratios as low as $R/R_A = 5$, that is, less than in MORBs; this is usually attributed to the incorporation of some recycled crustal material in the plume.

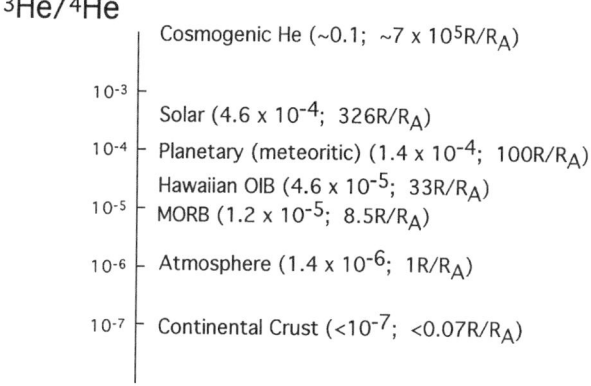

He isotopic ratios

³He/⁴He

- Cosmogenic He (~0.1; ~7 x 10⁵R/R$_A$)
- 10⁻³
- Solar (4.6 x 10⁻⁴; 326R/R$_A$)
- 10⁻⁴ — Planetary (meteoritic) (1.4 x 10⁻⁴; 100R/R$_A$)
- Hawaiian OIB (4.6 x 10⁻⁵; 33R/R$_A$)
- 10⁻⁵ — MORB (1.2 x 10⁻⁵; 8.5R/R$_A$)
- 10⁻⁶ — Atmosphere (1.4 x 10⁻⁶; 1R/R$_A$)
- 10⁻⁷ — Continental Crust (<10⁻⁷; <0.07R/R$_A$)

Figure 3.4. He isotopic compositions in terrestrial materials and some important components in the solar system.

Because there is no appreciable production of ^{3}He by radioactive processes in the mantle (Morrison and Pine, 1955), the ^{3}He observed in oceanic basalt glasses is considered to be of primordial origin, trapped within the primitive materials from which the Earth formed. As previously mentioned, the other isotope of helium, ^{4}He, is continually produced by radioactive decay of uranium and thorium. The variations in observed ^{3}He/^{4}He ratios in oceanic basalts therefore reflect the time-integrated effects of variations in the ratio of primordial helium to uranium and thorium [^{3}He/(U + Th)] brought about by the preferential loss (i.e., outgassing) of ^{3}He with respect to uranium and thorium in the mantle. The MORB-source region in the mantle, with its relatively low ^{3}He/^{4}He ratio, can be regarded as relatively "degassed" and as having a lower ^{3}He/(U + Th) ratio than the plume source, which has been less extensively degassed. The latter, which has retained a greater proportion of its primordial ^{3}He, has a higher ^{3}He/(U + Th) ratio and therefore in general has a higher ^{3}He/^{4}He ratio. Thus, the simple observation that helium isotopic ratios in MORBs are systematically lower than those seen in the Hawaiian OIBs suggests that the outgassing of the atmosphere has proceeded in a heterogeneous fashion, with the MORB source region (depleted mantle) being the primary source for the terrestrial atmosphere.

The ^{3}He/^{4}He ratio in continental-crustal rocks is very low, less than 10^{-7} (Figure 3.4). The ratio is dominated by the production of ^{4}He from decay of uranium and thorium in crustal rocks, but some ^{3}He also is produced by (n, α) reactions on lithium, where the neutrons are derived from uranium and thorium decay (Morrison and Pine, 1955). It should also be noted that at the very surface of the Earth, there is ^{3}He production by exposure to cosmic rays (Kurz, 1986), and this can be significant, especially at high altitude. However, in the global budget, the production of ^{3}He in the crust and by cosmic-ray exposure is negligible.

3.4.2. Neon

It is customary to present neon isotopic compositions in a three-isotope plot: the ^{20}Ne/^{22}Ne-versus-^{21}Ne/^{22}Ne diagram (Figure 3.2). Neon isotopic ratios that differ from the atmospheric values of ^{20}Ne/^{22}Ne = 9.8 and ^{21}Ne/^{22}Ne = 0.029 commonly have been reported from mantle-derived samples: MORB glasses (Poreda and Radicati di Brozolo, 1984; Sarda, Staudacher, and Allègre, 1988; Marty, 1989; Hiyagon et al., 1992); basaltic glasses from Loihi and Kilauea in Hawaii (Sarda et al., 1988; Honda et al., 1991, 1993b; Hiyagon et al., 1992), from Rocard seamount in the central Pacific, east of Tahiti (Staudacher and Allègre, 1989), and from Shona ridge in the South Atlantic (Moreira et al., 1995); back-arc-basin basalts from the Lau Basin (Honda et al., 1993c); ultramafic xenoliths from the islands of Réunion and Samoa (Staudacher, Sarda, and Allègre, 1990; Poreda and Farley, 1992); diamonds (Honda et al., 1987; Ozima and Zashu, 1988, 1991); gases from the

Yellowstone hotspot (Kennedy et al., 1985); CO_2 well gases (Phinney, Tennyson, and Frick, 1978; Caffee et al., 1988).

There are two particularly important data sets: results from MORB glasses and from Hawaiian basaltic glasses. It can be seen in Figure 3.2 that the neon isotopic signatures of the relatively highly degassed (MORB) mantle source and the less extensively degassed (OIB) mantle source are enriched in ^{20}Ne and ^{21}Ne relative to ^{22}Ne, as compared with atmospheric neon, and that they differ from one another. It is also important to point out that the neon data from xenolith samples from Samoa (South Pacific) and Réunion (Indian Ocean), both considered to be plume-related, plot between the Hawaiian and MORB data and that their helium isotopic ratios are slightly higher than the MORB value but lower than the Hawaiian value. We shall discuss these results in more detail in a later section.

3.4.3. Argon

Argon has three stable isotopes (Table 3.1), with ^{40}Ar by far the most important, comprising 99.6% of the argon in the Earth's atmosphere. Indeed, the high abundance of ^{40}Ar in the atmosphere, as compared with the other noble gases, led von Weizsäcker (1937) to infer that the ^{40}Ar had been generated from the decay of ^{40}K in the Earth. Subsequently, of course, that suggestion was confirmed, and the inventory of argon in the atmosphere, together with the estimated potassium contents of the silicate Earth, has provided important constraints on our models for the outgassing history of the Earth (cf. Ozima and Podosek, 1983). Thus, the ^{40}Ar currently in the atmosphere would require a minimum average potassium content of $\sim$115 ppm for the whole mantle, or $\sim$435 ppm in the upper mantle down to the 660-km discontinuity, if it is assumed that all the argon was derived from the upper mantle above that discontinuity. Current estimates of bulk potassium in the silicate Earth are around 250 ppm, based on estimated uranium abundances (20 ppb) (Zindler and Hart, 1986) and K/U $= 12,700$ (Jochum et al., 1983; McDonough and Sun, 1995; O'Neill and Palme, Chapter 1, this volume). Thus it is probable that even the lower mantle (i.e., below 660 km) has been significantly degassed to account for the ^{40}Ar found in the Earth's atmosphere. In this context, it is important to note that on the basis of mass-balance calculations relating to the partitioning of incompatible elements between the crust and the mantle, McCulloch and Bennett (Chapter 2, this volume) estimate the depleted mantle to be approximately $45 \pm 5\%$ of the mass of the total mantle. Because the upper mantle (above 660 km) accounts for only about 30% of the mass of the total mantle, this indicates that a significant portion of the lower mantle has also been depleted, consistent with the conclusion reached on the basis of the K/Ar system.

The high to very high $^{40}Ar/^{36}Ar$ ratios observed in argon extracted from MORB glasses, exceeding 20,000 in some cases (Staudacher et al., 1989), contrast with the $^{40}Ar/^{36}Ar$ ratio in the atmosphere of 295.5. As there is always the possibility of some mixing with atmospheric argon, the $^{40}Ar/^{36}Ar$ for the MORB source generally is thought to be >20,000. There is consensus that this high $^{40}Ar/^{36}Ar$ ratio is attributable to past outgassing of the MORB source, thus greatly reducing the amount of primordial ^{36}Ar; subsequent in situ production of ^{40}Ar from ^{40}K decay has resulted in the very high ratios observed at the present day (Hart et al., 1979, 1985; Allègre et al., 1983, 1987). This observation has given rise to the notion of a well-degassed upper mantle, or depleted upper mantle, from which the continental crust and the atmosphere have been derived (Figure 3.5). As discussed earlier, at least part of the lower mantle is regarded as much less degassed or undepleted

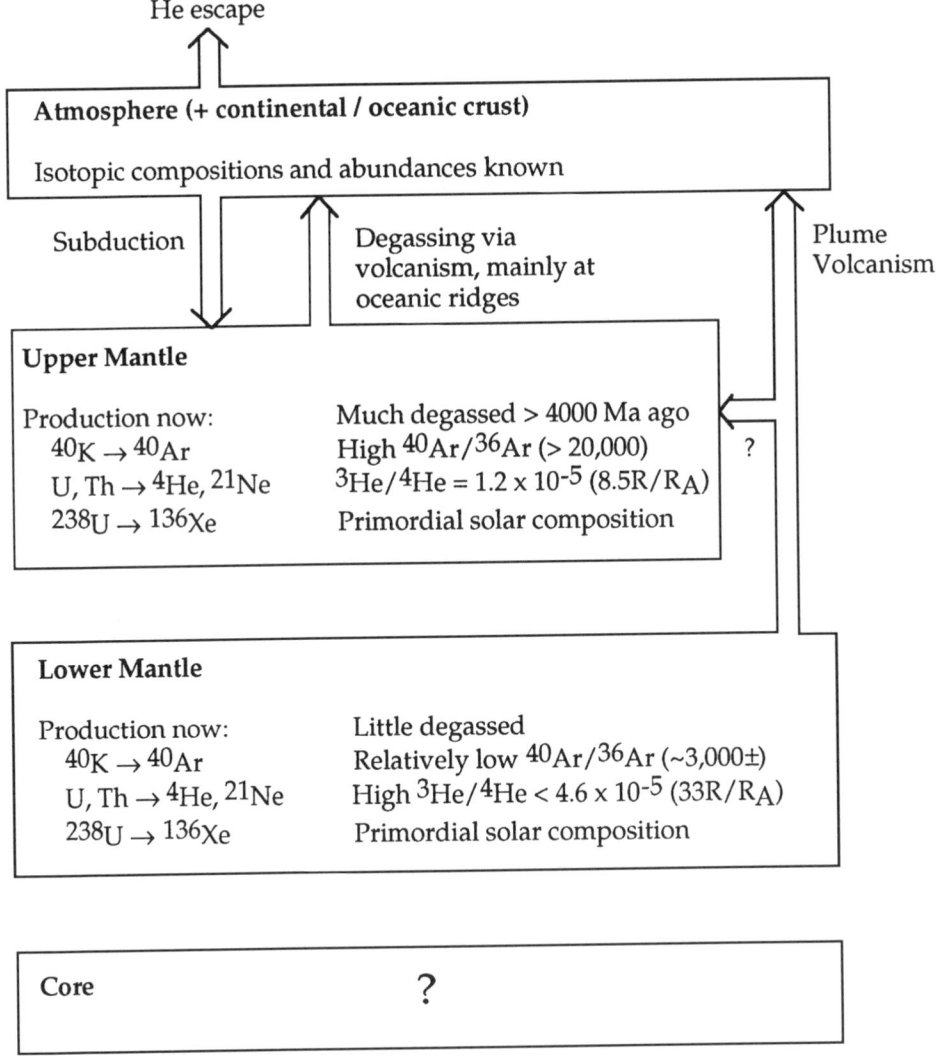

Figure 3.5. Postulated noble-gas reservoirs in the Earth, with some information as to their characteristics and their degassing paths and inventories. (Adapted from Porcelli and Wasserburg, 1995.)

and thus is predicted to have a much lower $^{40}Ar/^{36}Ar$ ratio at the present time than the depleted upper mantle. The discovery of $^{40}Ar/^{36}Ar$ ratios in chilled submarine basaltic glasses from Kilauea and Loihi, Hawaii, that are commonly <500 appeared to confirm that prediction, as these hotspot basalts are regarded as being derived from mantle plume material probably originating in the lower mantle. Although strong arguments have been presented by Patterson, Honda, and McDougall (1990) that the low observed $^{40}Ar/^{36}Ar$ values may in part simply be reflecting atmospheric argon contamination, it remains probable that the lower mantle has a substantially lower $^{40}Ar/^{36}Ar$ ratio than the upper mantle, perhaps ~3,000 (cf. Hiyagon et al., 1992) (Figure 3.5). Thus there is a concordance of evidence from both helium and argon for a layered mantle, an upper degassed part and a lower part that has been less extensively degassed.

As both ^{36}Ar and ^{38}Ar are regarded as primordial, their ratio may well be of considerable value in studies of the evolution of the Earth. In fact, the $^{38}Ar/^{36}Ar$ ratios measured in rocks from the mantle cover a very small range and are essentially indistinguishable from that found in the atmosphere. Interestingly, the $^{38}Ar/^{36}Ar$ ratios in the Earth also are very similar to the solar and planetary values (Table 3.2). The small differences that have been found thus far can be accounted for by minor isotopic fractionation (e.g., Krummenacher, 1970; Kaneoka, 1980).

3.4.4. Krypton

The krypton isotopic compositions in MORBs and OIBs are indistinguishable from atmospheric values within experimental uncertainties (e.g., Ozima, 1994). This might have been anticipated, because, as previously noted, the krypton isotopic composition of the Earth's atmosphere differs little from those of the solar and planetary (meteoritic) components, possible candidates for the Earth's primordial inventory. More accurate measurements of krypton isotopic compositions in mantle-derived samples may be able to identify the presence of solar or planetary krypton in the mantle.

3.4.5. Xenon

Xenon isotopic measurements of mantle-derived samples provide useful constraints on our models for the formation and evolution of the Earth's atmosphere. For example, variations in $^{129}Xe/^{130}Xe$ ratios for mantle-derived samples record isotopic events relating to the early history of the Earth. Note that ^{130}Xe is regarded as primordial and that it has not been augmented by fission or radiogenic products. Some samples show excess ^{129}Xe, relative to the atmospheric value; this is believed to be attributable to radioactive decay of the extinct nuclide ^{129}I (half-life of 17 million years), once present in the Earth. Excess ^{129}Xe has been observed in MORB glasses (Staudacher and Allègre, 1982; Allègre et al., 1983; Marty, 1989;

Hiyagon et al., 1992), ultramafic xenolith samples from the Samoan hotspot chain (Poreda and Farley, 1992), ancient diamonds from Zaire and Botswana (Ozima and Zashu, 1991), and CO_2 well gases from Harding County, New Mexico (Phinney et al., 1978), and Caroline, South Australia (Caffee et al., 1988). Some investigators have argued that the existence of ^{129}I-derived ^{129}Xe in MORB glasses in excess of that observed in the atmosphere requires that the Earth's atmosphere must have separated from the mantle before all the ^{129}I had decayed, that is, within about 100 million years of the formation of the Earth (Staudacher and Allègre, 1982; Allègre et al., 1983). That is another powerful argument in favour of early catastrophic outgassing of at least part of the Earth.

The excess in ^{129}Xe appears to be correlated with excesses in $^{131-136}Xe$ relative to atmospheric xenon. Measured $^{129}Xe/^{130}Xe$ and $^{136}Xe/^{130}Xe$ ratios for the MORBs, Samoan xenoliths, and diamonds are plotted in Figure 3.6. The excesses in $^{131-136}Xe$ are attributed to spontaneous fission-derived xenon, either from ^{238}U (half-life of 4,468 million years) or from an extinct nuclide, ^{244}Pu (half-life of 82 million years). However, because the differences in the fission xenon spectra produced from ^{238}U and ^{244}Pu are small, it is not possible to resolve whether the parent element is ^{238}U or ^{244}Pu or both. Thus, some investigators have argued that the uranium content in the mantle is sufficient to account for the fission-origin xenon observed in mantle-derived samples (e.g., Fisher, 1985; Allègre, Staudacher, and Sarda, 1987). On the other hand, it has been argued that ^{244}Pu was responsible for the fissiogenic xenon, because the data from MORBs (derived from the modern mantle) and from diamonds (derived from ancient mantle) lie on the same

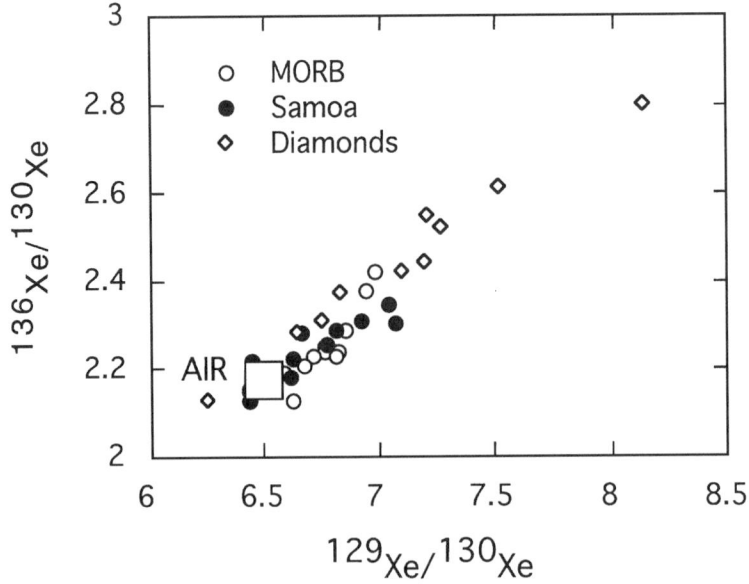

Figure 3.6. Data from MORBs, diamonds, and Samoan xenoliths plotted on a diagram of $^{136}Xe/^{130}Xe$ versus $^{129}Xe/^{130}Xe$. See text for data sources.

correlation line (Figure 3.6). Because fissiogenic xenon produced from ^{238}U accumulates over time in the mantle, the correlation between the samples derived from the modern mantle and the ancient mantle would be unlikely if ^{238}U were responsible for the fissiogenic xenon. Thus, it is argued that the correlation must have been established at an early stage in the history of the mantle by decay of ^{129}I and ^{244}Pu (Ozima and Zashu, 1991).

Xenon studies of the Caroline CO_2 well gas from South Australia showed excess ^{128}Xe, and to a lesser extent ^{126}Xe and ^{124}Xe, with respect to atmospheric xenon (Caffee et al., 1988). That was the first report of anomalies of the lighter xenon isotopes ^{124}Xe, ^{126}Xe, and ^{128}Xe, relative to atmospheric values, in samples presumed to have been derived from the mantle. Because there are no known nuclear reactions to produce all three of these xenon isotopes, the observed excesses of ^{128}Xe, ^{126}Xe, and ^{124}Xe are considered to reflect the presence of solar or planetary (meteoritic) xenon (Caffee et al., 1988). It will be very important to investigate further whether or not similar excesses of ^{128}Xe, ^{126}Xe, and ^{124}Xe can be detected in MORBs and in samples from hotspot volcanics.

3.5. Primordial Noble-Gas Components in the Earth
3.5.1. General Comments

Current views on the origin of the Earth favour the notion that it was formed by accretion from a hierarchy of planetesimals, rather than directly from nebular dust and gas (e.g., Wetherill, 1990). By analogy with the asteroids, the larger of these planetesimals are likely to have differentiated into metallic cores and silicate mantles, so that the Earth probably formed from a rather heterogeneous collection of material (O'Neill and Palme, Chapter 1, this volume). During assembly, as well as subsequently, the Earth differentiated into metallic core, silicate mantle, continental and oceanic crust, hydrosphere, and atmosphere. Today, the Earth remains a dynamic planet, with active convection in the mantle. Instabilities in upper and lower thermal boundary layers are providing the driving force for plate tectonics and mantle plumes (Griffiths and Turner, Chapter 4, this volume; Davies, Chapter 5, this volume). These processes cause considerable stirring and mixing in the mantle. Such activity is likely to have been much more vigorous in the past, owing to liberation of gravitational energy associated with accretion, and because of the much higher level of radioactive heating. An important and fundamental question is whether or not the vigorous convection, recycling, and crust formation have resulted in thorough homogenization of primordial components in the Earth since it formed. In fact, studies of various isotopic systems, especially Sm-Nd, Rb-Sr, U-Pb, and Re-Os, have provided much information relating to the broad questions about the Earth's evolution (e.g., Carson, 1994; Allègre and Schneider, 1994; McCulloch and Bennett, Chapter 2, this volume; Campbell, Chapter 6, this volume), and here we shall focus on the evidence derived from noble-gas geochemistry.

There remains considerable uncertainty as to the composition of the primordial noble-gas components in the Earth, despite the fact that the atmosphere generally is agreed to be a major reservoir of noble gases in the Earth, with the notable exception of helium, which is continually lost from the top of the atmosphere. As previously noted, noble gases in many mantle-derived samples have isotopic compositions differing significantly from those observed in the atmosphere. Some of these differences provide important clues to the degassing history of mantle reservoirs, and other differences yield information on the Earth's primordial components. This is best illustrated by the combined evidence from helium and neon isotopic ratios measured in mantle-derived samples. These findings provide very strong arguments in favour of a primordial noble-gas component in the Earth of solar composition. Because of the importance of such findings, we shall present an overview prior to discussing the implications for the Earth's evolution.

3.5.2. Helium and Neon Isotopic Compositions in Mantle-derived Samples

Craig and Lupton (1976) found neon isotopic compositions in several samples of submarine basaltic glasses and a gas sample from the Kilauea volcano, Hawaii, differing from the atmospheric neon composition by slight enrichments in ^{20}Ne and ^{21}Ne relative to ^{22}Ne. On that basis, they rather tentatively suggested that the Earth may contain a primordial component of solar composition.

Subsequently, many analyses have been made on mantle-derived samples, yielding isotopic compositions ranging from essentially atmospheric to varying, often systematic, enrichments in ^{20}Ne and ^{21}Ne. Well over 100 analyses are available showing these characteristics (Figure 3.2), including data from the following samples: MORBs; OIBs, especially from Hawaii; back-arc-basin basalts; mantle xenoliths; diamonds; additional gas samples from the Yellowstone, Wyoming, volcanic system; and some CO_2 samples from gas wells. In particular, Sarda et al. (1988) published good-quality neon isotopic analyses of gases trapped in MORB submarine glasses, later augmented by the findings of Hiyagon et al. (1992). These MORB data lie on a clear correlation line in the three-isotope neon diagram, passing through the atmospheric neon composition (Figure 3.7). Neon data from plume-related, dredged basaltic glasses from Loihi and Kilauea, Hawaii, also lie on a linear array (Figure 3.7); however, the slope is distinctly different from that of the MORB line. Both linear arrays have slopes readily distinguishable from that expected from single-stage mass fractionation of atmospheric neon by diffusion-related processes, shown as *mfl-a* in Figure 3.7. These linear arrays are interpreted as representing mixing lines between atmospheric neon (probably introduced to basalt magmas by interactions with seawater) and a mantle component, in each case having significantly greater ^{20}Ne/^{22}Ne and ^{21}Ne/^{22}Ne ratios. It is inferred that the mantle sources for MORBs and for the Hawaiian OIBs have distinctive compositions, further enriched in ^{21}Ne and ^{20}Ne, relative to atmospheric neon,

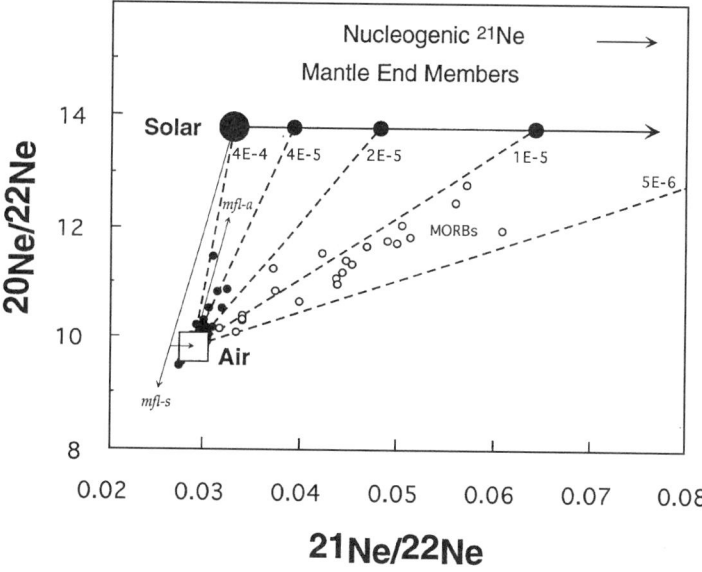

Figure 3.7. Three-isotope plots for neon, with hypothetical mixing lines (dashed), showing hypothetical neon end members in the mantle, comprising a mixture of solar, nucleogenic, and atmospheric neon. By adding radiogenic ^{4}He and corresponding nucleogenic ^{21}Ne, the mantle sources develop lower ^{3}He/^{4}He and higher ^{21}Ne/^{22}Ne ratios than the solar values. The calculated (model) ^{3}He/^{4}He ratios are indicated next to the mixing lines. Accordingly, the slope of a neon mixing line between a mantle end member and present-day atmosphere ('Air') becomes less steep as the ^{3}He/^{4}He ratio decreases. Small open and filled circles are data from MORBs and Hawaiian basalt glass samples, respectively. The mass-fractionation line for atmospheric neon is shown as *mfl-a*. The mass-fractionation line from the solar neon composition is labelled *mfl-s*. The present-day atmospheric neon isotopic composition can be explained by addition of nucleogenic neon (essentially pure ^{21}Ne) to isotopically fractionated solar neon.

as shown diagrammatically in Figure 3.7. Note that the highest ^{20}Ne/^{22}Ne ratios thus far measured in mantle-derived rocks approach, but do not exceed, the solar ^{20}Ne/^{22}Ne value of 13.8. The proposition, now widely accepted, is that the neon data provide strong evidence in favour of a primordial solar component in the mantle and that the mantle reservoirs have different compositions owing to differences in their proportions of ^{21}Ne, produced by nucleogenic processes in the source regions (Figure 3.7). This augmentation of ^{21}Ne is from the reactions ^{18}O (α, n) ^{21}Ne and ^{24}Mg (n, α) ^{21}Ne, the so-called Wetherill reactions (Wetherill, 1954), involving α-particles and secondarily produced neutrons associated with decay of uranium and thorium in the mantle. On this basis, as illustrated in Figure 3.7, it is inferred that the MORB-source region has a neon composition more enriched in nucleogenic ^{21}Ne than that of the Hawaiian OIB source, reflecting a higher $(U + Th)/Ne$ ratio in the depleted-mantle source compared with the OIB source. Thus, we suggest that the neon isotopic compositions in mantle-derived samples are reflecting a three-component system and that therefore we can express any measured neon isotopic composition as the sum of atmospheric, solar, and nucleogenic components.

Indeed, for each neon analysis, it is possible to calculate a ratio between nucleogenic ^{21}Ne and solar ^{22}Ne, designated $^{21}Ne^*/^{22}Ne_s$, particularly useful for comparison with helium isotopic ratios.

3.5.3. Solar Hypothesis

We noted earlier that MORB samples have a rather uniform $^3He/^4He$ ratio of $1.2 \pm 0.2 \times 10^{-5}(R/R_A = 8.5 \pm 1.4)$ and that the $^3He/^4He$ ratio in Hawaiian samples generally is much higher, in the range 2–$4.6 \times 10^{-5}(R/R_A = 14$–$33)$. These helium isotopic ratios are likely to closely reflect the ratios in the mantle source regions for these magmas, as the very low helium abundance in the atmosphere (5.2 ppm) means that contamination by atmospheric helium normally is not a problem. Thus, accepting the helium-isotope ratios at face value, there appears to be a correlation between the slope of the line in the neon three-isotope diagram and the observed $^3He/^4He$ ratio. The steeper the slope of the neon line, the higher the $^3He/^4He$ ratio. This link between neon and helium isotopic compositions provides very strong support for the idea of a primordial 'solar' component in the Earth. The link is expected because variations in the $^3He/^4He$ and $^{21}Ne/^{22}Ne$ ratios in the mantle arise from the time-integrated addition of radiogenic helium ($^4He^*$) and nucleogenic neon ($^{21}Ne^*$) having a unique $^{21}Ne^*/^4He^*$ production ratio of $\sim 1 \times 10^{-7}$ (Kyser and Rison, 1982), as a consequence of both being generated by decay of uranium and thorium present in the source regions. Thus, the in situ growth of $^4He^*$ progressively lowers the $^3He/^4He$ ratio with time, and the products of the decay that produces the helium also generate the $^{21}Ne^*$, as discussed earlier. This means that for a given neon-isotope correlation line, in principle we can predict the $^3He/^4He$ ratio for the same source.

If our hypothesis is correct, and if there were no uranium and thorium in the mantle, we would expect neon from the mantle-derived samples to lie on the mixing line Solar–Air in Figure 3.7, and we would predict a $^3He/^4He$ ratio of 4×10^{-4}, the solar value. But as the mantle contains some uranium and thorium, there is continuing incrementation of nucleogenic neon ($^{21}Ne^*$), as well as radiogenic helium ($^4He^*$), with progressive decreases in the $^3He/^4He$ ratio (Figure 3.7). As will be evident from Figure 3.7, the MORB data fall around the neon correlation line expected for a $^3He/^4He$ ratio of 1×10^{-5}, close to the ratio observed in MORBs, and a similar good correlation is found for Hawaiian OIBs.

As pointed out earlier, for samples with neon isotopic ratios differing significantly from that for atmospheric neon, we can calculate the $^{21}Ne^*$ and solar ^{22}Ne components for each sample. From these calculated $^{21}Ne^*/^{22}Ne_s$ ratios we can then calculate an expected or predicted $^3He/^4He$ ratio in the sample based upon the solar hypothesis (Honda et al., 1993a; Patterson et al., 1994). Figure 3.8 shows a plot of the predicted and the observed $^3He/^4He$ ratios, plotted in the order of increasing $^3He/^4He$. The ratio of calculated to observed values is close to 1, within a factor of about 2, indicating a good degree of concordance with the model. This correlation

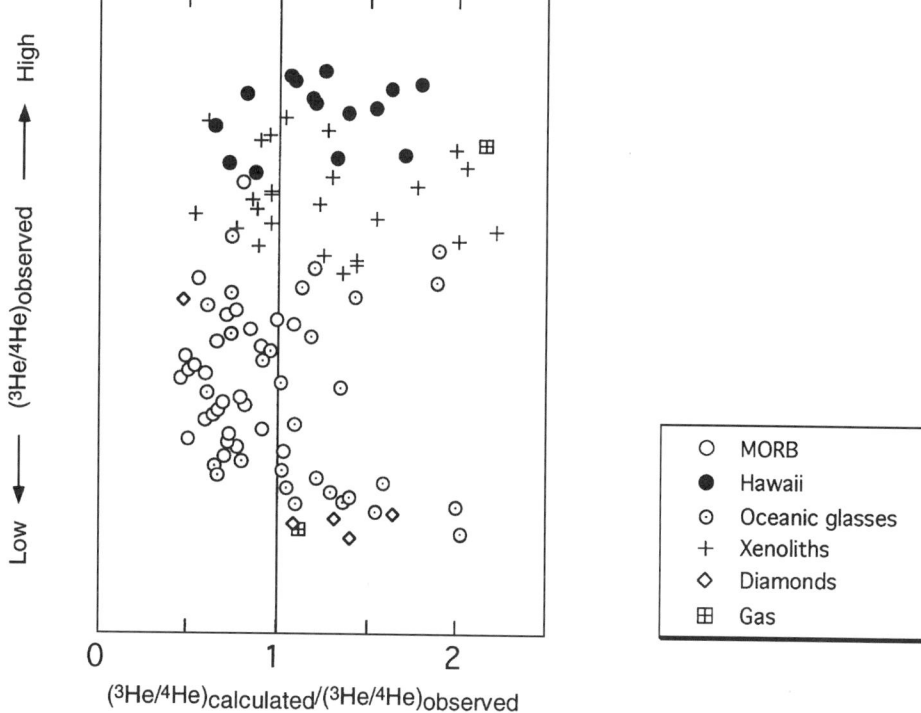

Figure 3.8. Plot of the ratio between calculated $^3He/^4He$ (based on the deconvolved solar-Ne/nucleogenic-Ne ratio) and the observed $^3He/^4He$ in mantle-derived samples. Data are shown in order of increasing $^3He/^4He$ ratios observed in the samples, ranging from 4.2×10^{-6} to 4.6×10^{-5}; the scale on the vertical axis is arbitrary.

of neon and helium isotopic ratios provides great confidence in the hypothesis that the Earth has a primordial solar component.

3.5.4. Solar Noble Gases in the Earth from Interplanetary Dust Particles?

An alternative scenario is that the solar noble gases in mantle-derived samples may have originated from a continuous input of interplanetary dust particles (IDPs), these having been introduced into the mantle sources by subduction (e.g., Anderson, 1993). High $^3He/^4He$ ratios ($>1 \times 10^{-4}$) and $^{20}Ne/^{22}Ne$ ratios (>12.0) have been observed in some deep-sea sediments (Amari and Ozima, 1988; Nier, Schlutter, and Brownlee, 1990), owing to the presence of IDPs. These IDPs are saturated with solar-wind implantations, so that they have very high concentrations of solar noble gases, as high as 0.1 cm^3 (STP) per gram of helium, for example. Thus, a very small input of IDPs (≤ 1 ppm by weight) produced the high helium- and neon-isotope ratios observed in some oceanic sediments. However, the idea of an IDP origin for solar-like helium and neon in the Earth is critically dependent on three assumptions: (1) a sufficient influx of IDPs to the Earth in the past, (2) resistance of IDPs to weathering, alteration, and diagenesis on the seafloor, and (3) sufficient retentivity of solar noble gases in IDPs during subduction to deliver solar-like noble gases into mantle sources. In regard to the first assumption, the absolute abundance of solar noble gases in deep-sea sediments seems to be much too low (Trull, 1994; Farley, 1995). Furthermore, in relation to the third assumption, recent diffusion experiments have shown that the solar helium and neon trapped in IDPs probably would have been lost within a few years at 500°C, and within 10^5 years at 200°C, indicating that diffusive loss from the subducting slab would have occurred at quite shallow depths (Hiyagon, 1994). Thus, the evidence against significant introduction of solar noble gases into the mantle via IDPs and subduction is so strong that we believe that the observations of solar-like helium and neon in mantle-derived samples can most reasonably be explained as primordial in origin, trapped within the Earth at a very early stage of its evolution.

Having established this relationship, we now need to examine the consequences, especially in relation to the evolution of the solid Earth and its atmosphere.

3.6. Reconciliation of Existing Evidence and Constraints on Earth Models

Brown (1952) recognized that the noble-gas abundances in the Earth's atmosphere are very substantially depleted with respect to solar abundances (Figure 3.1a). That led him to postulate that the Earth's atmosphere had not been generated by direct acquisition of a primary atmosphere from the surrounding solar nebula, but rather had derived from extensive outgassing of volatiles from within the solid Earth. Subsequently, several investigators suggested, particularly on the basis of significant differences in $^{40}Ar/^{36}Ar$ ratios between the atmosphere ($= 295.5$) and parts of the mantle ($>10,000$), that considerable outgassing must have occurred at an early stage following the Earth's formation; see, for example, the extensive reviews by Bernatowicz and Podosek (1978), Ozima and Podosek (1983), and Turner (1989). All models of the Earth's degassing based on consideration of the high $^{40}Ar/^{36}Ar$ ratios observed, especially in MORBs, postulate that extensive degassing of ^{36}Ar must have occurred from at least the upper mantle before there had been significant growth of radiogenic ^{40}Ar in the mantle, that is, within about 500 million years of the formation of the Earth. Similarly, as already discussed, the model of early catastrophic outgassing has been further supported by the discovery of excess ^{129}Xe in MORB samples.

Primordial volatiles are still continuously outgassing from the solid Earth, mainly from the mid-ocean-ridge system. Craig, Clarke, and Beg (1975) estimated the current global flux of primordial ^3He to be 4 atoms $\cdot$ cm$^{-2} \cdot$ s^{-1}, or $\sim 2.4 \times 10^7$ cm^3(STP) $\cdot$ yr^{-1}. Thus, the primordial solar ^{22}Ne flux can be estimated to be 6.7×10^6 cm^3(STP) $\cdot$ yr^{-1}, by using the solar $^3He/^{22}Ne$ abundance ratio of 3.6 (Benkert et al., 1993). The total amount of primordial ^{22}Ne outgassed from the Earth over the past 4.5 billion years, using the estimated present-day flux, would be about 3×10^{16} cm^3(STP), which compares with the actual abundance of ^{22}Ne in the atmosphere of 6.7×10^{18} cm^3(STP). Thus, outgassing of primordial noble gases from the Earth, based upon the current flux, has been insignificant compared with the total budget of noble gases in the atmosphere. This provides further strong evidence that outgassing occurred at a much greater rate earlier in the Earth's history, in keeping with the idea that the Earth was a much more active body in the past.

Allègre et al. (1983, 1987), Kaneoka (1983), Hart et al. (1985), and Zhang and Zindler (1989), among others, have contributed to the development of models for the evolution of the Earth's mantle and atmosphere. In these models, essentially three main noble-gas reservoirs generally are postulated: (1) an undegassed (or least degassed) lower mantle that has acted as a relatively closed system for 4.5 billion years and is sampled by mantle plumes, such as those seen in Hawaii, (2) the degassed mantle, the source for MORBs, and (3) the atmosphere plus, to a lesser extent, the continental crust. The present-day atmosphere is considered to have been formed by the nearly complete degassing of the upper (depleted) mantle, which now contains only a small proportion (<1%) (e.g., Allègre et al., 1987; Zhang and Zindler, 1989) of its original inventory of primordial noble gases, together with radiogenic noble gases. Such models imply that the undegassed mantle has elemental and isotopic ratios of the heavy noble gases (neon, argon, krypton, and xenon) similar to those of the atmosphere, provided that account is taken of the augmentation of some isotopes by radioactive decay or nucleogenic processes.

These models were consistent with many of the noble-gas data from mantle-derived samples (particularly from MORBs and submarine basalts from the Loihi seamount, Hawaii) available at that time. The ^{3}He/^{4}He ratios in Loihi basaltic glasses are as high as 4.6×10^{-5} ($R/R_A = 33$), consistently higher than the MORB value of 1.2×10^{-5} ($R/R_A = 8.5$). Similarly, whereas ^{40}Ar/^{36}Ar ratios in MORBs are higher than 10,000, ^{40}Ar/^{36}Ar ratios in Loihi samples typically are found to be only slightly higher than the atmospheric value of 295.5, ranging up to ~500. The low ^{40}Ar/^{36}Ar ratios observed in Loihi samples are close to the value calculated for a completely undegassed mantle reservoir (~400) (Allègre et al., 1987). Neon, krypton, and xenon isotopic ratios in Loihi samples were considered to be indistinguishable from atmospheric values, and the elemental abundances of neon, argon, krypton, and xenon in Loihi samples were found to be close to those in the atmosphere. These observations from MORBs and Loihi samples appeared to provide strong support for the models developed during the 1980s.

However, recent, more precise noble-gas data for mantle-derived samples, particularly from MORBs and Hawaiian samples, appear to conflict with these popular models. The discovery of 'solar' neon, both in MORBs and in Hawaiian samples (e.g., Sarda et al., 1988; Honda et al., 1991) strongly suggests that the neon-isotope compositions in the mantle are different from those in the atmosphere. As discussed earlier, the solar-like noble-gas signature can safely be regarded as primordial. Thus the discovery of solar-like neon in the mantle indicates that a closed-system approximation for the evolution of the Earth's atmosphere from the mantle – a basic premise of previous models – is incorrect. If we accept this powerful new argument, then how has the noble-gas composition of the present-day atmosphere been generated from an initially solar primordial composition?

Thus, an important conclusion arising from the recognition of primordial solar noble gases in the Earth is that the Earth's atmosphere probably has been an open system, at least in part, during its evolution. Generation of the present-day noble-gas composition of the atmosphere from a solar composition seems to have required selective removal of volatiles from the atmosphere and/or delivery of additional volatiles to the atmosphere during the Earth's early history. In the following paragraphs we shall examine these possibilities, in conjunction with current views on the formation of the Earth.

There appears to be a consensus from astrophysical calculations that the Earth grew to its present size within about 10–100 million years after the formation of the Sun. It is thought that during the final stage of its formation, the Earth was extensively melted as a consequence of the release of energy delivered by accreting planetesimals (e.g., Wetherill, 1990). Indeed, the separation of the Moon from the Earth is commonly postulated to have occurred as the result of collision of a Mars-size body with the Earth (Newsom and Jones, 1990); ultimately Ringwood (1986, 1990) became a strong supporter of this view. The energy released by such a giant impact is thought to have been sufficient to produce a magma ocean on the Earth (Benz, Slattery, and Cameron, 1986). As the solubilities of volatiles, including the noble gases, in silicate melts are extremely low, there can be little doubt that an impact responsible for the separation of the Moon from the Earth would have had profound consequences for the Earth's proto-atmosphere, probably causing its wholesale removal. Thus, the Earth's proto-atmosphere, considered to have a solar noble-gas composition, might have been generated at that time, but much of it was lost at the same time.

The postulated giant impact might have been responsible for modifying the solar noble-gas composition of the Earth's proto-atmosphere in the direction of its present-day composition. Hydrogen, a main constituent of the Earth's proto-atmosphere, would have escaped from the Earth's gravity field by virtue of the energy supplied by the impact of a Mars-size body (Benz and Cameron, 1990; Pepin, 1997). Such hydrodynamic flow might have carried with it heavier atmospheric constituents as well. A large flux of extreme ultraviolet (EUV) radiation from the young evolving Sun has been suggested as an alternative energy supply for inducing hydrodynamic escape. In any event, the process of hydrodynamic escape is considered to have been capable of causing isotopic fractionation of noble gases of solar composition to yield compositions like those of the present-day atmosphere (Hunten, Pepin, and Walker, 1987; Zahnle, Kasting, and Pollack, 1990; Pepin, 1991). Because lighter isotopes would have escaped more readily, the noble gases remaining in the atmosphere would have become enriched in their heavier isotopes.

Thus, during hydrodynamic escape, neon in the proto-atmosphere might well have changed its isotopic composition along the mass-fractionation line extending from the solar neon composition (shown as *mfl-s* in Figure 3.7). Under this hypothesis, when hydrodynamic escape ceased, the $^{20}Ne/^{22}Ne$ ratio in the atmosphere would have been essentially at the present-day value of 9.8, which is 29% lower than the value in solar neon. Owing to the addition of nucleogenic ^{21}Ne produced in the Earth over time, and outgassed to the atmosphere, the $^{21}Ne/^{22}Ne$ ratio in the atmosphere has increased progressively to the present value of 0.029. Theoretically, isotopic fractionation of solar argon, krypton, and xenon to yield atmosphere-like compositions would also have been possible (Pepin, 1991).

A further possible mechanism involved in modifying the solar noble-gas composition in the Earth's proto-atmosphere may have been the addition of volatiles of planetary (meteoritic) composition. Thus, during accretion, some of the volatiles may well have been derived from a late veneer of volatile-rich materials such as carbonaceous chondrites (Anders and Owen, 1977; Marty, 1989) or cometary material (Porcelli and Wasserburg, 1995; Owen and Bar-Nun, 1995). In support of that notion, it should be noted that the addition of a late-stage veneer is commonly postulated to account for the mantle abundances of the platinum-group elements (PGEs: Ru, Rh, Pd, Os, Ir, and Pt), S, Se, Te, C, and the heavy halides in the silicate Earth (O'Neill and Palme, Chapter 1, this volume). As indicated earlier (Figure 3.2), atmospheric neon lies between solar neon and planetary (meteoritic) neon in composition and may have been generated by mixing of these two distinctive primordial components. Under this scenario, the solar-like neon probably would have been degassed from the solid Earth, and the planetary (meteoritic) neon could have been derived from a late veneer of meteoritic or cometary material. Perhaps the abundance ratio of argon to neon in the atmosphere could also be explained by mixing of solar and planetary (meteoritic) components. However, that would require a rather ad hoc assumption that the non-radiogenic isotopic compositions of the heavier noble gases (argon, krypton, xenon) in the late-veneer material were close to atmospheric values.

In concluding this section, two important points need to be emphasized. First, the noble gases in the Earth's atmosphere are distinct in their isotopic and elemental compositions compared with solar noble gases. Therefore, because of the large elemental and isotopic differences between the two compositions, derivation of the Earth's atmospheric noble-gas composition from a primordial solar composition would have required the operation of rather complex processes. Second, it is clear that much of the Earth's present-day atmosphere was formed during a short period within a few hundred million years of the formation of the Sun. An important implication of the hypothesis of early degassing is that the postulated two-layer mantle structure, corresponding to the extensively degassed portion of the mantle and the less extensively degassed mantle, would have to have been established during that period. Degassing of the upper mantle usually is linked with the formation of the crust, hydrosphere, and atmosphere. Crust formation undoubtedly continued, albeit at slower rates, subsequently. Processes such as subduction result in atmospheric gases being transported into the mantle, possibly even into the less extensively degassed lower mantle, so that the noble-gas compositions in the present-day mantle have been determined by a variety of partly interrelated processes (Figure 3.5).

3.7. Summary and Conclusions

Noble-gas geochemistry has provided much useful and unique information concerning the origin and evolution of the Earth and its differentiation into core, mantle, crust, and atmosphere/hydrosphere. The atmosphere of the Earth is the major repository of the noble gases, reflecting a high degree of outgassing of the planet. Nevertheless, analyses of mantle-derived samples clearly show that outgassing of the noble gases from the Earth is continuing, albeit at very low rates compared with those that must have prevailed during the early history of the Earth.

There is a concordance of evidence from noble-gas geochemistry and from other isotopic systems that the Earth's crust, atmosphere, and hydrosphere were formed by differentiation from the upper mantle, possibly that part of the mantle above the 660-km seismic discontinuity, but also with a significant contribution from the lower mantle (McCulloch and Bennett, Chapter 2, this volume). The residual upper mantle is considered to be the source of MORB magmas, which have extremely high $^{40}Ar/^{36}Ar$ ratios, believed to average >20,000. Such high $^{40}Ar/^{36}Ar$ ratios provide compelling evidence for profound early degassing of that part of the mantle, certainly within the first 500 million years of Earth history. The presence of high $^{129}Xe/^{130}Xe$ ratios in MORBs, compared with atmospheric ratios, indicates that separation of the bulk of the atmosphere from the mantle by degassing probably occurred within the first 100 million years or so of Earth history. The remarkable coherence and homogeneity of the MORB source in the upper mantle is further reinforced by the concordance of helium and neon isotopic data, and this is also reflected in other isotopic systems. Thus, the upper-mantle source for MORB magmas is an important noble-gas reservoir, characterized by having been extensively degassed at an early stage in Earth history, related largely to the formation of the crust and atmosphere.

The source for plume-related lavas (OIBs, etc.) has contrasting noble-gas characteristics, indicative of a much less extensively degassed mantle reservoir, generally identified as the region below the depleted, degassed mantle. Whether or not that part of the mantle is homogeneous remains a moot point, as does the question of how much recycling has occurred owing to subduction. That relatively undegassed part of the mantle has commonly been postulated to have a noble-gas composition quite similar to that of the present-day atmosphere. That simple proposition needs to be modified, as the isotopic compositions of neon and argon, particularly in OIBs, commonly differ significantly from the atmospheric values.

The neon isotopic compositions found in the noble gases released from OIBs and MORBs differ in systematic ways from atmospheric values and from each other. They show remarkable correlations with helium isotopic compositions. The correlated neon and helium isotopic compositions provide powerful evidence in favour of the notion of a primordial solar component within the Earth, now a widely accepted proposition. However, if the whole Earth began with a solar noble-gas composition, then the derivation of the present-day atmosphere from the mantle by simple outgassing processes poses considerable problems. This dilemma can be resolved by postulating that elemental and isotopic fractionations occurred in the atmosphere as consequences of hydrodynamic-escape processes, associated with intensive EUV radiation from the Sun, and/or the rupture of the Moon from the Earth. Another possible scenario is that volatile-rich material accreting at a late stage in the Earth's formation may have had not a solar but a more planetary (meteoritic) noble-gas composition, so that the atmosphere may be a mixture of solar and planetary components. The helium and neon data are consistent with that hypothesis, and further work (especially on the heavier noble gases argon, krypton, and xenon) with mantle-derived samples should help to distinguish between the possibilities. Studies of this kind ultimately should lead to a fuller understanding of the Earth's origin and evolution, as well as better delineation of the kinds and compositions of the noble-gas reservoirs in the mantle.

Acknowledgements

We wish to thank Kurt Lambeck for his strong and continuing support for the noble-gas laboratory in the Research School of Earth Sciences, especially important during the initiation and setting-up phases when he was director of the school. Reviews of this manuscript by Ian Jackson, Geoff Davies, Geoff Taylor, M. Ozima, and especially John Stone were most helpful.

References

Allègre, C. J., and Schneider, S. H. 1994. The evolution of the Earth. *Sci. Am.* 271:44–51.
Allègre, C. J., Staudacher, T., and Sarda, P. 1987. Rare gas systematics: formation of the atmosphere, evolution and structure of the Earth's mantle. *Earth Planet. Sci. Lett.* 81:127–50.
Allègre, C. J., Staudacher, T., Sarda, P., and Kurz, M. 1983. Constraints on evolution of Earth's mantle from rare gas systematics. *Nature* 303:762–6.
Amari, S., and Ozima, M. 1988. Extra-terrestrial noble gases in deep sea sediments. *Geochim. Cosmochim. Acta* 52:1087–95.
Anders, E., and Owen, T. 1977. Mars and Earth: origin and abundance of volatiles. *Science* 198:453–65.
Anders, E., and Zinner, E. 1993. Interstellar grains in primitive meteorites: diamond, silicon carbide, and graphite. *Meteoritics* 28:490–514.
Anderson, D. L. 1993. Helium-3 from the mantle: primordial signal or cosmic dust? *Science* 261:170–6.

Benkert, J.-P., Baur, H., Signer, P., and Wieler, R. 1993. He, Ne, and Ar from the solar wind and solar energetic particles in lunar ilmenites and pyroxenes. *J. Geophys. Res.* 98:13147–62.

Benz, W., and Cameron, A. G. W. 1990. Terrestrial effects of the giant impact. In: *Origin of the Earth*, ed. H. E. Newsom and J. H. Jones, pp. 61–7. Oxford University Press.

Benz, W., Slattery, W. L., and Cameron, A. G. W. 1986. The origin of the Moon and the single-impact hypothesis. I. *Icarus* 66:515–35.

Bernatowicz, T. J., and Podosek, F. A. 1978. Nuclear components in the atmosphere. In: *Terrestrial Rare Gases*, ed. E. C. Alexander, Jr., and M. Ozima, pp. 99–135. Tokyo: Japan Scientific Society Press.

Brown, H. 1952. Rare gases and the formation of the Earth's atmosphere. In: *The Atmospheres of the Earth and Planets*, ed. G. P. Kuiper, pp. 260–8. University of Chicago Press.

Caffee, M. W., Hudson, G. B., Velsko, C., Alexander, E. C., Jr., Huss, G. R., and Chivas, A. R. 1988. Non-atmospheric noble gases from CO_2 well gases. *Lunar Planet. Sci.* 19:154–5.

Carson, R. W. 1994. Mechanisms of Earth differentiation: consequences for the chemical structure of the mantle. *Rev. Geophys.* 32:337–61.

Craig, H., Clarke, W. B., and Beg, M. A. 1975. Excess [3]He in deep water on the East Pacific Rise. *Earth Planet. Sci. Lett.* 26:125–32.

Craig, H., and Lupton, J. E. 1976. Primordial neon, helium, and hydrogen in oceanic basalts. *Earth Planet. Sci. Lett.* 31:369–85.

Farley, K. A. 1995. Cenozoic variations in the flux of interplanetary dust recorded by [3]He in a deep-sea sediment. *Nature* 376:153–6.

Fisher, D. E. 1985. Radiogenic rare gases and the evolutionary history of the depleted mantle. *J. Geophys. Res.* 90:1801–7.

Geiss, J., Buehler, F., Cerutti, H., Eberhardt, P., and Filleaux, C. H. 1972. Solar wind composition experiments. In: *Apollo 16 Preliminary Scientific Report*. SP-31514.1-14.10. NASA.

Göbel, R., Ott, U., and Begemann, F. 1978. On trapped noble gases in ureilites. *J. Geophys. Res.* 83:855–67.

Graham, D. W., Humphris, S. E., Jenkins, W. J., and Kurz, M. D. 1992. Helium isotope geochemistry of some volcanic rocks from Saint Helena. *Earth Planet. Sci. Lett.* 110:121–32.

Hart, R., Dymond, J., and Hogan, L. 1979. Preferential formation of the atmosphere–sialic crust system from the upper mantle. *Nature* 278:156–9.

Hart, R., Hogan, L., and Dymond, J. 1985. The closed-system approximation for evolution of argon and helium in the mantle, crust and atmosphere. *Chem. Geol. (Isotope Geosci.)* 52:45–73.

Hiyagon, H. 1994. Retention of solar helium and neon in IDPs in deep sea sediment. *Science* 263:1257–9.

Hiyagon, H., Ozima, M., Marty, B., Zashu, S., and Sakai, H. 1992. Noble gases in submarine glasses from mid-oceanic ridges and Loihi seamount: constraints on the early history of the Earth. *Geochim. Cosmochim. Acta* 56:1301–16.

Honda, M., McDougall, I., and Patterson, D. B. 1993a. Solar noble gases in the Earth: the systematics of helium-neon isotopes in mantle derived samples. *Lithos* 30:257–65.

Honda, M., McDougall, I., and Patterson, D. B., Doulgeris, A., and Clague, D. A. 1991. Possible solar noble gas component in Hawaiian basalts. *Nature* 349:149–51.

Honda, M., McDougall, I., Patterson, D. B., Doulgeris, A., and Clague, D. A. 1993b. Noble gases in submarine pillow basalt glasses from Loihi and Kilauea, Hawaii: a solar component in the Earth. *Geochim. Cosmochim. Acta* 57:859–74.

Honda, M., Patterson, D. B., McDougall, I., and Falloon, T. J. 1993c. Noble gases in submarine pillow basalt glasses from the Lau Basin. *Earth Planet. Sci. Lett.* 120:135–48.

Honda, M., Reynolds, J. H., Roedder, E., and Epstein, S. 1987. Noble gases in diamonds: occurrences of solarlike helium and neon. *J. Geophys. Res.* 92:12507–21.

Hunten, D. M., Pepin, R. O., and Walker, J. C. G. 1987. Mass fractionation in hydrodynamic escape. *Icarus* 69:532–49.

Jochum, K. P., Hofmann, A. W., Ito, E., Seufert, H. M., and White, W. M. 1983. K, U, and Th in mid-ocean ridge basalt glasses and heat production, K/U and K/Rb in the mantle. *Nature* 306:431–6.

Kaneoka, I. 1980. Rare gas isotopes and mass fractionation: an indicator of gas transport into or from a magma. *Earth Planet. Sci. Lett.* 48:284–92.

Kaneoka, I. 1983. Noble gas constraints on layered structure of the mantle. *Nature* 302:698–700.

Kennedy, B. M., Lynch, M. A., Reynolds, J. H., and Smith, S. P. 1985. Intensive sampling of noble gases in fluids at Yellowstone: I. Early overview of the data; regional pattern. *Geochim. Cosmochim. Acta* 49:1251–61.

Krummenacher, D. 1970. Isotopic composition of argon in modern surface volcanic rocks. *Earth Planet. Sci. Lett.* 8:109–17.

Kurz, M. D. 1986. Cosmogenic helium in a terrestrial rock. *Nature* 320:435–9.

Kurz, M. D. 1991. Noble gas isotopes in oceanic basalts: controversial constraints on mantle models. In: *Applications of Radiogenic Isotope Systems to Problems in Geology*, ed. L. Heaman and J. N. Ludden, pp. 259–86. Mineralogical Association of Canada Short Course Handbook, vol. 19.

Kurz, M. D., Jenkins, W. J., and Hart, S. R. 1982. Helium isotopic systematics of oceanic islands and mantle heterogeneity. *Nature* 297:43–6.

Kurz, M. D., Jenkins, W. J., Hart, S. R., and Clague, D. 1983. Helium isotopic variations in volcanic rocks from Loihi Seamount and the Island of Hawaii. *Earth Planet. Sci. Lett.* 66:388–406.

Kyser, T. K., and Rison, W. 1982. Systematics of rare gas isotopes in basic lavas and ultramafic xenoliths. *J. Geophys. Res.* 87:5611–30.

McDonough, W. F., and Sun, S.-s. 1995. The composition of the Earth. *Chem. Geol.* 120:223–53.

Marty, B. 1989. Neon and xenon isotopes in MORB: implications for the earth-atmosphere evolution. *Earth Planet. Sci. Lett.* 94:45–56.

Matsuda, J. (ed.). 1994. *Noble Gas Geochemistry and Cosmochemistry*. Tokyo: Terra Scientific Publishing.

Matsuda, J., Sudo, M., Ozima, M., Ito, K., Ohtaka, O., and Ito, E. 1993. Noble gas partitioning between metal and silicate under high pressures. *Science* 259:788–90.

Moreira, M., Staudacher, T., Sarda, P., Schilling, J.-G., and Allègre, C. J. 1995. A primitive plume neon component in MORB: the Shona ridge-anomaly, South Atlantic (51–52°S). *Earth Planet. Sci. Lett.* 133:367–77.

Morrison, P., and Pine, J. 1955. Radiogenic origin of the helium isotopes in rock. *Ann. N.Y. Acad. Sci.* 62:71–92.

Newsom, H. E., and Jones, J. H. (ed.). 1990. *Origin of the Earth*. Oxford University Press.

Niemann, H. B., Atreya, S. K., Carignan, G. R., Donahue, T. M., Haberman, J. A., Harpold, D. N., Hartle, R. E., Hunten, D. M., Kasprzak, W. T., Mahaffy, P. R., Owen, T. C., Spencer, N. W., and Way, S. H. 1996. The Galileo probe mass spectrometer: composition of Jupiter's atmosphere. *Science* 272:846–9.

Nier, A. O., Schlutter, D. J., and Brownlee, D. E. 1990. Helium and neon isotopes in deep Pacific Ocean sediments. *Geochim. Cosmochim. Acta* 54:173–82.

Owen, T., and Bar-Nun, A. 1995. Comets, impacts, and atmospheres. *Icarus* 116:215–26.

Ozima, M. 1994. Noble gas state in the mantle. *Rev. Geophys.* 32:405–26.

Ozima, M., and Podosek, F. A. 1983. *Noble Gas Geochemistry*. Cambridge University Press.

Ozima, M., and Zashu, S. 1988. Solar-type Ne in Zaire cubic diamonds. *Geochim. Cosmochim. Acta* 52:19–25.

Ozima, M., and Zashu, S. 1991. Noble gas state of the ancient mantle as deduced from noble gases in coated diamonds. *Earth Planet. Sci. Lett.* 105:13–27.

Patterson, D. B., Honda, M., and McDougall, I. 1990. Atmospheric contamination: a possible source for heavy noble gases in basalts from Loihi seamount, Hawaii. *Geophys. Res. Lett.* 17:705–8.

Patterson, D. B., Honda, M., and McDougall, I. 1994. Deconvolution of multiple components of neon and helium in mantle-derived samples. In: *Noble Gas Geochemistry and Cosmochemistry*, ed. J. Matsuda, pp. 179–89. Tokyo: Terra Scientific Publishing.

Pepin, P. O. 1967. Trapped neon in meteorites. *Earth Planet. Sci. Lett.* 2:13–8.

Pepin, R. O. 1991. On the origin and early evolution of terrestrial planet atmospheres and meteoritic volatiles. *Icarus* 92:2–79.

Pepin, R. O. 1997. Evolution of Earth's noble gases: consequences of assuming hydrodynamic loss driven by giant impact. *Icarus* 126:148–56.

Phinney, D., Tennyson, J., and Frick, U. 1978. Xenon in CO_2 well gas revisited. *J. Geophys. Res.* 83:2313–19.

Porcelli, D., and Wasserburg, G. J. 1995. Mass transfer of He, Ne, Ar, and Xe through a steady state upper mantle. *Geochim. Cosmochim. Acta* 59:4921–37.

Poreda, R., and Radicati di Brozolo, F. 1984. Neon isotope variations in Mid Atlantic Ridge basalts. *Earth Planet. Sci. Lett.* 69:277–89.

Poreda, R. J., and Farley, K. A. 1992. Rare gases in Samoan xenoliths. *Earth Planet. Sci. Lett.* 113:129–44.

Ringwood, A. E. 1986. Terrestrial origin of the Moon. *Nature* 322:323–8.

Ringwood, A. E. 1990. Earliest history of the Earth–Moon system. In: *Origin of the Earth*, ed. H. E. Newsom and J. H. Jones, pp. 101–34. Oxford University Press.

Rison, W., and Craig, H. 1983. Helium isotopes and mantle volatiles in Loihi Seamount and Hawaiian Islands basalts and xenoliths. *Earth Planet. Sci. Lett.* 66:407–26.

Sarda, P., Staudacher, T., and Allègre, C. J. 1988. Neon isotopes in submarine basalts. *Earth Planet. Sci. Lett.* 91:73–88.

Signer, P., and Suess, H. E. 1963. Rare gases in the Sun, in the atmosphere and in meteorites. In: *Earth Science and Meteorites*, ed. J. Geiss and E. D. Goldberg, pp. 241–72. New York: Wiley.

Staudacher, T., and Allègre, C. J. 1982. Terrestrial xenology. *Earth Planet. Sci. Lett.* 60:389–406.

Staudacher, T., and Allègre, C. J. 1989. Noble gases in glass samples from Tahiti: Teahitia, Rocard and Mehetia. *Earth Planet. Sci. Lett.* 93:210–22.

Staudacher, T., Sarda , P., and Allègre, C. J. 1990. Noble gas systematics of Réunion Island, Indian Ocean. *Chem. Geol.* 89:1–17.

Staudacher, T., Sarda, P., Richardson, S. H., Allègre, C. J., Sagna, I., and Dimitriev, L. V. 1989. Noble gases in basalt glasses from a Mid-Atlantic Ridge topographic high at 14°N: geodynamic consequences. *Earth Planet. Sci. Lett.* 96:119–33.

Stevenson, D. J. 1985. Partitioning of noble gases at extreme pressure within planets. *Lunar Planet. Sci.* 16:821–2.

Torgersen, T. 1989. Terrestrial helium degassing fluxes and the atmospheric helium budget: implications with respect to the degassing processes of continental crust. *Chem. Geol.* 79:1–14.

Trull, T. 1994. Influx and age constraints on recycled cosmic dust explanation for high $^3He/^4He$ ratios at hotspot volcanos. In: *Noble Gas Geochemistry and Cosmochemistry*, ed. J. Matsuda, pp. 77–88. Tokyo: Terra Scientific Publishing.

Turner, G. 1989. The outgassing history of the Earth's atmosphere. *J. Geol. Soc. London* 146:147–54.

von Weizsächer, C. F. 1937. Über die Möglichkeit eines dualen β^- Zerfalls von Kalium. *Phys. Zeitschr.* 38:623–4.

Wetherill, G. W. 1954. Variations in the isotopic abundances of neon and argon extracted from radioactive minerals. *Phys. Rev.* 96:679–83.

Wetherill, G. W. 1990. Formation of the Earth. *Annu. Rev. Earth Planet. Sci.* 18:205–56.

Wieler, R. 1994. "Q-gases" as "local" primordial noble gas component in primitive meteorites. In: *Noble Gas Geochemsitry and Cosmochemsitry*, ed. J. Matsuda, pp. 31–41. Tokyo: Terra Scientific Publishing.

Wieler, R., Anders, E., Baur, H., Lewis, R. S., and Signer, P. 1992. Characterisation of Q-gases and other noble gas components in the Murchison meteorite. *Geochim. Cosmochim. Acta* 56:2907–21.

Wieler, R., Baur, H., and Signer, P. 1986. Noble gases from solar energetic particles revealed by closed stepwise etching of lunar soil minerals. *Geochim. Cosmochim. Acta* 50:1997–2017.

Wieler, R., Etique, P., Signer, P., and Poupeau, G. 1983. Decrease of the solar flare/solar wind flux ratio in the past several aeons deduced from solar neon and tracks in lunar soil plagioclases. *J. Geophys. Res.* 88:A713–24.

Zahnle, K., Kasting, J. F., and Pollack, J. B. 1990. Mass fractionation of noble gases in diffusion-limited hydrodynamic hydrogen escape. *Icarus* 84:502–27.

Zhang, Y., and Zindler, A. 1989. Noble gas constraints on the evolution of the Earth's atmosphere. *J. Geophys. Res.* 94:13719–37.

Zindler, A., and Hart, S. R. 1986. Chemical geodynamics. *Annu. Rev. Earth Planet. Sci.* 14:493–571.

Part Two

Dynamics and Evolution of the Earth's Mantle

4

Understanding Mantle Dynamics through Mathematical Models and Laboratory Experiments

R. W. GRIFFITHS and J. S. TURNER

4.1. Introduction

Many geophysical and geological phenomena of the Earth's crust are now generally agreed to be consequences of thermal convection in the underlying mantle. Direct consequences include the relative motions of the continents, the spreading of the seafloor and formation of new crust, volcanism in its various tectonic settings, much of the Earth's seismic activity, and the magnitude of the observed heat flow through the surface. The flow of the mantle over geological time scales is driven by gravity acting on density differences that result from loss of heat from the Earth's surface and, to a lesser extent, from transfer of heat from the Earth's core to the mantle.

The current understanding of the nature of mantle convection and of the evolution of the Earth is based on a combination of observations made at the surface and deductions arrived at through application of the principles of physics. It depends on knowledge of the physical properties of mantle materials and an understanding of the macroscopic dynamical processes whereby those materials respond to the applied forces coupled to the thermodynamics of the system. This chapter is concerned with the latter part of the problem: the identification and study of various physical and dynamical phenomena in the mantle. In particular, we shall discuss the use of theoretical and experimental fluid dynamics as important contributors to the present state of understanding of the Earth, many aspects of which are outlined in this volume. Our approach is based on the belief that investigation of complex geophysical systems requires rigorous studies of individual processes, or a small number of interacting processes, and their sometimes complex physical consequences, rather than the use of models that attempt to take account of every complication that can be envisaged.

If this method is to be useful, it is essential that the most significant processes be studied and that application of the new insights to the Earth be justified by comparison with the effects of the physically plausible alternatives. The ultimate test of the theoretical and laboratory models must be quantitative comparison of their predictions with observations on the Earth. The strongest models are those that provide new and testable predictions. Many ideas that may at first glance seem qualitatively plausible can be eliminated in this way. The aim is to establish basic principles and to understand the essential processes underlying geophysical phenomena, thereby providing a foundation for further extension and refinement of models as new data become available.

Laboratory experiments have played an important part in answering questions about the behaviour of the mantle, particularly in bridging the gap between the kinematic 'plate-tectonic' description of the motions of the Earth's surface, on the one hand, and a predictive, dynamical understanding of the phenomena, on the other. Typically an investigation proceeds as follows: Observations, combined with prior physical knowledge and intuition, will give some insight into the relevant processes, and a model can then be set up in the laboratory, incorporating a limited number of parameters. Initial exploratory experiments often can demonstrate rather quickly whether or not the process envisaged can occur and what form the motions will take. Not infrequently even the qualitative behaviour provides new physical insights. The most important goals of such experiments are to identify the important dynamic parameters and those factors that can initially be neglected and to point the way toward an appropriate theoretical formulation.

Once the relevant physical parameters and the dominant force balances have been identified, it is possible to develop a predictive theory of the resulting motions. In all the examples given in this chapter, buoyancy forces due to compositional and/or temperature variations provide the driving energy, and viscous forces give rise to an opposing drag, with thermal diffusion playing a lesser (but often important) role. Often, dimensional arguments can take us a long way, with similarity solutions based on the governing equations providing the next step in understanding the temporal or spatial evolution. Such solutions assume that the dynamics remain unchanged and that the motions remain similar in structure (but not in magnitude or scale) through time. These solutions must be tested in quantitative experiments, which not only can provide a check on the original physical concept and choice of parameters but also can yield numerical values for the constants used in the theories. In some cases these constants can be evaluated in no other way. This concept of modelling is quite common in other fields of fluid mechanics, but less so in geophysics.

Returning to the important concept of dynamic similarity, we note that this can be achieved even when there are gross differences in scale or material properties. For example, a model flow in the laboratory in which both the viscosity and the scale are much smaller than in the Earth can be dynamically equivalent to a process occurring in the Earth, provided the two flows involve the same balance of forces. Furthermore, not all of the quantities of the system need to be in the same proportions as in the geophysical case: It is necessary only that the few important dimensionless ratios be similar, so that the systems will fall in the same dynamical regime.

4.2. Some Basic Assumptions and Deductions

Before treating a number of specific problems, we shall consider three basic concepts that explicitly or implicitly enter into every physically realistic discussion of convection in the mantle.

4.2.1. Rheology of the Mantle

Seismic and petrological evidence indicates that the bulk of the mantle is a crystalline solid. However, imposed stresses can produce irreversible deformation or creep. The two flow mechanisms considered most relevant to the mantle are *diffusion creep*, in which the strain rate is proportional to the stress, and *dislocation creep*, in which the strain rate is proportional to a higher power of the stress (Drury and Fitz Gerald, Chapter 11, this volume). Both these behaviours allow arbitrarily large strains, so that solids with these properties have no long-term strength. This ensures that in both cases an *effective viscosity* can be defined for mantle materials on geological timescales (although this 'viscosity' depends on the average stress level, if dislocation creep is appropriate). Hence the mantle is treated as a viscous fluid in analytical and numerical models of mantle convection, and laboratory experiments directly relevant to an understanding of mantle dynamics can be carried out with viscous fluids.

Regardless of the details of the rheology, the effective viscosity is strongly temperature-dependent. Assuming that diffusion creep is the mechanism by which deformation is accommodated, the viscosity η will be of the form

$$\eta = \eta_0 \exp(A T_M / T) \tag{1}$$

where T_M is the melting temperature. For a mantle of olivine, $A = 30$ at the pressures of interest, and $\eta_0 = 10^5$ Pa $\cdot$ s (Turcotte and Oxburgh, 1972; also see Stevenson and Turner, 1979). For $\eta = 10^{22}$ Pa $\cdot$ s (a mean value to the order of magnitude inferred from post-glacial uplift by Lambeck and Johnston, Chapter 10, this volume), $T = 0.77 T_M$, and η changes by an order of magnitude as T/T_M changes by only about 5%. We shall see later that the strong dependence of η on temperature ensures that it adjusts to a value that depends on the presence of mantle convection. That is, the value for this material property is determined (by way of the temperature, and within wide bounds set by the microscopic mechanics of the mantle material) by the dynamics and motions of the mantle. This conclusion contrasts with the view that whether or not mantle convection will occur is predetermined by the viscosity.

4.2.2. Inevitability of Convection in the Mantle

Following the argument put forward originally by Tozer (1965) and restated by Stevenson and Turner (1979), we shall consider the behaviour of the mantle when subjected to a purely vertical temperature gradient, and we begin by assuming that the physical properties are uniform. The stability of such a fluid layer, heated from below or cooled from above, is a classic problem in fluid mechanics, and we quote only the basic results. The onset of convection in this simplest approximation is governed entirely by the Rayleigh number Ra, which is essentially the ratio of the driving force (due to thermal buoyancy, and influenced by diffusion of heat) to the retarding force (due to diffusion of momentum by viscous stresses). For a fluid layer of depth H, with constant kinematic viscosity $\nu = \eta/\rho$ and thermal diffusivity κ,

$$Ra = g\alpha\beta H^4/\nu\kappa \qquad (2)$$

where g is the acceleration due to gravity, α is the coefficient of thermal expansion, and β is the difference between the actual overall temperature gradient (from top boundary to bottom boundary) and the adiabatic temperature gradient. If Ra exceeds a critical value Ra_c, of about 10^3 (the exact value depending on the boundary conditions), then convection will occur. For internal heating at a prescribed flux and cooling from the top boundary, the relevant Rayleigh number can still be defined as in equation (2), except that β is now the (horizontally averaged) superadiabatic temperature gradient that would be required for a conductive steady state given the imposed rate of heat generation.

Rather than try to evaluate Ra in the Earth using the currently poorly known values for the physical properties (β being the source of particularly large uncertainty), we can demonstrate the inevitability of mantle convection by an idealized thermal-evolution calculation based on the strong temperature dependence of viscosity [equation (1)]. Consider again a horizontal layer of thickness H, but now containing a uniformly distributed energy source, representing heating due to decay of radioactive elements. The bottom boundary is supposed to be insulated, and the top is fixed at $T = 0°C$. At time $t = 0$, we suppose that $T = T_0$ everywhere and that subsequently, but before convection occurs, the temperature distribution obeys the diffusion equation (with a source term included) – that is, the heat generated is transported only by conduction.

As discussed in more detail by Stevenson and Turner (1979), the scale and conductivity of the Earth are such that the heat generated cannot escape by conduction alone in the age t_E of the Earth. The diffusion length scale $l \sim (\kappa t)^{1/2}$ is a few hundred kilometres when $t = t_E$, so that a small body can lose most of its heat by conduction as it is generated. However, the much larger model Earth heats up, developing a temperature profile that is fixed at the surface, but with increasing temperature and temperature gradient at all depths. One can define a Rayleigh number at each depth z, based on a depth interval h centred on z (in which the temperature gradient changes by, say, a factor of 2) and the viscosity at the average temperature of that layer. As T increases, η, given by equation (1), rapidly decreases, and, for a great depth h, inevitably a time is reached when Ra (with h now used as the depth scale) near some depth z exceeds the critical value for convection to occur. The subsequent behaviour is for all regions eventually to become convective (except possibly the outermost, highly viscous layer, which is a boundary layer and will be discussed in more detail later). This follows from the fact that any nonconvecting region must continue to heat up, because the rate of conduction is too low to remove the heat generated, and so its viscosity must progressively decrease until the region takes part in the convection. Given the great depth of the mantle and the expected values of the constants in equation (1), a sufficiently low viscosity is reached at sub-solidus temperatures for convective heat transport to become possible before melting occurs at any depth.

4.2.3. Boundary Layers in Convection at High Rayleigh Numbers

The foregoing argument concentrates on the initiation of convection in the interior of a progressively heated mantle. It is clear that the eventual steady state must have a much greater rate of heat transport than can be achieved by conduction and that the corresponding Rayleigh number will be much larger than the critical value.

Two other points are useful in understanding the finite-amplitude flow in the Earth's mantle. Firstly, the viscosity v is very high, effectively infinite, relative to the thermal diffusivity κ, and so the viscous response to a perturbation will be instantaneous relative to the thermal response. Secondly, for large Rayleigh numbers, convective heat transport is much more important than conductive heat transport over most of the depth range (by the factor $uH/\kappa \sim 10^3$, where u is a typical flow velocity such as that of the tectonic plates). Conduction remains important, however, in the thin *boundary layers* through which heat is transported to and from the interior, and which in fact determine the magnitude of the flux that must be carried by the convection in the interior.

Some fundamental predictions can be made on the basis of dimensional reasoning, as follows: Suppose that the flux depends only on the material properties and on conditions very near the boundaries (i.e., the flux is independent of the total depth H). It follows from their definitions that the Rayleigh number Ra and the Nusselt number Nu [the ratio of the actual heat flux to the purely conductive flux down a linear (superadiabatic) temperature gradient between the two boundaries] are related by

$$Nu = cRa^{1/3} \tag{3}$$

because this is the only form that gives a flux independent of H. The constant $c \approx 0.1$, but can depend on the boundary conditions. A phenomenological theory due to Howard (1964) suggests that the conductive boundary layer is inherently unsteady, with cold (or hot) material breaking away intermittently. The mean thickness δ of the boundary layer is such that the Rayleigh number based on δ, Ra_δ say, is just critical ($\approx 10^3$). Thus,

$$Nu = H/\delta = (Ra/Ra_\delta)^{1/3} = 0.1Ra^{1/3} \tag{4}$$

in reasonable agreement with experiment (Turner, 1973a).

Equation (4) allows us to make crude estimates of Ra and η for the mantle. The measured temperature gradient near the Earth's surface is on the order of $20 \text{ K} \cdot \text{km}^{-1}$. Using a (poorly constrained) temperature of $3,500°\text{C}$ at the base of the mantle (Boehler, 1993), at a depth $H = 3,000$ km, an estimate of the overall temperature gradient through the mantle is $1.2 \text{ K} \cdot \text{km}^{-1}$. The conducting upper boundary layer, the lithosphere, is very thin compared with H, and equation (4) implies that $Nu > 10$, hence $Ra > 10^6$. Inserting the values for the depth and other properties[1] in equation (2), we deduce that the average viscosity is less than $\eta \sim 6 \times 10^{22}$ Pa $\cdot$ s. The average viscosity is thus determined by the heat flux and the efficiency of mantle convection. These conclusions, which are based on the assumption of uniform material properties, provide a first approximation to a description of the mantle. As will be seen later, there will be quantitative differences resulting from the temperature and pressure dependences of the viscosity and other material properties, but the basic conclusions remain unchanged.

The foregoing very robust general arguments show that the existence of a heat flux through a boundary of a convecting region inevitably implies that there will be an unstable conductive boundary layer. The two boundary layers at the top and bottom of the Earth's mantle are very different, however, and these differences will be the focus of subsequent sections. Because of the strong temperature dependence of the viscosity, the upper cold boundary layer will be stiff, and this property will affect the motions of the plates and of subducting slabs. If the plates are able to move and sink sufficiently rapidly, as is apparently the case for the present-day oceanic lithosphere, then they represent the unstable boundary layer. On the other hand, it is possible that a surface layer can be so viscous (or strong) that it will be stable and will not take part in the underlying convection, as has been suggested to be the case on Venus over the past 500 million years (Schaber et al., 1992; Solomatov and Moresi, 1996). The behaviour in a system with a very viscous, nonconvecting upper boundary layer (a problem that is also relevant in the dynamics of cooling magma chambers) has been addressed through laboratory experiments by Davaille and Jaupart (1993, 1994).

Because the Earth as a whole, including its core, is cooling, there will be a heat flux out of the core and into the base of the mantle, estimated to be on the order of 10% of the Earth's total surface heat flux (Davies, Chapter 5, this volume). The resulting boundary layer of hot, less dense, less viscous material behaves quite differently from the plates produced by surface cooling and can give rise to upwelling plumes (as discussed later). In addition, if there are any internal density interfaces in the mantle separating distinct convecting layers, then boundary layers must form on each side of such interfaces.

---

[1] The values substituted into equation (2) are $\alpha = 3 \times 10^{-5} \text{ K}^{-1}$, $\kappa = 10^{-6} \text{ m}^2 \cdot \text{s}^{-1}$, $\rho = 3 \times 10^3 \text{ kg} \cdot \text{m}^{-3}$, and $\beta = 0.9 \text{ K} \cdot \text{km}^{-1}$. Remember that by definition β is the difference between the overall temperature gradient over the whole depth, with or without convection, and the adiabatic gradient of $0.3 \text{ K} \cdot \text{km}^{-1}$. In the convecting region the gradient will, of course, be much closer to the adiabatic value.

Other questions that have begun to be addressed as a theoretical understanding of mantle convection has developed, particularly during the past decade, and as relevant observations and data have been identified, concern the extent to which the fluid ejected from each boundary layer mixes with the convecting interior and the effects of phase transformations on the convective flow. Can material arising at one or both boundary layers remain relatively unmixed until it reaches the opposite boundary? The answer is clearly yes in the case of upwelling plumes, which are believed to be the causes of surface phenomena such as chains of intraplate volcanoes, uplift of the seafloor surrounding hotspots, continental flood basalts, and oceanic plateaux (Davies, 1988a; Campbell, Chapter 6, this volume). In the opposite direction, the surface boundary layer (descending plates) could affect the dynamical processes at the core–mantle boundary in some significant way. In that case, each boundary layer can influence the chemical distributions at the other boundary (Campbell and Griffiths, 1992).

4.3. Upwelling Thermals and Plumes

We now turn to a discussion of models of specific convective processes in the mantle, starting from the core–mantle boundary (CMB) and working upwards. First we need to consider the implications of a heat flux through the CMB itself. It is also useful to keep in mind the application of these same concepts to convection arising at an internal interface, heated from below.

4.3.1. Initiation of Convection at the Base of the Mantle

There is a large density difference between the core and the mantle. The best estimates for the temperatures of the outer core and the lowermost mantle (the latter from extrapolation of the upper-mantle temperature adiabatically to the CMB) (Boehler, 1993) indicate that there is also a large temperature difference (approximately 1,300 K), so that there is a conductive heat flux from the core to the base of the mantle. This temperature drop must occur across a thermal boundary layer. There is direct seismic evidence for a spatially inhomogeneous boundary layer, the so-called D″ layer, above the CMB that in places is a few hundred kilometers thick. In this layer, the seismic-velocity-versus-depth gradients are anomalously small, but tomography has revealed large lateral variability (Kennett and van der Hilst, Chapter 8, this volume). Although there may be significant compositional variations within the D″ layer, it is likely that it also contains the thermal boundary layer.

Because of the strong temperature dependence of the effective viscosity, there will be a gradient of viscosity through the boundary layer at the bottom of the mantle, with a minimum at the CMB. This reduced viscosity will enhance the flow of the boundary-layer material into any region that has begun to break away from the boundary and convect upwards (as discussed in Section 4.2). Stacey and Loper (1983) developed an analysis of this lateral flow, assumed to be steady, and showed that it will be concentrated in a rheological boundary layer that will be much thinner than the thermal boundary layer and that the lateral flow can be replaced by a slow subsidence of the overlying mantle. Davies (1990a, 1993) combined heat-flux estimates (discussed later) with this theory to deduce the thicknesses of the two boundary layers. As a result of the viscosity variation, the temperature of the rising plume material will be strongly weighted towards the highest temperature in the thermal boundary layer. However, Griffiths and Campbell (1990) noted that the temperature of the source material for the plume might be much less than that of the core, because a thin, *gravitationally stable* conductive layer could persist between a partially miscible or reactive mantle and the much denser core. Such a dense stable layer (not to be confused with either the unstable boundary layer or the D'' layer) would support a large temperature drop without taking part in the boundary-layer convection.

In this picture, each plume draws boundary-layer material from a horizontal area determined only by the separation of unstable convective events. Presumably, if plumes are too far apart, perturbations of the boundary layer between plumes will grow to large amplitude, and a new plume will develop. There is as yet no prediction for that separation distance, and hence no prediction for the mean heat and buoyancy flux in each plume. However, we do know from theoretical stability arguments and a variety of experiments that the mean separation distance between plumes will not be related to the overall depth of the convecting layer.

Experiments have identified three basic forms of flow that may occur once convection has begun. During an initial period the flow is transient and forms a mushroom-shaped 'starting plume' consisting of a large head and narrow tail. Under some conditions the head may become cut off from the source boundary layer to form an isolated 'thermal'. Thermals are common in both experiments and numerical solutions of very viscous high-Rayleigh-number convection with uniform viscosity, forming when flow sweeps away the feeding conduit or when nearby instabilities on the boundary layer remove the supply of heat. However, it is unclear whether or not they can form in chaotic convection with large viscosity variations. The flow can reach a steady state (except for distortions due to surrounding motions) in which heat will be transported across the convecting layer by flow up a narrow conduit. Such a conduit can eventually shut down if neighbouring plumes carry away most of the heat flux. These limiting forms of convection are explored next.

4.3.2. Isolated Thermals

When a volume of buoyant fluid breaks away from the boundary layer, the resulting structure is known as a *thermal*, because of the superficial resemblance to the turbulent atmospheric thermals sought by birds and gliding enthusiasts to provide lift. During ascent of a thermal, heat can spread and warm up the surrounding cooler material (by conduction in the case of the extremely viscous mantle). However, as a result of the accompanying buoyancy, the heat will not be lost from the convecting region.

Consider first, for comparison, the case of a bubble of fluid for which the buoyancy is a consequence of an essentially nondiffusive property (such as a compositional difference in the mantle), in which case the volume of the less dense fluid will remain constant. It can be shown that the bubble will become spherical and that the velocity of rise U for a bubble of volume V and diameter D is given by the Stokes law:

$$U = \left(\frac{B}{2\pi D \eta_\infty} \right) f(\eta/\eta_\infty) \tag{5}$$

where $B = g \Delta \rho V$ is the total buoyancy, and η_∞ is the viscosity far from the bubble. This expresses the balance between the buoyancy force (generated by the action of gravity on the small density difference $\Delta \rho$) and the viscous drag. The last factor, which depends on the ratio of viscosities inside and outside the bubble, tends to $f = 1$ when this ratio is small. The outer viscosity always has the dominant effect on the rate of ascent.

Theory and laboratory experiments have shown that the same relation [equation (5)] applies to the rise of thermals in which the density difference $\Delta \rho = \rho_\infty \alpha \Delta T$ (where ρ_∞ is the environment density, and α is again the coefficient of thermal expansion) is due to a temperature difference ΔT, despite the effects of the conduction of heat (Griffiths, 1986a,b). Assuming that no heat is lost from a thermal during its ascent and that the thermal expansivity is constant, the conservation-of-heat constraint implies that the buoyancy B (where, in this case, $B = g\alpha \int \Delta T \, dV$) is conserved. This assumption is reasonable, because as heat diffuses outward into a thin boundary layer of thickness $\delta \sim (\kappa D/U)^{1/2}$ around the thermal, the newly heated layer will become buoyant (and less viscous) and will be drawn into the moving region, thus increasing its volume V. The inward volume flux due to this process of *thermal entrainment* is of order $dV/dt \sim UD\delta$, and the overall flow is characterized by a Rayleigh number $Ra_\mathrm{T} = B/\kappa \nu_\infty$, where ν_∞ is the kinematic viscosity of the environment.

A solution for self-similar flow can be derived using the aforementioned entrainment flux, conservation of buoyancy, and the velocity [equation (5)] (Griffiths, 1986a,b), and it predicts the diameter D and height of rise z (above a virtual source at the point $z = 0$, where $D = 0$ and $t = 0$) as functions of time t:

$$D = C Ra_\mathrm{T}^{1/4} (\kappa t)^{1/2} \tag{6}$$

and

$$z = (f/\pi C) Ra_\mathrm{T}^{3/4} (\kappa t)^{1/2} \tag{7}$$

where C is a constant of order unity. The value of this 'similarity constant' C could be predicted using numerical simulations capable of resolving the details of the flow within the boundary layer. More simply, it can be evaluated from experiments. Combination of (6) and (7) shows that the diameter increases linearly with height:

$$D = 2\varepsilon z \tag{8}$$

with a half-angle of spread $\varepsilon = (\pi C^2/2f)Ra_T^{-1/2}$, which is smaller for larger Rayleigh numbers. Hence the thermal enlarges less (before reaching a given height) for a larger temperature difference or smaller outer viscosity. The requirement that $\delta \ll D$ implies that the analysis applies to cases where $Ra_T \gg 1$. An additional internal source of heat from radioactivity can readily be included in the analysis (Griffiths, 1986c).

Experiments in which known volumes of heated viscous oil were injected into a cooler environment of the same oil showed that the behaviour is well described by equations (6)–(8). Fitting both equations (7) and (8) to the data, the similarity constant was found to be $C = 1.0 \pm 0.4$. It is important to emphasize that this laboratory value of C will be applicable to thermals in the mantle, provided the underlying assumptions are satisfied, and it allows predictions of ascent speed and plume properties. The model also allows predictions of the particle motions around these viscous, entraining thermals, and the shapes into which passive tracers are moulded by the flow compare well with those found in experiments (Griffiths, 1986b). At Ra_T greater than the moderate value of 200, the internal circulation forms a torus into which all the material originally in the thermal is eventually advected (Figure 4.1). The temperature distribution is not calculated and is not important for the overall evolution of the flow. However, because the entrainment process relies on conduction, it is clear that the heat is distributed more widely through the surrounding spherical volume.

4.3.3. Starting Plumes

When a steady flux of buoyancy is suddenly supplied at the base of a region of viscous fluid (by heating the boundary or by injecting hotter fluid), it will produce a nearly spherical volume of buoyant fluid that will grow slowly until it becomes large enough to leave the boundary. As such a spherical volume rises, it remains attached to the source by a cylindrical conduit through which buoyant fluid continues to flow, thus increasing the buoyancy and volume of the plume 'head'. This effect was first demonstrated by Whitehead and Luther (1975) using compositional buoyancy. Analysis of the important case of a starting plume driven by thermal buoyancy (Griffiths and Campbell, 1990) involved only a simple modification of the theoretical treatment for isolated thermals.

When the buoyancy is produced by heating, and the fluid has a viscosity that is highly temperature-dependent, flow in the conduit (i.e., the hot plume tail) is facilitated by the decrease in viscosity, whereas the motion of the head is again controlled by the larger outer viscosity. Conduction around the head again leads

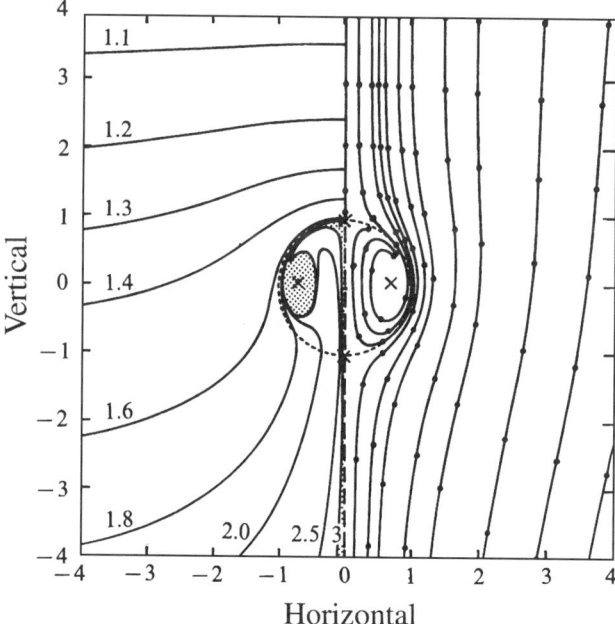

Horizontal

Figure 4.1. Particle paths (right) and deformation of material surfaces (left) near a thermal with Rayleigh number $Ra_T \approx 1{,}736$, relative to a frame of reference that is expanding with the diameter of the thermal. All fluid initially lying in a cone above the thermal (and bounded by a dividing stream surface) is eventually heated and entrained into the thermal. The material surfaces illustrate the large vertical displacement of surrounding fluid that does not form a part of the warm thermal. (From Griffiths, 1986b, with permission.)

to warming and entrainment of surrounding fluid. The foregoing description for isolated thermals can be modified to take into account the plume head's increasing buoyancy with time due to the source flux, and the increase in volume due to both the source flux and entrainment.

A satisfactory theoretical model can be based on the assumption that only the total buoyancy is relevant for the dynamics. Thus, an average temperature anomaly for the plume head can be defined that is less than the temperature of the source fluid arriving at the top of the plume and varies with time, providing a useful measure of the degree of entrainment and cooling. The evolution of the plume head is now governed by the heat-conservation relation

$$\Delta T V = q_0 \Delta T_0 (t - t_0) \tag{9}$$

where V is the volume, q_0 is the source volume flux, and ΔT_0 is the source temperature anomaly. For large times, when entrainment becomes important, the solutions for the diameter D, velocity U, and temperature anomaly ΔT of the head have the asymptotic forms

$$D \sim z^{3/5}, \qquad U \sim z^{1/5}, \qquad \Delta T / \Delta T_0 \sim z^{-1} \tag{10}$$

where the constants of proportionality are functions of the plume Rayleigh number, defined in this case by $Ra_p = g\alpha \Delta T_0 q_0^3 / \kappa^4 \nu_\infty$ (Griffiths, 1991). Note that in deriving the foregoing solution we did not need to make any specific assumptions about the form of the profiles for velocity or temperature either in the feeding conduit or in the plume head. The essential assumption is that these profiles remain similar as the flow develops; the use of a mean temperature does not require an assumption that the temperature is constant across the plume. But in each of the relations such as (10) there is a numerical multiplying factor or 'similarity constant' that depends on the real profiles and that has been evaluated experimentally. It might also be found through numerical models.

Experiments are again useful in testing the qualitative and quantitative forms of the theoretical solutions. These can be particularly simple in concept, because the behaviour of the head is insensitive to the source geometry, and so plumes arising from gravitational instabilities at a heated boundary layer can also be satisfactorily represented in a laboratory configuration involving a small inlet. A photograph of a hot starting plume in the interior of a laboratory tank is shown in Figure 4.2. This was produced by injecting hot, dyed syrup at a steady rate into the same (but cold and very viscous) syrup (Griffiths and Campbell, 1990). There was little cooling of the fluid flowing up the conduit until it arrived at the forward stagnation point of the rising head, where it met the resistance of the overlying fluid. There it spread laterally and axisymmetrically as a sheet, facilitating more efficient heat transfer to a boundary layer in the surrounding fluid, which thenceforth became part of the plume head. After the head had ascended a large distance, a continuous axisymmetric spiral of dyed material extended inward to a toroidal focus.

Although experiments have gone some way towards determining the similarity constant, there is still considerable uncertainty in its value (having to do with the rate of incorporation of external fluid into the rising plume head). Experiments with continuously fed plumes (Fitzpatrick, 1991) have shown that departures from self-similarity during the ascent (due to viscosity changes in the head, a finite volume in the conduit, temporal changes in the head shape, and side-wall effects) make it difficult to determine the constant to better than a factor of 2. However, given that uncertainty, the result is consistent with that for thermals ($C \approx 2$) and is sufficiently robust to allow some firm predictions about the scale and ascent rate of starting plumes, or large plume heads, in the mantle. The quantitative applications to the mantle (summarized in Section 4.4) are also consistent with recent computer modelling results (Davies, 1995a; Farnetani and Richards, 1995) and with a range of geophysical data. Modifications of the predicted plume behaviour in a mantle of power-law rheology have been computed by Weinberg (1997), indicating that plume heads can ascend more rapidly than predicted for a Newtonian mantle and can reach farther into the base of the lithosphere, but that the extent of entrainment and head size will not be greatly changed.

The distribution of source fluid in the plume head, as seen in Figure 4.2, does not parallel the temperature distribution, which we can safely assume will be much more smoothly distributed throughout the bulk of the plume head as a result of the

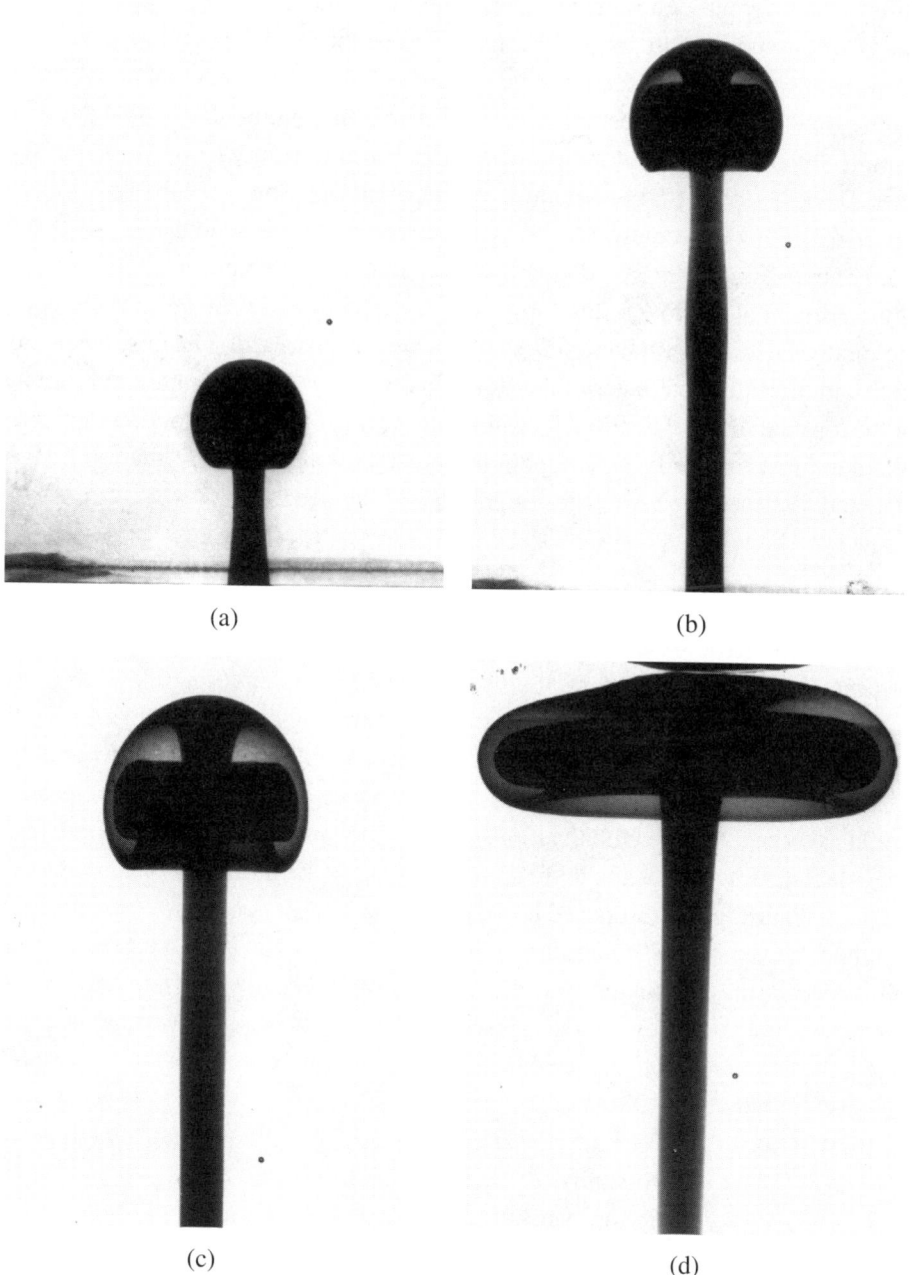

(a)   (b)

(c)   (d)

Figure 4.2. Photographs of a laboratory starting plume. Following its departure from the source region, it grows both by continuing addition from the source through the conduit and by entrainment, and eventually it spreads beneath the free surface. The source fluid is dyed. The temperature distribution is not seen. Note that it is not asserted that this experiment scales directly to mantle conditions; it is one of a series of experiments designed to test the theoretical similarity solution. However, it illustrates the predicted qualitative nature of newly forming mantle plumes; beyond that, it is necessary to use the theory referred to in Section 4.3 to make predictions for various mantle conditions. (From Griffiths and Campbell, 1990, with permission.)

nature of the thermal-entrainment process and continuing conduction within the head. Griffiths and Campbell (1990) concluded that the axial conduit and the radial outflow near the top of the head will be almost as hot as the source, that there may be some small remnant temperature maximum near the toroidal focus, and that the remainder of the source and entrained material in the head will be significantly cooler. Davies (1995a) and Farnetani and Richards (1995) have computed temperature distributions that confirm these ideas, in particular the conclusion that there are only small temperature gradients everywhere, except around the axial conduit and the horizontal outflow at the top of the head. They have shown that the coupling of advection and conduction is so effective at redistributing heat between the source and the entrained material that there is only a small temperature maximum near the toroidal focus of the flow. The hot outflow layer (and axial conduit) appears as the dominant feature in the temperature plots, covering much of the top of the plume and disappearing toward the outer edge of the head.

4.3.4. Long-Lived Plumes

When the head of a new plume reaches the top of the layer through which it is rising, and if the source flux is constant, the flow in the conduit delivering material from the source tends towards a steady state. If the surrounding fluid is otherwise at rest, the conduit will be vertical, and there will be little contamination by surrounding fluid from outside the bottom boundary layer, because a slow inflow from the surroundings will not mingle with the rapid vertical flow inside the conduit (Loper and Stacey, 1983). However, when there are larger-scale systematic motions in the surroundings, such as a superimposed horizontal shear flow or convective overturning, as might be driven by movement of tectonic plates, the conduit will be bent over in the direction of the horizontal flow. Skilbeck and Whitehead (1978) demonstrated this phenomenon and showed that when the angle of tilt from the vertical is large enough, the conduit becomes unstable and breaks up into a series of isolated blobs or diapirs[2] (Whitehead, 1988).

The relationship between the shear velocity and the tilt was examined, both theoretically and experimentally, by Richards and Griffiths (1988). They found that a satisfactory description can be given in terms of a vector addition of the horizontal advection velocity and the vertical Stokes velocity for each element of the conduit, in the form of equation (5), but with a different multiplicative constant. For example, with a linear shear profile and a conduit of fixed diameter and density, the conduit path would be parabolic. In an application to the mantle (which follows in Section 4.4), the plume source is taken as moving much more slowly than the surface plates, consistent with the apparent near-fixedness of oceanic hotspots relative to

---

[2] These authors suggested such conduit instability as an explanation for the formation of the discrete volcanic centres (islands) along chains of volcanic islands. However, this explanation seems unlikely to be correct, because more recent estimates of the buoyancy fluxes and temperature anomalies of plumes (see Section 4.4) indicate that the distance between diapirs would be much greater than the separation of islands.

one another (Morgan, 1971; Duncan and Richards, 1991) and with a plume origin deep in the lower mantle, where the convection velocities are believed to be lower by a factor of 10 or more because of an increase in viscosity with pressure (Gurnis and Davies, 1986b).

A consequence of the tilting of plume conduits is enhanced entrainment of material from the surroundings. While hot material flows upward along the conduit, each part of an inclined thermal-plume conduit must be rising through its surroundings (and continually displacing its surroundings upward) and therefore must contain a circulation in planes normal to the axis of the conduit – the quasi-two-dimensional equivalent of that shown in Figure 4.1 for the axisymmetric plume head. Surrounding material is again heated in a boundary layer around the rising cylindrical region and is drawn into it, thus increasing the volume flux in the conduit. The source material is concentrated into two cores, leaving a central strip that is relatively free of source fluid (which in the experiments is distinguished by dye) (Richards and Griffiths, 1989). Self-similar solutions for such flows (Griffiths and Campbell, 1991a) predict that entrainment will have a much greater effect on bent-over plumes when the temperature-dependent viscosity in the plume is allowed to increase with distance from the source because of entrainment and cooling, for in order to cope with the imposed buoyancy flux, the diameter of the conduit must then increase (with height) as the plume cools. Such solutions also predict that, as for starting plumes, the behaviour will be a function of the plume Rayleigh number $Ra_p = g\alpha\Delta T_0 q_0^3/\kappa^4 \nu_\infty$, where q_0 is the source volume flux. A plume with a greater buoyancy flux will be less tilted and will entrain a volume flux from the surroundings that will be smaller relative to the source volume flux (i.e., it will be less contaminated).

Experiments have been conducted to investigate the adjustment of plume conduits to a change in plate motion, such as the change in the velocity of the Pacific plate at 43 Ma (Griffiths and Richards, 1989), and, more recently, to examine the consequences of a plume rising beneath a spreading ridge (Feighner and Richards, 1995). In the first of these investigations it was concluded that the small radius of curvature of the bend in the Hawaiian–Emperor chain implies an upper limit of order 200 km on the extent of horizontal deflection of the underlying plume, a figure consistent with the prediction from the similarity solution for a plume having such a large buoyancy flux. In the second investigation, a source of compositional buoyancy was located a few centimetres under the surface of a tank of viscous fluid, vertically below the boundary between two plastic sheets running over rollers and in contact with the syrup. Drawing the sheets apart in a horizontal plane modelled the spreading at a ridge. When the plume reached the upper boundary, it first spread axisymmetrically and at the same rate as under a fixed plate. Once the spreading pool reached a certain size, a 'waist' developed at the 'ridge' and maintained a constant width W along the ridge as the buoyant material continued to spread laterally under the moving plates. The shape of the flattened region thus remained constant in a frame fixed in the ridge. The width W was found to be proportional to a length scale formed from the volume flux q, the separation velocity U_r, and the viscosity ratio η_r between the plume and the surrounding fluid, namely, $L_r = (q\eta_r/U_r)^{1/2}$. (The dependence on q and U_r follows from a simple dimensional argument, and the extra dependence on the viscosity ratio was determined empirically from the experiments.) Although these experiments with compositional buoyancy neglected any possible effects of thermal conduction out of the plume, the results scale reasonably well to observations, in particular to the isotopically 'enriched' footprint of the Iceland plume that extends a great distance along the Mid-Atlantic Ridge (Schilling, 1991).

4.4. Characteristics of Plumes in the Mantle

In order to apply the theoretical and laboratory results discussed earlier to predict plume velocities and sizes in the mantle, one must first make realistic estimates, from geophysical data, of the material properties (particularly the viscosity) and also of the temperature anomaly and heat flux at the source. Computer modelling requires similar inputs, and recent developments in that area are allowing valuable comparisons with the simple theories – though numerical modelling can go beyond analysis and laboratory experiments to provide detailed information on temperature distributions, partial melting, and the interactions of plumes with more realistic phase boundaries and lithospheres, in addition to facilitating investigations of the effects of nonlinear rheologies.

4.4.1. Plume Fluxes from Hotspot Tracks

One of the most important inputs to quantitative predictions for plumes in the mantle is the boundary condition on temperature or heat flux or both. The plume heat flux $F_H = q_0\rho c_p\Delta T_0$ [or, more precisely, the plume buoyancy flux $F_B = g\alpha F_H/c_p$, where c_p is the specific heat capacity, and q_0 is again the source volume flux, as in equation (9)] exerts the primary control on the plume flow. The temperature anomaly plays a lesser role through its influence on the viscosity difference and partial melting. The range of plume fluxes to be found in a convecting fluid with temperature-dependent viscosity is not yet well understood; it will be determined by the dynamics of the convection and will be related to the plume spacing. However, we can understand individual plumes by considering a single plume in isolation, in which case it is necessary to specify both a source temperature anomaly and buoyancy flux.[3]

---

[3] In most computer models of a convecting layer, only one of these is required, because the other is then determined by the coupling of conduction and convection of the bottom boundary layer. In numerical experiments (Olson et al., 1988) this was done by applying a temperature anomaly over a finite area of the bottom boundary. In the laboratory experiments described earlier, the temperature anomaly ΔT_0 and mass flux Q_0 from the source were prescribed independently, so that $F_B = g\alpha\Delta T_0 Q_0$.

Estimates of the plume buoyancy and heat fluxes in the mantle have been made using observations of their surface effects in oceanic settings, where the crust exerts relatively little masking effect compared with that of the thicker and more heterogeneous continental crust (Davies, 1988a; Sleep 1990). In making these estimates, it was assumed that a broad region of raised seafloor topography often associated with hotspot tracks is supported in isostatic balance by anomalously hot mantle material that has upwelled in a long-lived plume conduit and spread out under the lithosphere. The theoretical dependence of swell height on plume buoyancy is straightforward and has been confirmed in laboratory experiments (Olson and Nam, 1986; Griffiths, Gurnis, and Eitelberg, 1989). Thus the size of the hotspot swell can be combined with the velocity of the plate over the hotspot to obtain the rate of production of anomalous topography, which is then a measure of the buoyancy flux carried by the plume. For example, the existence of the Hawaiian swell, about 1,000 km wide and 1 km high, propagating across the Pacific plate at about 100 mm/yr, implies a flux (which Davies and Sleep expressed as a mass-deficit flux) in the plume of 2.3×10^{11} kg/yr. This mass-deficit flux is actually $\alpha \Delta T Q$, where Q and ΔT are the mass flux and temperature difference at any depth. This is related to the more physically meaningful buoyancy and heat fluxes through $F_B = g(\alpha \Delta T Q)$ and $F_H = (c_p/\alpha)(\alpha \Delta T Q)$, respectively. For the Hawaiian plume, $F_B \approx 8 \times 10^4 \, \mathrm{N \cdot s^{-1}}$ and $F_H \approx 3 \times 10^{11}$ W. Figure 4.3 shows the distribution

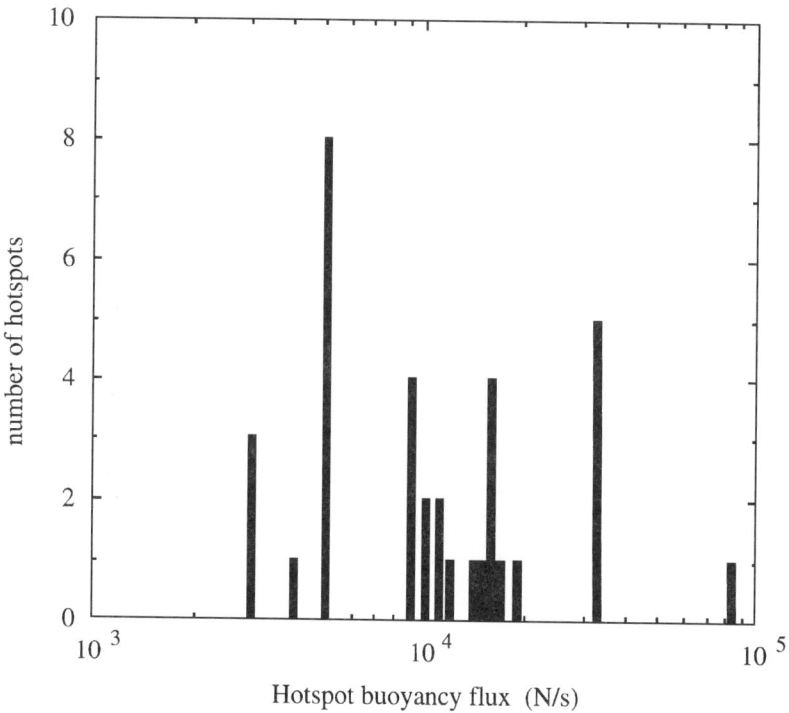

Figure 4.3. Plume buoyancy fluxes, adapted from the estimates of Sleep (1990) for the fluxes at 35 oceanic hotspots. (The buoyancy flux is Sleep's 'mass exchange flux' mutiplied by g.)

of buoyancy fluxes (adapted from Sleep, 1990) for 35 oceanic hotspots. Although there is considerable uncertainty in such estimates, they do indicate that plumes carry a range of fluxes and that the distribution is logarithmically centred about $10^4 \, \text{N} \cdot \text{s}^{-1}$.

An estimate of the volume or mass flux requires independent knowledge of the temperature anomaly, which is usually obtained from the petrology of erupted melts. However, the mass flux is not a conserved quantity, in that it, like the temperature, can vary with height along a plume. Nor is the mass flux well defined. Although the mass flux of hot material near the top of the plume (the flux that is relevant to melt production) for the Hawaiian example becomes $Q_{top} = F_B/(g\alpha\Delta T) \sim 3 \times 10^5 \, \text{kg} \cdot \text{s}^{-1}$ (assuming an average $\Delta T_{top} = 100$ K and no large-scale shearing), the movement of the lithospheric plate over the plume implies, as we have already explained, that the upper mantle is continuously being displaced by the plume and that there must be a *vertical* mass flux in the cooler surroundings. The upward mass flux relevant to overall motion and stirring in the mantle is then made up of both (1) the slow broad motion associated with ascent of the plume conduit in the presence of plate migration and (2) the relatively rapid piping of low-viscosity material upward through the narrow conduit. Later, in the context of stirring, we shall discuss experiments that have shown how the former of these two components leads to considerable vertical transport of material from the surroundings up toward the surface.

4.4.2. New Plumes and Flood Basalts

Predictions can be made for new plume heads by assuming that the rate of supply of buoyancy from the source boundary layer during the early stage in the life of a plume falls in the same range as the buoyancy fluxes derived for currently active hotspot tracks. In that case, a mantle viscosity of 10^{22} Pa · s implies that heads can grow as large as 400–600 km in diameter at the CMB before their ascent speed becomes sufficiently rapid to cause them to break away. Application of the complete form of equation (10) (Griffiths and Campbell, 1990) to the ensuing motions leads to the prediction of a further doubling of the diameter (and an increase in volume by an order of magnitude) as the plume head ascends through 2,800 km. Thus, plume heads that reach the lithosphere while still receiving a constant influx from their source regions can be predicted to be extremely large: 800–1,200 km in diameter. They will also have incorporated a volume of lower-mantle material comparable to the total volume supplied from the source, though the ratio of these two volumes will depend on the source flux. The head size, however, will be insensitive to the flux. The diameter D is instead dependent primarily on the mantle viscosity $(D \sim \eta_\infty^{1/5})$.

As a plume head approaches the upper boundary (the free surface in a laboratory tank, or the stiff lithosphere of the Earth), it must flatten and spread. Again, scaling

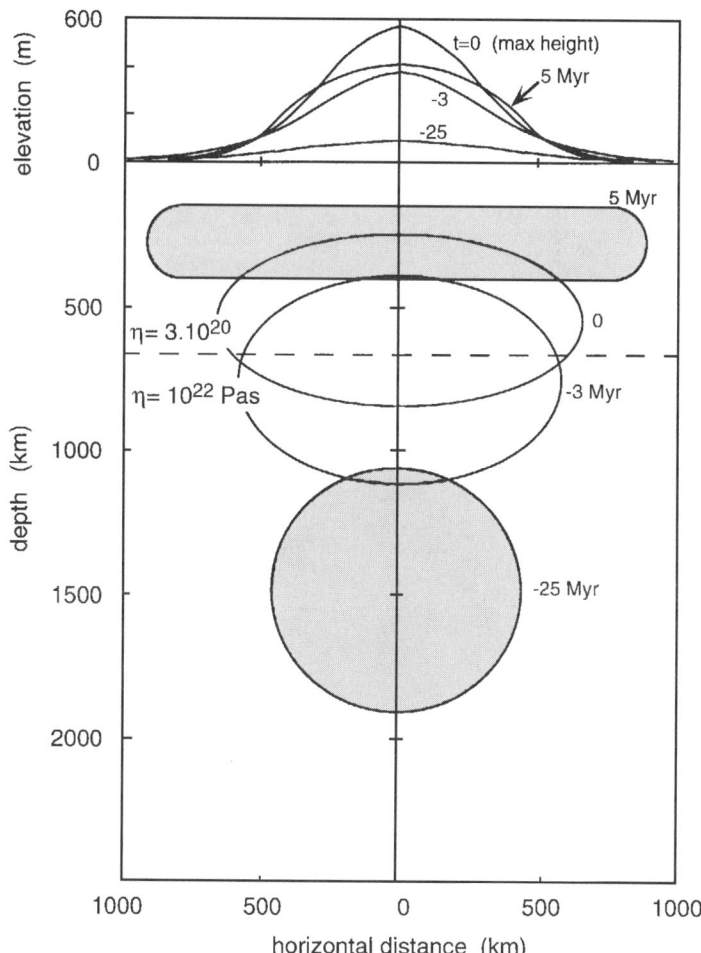

Figure 4.4. Diagrammatic summary of the predicted size of plume heads for plumes originating near the CMB (in this case assuming a source buoyancy flux of 3×10^4 N · s^{-1}) and the relationship of plume depth to surface uplift for the given lower- and upper-mantle viscosities. The horizontal extent is large compared with the depth of the upper mantle and is consistent with that of flood-basalt provinces. (From Griffiths and Campbell, 1991b, with permission.)

analysis and experiments such as that pictured in Figure 4.2 have helped to elucidate and quantify this process (Griffiths and Campbell, 1991b). The head is found to spread rapidly to twice its previous diameter, and much more slowly thereafter (according to $D \sim t^{1/4}$). Thus a spherical head 1,000 km in diameter should produce a pancake-shaped thermal anomaly about 2,000 km across at the base of the lithosphere. Figure 4.4 places the predicted size and position of a plume head in perspective relative to the depth of the upper mantle. These predictions provided the first suggestion that plume structures could have such enormous dimensions, although it should be remembered that the dimension given is the diameter of the equivalent sphere that would contain the plume-head buoyancy at an average temperature, whereas some of the head will actually be much cooler.

Campbell and Griffiths (1990) have argued that the chronology, tectonics, and geochemistry for at least two flood-basalt provinces believed to be attributable to plume heads are consistent with the 1,000–2,000-km scale. Similar head sizes and ascent times have been predicted by numerical experiments simulating mantle conditions (Davies, 1995a; Farnetani and Richards, 1995). Furthermore, continental flood volcanism is known to have been characterized by sudden onset, with most of the magmas erupted within a short period of 1–3 million years (Richards, Duncan, and Courtillot, 1989) and over a roughly equidimensional region 2,000–2,500 km across. Use of these comparisons to argue in the opposite direction provides evidence that at least those plume heads responsible for the major flood basalts had dimensions consistent only with an origin deep in the lower mantle and therefore most probably in the thermal boundary layer at the CMB. The apparent agreement between model and geophysical data also extends to the range of plume fluxes that could have led to flood basalts: The model predicts that plume heads created by buoyancy fluxes smaller than the minimum inferred from hotspot tracks (Figure 4.3) would have risen too slowly and might not have retained sufficient coherence within the convecting mantle to produce major volcanic episodes.

Other experiments (with and without an imposed stiff lithosphere) indicate that the axisymmetric surface uplift above a plume head will reach a maximum when the top of the plume described earlier is 50–100 km from the surface (or lithosphere), after only about 5 million years of significant uplift (Olson and Nam, 1986; Griffiths et al., 1989). Subsidence of the surface will follow over a period on the order of 100 million years because of continuing slow spreading of the thermal anomaly. On the Earth, the extent of subsidence will be enhanced if there is some lithospheric extension (which might be produced by the plume-generated uplift itself or by coincident extensional tectonics) (White and McKenzie, 1989; Campbell, Chapter 6, this volume). A similar process provides the best explanation for the subsidence, with age, of seafloor swells and volcanic-island chains: In that case, we suggest that plume material can continue to flow laterally long after it has been carried away from the original hotspot location by the plate motion.

4.4.3. Temperature of the Plume Source and Resulting Melts

Much of the geological significance of mantle plumes relates, of course, to the resulting volcanism and reworking of the Earth's crust. Hence it is vital to make progress towards a quantitative understanding, and predictive models, for the generation and composition of partial melts in dynamic plume environments. The model outlined earlier implies that the composition of plume melts will be influenced by the tendency of a plume to redistribute heat between its source and the entrained mantle materials before advecting at least some of this material through its solidus conditions. True mixing of any associated chemical and isotopic signatures may result from stirring if partial melts of varied compositions are produced and then ascend and mingle. We can expect that the bulk of melting will occur within the hottest plume material and at lowest pressures (shallowest depths). However, the predicted length scales and ascent speeds for plume heads strongly suggest that relatively little heating and melting are possible in the overlying lithosphere, unless extension is an important factor.

Campbell and Griffiths (1990) suggested that the compositions of the voluminous tholeiitic flood basalts are consistent with melting of material in the plume head, whose temperature may have been substantially lower than that in the conduit, whereas the compositions of picrites erupted in flood-basalt provinces (and of komatiites in Archaean examples) are consistent with melting at much higher temperatures, and could therefore represent melting of the hottest, least contaminated parts of the plume (Campbell, Chapter 6, this volume). In that view, the potential temperature anomaly in the plume source was inferred to be at least 300 K. If some cooling during flow up the conduit is allowed for, the source temperature anomaly might be as high as 400 K. On the other hand, the latest numerical modelling (Farnetani and Richards, 1995), which has begun to assess the potential for melting of source and entrained mantle material, suggests that only the hottest material undergoes melting. That suggestion depends on assumed source temperatures and on the parameterization of melting and thus requires further exploration. However, if correct, it would imply that melts came only from source material (enriched in incompatible elements) spreading near the top of a plume. That would require a different explanation for the ranges of compositions and apparent melting temperatures in flood basalts.

4.5. The Upper Boundary Layer and the Descent of Cold Slabs
4.5.1. The Lithosphere as an Active Boundary Layer

We turn now to the cooled upper boundary and the generation of the primary motions of mantle convection. The plume heat fluxes estimated from oceanic hotspot topography imply that only about 10% of the heat flux through the Earth's surface is supplied by mantle plumes, which carry heat from the core. The total flux at the surface is largely due to loss of internal heat and the heat generated by radioactive decay (O'Neill and Palme, Chapter 1, this volume; Davies, Chapter 5, this volume). That is, the mantle may be regarded as a layer of viscous fluid, largely internally heated, and cooled from above.

Early notions about mantle convection regarded plate tectonics as the surface reaction to an underlying pattern of convection occurring especially in the upper mantle. That view required the plates to be dragged along by a faster motion beneath. When the observations were compared with findings from laboratory studies with that picture in mind, it was puzzling that the width of the inferred convection cells was so much greater than their depth (often presumed to be that of the upper mantle), and that led to many investigations of the effects of variable fluid properties and different boundary conditions on the aspect ratio. A more consistent view is that the buoyancy forces acting on the colder, denser plates provide the primary driving mechanism for convection, at least under the present-day tectonic regime, so that subduction of lithospheric slabs is an active part of convection, not a secondary reaction to it. Those earlier questions about the horizontal scale are then readily answered (1) by considering that the convection may penetrate the full depth of the mantle and (2) by noting that the strength of the lithosphere (which can yield and break only at stresses greater than a few hundred MPa) can inhibit the initiation of subduction. In this view, cold material can break away from the upper boundary only where two plates meet, such that one plate can slide under the other. In the intervening regions (i.e., near mid-ocean ridges) there is a compensating, passive ascending flow. Thus, much of the geometry of convection in the mantle is determined by the pattern of the plates, though the development of criteria for the formation and size of these plates remains a major theoretical challenge.

The mechanism whereby subduction is initiated is not well understood, nor are the factors that determine the angle of subduction. However they are formed, it is clear that dense, highly viscous, two-dimensional descending plates or cold slabs are fundamental elements in any dynamical description of mantle convection. In the following sections we shall review experiments and related theories that have explored the behaviour of such slabs.

4.5.2. Interaction of Subducting Slabs with a Viscosity or Density Discontinuity

Seismic observations indicate that there are several abrupt changes in seismic wavespeed with increasing depth in the transition zone of the Earth's mantle. In particular, a major discontinuity has been identified at a depth of 660 km, but thus far there is no universally agreed interpretation of its dynamical significance. Jackson and Rigden (Chapter 9, this volume) present evidence to support the view that the seismic data are consistent with the behaviour to be expected of a mantle that is grossly uniform in chemical composition throughout. However, laboratory experiments conducted at high pressures and temperatures indicate that descending slabs undergo different series of phase transformations in the basaltic crustal layer and in the underlying peridotite (Irifune and Ringwood, 1987).

If there were a compositional step, that would imply a two-layer mantle, with each layer convecting separately, thus allowing mantle materials of different compositions to remain distinct over long periods. Much geophysical evidence and dynamical modelling, on the other hand (e.g., Davies, 1988a; Davies and Richards, 1992), are in conflict with the hypothesis of two-layer convection and point towards whole-mantle convection, though with an increase in viscosity below the 660-km discontinuity, allowing the deepest mantle to be partially isolated over long periods. There is also the issue whether or not phase changes in themselves can inhibit penetration of slabs (and plumes) through the phase boundary. That issue will not be addressed here, but see Jackson and Rigden (Chapter 9, this volume).

Leaving aside these questions about the mantle's properties, laboratory and numerical experiments can be used to gain an understanding of the dynamics of highly viscous descending slabs in the presence of a density or viscosity gradients. Here we shall cite several studies in which the viscosity changes were assumed to be concentrated across the 660-km discontinuity, although the evidence does not rule out a more gradual variation. Kincaid and Olson (1987) approached this problem by using moulded slabs of very cold (hence more dense and viscous) sucrose solution, laid on the surface in a tank containing two layers of less viscous sucrose solutions. One slab was introduced with a shallow-dipping bend in it; driven by its negative buoyancy, it was subducted under the edge of a second slab next to it. The far end of the subducting slab was either allowed to move freely or held fixed. The latter case corresponded to zero spreading speed on one side of a ridge, and the slab penetrated the upper layer at a shallow angle, with the subduction line moving back toward the ridge, and the slab being laid out along the interface. Those authors also found that the slab sank into the lower layer without distortion when it was very dense relative to the lower layer; there was limited penetration when the densities were nearly equal; and the slab was deflected at the interface when the slab was lighter than the lower layer. For the case of freely sinking slabs, and nearly vertical penetration, their observations showed behaviour similar to that predicted in a numerical model by Christensen and Yuen (1984).

In order to control independently a greater number of variables in the flow, Griffiths and Turner (1988a,b) used a different method of producing the subducting viscous slabs. Their experiments examined axisymmetric plumes and two-dimensional sheets of relatively viscous syrup (density ρ, kinematic viscosity ν) flowing out of a reservoir through a tube or slot and falling through an upper layer of glycerine (ρ_T, ν_T) onto a more viscous and denser bottom layer composed of mixtures of glycerine and syrup (ρ_B, ν_B). The system was characterized by the source–interface distance H relative to the sheet thickness d and the relative density difference $P = (\rho - \rho_B)/(\rho - \rho_T)$. For small dimensionless source–interface distances H/d, sheets much denser than the lower layer remained planar as they passed through the interface. When H was increased above a critical value H_c, the sheet buckled and folded; H_c increased with increasing P because of the decreasing resistance to penetration by the sheet. This behaviour is consistent with a theoretical result, due to Biot (1961), whereby folding of a sheet in longitudinal compression can occur only if the bending of the sheet and the induced motion of the outer, less viscous fluid causes less dissipation of energy than would be associated with a smooth thickening or spreading of the sheet. Thus H_c should be indefinitely large (i.e., the sheet will not fold at all) for comparable viscosities and will decrease as the sheet becomes much more viscous than its surroundings. Experimentally (with $P = 0$) it has been found that the dimensionless critical height

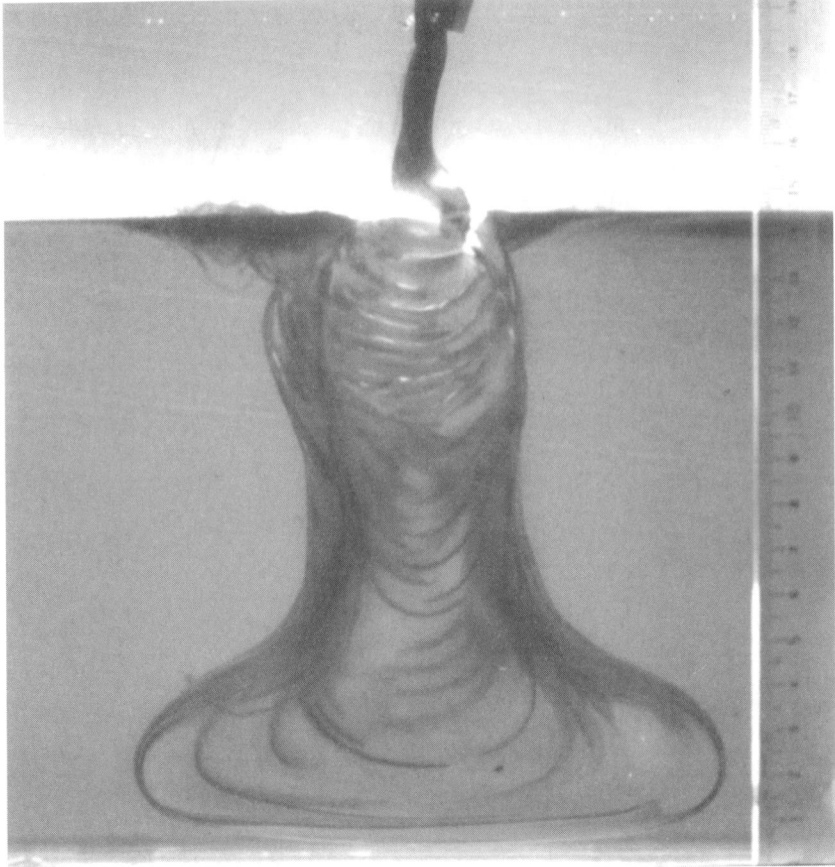

Figure 4.5. Photograph of a model of a folding slab descending through a two-layer mantle. A sheet of viscous syrup has folded on the interface between layers of glycerol and a glycerol/syrup mixture and has pulled less dense upper-layer fluid with it into the bottom layer. (From Griffiths and Turner, 1988a, with permission.)

H_c/d is approximately 4 for a viscosity ratio of 60 and approaches an asymptotic value of 1.7 above a viscosity ratio of about 10^4.

In the experiments, viscous drag gave rise to circulation cells in the upper layer, and upper-layer fluid was also pulled down through the interface (Figure 4.5). For large density increases across the interface (small positive P), the folding of a two-dimensional sheet leads to the incorporation of upper-layer fluid between the folds and the formation of a neutrally buoyant mixture that can spread along the interface, even when the sheet is denser than both layers. Because the upper-layer fluid is only trapped by the folds, rather than being thoroughly mixed with the plume fluid, given enough time the two fluids can separate again, with dense slab material sinking to the bottom. More dense folded sheets, on the other hand, cannot spread at the interface, but instead sink directly through the lower layer along with the trapped upper-layer material (Figure 4.5). The flux of entrained upper-layer material, compared with the flux in the slab itself, thus increases with increasing P.

Qualitatively similar behaviour was seen when the discontinuity was dominated by a viscosity step (and the density step was very small, $0.9 < P < 1$). Again, if the length of the viscous sheet (i.e., the depth of the upper layer) was large enough, it folded as it approached the discontinuity. The much more viscous lower layer reduced the vertical velocity and acted as a nearly rigid boundary. The folded sheet formed a composite plume that sank slowly through the lower layer. Although the viscosity increase in the mantle is attributed to the effect of pressure (either through a phase change or because of a positive activation volume for creep of the high-pressure mineral assemblage), the upper mantle will still be entrained (but in this case will not retain a low viscosity). The effect of an interface thus depends on the extent to which the buoyancy-driven descent of a slab is slowed by either density changes or viscosity changes in the surroundings.

One implication for our understanding of the mantle is that vertically descending slabs whose thicknesses are less than about 20% of the depth of the upper mantle (i.e., $d < 130$ km) are likely to fold near a depth of 660 km in the mantle if there is a rapid increase in viscosity at that level. That assumption has been used by Davies (1995a; Chapter 5, this volume) in a model of mantle evolution in order to predict when a slab will penetrate the 660-km discontinuity and when it will buckle and spread.

Laboratory experiments (Turner, 1973b) and later numerical modelling (Christensen and Yuen, 1984) have investigated the combined effects of compositional buoyancy and thermal buoyancy in the mantle for the case of a possible difference in compositional density between the upper mantle and lower mantle. It appears that cold viscous slabs can penetrate through a stable compositional-density interface when the thermal-buoyancy force exceeds the intrinsic stabilizing force, whereas there will be complete separation of the two layers when compositional differences between the layers dominate. For an intermediate range of conditions, the slab material may penetrate into the lower layer, thermally equilibrate there, and then rise again into the upper layer because of its intrinsic buoyancy. This is an example of the more general phenomena encompassed by the term 'double-diffusive convection', whereby two components with different molecular diffusivities have opposing effects on the density difference (Turner, 1985), and it is, in essence, the physical process proposed by Silver, Carlson, and Olson (1988) to reconcile the one- and two-layer models of mantle convection.

4.5.3. Effects of Trench Migration

Mantle convection and the resulting plate tectonics are particularly complex because of the mobility of the plate boundaries. The same buoyancy forces that drive the motions of the plates cause trenches to migrate, so that the locations of deep slabs were determined by the former positions of trenches. Whereas few laboratory experiments have been designed to address the causes and effects of trench migration, some recent experiments have begun to explore its implications for slab behaviour.

The earlier work with sheets falling from a stationary source was extended by using a slowly moving slit source from which high-viscosity syrup was released at a constant rate at the surface of a layer of less viscous and slightly less dense syrup (Griffiths, Hackney, and van der Hilst, 1995). A second layer of syrup beneath provided a density or viscosity discontinuity. These experiments showed that the ratio between the 'trench' migration speed and the buoyancy-driven sinking speed for the slab determined the angle of dip of the slab in the upper layer. In that layer the descent was steady when the trench speed was below a critical value, but at higher migration speeds the dip became oscillatory. The oscillatory behaviour is a result of the gravitational instability of a slab whose dip (determined by the coupling of trench migration speed and sinking velocity) is less than a critical angle from the horizontal.

When a model slab encountered a density or viscosity interface, and the migration speed was low, it was deflected, but continued sinking through the lower layer (at a smaller dip, measured from the horizontal, Figure 4.6). For higher trench migration speeds, it was laid out on the interface, producing a nearly horizontal portion of the slab. That horizontal portion was unstable and eventually sank into the lower layer at some distance behind the point of incidence of the slab onto the interface (Figure 4.6). The spacing of the resulting descending blobs of slab material in the lower layer was determined either by the period of oscillation of the slab in the upper layer (when the slab was unstable in the upper layer) or by the gravitational instability of the nearly horizontal portion of slab at the interface. A similar range of slab behaviours has been identified in numerical models of the mantle that have incorporated a viscosity step and endothermic phase changes in the sinking slab (Christensen, 1996). From such experiments, the behaviour can be summarized as a function primarily of two parameters: the viscosity contrast between the upper and lower layers (embodied in a ratio of sinking speeds), and the ratio of the migration speed to the buoyancy-driven sinking speed.

Scaling of these findings suggests that any of a number of regimes can occur under mantle conditions, so that the behaviour of slabs can vary and will depend on the trench migration speed. In particular, it is important to recognize that the implications of tomographic snapshots of present-day slab morphology are ambiguous, because they do not uniquely constrain the flow behaviour that created the observed morphology. Thus, patches of horizontal slabs detected near a depth of 660 km do not necessarily imply that those slabs will not penetrate into the lower mantle or that the mantle is not convecting as one single layer. The horizontal portions of slabs could have resulted from local and temporary deflections due to a decreased sinking speed for slabs in a more viscous lower mantle. Further evidence for this conclusion can be found in the structure across the transition zone in the mantle below the Fiji–Tonga arc, deduced by inversion of seismic travel-time data (van der Hilst and Seno, 1994; van der Hilst, 1995; Kennett and van der Hilst, Chapter 8,

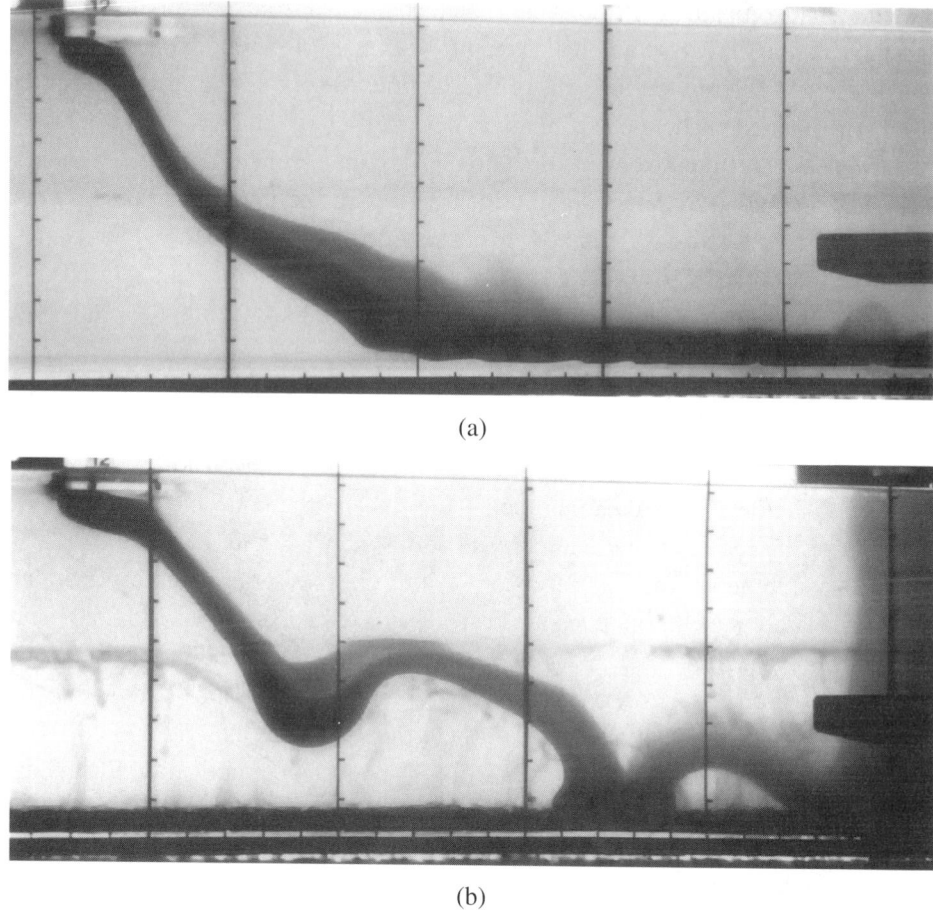

(a)

(b)

Figure 4.6. Photographs of laboratory models of subducting slabs impinging on a density-and-viscosity discontinuity from a slowly migrating two-dimensional source (trench), showing conditions under which (a) the slab is steady in the upper layer and passes through the discontinuity with only a reduction in the slab dip and (b) the slab dip is gravitationally unstable in the upper layer, so that the slab descends into the lower layer as a sequence of diapirs whose spacing is determined by the slab's frequency of oscillation in the upper layer. The discontinuity is at mid-depth, and the migration speed is constant. (From Griffiths et al., 1995, with permission.)

this volume). In the northern part of that arc there are significant changes in dip with depth: In the upper mantle it is about 50°; the slab is sub-horizontal in the transition zone; and the dip increases again in the lower mantle. At the southern end of the trench, the slab penetrates the 660-km boundary with no deflection. This is consistent with the laboratory data, given the high migration speed of the trench in the northern region and the much slower trench migration in the south.

4.6. The Heterogeneity and Stirring of the Mantle

Isotopic analyses of rocks from the mid-ocean ridges and oceanic hotspots have shown that there are chemical heterogeneities in the Earth's mantle and that material enriched in incompatible elements has remained distinct from the rest of the mantle for about 2×10^9 years. Samples of mantle with very different apparent isotopic ages were first interpreted in terms of two distinct 'reservoirs': a primitive region that had remained unchanged throughout the evolution of the Earth, and a depleted region that had resulted from fractionation during the formation of the continental crust. These ages were simplistically identified with the lower and upper layers, respectively, of a two-layer mantle. More extensive measurements now suggest that at least five source types are required to span the range of compositions observed in mid-ocean-ridge basalts and in ocean-island basalts, and there are even considerable variations within and between islands; see Zindler and Hart (1986) for a comprehensive review. Though the upper mantle, as sampled by mid-ocean-ridge basalts, is relatively more homogeneous chemically and isotopically than the mantle as a whole, the geochemical data give little reason to hypothesize a two-layer mantle. These data, and considerations of the convective transport of heat out of the lower mantle and core (Sections 4.2, 4.4, and 4.5), make it seem more plausible not to think in terms of several discrete large volumes of homogeneous material in the mantle, but instead to expect heterogeneities on many scales. More generally, it is essential to come to an understanding of the geometry and lifetimes of compositional heterogeneities in terms of the dynamics of convection, involving the production of heterogeneities by melting and chemical differentiation, and their removal by stirring. That was one motivation for the dynamical studies of plumes and other aspects of mantle convection discussed thus far.

4.6.1. Heterogeneity and the Convective Cycle

We have already discussed the fluxes associated with the ascent of plume heads originating near the CMB. As noted in Section 4.3, the action of viscous forces associated with the passage of a plume must drag upwards a large mass of material outside the buoyant plume head (Griffiths, 1986a,b). This results in a large mass transport, across any depth, of material having relatively uniform composition and potential temperature. The magnitude of such transport is difficult to define precisely, but it is an order of magnitude greater than the mass flux within the plume head (because velocities decay only slowly with distance from the axis of the diapir). From the estimate of one plume head ascending every 20 million years (based on geological data for the past 250 million years) (Hill et al., 1991, 1992), it is calculated that plume heads would fill a volume equal to that of the upper mantle with *warmed* material every 1,200 million years, whereas the total mass flux would imply a flushing time for the upper mantle on the order of 100 million years. Of course, much of this transport is of material from the top of the lower mantle, dragged across the phase boundary. This part of the mass flux will not influence the chemical or isotopic composition of the upper mantle if there is no significant difference in the top-most part of the lower mantle, as has been suggested by numerical convection experiments (Davies and Gurnis, 1986) indicating that the plate-driven overturning velocity decreases only slowly with depth through the mid-mantle. However, the magnitudes of the fluxes suggest that plume heads may have added significantly to motions and to the redistribution of heterogeneities in the mantle.

We have discussed the evidence that sufficiently large and dense plates can penetrate the 660-km discontinuity and continue to descend through the lower mantle. Like hot plumes, cold slabs do not transport only the material in the slabs themselves; as a result of viscous coupling, they also induce recirculation cells, pulling surrounding mantle down with them. Some of the downward-moving mantle will be cooled by the slabs, becoming denser than the surrounding mantle. A slab itself is also a layered structure, with a thin basaltic crust on its upper surface. If such a composite slab, with its load of cooled upper mantle, is still denser than the surrounding lower mantle when it reaches the core–mantle interface, it will spread out there and form an inhomogeneous boundary layer, which might be identified with the seismologically resolved D'' layer. The chemical properties of this layer would then reflect the near-surface processes that formed the slab and depleted the upper mantle of incompatible elements.

The heterogeneous material at the lower boundary may then contribute to the source of hot plumes. Indeed, Hofmann and White (1982) argued, on the basis of the enrichment of light rare-earth elements and other trace elements, that the source region for ocean-island (plume) basalts contains a large proportion of recycled oceanic crust. Campbell and Griffiths (1992) took this further, working from the apparent variations in the composition and temperature of plume melts with time, and outlined a possible, if somewhat speculative, scenario for the evolution of the mantle composition (Campbell, Chapter 6, this volume). This involves the onset of lithosphere subduction at about 4 Ga and subsequent formation of a depleted upper mantle and an enriched layer near the CMB. The enriched source could have heated up and been returned (in plumes) to the upper mantle to begin producing enriched ocean-island basalts at around 2.5 Ga, the time at which there appears to have been a change from depleted to enriched plume melts (and a coincident decrease in their maximum melting temperature).

Thus, in principle, we have closed the loop of the convective processes in the mantle. On the other hand, there remain many questions, particularly concerning the spacing of plumes at the CMB, the composition of the plume source region, and how plumes come to be so enriched, given chemical and isotopic heterogeneity expected of that region. Further studies are also needed to clarify the extent to which a compositionally layered slab and the upper-mantle material coupled to it stay together in the lower mantle and how slabs and plumes there are affected by convective motions driven by other slabs and plumes.

4.6.2. Stirring in Unsteady Convection

Apart from the cycling of material between the upper and lower regions of the convecting layer, the nature of stirring in the mantle is essentially dependent on the unsteadiness of flow patterns, which can give rise to chaotic particle paths and more rapid mingling of heterogeneities. From the geological evidence it is clear that both the convection driven by the surface cooling and that associated with upwelling plumes are far from steady over time scales of 100 million years and longer. The changes in the patterns of motion need not be associated with causal changes, because instabilities in the flow (arising especially in the boundary layers) can lead to spontaneously changing patterns even when the external boundary conditions are steady. This point was made by Tritton (1985) in relation to his extremely viscous laboratory models of convection, and it has also been demonstrated by a variety of numerical studies of convection at large Rayleigh numbers and infinite Prandtl number. Once we have unsteady flow, we can draw further on the findings from fundamental fluid-dynamics studies in order to understand the greatly increased mixing efficiency that occurs; see the review by Ottino et al. (1988). On the other hand, apart from the studies of plumes and slabs already discussed, most studies of stirring in the mantle have relied on numerical experiments in which neutrally buoyant marker particles have been followed for a long time (Davies, Chapter 5, this volume).

We shall next consider the question of unsteadiness and its effect on dispersion in the mantle. If we consider stagnation-point flows or chaotic velocity fields $\mathbf{v}$ characterized by pure straining motion ($\nabla \times \mathbf{v} = 0$), heterogeneities are stretched out into long tendrils at an exponential rate. Thus the average thickness of heterogeneities, which is relevant if one wishes to know the likelihood that a given piece of mantle will contain bands of subducted crust, decreases exponentially with time. On that basis, Kellogg and Turcotte (1986) and Hoffmann and McKenzie (1985) concluded that subducted lithosphere would be well mixed through the mantle in less than 500 million years. However, the calculations of Olson, Yuen, and Balsinger (1984), Gurnis and Davies (1986a), and Gurnis (1986) also monitored the sizes of the largest surviving blobs of different composition and found that in unsteady convection some blobs of labelled fluid persisted for much longer than was predicted for pure straining flow or even for simple shear flow ($\mathbf{v} \cdot \nabla \mathbf{v} = 0$, which leads to linear changes in dimensions with time). They concluded that some regions of the mantle could retain high concentrations of old material, persisting for 2 billion years, while rapid stirring on smaller scales occurred in other regions.

There are further complications, because different kinds of unsteadiness can yield very different rates of stirring (Christensen, 1989). For example, two-cell flows, with the boundary between cells slowly oscillating back and forth (Davies, 1988b), will give slow mixing: Blobs in some regions of the flow can be strained and stirred by simple shear and then occasionally 'unmixed' by a reversal of the sense of shear when the parcel is transferred to an adjacent cell. A second regime identified by Christensen (1989), fast stirring, occurs with an irregular motion of boundaries between cells, or imposed smaller-scale instabilities. Davies (1990b) has argued that this invokes a more random form of convection in the lower lithosphere, for which there is little observational evidence [but which was assumed by Kellogg and Turcotte (1986) in arriving at their estimates of fast mixing rates], and that the slow stirring regime is the appropriate one for the mantle. However, some modification of this conclusion may be needed when the effects of unsteady (and possibly random) convection from the bottom boundary layer (hot plumes) are added to those of the plate-driven convection.

In Section 4.3 we noted that plume heads can greatly perturb the streamlines of the plate-driven convection. This will add further tortuosity to particle paths and might induce faster stirring of any heterogeneities. R. W. Griffiths and M. Gurnis (unpublished data) have demonstrated a potentially larger role for stirring by steady, narrow plume conduits. In slow shearing flow or steady overturning cells, a large volume of mantle material is displaced vertically, sheared and twisted as it moves past a stationary plume conduit beneath a hotspot (Figure 4.7). Although

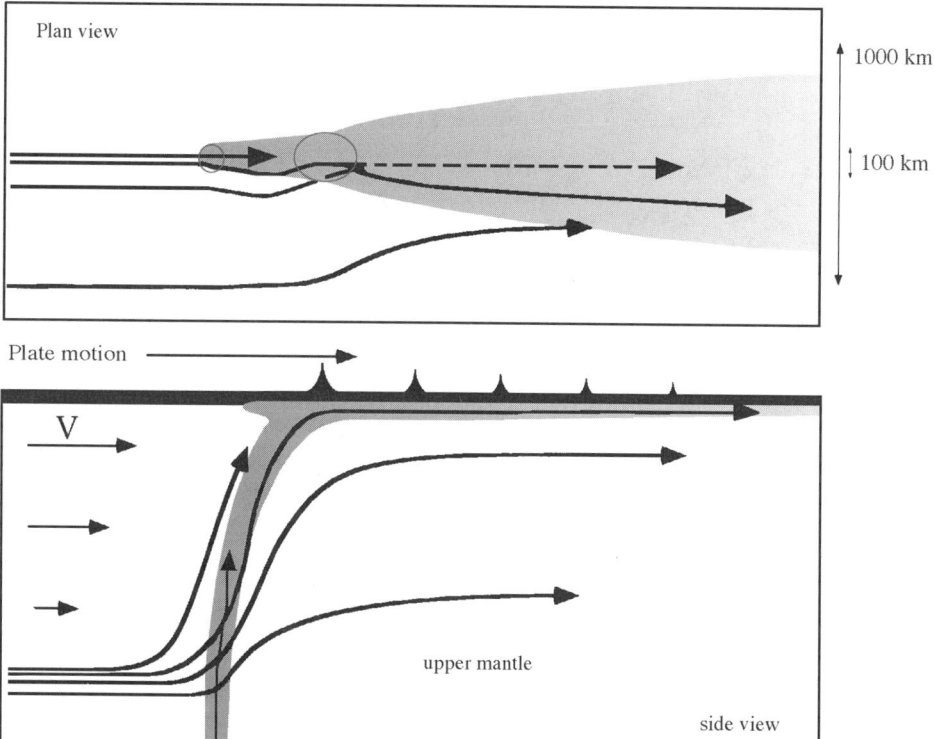

Figure 4.7. Sketch of the streamlines observed in experiments with shear flow about a steady plume conduit, showing the large stirring effect that could be caused as the mantle, driven by the motion of an overlying plate, passes around a stationary plume. In this experiment, the unidirectional shear flow was driven by an imposed surface velocity in a large circular tank. Similar results were obtained in other experiments in which the plume rose through an overturning convection cell (in a box with one end wall heated and the other cooled.)

the resulting effect on stirring has proved difficult to quantify, it is possible that the presence of a number of such plumes beneath large oceanic plates could produce a significant modification of the overall stirring regime.

4.6.3. Effects of Viscosity and Intrinsic Density Differences

We have noted the role of rheological contrasts in the dynamics of the upper and lower boundary layers of the mantle, in descending slabs, and in upwelling plumes. In addition, a large increase in viscosity with depth (pressure) can strongly modify the rate of stirring, so that the survival times for heterogeneities are considerably longer in the lower mantle than in the upper mantle (Gurnis and Davies, 1986b). Thus it is possible that a fraction of the material originally near the core–mantle interface may have persisted to the present. This conclusion is consistent with the higher ^{3}He/^{4}He ratios found in basalts, from oceanic hotspots, compared with those in mid-ocean-ridge basalts, which suggests that a more primitive, less degassed mantle (as well as recycled oceanic-crust material) is being sampled by plumes (McDougall and Honda, Chapter 3, this volume).

The driving force for the dominant motions of the mantle derives from density differences due to thermal expansion. However, it is possible that composition-related (intrinsic) density differences may also play a significant role in parts of the flow, and this should be considered in the context of stirring. The intrinsic buoyancy of the crust clearly causes the continents to float at the surface and may modify the buoyancy of subducted oceanic lithosphere. In general, it is necessary to consider the ratio between compositional density differences $\Delta\rho_i$ and those resulting from temperature differences $\rho\alpha\Delta T$ (we define the ratio $R_\rho = \Delta\rho_i/\rho\alpha\Delta T$) along with the *length scales of the compositional anomalies* and the magnitude of the viscous drag forces. If either R_ρ or the length scale of a heterogeneity is small, the intrinsic buoyancy will be small compared with both the viscous drag that it has to overcome and the thermal buoyancy forces that drive the large-scale flow. Gurnis (1986) explored the effect of an intrinsically more dense oceanic crust in numerical models of unsteady convection and concluded that the overall effect of intrinsic density differences on the settling of subducted material to the bottom of the convecting layer would be small. In fact, the results indicate that it would be significant only when R_ρ is so large as to imply a cool lithosphere made up entirely of oceanic crust.

Richards and Davies (1989) examined the gravitational segregation of a laminated sinking slab and showed that no appreciable segregation could occur before the material had penetrated deep into the lower mantle. Recent numerical calculations by Christensen and Hofmann (1994), on the other hand, have suggested that the oceanic-crust component may gravitationally segregate at the bottom of the convecting mantle, because of heating, and be stored there long enough (before being re-entrained into mantle plumes and brought to the surface) to develop the observed isotopic age and incompatible-element enrichment typical of hotspot basalts. Such segregation does appear to be essential if we are to explain the enrichment of plume melts in terms of source composition.

4.7. Synopsis

A physical understanding of complex phenomena can be greatly facilitated if individual physical and dynamical processes are first isolated and studied separately, often with guidance from studies of related flows in other contexts. For example, we have identified as components of mantle convection the generic processes of unsteady, high-Rayleigh-number convection driven by unstable boundary layers, individual plumes, and descending slabs, as well as the concepts of thermal entrainment and viscous entrainment. These can be complex processes in themselves, but they are much simpler individually than is the complete system encountered in the Earth. Though we have covered many of the most significant processes that transport material and heat about in the mantle, this has been done by considering them carefully one at a time, rather than attempting to put them all together from the start.

For each of these processes, an effective way to identify significant new phenomena and to elucidate the underlying physics is to carry out controlled laboratory experiments. Associated analytic models based on the new physical understanding and on simple scaling arguments can then lead to findings that can be used to make predictions about the behaviour under mantle conditions. This combination of experiment and theory can also guide the development of numerical models, starting with relatively simple, well-defined problems not too far removed from the conditions achievable in the laboratory, which can than be extended step-by-step to treat phenomena and parameter ranges that cannot be directly modelled in the laboratory. These include, in particular, phase changes and other pressure-dependent effects.

Finally, we mention a few of the outstanding questions concerning the mantle that need to be addressed in the near future through laboratory experiments and associated theory. Will we be able to make predictions about the spatial and temporal distributions of geochemical and isotopic compositions formed when a plume head ascends beneath a moving oceanic plate or at a mid-ocean ridge? Will we be able to predict the distribution and composition of melts in these cases? What determines the position and spacing of new plumes? Are the locations of plumes affected by remnants of subducted slabs lying near the CMB? How long do individual plumes remain active? Can we better describe the process of collection of old oceanic crust near the CMB and its sampling by plumes? What interactions can occur between plume formation in the mantle and convection in the core, and are these dynamical interactions influenced by miscibility or reaction at the CMB? What is the relationship of the temperature of the plume source to the temperatures of the lower mantle and outer core? Some of these problems have already been approached using ideas generated by laboratory models, and some represent combinations of processes that are already understood.

References

Biot, M. A. 1961. Theory of folding of statified viscoelastic media, and its implications in tectonics and orogenesis. *Bull. Geol. Soc. Am.* 72:1595–620.

Boehler, R. 1993. Temperatures in the Earth's core from melting-point measurements of iron at high static pressures. *Nature* 363:534–6.

Campbell, I. H., and Griffiths, R. W. 1990. Implications of mantle plume structure for the evolution of flood basalts. *Earth Planet. Sci. Lett.* 99:79–83.

Campbell, I. H., and Griffiths, R. W. 1992. The changing nature of mantle hotspots through time: implications for the geochemical evolution of the mantle. *J. Geol.* 92:497–523.

Christensen, U. R. 1989. Mixing by time-dependent convection. *Earth Planet. Sci. Lett.* 95:382–94.

Christensen, U. R. 1996. The influence of trench migration on slab penetration into the lower mantle. *Earth Planet. Sci. Lett.* 140:27–40.

Christensen, U. R., and Hofmann, A. W. 1994. Segregation of subducted oceanic crust in the convecting mantle. *J. Geophys. Res.* 99:19867–84.

Christensen, U. R., and Yuen, D. A. 1984. The interaction of a subducting lithospheric slab with a chemical or phase boundary. *J. Geophys. Res.* 89:4389–402.

Davaille, A., and Jaupart, C. 1993. Transsient high-Rayleigh-number thermal convection with large viscosity variations. *J. Fluid Mech.* 253:141–66.

Davaille, A., and Jaupart, C. 1994. Onset of thermal convection in fluids with temperature-dependent viscosity: application to the oceanic mantle. *J. Geophys. Res.* 99:19853–66.

Davies, G. F. 1988a. Ocean bathymetry and mantle convection: 1. Large scale flow and hotspots. *J. Geophys. Res.* 93:10467–80.

Davies, G. F. 1988b. Role of the lithosphere in mantle convection. *J. Geophys. Res.* 93:10451–66.

Davies, G. F. 1990a. Mantle plumes, mantle stirring and hotspot chemistry. *Earth Planet. Sci. Lett.* 99:94–109.

Davies, G. F. 1990b. Comment on 'Mixing by time-dependent convection' by U. Christensen. *Earth Planet. Sci. Lett.* 98:405–7.

Davies, G. F. 1993. Cooling the core and mantle by plume and plate flows. *Geophys. J. Int.* 115:132–46.

Davies, G. F. 1995a. Penetration of plates and plumes through the mantle transition zone. *Earth Planet. Sci. Lett.* 133:507–16.

Davies, G. F. 1995b. Punctuated tectonic evolution of the earth. *Earth Planet. Sci. Lett.* 136:363–79.

Davies, G. F., and Gurnis, M. 1986. Numerical study of high Rayleigh number convection in a medium with a depth-dependent viscosity. *Geophys. J. Royal Astron. Soc.* 85:523–41.

Davies, G. F., and Richards, M. A. 1992. Mantle convection. *J. Geol.* 100:151–206.

Duncan, R. A., and Richards, M. A. 1991. Hotspots, mantle plumes, flood basalts, and true polar wander. *Rev. Geophys.* 29:31–50.

Farnetani, C., and Richards, M. A. 1995. Thermal entrainment and melting in mantle plumes. *Earth Planet. Sci. Lett.* 136:251–67.

Feighner, M. A., and Richards, M. A. 1995. Plume–ridge and plume–plate interactions. *Earth Planet. Sci. Lett.* 129:171–82.

Fitzpatrick, M. F. 1991. The dynamics of viscous thermal plumes in the Earth's mantle. Honours thesis, The Australian National University.

Griffiths, R. W. 1986a. Thermals in extremely viscous fluids, including the effects of temperature dependent viscosity. *J. Fluid Mech.* 166:115–38.

Griffiths, R. W. 1986b. Particle motions induced by spherical convective elements in Stokes flow. *J. Fluid Mech.* 166:139–59.

Griffiths, R. W. 1986c. Dynamics of mantle thermals with constant buoyancy or anomalous internal heating. *Earth Planet. Sci. Lett.* 78:435–46.

Griffiths, R. W. 1991. Entrainment and stirring in viscous plumes. *Phys. Fluids* A3:1233–42.

Griffiths, R. W., and Campbell, I. H. 1990. Stirring and structure in mantle starting plumes. *Earth Planet. Sci. Lett.* 99:66–78.

Griffiths, R. W., and Campbell, I. H. 1991a. On the dynamics of long-lived plume conduits in the convecting mantle. *Earth Planet. Sci. Lett.* 103:214–27.

Griffiths, R. W., and Campbell, I. H. 1991b. The interaction of plumes with the Earth's surface and onset of small-scale convection. *J. Geophys. Res.* 96:18295–310.

Griffiths, R. W., Gurnis, M., and Eitelberg, G. 1989. Holographic measurements of surface topography in laboratory models of mantle hotspots. *Geophys. J.* 96:1–19.

Griffiths, R. W., Hackney, R. I., and van der Hilst, R. D. 1995. A laboratory investigation of the effects of trench migration on the descent of subducted slabs. *Earth Planet. Sci. Lett.* 133:1–17.

Griffiths, R. W., and Richards, M. A. 1989. The adjustment of mantle plumes to changes in plate motions. *Geophys. Res. Lett.* 16:437–40.

Griffiths, R. W., and Turner, J. S. 1988a. Folding of viscous plumes impinging on a density or viscosity interface. *Geophys. J.* 95:397–419.

Griffiths, R. W., and Turner, J. S. 1988b. Viscous entrainment by sinking plumes. *Earth Planet. Sci. Lett.* 90:467–77.

Gurnis, M. 1986. Stirring and mixing in the mantle by plate-scale flow: large persistent blobs and long tendrils coexist. *Geophys. Res. Lett.* 13:1474–7.

Gurnis, M., and Davies, G. F. 1986a. Mixing in numerical models of mantle convection incorporating plate kinematics. *J. Geophys. Res.* 91:6375–95.

Gurnis, M., and Davies, G. F. 1986b. The effect of depth-dependent viscosity on convective mixing in the mantle and the possible survival of primitive mantle. *Geophys. Res. Lett.* 13:541–4.

Hill, R. I., Campbell, I. H., Davies, G. F., and Griffiths, R. W. 1992. Mantle plumes and continental tectonics. *Science* 256:186–93.

Hill, R. I., Campbell, I. H., and Griffiths, R. W. 1991. Plume tectonics and the development of stable continental crust. *Exploration Geophys.* 22:185–8.

Hoffmann, N. R. A., and McKenzie, D. P. 1985. The destruction of chemical heterogeneities by differential fluid motions during mantle convection. *Geophys. J. Royal Astron. Soc.* 82:163–206.

Hofmann, A. W., and White, W. M. 1982. Mantle plumes from ancient oceanic crust. *Earth Planet. Sci. Lett.* 57:421–36.

Howard, L. N. 1964. Convection at high Rayleigh number. In *Proceedings of the 11th International Congress on Applied Mechanics*, ed. H. Görtler, pp. 1109–15. Berlin: Springer-Verlag.

Irifune, T., and Ringwood, A. E. 1987. Phase transformations in a harzburgite composition to 26 GPa: implications for dynamical behaviour of the subducting slab. *Earth Planet. Sci. Lett.* 86:365–76.

Kellogg, L. H., and Turcotte, D. L. 1986. Homogenization of the mantle by convective mixing and diffusion. *Earth Planet. Sci. Lett.* 81:371–8.

Kincaid, C., and Olson, P. 1987. An experimental study of subduction and slab migration. *J. Geophys. Res.* 92:13832–40.

Loper, D. E., and Stacey, F. D. 1983. The dynamical and thermal structure of deep mantle plumes. *Phys. Earth Planet. Int.* 33:304–17.

Morgan, W. J. 1971. Convection plumes in the lower mantle. *Nature* 230:42–3.

Olson, P. L., and Nam, I. S. 1986. Formation of seafloor swells by mantle plumes. *J. Geophys. Res.* 91:7181–91.

Olson, P. L., Schubert, G., Anderson, C., and Goldman, P. 1988. Plume formation and lithosphere erosion: a comparison of laboratory and numerical experiments. *J. Geophys. Res.* 93:15065–84.

Olson, P. L., Yuen, D. A., and Balsinger, D. 1984. Mixing of passive heterogeneities by mantle convection. *J. Geophys. Res.* 89:425–36.

Ottino, J. M., Leong, C. W., Rising, H., and Swanson, P. D. 1988. Morphological structures produced by mixing in chaotic flows. *Nature* 333:419–25.

Richards, M. A., and Davies, G. F. 1989. On the separation of relatively buoyant components from subducted lithosphere. *Geophys. Res. Lett.* 16:831–34.

Richards, M. A., Duncan, R. A., and Courtillot, V. E. 1989. Flood basalts and hotspot tracks: plume heads and tails. *Science* 246:103–7.

Richards, M. A., and Griffiths, R. W. 1988. Deflection of plumes by mantle shear flow. Experimental results and a simple theory. *Geophys. J.* 94:367–76.

Richards, M. A., and Griffiths, R. W. 1989. Thermal entrainment by deflected mantle plumes. *Nature* 342:900–2.

Schaber, G. G., Strom, R. G., Moore, H. J., Soderblom, L. A., Kirk, R. L., Dawson, D. J., Gaddis, L. R., Boyce, J. M., and Russell, J. 1992. Geology and distribution of impact craters on Venus: What are they telling us? *J. Geophys. Res.* 97:13257–301.

Schilling, J.-G. 1973. Icelandic mantle plume: geochemical evidence along Reykjanes Ridge. *Nature* 242:565–71.

Schilling, J.-G. 1991. Fluxes and excess temperatures of mantle plumes inferred from their interaction with migrating mid-ocean ridges. *Nature* 352:397–403.

Silver, P. G., Carlson, R. W., and Olson, P. 1988. Deep slabs, geochemical heterogeneity, and the large-scale structure of mantle convection: investigation of an enduring paradox. *Annu. Rev. Earth Planet. Sci.* 16:477–541.

Skilbeck, J. N., and Whitehead, J. A., Jr. 1978. Formation of discrete islands in linear island chains. *Nature* 272:499–501.

Sleep, N. H. 1990. Hotspots and mantle plumes: some phenomenology. *J. Geophys. Res.* 95:6715–36.

Solomatov, V. S., Moresi, L.-N. 1996. Stagnant lid convection on Venus. *J. Geophys. Res.* 101:4737–53.

Stacey, F. W., and Loper, D. E. 1983. The thermal boundary layer interpretation of D″ and its role as a plume source. *Phys. Earth Planet. Int.* 33:45–55.

Stevenson, D. J., and Turner, J. S. 1979. Fluid models of mantle convection. In: *The Earth: Its Origin, Structure and Evolution*, ed. M. W. McElhinny, pp. 227–63. London: Academic Press.

Tozer, D. C. 1965. Heat transfer and convection currents. *Phil. Trans. R. Soc. London* A258:252–71.

Tritton, D. J. 1985. Experiments on turbulence in geophysical fluid dynamics. II. Convection in very viscous fluid. In: *Turbulence and Predictability in Geophysical Fluid Dynamics and Climate Dynamics*, pp. 193–9. Bologna: Soc. Italiana di Fisica.

Turcotte, D. L., and Oxburgh, E. R. 1972. Mantle convection and the new global tectonics. *Annu. Rev. Fluid Mech.* 4:33–68.

Turner, J. S. 1973a. *Buoyancy Effects in Fluids*. Cambridge University Press.

Turner, J. S. 1973b. Convection in the mantle: a laboratory model with temperature–dependent viscosity. *Earth Planet. Sci. Lett.* 17:369–74.

Turner, J. S. 1985. Multicomponent convection. *Annu. Rev. Fluid Mech.* 17:11–44.

van der Hilst, R. D. 1995. Complex morphology of the subducted lithosphere in the mantle beneath the Tonga trench. *Nature* 374:154–7.

van der Hilst, R. D., and Seno, T. 1994. Effects of plate motion on the deep structure and penetration depth below the Izu-Bonin and Mariana island arcs. *Earth Planet. Sci. Lett.* 120:395–407.

Weinberg, R. F. 1997. Rise of starting plumes through mantle of temperature-, pressure-, and stress-dependent viscosity. *J. Geophys. Res.* 102:7613–23.

White, R., and McKenzie, D. 1989. Magmatism at rift zones: the generation of volcanic continental margins and flood basalts. *J. Geophys. Res.* 94:7685–729.

Whitehead, J. A., Jr. 1988. Fluid models of geological hotspots. *Annu. Rev. Fluid Mech.* 20:61–87.

Whitehead, J. A., Jr., and Luther, P. S. 1975. Dynamics of laboratory diapir and plume models. *J. Geophys. Res.* 80:705–17.

Zindler, A., and Hart, S. 1986. Chemical geodynamics. *Annu. Rev. Earth Planet. Sci.* 14:493–570.

5

Plates, Plumes, Mantle Convection, and Mantle Evolution

GEOFFREY F. DAVIES

5.1. Introduction

The major features of mantle convection can be deduced fairly directly from well-established observations and straightforward physics. The picture that emerges is that the tectonic plates are the dominant active component, whereas mantle plumes are an important secondary component. The plates comprise the cool, upper thermal boundary layer of the convecting mantle, and the cycle of creation, cooling, subduction, and reheating of plates is the dominant means by which heat is removed from the Earth's interior. Plumes arise mainly from a hot thermal boundary layer that is probably located at the base of the mantle, and they transport about 10% of the Earth's heat budget through the mantle to the base of the lithosphere. Only a small fraction of this plume heat escapes to the surface.

There are strong arguments against the mantle transition zone being a substantial long-term barrier to flow; in other words, the mantle convects as a single layer, to a good approximation. However, it seems to be close to a state in which phase-transformation effects and chemical-buoyancy effects in the transition zone would separate mantle convection into two layers, and this may have been the case in the past, either episodically or continuously. At present, some plates and plumes may not penetrate the transition zone immediately, but virtually all probably do on a timescale of the order of 10^8 years.

The plates are an integral part of mantle convection, being the main driving thermal boundary layer. The surface pattern of this convection, that is, the geometry of the plates, is not like that in 'normal' convecting fluids, because it is controlled by the essentially brittle rheology of the lithosphere, rather than by the ductile rheology of the underlying mantle.

This understanding of the present-day mantle is leading to better-defined hypotheses about the evolution of the Earth's thermal and tectonic regimes. It is plausible that in Precambrian times the mantle was episodically layered and has evolved to its present, essentially unlayered state within the past 1 billion years. It is also plausible that plate tectonics would not have worked when the mantle was more than about 50°C hotter than at present, which may have been the case as recently as 1.5 Ga. Nevertheless, the dynamical nature of earlier tectonic regimes is still unclear, and a smooth evolution of the Earth, with modern-style plate tectonics operating for the past 3 billion years or more, cannot be ruled out either. Whether plumes were more or less important than at present is also unclear. Plumes do not offer an alternative to plate tectonics, because they derive from a different thermal boundary layer.

The purpose of this chapter is to summarize the main arguments that have led to the foregoing conception of the present-day mantle and to suggest some implications for the mantle's thermal and chemical evolution and the tectonic evolution of the Earth. A much more detailed account of the arguments concerning the present-day mantle has been given by Davies and Richards (1992), but because the concern there was to address all of the many issues that had been raised, perhaps the simplicity of the main arguments was not so apparent in that paper. Griffiths and Turner (Chapter 4, this volume) have provided a more detailed discussion of some of the dynamics involved and of the contribution of laboratory experiments to an understanding of such dynamics. Here the emphasis is more on connecting the theory of dynamical behaviour with observations and developing an overall picture of mantle dynamics.

5.2. Convection

All thermal convection is driven by one or two thermal boundary layers (Griffiths and Turner, Chapter 4, this volume). *Convection* is the fluid motion that occurs when the buoyancy (positive or negative) of the thermal boundary layer causes it to detach from the fluid boundary and rise or fall through the interior of the fluid layer. It is necessary to state this basic point in order to steer us past some misconceptions that are still rather commonly encountered in the geological community. Thus, mantle convection need not be regarded as something mysterious and arbitrary that goes on 'down there' below the plates. If there is a mode of convection present, there must be a thermal boundary layer that is driving it, and this limits the possibilities in important ways. We shall see later that the question of whether the plates are active or passive participants in mantle convection can be readily resolved, as can the question of how mantle convection can fit in with the peculiar geometry of the plates.

Textbook examples of convection usually show the case of a layer of fluid heated from below and cooled from above. In this case there is a hot thermal boundary layer at the bottom and a cool thermal boundary layer at the top (Figure 5.1a). However, it is possible to have, for example, a fluid layer heated from within and cooled from above, in which case there will be no hot thermal boundary at the bottom, only the cool one at the top (Figure 5.1b). The mantle seems to be rather close to this condition, as we shall see. The point here is that each thermal boundary layer can drive a distinctive pattern of convection. Again, this is true of the mantle, because

(a)

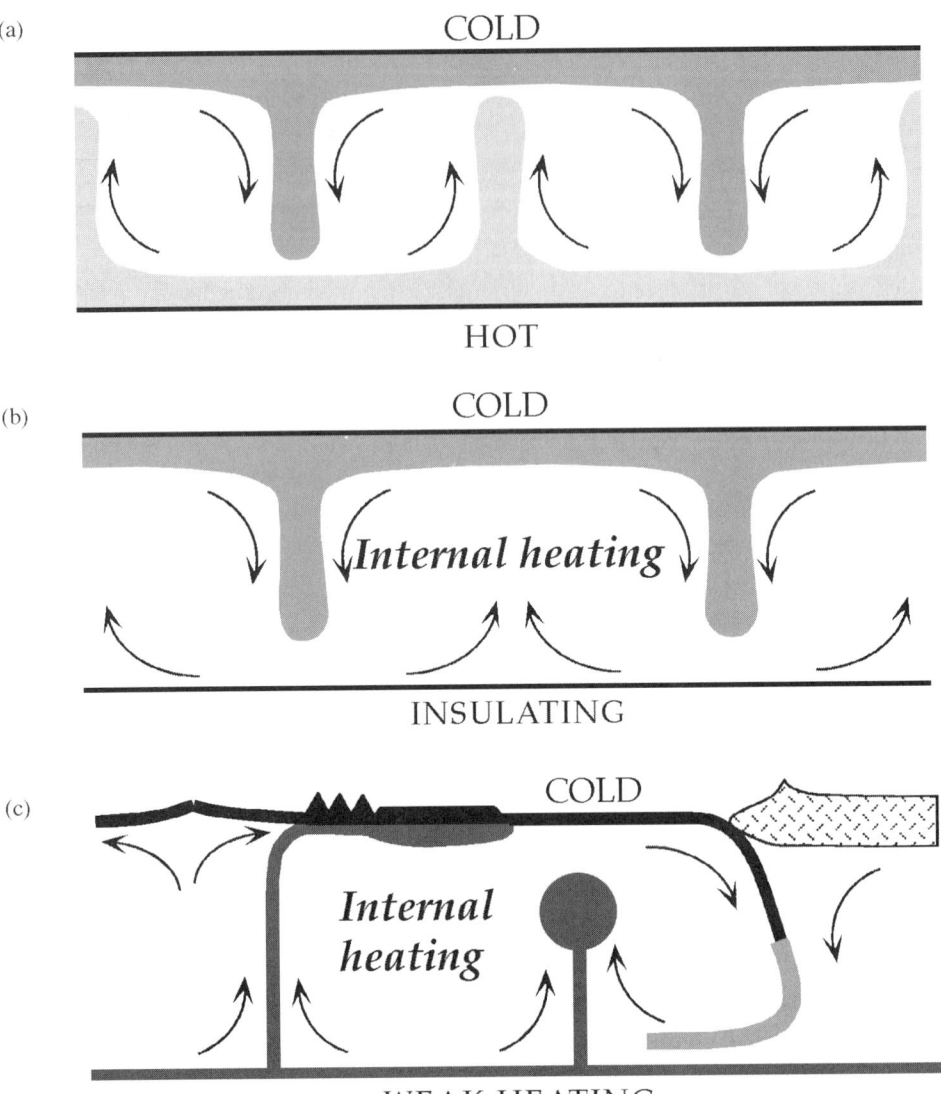

Figure 5.1. Sketches of different types of convection. (a) 'Textbook' convection, with a fluid layer heated from below and cooled from above. Two thermal boundary layers exist, and each drives flow, though in a co-operative pattern. (b) Internally heated convection cooled from above. There is a thermal boundary layer only at the top. (c) Mantle convection. The fluid is cooled at the top, but heated mainly from within and weakly from below. As well, the temperature dependence of the rheology causes the thermal boundary layers to behave differently, and the flows they drive are to some extent independent, with columnar plumes punching through big cells driven by the plates.

of the different physical properties of the two boundary layers: we can think of a plate mode of convection and a plume mode of convection. In the mantle, these two modes seem to operate with a considerable degree of independence (Figure 5.1c). This is in contrast to the usual textbook case (Figure 5.1a), in which the two thermal boundary layers cooperate in driving a single pattern of flow.

(a) Top:

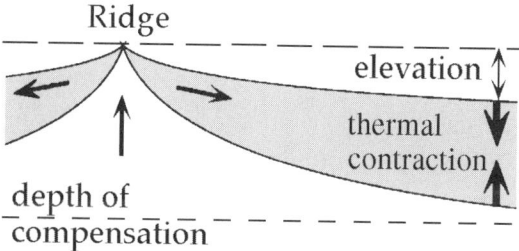

(b) Bottom:

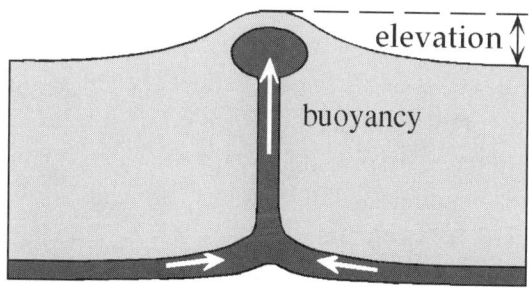

Figure 5.2. Top: Sketch of the topography due to the cool upper thermal boundary layer of the mantle. Bottom: The topography due to the hot lower thermal boundary layer of the mantle; the top surface is affected only when the buoyant fluid from the bottom boundary layer approaches the top surface.

One more point will be made here because it is central to the case to be developed. Convection in a fluid is driven by the buoyancy of the thermal boundary layers, and that same buoyancy will deflect the top surface of the fluid when the buoyant material is close to that surface; in other words, mantle convection generates topography at the Earth's surface. The top thermal boundary layer generates topography mainly because of variations in its thickness. Fluid from the bottom thermal boundary layer can generate significant topography only when it rises to the top of the layer (Figure 5.2). We shall see that the Earth's topography is the principal observation that constrains our understanding of the form of mantle convection.

5.3. The Lithosphere

The lithosphere is defined as the strong outer layer of the mantle that is effectively rigid on geological timescales. Because its strength derives from the fact that it is cooler than the underlying mantle, it coincides substantially with the cool thermal boundary layer at the top of the mantle. (A thermal boundary layer, in this context, is the region where the temperature changes from its value at the surface to its value in the interior of the fluid.) The thickness of the thermal boundary layer is quite variable (Figure 5.1c). In oceanic regions, it varies from about 10 km at spreading centres to over 100 km under old seafloor. In continental areas it may be over 200 km thick under the oldest crust. The continental crust is a compositional heterogeneity that is incorporated into the thermal boundary layer, and there is evidence that the sub-crustal portion in continental regions is also compositionally distinct.

The lowest part of the lithosphere is the warmest, and its properties grade into those of the underlying mantle. Thus there is a gradational transition from the characteristics of high strength and brittle failure in the cool upper part to ductility deeper down, and the lowest 20 km or so of the thermal boundary layer may be somewhat mobile, depending on circumstances, and not strictly part of the lithosphere. (The sub-lithospheric mantle is often called the 'asthenosphere', meaning the weak layer, but I avoid this term here because it has acquired some narrower meanings identifying it with the low-viscosity zone in the uppermost mantle, or with a zone of low seismic wave speeds, or sometimes with all of the upper mantle.)

A crucial property of the Earth's lithosphere is that it is broken into pieces. This allows the pieces to move, despite their 'rigidity'. This concept of moving, rigid pieces or plates is the central idea in Wilson's formulation of the theory of plate tectonics (Wilson, 1965).

These special mechanical properties of the lithosphere result from the temperature dependence of the rheology of silicates. Thus at high temperatures, silicates are ductile and approximately viscous ('viscous' implies a linear relationship between stress and strain rate), and the viscosity is strongly temperature-dependent, decreasing approximately by a factor of 5 for each $100°C$ increase in temperature (Drury and Fitz Gerald, Chapter 11, this volume). On the other hand, at low temperatures silicates are brittle–elastic, a fact well known to structural geologists and seismologists. Reflecting this, the boundaries between plates can be recognized as corresponding to one of the three standard types of faults: normal, reverse, or strike-slip.

The brittle behaviour of the lithosphere provides the clue also to the angular shapes, varied sizes, and kinematics of plates. Normally (i.e., when no new plate boundaries form), seafloor spreading at mid-ocean ridges is symmetric (i.e., equal amounts of new plate are formed on the two sides of the spreading centre), but subduction zones consume material asymmetrically (only one plate goes down). This means that plates may grow or shrink, depending on the relative spreading and subduction rates. Magnetic anomalies record the growth of the Pacific plate from a small plate in the Mesozoic, and a corresponding decrease in the size of the Farallon plate is inferred. The asymmetry of subduction is a straightforward consequence of this type of boundary being a reverse fault: gravity prevents the overthrusting plate from going up, so the only way in which continuing convergence can be accommodated is for the underthrusting plate to keep going down. The observed symmetry of spreading centres requires for its explanation the plausible assumption that the plates will separate at the weakest place: the centre of the most recent upwelling or intrusion.

5.4. Plate-Scale Flow

We have inferred an upper thermal boundary layer with distinctive and unusual mechanical properties: it is very stiff, but broken into pieces that are mobile. Referring back to our statement that convection is driven by thermal boundary layers, we next ask what form of convection this boundary layer will drive. The way in which the negatively buoyant lithosphere detaches from the Earth's surface is by subduction. It is at subduction zones that plates sink into the mantle and drive mantle circulation: the subducted plate drags viscous mantle along with it, and it also drags the rest of the attached surface plate along behind it. The resulting flow is a *circulation*, that is, it returns upward in regions away from subduction zones (Figure 5.1b). This upward flow will reach the surface where two plates are pulling apart at a spreading centre. The total circulation driven by the sinking plates can be called the plate-scale flow.

Several important points can be made here. Firstly, in the absence of another boundary layer, the upward flow is passive, meaning that it is not driven by a positive buoyancy of its own: it is normal mantle at normal temperature. We shall see later that there is clear evidence that this is usually the case at spreading centres.

It is then easier to understand how mantle flow can accommodate to the relative motions and irregular geometry of spreading centres, with ridge segments disconnected and offset from each other by transform faults. Because the flow is passive, it ascends only where there is a spreading segment, and doesn't ascend under a transform segment. Whether or not the spreading centres are moving relative to some other feature (such as a subduction zone) is immaterial: each spreading centre merely pulls up local mantle, wherever it happens to be.

It has been suggested that there is a pervasive 'small-scale' convection driven by the lowermost, softest part of the boundary layer (Parsons and McKenzie, 1978). If that were the case (there is no clear evidence for it), it would have to be a relatively minor mode of convection, because most of the negative buoyancy of the boundary layer is trapped by the strength of the plates until it is released into the mantle at a subduction zone.

It is clear from this approach that the plates are an active component of mantle convection. This does not preclude other components that may influence plates, but we shall see that their influence seems to be secondary. Nor does it decide which of various proposed forces acting on plates at the surface (such as 'slab pull' or 'ridge push') is dominant: Each of these is a result of the existence of the thermal boundary layer, and the more general point being made here is that it is the thermal boundary layer that is active. Nor is it necessarily inconsistent with the relative motions of plates that are not subducting (such as the African, South American, and Antarctic plates), because these may be driven secondarily by mantle motion driven by subduction of other plates, as well as by the weaker internal ridge-push force. The cause of the motion of these plates is not clear-cut, however, because plumes may play some role.

The driving force due to subducted lithosphere does not cease when it becomes aseismic at depth: its negative buoyancy is undiminished, even though it smears out by thermal diffusion, and it continues to act until it reaches the base of the convecting layer. Thus, there must still be downwelling under western North America because of the subduction of the Farallon plate there until relatively recently, even though the subducted Farallon plate is not seismically active.

5.5. Plumes

There is a robust argument, which I shall now present, that if the mantle has been cooling, then there must be thermal plumes. Suppose the Earth was very hot soon after its formation, because of trapped gravitational energy of accretion, or, more specifically, because of the impact of a Mars-size body, and that there was no temperature difference between the mantle and the core. The mantle would have cooled by convection, and as it did a temperature difference between the mantle and the core would have developed. Heat would then have been conducted out from the core, forming a hot thermal boundary layer at the base of the mantle. This thermal boundary layer would have had a lower viscosity, and both theory and experiment indicate that it would have generated buoyant upwellings that would have been columnar in form (rather than sheets) (Whitehead and Luther, 1975). In other words, mantle plumes would have developed.

A new upwelling advances as a large spherical 'head', because a great amount of buoyancy is required to push through the higher-viscosity surrounding mantle. Once a path is made, the lower-viscosity material from the boundary layer requires only a narrow conduit through which to flow after the plume head. The result is the characteristic head-and-tail structure of a new plume (Whitehead and Luther, 1975; Griffiths and Campbell, 1990). The head grows by addition of material from the tail and by thermal entrainment. After the head reaches the top of the layer, the tail may continue for some time. The dynamics of plumes are described in more detail by Griffiths and Turner (Chapter 4, this volume).

We can thus deduce the probable existence of thermal mantle plumes on the basis of experiments, theory, the temperature dependence of mantle viscosity, and the likelihood that the mantle has cooled relative to the core. If there is, or was, a barrier to flow in the mantle transition zone, then we could also expect plumes to arise at that interface. We shall see that the existence of mantle plumes can also be inferred rather directly from observations.

5.6. Signatures of Plate-Scale Flow

We have deduced the existence of plate-scale flow from the existence of a cool thermal boundary layer at the surface of the Earth, from its special mechanical properties, and from the fact that it is broken into plates that are moving. We can deduce important quantitative information about this flow from the observed distributions of topography and heat flux at the Earth's surface.

5.6.1. Mid-Ocean-Ridge Topography

The general types of topography to be expected from mantle-type convection are sketched in Figure 5.2. There are two types: the mid-ocean-ridge topography (due to the upper, cool thermal boundary layer) and swells (due to the ascent of buoyant material from a hot boundary layer at depth). Mid-ocean-ridge topography has a special and relatively simple form in the mantle convection system, and it is well explained by a simple physical theory that we shall now examine.

As a plate moves away from a spreading centre, it cools by conduction of heat to the Earth's surface. As a result, it becomes thicker, because it is the cooled part that has the strength to behave as a stiff plate. Heat conduction is a thermal-diffusion process, and diffusion processes have the general property that the distance over which a diffusing quantity extends is proportional to the square root of the time elapsed since the diffusion began. This can be understood qualitatively by noting that in order for the thickness of the cool layer to double, twice as much heat will have to be removed, but in the process the thermal gradient will be halved, so the rate of heat removal by conduction will be halved. Consequently it will take *four* times as long for the layer thickness to double as it did to reach its present thickness. A rigorous mathematical analysis leads to the same conclusion, which can be expressed very generally in the form

$$t = az^2/\kappa \tag{1}$$

where t is time, z is depth, κ is the thermal diffusivity, and a is a dimensionless constant that will be of the order of 1 if representative values of z and t are used. Note that $\kappa = K/\rho C_P$, where K is the thermal conductivity, ρ is density, and C_P is specific heat.

Typically, for rocks, κ is about 10^{-6} m^2/s, or 30 km^2 per million years, and for this problem, $a = 0.25$ (Turcotte and Schubert, 1982). These values predict that it takes about 80 million years for the lithosphere to reach 100 km in thickness, quite good approximations to the age and thickness of the older oceanic lithosphere.

As lithosphere thickens, it undergoes thermal contraction, while remaining in isostatic balance with the younger and older parts of the plates. The thermal contraction is therefore accommodated by subsidence of the top surface of the lithosphere relative to its level at the spreading centre. For upper-mantle rocks, the volume coefficient of thermal expansion, α, is about 3×10^{-5} °C^{-1}. The temperature of the upper mantle is about 1,300°C, so the average temperature drop, ΔT, within the lithosphere material since it was under the spreading centre is about 650°C. The increase in depth, Δd, to the top of the lithosphere (the seafloor), when the lithosphere thickness is z, is then

$$\Delta d = \alpha \Delta T (\rho/\Delta\rho)z = \alpha \Delta T (\rho/\Delta\rho)(\kappa t/a)^{1/2} \tag{2}$$

where ρ is the density of the upper mantle, $\Delta\rho$ is the density contrast between the upper mantle and seawater, and the factor $\rho/\Delta\rho$ takes account of the isostatic balance under seawater. With $\rho = 3{,}300 \, \text{kg} \cdot \text{m}^{-3}$, this gives $\Delta d = 3 \, \text{km}$ of subsidence for $z = 100 \, \text{km}$, a good approximation to the change in depth of the ocean between a spreading centre and the old seafloor.

One of the remarkable early discoveries that followed the recognition of seafloor spreading was that the increase in the depth of the seafloor is indeed proportional to the square root of its age (Sclater and Francheteau, 1970). We see that this is explained by a remarkably simple theory: it is due to the conductive cooling and thermal contraction of a plate as it moves away from a spreading centre.

Now let us adopt a different perspective. The mid-ocean-ridges stand high relative to old seafloor because of the thermal contraction just considered. We have actually fully explained the existence of the mid-ocean ridges. We do not need to invoke any uplift mechanism at ridges to explain their presence. We have explained them by assuming, implicitly, that the mantle underlying the plates is passive. Specifically, *no extra buoyant upwelling is required under the spreading centres to explain their relative elevation*. In fact, any such extra buoyancy would destroy the agreement between theory and observation.

In other words, the agreement, to first order, between our simple cooling-plate model and the observed seafloor topography implies that the mantle under the plates is relatively passive. There are deviations from the simple square-root-of-age depth dependence, and we shall see that they tell us some important things about mantle dynamics, but they are secondary, with amplitudes of no more than 1 km.

The two dominant features of the Earth's topography are the continents and the mid-ocean-ridge system. The elevation of the continents relative to the seafloor is explained by their lower density and consequent relative buoyancy. The mid-ocean-ridge system is explained by a near-surface process: the thickening of the oceanic lithosphere by thermal conduction, and its consequent thermal contraction. The oceanic lithosphere, as we have seen, is a thermal boundary layer that drives what we have called the plate-scale mode of mantle convection. Thus the dominant dynamic topography of the Earth is a direct expression of the plate-scale mode of mantle convection.

5.6.2. Heat Flow through the Seafloor

It is observed that heat flux through the seafloor declines with age in proportion to $(\text{age})^{-1/2}$, to a good approximation (Sclater, Jaupart, and Galson, 1980). As a plate thickens with age according to equation (1), the heat flux through its surface is

$$q = K \, \Delta T \, (b/\kappa t)^{1/2} \tag{3}$$

where b is a dimensionless constant. This form can be obtained directly from equation (1). A rigorous analysis (Turcotte and Schubert, 1982) gives $b = 1/\pi$.

With $\Delta T = 1,300°C$, $t = 100$ million years, and the other quantities as before, this yields $q = 40$ mW/m^2, a good approximation to the observed heat flux through old ocean floor. Thus the model of lithosphere thickening by conduction explains the observed magnitude and age dependence of the heat flux through the seafloor, as well as the observed subsidence of the seafloor with age.

Using this result, the total heat flow through the seafloor has been estimated to be 3×10^{13} W (Sclater et al., 1980), which is about 85% of the heat emerging from the mantle (Davies and Richards, 1992). This implies directly that the cycle of plate formation at a spreading centre, cooling at the surface, subduction, and reheating in the mantle is the dominant process by which heat is removed from the Earth's interior. In other words, the plate-scale flow is the dominant mode of mantle convection.

5.7. Plume Signatures
5.7.1. Plume Topography

Around many volcanic hotspots there are broad swells, each typically 0.5–1 km high and up to 1,000 km in diameter (Crough, 1983). [I shall use the term 'hotspot' here to refer to isolated surface volcanic centres. Wilson (1963) originally used the term to refer to a hypothetical isolated source of melting in the mantle, but that theory has been supplanted by the plume theory described here.] These swells are clear deviations from the square-root-of-age subsidence just described. The swells cannot be explained by excess thickness of the oceanic crust, and they are too broad to be supported by elastic flexure of the lithosphere. The clearest example is around Hawaii. Recent seismic evidence has shown that there is little thinning of the lithosphere under the Hawaiian swell (Woods, Leveque, and Okal, 1991), and there is no clear heat-flow anomaly (Von Herzen et al., 1989), which also indicates normal lithosphere thickness. The only viable explanation is the presence of buoyant mantle under the lithosphere. The occurrence of active volcanism at Hawaii, with progressively older extinct volcanism extending to the northwest along the volcanic chain (Clague and Dalrymple, 1989), implies the existence of a long-lived column of rising, buoyant mantle under Hawaii, that is, of a mantle plume (Morgan, 1971). The swell can be explained as due to the spreading of the buoyant plume material under the lithosphere.

The buoyancy and anomalous melting of a plume might, in principle, be attributable either to the plume being hot or to it containing excess volatile components (H_2O and CO_2) (Green and Falloon, Chapter 7, this volume). We have seen that thermal plumes are to be expected in a cooling Earth, so it is straightforward to identify the buoyant columns inferred in the preceding paragraph with such thermal plumes. It is plausible that plumes may also contain excess or variable amounts of volatiles, which could exert important influences. However, in the absence of a plausible dynamical scheme to account for the origin of long-lived, slow-moving 'wet spots' or wet plumes, and in the absence of compelling geochemical evidence, the extreme case of a plume being wet but not hot will not be further considered here.

Hotspot swells comprise a second type of topography that can be associated with a mode of mantle convection. In this case the associated thermal boundary layer is a hot one at depth. We might suspect, from the fact that hotspot swells are a much less prominent form of dynamic topography than mid-ocean ridges, that the plume mode of convection is secondary in comparison with the plate mode.

5.7.2. Plume Heat Flow

Davies (1988) pointed out that the size of a hotspot swell can be used to estimate the flux of buoyancy up a plume. In the Hawaiian case, the swell must be extended to the southeast, as the Pacific plate moves to the northwestward over the plume at about $100 \text{ mm} \cdot \text{yr}^{-1}$. The weight of the new swell topography created in one year (1 km high by 1,000 km across the chain, and 100 mm along the chain) balances the buoyancy of the plume material arriving each year. There is then a simple conversion from buoyancy flux B to heat flow Q:

$$Q = C_P B / \alpha g \qquad (4)$$

Davies (1988) and Sleep (1990) used this approach to deduce that the Hawaiian plume (which is the strongest) carries about 1% of the Earth's heat budget, while all plumes together carry about 6%.

Morgan (1981) noted that several hotspot tracks emerge from flood-basalt provinces, and he proposed that the flood basalts were due to the arrival of a plume head, with the emerging hotspot track being due to the following plume tail (Griffiths and Turner, Chapter 4, this volume). This idea has now gained considerable support (Richards, Duncan, and Courtillot, 1989; Campbell and Griffiths, 1990). Hill et al. (1992) used this inference and the observed frequency of flood-basalt eruptions to estimate that plume heads transport about 50% as much heat as plume tails.

Thus plume heads and tails together carry about 10% of the Earth's heat budget through the mantle. Note that this is heat carried *through* the mantle to the base of the lithosphere. The Hawaiian example illustrates that if the plume arrives under thick lithosphere, little of its excess heat may escape to the surface. Much of it may be stirred back into the mantle.

5.8. A Barrier at a Depth of 660 km?

There has been some question whether or not the mantle is divided into two separately convecting layers at a depth of 660 km, where seismic wavespeeds and density increase discontinuously. We shall examine three kinds of evidence: seafloor topography, seismic tomography, and numerical modelling. Many other arguments, including an important argument from the gravity field, were discussed by Davies and Richards (1992), and they judged there that the evidence either supports the conclusion suggested here or is not inconsistent with it.

5.8.1. The Topographic Constraint

The most direct and most robust constraint again comes from seafloor topography and the Earth's heat budget. The upper mantle (or, more accurately, the source for mid-ocean-ridge basalts) is strongly depleted of the main radioactive heat-producing elements (K, U, and Th). If the observed concentrations of these elements are integrated over a depth of 660 km, they are sufficient to produce only about 2% of the heat emerging from the mantle (Jochum et al., 1986; Davies and Richards, 1992). Approximately another 6% can be accounted for by the slow cooling of the upper mantle, inferred from models of the thermal evolution of the mantle. This means that most of the heat emerging from the mantle must be coming from deeper than 660 km.

The heat must be transported in one of two ways. If there is no barrier to convective flow at 660 km, then it can be advected. That is, it can be transported in the normal way for a convecting fluid via descending cool material and/or ascending warm material. If there is a barrier to flow, then the heat must be conducted. That requires a steep temperature gradient, or, in other words, a thermal boundary layer. In fact, there would be two thermal boundary layers, one above and one below the interface. The resulting hot boundary layer at the base of the upper mantle would give rise to upwellings, probably in the form of plumes, for the reasons discussed earlier. When these reached the base of the lithosphere, they would lift it and thus generate topography.

Now we can see that if there is a barrier to flow at 660 km, then there should be upwellings arising from that depth and carrying about 90% of the heat emerging from the mantle, and generating proportionate uplifts. Simple estimates suggest, and numerical calculations (Davies, 1988) confirm, that if the upwellings carried most of the mantle's heat budget, then the resulting uplifts should be comparable in amplitude and extent to the mid-ocean-ridge topography, though different in form (Figure 5.2).

It is clear that there is no such topography on the seafloor. The observed deviations from the square-root-of-age topography expected from the cooling plates are of two forms. There are the hotspot swells, which, as we have seen, correspond to plumes transporting about 10% of the mantle's heat budget. There are also broad, low-amplitude variations whose origin is debated, but which in any case correspond to only a few percent of the mantle's heat budget; these will be discussed later. In concise terms, the seafloor topography is dominated by the mid-ocean-ridge system, and other types of topography are secondary.

Correspondingly, mantle heat transport is dominated by the plates, and the plumes or other upwellings are secondary. Thus there cannot be a substantial barrier to flow at a depth of 660 km. The plumes, or most of them, must be coming from below the region where most of the mantle's heat is being generated by radioactivity or is being released by slow cooling. The most straightforward inference is that the plumes originate at the base of the mantle and that they transport the heat that is conducted from the core into the base of the mantle.

One point may need clarification here. The lower thermal boundary layer and its associated plumes transport heat *into* and *through* the interior of the convecting layer, whereas the upper thermal boundary layer and its associated downwellings (subducted lithosphere) transport heat *out of* the convecting layer. Thus the percentages of heat transport do not have to add up to 100%. In a purely bottom-heated case, they would add to 200%, but they really represent two different stages of the heat transport. It follows, as Stacey and Loper (1984) first pointed out, that plates cool the mantle, and plumes cool the core.

5.8.2. Seismic Tomography

Several seismic-tomography studies of subduction zones, as reviewed by Kennett and van der Hilst (Chapter 8, this volume), have revealed that some of the subducted lithosphere seems to have been deflected horizontally above a depth of 660 km, whereas in other places it seems to have penetrated directly into the lower mantle. This conclusion seems likely to survive, although there are some important questions about these studies still to be clarified. For example, the mantle transition zone, where seismic wavespeeds undergo two or more discontinuous increases with depth, is the most difficult place in which to attempt to detect sub-horizontal lithospheric remnants. It will be essential to have a robust reference model before reliable detections can be claimed (van der Hilst et al., 1991). Van der Hilst and others (van der Hilst and Seno, 1993; van der Hilst, 1995; Kennett and van der Hilst, Chapter 8, this volume) have shown that at least two cases of horizontally deflected slabs can be explained by the kinematics of trench migration at the surface. In the case of Tonga, the horizontal deflection was temporary, with the slab eventually penetrating into the lower mantle. These studies and that of Grand (1994) have presented strong cases for some subducted lithosphere having penetrated into the lower mantle.

5.8.3. Numerical Modelling

There have been many numerical models that have attempted to address the effect on mantle convection of a pressure-induced phase transformation with a negative Clapeyron slope. The Clapeyron slope β is the temperature derivative of the transition pressure P_{tr}: $\beta = dP_{tr}/dT$. Several authors have concluded that convection would be separated into two layers, but others have reached the opposite conclusion.

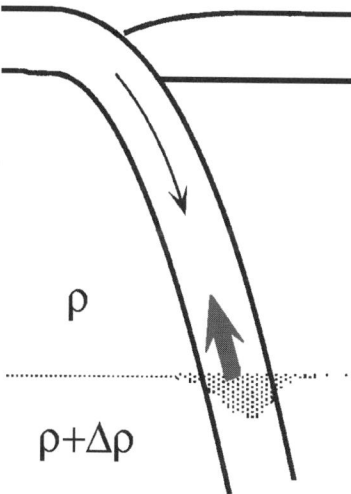

Figure 5.3. Sketch of the deflection of a phase-transformation boundary with a negative Clapeyron slope by cool, descending, subducted lithosphere. Buoyancy is induced in the slab where the phase transformation occurs, at greater depth (shaded region), because the material is less dense than the adjacent material at the same depth. This buoyancy opposes the flow of the slab through the phase boundary.

The mechanism usually invoked to explain layering arises from the fact that a transformation with a negative Clapeyron slope would yield buoyancy-forces that would oppose convective penetration of the phase transformation zone. The pressure-induced transformation of Mg_2SiO_4 in the spinel structure (ringwoodite) into $MgSiO_3$ (perovskite structure) plus MgO is thought to have a Clapeyron slope of about -2 to -3 MPa/K (Bina and Helffrich, 1994) and to be the main transformation accounting for the 660-km seismic discontinuity. If β is negative, then in a cold descending flow, such as subducting lithosphere, the mantle material will transform at a higher pressure than the surrounding warmer material (Figure 5.3). This means that the lower-density low-pressure phase will exist at the same depth as the adjacent higher-density high-pressure phase, and consequently the cool, untransformed material will be buoyant. That buoyancy will oppose the negative buoyancy driving the descending flow. If it is sufficiently strong, it can block the convective flow and deflect it sideways. A rising warmer flow will also be inhibited by this effect.

It remains unclear how important this effect might be in the mantle. There is significant uncertainty about the value of the Clapeyron slope for the spinel–perovskite transformation. Potentially more important are the effects of transformations in the pyroxene-stoichiometry component of the mantle at similar depths, which may cancel much of the effect of the olivine-stoichiometry component. The compositional layering of subducted lithosphere may also play a significant role, and it may have been particularly important in the early Earth, when the mantle probably was hotter. These uncertainties and other possibilities will be discussed in more detail by Jackson and Rigden (Chapter 9, this volume), and the early evolution of the mantle will be discussed later in this chapter. At this stage it seems that the net effect of thermal deflections of phase boundaries could be anywhere from zero to the equivalent of $\beta = -3$ MPa/K, and the role of chemical deflections of phase boundaries has hardly been explored.

The diversity of outcomes from numerical modelling can be attributed to two main factors: whether the model is two-dimensional or three-dimensional, and whether the viscosity is constant or the models include something resembling stiff lithospheric plates. With constant viscosity and two-dimensionality, transient layering can be induced with $\beta = -2.5$ MPa/K. The behaviour is illustrated in Figure 5.4: layering can persist for some time, but eventually the upper layer becomes cool and dense enough that it breaks through, and there is a catastrophic overturn in which the upper, cool layer drains to the base of the lower layer and is replaced by hot material from the lower layer.

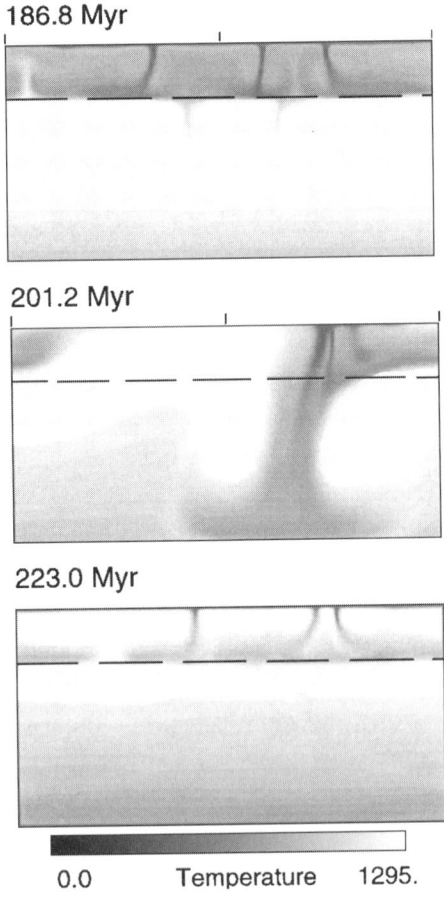

186.8 Myr

201.2 Myr

223.0 Myr

0.0   Temperature   1295.

Figure 5.4. Sequence from a numerical convection model with constant viscosity showing transient layering and an overturn event. The buoyancy from a phase transformation with a Clapeyron slope of -3 MPa/°C is included at the depth of the dashed line.

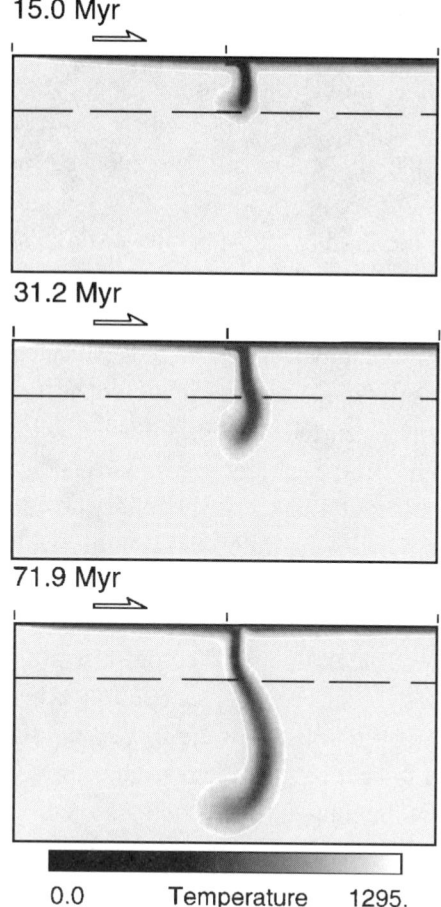

15.0 Myr

31.2 Myr

71.9 Myr

0.0　　　Temperature　　1295.

Figure 5.5. Sequence from a numerical convection model with stiff, but mobile, plates. The viscosity is temperature-dependent, with a maximum viscosity at the cold surface 200 times the viscosity of the interior. Also, the viscosity is reduced locally at each end of the mobile plate, allowing it to move. Phase-transformation buoyancy with a Clapeyron slope of -3 MPa/$°$C is included. The thicker, stronger slab is able to penetrate the phase barrier, whereas the constant-viscosity downwellings of Figure 5.4 did not.

For three-dimensional models with constant viscosity it is found that break-throughs are more frequent, more localized, and less catastrophic (Tackley et al., 1993). Apparently this is because columnar upwellings or downwellings can penetrate more readily than the sheet upwellings and downwellings implicit in two-dimensional models. Nevertheless, significant thermal layering can develop in three-dimensional models.

Constructing a model to allow for variable viscosity is technically difficult, but it is essential for simulating lithospheric plates, and it has important consequences. This is illustrated in Figure 5.5, which shows a two-dimensional model that includes an approximation to a stiff, mobile plate. Such a subducting plate can directly penetrate a phase barrier with $\beta = -3$ MPa/K. A value of β at least as negative as -4

MPa/K is required to prevent penetration of such a plate. Ascending plume heads also can penetrate with $\beta = -3$ MPa/K, whereas plume tails may be blocked with $\beta = -2.5$ MPa/K, a value similar to that found for constant-viscosity convection. Even when a slab was blocked, for the model with $\beta = -4$ MPa/K a delayed break-through was observed that allowed most of the blocked material to pass through the phase barrier.

As discussed earlier, the Earth's lithosphere is distinguished by being both stiff and broken into mobile pieces. Both aspects play important roles in the model findings just described. The stiffness means that more of the weight of the descending column of cold fluid is brought to bear against the resistance of the phase boundary than in a constant-viscosity fluid. The combination of stiffness and mobility means that the boundary layer at the surface becomes much thicker before it detaches and descends into the fluid interior. Such thickness inhibits buckling of the subducted slab, and hence promotes penetration. If the boundary layer is not stiff, it drips away more often and is thinner and more deformable and less likely to penetrate. If the boundary layer is stiff but is not broken into pieces that can move, only the lower-most part of the boundary layer drips away, and again these drips are thin and soft.

The temperature dependence of mantle rheology also affects the ability of plumes to penetrate a phase barrier, because it gives rise to the head-and-tail structure of plumes. The resulting large plume heads are more able to penetrate because they bring greater buoyancy to bear against the resistance: one can accurately think of the 'hydrostatic head' of the plume head as being large because the plume head is large. This does not apply to narrow plume tails, because the viscous resistance of the plume walls locally balances and neutralizes the plume buoyancy.

5.8.4. Summary

A consistent view can be derived from these arguments. It is that plates (and plumes?) may be temporarily prevented from penetrating the 660-km discontinuity, but that ultimately most or all of this material will penetrate. Evidence of transient blockage is provided by seismic tomography, which shows some slabs penetrating and others apparently being blocked, and by numerical models, which show that reasonable simulations of subducted slabs and plumes are likely to penetrate, but may be temporarily blocked. Laboratory models of descending sheets interacting with density and viscosity interfaces (Griffiths and Turner, Chapter 4, this volume) support this conclusion.

Penetration over the long term is supported by the seafloor-topography constraint: this constraint requires only that no significant temperature difference between the upper mantle and lower mantle should accumulate. Such a difference would require complete blockage for times on the order of 100 million years or more. Localized and temporary blockages lasting no more than tens of millions of years are ample to explain the tomographic images, are consistent with the numerical models, and would not result in substantial thermal stratification. Thus there is no inconsistency between these inferences.

It is worth noting that both seismic tomography and numerical modelling indicate that the mantle, at present, seems to be quite close to the conditions in which layering would occur. This may not be a coincidence, as we shall see later.

5.9. Other Internal Layering and Heterogeneity

There is evidence for different seismic properties within the lowest few hundred kilometres of the mantle. There are indications that the viscosity of the mantle increases with depth. There has been controversy and some confusion about whether or not there is or can be additional stratification of minor and trace chemical components. Each of these aspects will now be discussed briefly.

5.9.1. The Bottom of the Mantle (D″ Region)

The D″ layer was defined by Bullen (1965) as a layer at the bottom of the mantle with a thickness of about 200 km. Currently it is viewed as being a zone of seismic heterogeneity on both the small scale (tens of kilometres) and the large scale (thousands of kilometres) (Kennett and van der Hilst, Chapter 8, this volume). The large-scale heterogeneity includes a sharp interface that is not consistently detected and may not be developed everywhere, as well as large variations in the thickness of the zone.

The D″ layer has been interpreted as being either a thermal boundary layer or a chemically distinct zone, or both. However, if the plume flux deduced earlier represents the heat flux through the base of the mantle, it is unlikely that the thermal gradient in the boundary layer is sufficient to account for the seismological properties (Davies, 1990), and it could not alone account for a sharp interface. There is most likely a thermal boundary layer there, as argued earlier, but it is plausible that the seismic structures are due mainly to compositional effects (Griffiths and Turner, Chapter 4, this volume).

If the D″ region is due to a zone of material with slightly greater density, then its large-scale variability is readily explained, because it would be swept toward regions of mantle upwelling, where it would have greatest thickness, and it might be thin or absent under regions of downwelling (Davies and Gurnis, 1986). Hofmann and White (1982) proposed that such a layer might comprise subducted oceanic crust that had segregated and sunk to the bottom of the mantle.

5.9.2. Viscosity Stratification

Modelling of both post-glacial rebound (Mitrovica, 1996; Lambeck and Johnston, Chapter 10, this volume) and the gravity field over subduction zones (Richards and Hager, 1988) indicates that the viscosity of the mantle increases with depth. Mitrovica (1996) has demonstrated that the post-glacial rebound for sites near the centres of the Fennoscandian and North American ice sheets constrains the average viscosity of the mantle, to a depth of about 2,000 km, to be close to 10^{21} Pa $\cdot$ s. He has also shown that such a constraint is not inconsistent with a viscosity increase through the transition zone by a factor of 10–20, specifically from about 3×10^{20} Pa $\cdot$ s in the upper mantle to about 6×10^{21} Pa $\cdot$ s in the middle mantle (700–2,000 km).

A comparable increase in viscosity through the transition zone, by one or two orders of magnitude, has been inferred by Richards and Hager (1988) on the basis of positive geoid anomalies over subduction zones. The viscosity in the deepest mantle (depths of 2,000–3,000 km) is not strongly constrained, but values of at least 10^{22} Pa $\cdot$ s are plausibly consistent with the available constraints (Lambeck and Johnston, Chapter 10, this volume; Mitrovica, 1996). Such high values near the base of the mantle could explain the low horizontal velocities of hotspots, because horizontal convective velocities in the deep mantle would be about an order of magnitude lower (Gurnis and Davies 1986b), so that plume sources would move only slowly.

5.9.3. Chemical Stratification and Heterogeneity

Even if the mantle convects basically as a single layer, there still are several reasons that there might be significant chemical heterogeneity, including stratification, in the mantle. We need to distinguish distinct layering from continuous gradation, and these vertical heterogeneities from more general heterogeneity, which is the variation of composition in three dimensions.

We also need to distinguish between mixing and stirring. Stirring is the intermingling of compositionally distinct components, whereas mixing is the homogenization of distinct components down to the molecular level, thus forming an intermediate composition. In the mantle, solid-state diffusion is so slow that homogenization in the solid state occurs over length scales only on the order of metres after billions of years. However, effective homogenization on larger scales may occur during the generation and extraction of melt, depending on the degree of melting and the dynamics of melt migration and extraction.

The main dispute about mantle layering has concerned whether or not there is a barrier to flow in the transition zone at a depth near 660 km. If there is not, as suggested here, there may be other layering, particularly in the D″ region. Even if that is the only layering, there still could be other chemical differences. Firstly, convection does not instantly destroy heterogeneity. The overturn time for whole-mantle convection is on the order of several hundred million years, and it may take 10 or more overturns to remove most of the heterogeneity. Secondly, the stirring rates will be slower in the lower mantle if it has a higher viscosity, so the lower mantle may be more heterogeneous. Thirdly, extraction of incompatible elements takes place at the top of the mantle, so the upper mantle may be more depleted (and more homogeneous) than the lower mantle. Davies (1990) argued that these three factors might explain much of the observed isotopic heterogeneity of the mantle, which has an apparent age scale of about 2 billion years. Fourthly, although the transition zone may not strongly inhibit flow between the upper mantle and lower mantle, it may reduce the flow sufficiently for some chemical differences to accumulate; the degree to which that can happen needs quantitative evaluation, because heat transport cannot be greatly inhibited, according to the arguments of the preceding section.

The preceding four factors apply to passive heterogeneities, that is, to heterogeneities that do not affect the buoyancy of the material, such as variations in trace-element abundances. A fifth factor that does involve buoyancy and major-element compositions is the possibility of gravitational segregation of the oceanic-crust component of subducted lithosphere, which has a density significantly different from that of the refractory harzburgite layer of oceanic lithosphere and from that of normal mantle (Anderson and Bass, 1986; Ringwood and Irifune, 1988; Ringwood, 1991). Gurnis (1986) concluded from numerical models that such segregation would be inefficient, but might be significant over the whole depth of the mantle (see also Griffiths and Turner, Chapter 4, this volume). Christensen and Hofmann (1994) pointed out, however, that segregation would be more efficient within the thermal boundary layer at the base of the mantle, where the lithosphere would become hotter and softer. That could yield a distinct layer with a high proportion of old crust, as proposed by Hofmann and White (1982), as well as the continuous gradation in average crustal proportions envisaged by Gurnis (1986) and by Davies (1990). Such a layer would provide a plausible explanation for the seismic D″ zone.

Stratification involving buoyancy would affect the dynamics of convection to some degree. In the extreme case of a large density difference, such as between the mantle and core, there is complete separation of flow, and no clear chemical signature of the deeper layer has been discerned in the upper layer. At the other extreme of a very small density difference, flow would be little affected, and any stratification would tend eventually to be removed by the flow, except for the kinematic delay involved in stirring through the whole depth of a viscously stratified mantle, as discussed earlier. In the intermediate range that we are distinguishing here, there would be transport by entrainment across any interface or compositional gradient (Olson, 1984; Griffiths and Turner, Chapter 4, this volume). The viability of such a stratified model of the mantle must then be evaluated quantitatively in terms of the compositions and heterogeneities of the layers in relation to the entrainment flux and the flux due to any renewing mechanism, as well as in relation to observations of chemical heterogeneity and the physical constraints discussed earlier.

At this stage it seems that a distinct layer at the base of the mantle is plausible, with this layer being renewed by addition of subducted oceanic crust, and diminished by entrainment into plumes. Some modest gradation in major-element composition and density through the deeper mantle may also be possible. Substantial changes in composition and hence density within the transition zone seem unlikely for the reasons given earlier.

Substantial gradation in trace-element composition and degree of heterogeneity between the top and bottom of the mantle is dynamically plausible. Observational support for this is provided by the distinctive compositional ranges of plume melting products compared with mid-ocean-ridge basalts (Campbell, Chapter 6, this volume; Davies, 1990); mid-ocean-ridge basalts sample the uppermost mantle, whereas plumes probably sample the bottom of the mantle. In general, mid-ocean-ridge basalts are more depleted of incompatible elements and are more homogeneous, whereas plumes are less depleted and show more heterogeneity, particularly between different plumes. This would be consistent with the uppermost mantle being more depleted and homogenized, and the lowermost mantle being more heterogeneous and less depleted, and possibly involving a distinct layer or zone with a different major-element composition, due, for example, to a higher proportion of subducted oceanic crust.

5.10. Thermal Evolution of the Mantle

The age of the Earth, the nature of its tectonic engine and its tectonic evolution, and the nature of its internal thermal regime and its thermal evolution were at one time major puzzles and the topics of protracted and famous controversies in geology. These aspects are all related, through heat and radioactivity. The discovery of radioactivity and the recognition that it could be a substantial heat source within the Earth allowed the resolution of the controversy about the age of the Earth between Lord Kelvin and prominent geologists, Kelvin's case having been based on arguments about how long it would take for the Earth to cool. Later, an understanding of radioactivity was used to estimate the age of the Earth more directly.

However, it was not until there was widespread acceptance of the concepts of continental drift, plate tectonics, and mantle convection in the late 1960s that a reasonable picture of the Earth's internal thermal regime could be constructed. The amount of heat observed to be emerging from the interior could not be satisfactorily explained until convection came to be seen as the principal means of heat transport. Fisher (1881), a century earlier, not only had proposed mantle convection as a tectonic agent (as discussed later) but also had noted that it would go far towards resolving the argument between Lord Kelvin and the geologists by making available a much greater reservoir of heat and thus increasing the time required for the surface heat flux to fall to the present value. A lively account of some of these controversies has been given by Hallam (1973).

It was not until about 1980 that a simple relationship between temperature and convective heat transport became appreciated by geophysicists, but it was then applied essentially independently by a number of groups to calculate the thermal

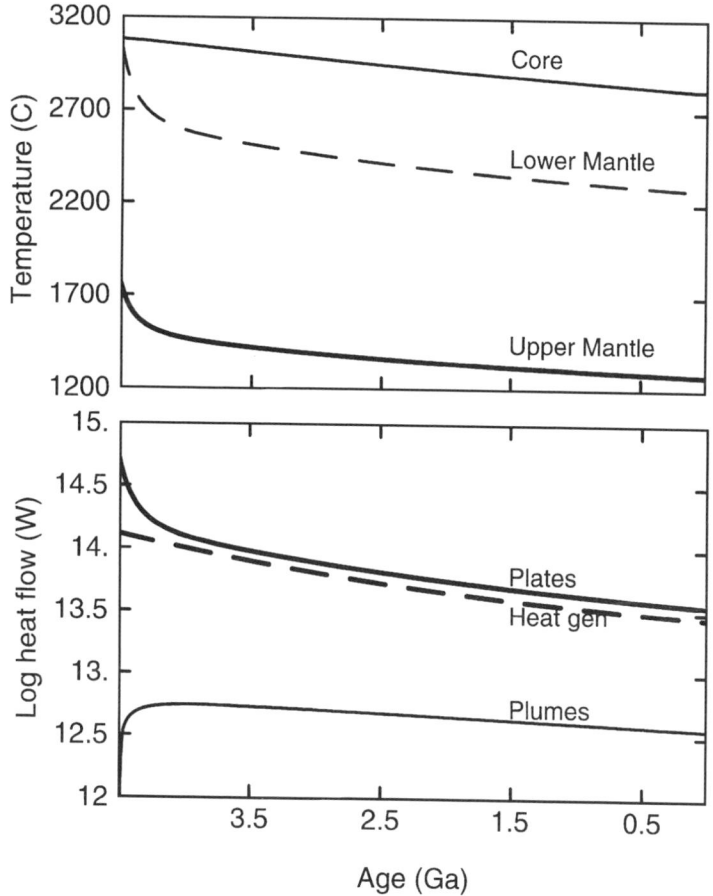

Figure 5.6. Thermal evolution of the mantle and core from a model without phase-barrier effects, and assuming whole-mantle convection. The 'Lower Mantle' temperature refers to the bottom of the mantle, just above the thermal boundary layer. 'Plates' indicates heat loss from the mantle to the Earth's surface; 'Heat gen' indicates radioactive heating; 'Plumes' indicates plume heat flow (from the core into the mantle).

evolution of the interior. (Turcotte, Cooke, and Willemann, 1979; Davies, 1980; Schubert, Stevenson, and Cassen, 1980; Stacey, 1980). Another key ingredient in these analyses was the known strong temperature dependence of mantle viscosity. Such calculations showed that after an initial transient lasting perhaps 500 million years, the temperature of the mantle would have been controlled by the slow decay of the long-lived radioactive heat sources (K, U, and Th) and would have declined by no more than 200–300°C over the past 4 billion years. An example is shown in Figure 5.6.

The possibility that phase transformations in the transition zone might affect mantle convection was recognized even before plate tectonics became established (Verhoogen, 1965) and was further developed early in the plate-tectonics era (Ringwood, 1975; Schubert, Yuen, and Turcotte, 1975), though a clear answer

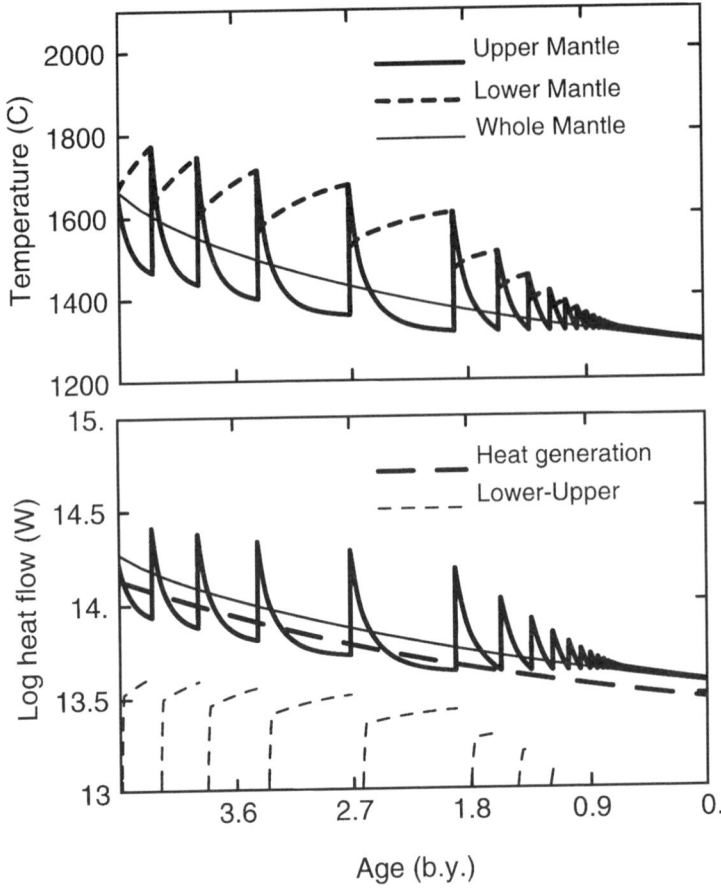

Figure 5.7. Thermal evolution of the mantle with a phase barrier between the upper mantle and lower mantle. The spikes correspond to mantle overturns, when the upper mantle is replaced by hot material from the lower mantle, and the lower-mantle temperature is reset to be the mean of the remaining hot lower-mantle material and the foundered cool material from the upper mantle (as depicted in Figure 5.4). A smooth curve for a whole-mantle model is included for comparison. Here the 'Lower Mantle' temperature is that for the top of the lower mantle. The 'Lower-Upper' curve is the heat flow passing between the mantle layers. The core is not included explicitly in this model.

was slow to emerge (Christensen and Yuen, 1984; Davies, 1995a). However, a further implication was not demonstrated until 1991, when Machetel and Weber (1991) showed that a phase transformation in a simplified convection model induced transient layering, as described earlier. This raised the possibility that the thermal evolution of the Earth has been episodic.

This possibility is only now beginning to be explored quantitatively, but potentially important findings have already emerged. An example is shown in Figure 5.7 (Davies, 1995b). In this model, episodic layering occurs until about 1 Ga, after which whole-mantle convection takes over – consistent with the arguments given earlier against layering of the present-day mantle. During the earlier layered episodes, the temperatures of the upper mantle and lower mantle diverge because the plate-scale flow cools only the upper mantle, while radioactivity is assumed to heat both layers. The layered episodes are interrupted by overturns in which the cooler upper-mantle material drains to the bottom of the mantle, to be replaced by hot material from the lower mantle. For simplicity, the lower mantle is represented in these calculations by its new average temperature.

Before 2.7 Ga, overturns are triggered by an instability at the interface between the layers, which develops when the temperature difference between the layers reaches a critical value (determined by a local Rayleigh number). These overturns occur at increasingly long intervals because of the decline of radioactive heating.

From about 2 Ga onwards in this model, overturns are triggered by plates sinking from the surface and punching through the phase-transformation barrier. They do not do so earlier because the upper mantle is too hot: its viscosity is lower, it convects faster, and the plates move faster and are younger and thinner when they arrive at a subduction zone. Because they are thinner, they can buckle more easily and exert less force on the phase barrier. Overturns triggered in this way occur at decreasing intervals until the lower-mantle temperature is reduced essentially to the upper-mantle temperature. Thereafter, whole-mantle convection is assumed.

This model is neither definitive nor unique. Its significance lies in demonstrating two dynamical phenomena and in some of its resemblances to important aspects of the Earth's tectonic evolution. The dynamical phenomena are its episodic layering, which indeed seems to be quantitatively plausible with reasonable parameter values, and the billion-year phases of behaviour within which the style or presence of overturns is distinctive. The tectonic implications will be discussed in the next section.

Some key aspects of our current understanding of the thermal evolution of the Earth's interior are as follows: The main regulation is by the mantle, with the liquid core acting as a secondary heat reservoir (Davies, 1993b). The temperature of the mantle is determined by a near-balance between radioactive heating and convective cooling (by the plate-scale flow). Changes in the required rate of convective cooling can be accommodated by relatively small changes in mantle temperature, because of the strong temperature dependence of mantle rheology: Increasing the temperature by $100°C$ will decrease the viscosity by about a factor of 5, allowing much faster convection (see also Griffiths and Turner, Chapter 4, this volume). Phase transformations provide a plausible mechanism by which the thermal evolution might be episodic. Without this, the evolution necessarily would be very smooth.

5.11. Tectonic Evolution of the Earth

The beginnings of our modern understanding of the nature of the tectonic engine can be seen in Fisher's proposal (Fisher, 1881) that convection currents in the mantle rise under the oceans and descend under the continents, though he assumed the mantle to be liquid (just prior to presentation of the evidence from instrumental seismology that it is solid). The theory of continental drift, which is a kinematic theory that describes motions, but not the forces that drive them, raised the question of the driving mechanism and led to Holmes's proposal that continents ride on convection currents (Holmes, 1931). The development of the theory of plate tectonics (Wilson, 1965) is deservedly recognized as having promoted a great (though not complete) unification in geology. However, plate tectonics is also a kinematic theory. The theory of mantle convection has the potential to provide the long-sought dynamical theory of the driving forces for both plate tectonics and the subsequently recognized mantle plumes (Morgan, 1971). It is reasonable to claim that the main elements of that theory are now in place.

The acceptance of plate tectonics as the dominant tectonic mode in the present-day Earth soon raised the question whether or not it had been operative throughout geological history. The early opinion was predominantly that it had not, because it was known that the tectonic pattern for Archaean terranes is distinctly different from that for typical Phanerozoic terranes, notably by the absence of the great elongated mountain belts (or their 'geosyncline' remnants) associated with subduction zones. More recently, opinion has swung the other way, with interpretations of Proterozoic and Archaean tectonics and geochemistry commonly being made in a plate-tectonics framework (Nutman et al., 1996). Often these interpretations have involved an explicit or implicit assumption that plates or continents were smaller then, to account for the smaller scale of tectonic elements (Condie, 1995).

There are good reasons to regard these questions concerning the nature of Archaean tectonics and the tectonic evolution of the Earth as yet to be answered unequivocally. Davies (1992) has noted that it is not clear that plates could have been subducted at a sufficient rate to keep the Earth from heating up if the mantle had been as little as $50°C$ hotter than at present. In this scenario, there are two effects of a higher mantle temperature, both working in the same direction. Firstly, melting at mid-ocean ridges would be greater, the oceanic crust would be thicker, and the oceanic lithosphere would be more buoyant. Secondly, the mantle viscosity would be lower, and plates would move faster and thus would be younger and thinner and have less negative thermal buoyancy when they reached a subduction zone. Above a certain critical temperature, there might therefore have been a paradoxical condition in which plates arrived at the location of a potential subduction zone still positively buoyant and unable to be subducted. At higher temperatures, subducting plates still could have existed, but they would have had to move more slowly in order to age for longer periods, and consequently would not have removed heat at a sufficient rate. In a model of steady thermal evolution (Figure 5.6), the critical condition would have applied about 1.5 billion years ago, whereas in a model of episodic thermal evolution, normal subduction might have been possible between overturns as far back as the late Archaean (Figure 5.7). It is possible to conjecture that transformation of the thicker basaltic oceanic crust to denser eclogite might extend the range of conditions under which subduction could occur, but it would

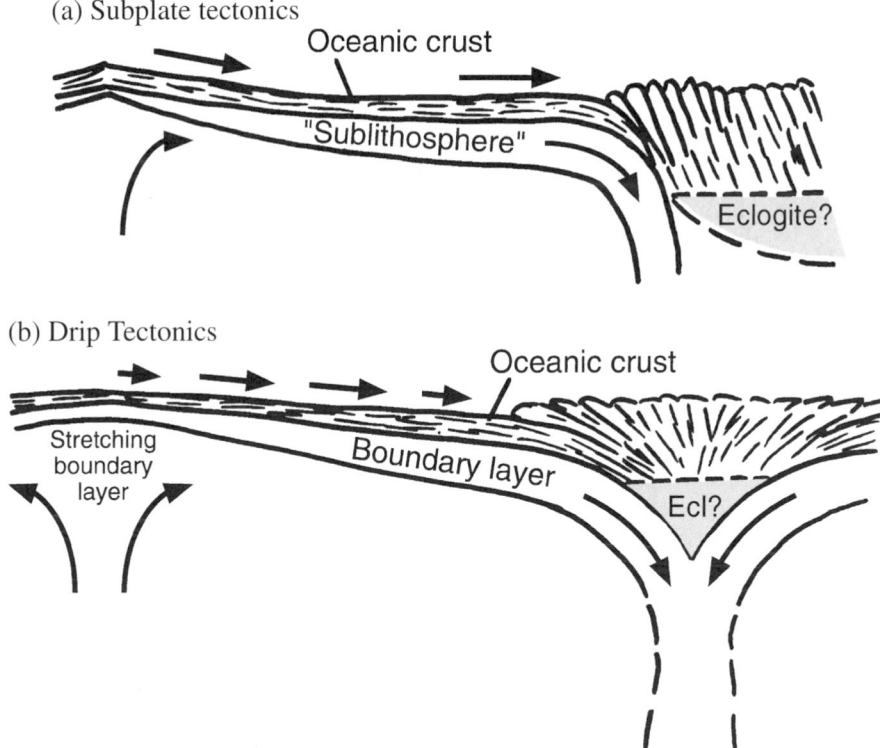

Figure 5.8. Conjectured tectonic modes that may have operated if plates were too buoyant to subduct. (a) Plate-like convection, but with the buoyant oceanic crust delaminating and remaining at the surface. (b) A non-plate mode in which the lithosphere is too soft to be rigid, the descending flow is symmetric, and the mafic crust remains at the surface. In both cases it is possible that the mafic crust might founder if its root transforms to eclogite.

need to be demonstrated quantitatively that subduction could occur at a sufficient rate, and this has not yet been done.

A more general point to be made here is that the behaviour of the top thermal boundary layer of the mantle may be strongly affected by its composition and mineralogy. Alternatives to standard plate tectonics can be envisaged in which the buoyant oceanic crust is scraped off, and only the mantle part of the lithosphere is subducted (Figure 5.8a), or in which the lithosphere is too thin for brittle behaviour to dominate, the boundary layer is more fluid, descending flow is more symmetric, and again the crust may remain at the surface (Figure 5.8b). In either of these cases, some or all of the oceanic crust might ultimately sink, if its roots were to transform to eclogite. However the effect still would be that the coldest part of the lithosphere would be prevented from sinking, temporarily or over the longer term, and this would reduce the ability of such modes of behaviour to cool the mantle (Davies, 1993a,b).

There is an important parallel here with the role of the phase transformations in the transition zone. We discussed earlier how the effect of temperature on the depth of transformations might prevent subducted lithosphere from penetrating. The compositional variation within subducted lithosphere may also be important. Ringwood and Irifune (1988) summarized the experimental evidence that differences in the depths of transformations between subducted crust, depleted mantle, and normal mantle would result in a net buoyancy at depths near 660 km, and they proposed that such buoyancy would prevent the penetration of lithosphere into the lower mantle (Jackson and Rigden, Chapter 9, this volume). This could be a contributing factor in the apparent hesitation of some slabs to penetrate, though quantitative estimates of the dynamical effect indicate that such buoyancy is unlikely to cause complete blocking of slab penetration in the present-day mantle (Davies, 1995a). However, this effect could have been much more important in the past, if slabs were thinner and had a thicker oceanic-crust component.

Thus we can argue more generally that compositional and phase differences may have had major influences on mantle dynamics in the past. The oceanic-crust component of the lithosphere, in particular, may have controlled the viability of plate tectonics and may have influenced or controlled the ability of subducted lithosphere to penetrate the transition zone.

Discussions of alternatives to plate tectonics, either for the early Earth or for other planets, particularly Venus, have often suggested plumes as an alternative. This is not an accurate conception of the role of plumes. As was explained earlier, plates and plumes are expressions of different thermal boundary layers. The top boundary layer exists because heat conducts out of the fluid. The bottom boundary layer exists because heat conducts into the fluid. Their relative strengths and behaviours should be considered independently, even though they may interact to a greater or lesser degree.

Thus, if the top boundary layer does not behave as plates, it must behave in some other way (such as suggested in Figure 5.8), and this dynamical behaviour will control the efficiency of heat removal. Conversely, the rate of heat removal presumably must be at least as great as the rate of radioactive heat generation. This may seem a trivial point, but the modes of behaviour sketched in Figure 5.8 may actually remove heat less efficiently at higher temperatures, as noted earlier. Thus, these modes may look attractive from the point of view of tectonics, but at face value they are not thermally viable, because the mantle could not have cooled through these modes to get to the more efficient plate mode. The point here is not that there is no prospect of finding a viable alternative to plate tectonics, but that any alternative must be both tectonically and thermally viable.

Major mantle overturns of the kind shown in Figure 5.7 and illustrated in Figure 5.4 would have had dramatic geological consequences (Davies, 1995b). They would have produced large amounts of mafic mantle melt and would have caused tectonic and thermal reworking of pre-existing crust within a short period, on the order of 10 million years. They might account for some of the apparent pulses of crust formation and consolidation in the Archaean and Proterozoic (e.g.,

McCulloch and Bennett, Chapter 2, this volume). They might have caused major extinctions, even of primitive life, and strongly affected the composition of the atmosphere. Mantle overturns are dynamically distinct from the actions of plates and plumes and should be added to the repertoire of hypotheses considered by geologists.

Turning to plumes, the question whether plumes were more or less important in the past (or in Venus) depends on the strength of the thermal boundary layer at the bottom of the mantle, which depends on the relative temperatures of the mantle and core. The possibilities here are still open. Thus, in Figure 5.6, the temperature difference between mantle and core is fairly constant over much of the Earth's history, and the plume flux does not vary much. In Figure 5.7, the lower mantle is much hotter early in the Earth's history, and this implies that plumes would, on average, have been less active in the past, approaching their present-day strength only in the past 1 billion years or so. On the other hand, Campbell and Griffiths (1992; Campbell, Chapter 6, this volume) have suggested that subducted oceanic crust may have begun to accumulate at the base of the mantle near the end of the Archaean. This would partially have insulated the core from the mantle. In that scenario, plumes would have been hotter and more numerous early in the Earth's history, becoming more muted later.

To summarize, the nature of the tectonic regime during earlier times in the Earth's history still is quite unclear. Plate tectonics may or may not have been operative, and whether the alternative behaviour of the upper thermal boundary layer was a modified form of plate tectonics or something substantially different is not clear. Plumes may have been less important, about equally important, or more important in the past. Episodic layering of the mantle, alternating with dramatic overturns, may have occurred in the past, but that possibility is only beginning to be quantitatively explored. Each of those processes may have been strongly affected by compositional effects and phase changes.

On the other hand, the basic concepts of convection and our reasonably good understanding of the present-day mantle allows us to define the possibilities more sharply, and these are now beginning to be quantified, so that physical, chemical, and geological tests should emerge. For example, long-term layering or episodic layering would have a substantial influence on the geochemical evolution of the mantle and on the survival of heterogeneities within it, and the magmatic and geological effects of a major mantle overturn might be so dramatic (Davies, 1995b) that they could be readily detectable in preserved crust.

References

Anderson, D. L., and Bass, J. D. 1986. The transition region of the earth's upper mantle. *Nature* 320:321–8.

Bina, C. R., and Helffrich, G. 1994. Phase transition Clapeyron slopes and transition zone seismic discontinuity topography. *J. Geophys. Res.* 99:15853–60.

Bullen, K. E. 1965. *An Introduction to the Theory of Seismology.* Cambridge University Press.

Campbell, I. H., and Griffiths, R. W. 1990. Implications of mantle plume structure for the evolution of flood basalts. *Earth Planet. Sci. Lett.* 99:79–83.

Campbell, I. H., and Griffiths, R. W. 1992. The changing nature of mantle hotspots through time: implications for the chemical evolution of the mantle. *J. Geol.* 92:497–523.

Christensen, U. R., and Hofmann, A. W. 1994. Segregation of subducted oceanic crust in the convecting mantle. *J. Geophys. Res.* 99:19867–84.

Christensen, U. R., and Yuen, D. A. 1984. The interaction of a subducting lithospheric slab with a chemical or phase boundary. *J. Geophys. Res.* 89:4389–402.

Clague, D. A., and Dalrymple, G. B. 1989. Tectonics, geochronology and origin of the Hawaiian–Emperor volcanic chain. In: *The Eastern Pacific Ocean and Hawaii,* ed. E. L. Winterer, D. M. Hussong, and R. W. Decker, pp. 188–217. Boulder: Geological Society of America.

Condie, K. C. 1995. Episodic ages of greenstones: a key to mantle dynamics? *Geophys. Res. Lett.* 22:2215–18.

Crough, T. S. 1983. Hotspot swells. *Annu. Rev. Earth Planet. Sci.* 11:165–93.

Davies, G. F. 1980. Thermal histories of convective earth models and constraints on radiogenic heat production in the earth. *J. Geophys. Res.* 85:2517–30.

Davies, G. F. 1988. Ocean bathymetry and mantle convection. 1. Large-scale flow and hotspots. *J. Geophys. Res.* 93:10467–80.

Davies, G. F. 1990. Mantle plumes, mantle stirring and hotspot chemistry. *Earth Planet. Sci. Lett.* 99:94–109.

Davies, G. F. 1992. On the emergence of plate tectonics. *Geology* 20:963–6.

Davies, G. F. 1993a. Conjectures on the thermal and tectonic evolution of the earth. *Lithos* 30:281–9.

Davies, G. F. 1993b. Cooling the core and mantle by plume and plate flows. *Geophys. J. Int.* 115:132–46.

Davies, G. F. 1995a. Penetration of plates and plumes through the mantle transition zone. *Earth Planet. Sci. Lett.* 133:507–16.

Davies, G. F. 1995b. Punctuated tectonic evolution of the earth. *Earth Planet. Sci. Lett.* 36:363–80.

Davies, G. F., and Gurnis, M. 1986. Interaction of mantle dregs with convection: lateral heterogeneity at the core–mantle boundary. *Geophys. Res. Lett.* 13:1517–20.

Davies, G. F., and Richards, M. A. 1992. Mantle convection. *J. Geol.* 100:151–206.

Fisher, O. 1881. *Physics of the Earth's Crust.* London: Murray.

Grand, S. P. 1994. Mantle shear structure beneath the Americas and surrounding oceans. *J. Geophys. Res.* 99:11591–621.

Griffiths, R. W., and Campbell, I. H. 1990. Stirring and structure in mantle plumes. *Earth Planet. Sci. Lett.* 99:66–78.

Gurnis, M. 1986. The effect of chemical density differences on convective mixing in the Earth's mantle. *J. Geophys. Res.* 91:11407–19.

Gurnis, M., and Davies, G. F. 1986a. The effect of depth-dependent viscosity on convective mixing in the mantle and the possible survival of primitive mantle. *Geophys. Res. Lett.* 13:541–4.

Gurnis, M., and Davies, G. F. 1986b. Numerical models of high Rayleigh number convection in a medium with depth-dependent viscosity. *Geophys. J. Royal. Astron. Soc.* 85:523–41.

Hallam, A. 1973. *A Revolution in the Earth Sciences.* Oxford: Clarendon Press.

Hill, R. I., Campbell, I. H., Davies, G. F., and Griffiths, R. W. 1992. Mantle plumes and continental tectonics. *Science* 256:186–93.

Hofmann, A. W., and White, W. M. 1982. Mantle plumes from ancient oceanic crust. *Earth Planet. Sci. Lett.* 57:421–36.

Holmes, A. 1931. Radioactivity and earth movements. *Geol. Soc. Glasgow, Trans.* 18:559–606.

Jochum, K. P., Hofmann, A. W., Ito, E., Seufert, H. M., and White, W. M. 1986. K, U, and Th in mid-ocean ridge basalt glasses and heat production. *Nature* 306:431–6.

Machetel, P., and Weber, P. 1991. Intermittent layered convection in a model mantle with an endothermic phase change at 670 km. *Nature* 350:55–7.

Mitrovica, J. X. 1996. Haskell [1935] revisited. *J. Geophys. Res.* 101:555–69.

Morgan, W. J. 1971. Convection plumes in the lower mantle. *Nature* 230:42–3.

Morgan, W. J. 1981. Hotspot tracks and the opening of the Atlantic and Indian Oceans. In: *The Sea,* vol. 7, ed. C. Emiliani, pp. 443–87. New York: Wiley.

Nutman, A. P., McGregor, V. R., Friend, C. L., Bennett, V. C., and Kinney, P. D. 1996. The Itsaq gneiss complex of southern West Greenland; the world's most extensive record of early (3,900–3,600 Ma) crustal evolution. *Precambrian Research* 89:1–40.

Olson, P. 1984. An experimental approach to thermal convection in a two-layered mantle. *J. Geophys. Res.* 89:11293–301.

Parsons, B., and McKenzie, D. P. 1978. Mantle convection and the thermal structure of the plates. *J. Geophys. Res.* 83:4485–96.

Richards, M. A., Duncan, R. A., and Courtillot, V. E. 1989. Flood basalts and hot-spot tracks: plume heads and tails. *Science* 246:103–7.

Richards, M. A., and Hager, B. H. 1988. The earth's geoid and the large-scale structure of mantle convection. In: *Physics of the Planets*, ed. S. K. Runcorn, pp. 247–72. New York: Wiley.

Ringwood, A. E. 1975. *Composition and Petrology of the Earth's Mantle*. New York: McGraw-Hill.

Ringwood, A. E. 1991. Phase transformations and their bearing on the constitution and dynamics of the mantle. *Geochim. Cosmochim. Acta* 55:2083–110.

Ringwood, A. E., and Irifune, T. 1988. Nature of the 650-km discontinuity: implications for mantle dynamics and differentiation. *Nature* 331:131–6.

Schubert, G., Stevenson, D., and Cassen, P. 1980. Whole planet cooling and radiogenic heat source contents of the Earth and Moon. *J. Geophys. Res.* 85:2531–8.

Schubert, G., Yuen, D. A., and Turcotte, D. L. 1975. Role of phase transitions in a dynamic mantle. *Geophys. J. Royal Astron. Soc.* 42:705–35.

Sclater, J. G., and Francheteau, J. 1970. The implications of terrestrial heat flow observations on current tectonic and geochemical models of the crust and upper mantle of the Earth. *Geophys. J. Royal Astron. Soc.* 20:509–42.

Sclater, J. G., Jaupart, C., and Galson, D. 1980. The heat flow through the oceanic and continental crust and the heat loss of the earth. *Rev. Geophys.* 18:269–312.

Sleep, N. H. 1990. Hotspots and mantle plumes: some phenomenology. *J. Geophys. Res.* 95:6715–36.

Stacey, F. W. 1980. The cooling earth: a reappraisal. *Phys. Earth Planet. Int.* 22:89–96.

Stacey, F. W., and Loper, D. E. 1984. Thermal histories of the core and mantle. *Phys. Earth Planet. Int.* 36:99–115.

Tackley, P. J., Stevenson, D. J., Glatzmaier, G. A., and Schubert, G. 1993. Effects of an endothermic phase transition at 670 km depth in a spherical model of convection in the earth's mantle. *Nature* 361:699–704.

Turcotte, D. L., Cooke, F. A., and Willemann, R. J. 1979. Parameterized convection within the Moon and terrestrial planets. In: *Proceedings of the 10th Lunar and Planetary Science Conference*, pp. 2375–92.

Turcotte, D. L., and Schubert, G. 1982. *Geodynamics: Applications of Continuum Physics to Geological Problems*. New York: Wiley.

van der Hilst, R. 1995. Complex morphology of subducted lithosphere in the mantle beneath the Tonga trench. *Nature* 374:154–7.

van der Hilst, R., Engdahl, R., Spakman, W., and Nolet, G. 1991. Tomographic imaging of subducted lithosphere below northwest Pacific island arcs. *Nature* 353:37–43.

van der Hilst, R., and Seno, T. 1993. Effects of relative plate motion on the deep structure and penetration depth of slabs below the Izu-Bonin and Mariana island arcs. *Earth Planet. Sci. Lett.* 120:395–407.

Verhoogen, J. 1965. Phase changes and convection in the earth's mantle. *Phil. Trans. R. Soc. London* A258:276–83.

Von Herzen, R. P., Cordery, M. J., Detrick, R. S., and Fang, C. 1989. Heat flow and thermal origin of hotspot swells: the Hawaiian swell revisited. *J. Geophys. Res.* 94:13783–99.

Whitehead, J. A., and Luther, D. S. 1975. Dynamics of laboratory diapir and plume models. *J. Geophys. Res.* 80:705–17.

Wilson, J. T. 1963. Evidence from islands on the spreading of the ocean floor. *Nature* 197:536–8.

Wilson, J. T. 1965. A new class of faults and their bearing on continental drift. *Nature* 207:343–7.

Woods, M. T., Leveque, J.-J., and Okal, E. A. 1991. Two-station measurements of Rayleigh wave group velocity along the Hawaiian swell. *Geophys. Res. Lett.* 18:105–8.

6

The Mantle's Chemical Structure: Insights from the Melting Products of Mantle Plumes

I. H. CAMPBELL

6.1. Introduction

Mantle convection can be divided into two distinct components, each driven by one of the mantle's boundary layers (Davies and Richards, 1992; Davies, Chapter 5, this volume). The outer boundary layer is cold. Oceanic lithosphere forms at the mid-ocean ridges, cools over a period of time so that it eventually becomes denser than the underlying mantle, and sinks back into the mantle at subduction zones. The plate motions associated with subduction create low-pressure zones at mid-ocean ridges that result in passive upwellings of upper mantle beneath the ridges. As this hot mantle rises, it undergoes decompressional melting to produce mid-ocean-ridge basalts (MORBs). Because the physics of the process is well understood and because the chemistry of MORBs is well documented, the chemistry of the upper mantle is well known. It is relatively homogeneous and has been depleted of the highly incompatible elements, including the light rare-earth elements, and therefore has a low $^{87}Sr/^{86}Sr$ ratio and a high $^{143}Nd/^{144}Nd$ ratio (e.g., McCulloch and Bennett, Chapter 2, this volume).

In order to unravel the chemical structure of the deep mantle it is necessary to understand the dynamics of the mantle's hot lower boundary layer. Long-lived plumes require a replenishing source of buoyancy and therefore must originate from a thermal boundary layer deep within the mantle. Their melting products provide samples of the thermal boundary layer from which they originate, as well as samples of any overlying material that is entrained into the plumes as they ascend through the mantle. However, before the chemistry of these melting products can be interpreted, it is necessary to understand the dynamics of mantle plumes and to know the depth of the boundary layer from which they originate. Mantle plumes could originate from one of two boundary layers: the core–mantle boundary or the upper-mantle–lower-mantle boundary. In this discussion, plumes will be assumed to originate from the core–mantle boundary. The implications of a source in the upper-mantle–lower-mantle boundary layer will be discussed where appropriate.

6.2. The Fluid Dynamics of Mantle Plumes

The dynamics of mantle plumes is reviewed at length by Griffiths and Turner (Chapter 4, this volume) and will be covered only briefly here. The core is thought to be hotter than the overlying mantle (Boehler, Chopelas, and Zerr, 1995). Heat conducted out of the core warms the mantle in the boundary layer above the core and lowers its density. The material in this boundary layer is lighter than the overlying mantle material and it begins to rise, but before it can rise at a significant rate it must acquire enough thermal buoyancy to overcome the viscosity of the overlying mantle that opposes its rise. As a consequence, a new plume will have a large head, followed by a relatively small tail (Whitehead and Luther, 1975; Richards, Duncan, and Courtillot, 1989). The tail, or feeder conduit, is narrower because the hot, low-viscosity material flowing up the existing pathway can rise faster than the head, which must displace cooler, higher-viscosity mantle. As the plume head rises through the mantle it grows, and it does so for two reasons (Griffiths and Campbell, 1990). Firstly, the more rapidly flowing material in the plume tail feeds a continuous flux of new material into the head. When this material reaches the stagnation point at the top of the plume, it flows radially, giving the head its characteristic doughnut structure. Secondly, the head grows by entrainment. As a plume head ascends through the mantle, heat is conducted from the head and tail of the plume into the adjacent mantle. The temperature of the boundary layer surrounding the plume increases, and its density falls, so that it begins to rise with the plume. This material then becomes part of the plume and is stirred in with the source material by a recirculating flow within the head. The head is therefore a mixture of material from the hot source and material from the cooler mantle through which the head passes (Figure 6.1).

Griffiths and Campbell (1990) have argued that the material entrained into the head of a plume is entirely from the lower mantle. In fact, a detailed computer study of the entrainment process (G. Davies, personal communication, 1996) has shown that most of the entrained material comes from the bottom of the lower mantle. The explanation for that finding lies in the interplay between the mechanism by which a plume head rises and the nature of the entrainment process. Consider, initially, a plume head that rises without entrainment, as would be the case if its ascent were driven by compositional buoyancy. Imagine that the mantle through which it is rising is divided by marker horizons 10, 100, 200, and 500 km above the base and 660 km from the top, as shown in Figure 6.2. The head rises by lifting and thinning the overlying mantle. When the plume head reaches the top of its ascent in the upper mantle, it has, pushed before it and draped over its sides, a considerable volume of material that originated in the lower mantle. All of the imaginary layers in this material will be present in the sheath, but their thicknesses will be greatly reduced, especially the lower ones, which relative to the scale of the mantle, will have become very thin. In a thermal plume head, the entrainment process is superimposed on the continual thinning of the lower-mantle sheath. Calculations based on equation 7 of

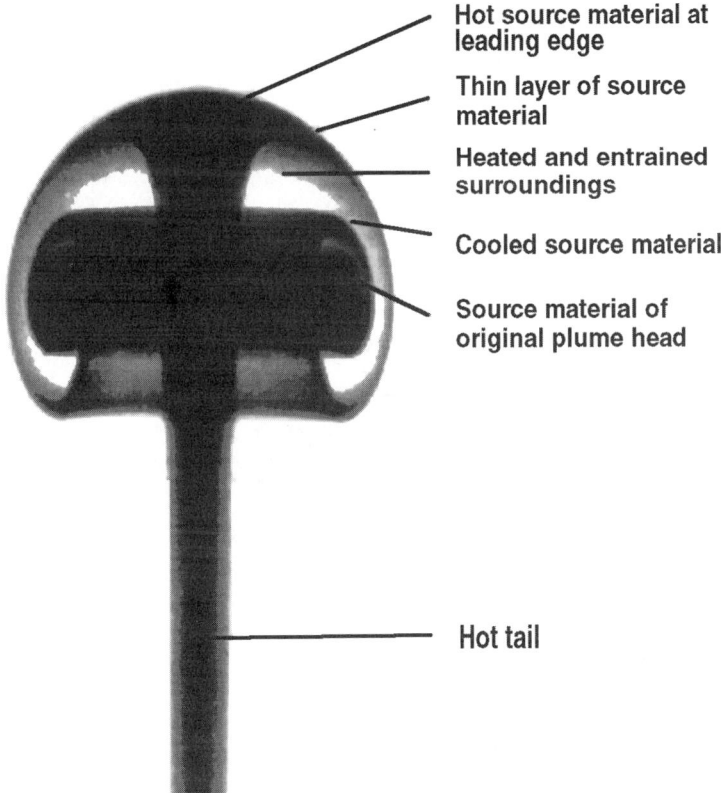

Hot source material at leading edge

Thin layer of source material

Heated and entrained surroundings

Cooled source material

Source material of original plume head

Hot tail

Figure 6.1. Photograph of a laboratory model of a starting thermal plume. Conduction of heat during the rise of the plume causes entrainment of cooler material and smoothing of the temperature gradients within the head, where the mean temperature is intermediate between that of the source and that of the ambient mantle. The temperature in the plume tail (including the central axis of the head) and along the leading edge of the head is appreciably greater than the average temperature of the head and approaches the temperature of the material in the plume source region.

Griffiths and Campbell (1990) suggest that the thickness of the thermal boundary layer that surrounds the head lies within the range 5–10 km, being thinnest at the top of the head and thickest at the bottom. Material within this boundary layer is swept into the base of the head by its recirculating flow and in that way is removed from the sheath. When an imaginary layer adjacent to the head is reduced in thickness to less than about 5 km, it begins to be heated and entrained. The findings of Davies, referred to earlier, show that only material that originates within about 500 km of the bottom of the mantle thins sufficiently to be incorporated into the plume head (Figure 6.2). As a consequence, a plume head can entrain no material from the upper mantle.

Computer models of the ascent of plume heads show that they 'neck' and become distorted when they pass through the viscosity change that distinguishes the upper mantle from the lower mantle (Richards and Hager, 1988; Lambeck and Johnston,

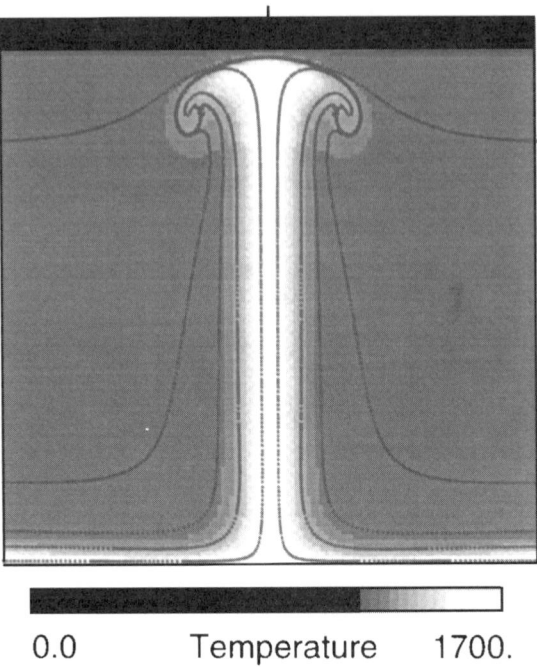

0.0   Temperature   1700.

Figure 6.2. A thermal plume with tracers showing the source of material entrained into the plume (G. F. Davies, personal communication). The model was started with lines of tracers at heights of 10, 100, 200, and 500 km above the base and at a depth of 660 km from the top. The material entrained into the head has come from the lowest 500 km, and mostly from the lowest 200 km. No upper-mantle material is entrained.

Chapter 10, this volume). However, this is a second-order effect, and a head will recover most of its spherical shape by the time it reaches the top of the mantle (Farnetani and Richards, 1994).

The principal factor controlling the size of a plume head is the distance it rises through the mantle. A plume head originating at the core–mantle boundary attains a diameter of about 1,000 km by the time it reaches the top of the mantle. There its ascent is arrested by interaction with the lithosphere, and it flattens to form a disk approximately 2,000 km across and 180 km deep. This contrasts with a calculated plume-head diameter of about 250 km, and a flattened-disk diameter of about 500 km, for a plume originating from the upper-mantle–lower-mantle boundary (Campbell and Griffiths, 1990; Griffiths and Campbell, 1990).

Depending on the dynamics of the source region, flow in a conduit may continue long after the plume head has flattened itself against the oceanic or continental lithosphere. If the conduit is tilted sideways by motion of the overlying plate or by convection within the mantle, it will rise as a tilted plume and entrain surrounding mantle as it ascends (Griffiths and Richards, 1989). The amount of entrainment into the tail decreases as the velocity of the overlying plate decreases and as the plume's buoyancy flux increases, the latter being the more important (Griffiths and

Campbell, 1991a). As a consequence of these complications, the basaltic material erupted above a plume conduit may be the product of melting mantle that is a mixture of the plume source material and material entrained into the plume during ascent.

6.3. The Melting Products of Mantle Plumes

Continental flood basalts, oceanic plateaus, and chains of volcanoes such as the Walvis Ridge, Chagos-Laccadive Ridge, and Emperor Seamounts are interpreted to be the melting products of mantle plumes (Morgan, 1981). The characteristic features of flood basalts are as follows:

1. They are the first eruptive products of an extended phase of volcanic activity.
2. The magma erupted is dominantly tholeiitic basalt, with lesser amounts of picrite and alkali basalt.
3. The erupted volumes are enormous, and the areas covered typically are equidimensional and 2,000–2,500 km across.
4. The timescale for the main phase of volcanic activity is short, normally less than 1 million years.
5. Eruption of basalt is preceded by about 1,000 m of uplift covering an areal extent similar to that covered by the basalts.

Oceanic plateaus, such as the Ontong-Java Plateau and Kerguelen Plateau, are the oceanic equivalents of continental flood basalts. Their erupted volumes are enormous, being 5–10 times greater than the volumes of the largest continental flood basalts (Coffin and Eldholm, 1994). Both continental flood basalts and oceanic plateaus are commonly connected to the current positions of associated hotspots by long narrow chains of volcanoes that typically are 200–300 km wide. Examples include the Walvis Ridge (Paraná) and 90 East Ridge (Kerguelen). The Mascarene Plateau-Chagos-Laccadive Ridge, connecting the Réunion hotspot to the Deccan Traps, is shown in Figure 6.3.

6.4. Hypotheses for the Origin of Flood Basalts

There are two principal hypotheses that seek to explain the relationship between mantle plumes and flood basalts: the Campbell and Griffiths (1990) plume-head hypothesis and the White and McKenzie (1989) extension hypothesis.

6.4.1. The Plume-Head Hypothesis

The Campbell and Griffiths (1990) hypothesis is illustrated in Figure 6.4: A large plume head, originating at the core–mantle boundary, rises beneath the lithosphere

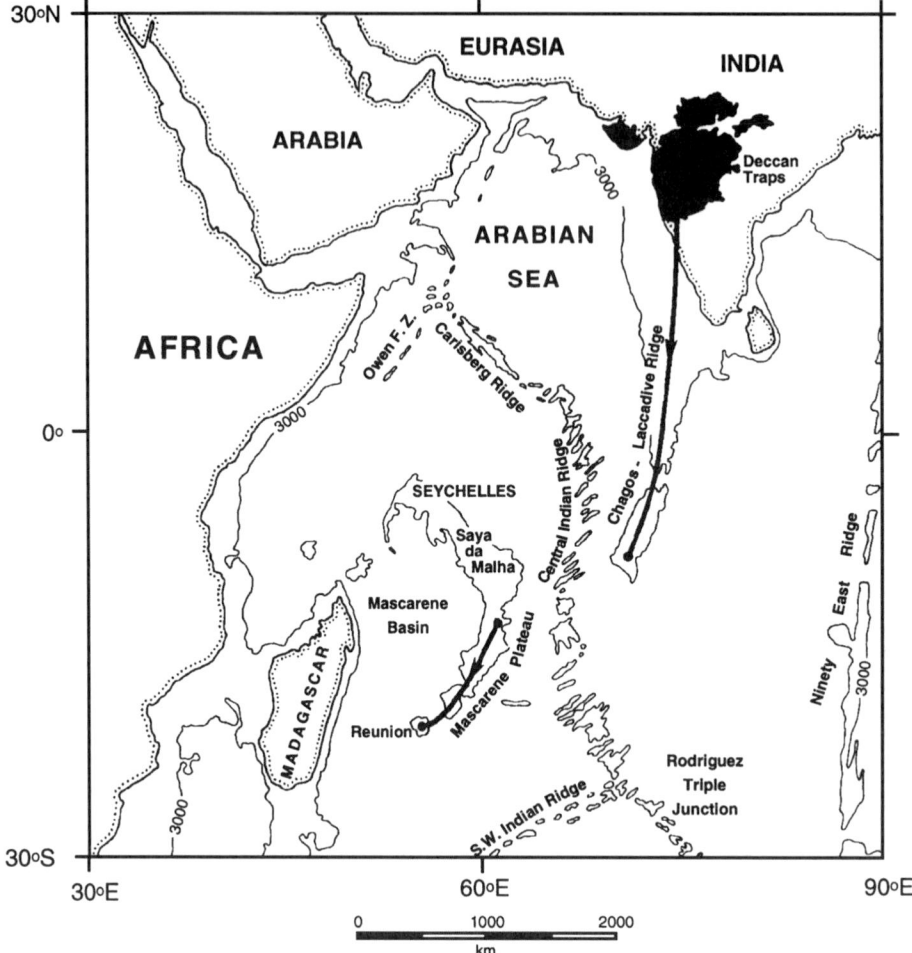

Figure 6.3. Geography of the western Indian Ocean, showing the distribution of volcanic rocks associated with the Réunion-Deccan plume. The Saya da Malha was part of the Deccan Traps prior to its separation, caused by spreading on the Central Indian Ridge. Note that the Deccan Traps and Saya da Malha are connected to Réunion Island, the current position of the hotspot, by the 200–300 km wide volcanic chains of the Mascarene Plateau and the Chagos-Laccadive Ridge. (Adapted from White and McKenzie, 1989.)

and flattens to form a disk about 2,000–2,500 km across when it reaches the top of its ascent, leading to an uplift of 500–1,000 m, followed by volcanism. Melting of the plume head occurs as a consequence of adiabatic decompression and takes place when it reaches the top of the upper mantle. Melting continues as long as the head continues to rise and flatten. The timescale for the main eruptive phase is the time required for the final stage of flattening of the plume head, which is directly proportional to the viscosity of the surrounding mantle (Griffiths and Campbell, 1991b). A timescale of 1 million years implies that the viscosity at the *top* of the upper mantle is 5×10^{19} Pa · s, almost an order of magnitude lower than the estimates for the *average* viscosity of the upper mantle, based on

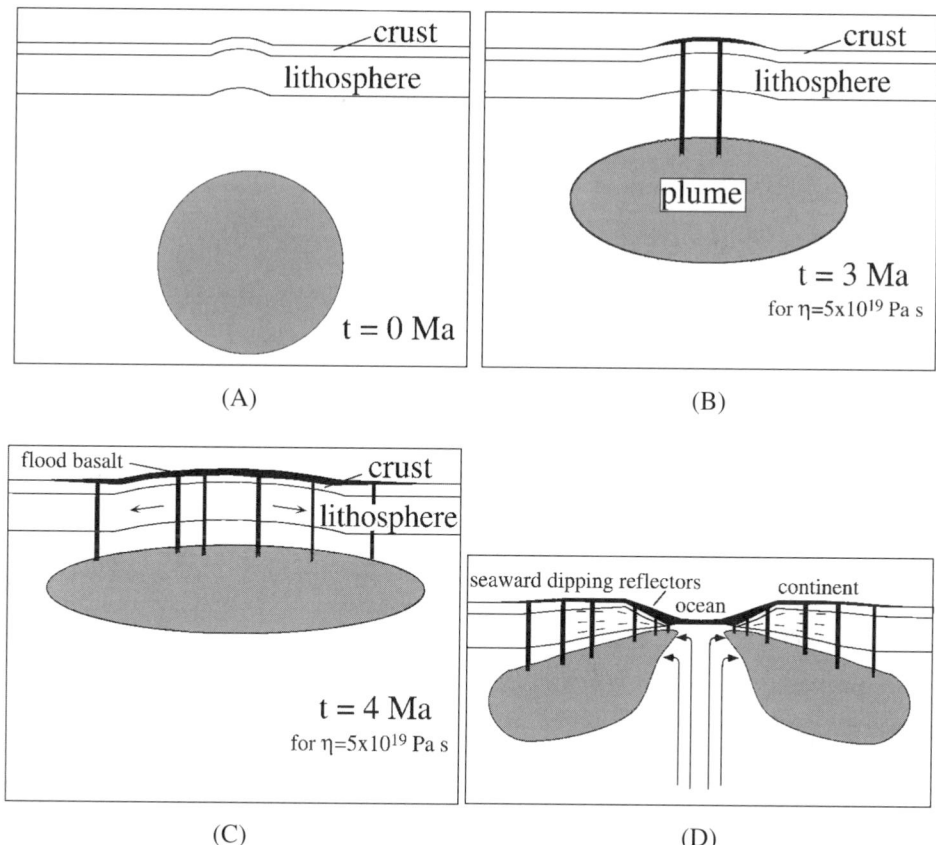

Figure 6.4. The Campbell and Griffiths hypothesis for formation of flood basalts. (A) A 1,000-km-diameter plume head rises beneath continental crust. (B) As it reaches the top of its ascent, the head begins to flatten and melt, leading to eruption of basalt at the surface. (C) Melting of the plume results from adiabatic decompression, and melting continues as long as the head continues to rise and flatten. The arrival of the plume head also leads to surface uplift, placing the lithosphere under tension, as indicated by the arrows. The final diameter of the flattened disk is about twice that of the plume head (i.e., ~2,000 km). (D) The tension induced by the plume head may lead to runaway extension and to the formation of a new ocean basin. Where that happens, it will lead to a second phase of decompressional melting that may be comparable in volume to the first. The plume tail and the structure of the plume head have been omitted to simplify the diagram.

glacial rebound analyses (Lambeck and Johnston, Chapter 10, this volume). The amount of volcanism produced depends on the height to which the top of the plume rises, which in turn is controlled by the local properties of the lithosphere, especially the topography of its lower boundary. In the Campbell and Griffiths model, the areal extent of volcanism (about 2,500 × 2,500 km) is largely controlled by the dimensions of the flattened plume head. Melting will start where the temperature is highest, that is, at the plume axis and along the hot leading edge at the top of the plume. These high-temperature parts of the plume can melt to produce high-MgO magmas such as komatiites and picrites. Later, as the plume head continues to rise and flatten, the cooler entrained mantle material below the leading edge of the plume may start to melt if it rises to a sufficiently low pressure (i.e., shallow depth).

The uplift associated with the arrival of a plume head places the lithosphere under tension, but the stresses produced are, by themselves, too small to produce runaway extension leading to the formation of a new ocean basin. However, normally there are additional subduction-related stresses acting on a plate, and where these stresses are sufficiently large, the arrival of a new plume may initiate extension by (1) providing an additional tensional force and (2) heating and weakening the overlying lithosphere. Because heat is transmitted from plumes to the overlying lithosphere by conduction, there is a delay between the arrival of a new plume and the onset of extension (Hill, 1991). The length of that delay will depend on the strength of the lithosphere prior to the arrival of the plume and on the magnitude of the subduction-related forces acting on the plate. The observed time interval between volcanism and runaway extension varies between 1 and 20 million years (Hill, 1991).

The timescale for cooling of a plume head is hundreds of millions of years. If the main phase of volcanism is followed by a period of extension and lithospheric thinning, the associated adiabatic decompression may lead to a second phase of volcanic activity, by the White and McKenzie mechanism described next. If extension results in the formation of a new ocean basin, that second phase of volcanic activity may be as significant as the first. The seaward-dipping reflectors off the east coast of Greenland and those associated with the Deccan, Paraná, and Karoo (White and McKenzie, 1989) are possible examples of this type of volcanism.

6.4.2. The Extension Hypothesis

White and McKenzie (1989) have suggested a steady-state model for flood basalts in which the thermal anomaly required to produce flood volcanism is built up over a long period of time by what is essentially a plume tail rising under stable lithosphere prior to the onset of volcanism. Volcanism is triggered by lithospheric extension, which allows the anomalous hot mantle below the lithosphere to rise adiabatically and melt. The White and McKenzie (1989) starting plume does include a head, but its calculated diameter is only about 200 km compared with 50–75 km for a plume tail. The arrival of a plume head of this size, with a volume only about 2% of that in the model proposed by Griffiths and Campbell (1990), will have little influence on the rate of volcanic eruption from the plume.

The sequence of events predicted by the White and McKenzie (1989) hypothesis is shown in Figure 6.5. A plume produces an accumulation of hot mantle beneath a stable lithosphere over an *extended* period of time. This results in uplift at the surface, but no volcanism, because the plume is unable to penetrate the lithosphere sufficiently to allow melting. Volcanism is triggered by lithospheric extension, which causes the hot mantle below the lithosphere to rise adiabatically and melt. For significant melting to occur, the lithosphere must thin by at least 50% (i.e.,

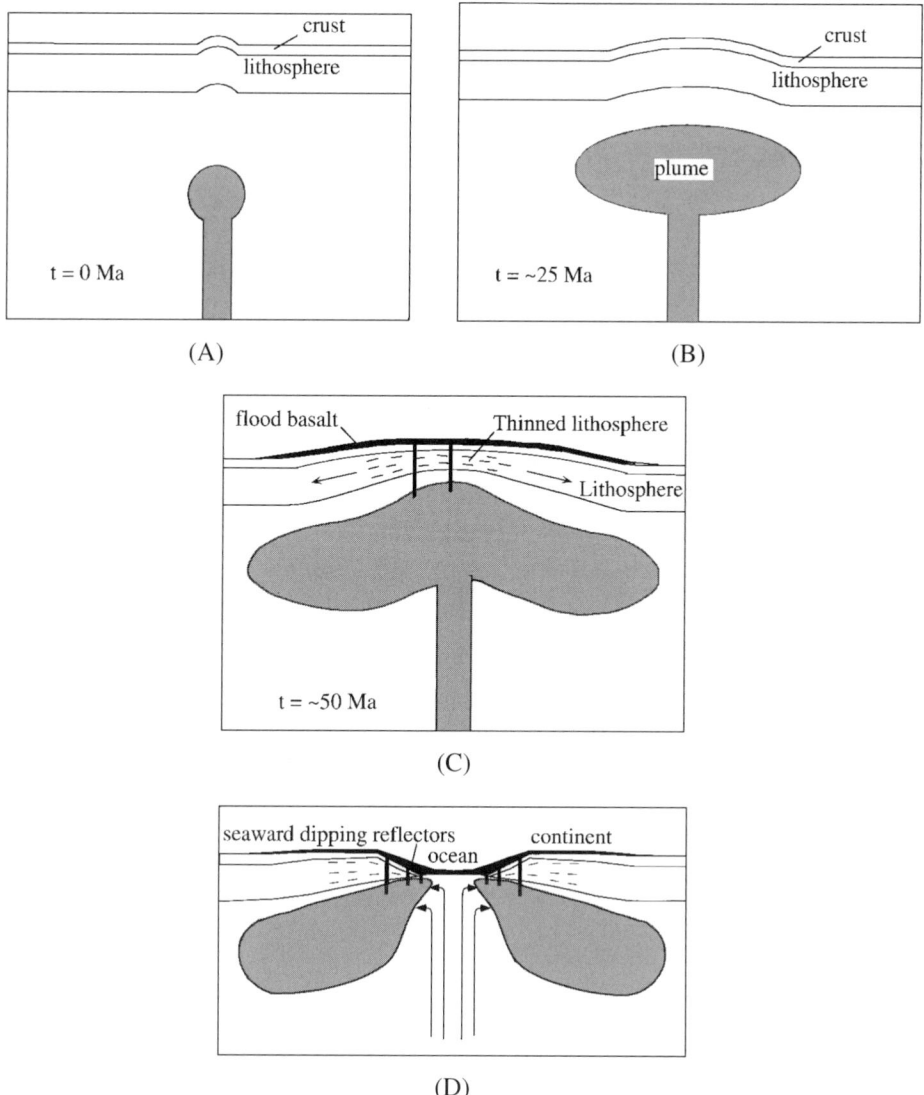

Figure 6.5. The White and McKenzie hypothesis for formation of flood basalts. (A) A new plume with a head diameter of about 200 km ascends beneath the continental lithosphere, leading to surface uplift, but no volcanism. (B) The mushroom head of the plume expands as material continues to flow up the tail of the plume into the head. This leads to further surface uplift, but still no volcanism. (C) Volcanism is triggered by extension when the stretching factor β exceeds about 2. (D) Extension may lead to the formation of a new ocean basin.

a stretching factor $\beta > 2$), and this can lead to runaway extension, continental breakup, and the formation of a new ocean basin.

There are three key differences between the White and McKenzie model for flood volcanism and the Campbell and Griffiths model. Firstly, White and McKenzie assume that plumes originate from the upper-mantle–lower-mantle boundary, whereas Campbell and Griffiths assume that they originate from the core–mantle boundary. Secondly, White and McKenzie ignore the possibility that plume heads may entrain mantle material during their ascent. This is not a serious problem if plumes originate from the upper-mantle–lower-mantle boundary, because the amount of entrainment would be small, but it is a serious problem if they originate from the core–mantle boundary. Thirdly, the White and McKenzie hypothesis requires extension to precede volcanism, whereas the Campbell and Griffiths hypothesis does not.

6.5. An Evaluation of the Plume-Head and Extension Hypotheses

The two hypotheses for mantle plumes make contrasting predictions regarding (1) the incubation time required to form a flood-basalt province and (2) the timing of extension relative to volcanism. These predictions can be tested against observations.

6.5.1. Incubation Time

White and McKenzie (1989) do not specify the size of the thermal anomaly they require nor the time needed to produce it. An estimate of the incubation time required to form the mantle thermal anomaly can be made on the basis of the eruption rate of the Iceland plume. Iceland is chosen because its plume ascends beneath a spreading centre, so that the amount of melt produced is the maximum possible. The surface area of the Iceland Plateau is 1.7×10^5 km^2, and the excess thickness of the crust that can be attributed to the plume is 25 km (White et al., 1996). Therefore the volume of basalt generated from the plume has been approximately 4×10^6 km^3. The minimum time required for formation of the Iceland Plateau would have been 15 million years, the age of the oldest rocks exposed on Iceland (McDougall, Kristiansson, and Saemundsson, 1984). A maximum time of about 40 million years can be estimated from the age of the oldest seafloor covered by the plateau. Because most of the Iceland Plateau is below sea level, and only the youngest rocks at the top of the plateau are available for dating, the time taken to form the plateau is likely to have been closer to the maximum age than to the minimum. It is therefore assumed that the time required to form the Iceland Plateau was about 30 million years, which gives an eruption rate of 0.13 km^3/yr. At that rate, a plume with the strength of the Iceland plume, ascending beneath a spreading centre, would require a minimum of 20 million years to produce a flood-basalt province 2,000 km across and 1 km thick (3×10^6 km^3). Alternatively, the same volume of melt could have been produced if hot mantle material from the plume had accumulated below a thick, stable lithospheric lid that had then ruptured abruptly, in the manner described by White and McKenzie (1989). Obviously, the minimum incubation required for formation of a flood-basalt province by that mechanism, assuming that hot mantle material would have melted with the same high efficiency as in a plume rising below a spreading centre, would be 20 million years. However, that high level of efficiency would be unlikely to be achieved by the White and

McKenzie (1989) mechanism. If plume material incubates beneath a continental lithosphere and is then drawn into a zone of decompressional melting during a period of extension, much of the hot plume material can be expected to remain under the continental lithosphere (Figure 6.4D), rather than being drawn into the zone of melting. This approach therefore gives a very conservative estimate for the required incubation time. It is interesting to compare the eruption rate for the Iceland plume of $0.13 \text{ km}^3/\text{yr}$ with the plume volume flux of $1.5 \text{ km}^3/\text{yr}$ (Sleep, 1990), based on a plume excess temperature of $300°C$; the implied average percentage of partial melting of material rising in the mantle plume is thus about 8%.

Incubation times can also be estimated from the volume fluxes for modern plumes. The flux for a typical strong plume, such as Pitcairn or Tahiti, is $3.0 \text{ km}^3/\text{yr}$ (Sleep, 1990), for a temperature excess of $300°C$. If it is assumed that all of the accumulated plume material undergoes an average of 10% partial melting during extension, then the incubation time required for a flood basalt with an erupted volume of $3 \times 10^6 \text{ km}^3$ will be 10 million years. For average plumes, such as Iceland and Tristan da Cunha, with volume fluxes of about $2 \text{ km}^3/\text{yr}$, the minimum time required will be 15 million years. As already noted, during the rifting of continents, some of the hot plume material may remain trapped under the continental lithosphere (Figure 6.4D). Furthermore, when melting takes place beneath thinned lithosphere, rather than beneath an oceanic spreading centre, the percentage of partial melting will be appreciably less than 10%. For example, the Hawaiian and Réunion plumes, which ascend beneath oceanic lithosphere, have plume fluxes of 9.0 and $2.0 \text{ km}^3/\text{yr}$, respectively (Sleep, 1990), compared with eruption rates of only 0.03 and $0.005 \text{ km}^3/\text{yr}$ (Clague and Dalrymple, 1989; I. H. Campbell and G. F. Davies, unpublished data). The average percentage of partial melting for the material rising in both plumes is therefore less than 0.5%. If continental flood basalts erupt through thinned continental lithosphere, rather than at an oceanic spreading centre, the average percentage of partial melting will be well below 10%. Halving the percentage of partial melting will double the incubation time to 20–30 million years, depending on the plume flux. I therefore suggest that for the White and McKenzie (1989) mechanism to be viable, the minimum plume incubation time required would be 30 million years, and 50–100 million years would be a more realistic estimate.

There are geological observations that constrain the incubation times available for the formation of two flood-basalt provinces (the Deccan Traps and Wrangellia), and these times are inconsistent with the White and McKenzie hypothesis.

In the case of the Deccan Traps, the Indian plate was moving at 20 cm per year, or 200 km per million years, at the time of eruption of the Deccan lavas (Patriat and Achache, 1984). Because the north–south dimension (the direction of plate movement) of the traps is less than 2,000 km, the time available for a pool of hot mantle material to have accumulated beneath the Indian plate would have been less than 10 million years, well short of the tens of millions of years required by the White and McKenzie hypothesis.

An incubation time can also be determined from the duration of the uplift event that preceded flood volcanism. Such a timescale is best illustrated by the Wrangellia province (Richards et al., 1991), where the stratigraphic sequence is radiolarian chert overlain by 3,000–6,000 m of subaerial basalts and pillow basalts that are in turn overlain by limestone. The cherts must have formed below the carbonate-compensation depth of 1,000 m. Furthermore, the total timescale for uplift and volcanism, based on the fossil evidence, was less than 5 million years, so that volcanism was preceded by uplift of at least 1,000 m in less than 5 million years (Richards et al., 1991). The maximum timescale for plume incubation prior to volcanism was therefore 5 million years.

6.5.2. Timing of Extension Relative to Volcanism

The timing of extension relative to volcanism provides an unambiguous test of the two hypotheses. In the White and McKenzie model, extension causes volcanism and must therefore precede it. The plume-head hypothesis does not require extension to precede volcanism, *although it does not preclude this possibility*. The relationship between extension and volcanism has been documented for the Deccan, the Columbia River basalt, and the Paraná. In each case the main phase of continental volcanic activity preceded the onset of significant extension.

The best evidence comes from the work of Hooper (1990) on the Western Ghats of the Deccan Traps, where an exposed 3,000-m-thick section consists entirely of the type of tholeiitic basalts typical of the traps. Associated with the eruption of the Deccan magmas were several suites of dikes, as shown in Figure 6.6. Large dikes, with tholeiitic compositions similar to those of the overlying Western Ghats, outcrop on the coastal plain below the Ghats. These dikes decrease in number up-section, indicating that they were emplaced at the time of eruption of the Western Ghats. Thus their field relationships and chemistry are consistent with the suggestion of Hooper (1990) that they acted as feeders to the Western Ghats. The orientations of these dikes are random, implying that the crust was not subject to a uniform tensile stress during the main eruptive phase of the Deccan Traps.

There is a second suite of dikes, cutting the uppermost subgroup of the Ghats near the Panvel Flexure and trending parallel to the west coast (Figure 6.6), whose members clearly are younger than the main phase of tholeiitic volcanism. They consist of rocks of both tholeiitic and lamprophyric composition and, according to Hooper (1990), may have acted as feeders to small volumes of late alkalic and intermediate flows that lie above normal Deccan tholeiites on Bombay Island. The dikes are parallel to horst-graben structures off the coast of India and have been linked to east–west extension that was associated with the separation of the Indian and Seychelle plates. They can be explained by the extension hypothesis of White and McKenzie (1989), but the main phase of Deccan volcanism cannot.

Eruption of the main tholeiitic phase of the Columbia River basalts, between

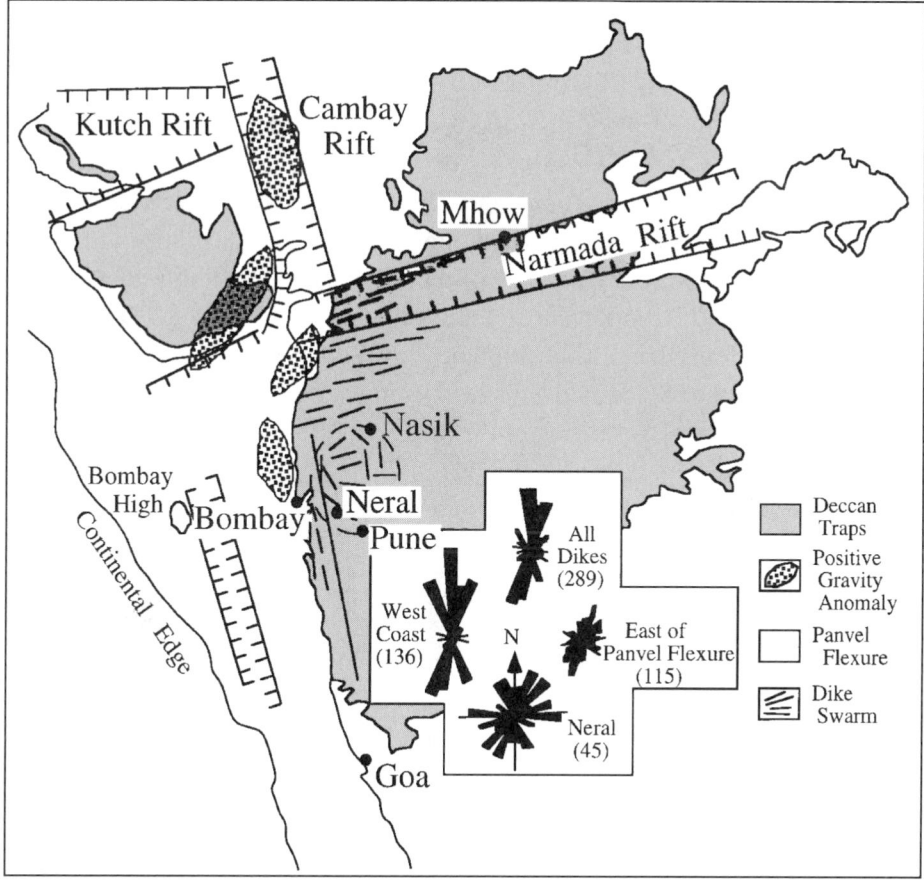

Figure 6.6. Sketch map of the Deccan Traps, showing the orientations of dikes. The feeder dikes to the main phase of flood volcanism in the Neral region have random orientations and cannot have been produced during a period of extension. A suite of dikes parallel to the west coast of India probably were produced by the extension associated with the separation of the Indian and Seychelle plates. They are perpendicular to the direction of extension, but cut the main phase of tholeiitic volcanism and must therefore be younger. (Adapted from Hooper, 1990.)

17 and 15 Ma, was accompanied by minor east–west extension, strong enough to produce north–south-oriented feeder dikes, but not strong enough to produce detectable crustal thinning. Hooper (1990) estimated the amount of extension to have been less than 1%. Significant extension (about 20%) began at approximately 15 Ma, *following* the eruption of the voluminous Grande Ronde Basalts, and it was confined to the southern half of the field. It was accompanied by a change in volcanism from massive flows of tholeiitic sheets to small-volume flows of olivine tholeiite, calc-alkalic to alkalic, and silicic volcanics similar to the volcanics of the extensional Basin and Range Province to the south.

There are three sets of dikes associated with the Paraná volcanics: a small set of dikes west of Rio de Janeiro, the Etendeka dikes, and the west–northwest-trending dikes of the Ponta Grossa and eastern Paraguay regions (Turner et al., 1994). Together these three suites form a triple junction centred on the South American coast near Curitiba. Two of the suites, the Etendeka and the Rio de Janeiro dikes, are parallel to the coast, but they lie outside the volcanic field, making it unlikely that they acted as feeders to the main phase of Paraná volcanism. The Ponta Grossa and eastern Paraguay dikes are much more likely candidates. Their location at the centre of the Paraná, broad aereal extent, timing, and chemical affinity with the Pitanga and Paranasanema magma types of the Paraná all support this interpretation (Turner et al., 1994). The orientation of the Ponta Grossa dikes requires northeast–southwest extension at the time of eruption of the Paraná basalts, at a right angle to the tensional direction associated with the opening of the South Atlantic and 10–15 million years before the formation of the first oceanic crust in the central Atlantic at 120 Ma (Guiraund and Maurin, 1992).

Two generalizations can be drawn from the foregoing discussion of relationships between dikes and flood volcanism: Firstly, in each case, the main phase of volcanism preceded significant extension. Secondly, where the products of flood volcanism are accompanied by a suite of dikes oriented perpendicular to the direction of extension (normally parallel to the coast) and their temporal relationship is known, the dikes postdate the main phase of continental volcanism.

The White and McKenzie hypothesis, therefore, fails both tests. The field evidence shows that the main phase of volcanism preceded extension, at least for the Deccan, Paraná, and Columbia River basalts. Furthermore, the incubation time of less than 5 million years for the Wrangellia Province would have been too short for the buildup of hot mantle material required by the White and McKenzie hypothesis.

6.6. Refinements to the Plume-Head Hypothesis

An apparent weakness of the plume-head hypothesis is that a head of pyrolite composition is unlikely to undergo extensive melting if it ascends beneath thick continental lithosphere. This point is illustrated in Figure 6.7a, which shows the results of a series one-dimensional melting calculations for a rising plume head, based on the approach of McKenzie and Bickle (1988). The head is modelled as a uniform disk of hot mantle that rises adiabatically and undergoes decompressional melting until its ascent is arrested by the base of the lithosphere. The thickness of a layer of melt produced at the surface has been calculated as a function of plume temperature and lithosphere thickness (Cordery et al., 1997). If it is assumed that the plume head is a mixture of hot mantle material from the plume source and cooler entrained material, so that its average potential temperature is 100°C greater than that of the upper mantle (i.e., 1,380°C), no melting will occur in the model plume head composed of pyrolite if its ascent is stopped by the base of the lithosphere, at a depth of 100 km. Farnetani and Richards (1994) recognized this problem and suggested that plume temperatures must be at least 350°C greater than that of the

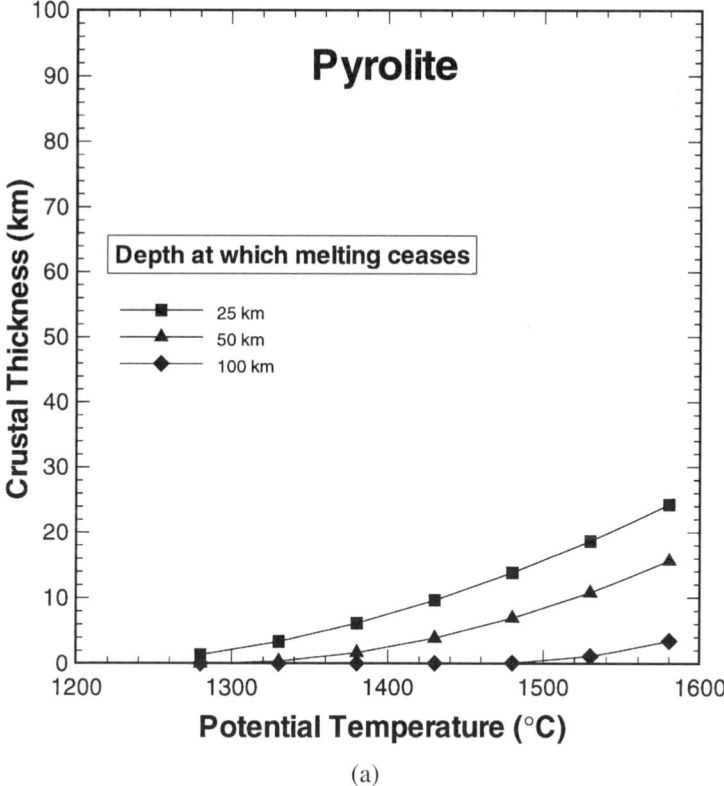

Figure 6.7a. One-dimensional decompressional melting of a plume head consisting of pyrolite, based on the approach of McKenzie and Bickle (1988). The diagram gives the thickness of the basaltic layer (representing the amount of melt) as a function of the potential temperature of the plume head, and the depth corresponding to the base of the lithosphere, at which melting ceases.

upper mantle for melting to occur beneath the continental lithosphere without the aid of extension. Their hypothesis predicts that flood volcanism should be associated with 2–4 km of uplift and that the primary magmas should be picrites, with 15–20% MgO. Although picrites are a common feature of flood basalts, they generally are a minor component, confined to the early stages of volcanism. Farnetani and Richards (1994) attributed the scarcity of picrites to fractional crystallization at the crust–mantle boundary, but the amount of fractionation required seems excessive, and the restriction of picrites to the early volcanics is not explained. A more serious problem is the magnitude of the uplift predicted by their hypothesis. Most flood basalts are associated with about 1 km of uplift, not 2–4 km as predicted by Farnetani and Richards (1994). As the amount of uplift is directly related to plume temperature, lower average plume-head temperatures are implied.

The simple analysis presented earlier, suggesting that a plume head cannot melt if it ascends beneath continental lithosphere, ignores three factors that may have important influences on the amount of melt generated by a rising plume head: (1) the role of eclogite in lowering the melting temperature of the plumes, (2) the thermal

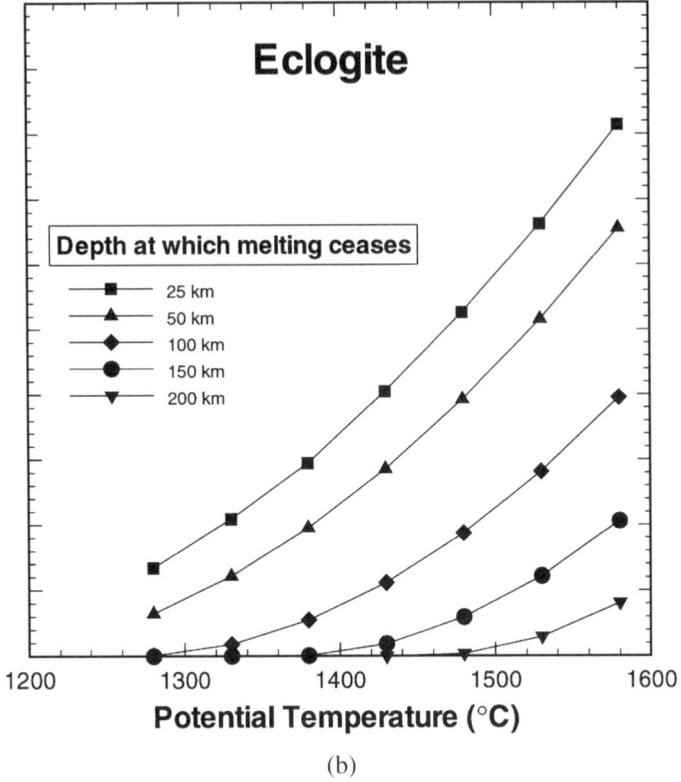

(b)

Figure 6.7b. One-dimensional decompressional melting for a model plume head consisting solely of eclogite. (Adapted from Cordery et al., in press.)

structure of plume heads, and (3) the possibility that secondary instabilities may develop in plume heads. Because the role of eclogite in the plume source region is essential to the arguments developed in this chapter, I shall begin by discussing the case for eclogite in the mantle source of ocean-island basalt (OIB).

6.6.1. The Case for Eclogite in OIB-type Mantle

The geochemical characteristics of basalts derived from OIB-type mantle (OIBs and flood basalts) that distinguish them from basalts derived from MORB-type mantle are (1) higher concentrations of incompatible elements, (2) higher Rb/Sr and Nd/Sm ratios, (3) higher $^{87}Sr/^{86}Sr$ ratios, but lower $^{143}Nd/^{144}Nd$ ratios, (4) $^{187}Os/^{188}Os$ ratios that are higher than that for the bulk Earth and up to 20% more radiogenic than those for unaltered MORBs, and (5) higher FeO contents at a given SiO_2 content (Figure 6.8). Furthermore, in $^{87}Sr/^{86}Sr–^{143}Nd/^{144}Nd$ space, the field for plume-related magmas is distinct from, and more dispersed than, the field for MORBs. Flood basalts and OIBs with these chemical and isotopic characteristics cannot be produced by melting pyrolite. A source region of different composition is required.

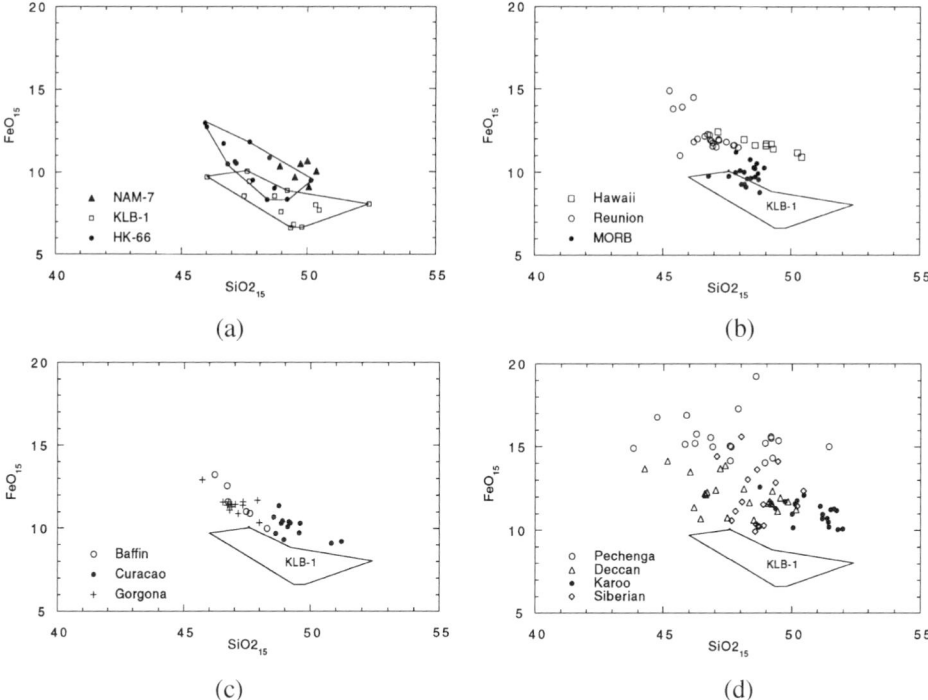

Figure 6.8. A plot of FeO against SiO_2 for selected experimental melts, MORBs, and plume-related basalts and picrites. Following Scarrow and Cox (1995), all samples have been normalized to an MgO content of 15% by incremental addition or subtraction of olivine with a composition defined by the Roeder and Emsley (1970) relationship. Only samples with MgO contents above 8% are included in this diagram. (a) Experimental melts produced by melting a natural peridotite (KLB-1) whose composition is similar to that of pyrolite, an SiO_2-FeO-rich peridotite (HK-66), and a MORB-type tholeiitic basalt (NAM-7). (b) Typical MORBs and basalts and picrites from Hawaii and Réunion as representatives of OIBs. Each Hawaiian datum is the average of a large number of samples from a volcanic centre (Hauri, 1996). (c) Depleted picrites from Baffin Bay, Gorgona, and Curaçao. (d) Typical flood basalts. Data from Bender et al. (1984), Kay et al. (1970), Hirose and Kushiro (1993), Yasuda et al. (1994), Kerr et al. (1996a,b), Echeverria (1982), Ludden (1978), Mahoney (1988), Beane et al. (1986), Ghose (1976), Lightfoot et al. (1990, 1993), Cox and Hawkesworth (1985), Mahoney et al. (1985), Krishnamurphy and Cox (1977), Clarke (1970), Aitken and Echeverria (1984), Bristow (1984), Ellam and Cox (1989, 1991), and Gill et al. (1988).

During the 1970s, the high incompatible-element concentrations in OIBs were attributed to very low degrees of partial melting. However, this hypothesis is inconsistent with the evidence (Hofmann and White, 1982). OIBs are associated with the melting of anomalously hot mantle and with the formation of thickened oceanic crust (e.g., Iceland). That is, OIBs appear to require high degrees of partial melting, rather than low. A second characteristic of most OIBs is that they have Rb/Sr and Nd/Sm ratios well above the chondritic value (i.e., they are enriched in light rare-earth elements, LREEs), whereas their $^{87}Sr/^{86}Sr$ and $^{143}Nd/^{144}Nd$ values indicate that the time-integrated Rb/Sr and Nd/Sm ratios in the mantle source regions are below the bulk-Earth values. Hofmann and White (1982) suggested that these characteristics of the OIB source, together with some aspects of the U/Pb and Th/Pb systematics, could be explained if the OIB source region contained a significant component of recycled oceanic crust. This concept is strongly supported by the recent Os isotopic evidence. The presence of recycled oceanic crust in the OIB source region could explain the high $^{187}Os/^{188}Os$ ratios in OIBs provided the recycling time was at least 1.0 billion years, because oceanic crust has a Re/Os ratio that is well above the value for the bulk Earth, as reviewed by Hauri and Hart (1993).

The principal advantages of the recycled-oceanic-crust hypothesis are that (1) high incompatible-element concentrations in OIBs can be reconciled with high degrees of partial melting, because the concentrations of the incompatible elements are 10 times as high in basalt as in pyrolite, and (2) garnet and clinopyroxene remain in the residue of an eclogite source region, so that the incompatible elements are fractionated even at high degrees of partial melting. This can explain the apparent paradox of Nd and Sr isotopes in OIBs: the production of high Rb/Sr and Nd/Sm ratios in basalts produced by the melting of a source region whose ratios of these elements are below the bulk-Earth values. The recycled oceanic crust would, of course, invert to eclogite at the pressure at which plumes melt in the upper mantle.

The source region for OIBs, however, cannot be pure eclogite. Eclogite melts to produce quartz-saturated tholeiites, whereas most OIBs are SiO_2-undersaturated. Furthermore, 100% melting of eclogite would produce basalt with a maximum MgO content of 7–8%. Melting of an eclogite source region, therefore, could not produce the picrites that are an important feature of many flood-basalt provinces. These problems can be overcome if the OIB source region is a mixture of eclogite and pyrolite.

The high FeO-SiO_2 contents of OIBs and flood basalts, seen in Figure 6.8, are also consistent with the eclogite hypothesis. The inverse relationship between [FeO] and [SiO_2] for the melts of KLB-1 and HK-66 (Figure 6.8a) is due to the influence of pressure on the stabilities of olivine and orthopyroxene. Raising the pressure increases the stability field for orthopyroxene relative to olivine in the crystalline residuum, leading to FeO-enriched, SiO_2-deficient melts. Note that the FeO content at a given [SiO_2] for melts produced from the basalt NAM-7 is greater than those produced by melting HK-66, which in turn are greater than those produced by melting KLB-1. Figure 6.8b shows typical MORBs and selected OIBs (Hawaii and Réunion), Figure 6.8c shows depleted picrites from Baffin Bay, Gorgona, and Curaçao, and Figure 6.8d shows some representative flood basalts. The MORBs lie within or close to the KLB-1 field (Figure 6.8b), whereas all of the plume-related magmas lie above that field. Although there is some overlap between the fields, the FeO content of flood basalt at a given SiO_2 is greater than those for OIBs, which in turn are greater than those for the depleted picrites. Even if the Fe-rich Pechenga picrites are ignored, the average difference in FeO content at a given SiO_2 content between MORBs and flood basalts is an impressive 2–3%.

The high FeO content in OIBs and flood basalts compared with MORBs cannot be explained by high pressure. If pressure were responsible for the difference, plume-related magmas should lie on the high-FeO extension of the KLB-1 field. That is, high FeO should be accompanied by low SiO_2. The correlation between high FeO and high SiO_2 requires OIBs and flood basalts to form by melting of a source composition that is richer in FeO and/or SiO_2 than pyrolite (Scarrow and Cox, 1995).

The presence of eclogite in the plume source region can explain the high FeO-SiO_2 content of plume basalts. Two factors contribute to that difference: (1) Eclogite has higher FeO and SiO_2 contents than pyrolite (10.5% of total iron as FeO and 50% SiO_2 in eclogite, compared with 8% and 45%, respectively, in pyrolite). As might be expected, raising the FeO-SiO_2 concentration of the parental material increases the SiO_2-FeO content of the basalts produced by melting it (Figure 6.8a). (2) If eclogite is recycled MORB, it will include a large proportion of oxidized oceanic crust. Dehydration during subduction of a slab would, of course, have lowered its water content and changed its fO_2. The extent of such changes is uncertain, but the evidence suggests that the OIB source is more oxidized (about one log unit higher) (Ballhaus, Berry, and Green, 1990) and has a higher water content (typical OIBs have about 0.5 wt% water, compared with about 0.2 wt% for MORBs) (Muenow et al., 1990) than the MORB source. Because Fe^{3+} acts as a moderately incompatible element during partial melting, the higher fO_2 of the OIB source region will result in the production of melts with elevated Fe_2O_3 contents and therefore higher total Fe. The very high FeO content of the Pechenga picrites requires a source with a very high fO_2 or unusually high Fe content.

The picrites of Gorgona, Baffin Bay, and Curaçao have lower [FeO] than the OIBs and flood basalts (Figure 6.8). These picrites are also depleted of incompatible elements, indicating derivation from a depleted-mantle source. Plume picrites therefore show a correlation between high FeO at a given SiO_2 and enrichment in incompatible elements. I suggest that this difference is due to the presence or absence of eclogite in the plume source region. Plumes that contain a high percentage of eclogite melt to produce FeO-SiO_2-rich basalts that are LREE-enriched, whereas plumes that contain little or no eclogite produce LREE-depleted basalts with lower FeO-SiO_2 contents.

A related feature of flood basalts that can be explained by eclogite in the plume source is their low MgO content, reflected in low Mg/Fe ratios. Flood basalts commonly have MgO contents below 8% and Mg/Fe ratios below the value expected for a melt in equilibrium with pyrolite (i.e., with an olivine of composition Fo_{89}). These properties of the basalts have been attributed to the magmas having undergone extensive fractionation en route to the surface (e.g., Cox, 1980). However, if the plume source contains eclogite, the amount of fractionation required to produce basalts with low MgO and low Mg/Fe will be reduced. Furthermore, 'primitive' flood basalts and OIBs are not constrained to be in equilibrium with olivine Fo_{89}. It is possible that some OIBs and flood basalts with 6–8% MgO are primary magmas.

6.6.2. The Role of Eclogite in Lowering the Melting
Temperature of Plumes

The presence of eclogite in a plume might have a profound influence on the temperature at which melting begins (Campbell, Cordery, and Davies, 1995). This point is dramatically illustrated in Figure 6.9, which shows that the dry eclogite solidus is 200°C below the pyrolite solidus at depths between 100 and 150 km (Yasuda, Fujii, and Kurita, 1995), the likely depth range for melting in a plume ascending beneath continental crust. In fact, a plume consisting of pure eclogite that is 200°C hotter than the upper mantle would begin melting at a depth of 200 km, and if no melt escaped from the plume, it would be completely molten by the time it reached a depth of 130 km. The depth at which melting starts will be further increased if the OIB source region contains water at 500–1,000 ppm, as suggested by the 0.5% water content of OIBs. This amount of water lowers the pyrolite solidus by about 50°C (Hirose and Kawamoto, 1995; Green and Falloon, Chapter 7, this volume), but its influence on the eclogite solidus is unknown.

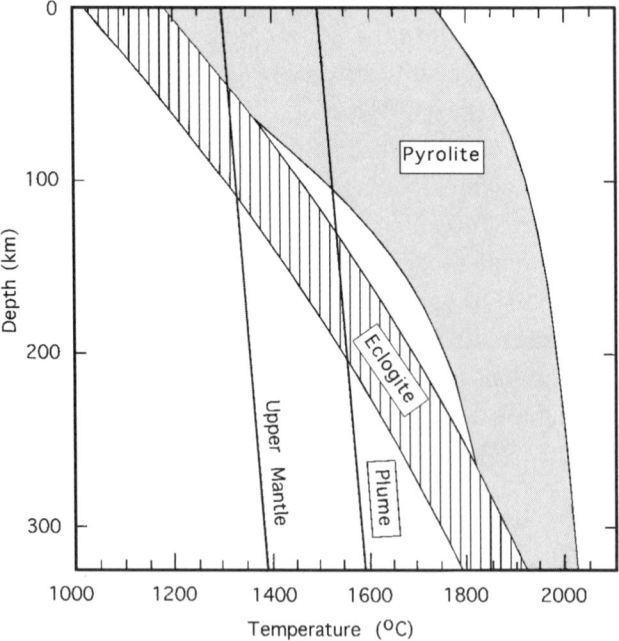

Figure 6.9. The solidus and liquidus temperatures for pyrolite and eclogite as functions of depth. The adiabatic temperature paths for the upper mantle (potential temperature, $T_p = 1,300°C$) and for a plume with a temperature excess of 200°C ($T_p = 1,500°C$) are also shown. Eclogite solidus and liquidus data from Yasuda et al. (1994). (Adapted from Cordery et al., in press.)

If the mantle were enriched in eclogite, however, that would not necessarily lower its melting point. O'Hara (1965) has shown that there is a thermal divide between model mantle ('peridotite') and 'eclogite' compositions in the synthetic system $CaO-Al_2O_3-MgO-SiO_2$. In this system, the addition of a small amount of 'eclogite' to the model mantle does not change its melting point. Only if sufficient 'eclogite' is added to cross the thermal divide does the eclogite solidus become relevant. In natural systems, which include FeO and SiO_2, the addition of a small amount of eclogite to a pyrolite source region can be expected to fertilise the mantle and lower its solidus temperature. Nevertheless, the conclusion that the addition of a small amount of eclogite to pyrolite has little influence on the melting point of the mixture is likely to remain valid. If eclogite were to play a role in lowering the melting temperature for plumes, the percentage of eclogite in the source region would have to be large, probably in excess of 50%, in order to take advantage of the low temperature of the eclogite solidus.

The compositions of the first melts produced by partial melting of eclogite pose an additional complication. Green and Ringwood (1968) have shown that the first melts are andesitic, with an SiO_2 content of 62%, far higher than the SiO_2 content of a typical flood basalt. There are two possible solutions to this problem. Firstly, if the melt is produced by melting an eclogite-pyrolite hybrid, its SiO_2 content will be buffered by the presence of olivine and therefore will be lower than that observed in the Green and Ringwood experiments. Secondly, if the region of melting consists of pure eclogite, the SiO_2 content of the melt will decrease as the melt fraction increases. For example, the melts reported in the eclogite-melting experiments of Yasuda et al. (1994), produced at *high* degrees of partial melting, had SiO_2 contents between 50.8% and 52.4% SiO_2 and compositions that were remarkably similar to those of flood basalts (Figure 6.8a, NAM-7). It is apparent from Figure 6.9 that very high degrees of partial melting are possible in plumes or parts of plumes that consist dominantly of eclogite.

As already argued, plumes are unlikely to consist of pure eclogite. Recycled oceanic lithosphere consists of a mixture of former basalt (plus gabbro), depleted harzburgite, and pyrolite. These components were formed by the melting of MORB-type mantle, and if on recycling through the mantle they were to become thoroughly mixed, the average composition of the mixture would again be MORB-type mantle. The solidus of that mixture would be the same as for pyrolite (except on a local scale), and the melts produced would have the same isotopic composition as MORB. Well-mixed, recycled oceanic lithosphere cannot be the source of OIBs. If recycled oceanic lithosphere were to play a role in the formation of OIBs, the basaltic component would have to be concentrated relative to the refractory harzburgite component.

There are two possible mechanisms. The first is that the two components might be separated gravitationally (Christensen and Hofmann, 1994). The basalt component would be richer in Fe, Al, and Si than the depleted harzburgite component and, at upper-mantle pressures, would transform to eclogite, which is appreciably denser. The relative densities of basalt and harzburgite at core–mantle boundary pressures are not known, but a recent assessment by S. E. Kesson and others at The Australian National University suggests that the basaltic component would again be denser. Isotopic evidence requires the OIB source material to have been isolated from MORB-type mantle for 1.5–2.0 billion years (Chase, 1981), and that can be taken to be the recycling time for oceanic lithosphere. Much of that time, say 0.5–1.0 billion years, would be spent within the lower thermal boundary layer while the initially cool subducted lithosphere acquired enough heat and buoyancy to ascend in a mantle plume. It would be during this period, when the two components would be hottest and would have their lowest viscosities, that they would be most likely to undergo gravitational separation (Christensen and Hofmann, 1994). If the basaltic component were the denser, it would sink to the bottom of the boundary layer, where it would be preferentially sampled by plumes.

The second possibility is that the basalt and harzburgite components might not separate, but rather fail to mix thoroughly during the recycling process. Whether or not this process could effectively isolate the basaltic component in the plume source region would depend on the scale of mixing. Consider a plume consisting of lenses of eclogite surrounded by pyrolite or harzburgite. Even at plume temperatures, melting would be confined to the eclogite lenses because of their lower melting temperatures. The melt produced would be buoyant and would rise through the eclogite to the contact with the overlying harzburgite and/or pyrolite. Such a melt would, of course, be out of equilibrium with its new environment, and it would react with the mineral phases present, consuming heat and part of the melt in the process and producing a narrow zone of refractory pyroxene and garnet. Because of its high solidus temperature, this material might act as a cap that would impede the ascending melt, which would collect below it until the melt acquired enough buoyancy to break the cap and escape to the surface (Figure 6.10). It is suggested that trapping of melt below that cap might allow the eclogite lenses to continue melting during adiabatic ascent and reach the high degrees of partial melting required for the formation of continental flood basalts. If the thickness of the eclogite lenses were a few metres or less, it would be unlikely that the melt produced would be able to acquire enough buoyancy to escape from the mantle. If, however, the thickness of the eclogite lens were a few tens of metres or more, a much larger volume of melt would be produced, and it might acquire enough buoyancy to escape from the mantle, although extensive interaction with the adjacent pyrolite and harzburgite would be inevitable.

The likely length scale for layering in recycled oceanic lithosphere is difficult to assess. Prior to subduction, the thicknesses of the basaltic and harzburgitic layers are approximately 7 and 50 km, respectively. However, during flow from the boundary layer into the plume and during flow up the plume, both layers would be thinned by stretching. If stretching reduced the thickness of the basaltic layers to a few metres, melting might occur on a local scale, but it would be unlikely that the melt would

Pyrolite

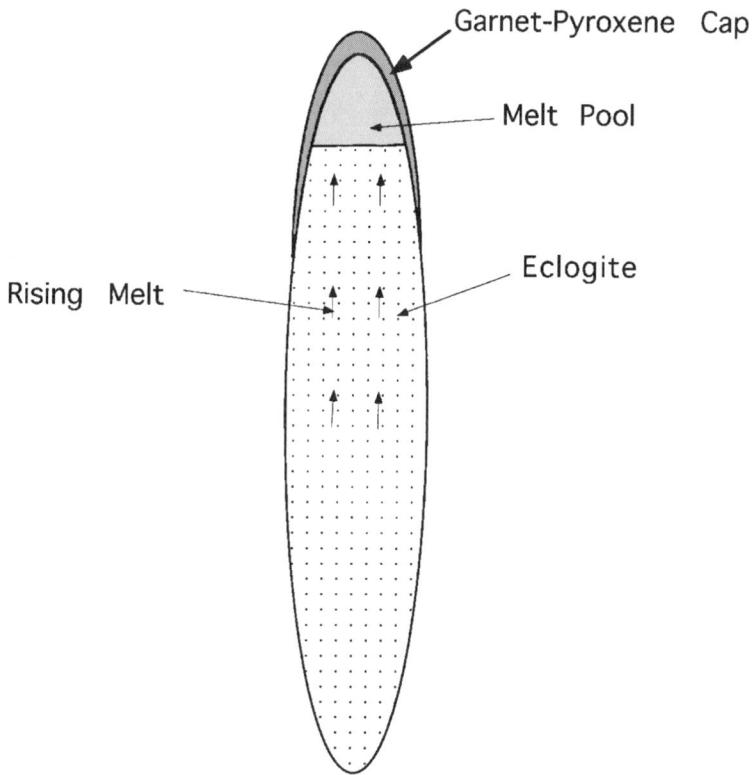

Figure 6.10. Sketch of a lens of eclogite surrounded by pyrolite (or harzburgite). In a rising plume, the eclogite will melt to produce a magma that will rise to the top of the lens, where it will react with the pyrolite, changing its composition and producing a refractory pyroxene-garnet cap. Later melts may accumulate under that cap.

be able to escape from the mantle. The critical requirement for the basaltic layers, if they are to play a significant role in the formation of flood basalts, is that they be large enough to generate melt with sufficient buoyancy to escape from the mantle.

A change in plume composition from pyrolite to eclogite could have a marked influence on the amount of melt produced. It was shown previously, in connection with Figure 6.7a, that a plume composed of pyrolite, with a temperature excess of 100°C, would produce no melt if its ascent was arrested by the lithosphere at a depth of 100 km. Had the plume been composed of eclogite, a layer of basalt 5 km thick would have been produced. It is not possible to predict quantitatively the amount of magma that would be produced by the melting of a hybrid plume head consisting of pyrolite and eclogite, because no data exist for the melting relationships for such a mixture. A simple linear relationship has been assumed which takes no account of the latest heat consumed by the reaction between basalt and pyrolite, and therefore provides, what is strictly, a maximum value for the amount of melt produced. It might be appropriate for a plume containing eclogite in large lenses. If the same temperature and height of rise are assumed, a plume head consisting of 30% eclogite and 70% pyrolite would produce a layer of melt about 2 km thick (Figure 6.7b). That is comparable to the thicknesses of basalts in typical flood-basalt provinces, which commonly are 1–2 km. If the source region is assumed to contain 10% eclogite in thick layers, melting should produce about 500 m of basalt.

6.6.3. The Thermal Structure of Plume Heads

Entrainment of lower-mantle material into a plume head will cool the head, so that its average temperature will be well below the temperature of hot material in the plume tail (Griffiths and Campbell, 1990). Although the average temperature of the plume head will be lowered by this process, the temperature at the top of the plume head, especially near its centre, will be little affected. This is because the upper portion of the head will be continually fed by hot mantle flowing up the tail. The effect on the thermal structure of the head is illustrated in Figure 6.11. If the material in the plume tail is 300°C hotter than normal mantle material, the average temperature of the top of the plume head will be about 200°C hotter, although there will be a thermal gradient from the centre to the margin (Figure 6.11).

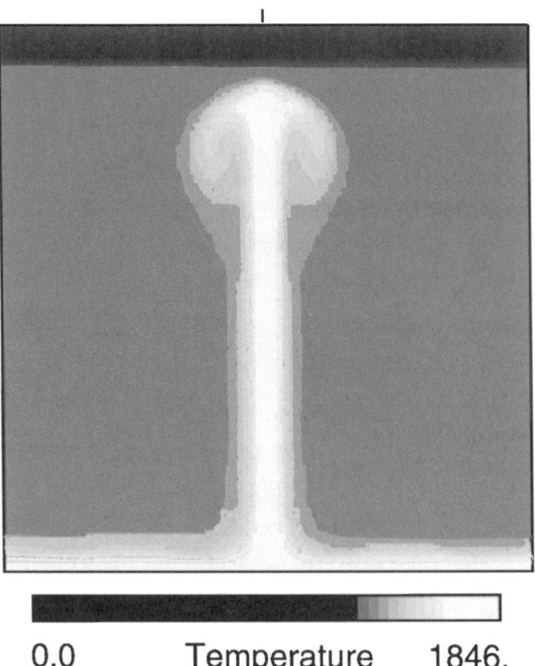

0.0   Temperature   1846.

Figure 6.11. Thermal structure in a plume, from a numerical model (G. F. Davies, personal communication). The model is of a cylindrical volume of fluid with radius equal to the depth (only half the radius is shown), and the background viscosity is 10^{22} Pa · s. The bottom boundary has an excess temperature of 430°C, yielding an excess plume-tail temperature of about 300°C, with a viscosity less than that of the background by a factor of about 120.

If the same height of rise is assumed as before (i.e., the top of the plume reaches a minimum depth of 100 km), then raising the temperature at the top of the head to 200°C above that of the upper mantle still would produce no melt during adiabatic decompression if the head were composed only of pyrolite. However, if it contained 10% eclogite in thick lenses, it would produce up to 2 km of melt, and up to 6 km if it contained 30% eclogite (Figure 6.7b).

6.6.4. Secondary Instability

The ascent of a plume head will be arrested soon after it encounters the mantle lithosphere. In order to rise farther, the head will have to displace the overlying mantle, which must flow sideways, out of the path of the rising head. Because the lithosphere that must be displaced is cold and stiff and because the distance it must flow is large, it is unlikely that a plume head could penetrate more than about 20 km into the mantle lithosphere by this process (Arndt and Christensen, 1992). However, detailed laboratory studies of the interaction between a rising sphere of buoyant fluid and an upper thermal boundary layer (Griffiths and Campbell, 1991b) have revealed a secondary instability that develops at the top of the sphere. Small secondary plumes of dense fluid in the boundary layer sink into the sphere, with a complementary flow of hot plume material moving in the opposite direction. If that type of secondary instability should develop during the interaction between a plume head and the mantle lithosphere, the head might penetrate farther into the lithosphere, because the lithosphere would be thinned by small-scale vertical flow, rather than by large-scale horizontal flow. The distance that the top of a head could penetrate via this mechanism is uncertain, but 75 km from the surface seems possible. If the top of the plume head were composed of pyrolite and were 200°C hotter than the upper mantle, it would produce a layer of basalt 3 km thick if it reached a depth of 75 km, 5 km of basalt if it were made up of pyrolite plus 10% eclogite, and 12 km of basalt if it contained pyrolite plus 30% eclogite (Figure 6.7).

The relative importance of the factors discussed earlier for the production of flood basalts is uncertain and will vary from case to case. The calculations are semiquantitative at best and are presented only to show that a plume head can produce a flood-basalt province without extension. However, it is important to remember that melting will be confined to the low-pressure region at the top of the head, where temperatures are highest. This part of the plume will be composed of material from the original source region (OIB mantle) and therefore will be likely to have a high fraction of eclogite. Thus, all of the factors that aid melting come together at the top of a plume head: (1) low pressure, (2) high temperature, and (3) a high concentration of low-melting-point eclogite. A more detailed analysis by Cordery et al. (1997) suggests that the role of eclogite in lowering the melting point of the mantle material at the top of a plume is essential for the formation of flood basalts if, as the field evidence suggests, volcanism precedes rifting.

6.7. The Chemistry of the Boundary-Layer Source for Plumes

Because plume heads and tails derive material from different depths, their melting products provide information about different levels in the mantle. We shall begin by discussing plume tails (including the hot axis of the plume head), because their melting mechanism is thought to be well understood, and the interpretation is less controversial.

6.7.1. How Plumes Sample a Boundary Layer

As already noted, mantle plumes must originate from a thermal boundary layer in the mantle, which (we continue to assume) is the core–mantle boundary. Because the core is hotter than the mantle, heat is conducted from the core into the overlying mantle, leading to the formation of a thermal boundary layer 100–200 km thick (Davies, 1990). The temperature step across this boundary layer is poorly constrained, but probably is 300–400°C for the modern mantle and possibly was as high as 600–700°C for the Archaean mantle (Griffiths and Campbell, 1991a). The viscosity of the mantle is strongly temperature-dependent and decreases by about three orders of magnitude for a 400°C increase in temperature. As a consequence, the viscosity of the mantle immediately above the core is much less than that of the mantle outside the boundary layer. This difference could have been as much as four orders of magnitude during the Archaean. Because the rate of flow of the mantle is a function of its viscosity, material within 10–15 km of the core will dominate the lateral flow within the boundary layer that feeds a plume (Davies, 1990) (Figure 6.12a). As this material is withdrawn from the bottom of the boundary layer, it is replaced by overlying material (Figure 6.12a). This extends the width of the zone sampled by a plume, and it can lead to a change in the composition of the material flowing up the plume over time if the composition of the mantle in the boundary layer is zoned vertically.

Lateral compositional variations within the source region can also influence the composition of the mantle sampled by a plume. Consider a source region consisting dominantly of C-type mantle, but with a pool of A-type mantle on one side of the plume, close to the plume axis, and a pool of B-type mantle a little farther from the axis on the other side (Figure 6.12b). Initially, an axisymmetric plume will sample only material close to the plume axis (i.e., C-type mantle). Over time, the radius of influence of the plume will increase, so that firstly A-type mantle and then B-type mantle will be drawn into it. As these different mantle types enter the plume they will be stretched and smeared during flow across the boundary layer and up the tail of the plume. Further stretching may occur during the lateral spreading that will take place when the ascent of the plume is arrested by interaction with the lithosphere. Note that this is smearing, not mixing at the molecular scale. However, if stretching reduces the length scale for the various mantle types, so that it is small compared

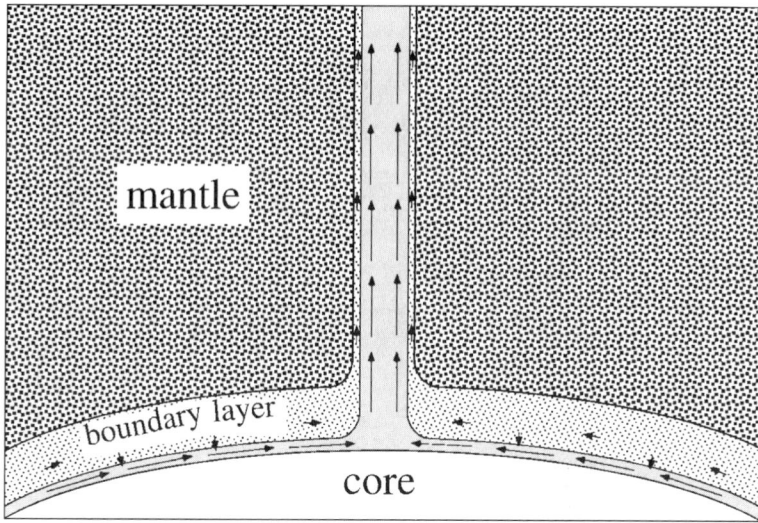

Figure 6.12a. Flow into a plume tail from a boundary layer above the core. The flow rate is greatest in the mantle adjacent to the core, where temperatures are highest and viscosities are lowest. As a consequence, most of the material flowing into the plume comes from within 10–15 km of the core. The vertical arrows in the boundary layer are a reminder that as material is drained from the layer adjacent to the core, it is replaced by overlying mantle.

with the length scale for the zone of melting in the plume, then for geochemical purposes the various mantle types can be regarded as mixed. In the example under consideration, the first melts issuing from the plume will show geochemical characteristics reflecting a C-type mantle source, then C-type plus B-type mantle in varying proportions, and finally all three types. The mechanism by which plumes sample the boundary layer can also lead to lateral compositional zoning within the plume tail. In that case, the left-hand side of the tail will be enriched in A-type mantle, and the right-hand side in B-type mantle. Sampling of a heterogeneous boundary layer, like that illustrated in Figure 6.12b, is the likely explanation for the subtle chemical differences between the two lines of volcanoes that make up the Hawaiian chain (Frey and Rhodes, 1993). For the same reason, there is every likelihood that basalts erupting on one side of a flood-basalt province or oceanic plateau will show distinct or subtle chemical differences from those erupting on the other.

6.7.2. Plume Tails

A most important point, in the present context, is that plume tails sample mainly the bottom of the boundary layer. If plumes originate from the core–mantle boundary, then OIBs and the lower volcanics from the central parts of flood-basalt provinces (which are the melting products of plume tails) can provide information about the mantle that immediately overlies the core (Campbell and Griffiths, 1992). Those melting products include alkali basalts, tholeiites, and picrites. In principle, the

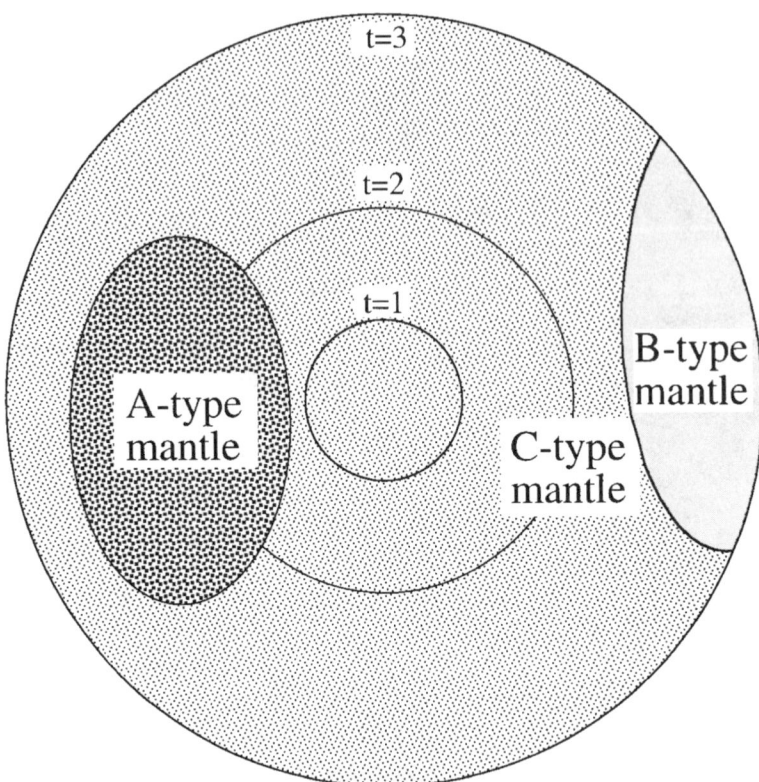

Figure 6.12b. Plan view of the mantle sampled by a plume at different times. At time $t = 1$ the plume will sample only mantle type C; at $t = 2$ it will sample types C and A, and at $t = 3$, types C, A, and B.

chemistry of any of these rock types can be inverted to provide information about the mantle that melted to produce them. However, where the aim of studying OIBs is to identify the nature of the plume source region, picrites are preferred over tholeiites, which are, in turn, preferred over alkali basalts. There are two reasons for this. Firstly, alkali basalts form by low degrees of partial melting and constitute only a small fraction of OIB volcanoes. They sample only a small volume of mantle and are not necessarily representative of the main eruptive phase. Picrites, which form by the highest degree of partial melting, are likely to sample the largest volume of mantle. Secondly, plume tails can entrain surrounding mantle material if they become tilted as they rise through the mantle. Where that happens, material from the plume source becomes diluted by mixing with material from the upper mantle and/or lower mantle. Extracting information about the plume source region from magmas produced by melting of such hybrid material is complicated, because the amount of entrainment and the composition of the entrained material are poorly constrained. Picrites and komatiites, which represent the least-contaminated samples of plume material, therefore provide the best indications of the chemical composition of the mantle's lower boundary layer (Campbell and Griffiths, 1992).

6.7.3. Identifying the Melting Products of Plume Tails

Plume-related volcanic rocks are found throughout the geological record. If the melting products of plume tails can be recognized with confidence, they can be used to characterize the mantle at the core–mantle boundary and to show how it has varied through geological time. For the modern Earth and for Phanerozoic periods this can normally be done by plate reconstructions that correlate volcanic rocks with a given hotspot track, but this is not possible for Proterozoic and Archaean volcanic rocks, and a different criterion is required.

Temperature variations within the convecting mantle, away from subduction zones, boundary layers, and hotspots, are thought to be small and within a few tens of degrees of the adiabatic gradient (Davies, 1988). McKenzie and Bickle (1988) suggested that the potential temperature (T_p, the temperature after adiabatic decompression to 1 atm) of the modern upper mantle is 1,280°C and that MORBs are formed by melting of mantle of this temperature during lithospheric stretching. Using this approach, they have been able to model the thickness and composition of MORBs with surprising accuracy and have shown that the maximum MgO content of basalts produced in this way is about 11 wt%. Klein and Langmuir (1987) reached similar conclusions, using a T_p slightly in excess of 1,300°C. Because the liquidus temperature of a dry magma is a function of MgO content, basalts with MgO contents well above 11 wt% must have come from anomalously hot zones within the mantle, that is, from mantle plumes. This provides a simple criterion for the recognition of plume activity and an additional reason to concentrate on picrites and komatiites. It follows that Precambrian volcanic associations that include magmas with high liquidus temperatures, such as picrites and komatiites, were produced by mantle plumes. For the modern mantle, Campbell and Griffiths (1992) have proposed that an [MgO] threshold of 14 wt% be used for identification of plume-related magmas, and 18 wt% for the hotter Archaean mantle.

It is important to remember that this criterion for identifying plume-derived magmas applies only to anhydrous melting. Rare picrites that form above subduction zones, the 'cold zones' in the mantle convection cycle, probably are produced by water-induced melting of the mantle wedge above a subducted slab. However, these magmas are readily identified by their calc-alkaline geochemical characteristics (e.g., low Ni and Nb/La ratio) (Campbell and Griffiths, 1993).

6.7.4. Classification of Mantle Source Regions

Basalts derived from the two principal mantle reservoirs recognized by geochemists, OIB-type mantle and MORB-type mantle, can be classified on the basis of their Nd isotopes and incompatible-trace-element characteristics. Basalts produced by melting of MORB-type mantle are depleted of the highly incompatible elements and have ε_{Nd} values that lie close to the evolution curve for the upper mantle at the time of crystallization (see McCulloch and Bennett, Chapter 2, this volume, for definitions of ε_{Nd} and the evolution curve for Nd). Those that show enrichment in highly incompatible elements, with ε_{Nd} values below the expected range for the depleted mantle at the time of crystallization, are classified as having come from enriched or OIB-type mantle.

The classification of mantle source regions as depleted or enriched on the basis of trace-element signatures in picrites and komatiites suffers from two potential limitations when applied to continental flood basalts. Firstly, the continental crust is enriched in highly incompatible elements and has a high $^{87}Sr/^{86}Sr$ ratio and a low $^{143}Nd/^{144}Nd$ ratio. A basaltic magma that was derived from a depleted-mantle source, but acquired high concentrations of incompatible elements through crustal contamination, could therefore have a geochemical signature similar to those of basalts derived from enriched mantle. However, the continental crust has two characteristics that are not shared by the OIB mantle source: high silica content and low Nb/La. To guard against the possibility of crustal contamination, in our assessments we omit all samples with anomalously high silica contents at a given MgO value and all samples that are enriched in highly incompatible elements and simultaneously display Nb/La ratios less than 80% of the chondritic value (Campbell and Griffiths, 1992).

The second factor that complicates interpretation of the chemistry of continental flood basalts is the suggestion that they form by melting of the continental mantle lithosphere or that they are contaminated by melts derived from that source. The first of these possibilities seems unlikely. The continental mantle lithosphere forms part of the cold boundary layer at the top of the mantle, and thus it is the part of the mantle that is least likely to melt (Campbell and Griffiths, 1990). It can be melted by uplift and stretching only if it also gains heat by conduction from a hot plume (White and McKenzie, 1989). Models of plume–lithosphere interaction (Arndt and Christensen, 1992) and simple conductive calculations (Campbell and Griffiths, 1992) suggest that melting of the continental mantle lithosphere is unlikely to be significant on the timescale of continental flood volcanism.

Gallagher and Hawkesworth (1992) have countered these arguments by suggesting that the continental mantle lithosphere can melt if it contains sufficient water. However, most flood basalts have erupted sub-aerially, and if their volatile content were high, pyroclastics should have been the dominant form of volcanism. With the notable exception of the Siberian Traps, pyroclastics are conspicuously absent from flood basalts. Furthermore, the mineralogy of the continental mantle lithosphere, as sampled by mantle nodules in alkali basalts and kimberlites, is dominated by olivine, pyroxene, garnet, and spinel. Amphibole, although known, is rare. Mantle nodules provide little support for the concept of a hydrous continental mantle lithosphere. The geochemical argument for lithospheric mantle contamination is based on two observations. Firstly, low-Ti basalts that are an important subgroup of the Gondwana flood basalts (Paraná, Karoo, Ferrar, and Tasmanian dolerites) have geochemical and isotopic characteristics that lie outside the field for OIBs. Secondly,

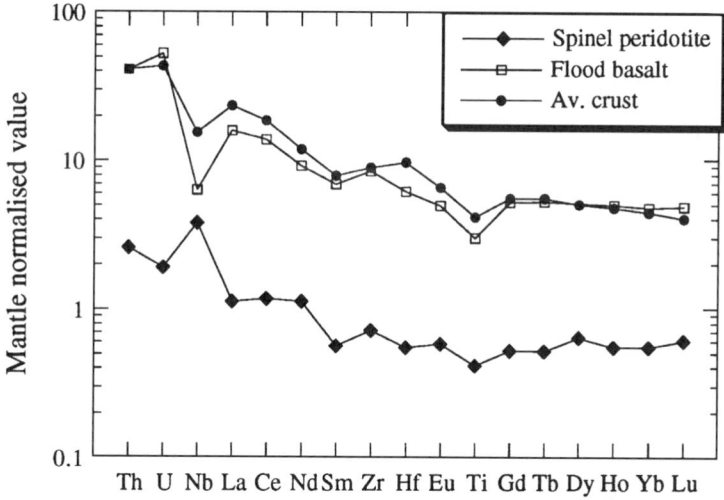

Figure 6.13. Primitive-mantle-normalized trace-element abundance pattern for the continental mantle lithosphere, as represented by spinel peridotites, compared with the pattern for a typical Nb-depleted flood basalt. (Adapted from McDonough, 1990.)

the distribution of these basalts appears to be geographically controlled and linked to Archaean cratons (Gibson et al., 1995). However, the suggestion of Gibson et al. (1995) that low-Ti basalts derive their geochemical characteristics from the continental mantle lithosphere is not without its problems. A characteristic feature of continental flood basalts that lie outside the OIB field is that they have Nb/La ratios that are well below the chondritic value. The average and median Nb/La ratios for the continental mantle lithosphere, as sampled by spinel peridotites, are 1.8 and 3.5, respectively, as compared with 1.0 for the primitive mantle (McDonough, 1990). Although there are a few examples of garnet-phlogopite-peridotite with low Nb/La ratios (e.g., sample JJG 319 of Cox et al., 1984) that might constitute suitable source material, such samples are rare. On average, the continental mantle lithosphere is enriched, not depleted, in Nb relative to La (Figure 6.13). It is therefore an unlikely source for low-Ti basalts.

The problem of the origin of low-Nb/La flood basalts has been circumvented in this study. By omitting all samples enriched in highly incompatible elements and with low Nb/La ratios, we are rejecting not only any samples that may have been contaminated by continental crust but also those samples that some geochemists believe might contain a contribution from the continental mantle lithosphere. However, application of the Nb/La-ratio criterion to picrites and komatiites leads to the rejection of two suites of picrites, those of the Keweenawan and Karoo, that clearly are associated with mantle plumes. In the case of the Keweenawan, the low Nb/La ratios probably are due to crustal contamination, and if that is the case, these samples have been correctly rejected from this study. However, Ellam and Cox (1989, 1991) and others have argued convincingly that the low Nb/La ratios of the Karoo picrites are of mantle origin. Whether the Karoo picrites derived their isotopic and trace-element geochemical characteristics from the plume source (Campbell and Griffiths, 1992) or as a consequence of contamination from lithospheric mantle (Ellam and Cox, 1989, 1991) has important implications for the Dupal anomaly and for the range of compositions that may be present in the plume source region (Campbell and Griffiths, 1992), but it is not relevant to the principal arguments developed in this chapter.

6.7.5. *The Geochemistry of Picrites and Komatiites through Time*

Mantle-normalized incompatible-trace-element abundance patterns for komatiites and picrites of various ages are plotted in Figure 6.14. Komatiites older than 3.4

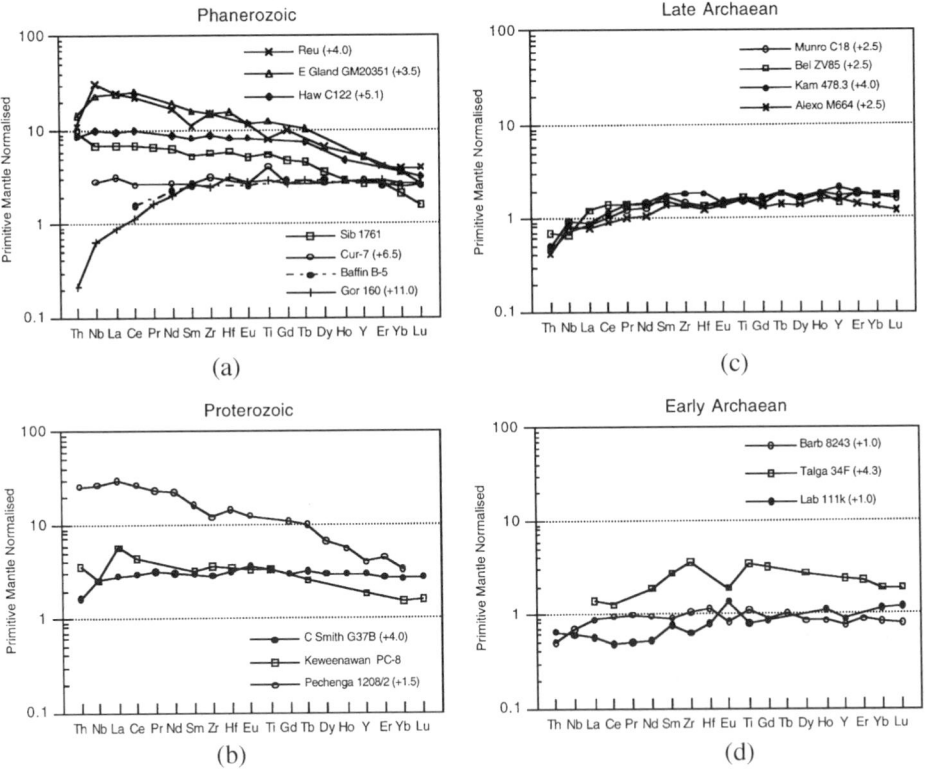

Figure 6.14. Mantle-normalized incompatible-trace-element patterns for picrites and komatiites of various ages. (a) Phanerozoic: 1761, picrite from the Siberian Traps (Lightfoot et al., 1990); C122, Hawaiian picrite (Chen and Frey, 1985); Reu, Réunion parent magma (Fisk et al., 1988); Gor 160, Gorgona komatiites (Jochum et al., 1991); GM20351, eastern Greenland picritic basalt (Gill et al., 1988); B-5, Baffin Bay (O'Nions and Clarke, 1972). (b) Proterozoic: a Pechenga ferro-picrite (Hanski and Smolkin, 1989); G37B, Cape Smith komatiite (Jochum et al., 1991); PC-8, Keweenawan picrites (Klewin and Berg, 1991). (c) Late Archaean: C18, Abitibi komatiite (Jochum et al., 1991); Kam 478.3, Yilgarn komatiite (Jochum et al., 1991); ZV85, Belingwe komatiite (Jochum et al., 1991). (d) Early Archaean: Barb 8243, Barberton komatiite (Jochum et al., 1991); 34F, Pilbara komatiite (Gruau et al., 1987); Lab 111k, Labrador komatiite (Collerson et al., 1991).

billion years are of two types. Those from Barberton and Labrador show weak depletion of highly incompatible elements and have low positive ε_{Nd} values (~ 1), whereas those from Pilbara show stronger depletion of highly incompatible elements and have an ε_{Nd} value of +4.3. All 2.7-Ga komatiites are depleted of highly incompatible elements and have ε_{Nd} values of 2.5 or greater.

There are only three examples of Proterozoic picrites for which high-quality geochemical data are available: Pechenga, Cape Smith, and the Keweenawan. The 2.0-Ga Pechenga picrites are strongly enriched in highly incompatible elements and have a low positive ε_{Nd} value, whereas the 1.9-Ga picrites of the Cape Smith Chukotat Suite are depleted of their incompatible elements and have moderate positive ε_{Nd} values. The Keweenawan picrites, which have an age of 1.1 Ga, are enriched in highly incompatible elements but have a pronounced Nb anomaly and are therefore excluded from consideration, for the reasons given earlier. Although the number of examples of Proterozoic picrites is limited, the available evidence indicates that Proterozoic plumes tapped sources in both enriched and depleted mantle. The Pechenga picrites provide clear evidence of an enriched OIB-type mantle source by 2.0 Ga, and this conclusion is supported by the occurrence of alkali basalts in the Povungnituk group, forming part of the 1.9-Ga Cape Smith volcanics.

Most Phanerozoic picrites are enriched in highly incompatible elements and show ε_{Nd} values well below the modern MORB value. However, two examples, from Gorgona and Baffin Bay, show depleted MORB-like geochemical characteristics, and a third, from Curaçao, has flat REE patterns. The Gorgona and Curaçao picrites have ε_{Nd} values of +11.0 and +6, respectively, but there are no Nd data available for Baffin Bay.

From the data presented in Figure 6.14 it is inferred that the evolution of material at the core–mantle boundary has been as follows: It consisted of a mixture of weakly depleted and depleted mantle at 3.5 Ga, was dominated by depleted mantle at 2.7 Ga, and consisted mainly of OIB-type mantle, with minor amounts of depleted mantle, between 0.5 Ga and the present. Its nature during the Proterozoic is less well constrained, but the data suggest that both OIB-type mantle and depleted mantle were involved. With the exception of a minor occurrence of a 2.7-Ga alkali basalt at Norseman, Western Australia (McCuaig, Kerrich, and Xie, 1994), OIB-type magmas are unknown from the Archaean. An interpretation consistent with the available data is that OIB-type mantle has become an increasingly important component in the plume source region beginning at the end of the Archaean.

It is important to note that the foregoing conclusions regarding the nature of the plume source and its secular variations are independent of the fluid dynamical model used for mantle plumes (i.e., White and McKenzie, 1989; Griffiths and Campbell, 1990). The only assumption required is that picrites and komatiites are produced by melting of the hottest parts of plumes, which represent material sampled from the bottom (or hottest part) of the boundary layer. Because this assumption is based on sound fluid dynamical principles, the conclusions should be robust.

What is controversial is the location of the boundary layer. As already noted, there are two possibilities: the core–mantle boundary and the upper-mantle–lower-mantle boundary. There are several lines of evidence that point to the core–mantle boundary as the most likely source for mantle plumes (Davies and Richards, 1992). These include (1) the fixed positions of hotspots relative to each other, (2) the close agreement between the predicted size for the head of a plume originating at the core–mantle boundary and the observed dimensions of flood-basalt provinces (Griffiths and Campbell, 1990), (3) the low percentage of the mantle's heat budget carried in mantle plumes (Davies, 1988), and (4) recent tomographic studies, using both P and S waves, showing clear images of slabs penetrating through the 660-km discontinuity (e.g., Kennett and van der Hilst, Chapter 8, this volume). On the other hand, geochemical evidence, especially that based on Ne, Ar, and He isotopes, has been used to argue that the mantle convects in two layers, with plumes originating from the 660-km discontinuity (O'Nions and Tolstikhin, 1996). In the interest of presenting a balanced view, both possibilities are considered. Figure 6.15 illustrates the chemical evolution of the plume source through time, assuming the source to be the core–mantle boundary. In Figure 6.16 it is assumed that plumes originate from the upper-mantle–lower-mantle boundary.

6.8. The Chemistry of the Lower Mantle
6.8.1. Melting of Plume Heads

If plume heads originate from the core–mantle boundary and entrain material as they rise through the mantle, then plume heads are mixtures of plume source material and material from the lower mantle. Their melting products, therefore, should provide information about the lower mantle. Before discussing the implications of this possibility, it is necessary to describe in greater detail the nature of the melting process. For simplicity it will be assumed that the material in the thermal boundary layer from which a plume originates is OIB-type mantle, although it is recognized that such is not always the case. The plume head will consist of a layer of eclogite-rich, OIB-type mantle at the top, underlain by entrained material from the lower mantle (Figure 6.1).

Campbell and Griffiths (1990) have considered the likely sequence of events that occurs when a plume head rises to the top of the mantle and flattens to form a disk. As already noted, melting results from adiabatic decompression, so that the first material to melt should be the OIB-rich layer at the top of the plume head. That is where the temperature of the head is highest, and where the pressure and solidus are lowest. As the head continues to rise and flatten, the underlying layer of entrained lower-mantle material may also enter the melt zone. If that happens, the first melts produced by a plume head will have geochemical characteristics consistent with derivation from an OIB-rich source, whereas later magmas will show evidence of derivation from a mixed OIB–lower-mantle source. However, if

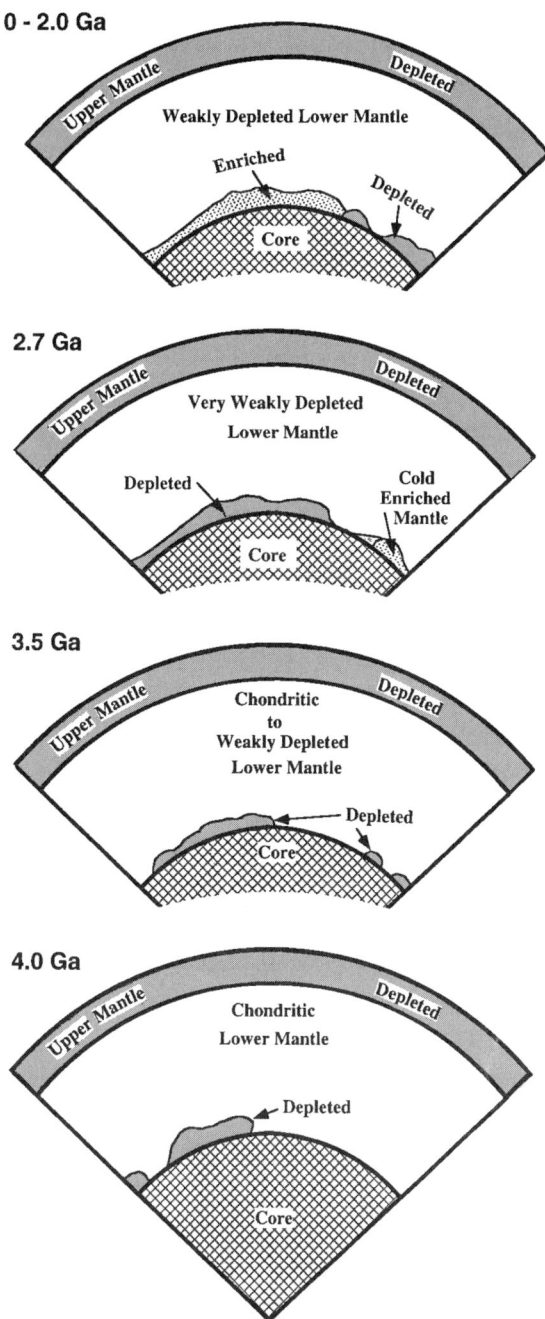

Figure 6.15. Sketches of sections through the mantle showing the inferred chemical structure of the mantle at selected times, assuming that plumes originate from the core–mantle boundary. The early lower mantle is assumed to have had chondritic nonvolatile-trace-element ratios and to have gradually evolved to a weakly depleted state at present. The boundary between the upper mantle and lower mantle is shown as being sharp, for clarity, but it probably is gradational in reality. (Adapted from Campbell and Griffiths, 1992.)

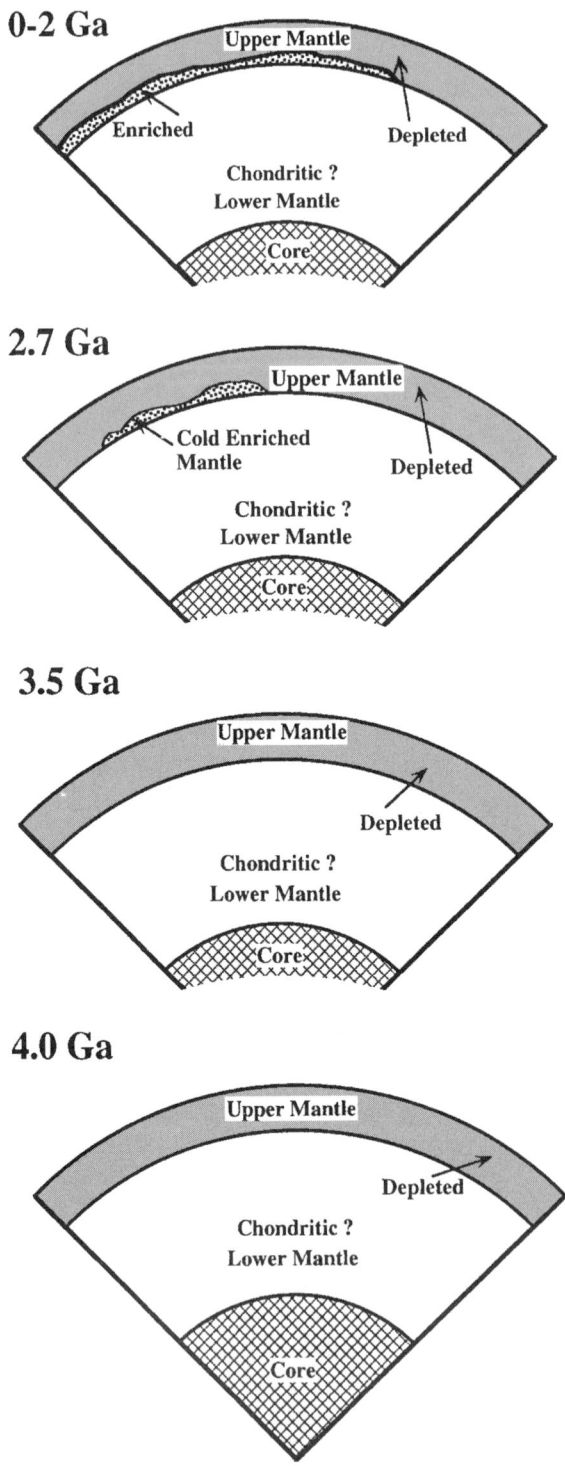

Figure 6.16. Sketches of sections through the mantle showing its inferred chemical structures at various times, assuming that plumes originate from the upper-mantle–lower-mantle boundary. In this scenario, the chemistry of the lower mantle remains enigmatic, but is probably chondritic.

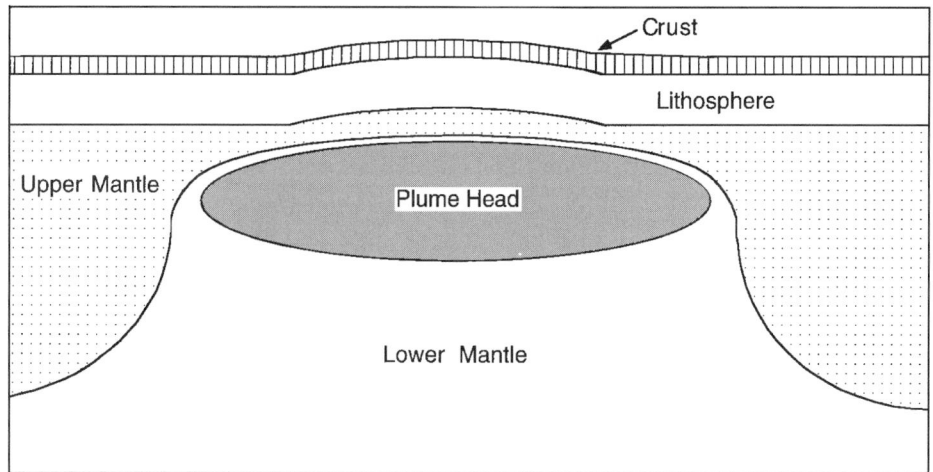

Figure 6.17. Sketch of a plume head at the top of its ascent, showing a thin layer from the lower mantle pushed in front of the head. The lower 3 km (approximately) of this layer is heated by the plume head and may take part in the melting process.

the entrained layer consists of pyrolite with a T_p about 100°C greater than that of normal upper mantle, it will be able to melt only if the layer ascends to a depth of less than approximately 75 km (Figure 6.7a). That may happen when a plume head ascends beneath young oceanic lithosphere, but it is less likely for plumes ascending beneath continental lithosphere unless the lithosphere is unusually thin or is thinned by a secondary instability (Griffiths and Campbell, 1991b) or by extension. That is, material entrained into the plume head may play an important role in the formation of oceanic plateaus, but is less likely to do so in the case of continental flood basalts.

There is an additional feature of the entrainment process, relevant to the melting of plume heads, that was not explicitly considered by Campbell and Griffiths (1990). A plume head rises by thinning the overlying mantle and pushing before it a continually thinning layer of lower mantle, as shown in Figure 6.17. As a consequence, when the ascent of the head is arrested by its interaction with the lithosphere, the head is overlain first by a layer of former lower-mantle material and then by a layer of upper-mantle material that will lie wedged between the top of the plume and the base of the lithosphere. The lower 3 km or so of that wedge, which will consist entirely or dominantly of material that originated in the lower mantle, will lie within the thermal boundary layer of the plume. It will be heated to temperatures comparable to that of the entrained mantle material in the centre of the plume. Material within the upper boundary layer is, however, more likely to contribute to flood basalts, for two reasons: Firstly, mantle material in the upper boundary layer will be at a lower pressure than the entrained mantle material within the plume, and if the materials have similar compositions, it will have a lower solidus temperature. Secondly, melts generated within the plume will have to pass through the boundary layer above the plume and therefore may become contaminated by melts derived from it.

On the other hand, if the White and McKenzie (1989) model for flood basalts is correct, plume heads originate within the upper mantle and entrain little material as they rise through the mantle. Under these circumstances, flood basalts and oceanic plateaus should be formed by melting mantle material of the same composition as ocean-island chains (i.e. plume tails). Studies of the chemistry of flood basalts and oceanic plateaus would then tell us little about the mantle that could not be learnt by studying ocean-island chains. Furthermore, if plumes originate from the upper-mantle–lower-mantle boundary, they provide no information about the lower mantle.

6.8.2. The Nature of the Lower Mantle

The plume-head hypothesis for flood basalts and oceanic plateaus predicts that basalts were formed by the melting of either mantle material from the plume's ultimate source at the core–mantle boundary or a mixture of that material and entrained lower-mantle material. The chemistry of flood basalts and oceanic-plateau basalts therefore has the potential to provide information about the lower mantle. Such chemistry can best be studied when the basalts are associated with komatiites or picrites, so that the chemistry of the high-temperature magmas that produced them can be used to identify the plume-source end member.

A study of basalt-picrite and basalt-komatiite associations in flood basalts and oceanic plateaus by Campbell and Griffiths (1992, 1993) has shown that the basalts generally are less extreme in their geochemical signatures than the associated komatiites and picrites (Figure 6.18). Picrites that are highly enriched in light REEs from flood-basalt provinces, such as the Karoo, Deccan Traps, and Siberian Traps, are associated with basalts that generally have flatter REE patterns (if samples contaminated by continental crust are excluded). For basalts from Gorgona Island, which is believed to be a fragment of the Caribbean oceanic plateau produced by the Galapagos starting plume (Richards et al., 1989; Storey et al., 1991; Hill, 1993), the situation is reversed. The komatiites are strongly LREE-depleted, with ε_{Nd} values of +11, whereas the associated basalts have flat REE patterns and ε_{Nd} values of +6. The picrite-basalt association of Curaçao, which also forms part of the Caribbean plateau, provides an intermediate example. Both the picrites and basalts have flat REE patterns, with ε_{Nd} values of +6.

The geochemical characteristics of Archaean basalt-komatiite associations, such as the Abitibi and Norseman-Wiluna greenstone belts, are similar to those for Gorgona; the komatiites are more depleted in LREEs and have higher ε_{Nd} values than the associated basalts (Campbell et al., 1989; Storey et al., 1991). Examples of REE patterns for basalt-picrite (komatiite) associations from the Siberian Traps, Gorgona, and Kambalda are illustrated in Figure 6.18. From these relationships,

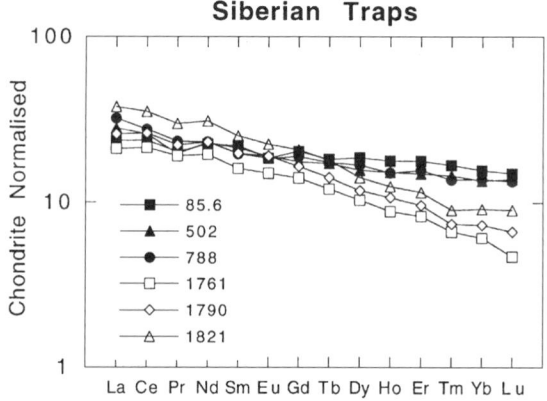

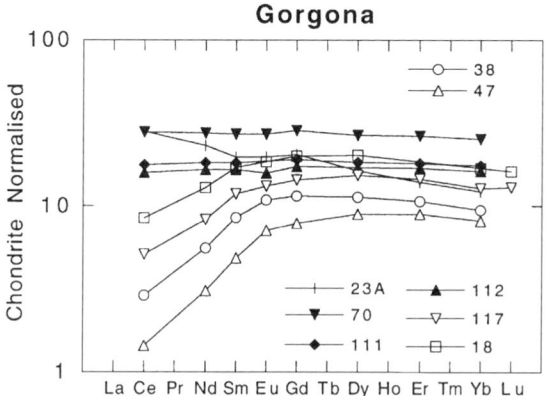

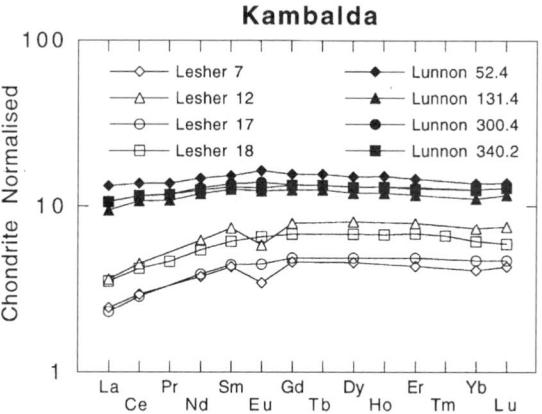

Figure 6.18. Chondrite-normalized REE patterns for komatiites (or picrites) (open symbols) and tholeiitic basalts (solid symbols) from Gorgona, the Siberian Traps, and Kambalda, Yilgarn Block, Western Australia. Data from Aitken and Echeverria (1984), Lightfoot et al. (1990), Lesher and Arndt (1995), and I. H. Campbell and P. J. Sylvester (unpublished data).

Campbell and Griffiths (1992) concluded that if continental flood basalts and oceanic plateaus were produced by the melting of plume heads that were mixtures of material from the ultimate plume source and material from the lower mantle, then the latter should have geochemical characteristics less extreme than those for OIB-type or MORB-type mantle. This does not mean that the lower mantle had chondritic non-volatile-trace-element ratios, but rather that it was neither as depleted of highly incompatible elements as MORB-type mantle nor as enriched (presumably in eclogite) as OIB-type mantle.

The chemistry of oceanic plateaus, such as the Ontong-Java Plateau, the Caribbean Plateau, and the Kerguelen–Broken Ridge, can be used to place further constraints on the nature of the lower mantle, even where picrites are absent. Oceanic plateau basalts have two advantages over continental flood basalts. Firstly, they do not erupt through continental crust and therefore are free from crustal contamination. Secondly, they erupt through thin oceanic lithosphere. As a consequence, the plume rises to shallower levels, and more of the head becomes involved in the melting process. If plateau basalts are interpreted in terms of the plume-head model, the early melts that form the bottom of a plateau should come dominantly from the layer of source mantle at the top of the plume (Figure 6.1) and/or the thermal boundary layer of former lower-mantle material pushed ahead of the plume. As the head rises and flattens against the lithosphere, more of the underlying layer of entrained material enters the melt zone, until, finally, melt from this layer may become the dominant component in the plume's volcanic output. Because the material entrained into a plume head is predicted to be largely from the lower mantle, these late melts should be the products of the melting of a mixture of material from the core–mantle boundary and material from the lower mantle, but dominated by the latter. It is these late basalts, which form the upper sequences of the oceanic plateaus, that have been sampled by the ocean dredging and drilling programs. Samples from the Kerguelen Plateau are highly enriched in LREEs. If there was a lower-mantle component in the source region for these basalts, it has been masked by the strong OIB-type (enriched) geochemical signature that is characteristic of the Kerguelen volcanics. An additional complication is that plate movement has returned the Kerguelen Plateau to the Kerguelen hotspot, so that much of the plateau is overlain by younger, OIB-type volcanics. For these reasons, samples from Kerguelen are unsuitable for evaluating the composition of the lower mantle.

The basalts from Ontong-Java, Malaita Island, the Naura basin, and the Manihiki Plateau, which formed part of the dismembered Ontong-Java Plateau (Coffin and Eldholm, 1994), and those from Curaçao and Gorgona, which form part of the Caribbean Plateau (Kerr et al., 1996a,b), provide more suitable material. Most of the basalts from these locations have flat REE patterns: ε_{Nd}, +3 to +7.5; $^{87}Sr/^{86}Sr$, 0.7030–0.7044; $^{206}Pb/^{204}Pb$, 18.2–19.2; $^{207}Pb/^{204}Pb$, 15.48–15.59; $^{208}Pb/^{204}Pb$, 38.2–39.0 (Mahoney et al., 1993; Kerr et al., 1996a). These are not the expected geochemical characteristics of basalts produced by the melting of OIB- or MORB-type mantle sources. Campbell and Griffiths (1992, 1993) suggested that these magmas had been produced by the melting of former lower-mantle material or by the melting of a hybrid source in which lower-mantle material was a significant component. A lower mantle with that composition would satisfy the earlier suggestion that it should be geochemically less extreme than OIB- or MORB-type mantle. The Ontong-Java Plateau and Caribbean Plateau are both large volcanic provinces, suggesting that they were derived from a major mantle reservoir. The implication is that there is a significant component in the lower mantle with the characteristics listed earlier. Furthermore, the composition inferred for the lower mantle on the basis of these oceanic plateau basalts is remarkably similar to FOZO, identified by Hart et al. (1992) as the common mantle component entrained into hotspots (plume tails) that has the following attributes: ε_{Nd}, +4 to +8; $^{87}Sr/^{86}Sr$, 0.703–0.704; $^{206}Pb/^{204}Pb$, 18.5–19.5; $^{207}Pb/^{204}Pb$, 15.5–15.65; $^{208}Pb/^{204}Pb$, 38.8–39.3.

6.8.3. A Geochemical Test for the Source of Mantle Plumes

Hofmann et al. (1986) have shown that U and Nb have similar partition coefficients during the partial-melting processes that produce OIBs and MORBs, so that the Nb/U ratios in these rock types should be essentially the same as in their mantle source region. However, during extraction of the continental crust from the upper mantle, U was fractionated from Nb, so that the upper mantle has a Nb/U ratio of about 48, compared with about 32 for primitive (chondritic) mantle (Figure 6.19). The OIB source has an Nb/U ratio similar to that of MORB-type mantle, but a higher Th/U ratio. As a consequence, OIBs and MORBs occupy different fields on a plot of Nb/Th against Nb/U. The lower mantle, which should have been less affected by the extraction of the continental crust, is expected to have Nb/Th and Nb/U ratios lying close to the values for the primitive mantle. Mass-balance calculations suggest

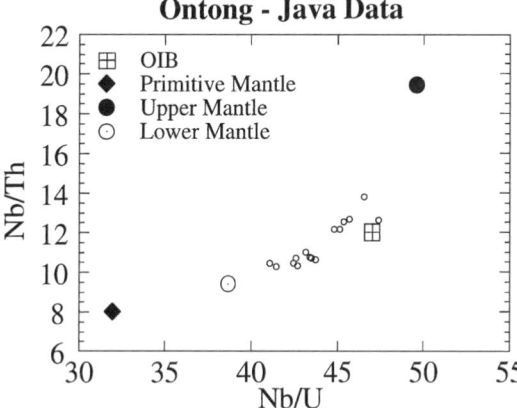

Figure 6.19. A plot of Nb/Th against Nb/U for a suite of basalts from Malaita Island, part of the Ontong-Java Plateau. Note that these samples lie on a mixing line between the OIB and lower-mantle compositions, as predicted by the plume-head hypothesis. (Adapted from Campbell et al., in press.)

that the incompatible trace elements in the continental crust could not have been extracted entirely from the upper mantle, so that some shift in the Nb/U and Nb/Th ratios away from the primitive-mantle values is to be expected. If the most recent estimates for the concentrations of Nb and Th in the continental crust and primitive mantle are used (Rudnick and Fountain, 1995; McDonough and Sun, 1995), the calculated Nb/U and Nb/Th ratios for the lower mantle are 39 and 9.5, respectively. These values are, of course, estimates that involve a degree of uncertainty. However, the lower mantle should have the same Th/U ratio as the primitive Earth (i.e., 4.2) (Allègre, Dupre, and Lewin, 1986), in which case the average lower-mantle value should lie between the values for the primitive mantle and OIB-type mantle (Th/U = 4.0) on a plot of Nb/Th versus Nb/U. Plots of Nb/Th against Nb/U therefore provide a simple and unambiguous procedure for classifying basalts as originating from MORB-type reservoirs, OIB-type reservoirs, or lower-mantle reservoirs.

If the Campbell and Griffiths (1990, 1992) plume-head hypothesis is correct, the basalts from the top of the Ontong-Java Plateau should display a clearly recognizable lower-mantle signature. Figure 6.18 is a plot of Nb/Th against Nb/U for a suite of basalts from Malaita Island, part of the Ontong-Java Plateau. Notice that the data lie on a mixing line between OIB-type mantle and lower mantle, as predicted (Campbell et al., in press).

6.9. A Model for the Evolving Chemical Structure of the Mantle

A model for the evolving chemical structure of the mantle, based on the foregoing analysis and on the assumption that plumes originate from the core–mantle boundary, is summarized in Figure 6.15. The thermal boundary layer above the core, as sampled by picrites and komatiites, consisted of a mixture of depleted mantle and chondritic mantle between 3.8 and 3.5 Ga. By 2.7 Ga, depleted MORB-like mantle was the dominant component in this region. It may also have contained a component of enriched, OIB-type mantle, but if it did, this enriched material was too cold (and therefore too dense) to have contributed to Archaean mantle plumes. During the Proterozoic, OIB-type mantle became increasingly important in the boundary layer, eventually becoming the dominant component in the Phanerozoic. However, the depleted picrites of Gorgona and Baffin Bay provide evidence of some MORB-like mantle in the modern plume source.

Our knowledge of the modern upper mantle is constrained by the compositions of MORBs, which, away from plumes, show remarkably uniform geochemical characteristics. As already noted, the upper mantle must be well mixed and depleted of incompatible trace elements, with low $^{87}Sr/^{86}Sr$ ratios, high $^{143}Nd/^{144}Nd$ ratios, and so forth. The Archaean upper mantle also appears to have had depleted geochemical characteristics. The oldest continental crust consists of gneisses and volcanic rocks, most of which have positive ε_{Nd} values (Hamilton et al., 1983; Gruau et al., 1987; Collerson et al., 1991; Galer and Goldstein, 1991; Bennett, Nutman, and McCulloch, 1993; McCulloch and Bennett, Chapter 2, this volume). If the Archaean crust was extracted from the upper mantle, the Archaean upper mantle must have had geochemical characteristics similar to those of the modern MORB-type mantle. This conclusion is supported by recent Lu-Hf isotopic studies of 3.8-Ga zircons from West Greenland, although the degree of depletion inferred from Hf-Lu systematics is somewhat less than that from Sm-Nd (Vervoort et al., 1996).

The evidence from continental flood basalts, oceanic plateaus, and OIBs is that the modern lower mantle is weakly depleted of incompatible elements, with an average ε_{Nd} value of +4 to +6. It is, however, unlikely to be homogeneous and probably includes other components, such as fragments of enriched mantle in the form of basaltic crust within subducted lithosphere sinking towards the core–mantle boundary.

The composition of the lower mantle is, of course, expected to have changed through time. It is widely assumed by geochemists that the primitive mantle had chondritic ratios of the nonvolatile incompatible trace elements. The 3.5-Ga Barberton komatiites, with chondritic REE patterns and an ε_{Nd} of +1, are interpreted to have originated from a source region that contained a high proportion of that primitive mantle. I suggest that the lower mantle has evolved to its present composition by mixing with material from the upper mantle that has slowly leaked into the lower mantle, some being dragged down by descending slabs, and some descending as the counterflow to a potentially large mass flux dragged upwards by rising plumes (Griffiths and Turner, Chapter 4, this volume).

An alternative interpretation, which assumes that plumes originate from the upper-mantle–lower-mantle boundary, is illustrated in Figure 6.16. The principal differences between this interpretation and the preceding one are (1) that the inferred composition of the plume source region is transferred from the core–mantle boundary to the bottom of the upper mantle, and (2) the geochemistry of the lower mantle cannot be constrained by observation, but can be expected to have remained isolated and therefore essentially chondritic throughout the Earth's history.

6.10. Discussion

There are two essential features of my preferred chemical structure for the mantle, as illustrated in Figures 6.15 and 6.16, that require explanation: (1) How did the boundary-layer source region for plumes become depleted during the Archaean? (2) Why was this material largely replaced by enriched, OIB-type mantle during the Proterozoic?

6.10.1. The OIB Convective Cycle

The explanation for the absence of OIB-type magmas during the Archaean is more straightforward and will be considered first. If the widely held view that OIB-type mantle includes an essential component of modified oceanic lithosphere is correct

(Chase, 1981; Hofmann and White, 1982; Ringwood, 1982), then the absence of OIB-type mantle in Archaean plumes can be attributed to the time taken for early oceanic lithosphere to cool, sink back into the mantle, become heated in a boundary layer, and rise back to the surface as a plume. If subduction started at about 3.8–4.0 Ga (the time of formation and stabilization of the first continental crust) (McCulloch and Bennett, Chapter 2, this volume) and the changeover from 'depleted' to 'enriched' mantle plumes occurred between 2.0 and 2.7 Ga, then the timescale for the completion of the first mantle convective cycle that produced OIB-type plumes was 1.1–2.0 Ga. That timescale is similar to the time required to form the observed isotopic anomalies in modern OIB-type basalts (Chase, 1981). This explanation is valid whether plumes originate from the core–mantle boundary or from the upper-mantle–lower-mantle boundary, although a long recirculation time seems less likely through the low-viscosity upper mantle.

6.10.2. The Depleted-Mantle Convective Cycle

The nature of the depleted-mantle convective cycle is more problematical. At first glance the upper-mantle–lower-mantle boundary appears to be the most likely source for depleted Archaean plumes. As discussed earlier in this chapter, the formation of the continental crust appears to have depleted the upper mantle of incompatible elements at some time prior to 4.0 Ga. If Archaean plumes originated from the upper-mantle–lower-mantle boundary, they should have the depleted geochemistry of MORB-type mantle, as observed. Once subduction started, the depleted mantle in the boundary layer would have been gradually replaced by mantle enriched in former basaltic crust, which made its first appearance in plumes during the early Proterozoic.

There are two problems with that explanation. Firstly, it ignores the accumulating geophysical and geochemical evidence that plumes originate at the core–mantle boundary. Secondly, it fails to explain why the basalts associated with Archaean komatiites have ε_{Nd} values of $+2$, compared with $+4$ to $+5$ for the komatiites. If the bottom of the upper mantle has an ε_{Nd} of $+4$, and if the upper mantle has been well mixed by convection, there is no obvious explanation for the basalts having an ε_{Nd} value of $+2$. Bearing in mind that basalts must form by melting at shallower depths than komatiites, the problem is to explain why the top of the upper mantle should have an ε_{Nd} value of $+2$, while the bottom is $+4$.

A mechanism by which depleted material could have accumulated at the base of the mantle has been suggested by Campbell and Griffiths (1992, 1993). They argued that the primitive Earth was hotter than the modern Earth, and that resulted in the formation of a thick outer basaltic shell. Because the basaltic crust was lighter than the underlying mantle, it stabilized the lithosphere and made it more resistant to subduction (Sleep and Windley, 1982). Campbell and Griffiths suggested that the commencement of subduction of the pre-Archaean lithosphere coincided with the beginning of formation of continental crust. Equivalently, the timescale for subduction of the pre-Archaean lithosphere would have been about 500 million years, the time difference between the formation of the Earth and the first appearance of continental crust. This figure is reasonable given that sections of the modern oceanic crust have survived for up to 200 million years. This long timescale for initiation of subduction implies that the main mechanism for heat loss from the pre-Archaean mantle was conduction through the stable lithosphere. This heat loss would have resulted in the development of a cool unstable boundary layer below the stable upper layer. If the timescale for which this boundary layer remained stable was less than the timescale for initiation of subduction, mantle convection would have occurred beneath a thin outer buoyant shell (Figure 6.20). Calculations by Campbell and Griffiths (1992) suggest that the timescale for the instability of the boundary layer was about 100–200 million years. Dripping of cold plumes from the thermal boundary layer beneath the stable Archaean lithosphere is therefore believed to have been the dominant form of downward convection in the Earth's early mantle.

Cold plumes are fundamentally different from hot plumes. The material in sinking plumes is colder than the surrounding mantle, and it is also more viscous, so that material in the plume tail will not sink faster than material in the plume head. Consequently, a cold sinking plume will not develop a large head (Olson and

Pre-Archaean Upper Boundary Layer

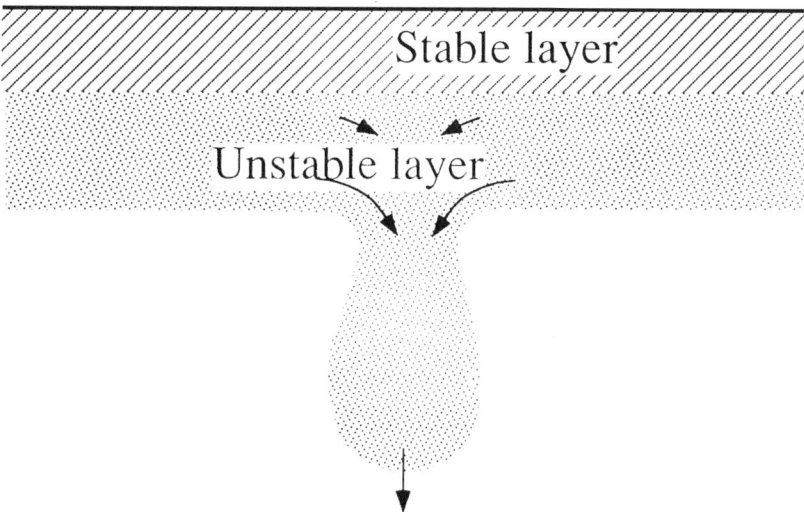

Figure 6.20. Sketch of a section through a pre-Archaean upper boundary layer, showing a stable plate, an unstable boundary layer below, and the formation of a drip. (Adapted from Campbell and Griffiths, 1992.)

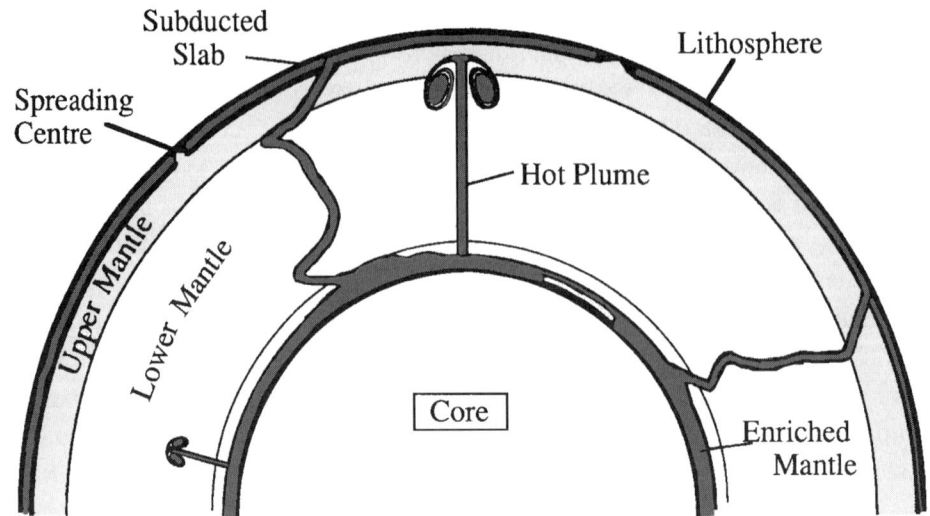

POST - ARCHAEAN   0 - 2.0 Ga

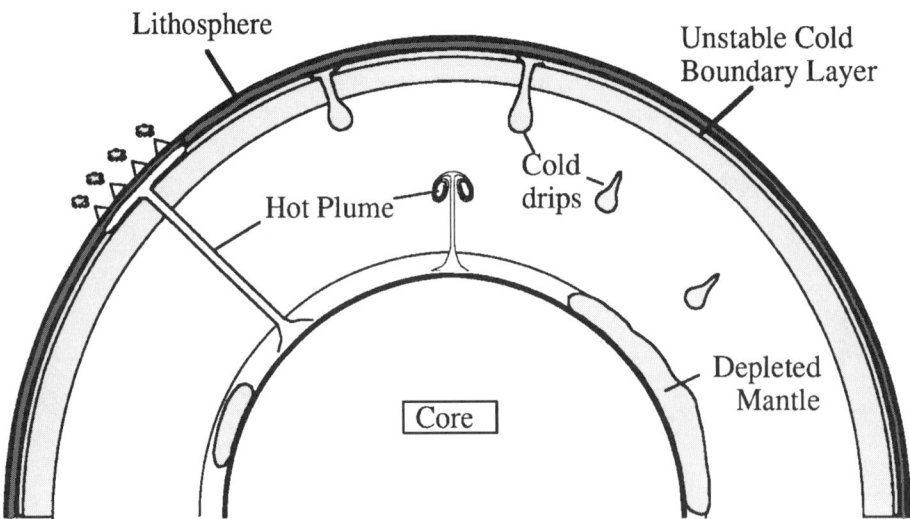

PRE-ARCHAEAN   4.0 - 4.5 Ga

Figure 6.21. Sketches of sections through the mantle showing the postulated convection style dominant in the pre-Archaean and post-Archaean mantle. Some of the cold plumes are shown as being continuous, but they may well have formed discontinuous drips in the Archaean mantle. (Adapted from Campbell and Griffiths, 1992.)

Singer, 1985) like that of a hot rising plume, and there will be no entrainment of the surrounding material into the descending plume (Figure 6.21). Hence, in the absence of a large compositional-density stratification within the mantle, the original material taken from the upper cold boundary layer could sink through the mantle whilst remaining compact and undiluted. These descending plumes therefore provided a mechanism for transferring material from the extreme top of the asthenosphere to the base of the mantle, with minimal contamination en route. If the upper mantle developed its depleted geochemical characteristics early in the history of the Earth, cold plumes might have delivered a flux of depleted material into the deep mantle, where it could have accumulated in the thermal boundary layer above the core. That layer of depleted mantle could later have given rise to Archaean plumes with depleted, MORB-like geochemistry.

6.11. Conclusion

The chemistry of picrites and komatiites of different ages can be used to constrain the geochemical characteristics of the ultimate source region for plumes, believed to lie at the core–mantle boundary, and to show how it has varied through time. During the Archaean, this region was dominated by depleted, MORB-like mantle. This type of mantle was gradually replaced by enriched, OIB-type mantle during the Proterozoic, so that the modern plume source consists mainly of OIB-type mantle. If the source of plumes is the thermal boundary layer that overlies the core, these conclusions apply to D''. If plumes originate from the upper-mantle–lower-mantle boundary, they apply to the transition zone at the base of the upper mantle. The enriched character of OIB-type mantle is generally attributed to the presence of former oceanic crust in the plume source region. If this hypothesis is correct, the absence of enriched geochemical signatures in Archaean plume-related magmas can be interpreted as a reflection of the time lag required for oceanic lithosphere to cool and become denser than the underlying mantle, sink back into the mantle, become heated in a thermal boundary layer, and return in plumes.

If plumes originate from the core–mantle boundary, the heads of new plumes will entrain lower-mantle material as they ascend, and their melting products, flood basalts and oceanic plateaus, can be used to constrain the chemistry of the lower mantle. This approach indicates that the lower mantle is depleted of incompatible elements, but less so than MORB-type mantle, with an average ε_{Nd} value of $+4$ to $+6$.

References

Aitken, B. G., and Echeverria, L. 1984. Petrology and geochemistry of komatiites and tholeiites from Gorgona Island, Colombia. *Contrib. Min. Pet.* 86:94–105.

Allègre, C. J., Dupre, B., and Lewin, E. 1986. Thorium/uranium ratio of the Earth. *Chem. Geol.* 56:219–27.

Arndt, N. T., and Christensen, U. 1992. The role of lithospheric mantle in continental flood volcanism: thermal and geochemical constraints. *J. Geophys. Res.* 97:10967–81.

Ballhaus, C., Berry, R. F., and Green, D. H. 1990. Oxygen fugacity controls in the Earth's upper mantle. *Nature* 348:437–40.

Beane, J. E., Turner, C. A., Hopper, P. R., Subbarao, K. V., and Walsh, J. N. 1986. Stratigraphy, composition and form of the Deccan Basalts, Western Ghats, India. *Bull. Volcanol.* 48:61–83.

Bender, J. F., Langmuir, C. H., and Hanson, C. N. 1984. Petrogenesis of basalt glasses from the Tamayo Region, East Pacific Rise. *J. Petrol.* 25:213–54.

Bennett, V. C., Nutman, A. P., and McCulloch, M. T. 1993. Nd isotopic evidence for transient, highly depleted mantle reservoirs in the early history of the Earth. *Earth Planet. Sci. Lett.* 119:299–317.

Boehler, R., Chopelas, A., and Zerr, A. 1995. Temperature and chemistry of the core–mantle boundary. *Chem. Geol.* 120:199–205.

Bristow, J. W. 1984. Picritic rocks of north Lebombo and south-east Zimbabwe. *Geol. Soc. S. Afr. Spec. Publ.* 13:105–23.

Campbell, I. H., Babbs, T. L., and Saunders, A. D. In press. Nb : U : Th ratios in Outong Java basalts: Implications for the source of mantle plumes. *Science.*

Campbell, I. H., Cordery, M. J., and Davies, G. 1995. The relationship between mantle plumes and continental flood basalts. In: *Proceedings of the International Field Conference and Symposium on Petrology and Metallogeny of Volcanic and Intrusive Rocks of the Midcontinent Rift System*, pp. 23–4.

Campbell, I. H., and Griffiths, R. W. 1990. Implications of mantle plume structure for the evolution of flood basalts. *Earth Planet. Sci. Lett.* 99:79–93.

Campbell, I. H., and Griffiths, R. W. 1992. The changing nature of mantle hotspots through time: implications for the geochemical evolution of the mantle. *J. Geol.* 92:497–523.

Campbell, I. H., and Griffiths, R. W. 1993. The evolution of the mantle's chemical structure. *Lithos* 30:389–99.

Campbell, I. H., Griffiths, R. W., and Hill, R. I. 1989. Melting in an Archaean mantle plume: heads it's basalts, tails it's komatiities. *Nature* 339:697–9.

Chase, C. G. 1981. Oceanic island Pb: two-stage histories and mantle evolution. *Earth Planet. Sci. Lett.* 52:277–84.

Chen, C.-Y., and Frey, F. A. 1985. Trace element and isotopic geochemistry of lavas from Haleakala volcano east Maui, Hawaii: implications for the origin of Hawaiian basalts. *J. Geophys. Res.* 90:8743–68.

Christensen, U. R., and Hofmann, A. W. 1994. Segregation of subducted oceanic crust in the convecting mantle. *J. Geophys. Res.* 99:19867–84.

Clague, D. A., and Dalrymple, G. B. 1989. Tectonics, geochronology and origin of the Hawaiian–Emperor volcanic chain. In: *The Geology of North America.* Vol. N: *The Eastern Pacific Ocean and Hawaii*, ed. E. L. Winterer, D. M. Hussong, and R. W. Decker, pp. 188–217. Boulder: Geological Society of America.

Clarke, D. B. 1970. Tertiary basalts of Baffin Bay: possible primary magma from the mantle. *Contrib. Min. Pet.* 25:203–24.

Coffin, M. A., and Eldholm, O. 1994. Large igneous provinces: crustal structure, dimensions and external consequences. *Rev. Geophys.* 32:1–36.

Collerson, K. D., Campbell, L. M., Weaver, B. L., and Palacz, Z. A. 1991. Evidence for extreme fractionation in early Archaean ultramafic rocks from northern Labrador. *Nature* 349:209–14.

Cordery, M. J., Davies, G., and Campbell, I. H. 1997. A numerical study on the genesis of continental flood basalts on oceanic plateaux from eclogite-bearing mantle plumes. *J. Geophys. Res.* 102:20179–197.

Cox, K. G. 1980. A model for flood basalt volcanism. *J. Petrol.* 21:629–50.

Cox, K. G., Duncan, A. R., Bristow, J. W., Taylor, S. R., and Erlank, A. J. 1984. Petrogenesis of the basic rocks of the Lebombo. *Geol. Soc. S. Afr. Spec. Publ.* 13:149–69.

Cox, K. G., and Hawkesworth, C. J. 1985. Geochemical stratigraphy of the Deccan Traps at Mahabaleshwar, Western Ghats, India, with implication for open system magmatic processes. *J. Petrol.* 26:355–77.

Davies, G. F. 1988. Ocean bathymetry and mantle convection. 1. Large-scale flow and hotspots. *J. Geophys. Res.* 93:10467–80.

Davies, G. F. 1990. Mantle plumes, mantle stirring and hotspot chemistry. *Earth Planet. Sci. Lett.* 99:94–109.

Davies, G. F., and Richards, M. A. 1992. Mantle convection. *J. Geol.* 100:151–206.

Echeverria, L. M. 1982. Komatiites from Gorgona Island, Colombia. In: *Komatiites*, ed. N. T. Arndt and E. G. Nisbet, pp. 199–209. London: Allen & Unwin.

Ellam, R. M., and Cox, K. G. 1989. A Proterozoic lithospheric source for Karoo magmatism: evidence from the Nuanetsi picrites. *Earth Planet. Sci. Lett.* 92:207–18.

Ellam, R. M., and Cox, K. G. 1991. An interpretation of Karoo picrite basalts in terms of interaction between asthenospheric magmas and the mantle lithosphere. *Earth Planet. Sci. Lett.* 105:330–42.

Farnetani, C. G., and Richards, M. A. 1994. Numerical investigations of the mantle plume initiation model for flood basalt events. *J. Geophys. Res.* 99:13813–34.

Fisk, M. R., Upton, B. G. J., and Ford, C. E. 1988. Geochemistry and experimental study of the genesis of magmas of Réunion Island, Indian Ocean. *J. Geophys. Res.* 93:4933–50.

Frey, F. M., and Rhodes, J. M. 1993. Intershield geochemical differences among Hawaiian volcanos: implications for source compositions, melting processes and magma ascent paths. *Phil. Trans. R. Soc. London* A342:121–36.

Galer, S. J. G., and Goldstein, S. C. 1991. Early mantle differentiation and its thermal consequences. *Geochim. Cosmochim. Acta* 55:227–39.

Gallagher, K., and Hawkesworth, C. 1992. Dehydration melting and the generation of continental flood basalts. *Nature* 358:57–9.

Ghose, N. C. 1976. Composition and origin of Deccan basalts. *Lithos* 9:65–73.

Gibson, S. A., Thompson, R. N., Dickin, A. P., and Leonardos, O. H. 1995. High-Ti and low-Ti mafic potassic magmas: key to plume–lithosphere interactions and continental flood-basalt genesis. *Earth Planet. Sci. Lett.* 136:149–65.

Gill, R. C. O., Nielsen, C. K., Brooks, C. K., and Ingram, G. A. 1988. Tertiary volcanism in the Kangerdlugssuaq region, E. Greenland: trace-element geochemistry of the Lower Basalts and tholeiitic dike swarms. In: *Early Tertiary Volcanism and the Opening of the NE Atlantic*, ed. A. C. Morton and L. M. Parson, pp. 161–79. Geological Society Special Publication 39. London: Geological Society.

Green, T. H., and Ringwood, A. E. 1968. Genesis of the calc-alkaline igneous rock suite. *Contrib. Min. Pet.* 18:105–62.

Griffiths, R. W., and Campbell, I. H. 1990. Stirring and structure in mantle starting plumes. *Earth Planet. Sci. Lett.* 99:66–78.

Griffiths, R. W., and Campbell, I. H. 1991a. On the dynamics of long-lived plume conduits in the convecting mantle. *Earth Planet. Sci. Lett.* 103:214–27.

Griffiths, R. W., and Campbell, I. H. 1991b. Interaction of mantle plume heads with the Earth's surface and onset of small-scale convection. *J. Geophys. Res.* 96:18295–310.

Griffiths, R. W., and Richards, M. A. 1989. The adjustment of mantle plumes to changes in plate motion. *Geophys. Res. Lett.* 16:437–40.

Gruau, G., Jahn, B. M., Glikson, A. Y., Davy, R., Hickman, A. H., and Chauvel, C. 1987. Age of the Archean Talga-Talga Subgroup, Pilbara Block, Western Australia, and early evolution of the mantle: new Sm-Nd isotopic evidence. *Earth Planet. Sci. Lett.* 85:105–16.

Guiraund, R., and Maurin, J. C. 1992. Early Cretaceous rifts of western and central Africa – an overview. *Tectonophysics* 23:153–68.

Hamilton, P. J., O'Nions, R. K., Bridgewater, D., and Nutman, A. 1983. Sm-Nd studies of Archean metasediments and metavolcanics from West Greenland and their implications. *Earth Planet. Sci. Lett.* 62:263–72.

Hanski, E. J., and Smolkin, V. F. 1989. Pechenga ferropicrites and other early Proterozoic picrites in the eastern part of the Baltic shield. *Precambrian Research* 45:63–82.

Hart, S. R., Hauri, E. H., Oschmann, L. A., and Whitehead, J. A. 1992. Mantle plumes and entrainment: the isotopic evidence. *Science* 256:517–20.

Hauri, E. H. 1996. Major-element variability in the Hawaiian mantle plume. *Nature* 382:415–19.

Hauri, E. H., and Hart, S. R. 1993. Re-Os isotope systematics of HIMU and EMII oceanic island basalts from the south Pacific Ocean. *Earth Planet. Sci. Lett.* 114:353–71.

Hill, R. I. 1991. Starting plumes and continental break-up. *Earth Planet. Sci. Lett.* 104:398–416.

Hill, R. I. 1993. Mantle plumes and continental tectonics. *Lithos* 30:193–206.

Hirose, H., and Kawamoto, T. 1995. Hydrous partial melting of lherzolite at 1 GPa: the effect of H_2O on the genesis of basaltic magmas. *Earth Planet. Sci. Lett.* 133:463–73.

Hirose, K., and Kushiro, I. 1993. Partial melting of dry peridotites at high pressures: determination of compositions of melts segregated from peridotite using aggregates of diamonds. *Earth Planet. Sci. Lett.* 114:477–89.

Hofmann, A. W., Jochum, K. P., Seufert, M., and White, W. M. 1986. Nb and Pb in oceanic basalts: new constraints on mantle evolution. *Earth Planet Sci. Lett.* 79:33–45.

Hofmann, A. W., and White, W. M. 1982. Mantle plumes from ancient oceanic crust. *Earth Planet. Sci. Lett.* 57:421–36.

Honda, M., McDougall, I., and Patterson, D. 1993. The systematics of helium-neon isotopes in mantle derived samples. *Lithos* 30:257–66.

Hooper, P. R. 1990. The timing of crustal extension and the eruption of continental flood basalts. *Nature* 345:246–9.

Jochum, K. P., Arndt, N. T., and Hofmann, A. W. 1991. Nb-Th-La in komatiites and basalts: constraints on komatiite petrogenesis and mantle evolution. *Earth Planet. Sci. Lett.* 107:272–89.

Kay, R., Hubbard, N. J., and Gast, P. W. 1970. Chemical characteristics and origin of oceanic ridge volcanic rocks. *J. Geophys. Res.* 75:1585–613.

Kerr, A. C., Marriner, G. F., Arndt, N. T., Tarney, J., Nivia, A., Saunders, A. D., and Duncan, R. A. 1996a. The petrogenesis of Gorgona komatiites, picrites and basalts: new field, petrographic and geochemical constraints. *Lithos* 37:245–60.

Kerr, A. C., Tarney, J., and Marriner, G. F. 1996b. The geochemistry and petrogenesis of the late-Cretaceous picrites and basalts of Curacao, Netherlands Antilles. *Contrib. Min. Pet.* 124:29–43.

Klein, E. M., and Langmuir, C. H. 1987. Global correlations of ocean ridge basalt chemistry with axial depth and crustal thickness. *J. Geophys. Res.* 92:8089–115.

Klewin, K. W., and Berg, J. H. 1991. Petrology of the Keweenawan Mamainse point lavas, Ontario: petrogenesis and continental rift evolution. *J. Geophys. Res.* 96:457–74.

Krishnamurphy, P., and Cox, K. G. 1977. Picritic basalts and related lavas from the Deccan Traps of western India. *Contrib. Min. Pet.* 62:53–75.

Lesher, C. M., and Arndt, N. T. 1995. REE and Nd isotope geochemistry, petrogenesis and volcanic evolution of contaminated komatiites at Kambalda, Western Australia. *Lithos* 34:127–57.

Lightfoot, P. C., Hawkesworth, C. J., Hergt, J., Naldrett, A. J., Gorbachev, N. S., Fedorenko, V. A., and Doherty, W. 1993. Remobilisation of the continental lithosphere by a mantle plume: major-, trace-element, and Sr-, Nd- and Pb-isotope evidence from picritic and tholeiitic lavas of the Noril'sk District, Siberian Traps, Russia. *Contrib. Min. Pet.* 114:171–88.

Lightfoot, P. C., Naldrett, A. J., Gorbachev, N. S., Doherty, W., and Fedorenko, V. A. 1990. Geochemistry of the Siberian Trap of the Noril'sk area, USSR, with implications for the relative contributions of crust and mantle to flood basalt magmatism. *Contrib. Min. Pet.* 104:631–44.

Ludden, J. N. 1978. Magmatic evolution of the basaltic shield volcanos of Reunion Island. *J. Volcanol. Geotherm. Res.* 4:171–98.

McCuaig, T. C., Kerrich, R., and Xie, Q. 1994. Phosphorous and high field strength element anomalies in Archean high-magnesian magmas as possible indicators of source mineralogy and depth. *Earth Planet. Sci. Lett.* 124:221–39.

McDonough, W. F. 1990. Constraints on the composition of the continental lithospheric mantle. *Earth Planet. Sci. Lett.* 101:1–18.

McDonough, W. F., and Sun, S.-S. 1995. The composition of the Earth. *Chem. Geol.* 120:223–53.

McDougall, I., Kristjansson, L., and Saemundsson, K. 1984. Magnetostratigraphy and geochronology of northwest Iceland. *J. Geophys. Res.* 89:7029–60.

McKenzie, D., and Bickle, M. H. 1988. The volume and composition of melt generated by extension of the lithosphere. *J. Petrol.* 29:625–79.

Mahoney, J. J. 1988. Deccan Traps. In: *Continental Flood Basalts*, ed. J. D. Macdougall, pp. 151–94. Amsterdam: Kluwer.

Mahoney, J. J., Macdougall, J. D., Lugmair, G. W., Gopalan, K., and Krishnamurphy, P. 1985. Origin of contemporaneous tholeiitic and K-rich alkalic lavas: a case study from the northern Deccan Plateau, India. *Earth Planet. Sci. Lett.* 72:39–53.

Mahoney, J. J., Storey, M., Duncan, R. A., Spencer, K. J., and Pringle, M. 1993. Geochemistry and geochronology of leg 130 basement lavas: nature and origin of the Ontong Java Plateau. In: *Proceedings of the Ocean Drilling Program, Scientific Results*, vol. 130, pp. 3–22. College Station, TX: Ocean Drilling Program.

Morgan, W. J. 1981. Hotspot tracks and the opening of the Atlantic and Indian Ocean. In: *The Sea*, vol. 7, ed. C. Emiliani, pp. 443–87. New York, Wiley.

Muenow, D. W., Garcia, M. O., Aggrey, K. E., Bednare, U., and Schmincke, H. U. 1990. Volatiles in submarine glasses as a discriminant of tectonic origin; application of the Troodos ophiolite. *Nature* 343:159–61.

O'Hara, M. J. 1965. Primary magmas and the origin of basalts. *Scot. J. Geol.* 1:19–40.

Olson, P., and Singer, H. A. 1985. Creeping plumes. *J. Fluid Mech.* 158:511–31.

O'Nions, R. K., and Clarke, D. B. 1972. Comparative trace element geochemistry of Tertiary basalts from Baffin Bay. *Earth Planet. Sci. Lett.* 15:436–46.

O'Nions, R. K., and Tolstikhin, I. N. 1996. Limits on the mass flux between lower and upper mantle and stability of layering. *Earth Planet. Sci. Lett.* 139:213–22.

Patriat, P., and Achache, J. 1984. India–Eurasia collision chronology has implications for crustal shortening and driving mechanism of plates. *Nature* 311:615–21.

Richards, M. A., Duncan, R. A., and Courtillot, V. E. 1989. Flood basalts and hotspot tracks; plume head and tails. *Science* 246:103–7.

Richards, M. A., and Hager, B. H. 1988. The Earth's geoid and the large-scale structure of mantle convection. In: *Physics of the Planet*, ed. S. K. Runcorn, pp. 247–72. New York: Wiley.

Richards, M. A., Jones, D. L., Duncan, R. A., and De Paolo, D. J. 1991. A mantle plume initiation model for the Wrangellia flood basalt and other oceanic plateaus. *Science* 254:263–7.

Ringwood, A. E. 1979. *Origin of the Earth and Moon*. Berlin: Springer-Verlag.

Ringwood, A. E. 1982. Phase transformations and differentiation in subducted lithosphere: implications for mantle dynamics, basalt petrogenesis, and crustal evolution. *J. Geol.* 90:611–43.

Roeder, P. L., and Emsley, R. F. 1970. Olivine–liquid equilibrium. *Contrib. Min. Pet.* 29:225–82.

Rudnick, R. L., and Fountain, D. M. 1995. Nature and composition of the continental crust: a lower crustal perspective. *Rev. Geophys.* 33:267–310.

Scarrow, J. H., and Cox, K. G. 1995. Basalts generated by decompressive adiabatic melting of a mantle plume: a case study from Isle of Skye, N. W. Scotland. *J. Petrol.* 36:3–22.

Sleep, N. H. 1990. Hotspots and mantle plumes: some phenomenology. *J. Geophys. Res.* 95:6715–36.

Sleep, N. H., and Windley, B. F. 1982. Archaean plate tectonics: constraints and inferences. *J. Geol.* 90:363–79.

Storey, M., Mahoney, J. J., Kroenke, L. W., and Saunders, A. B. 1991. Are oceanic plateaus sites of komatiite formation? *Geology* 19:376–9.

Sun, S.-S., and Nesbitt, R. W. 1978. Petrogenesis of Archaean ultrabasic and basic volcanics: evidence from rare earth elements. *Contrib. Min. Pet.* 65:301–25.

Turner, S., Regelous, M., Kelley, S., Hawkesworth, C., and Mantovani, M. 1994. Magmatism and continental break-up in the South Atlantic: high precision ^{40}Ar-^{39}Ar geochronology. *Earth Planet. Sci. Lett.* 121:333–48.

Vervoort, J. D., Patchett, P. J., Gehrels, G. E., and Nutman, A. P. 1996. Constraints on early Earth differentiation from hafnium and neodymium isotopes. *Nature* 379: 624–7.

White, R. S., McBride, J. H., Maguire, R. K. H., Brandsdottir, B., Meke, W., Minshull, T. A., Richardson, K. R., Smallwood, J. R., Staples, R. K., and the FIRE Working Group. 1996. Seismic images of crust beneath Iceland contribute to long-standing debate. *EOS, Trans. AGU* 77:197–9.

White, R., and McKenzie, D. 1989. Magmatism at rift zones: the generation of volcanic continental margins and flood basalts. *J. Geophys. Res.* 94:7685–729.

Whitehead, J. A., and Luther, D. S. 1975. Dynamics of laboratory diapir and plume models. *J. Geophys. Res.* 80:705–17.

Yasuda, A., Fujii, T., and Kurita, K. 1994. Melting phase relations of an anhydrous mid-ocean ridge basalt from 3 to 20 GPa: implications for the behaviour of subducted oceanic crust in the mantle. *J. Geophys. Res.* 99:9401–14.

7

Pyrolite: A Ringwood Concept and Its Current Expression

DAVID H. GREEN and TREVOR J. FALLOON

7.1. Introduction

'Pyrolite' was chosen by Ringwood (1962a,b) as the name for a model chemical composition for the mantle, consisting predominantly of pyroxene and olivine. Subsequently, the pyrolite concept has been quantified, and assessment of model pyrolite compositions has been a major theme in experimental studies at The Australian National University and the University of Tasmania. Such experimental studies have been designed to achieve an understanding of the mineralogy of the upper mantle, the transition zone, and the lower mantle as a function of pressure and temperature and to investigate the melting behaviour and melt products of the upper mantle to depths of approximately 150 km. The purpose of this chapter is to review the experimental studies of pyrolite compositions relevant to the Earth's upper mantle and to summarize the applications of such studies.

We shall begin with the definition of pyrolite in its historical context before presenting a review of the experimental studies (both published and ongoing) that have been performed on a range of different pyrolite compositions. These experimental studies have emphasized the very important influence of volatile $(C + H + O)$ composition on the solidus position (the pressure–temperature conditions at which melt first appears). The experimental findings have led directly to models for the upper mantle that predict distinctive sub-solidus mineralogies and melting characteristics in particular tectonic environments. In the final part of this chapter we shall consider the petrological constraints on magma genesis based on the pyrolite concept and the findings from the experimental studies.

The petrological framework presented differs from that used in most models of magma genesis by the introduction of two very important concepts: the *petrological lithosphere* and the *incipient-melting regime*. The concept of the petrological lithosphere derives from experimental observations of the stability field for amphibole (magnesian pargasite) and the marked depression of the pyrolite solidus at about 95 km. The term 'incipient-melting regime' denotes the pressure–temperature field

311

between the pyrolite-$(C + H + O)$ solidus and the $(C + H)$-free solidus, in which the amount of melt present is determined primarily by the $(C + H + O)$ content. Our interpretation of magma genesis at mid-ocean ridges emphasizes dynamic, three-dimensional upwelling, with significant melt retention during adiabatic ascent in the mantle before segregation and eruption of primary mid-ocean-ridge basaltic (MORB) magmas. We shall present evidence that primary MORB magmas must have MgO contents of 13–16 wt%, requiring mantle potential temperatures (T_p) of 1,380–1,430°C, significantly higher than that currently assumed in some models of MORB petrogenesis. The observed variations in incompatible elements and isotopes for MORBs and for intraplate basalts are explained with reference to processes leading to mantle heterogeneity and *wall-rock reaction* or *thermal entrainment* during dynamic upwelling.

Both compositional and thermal buoyancies are invoked in a model of 'hotspot' or intraplate plume magmatism in which there is an important role for $(C + H + O)$ fluids and associated redox reactions as a mechanism for initiating melting. The model does not require that large temperature variations be associated with mantle plumes.

Finally, the pyrolite concept is used to present a petrological framework for convergent-margin magmatism. This framework emphasizes the complexities of melting in the subducted slab and overlying mantle wedge and predicts a role for dolomitic carbonatite melt within the mantle wedge. Both H_2O-rich fluid and carbonatitic melt can be vehicles for metasomatism and melt fluxing in the mantle wedge overlying the subducted slab.

7.2. The Pyrolite Concept
7.2.1. The Historical Development of the Pyrolite Concept

In the late 1950s there was a lively debate on the nature of the upper mantle and of the crust–mantle boundary or Mohorovicic discontinuity ('Moho'). Ringwood, working largely with Earth models based on meteorite compositions (a chondritic Earth model), considered that the Earth's mantle is olivine-rich and of peridotitic composition. A complementary argument involved the predicted phase transformations in olivine (to spinel structure) and orthopyroxene (to Mg_2SiO_4 spinel $+$ stishovite or to other polymorphs) at the high pressures of the transition zone in the mantle. The argument for a peridotitic upper mantle was countered by others who argued for a basaltic oceanic crust and lower continental crust and an explanation for the Moho in terms of the basalt(gabbro)-to-eclogite reaction.

Studies of large ultramafic bodies in young orogenic belts (ophiolite complexes) revealed large volumes of harzburgite, but also significant dunite, wehrlite, pyroxenite, troctolite, and gabbro. Peridotitic rocks are commonly layered, often strongly deformed, and extremely refractory in chemical composition (i.e., having lost a low-melting-point or fusible fraction). Although there are obvious cogenetic relationships among the various rock types of orogenic layered ultramafic complexes, arbitrary assumptions concerning the relative abundances of the diverse rock types are required in order to construct an average mantle composition. An important paper by Ross, Foster, and Myers (1954) documented the mineralogy of peridotite xenoliths, which are inclusions of coarse-grained olivine (Fo_{89-91}), orthopyroxene, clinopyroxene, and spinel occurring in silica-undersaturated basalts, basanites, olivine nephelinites, and so forth, found worldwide in intraplate settings. In contrast to the peridotites in layered ultramafic complexes in ophiolite suites, both the orthopyroxene and clinopyroxene in the peridotite xenoliths have high Al_2O_3 contents (usually >3 wt%). Similarly, the spinel present is aluminous, commonly with more than 40 wt% Al_2O_3. In bulk chemical composition, these peridotite xenoliths range from clinopyroxene-bearing harzburgite (with 1–2 wt% Al_2O_3 and <1 wt% CaO) to lherzolite (with 15–20% clinopyroxene and 3–4 wt% Al_2O_3 and CaO) (see Figure 7.10).

The combined field, petrographic, and mineralogical study of the Lizard Peridotite in Cornwall (Green, 1964a–c) provided an important linkage between orogenic peridotites and peridotite xenoliths. The composition and mineralogy of the inner core of Lizard Peridotite is coarse-grained spinel lherzolite, characterized by olivine (Fo_{89}), aluminous orthopyroxene, aluminous clinopyroxene, and aluminous spinel (i.e., the same mineralogy as the peridotite xenoliths in basalt). The coarse-grained lherzolite core passes transitionally to fine-grained, foliated, plagioclase lherzolite at the deformed and recrystallized margins of the body, where pyroxenes show decreased Al_2O_3 contents, and relict chrome-rich spinel is rimmed by plagioclase. The bulk compositions (2–4 wt% CaO and Al_2O_3) of the marginal rocks match those of the lherzolite core. The reactions of spinel + aluminous pyroxene lherzolite to plagioclase lherzolite were interpreted as decompression reactions from an initial crystallization at high pressure and high temperature in the mantle to a low-pressure, high-temperature crustal emplacement, possibly in a Red Sea–style rift or back-arc-basin environment. The Lizard Peridotite thus represents a relatively large mantle sample, emplaced as a high-temperature mantle upwelling or diapir. Significantly, the 'mantle sample' in this case is lherzolitic, with olivine of Fo_{89} composition and containing 3-4% CaO and Al_2O_3 in the coarse-grained spinel lherzolite.

The multiple themes we have briefly summarized led directly to the concept of 'pyrolite' (Ringwood, 1962a,b; Green and Ringwood, 1963). Evidence from natural 'mantle samples' indicated that there is a spectrum of compositions from highly refractory harzburgite and dunite (with olivine commonly >Fo_{91}) to lherzolite (with olivine of Fo_{89-90}, and containing 3–4 wt% CaO and Al_2O_3). The natural samples of lherzolite demonstrate three anhydrous mineral assemblages (i.e., plagioclase lherzolite, aluminous spinel + aluminous pyroxene lherzolite, and garnet lherzolite) stable at high temperatures and representing increasing pressures of crystallization from crust to mantle. A fourth, high-temperature assemblage, pargasite lherzolite, also occurs and contains up to about 30% amphibole (pargasite), and about 0.4 wt% H_2O.

The spectrum of chemical compositions of natural mantle samples was conceived as being related by one or more partial-melting events yielding magmas of generally basaltic character. Mantle-derived magmas include olivine-rich and silica-undersaturated magmas (i.e., kimberlites, olivine melilitites, olivine nephelinites, olivine basanites, and alkali olivine basalts) that host dense peridotite xenoliths (Frey, Green, and Roy, 1978). The minor-element contents (particularly P_2O_5, K_2O, Na_2O, and TiO_2) of these magmas suggest increasing degrees of melting through the foregoing sequence, with alkali olivine basalts representing the highest degree of melting. Although tholeiitic magmas do not contain mantle xenoliths, their relative abundance in ocean-floor sampling, in the eruptive sequence of the Hawaiian Islands, and in continental flood-basalt provinces all argue that tholeiitic basalts also come from primitive, mantle-derived magmas with lower contents of minor elements than are found in alkali olivine basalts and formed by greater degrees of melting of the mantle source (Green and Ringwood, 1967a; Frey, et al., 1978).

Thus, the rationale for the pyrolite model was to combine a mantle-derived magma (basalt) with refractory residue (harzburgite or dunite) in proportions such that the resultant model mantle or 'pyrolite' composition would contain 3–4 wt% CaO, Al_2O_3, and olivine of approximately Fo_{89} composition. The composition of the least refractory or most fertile of the high-pressure mantle sample suite was thus employed as a major constraint on the mantle composition. The basalt or melt component determined the concentrations of Na_2O, K_2O, P_2O_5, TiO_2, and all 'incompatible' trace elements in the model source. Conversely, the FeO, MgO, SiO_2, Cr_2O_3, and NiO contents of the model source reflected both the model residue and the magma chosen, but were most sensitive to the olivine and enstatite compositions and their relative proportions and to the abundance of chromite in the residue. It is thus no surprise that the pyrolite model and the mantle compositions deduced from the least refractory of mantle xenolith suites are convergent in major and minor elements (O'Neill and Palme, Chapter 1, this volume).

The name 'pyrolite' was coined deliberately to emphasize that this model composition was not a rock type but was a model-dependent composition, dominated by olivine > pyroxenes in its mineralogy. Since the original definition, the term 'pyrolite' has been used for theoretical model compositions, particularly those used in experimental studies (e.g., 'Hawaiian pyrolite', 'MORB pyrolite'), whereas the terms 'lherzolite' and 'peridotite' (e.g., 'Tinaquillo lherzolite', 'peridotite KLB-1') (Takahashi, 1986) have been used for any natural peridotite compositions employed in experimental studies. There have been many studies of peridotites from diverse natural situations. The original conclusions remain valid:

1. Orogenic peridotites (ophiolitic peridotites) are extremely heterogeneous, with both igneous and tectonic layering. Their mineral assemblages are characteristically low-pressure, within the olivine + plagioclase stability field. Refractory harzburgite is common, with about 80% modal olivine (Fo_{91-92}), enstatite, and minor chromite. Geochemical studies, including isotopic analyses, have

emphasized that the harzburgitic rocks are refractory residues from basaltic extraction and are not themselves appropriate as source compositions for basaltic magmas.

2. Moderate-pressure (1–2.5 GPa), high-temperature lherzolites, characterized by mineralogy of olivine + aluminous pyroxenes + aluminous spinel, occur within some orogenic peridotites and are characteristic of mantle xenolith suites in primitive, nepheline-normative, and olivine-rich magmas in intraplate settings ('rift', 'hotspot', 'ocean island', 'plateau basalt'). Petrological, mineralogical, and geochemical studies have revealed a spectrum of compositions, from harzburgite to lherzolite, with the most fertile samples (i.e., those compositions richest in the low-melting-point or fusible fraction) containing 3–4 wt% CaO and Al_2O_3 and around 55–60% modal olivine (Fo_{89}). Trace-element and isotopic studies of particular peridotite suites have demonstrated complex histories, including residue from melt extraction (Nickel and Green, 1984), overprinting by metasomatism, and variable interaction with host magma during transport (Frey and Green, 1974). Metasomatism by addition of or reaction with fluid (H_2O-rich), by carbonatite melt, or by silicate melt addition has been demonstrated (O'Reilly and Griffin, 1988; Yaxley, Crawford, and Green, 1991; Rudnick, McDonough, and Chappell, 1993). Rarely, xenolith suites include spinel + garnet lherzolites, the five-phase assemblage stable for these compositions at 2.0 ± 0.2 GPa at a temperature of 1,000°C (Green and Ringwood, 1967b).

3. Peridotite xenoliths in kimberlite, and some orogenic peridotites associated with eclogites and other high-pressure metamorphic rocks, include garnet lherzolites and garnet harzburgites. Examples of associated meta-rodingites and enstatite-rich harzburgites provide convincing evidence of subduction of ophiolites, including serpentinite, and dehydration under high-pressure, moderate-temperature conditions (Evans and Trommsdorff, 1978). Garnet lherzolites in orogenic peridotites are heterogeneous and have complex pre-histories, which in some cases, at least, include subduction of near-surface precursors.

   Garnet-lherzolite xenoliths in kimberlites are also complex, residual to fertile in geochemical characteristics, and individual suites show evidence for metasomatism by potassic fluids or melts or by carbonatite melts (Rudnick et al., 1993). The significance of cataclastic (sheared garnet lherzolite) textures remains debatable. An important observation from South African suites is that the most refractory end members are harzburgites with olivine ($\sim Fo_{93}$) and abnormal enstatite : olivine modal proportions (>35% enstatite) (Boyd, 1989). The garnet-lherzolite-to-harzburgite xenolith suite from kimberlites is attributed to cool, deep lithosphere, formed as refractory residual compositions from Archaean/Proterozoic magmatism (and possibly subduction, etc.), with superimposed metasomatic and metamorphic overprints.

The natural mantle samples provide evidence of peridotite samples derived from a considerable depth/pressure range and consistently point to the 'least refractory'

or 'most fertile' compositions as being lherzolite, with about 60% olivine (Fo$_{89}$) and
3–4 wt% CaO and Al$_2$O$_3$. Such samples, however, are inferred to have been derived
from the Earth's lithosphere. The processes of magma formation at mid-ocean
ridges or in intraplate basaltic provinces imply upwelling of asthenospheric mantle.
Magma source regions are thus generally inferred to lie beneath the lithosphere
(and certainly at depths greater than the depths of sampling of xenolith suites). The
'pyrolite' concept addresses this dilemma by combining the magma component
with an estimate of the residue composition. Consistency requires that the inferred
highest-degree melt be combined with the most refractory residue in proportions
to yield 3–4% CaO and Al$_2$O$_3$ in the model pyrolite. This is the rationale for
calculation of 'Hawaiian pyrolite' and 'MORB pyrolite' (Table 7.1).

It is important to note that the Hawaiian-pyrolite and MORB-pyrolite compositions are attempts at estimating the 'asthenosphere' composition beneath Hawaii
and beneath the Atlantic mid-ocean ridge. As will become apparent later, the asthenospheric source in both cases may normally exist with a small incipient-melt
fraction present under upper-mantle conditions. If there were some means for direct sampling of the inferred Hawaiian-pyrolite composition at 100–150 km beneath
Hawaii, it would be present as garnet lherzolite, with a very small (<2%?) melt fraction of olivine-melilitite or olivine-nephelinite composition. The same composition,
upwelling to about 60 km, is represented by olivine + aluminous orthopyroxene,
together with about 30% tholeiitic picrite closely similar to the 1959–60 Kilauea
Iki picrite (Table 7.1A, Figure 7.1).

In a similar way, the MORB pyrolite at 100–150 km beneath the mid-ocean
ridge probably would contain less than a 1% melt of olivine nephelinite or olivine-
rich basanite composition in a garnet-lherzolite residue. The same composition
upwelling to a depth of about 60 km would contain 15–20% melt, with olivine, orthopyroxene, and trace clinopyroxene as residual phases. Thus the emphasis in the
'pyrolite model' is on a source containing crystals and liquid, both varying in character and proportions as functions of pressure and temperature. Pyrolite can 'freeze'
(as discussed later), but its common fate is to separate into residue (lithosphere) and
liquid (basalt, carbonatite) that will migrate or be inhomogeneously redistributed
during the dynamic complexities of plate movement, rifting, and subduction.

7.2.2. The Calculation of Distinctive Pyrolite Compositions

The initial pyrolite composition (pyrolite I of Green and Ringwood, 1963) combined a relatively primitive (Mg# 73)[1] basaltic composition with refractory dunite
in proportions 1 : 3 to yield a composition with 2.7 wt% CaO and 4.0 wt% Al$_2$O$_3$,
but with a low pyroxene : olivine ratio (and a low pyroxene : Al$_2$O$_3$ ratio). In experimental studies (Green and Ringwood, 1967b) this composition was found to retain
(Al + Cr) spinel up to the anhydrous solidus, with garnet present at the solidus at

---

[1] Mg# is the atomic ratio 100Mg/(Mg + Fe^{2+}).

Table 7.1A. *Calculation of the composition of Hawaiian pyrolite from the Kilauea Iki parent composition and liquidus-phase compositions[a]*

Compound	54.5% olivine (Mg# 90.5)	+	15.0% orthopyroxene (Mg# 91)	+	0.5% spinel (Mg# 64)	+	30% melt (Mg# 74.1)	=	100% source (Mg# 87.9)	(70%) residue (Mg# 90.5)
SiO$_2$	40.9		56.4		—		47.8		45.1	44.0
TiO$_2$	—		0.25		1.2		2.1		0.7	0.05
Al$_2$O$_3$	—		2.3		19.2		10.5		3.6	0.6
Fe$_2$O$_3$	—		—		—		1.8		0.6	—
FeO	9.2		5.8		17.8		10.0		9.0	8.5
MnO	0.1		0.1		0.1		0.1		0.1	0.1
MgO	49.1		32.7		17.8		16.0		36.6	45.4
CaO	0.24		2.3		—		9.3		3.3	0.7
Na$_2$O	—		—		—		1.8		0.5	—
K$_2$O	—		—		—		0.43		0.13	—
P$_2$O$_5$	—		—		—		0.20		0.06	—
NiO	0.50		—		—		0.10		0.30	0.4
Cr$_2$O$_3$	—		0.3		43.5		0.13		0.30	0.4

[a]Data from Eggins (1992a).
Note: This revised composition is essentially identical with the Hawaiian pyrolite of Table 7.2.

Table 7.1B. *Calculation of the composition of MORB pyrolite*[a]

Compound	57.5% olivine (Mg# 91.6)	+ 18.0% orthopyroxene (Mg# 92.1)	+ 0.5% spinel (Mg# 64)	+ 24% melt (Mg# 79.0)	= 100% source (Mg# 90.1)	(76%) residue (Mg# 91.6)
SiO_2	40.9	54.8	—	48.3	44.7	43.9
TiO_2	—	<0.1	1.2	0.6	0.17	0.04
Al_2O_3	—	5.5	19.2	13.2	4.4	1.4
FeO	8.2	4.9	17.8	7.9	7.6	7.5
MnO	—	—	0.1	0.1	0.1	0.1
MgO	50.4	32.0	17.8	16.7	38.6	45.9
CaO	0.4	2.8	—	10.9	3.4	0.95
Na_2O	—	—	—	1.65	0.4	—
K_2O	—	—	—	0.01	0.0	—
NiO	0.40	0.05	—	0.08	0.3	0.3
Cr_2O_3	—	1.2	43.5	0.06	0.4	0.6

[a]Data from Green et al. (1979).

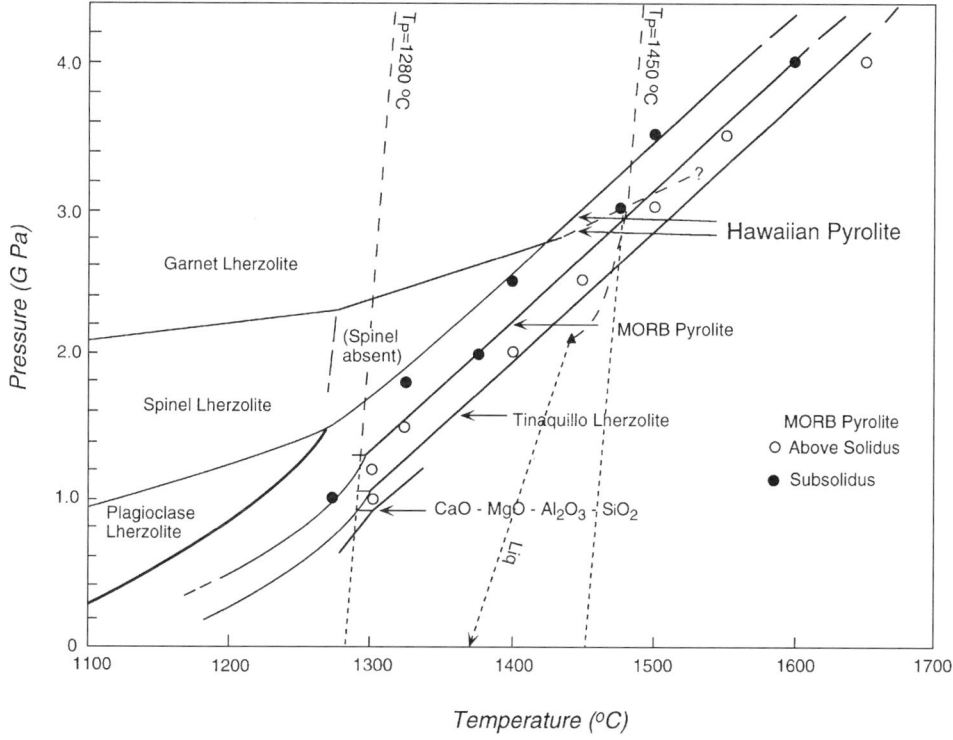

Figure 7.1. Experimentally determined solidi for Hawaiian-pyrolite, MORB-pyrolite, and Tinaquillo-lherzolite compositions and stability fields for plagioclase, spinel, and garnet-lherzolite mineralogy for the Hawaiian-pyrolite composition. Experimental points for MORB pyrolite are shown. Solidi for other compositions are from sources cited in the text. Adiabats for mantle potential temperatures of 1,280°C and 1,450°C are shown, with a sub-solidus adiabatic gradient of 10°C/GPa (Birch, 1952) and a liquid adiabat of 30°C/GPa (McKenzie and Bickle, 1988). The solid triangle is the inferred depth for magma segregation of MORB picrite (2 GPa, 1,430°C) from Green et al. (1979).

$P = 2.5$ GPa. A second composition (pyrolite II) was calculated by using harzburgite rather than dunite as the residue component (Green and Ringwood, 1967b). A more appropriate estimate for pyrolite was made after experimental studies on a Hawaiian tholeiitic picrite had established that its liquidus phases were olivine (Fo_{90}) + aluminous pyroxenes at $P = 1.35$ GPa and $T = 1,350$°C (Green and Ringwood, 1967a). Thus this composition matched a key criterion[2] for a primary melt from mantle lherzolite/harzburgite. Ringwood (1966) selected a combination of refractory harzburgite and Kilauea Iki olivine tholeiite (but not the parental 1959–60 Kilauea Iki picrite) to calculate the pyrolite III composition, later commonly referred to as 'Hawaiian pyrolite'. It is possible to refine the Hawaiian-pyrolite model composition further by using the experimental studies of Eggins (1992a).

---

[2] The criterion is that a primary melt from mantle lherzoliter will be saturated in olivine ($\geq Fo_{89}$) + orthopyroxene (and possibly with other lherzolite phases) at some pressure and temperature (i.e., conditions at which the primary magma segregated from the residual mantle).

Eggins used improved estimates for the parental composition (tholeiitic picrite) of the 1959–60 Kilauea Iki eruption and established olivine + orthopyroxene saturation (olivine Mg# 90.5) at 2 GPa and 1,460°C for the anhydrous composition. In Table 7.1A, olivine, orthopyroxene, spinel, and liquid are combined in the ratios 54.5 : 15 : 0.5 : 30 to yield a model source composition. The new composition is barely distinguishable from the earlier calculation and retains the distinctive features of high TiO_2, Na_2O, and K_2O contents for chosen CaO and Al_2O_3 contents. The high TiO_2 content is important in determining pargasite stability and a subsolidus role for ilmenite at pressures beyond those for pargasite stability. It is worth emphasizing that although the choice of the proportions of olivine, orthopyroxene, and melt to produce a lherzolite with about 3–4 wt% CaO + Al_2O_3 is arbitrary, the experimental observation from Kilauea Iki picrite that only olivine and orthopyroxene are liquidus phases at 2 GPa (Eggins, 1992a) is the reason that the relative CaO, Al_2O_3, TiO_2, K_2O, and Na_2O contents are distinctive and differ from those for MORB pyrolite and Tinaquillo lherzolite.

The experimental studies of Hawaiian pyrolite have been extensive and are considered particularly relevant for intraplate settings. In addition to the major phases, this composition crystallizes ilmenite, phlogopite, and, in some experiments, apatite – minor phases in the mantle, but significant for some geochemical and petrologic studies. Characteristically, intraplate magmas, including hotspot basalts, have high TiO_2 contents and higher K_2O/Na_2O, Na_2O/CaO, and TiO_2/Al_2O_3 ratios than 'normal' MORBs (N-MORBs) or back-arc-basin basalts.

The MORB-pyrolite composition (Table 7.1B) was calculated (Green, Hibberson, and Jaques, 1979) after experimental studies of a primitive olivine tholeiite (DSDP 3-18) established *P–T* conditions for olivine + orthopyroxene saturation in a picritic composition (DSDP 3-18 + 17% olivine) at 1.8–2.0 GPa and $T = 1,430°C$. The calculation of the MORB-pyrolite model composition again used the 3–4 wt% CaO and Al_2O_3 and the Fo_{89-90} constraints on the model source composition. It should be noted that in addition to lower TiO_2, Na_2O, and K_2O contents than in Hawaiian pyrolite, the MORB pyrolite has a lower CaO/Al_2O_3 ratio (Table 7.2).

Extensive experimental studies have also been carried out on a composition called Tinaquillo lherzolite. This composition (Table 7.2) is very similar in terms of major elements (SiO_2, MgO, FeO, Al_2O_3, CaO) to Hawaiian pyrolite, MORB pyrolite, and the whole group of 'fertile' natural lherzolites (particularly the Lizard Peridotite). However, it is more depleted of TiO_2 and Na_2O and is strongly depleted of incompatible trace elements (K_2O, light rare-earth elements, etc.). The composition thus completes a spectrum from incompatible-element-enriched mantle (or hotspot mantle), through normal upper mantle (MORB mantle), to incompatible-element-depleted mantle (Tinaquillo lherzolite). One further model composition, based on an estimate of a source composition for K-enriched Tertiary basanites from the North Hessian Depression (designated the NHD Peridotite) was studied specifically to

Table 7.2. *Model compositions for the upper-mantle sources of basalts and olivine-depleted compositions, both used for experimental studies*

Compound	HPY[a]	HPY-40[b]	MPY[c]	MPY-40[d]	TQ[e]	TQ-40[f]
SiO_2	45.20	47.90	44.74	47.15	44.95	47.50
TiO_2	0.71	1.18	0.17	0.28	0.08	0.13
Al_2O_3	3.54	5.91	4.37	7.28	3.22	5.35
FeO	8.47	8.81	7.55	7.27	7.66	7.51
MnO	0.14	0.13	0.11	0.12	0.14	0.18
MgO	37.50	28.80	39.57	30.57	40.03	32.80
CaO	3.08	5.14	3.38	5.63	2.99	4.97
Na_2O	0.57	0.95	0.40	0.66	0.18	0.30
K_2O	0.13	0.22	0.00	0.00	0.02	0.03
P_2O_5	0.04	0.06	0.00	0.00	0.01	0.02
NiO	0.20	0.13	0.26	0.29	0.26	0.43
Cr_2O_3	0.43	0.72	0.45	0.75	0.45	0.75
Mg#	89.0	85.5	90.0	88.0	90.0	89.0

[a] HPY, Hawaiian pyrolite.
[b] HPY-40, Hawaiian pyrolite minus 40% olivine ($Mg_{91.6}Fe_{8.1}Ni_{0.2}Mn_{0.1}$).
[c] MPY, MORB pyrolite.
[d] MPY-40, MORB pyrolite minus 40% olivine ($Mg_{91.6}Fe_{8.1}Ni_{0.2}Mn_{0.1}$).
[e] TQ, Tinaquillo lherzolite.
[f] TQ-40, Tinaquillo lherzolite minus 40% olivine ($Mg_{91.9}Fe_{8.0}Mn_{0.1}$).

explore the stability of coexisting phlogopite and pargasite. This study determined the maximum K solubility in pargasite within its stability field and the melting relationships of pargasite and phlogopite-bearing peridotite (Mengel and Green, 1989).

7.3. Experimental Studies of Pyrolite Compositions
7.3.1. Experimental Methodology

A distinctive feature of the experimental studies performed on pyrolite compositions has been the use of sintered-oxide starting materials, rather than crushed natural minerals (e.g., Takahashi, 1986; Baker et al., 1995; Hirose and Kushiro, 1993; Kushiro, 1996). As will be briefly discussed later, the nature of the starting materials can have significant effects on the products of experimental studies. The broad features of the experimental studies that used the pyrolite compositions, as reviewed here, are consistent with experimental studies by other workers who used natural minerals, although there are discrepancies in the details (e.g., Falloon and Green, 1987; Falloon et al.,1996).

For the experimental studies, sintered-oxide mixes of pyrolite and model peridotite compositions are prepared from analytical reagent-grade chemicals by an initial firing (in air at 1,100°C) of an iron-free composition or a composition with only part of the total iron added. An initial high-temperature firing produces an olivine + pyroxenes + plagioclase + magnetite/chromite assemblage of very fine grains ($<5\,\mu$m). Metallic iron is then added, and each sample is fired in an evacuated silica tube at about 950°C to reduce the initial Fe_2O_3 to FeO. Alternatively, iron is added as fayalite. Microanalysis of the starting materials confirms very low Fe^{3+}/Fe^{2+} ratios, and electron-microprobe analysis of olivines in the experimental run products confirms the expected forsterite/fayalite ratios.

Because olivine comprises more than 50%, modally, of all high-temperature sub-solidus assemblages of pyrolite compositions at pressures less than 5 GPa, it is possible to subtract olivine from the bulk composition without affecting the sub-solidus phase relationships (provided that the olivine composition extracted matches that of the sub-solidus assemblage). The extraction of olivine aids in crystal growth and in the analysis of other phases by increasing their proportions, and also increases the degree of partial melting at given pressure and temperature. Provided that olivine remains a residual phase, the only effect that extraction of olivine has on melting relations is to *increase* the rate of change of the percentage melting as a function of temperature above the solidus. This increases the rate of change in the composition (Mg#) of residual olivine and other phases. By extracting olivine of Fo_{91} composition, rather than olivine with the Mg# of the bulk composition, the effect is to derive more Fe-rich liquids and residual olivine for experiments conducted below the temperature at which the residual olivine is Fo_{91}, and more Mg-rich liquids and residual olivine for experiments above that temperature. The effects are not large for up to 30–40% melting of Hawaiian pyrolite, that is, 50–67% melting of HPY-40 (Hawaiian pyrolite minus 40% olivine).

The use of sintered-oxide mixes also allows synthesis of starting compositions that have small and constant water contents, by means of synthesis of pargasite lherzolite at 1.5 GPa and 925°C under water-saturated conditions. The product contains a fixed modal proportion of pargasite of known composition. Excess water is driven off by crushing and firing at 200–300°C, and the starting material of pargasite lherzolite can be used to define its dehydration solidus as a function of pressure and temperature or as a function of the C – H fluid composition (Green, 1973b; Wallace and Green, 1988; Falloon and Green, 1990).

At a relatively early stage in the experimental studies of peridotite, consideration was given to the possibility of using crushed natural lherzolite as starting material – a practice used in some other experimental laboratories (Hirose and Kushiro, 1993; Baker et al., 1995). The following observations and difficulties led to abandonment of that approach:

1. The crushing of natural peridotite to a very fine ($<10\,\mu$m) and uniform grain size proved difficult, and prolonged crushing and grinding introduced contamination from the medium used: SiO_2 (agate), Al_2O_3 (corundum), WC (tungsten carbide), Fe (hardened steel).

2. The processes of mineral separation, fine crushing, and elutriation of individual minerals and recombination of sized fractions of crushed minerals were able to eliminate most, if not all, of the contamination problem, but there still remained a range of grain sizes. In addition, care had to be taken with the crushing and drying to eliminate all absorbed water or organic (alcohol, acetone) components.

3. Experiments with natural minerals failed to achieve homogeneity of phase compositions under sub-solidus conditions. In the presence of melt, minerals readjusted their compositions on grain boundaries, but spinel, garnet, orthopyroxene, and clinopyroxene characteristically remained compositionally zoned, requiring the selection of 'equilibrium compositions' for the experiments and creating uncertainty about the reacting composition for the run, so that mass-balance criteria could not be applied to test the reliability of run products. Olivine, unlike other phases, readjusted its Mg# throughout, but anomalous Fe/Mg partitioning between olivine and pyroxenes persisted because of the lack of equilibration of the latter.

4. The use of high-pressure natural assemblages, including aluminous pyroxenes, means that such phases are placed metastably at P–T conditions differing from those of their initial crystallization. Recent experimental studies on single phases have shown 'early partial melting' (Doukhan et al., 1993), manifested as small internal melt pools that are enriched in components rejected by the pyroxenes at the new pressure and temperature. Such disequilibrium melting is a complexity to be avoided and is a further factor in producing local reactions and reaction products, as opposed to a uniform, equilibrated run product.

Sintered-oxide experimental mixes react from low-pressure assemblages to high-pressure assemblages. They experience crystal growth from the initial, very fine grain size and characteristically produce slightly porphyroblastic textures because of the presence of orthopyroxene laths. They also yield strongly poikiloblastic garnet in sub-solidus experiments close to the 'garnet-in' boundary (contrasting with the abundant small sub-hedral garnets seen at higher pressures) (Green and Ringwood, 1967b, 1970). In defining sub-solidus reactions, such as the disappearance of plagioclase or the appearance of garnet with increasing pressure, it is possible for a low-pressure assemblage to metastably persist to higher pressures. Reversal experiments (growing the low-pressure assemblage from a previously crystallized high-pressure assemblage, or determining the direction of reaction from mixtures of low- and high-pressure assemblages) are desirable to confirm the reaction boundaries established by synthesis from sintered-oxide mixes or from glasses (e.g., Green and Hibberson, 1970a; Green and Ringwood, 1970). A recent detailed study of pargasite stability in MORB pyrolite has shown that reliable phase compositions can be determined at 925°C (in the presence of excess H_2O-rich fluid), and excellent modal abundances and mass balance can be achieved using sintered-oxide mixes and modern electron-microprobe techniques (Niida and Green, in press).

7.3.2. Sub-solidus Mineralogy of Pyrolite as a Function of P–T to a Depth of 150 km

Experimental study of HPY-40 (Hawaiian pyrolite minus 40% olivine) established the stability fields for plagioclase lherzolite, spinel + aluminous pyroxene lherzolite, and garnet lherzolite (Figure 7.1). A consequence of the high proportion of pyroxene relative to trivalent oxides that are not coupled to alkalis [i.e., $(Al_2O_3 + Cr_2O_3) - Na_2O$] in the Hawaiian-pyrolite composition is the elimination of spinel near the solidus at pressures greater than about 1.5 GPa, giving assemblages of olivine + aluminous pyroxenes only.

In the system $CaO + MgO + Al_2O_3 + SiO_2$ (CMAS), the plagioclase-lherzolite-to-spinel-lherzolite reaction is univariant and is simply represented as

$$CaAl_2Si_2O_8 + 2Mg_2SiO_4 \rightarrow MgAl_2O_4 + CaMgSi_2O_6 + 2MgSiO_3$$
anorthite   forsterite   spinel   diopside   enstatite

The reaction is made complex by the effect of increasing Ca-Tschermak ($CaAl_2SiO_6$) solid solution in both orthopyroxene and clinopyroxene with increasing temperature. This is most significant in the CaO-MgO-FeO-Al_2O_3-SiO_2 system; it has decreasing effect as plagioclase becomes more sodic and is absent from the Na_2O-MgO-FeO-Al_2O_3-SiO_2 system. Similarly, the univariant boundary in the CaO-MgO-Al_2O_3-SiO_2 system becomes divariant with addition of any one of the components FeO, Cr_2O_3, or Na_2O. Because we can assume a mantle Mg# of 90 ± 2, the Fe effect is small. The addition of Cr_2O_3 introduces Cr-rich spinel as a fifth but very minor phase in the plagioclase-lherzolite field. The characteristically extremely small size of spinel grains and the presence of compositional zoning or variation (e.g., Cr-rich spinels entirely enclosed in olivine or orthopyroxene) have precluded detailed study of the change from chromite to aluminous spinel through the plagioclase-lherzolite-to-spinel-lherzolite transition. Plagioclase becomes more sodic through the transition, but this is also moderated by the diopside content (acting as host to jadeite solid solution). Studies of Tinaquillo lherzolite, MORB pyrolite, and Hawaiian pyrolite have empirically confirmed these predictions, so that plagioclase is stable at the solidus to about 1.5 GPa in Hawaiian pyrolite, but to only about 1.2 GPa in MORB pyrolite. The very calcic plagioclase of Tinaquillo lherzolite produces a minimal 'cusp' at the solidus, and plagioclase disappearance at the lowest pressure (Figure 7.1).

Although the spinel-lherzolite-to-garnet-lherzolite reaction is a univariant boundary in the MgO-Al_2O_3-SiO_2 system, the boundary is made complex by two competing reactions:

$$MgAl_2O_4 + nMg_2Si_2O_6 \rightarrow Mg_2SiO_4 + (n-1)Mg_2Si_2O_6 \cdot MgAl_2SiO_6$$

    spinel   enstatite   olivine   aluminous enstatite

$$MgAl_2O_4 + 2Mg_2Si_2O_6 \rightarrow Mg_2SiO_4 + Mg_3Al_2Si_3O_{12}$$

       spinel   enstatite   olivine   pyrope

Because the first reaction is strongly temperature-dependent and the second reaction is both pressure- and temperature-dependent, selection of bulk compositions (e.g., low spinel : orthopyroxene) can produce a boundary between spinel harzburgite and garnet harzburgite at low temperatures and a boundary between spinel-free harzburgite (with aluminous orthopyroxene) and garnet harzburgite at high temperatures (Figure 7.1).

In systems more complex than $MgO-Al_2O_3-SiO_2$ the spinel-lherzolite-to-garnet-lherzolite reaction is a multivariant field that widens with increasing Cr_2O_3 content of the lherzolite and is displaced to higher pressures (for the incoming of garnet) by higher $Cr/(Cr+Al)$ ratios of the sub-solidus spinel. The latter ratio is not simply determined by the $Cr/(Cr+Al)$ ratio of the bulk composition but is also a function of the diopside and enstatite contents and the solubilities of $(Cr+Al)$ (coupled with Na and as Tschermak-type molecules) in the pyroxenes. In the absence of the detailed studies that would be necessary to model these complex interactions, the experimental studies of Hawaiian pyrolite, MORB pyrolite, and Tinaquillo lherzolite show only minor differences in the pressures for garnet appearance at $1{,}200°C$ or at the respective solidi.

7.4. Pyrolite Solidi and Melting Relationships
7.4.1. Pyrolite: (C + H)-absent

The solidi for Hawaiian pyrolite, MORB pyrolite, and Tinaquillo lherzolite have been experimentally determined for $(C+H)$-absent compositions (Figure 7.1). There are significant differences in solidus temperatures at a given pressure, and thus the differences in pressure for the intersection of a mantle adiabat with these solidi can be greater than 0.7 GPa (i.e., >20 km of depth). It is inappropriate to eliminate these differences as being due to experimental uncertainty in an attempt to estimate a single 'mantle solidus'.

The lower solidus temperature for Hawaiian pyrolite is in part a consequence of the K_2O, P_2O_5, and TiO_2 contents and their roles both in solid solutions (plagioclase, garnet, pyroxenes) and in forming separate but very minor phases (K-feldspar, apatite, ilmenite). Insofar as these elements are present in basalts and thus in their source regions and melting regimes, they will lower the solidi, and the melt fraction formed will be related to the modal abundances of these minor components. If these minor components are present in concentrations that may be contained within the major phases (olivine, orthopyroxene, and clinopyroxene, ±plagioclase,

±spinel, ±garnet) then the solidi are directly related to compositional differences among coexisting minerals. This is well illustrated by the behaviour of Na_2O and plagioclase. In the plagioclase-lherzolite field, the compositions of near-solidus plagioclase at about 1 GPa are An_{50} in Hawaiian pyrolite, An_{65} in MORB pyrolite, and An_{75} in Tinaquillo lherzolite – the differences in the 'plagioclase-out' boundaries (sub-solidus) and the temperatures of the solidi directly reflect these different plagioclase compositions (and consequently determine the position and magnitude of the cusp at the solidus).

Recent experimental studies, including those of the systems olivine (Fo_{90}) + jadeite + quartz and olivine (Fo_{90}) + $(CaAl_2SiO_6$-$NaAlSi_2O_6)$ + quartz, and these same systems with the addition of diopside, have explored the lherzolite minimum-melt composition at 1 GPa. These results are summarized in Figure 7.2, which traces the compositions of melts from a near-solidus melt in Hawaiian pyrolite to high-degree melts with residual olivine + orthopyroxene only. More extreme compositions (e.g., albite-bearing harzburgite) have also been investigated. For sodium-rich bulk compositions, minimum melts (at 1,220–1,230°C) are quite SiO_2-rich, but are very sodium-rich and are (nepheline + olivine)-normative (i.e., they are 'phonolitic' or 'trachytic' liquids). For Na-free systems, the minimum melt (at 1,310°C) is (olivine + hypersthene)-normative [note that this is the olivine + orthopyroxene + clinopyroxene + spinel eutectic in the CaO-MgO-FeO-Al_2O_3-SiO_2 system, as the reaction olivine (Fo_{90}) + anorthite $\leftrightarrow$ spinel + pyroxenes intersects the solidus just below 1 GPa].

The most sodic liquids in Figure 7.2 are formed only in extreme peridotite compositions, albite-bearing harzburgite, or lherzolite. However, 'fertile' compositions such as Hawaiian pyrolite (normative plagioclase is An_{55}) produce initial melts at about 1,240°C, in equilibrium with plagioclase of about An_{60} composition. These minimum melts are nepheline-normative. In MORB pyrolite, the normative plagioclase is An_{74}, and the solidus at 1 GPa is at about 1,275°C. Initial melts derived from sub-solidus An_{65} plagioclase are very close to the critical saturation boundary (i.e., with a very small content of either normative nepheline or normative hypersthene). For Tinaquillo lherzolite, near-solidus melts at 1 GPa are olivine- and hypersthene-normative, and the residual plagioclase is more calcic than An_{75}. The experimental data at 1 GPa is sufficient to define the minimum melt compositions for a wide spectrum of plausible peridotite compositions. The data of Figure 7.2 constrain melting models based on accumulation of melt from a large melting volume by flow through porous media. If the pooled melt equilibrates at 1 GPa with the crystalline matrix of olivine + orthopyroxene + clinopyroxene ± plagioclase ± spinel, then its major- and minor-element compositions will lie on the cotectics of Figure 7.2.

At pressures above 1.5 GPa, the three experimentally determined solidi retain their relative positions (Figure 7.1), but with smaller temperature differences than near 1 GPa. Apart from the effects of minor phases (ilmenite, apatite, K-feldspar), significant differences between the three compositions will be expressed in the

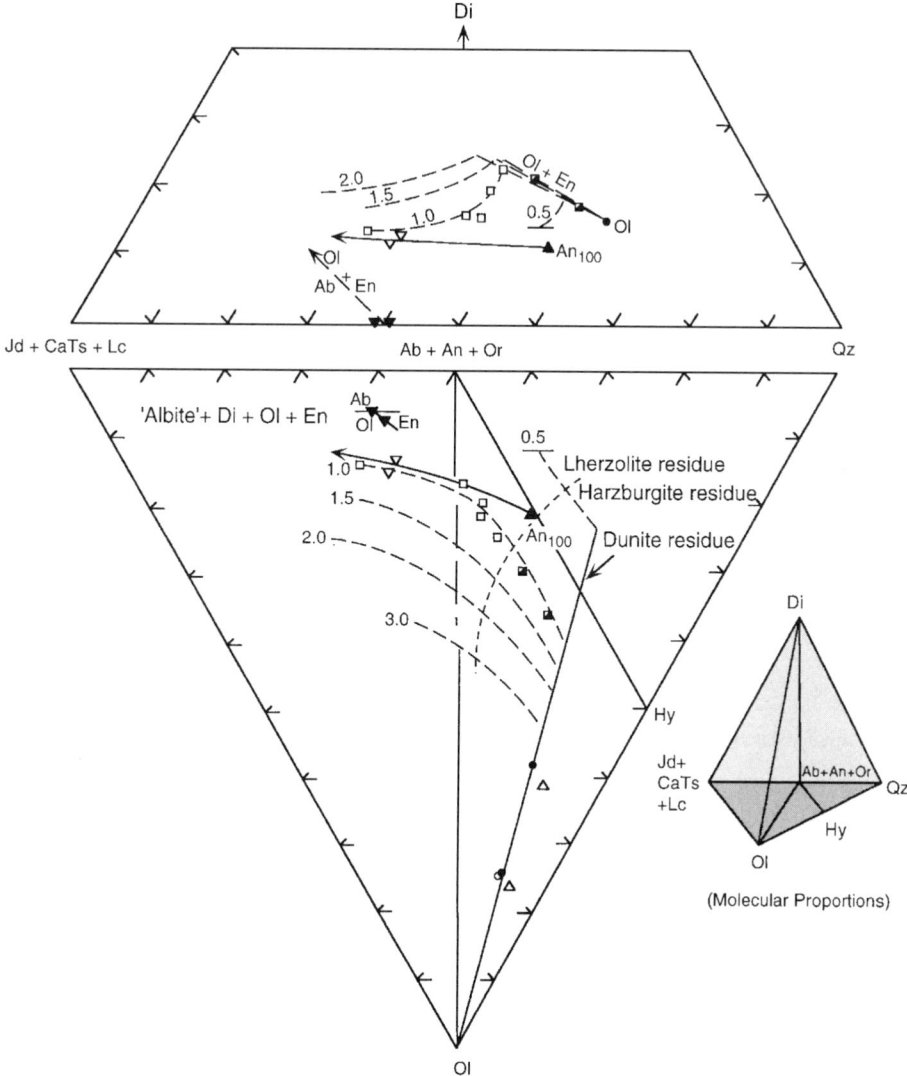

Figure 7.2. Projection of liquid compositions resulting from the melting of pyrolite onto the rear [Di, (Jd + CaTs + Lc), Qz] and lower [Ol, (Jd + CaTs + Lc), Qz] faces of the basalt tetrahedron (inset). Near-solidus, low-degree melts lie to the left of the figures; with increasing degrees of melting (i.e., increasing T at constant P), liquid compositions move to the right with the lherzolite residue (Ol + Opx + Cpx ± spinel) until diopside is eliminated, and then follow the Ol + Opx cotectic (harzburgite residue) to the olivine-control line (dunite residue). Melting paths at 0.5, 1.0, 1.5, 2.0, and 3.0 GPa are shown by dashed lines for the Hawaiian-pyrolite composition. The solid line at 1 GPa is the locus of melts at four-phase saturation with Ol + En + Di + plagioclase. This shifts from strongly nepheline-normative compositions, where the residual plagioclase is albite-rich, to olivine-normative compositions very close to the Hy-Plag join for sodium-free compositions. [Note that anorthite disappears at the solidus at 0.92 GPa (Figure. 7.1), so that the five-phase eutectic at 1 GPa in CaO-MgO-Al$_2$O$_3$-SiO$_2$ is actually Ol + En + Di + Sp + liquid (L).] The position of the eutectic Ol + En + Ab + L in the Ca-free system is also shown and provides a link between Figures 7.2 and 7.3. Filled circle, Hawaiian-pyrolite (HPY) and HPY-40 (HPY minus 40% olivine) compositions; open circle, MORB pyrolite (MPY); open triangle, Tinaquillo-lherzolite (TQ) and TQ-40 (TQ minus 40% olivine) compositions; open square,

clinopyroxene (cpx) composition. With $(Na/Ca)_{liquid} > (Na/Ca)_{cpx}$ at low pressures, the sequence of solidus temperatures for Hawaiian pyrolite < MORB pyrolite < Tinaquillo lherzolite is to be expected. The differences in solidus temperatures will also decrease if $(Na/Ca)_{liquid}$ approaches $(Na/Ca)_{cpx}$ at $P > 5$ GPa.

The experimental studies on peridotite melting show that with increasing pressure, melts become increasingly olivine-normative (picritic), and near-solidus melts become more silica-undersaturated (i.e., more nepheline-normative) (Figure 7.2).

The appearance of pyrope-rich garnet at the solidus and its presence as a residual phase are often considered important in modifying or controlling trace-element concentrations, particularly in retaining heavy rare-earth elements (so that the coexisting liquid will be enriched in light rare-earth elements) (Schilling, 1971; Leeman et al., 1980; Feigenson, 1986; Frey and Roden, 1987). It has also been argued that residual garnet fractionates Zr from Hf, and U from Th (Salters and Hart, 1989; La Tourrette, Kennedy, and Wasserburg, 1993). If garnet is to play these roles under anhydrous melting conditions, then the phase relations for the pyrolite compositions and the Tinaquillo lherzolite provide important constraints. Firstly, garnet can play no role at or above the solidus at pressures less than 2.8 GPa (Figure 7.1), (i.e., at depths <90 km) if the mantle is (C + H)-free. Secondly, when it first appears at the solidus, garnet is present in very small amounts because of the high Al_2O_3 solubility in pyroxenes, and the proportion of garnet increases with pressure as the Al_2O_3 solubility in pyroxenes decreases. The proportion of garnet also decreases rapidly above the solidus, so that the opportunity for garnet to influence trace-element concentrations in melts or residues is quite limited unless the depths for melt separation in a (C + H)-free mantle are considerably greater than 90 km.

7.4.2. Melting of Pyrolite in the Presence of (C + H + O) Fluids

At pressures of 2 GPa and higher and temperatures greater than 800°C, graphite (or, at higher pressures, diamond) is stable with fluid that approaches H_2O in composition and has extremely small contents of either CH_4 or CO_2 – this is often referred to as the graphite–water maximum for activity of H_2O (Green, Falloon, and Taylor, 1987). At oxygen activities higher than that defined by the graphite–water maximum, graphite is stable with $(CO_2 + H_2O)$ fluid over a small interval of fO_2, and at still higher oxygen activities graphite or diamond is unstable. A further

Figure 7.2 (*cont.*). liquid from Hawaiian pyrolite at 1 GPa, lherzolite residue; half-filled square, liquid from Hawaiian pyrolite at 1 GPa, harzburgite residue; open inverted triangle, plagioclase (An_{45}), lherzolite residue (Falloon et al., 1996); filled inverted triangle, experiments in the systems Jd + Fo + Qz and Jd + Di + Fo + Qz; filled triangle, experiment in CaO-MgO-Al_2O_3-SiO_2.

complication arises from the carbonation reactions

$$\text{olivine} + CO_2 \leftrightarrow \text{enstatite} + \text{magnesite}$$
$$\text{olivine} + \text{diopside} + CO_2 \leftrightarrow \text{dolomite} + \text{enstatite}$$

These reactions eliminate CO_2-rich fluids in favour of sub-solidus carbonate at pressures greater than 2 GPa at 1,000°C. (Note that these reactions move to higher pressures with increasing H_2O in a mixed $CO_2 + H_2O$ fluid, Figure 7.5.) At oxygen fugacities below the graphite–water maximum, graphite may coexist with $(CH_4 + H_2O)$ fluids. These comments signal the potential complexity of peridotite $+ (C + H + O)$ relationships. Our experimental approach has been to examine peridotite $+ H_2O$ (fO_2 graphite/water), peridotite $+ CO_2$, peridotite $+ (CO_2 + H_2O)$, and peridotite $+ (CH_4 + H_2O)$.

A parallel approach has been to examine the system nepheline $+$ forsterite $+$ quartz ($NaAlSiO_4 + Mg_2SiO_4 + SiO_2$) and its interactions with $(C + H + O)$ fluids (Green et al., 1987; Gupta, Green, and Taylor, 1987; Taylor and Green, 1987). This system has been studied for the purpose of clarifying the effect of pressure on the olivine-orthopyroxene cotectic (Figure 7.3a). Also, the composition of the minimum melt at the Na_2O-rich end of the olivine $+$ orthopyroxene cotectic is a simple-system analogue for near-solidus melting of lherzolite compositions (Figure 7.2). The presence of $C + H + O$ fluids at high-pressures strongly depresses the solidus in $Ne + Fo + Qz$ and also displaces the position of the olivine $+$ orthopyroxene cotectic. Infrared spectroscopy of quenched glasses permits the identification of solution mechanisms for carbon and hydrogen (e.g., Taylor and Green, 1987; Green et al., 1987). At high pressures (>2.0 GPa), CO_2 is very soluble in alkali-rich and 'basaltic' melts and polymerizes the melt structure, expanding the liquidus field of orthopyroxene at the expense of olivine. In contrast, H_2O solubility in alkali-rich and 'basaltic' melts also increases with pressure but, at least to 3 GPa, solution of $(OH)^-$ in the liquids has the effect of breaking chain structures (depolymerizing) in the melt. Increased $(OH)^-$ solubility expands the field of olivine at the expense of orthopyroxene (Figure 7.3b). Used as an analogue for melting in complex peridotite, the simple-system studies predict that oxidized and CO_2-rich fluids in the upper mantle (~ 90 km) will result in very undersaturated melts at low degrees of melting (i.e., in the Ne-Jd-Fo field of Figure 7.3b). The effect of H_2O-rich fluids is to produce more silica-rich melts from a 'peridotite' than would be produced in the absence of a fluid, although at 2.8 GPa liquids in the $Ne + Fo + Qz$ systems remain strongly silica-undersaturated (i.e., on the Ne side of the Ab-Fo join).

In the $C + H + O$ system at lower oxygen fugacity (fO_2) and high pressures, fluids become mixtures of $H_2O + CH_4$, or $CH_4 + H_2$ at fO_2 values below the iron $+$ wüstite buffer ($fO_2 < IW$).[3] Experiments have located the olivine $+$ enstatite cotectic (Figure 7.3b) in $Ne + Fo + Qz$ at 2.8 GPa under conditions of $fO_2 = IW + 1$ log unit (at which the fluid is $\sim 1 : 1$ in $CH_4 : H_2O$ molecular proportions) and at

---

[3] IW is the oxygen fugacity (in log units) of the iron/wüstite oxygen buffer.

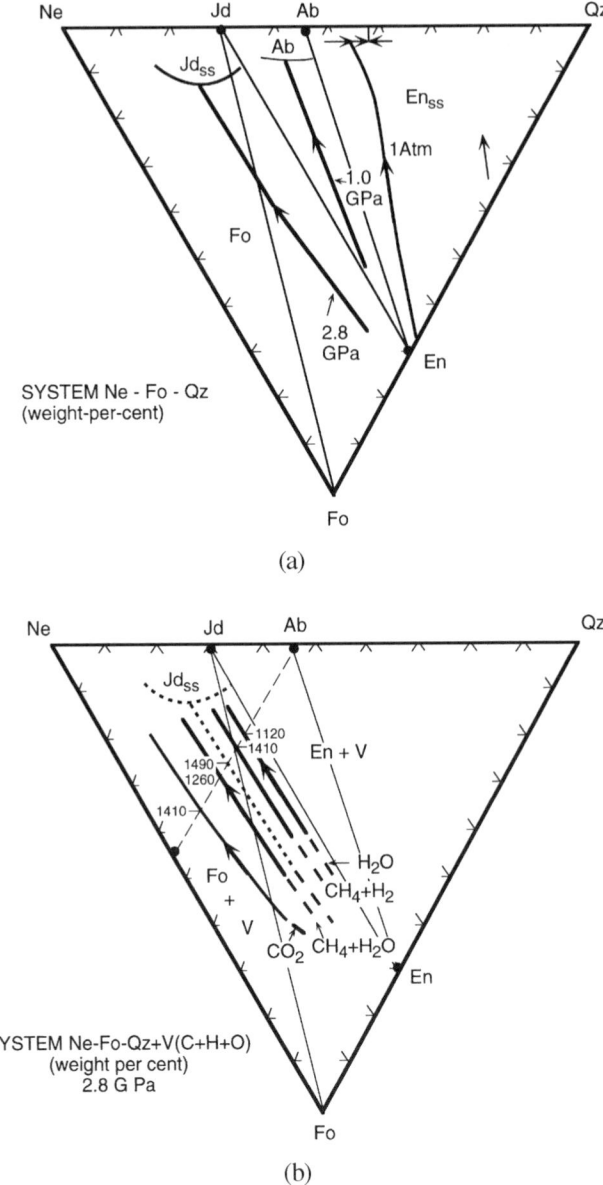

(a)

(b)

Figure 7.3. Partial phase diagrams for the system nepheline + forsterite + quartz as a simple system analogue for the base of the basalt tetrahedron (Figure 7.2). [Note that this figure is plotted in wt%, not mol% (Figure 7.2).] (a) Experimentally determined olivine + enstatite cotectics (cf. Figure. 7.2) at 1 atm and 1.0 GPa and 2.8 GPa, showing the strong pressure effect that moves the minimum melting composition to an undersaturated (nepheline-normative) composition. The arrows indicate the high-temperature-to-low-temperature directions along the cotectics. (b) Experimentally determined shifts in the Ol + En cotectic at 2.8 GPa due to saturation in H_2O, $CH_4 + H_2$ [$fO_2 \sim$ IW− (3–4)], $CH_4 + H_2O$ ($fO_2 =$ IW + 1), and CO_2 [$fO_2 >$ IW+ (3–4)]. The figures 1120, 1410, etc., denote liquidus temperatures (in °C) for the Ol + En cotectic at its intersection with the join of Ab to $Ne_{55}Ol_{45}$. The dashed line is the (C + H + O)-free cotectic from Figure 7.3a.

$fO_2 < IW - 4$ log units (fluid is $CH_4 > H_2 \gg H_2O$) (Taylor and Green, 1987). It is important to note from Figure 7.3b that at fO_2 values of about 1 log unit above the iron-wüstite buffer, the possible fluid phase is approximately a $1:1$ mixture of $CH_4 : H_2O$, but both carbon and hydrogen remain sufficiently soluble [as $(CO_3)^{2-}$ and $(OH)^-$] in low-SiO_2, alkali-rich melts to depress melting temperatures by about $230°C$. The olivine + orthopyroxene cotectic remains in the field of strong silica undersaturation (i.e., liquids are nepheline-normative) (Figure 7.3b).

7.4.3. Pyrolite + H₂O

7.4.3.1. Water-saturated Experiments

The water-saturated solidi for Hawaiian pyrolite and Tinaquillo lherzolite have been determined and occur at similar temperatures (Green, 1973a,b; Wallace and Green 1991). The solidi are depressed dramatically below the $(C + H)$-free solidus, and the solidus temperature decreases with increasing pressure (water pressure) up to about 1.5 GPa, reaching a minimum of about $975°C$ (Figure 7.4). At higher pressures, dT/dP is positive, and at 3 GPa the solidus lies between $1,000°C$ and $1,030°C$, approximately $450°C$ below that for the $(C + H)$-free solidus. Pargasitic amphibole is present below the solidus up to approximately 3 GPa, and its composition and modal abundance are dependent on both bulk composition and pressure. In Hawaiian pyrolite, the amphibole is more potassic, more sodic, more titaniferous, and more abundant than in either MORB pyrolite or Tinaquillo lherzolite. In Tinaquillo lherzolite, sub-solidus pargasite breaks down with increasing pressure at about 2.6 GPa and $1,000°C$; in MORB pyrolite it persists to 2.8 GPa, and in Hawaiian pyrolite it persists to 2.9 GPa (Figure 7.4) in the system with H_2O only.

In the study of MORB pyrolite, all minerals observed in experiments at $925°C$ could be analyzed, thereby demonstrating the compositional change in pargasite as a function of pressure. In parallel, there was an initial increase in the modal abundance of pargasite to 1.5 GPa, and then a progressive decrease to its disappearance at 2.8 GPa (Niida and Green, in press).

The water-saturated melting behaviour of Hawaiian pyrolite was investigated at 1 and 2 GPa, using the reversal approach to resolve problems produced by quench modification of melts (Green, 1976). At 1 GPa, liquids in equilibrium with lherzolite residue at $1,050°C$ are silica-rich and hypersthene- and quartz-normative. The compositions of melts at lower temperatures and lower degrees of melting have not yet been determined. With increasing temperature, liquids remain quartz-normative until olivine is the only residual phase. At 2 GPa, a liquid in equilibrium with lherzolite residue at $1,100°C$ is almost exactly silica-saturated (i.e., with neither normative quartz nor olivine). At higher temperatures, liquids with lherzolitic residue become olivine-normative.

The water-saturated melting experiments demonstrated an expansion of the liquidus field for olivine at the expense of pyroxenes, as in the simple-system analogue

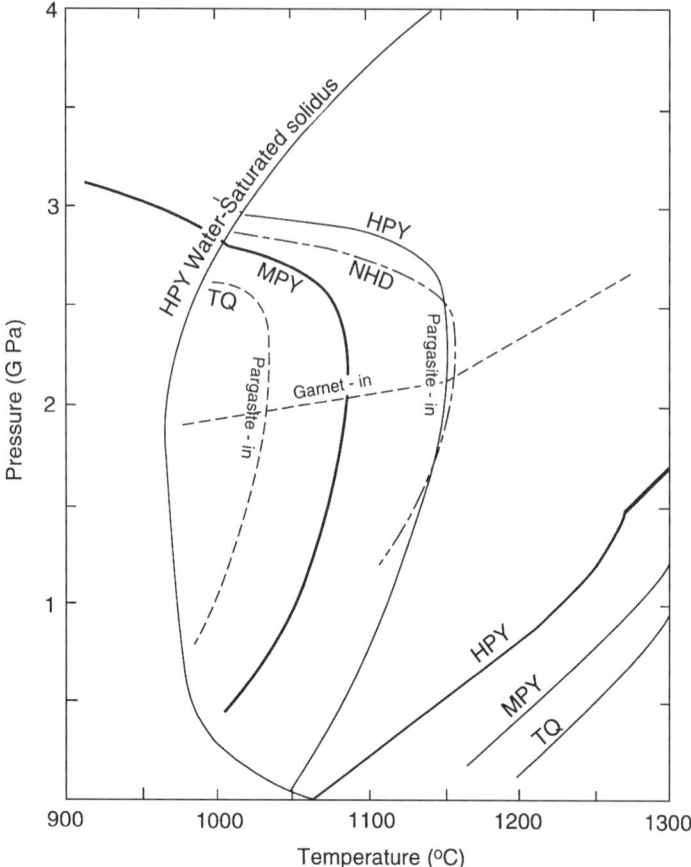

Figure 7.4. Experimentally determined dehydration solidi for Hawaiian pyrolite, NHD peridotite (Mengel and Green, 1986), MORB pyrolite, and Tinaquillo lherzolite. The dehydration solidi mark the beginning of melting (and also the disappearance of pargasite) for compositions in which small amounts of water are contained in sub-solidus pargasite (i.e., melting of pargasite lherzolite). The position of the water-saturated solidus (i.e., for H_2O content greater than can be contained in pargasite) and that for the incoming of garnet (both below and above the dehydration solidus) for Hawaiian pyrolite are also shown.

(Figure 7.3b). In the Earth's mantle, water-saturated melting at the solidus will occur if there is an influx of water in excess of that which can be held in pargasite at $P < 3$ GPa. In alternative scenarios at which mantle temperatures are above the water-saturated solidus, but there is limited availability of water, the water 'content' of the source becomes the primary determinant of the degree of melting at a given $P–T$. In this case, water-undersaturated melts can form, and their compositions will lie between the water-saturated and $(C + H)$-free partial-melt compositions, if comparisons are made for similar degrees of melting.

Although the compositions of water-saturated melts for pyrolite at pressures of 3 GPa or higher have not yet been determined, it is inferred that they will be olivine-normative, nepheline-normative at low melt fractions, but becoming hypersthene-normative at higher melt fractions (cf. Figure 7.3b). In addition, the compositions of near-solidus (975–1,035°C) melts for water-saturated melting at 0.5–3 GPa have not been determined, but such melts will have high contents of Na_2O, K_2O, and other incompatible elements, and this will make them olivine-normative and nepheline-normative at pressures of 1 GPa and higher.

7.4.3.2. *Water-undersaturated Melting or Dehydration Melting*

In a rock in which a hydrous phase is stable to temperatures at or above the water-saturated solidus there is a melting regime in which the solidus temperature is determined by breakdown of the hydrous phase along cotectic melting surfaces with the other sub-solidus minerals. Because water usually is strongly partitioned into the melt, the hydrous phase melts incongruently over a small temperature interval, with all of the water entering the melt. No fluid phase is produced, and the degree of melting is directly related to the amount of the hydrous phase present below the solidus. The water content of the melt, on the other hand, is primarily a function of the pressure and temperature of melting and is little affected either by the amount of hydrous phase present or by minor compositional variations of the melt phase.

Pargasite and phlogopite are hydrous phases that are stable to the solidus of fertile lherzolite at pressures of the uppermost mantle. The experimental determinations of dehydration solidi for four lherzolite compositions are presented in Figure 7.4. The experimental approach in each case was to synthesize the amphibole-bearing lherzolite composition at 1.5 GPa and 900–950°C under water-saturated conditions, in batches of 100–150 mg. This material was crushed and dried at 250–300°C, after a fragment was retained for mineral analysis. The modal abundances of amphibole and the compositions of the amphibole were different for each of the bulk compositions, and amphibole was most abundant and most titaniferous in the Hawaiian-pyrolite composition (Table 7.3).

In the NHD-peridotite composition, phlogopite is stable below the amphibole dehydration solidus to about 2.5 GPa and coexists with K_2O-saturated pargasite. Phlogopite is also stable above the dehydration solidus above 2.5 GPa in NHD peridotite and continues as a stable phase within the melting interval at pressures greater than 2.9 GPa, where amphibole breaks down, and up to pressures of at least 3.5 GPa. In Hawaiian pyrolite, phlogopite is absent within the pargasite stability field, but it appears as a residual phase at $P > 3$ GPa over a limited temperature interval above the water-saturated solidus (because of water released from pargasite breakdown) (Green, 1973b).

The relative $P–T$ positions of the dehydration solidi in Figure 7.4 show an apparent paradox in that the Tinaquillo peridotite has the lowest solidus temperature at pressures to 3 GPa, and Hawaiian pyrolite has the highest solidus temperatures, that is, the converse of the case for the $(C + H)$-free solidi. The reason for the paradox lies in the temperature-dependent stability of pargasite: Na_2O-, K_2O-, and TiO_2-rich pargasite (in Hawaiian pyrolite) is stable to a higher temperature than is

Table 7.3. *Comparison of pargasite compositions near the upper stability limit in three lherzolite compositions*[a]

Component	HPY[b] (2.0 GPa, 1,100°C)	MPY[c] (2.0 GPa, 1,050°C)	TQ[d] (2.0 GPa 1,000°C)
SiO_2	44.0	43.9	44.1
TiO_2	2.4	0.9	0.7
Al_2O_3	13.0	15.1	14.7
FeO	4.7	4.0	3.7
MgO	18.4	19.3	19.0
CaO	10.2	10.6	11.2
Na_2O	3.2	3.4	2.9
K_2O	0.8	0.0	0.6
Cr_2O_3	1.5	0.9	1.2
Mg#	87.5	89.6	90.2

[a] See Figure 7.4; this comparison illustrates the higher TiO_2, K_2O, and Na_2O contents of the pargasites that are stable (at $fH_2O < 1$) to higher temperatures in more fertile compositions.
[b] Hawaiian pyrolite data from Wallace and Green (1991).
[c] MORB pyrolite data from Niida and Green (in press).
[d] Tinaquillo lherzolite data from Wallace and Green (1991).

the low-Na_2O, low-TiO_2 pargasite formed in Tinaquillo lherzolite at 1.5 GPa and 1,000°C (Table 7.3). In each case, pargasite breakdown is accompanied by melting, but the amount of melt at the solidus is predicted to be lowest for the Tinaquillo lherzolite and highest for the Hawaiian pyrolite. These differences reflect the different abundances of amphibole at the solidi and thus different water contents at the dehydration solidi. Similarly, the amount of melt formed decreases with increasing pressure, particularly at pressures greater than 1.5 GPa. The reasons for change in the percentage of melting at the solidus are the changing sub-solidus composition and abundance of amphibole and the fact that the amphibole-breakdown curve is intersecting liquidus surfaces for basaltic melts with different water contents. For example, depression of the liquidus of an olivine-normative basalt at 1 GPa from the temperature of the anhydrous peridotite solidus at 1,250°C to that of the dehydration solidus at 1,100°C requires about 1.5 wt% H_2O in the basalt. With 0.3% H_2O in pargasite lherzolite, that implies 20% melting. However, at 2.5 GPa, depression of the liquidus of an olivine-normative basalt from the temperature of anhydrous solidus at 1,450°C to that of the dehydration solidus at 1,130°C requires about 7.5 wt% H_2O. At this pressure, there is only 0.15% H_2O in the pargasite-lherzolite source, and breakdown of the more sodic pargasite yields only 2% melt at the dehydration solidus.

The inverse dependence of dehydration solidi on 'fertility' in lherzolite compositions has significant implications for melting behaviour under conditions of lithosphere heating or upwelling. If the lithosphere is initially inhomogeneous (possibly because of variable melt extraction) and contains pargasite, then either heating or upwelling could produce partial melting in the more chemically depleted volumes, while Na_2O-, K_2O-, and TiO_2-rich volumes remain sub-solidus. The melt component thus produced would then be capable of reacting with any neighbouring fertile volumes to precipitate more Na_2O-, K_2O-, and TiO_2-rich pargasite. The process is one of local metasomatism in which pargasite-rich layers/volumes will form, intercalated with anhydrous lherzolitic or harzburgitic layers. This metasomatism via incipient-melt migration will impose incompatible-element signatures on both residue and enriched layer, the signature in the latter being that of an incompatible-element-enriched basalt. A random sampling of lithosphere affected by metasomatism of this nature, such as xenoliths in basaltic magmas, will provide cryptically metasomatized (pargasite-free) and modally metasomatized (pargasite-rich) lherzolites. The process is one of the mechanisms responsible for creating inhomogeneity in the lithosphere and providing an explanation for the decoupled major-element and incompatible-element abundances in spinel-lherzolite xenoliths and in high-pressure, high-temperature spinel-lherzolite bodies (Frey and Green, 1974).

7.4.4. Pyrolite + CO₂

The melting behaviour of peridotitic compositions with CO_2 (Falloon and Green, 1989) is relevant only at fO_2 conditions at or higher than the olivine-magnesite-orthopyroxene-graphite buffer. The melting behaviour is divided into three fields. At pressures below those for the carbonation reaction olivine + diopside + $CO_2 \rightarrow$ enstatite + dolomite, CO_2 has little effect on the peridotite other than a slight and increasing solidus depression due to small increases in the solubility of CO_2 in basaltic melts as pressures increase. At pressures approaching the carbonation reaction, the solidus is abruptly depressed, and the melt formed is carbonatitic (dolomitic), with less than 5 wt% SiO_2. The solidus will be vapour-undersaturated provided that the amount of CO_2 available is insufficient to eliminate olivine by the carbonation reactions. At higher pressures, the sub-solidus reaction dolomite + enstatite $\rightarrow$ diopside + magnesite is encountered, and the carbonatitic liquids at the solidus probably become more magnesitic.

The solidus for Hawaiian pyrolite + CO_2 is illustrated in Figure 7.5a. There is an abrupt depression of the solidus at about 1.8 GPa, but it is to be noted that the minimum solidus temperature at about 2.5 GPa is some 50–100°C higher than that for Hawaiian pyrolite + H_2O.

7.4.5. Pyrolite + (CO₂ + H₂O)
7.4.5.1. Fluid-saturated Experiments

Two different regimes have been investigated experimentally: The first was (CO_2 + H_2O)-fluid-saturated, with small fixed H_2O and CO_2 contents (Falloon and Green,

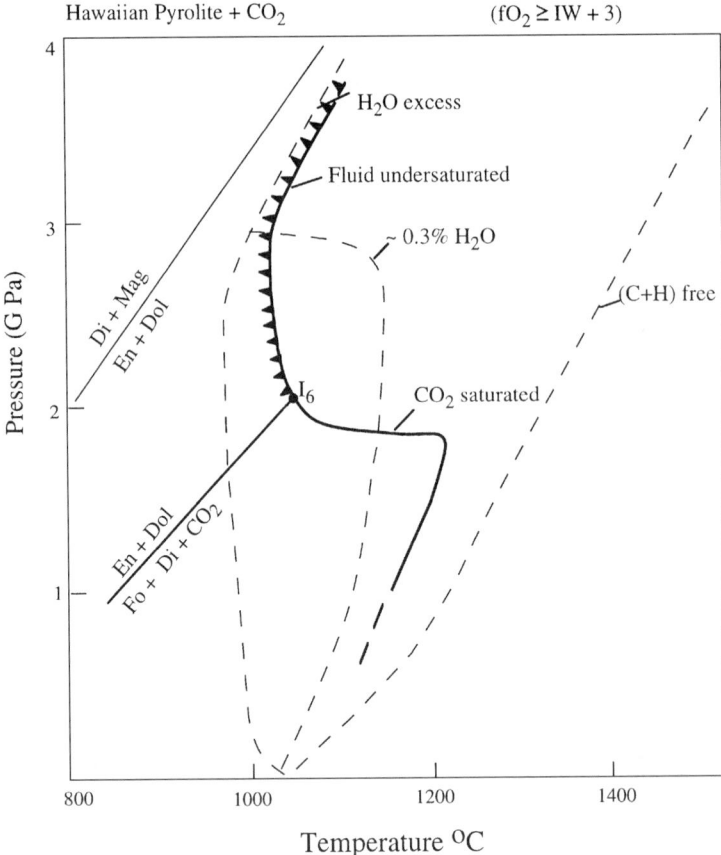

Figure 7.5a. The three parts (a, b, and c) of Figure 7.5 show the solidi for Hawaiian pyrolite + (C + H + O) compared with the (C + H)-free solidus from Figure 7.1 and the H_2O-saturated and H_2O-undersaturated solidi from Figure 7.4. These three solidi are shown as short dashed lines. Figure 7.5a shows the experimentally determined solidus for Hawaiian pyrolite + CO_2 (heavy solid line). The solidus is CO_2-saturated below the intersection of the carbonation reaction with the solidus at I_6 (Wyllie, 1978, 1987). At higher pressures, dolomite is present as a sub-solidus phase, the solidus is fluid-undersaturated, and the melt phase is sodic dolomitic carbonatite.

1990). The fluid-saturated experiments (Figure 7.5b) did not extend across the full range of H_2O and CO_2 concentrations, but were fluid-saturated principally because the amount of water added exceeded the amount that can be accommodated in pargasite. As in the pyrolite + CO_2 experiments, the CO_2 contents were insufficient to eliminate either diopside or olivine from the sub-solidus mineral assemblage via the carbonation reaction. Thus, all the experiments at pressures greater than those for the carbonation reactions were on carbonate-bearing lherzolite.

The fluid is CO_2-rich below the carbonation reaction, as most of the water remains in pargasite for sub-solidus conditions. The pargasite breakdown and the solidus are essentially coincident up to 1.5 GPa, lying between the pyrolite + H_2O dehydration solidus and the water-saturated-pyrolite solidus. As pressures approach those for

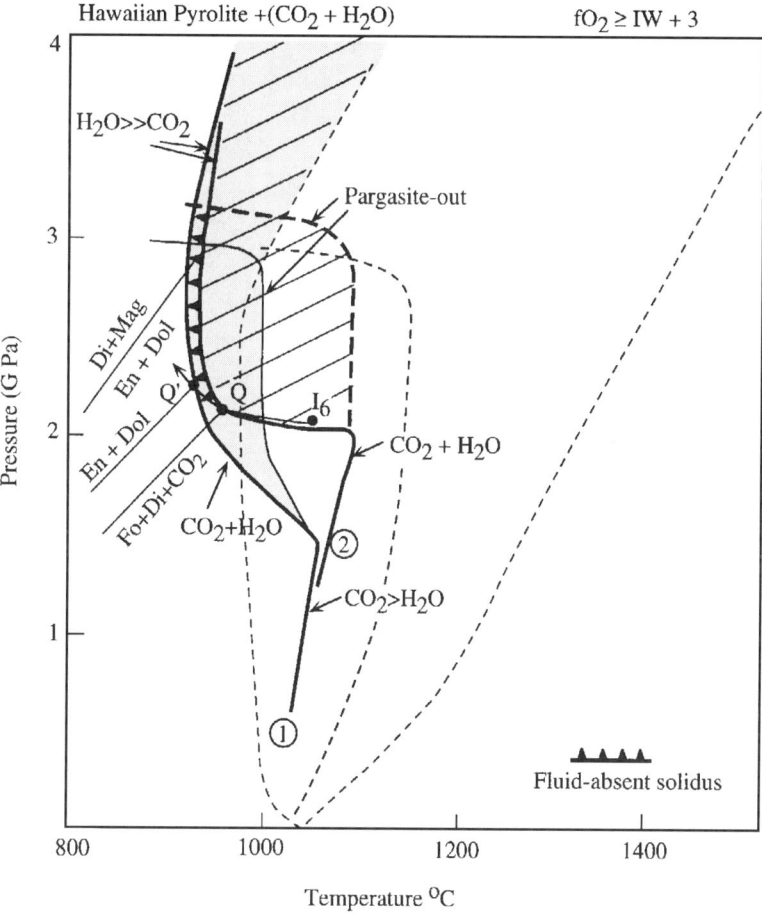

Figure 7.5b. The circled number 1 shows the solidus for Hawaiian pyrolite + $(CO_2 + H_2O)$ for fluid-saturated conditions. The nature of the fluid changes from $CO_2 > H_2O$ at pressures below the carbonation reaction $Fo + Di + CO_2 \rightarrow En + Dol$ to $H_2O \gg CO_2$ at high pressure. Pargasitic amphibole is present below the solidus to about 2.9 GPa and also persists above the solidus to about 1,020°C at 2.5 GPa. The melt at the solidus at pressures greater than 1.5 GPa is carbonatite, and the P–T field for carbonatite melt is shown by shading. At temperatures above the pargasite-out boundary, the melt is a silicate melt with dissolved $(CO_3)^{2-}$ and $(OH)^-$. The circled number 2 shows the solidus for Hawaiian pyrolite with a water content of $0 < H_2O < 0.4$ wt% and a small CO_2 (carbonate) content. The field for carbonatite melt is shown by diagonal shading, and between 2 GPa and 3.2 GPa a sodic dolomitic carbonatite melt coexists with residual pargasite. At less than 2.0 GPa, the sub-solidus assemblage is spinel + pargasite lherzolite, the solidus is fluid saturated, and the melt at the solidus is a hydrous silicate melt with low $(CO_3)^{2-}$ content. At pressures greater than 3.2 GPa, the solidus is also fluid-saturated, but pargasite is unstable below the solidus, the fluid is H_2O-rich, and the melt at the solidus is a carbonatite with about 3 wt% SiO_2. Between 2 GPa and 3.2 GPa, the carbonatite solidus (solid line with 'teeth') is fluid-undersaturated, and fluid-undersaturated silicate melt forms only at temperatures above the pargasite-out boundary. The carbonation reaction $Fo + Di + CO_2 \rightarrow En + Dol$ is displaced to higher pressures with increased fH_2O in the $(CO_2 + H_2O)$-fluid, displacing the intersection of this reaction with the solidus from Q to Q' with increasing $H_2O : CO_2$ in the low-pressure fluid phase.

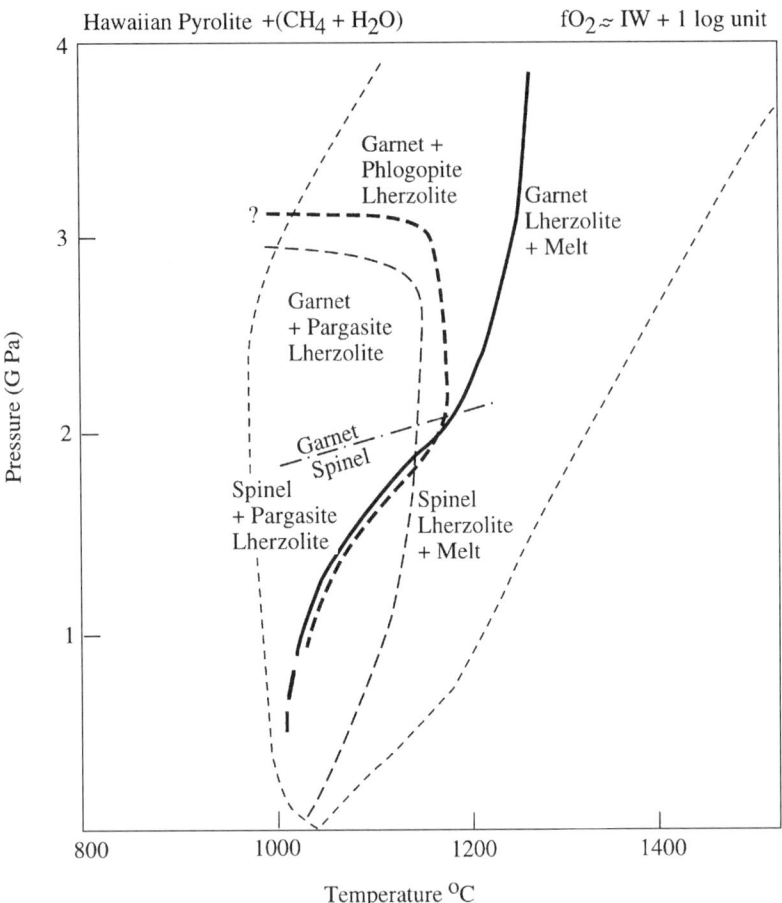

Figure 7.5c. Solidus (solid line) for Hawaiian pyrolite $+ (CH_4 + H_2O)$ at $fO_2 \sim IW + 1$ log unit. Fluid is present at the solidus, and the melt at the solidus is a silicate melt with dissolved $(OH)^-$ and with very low dissolved $(CO_3)^{2-}$ at low pressure, but with increasing dissolved $(CO_3)^{2-}$ at pressures greater than 2 GPa. Pargasite is stable to the solidus at $P < 2$ GPa, but it breaks down at $1,175°C$ to garnet lherzolite + phlogopite + fluid $(CH_4 + H_2O)$ at pressures greater than 2 GPa, and it is unstable at pressures greater than 3.2 GPa even at low temperatures.

the carbonation reaction, the solidus is markedly depressed and lies at temperatures below those for both the CO_2-saturated and H_2O-saturated solidi (Figure 7.5b). Along this part of the solidus, both carbonate and pargasite are stable below the solidus, and the fluid present is H_2O-rich. Pargasite persists for approximately 70–80°C above the solidus at 2.5 GPa. The melt formed at the solidus is carbonatitic (sodic dolomitic carbonatite) and coexists with both a water-rich fluid and pargasite-bearing lherzolite. These observations are particularly important in considering the subduction regime, where models for the subducted-slab/mantle-wedge interactions suggest P–T and fO_2 conditions in the mantle wedge for which a carbonatite melt and water-rich fluid field can be expected to result from release of water and CO_2 from the subducted slab (Figure 7.11).

Pargasite is not stable at pressures greater than 3 GPa in mantle lherzolite, and the sub-solidus carbonate is magnesite. Fluids remain H_2O-rich, and the minimum melts are carbonatitic. Phlogopite was not observed in the fluid-saturated sub-solidus assemblages nor in the fluid + carbonatite melt + garnet-lherzolite assemblages, and microprobe observations suggested concentration of K in the H_2O-rich fluid phase at $P \geq 3$ GPa. The H_2O-rich fluid phase produced very small quench aggregates that could not be analyzed (Wallace and Green, 1988; Falloon and Green, 1989, 1990). At temperatures above the fluid-saturated silicate solidus, both within the pargasite stability field and at higher pressures, there is no evidence in the partially molten lherzolite of a two-phase liquid field, and it is therefore assumed that the carbonatite melt and $(CO_3)^{2-}$-rich silicate melt (olivine melilitite or olivine leucitite) are miscible.

7.4.5.2. Fluid-undersaturated Conditions

This experimental study (Green and Wallace, 1988; Wallace and Green, 1988) was designed to investigate the melting behaviour of mantle peridotite for conditions matching those of the pyrolite dehydration melting experiments, but with addition of a small amount of carbonate. At pressures below those for the carbonation reaction, a CO_2-rich fluid was present, and the solidus was depressed slightly below the dehydration solidus for Hawaiian pyrolite + H_2O (Figure 7.5b). Melt compositions and melt percentages are expected to be very similar to those in Hawaiian pyrolite + H_2O, because although the melts will be fluid-saturated, the fluid will be dominantly CO_2, and at pressures less than 1.5 GPa the CO_2 will have low solubility in the melt, particularly relative to water.

At pressures approaching the solidus/carbonation-reaction intersection, the solidus is depressed to about 935°C. At higher pressures (beyond Q' in Figure 7.5b), fluid is absent along the solidus. Remarkably, the upper temperature limit for pargasite remains at about 1,100°C and does not follow the solidus depression. The experimental results show that water partitions between pargasite and the carbonatite melt, implying a relatively low solubility of about 2–3% in the latter. The composition of the melt at 2.2 GPa and 1,020°C was determined as sodic dolomitic carbonatite, and the compositions of all coexisting pargasite-lherzolite phases were determined. The carbonatite melt contains about 3% SiO_2. The silicate melting regime is encountered at the upper stability limit of pargasite, and at 2.6 GPa this is 165°C above the carbonatite solidus (Figure 7.5b). At $P > 3$ GPa, carbonatite melt also occurs at temperatures of about 940°C, some 120–150°C below the water-saturated silicate solidus. The silicate-melt solidus coincides with pargasite breakdown at pressures below 3 GPa and with the water-saturated solidus at higher pressures. Carbonatite melt dissolves in the more voluminous silicate melt at $T > 1,100°C$ for pressures below 3 GPa, and there also appears to be a single-melt phase at $P > 3$ GPa and $T > 1,050°C$.

7.4.6. *Pyrolite* + (C + H + O) *Fluids at Low Oxygen Fugacities*

The evidence from mantle samples and from primitive magmas indicates that the Earth's mantle is inhomogeneous in its oxygen fugacity (Christie, Carmichael, and Langmuir, 1986; Ballhaus, Berry, and Green, 1990, 1991). Primitive magmas produced at convergent margins are more oxidized than MORBs or intraplate basalts, an observation consistent with recycling of oxidized oceanic crust and lithosphere into the mantle via subduction. Intraplate (including hotspot) primary magmas apparently are more oxidized than MORBs, although that may in part be a consequence of volatile $CH_4 + CO_2 + H_2O$ degassing at intermediate to low pressures, rather than a primary source characteristic. Both MORBs and ocean-floor peridotites provide evidence for log fO_2 = IW + 1–2 log units. There have also been arguments that mantle fO_2 decreases with increasing pressure (O'Neill et al., 1993; Ballhaus and Frost, 1994). With respect to the (C + H + O) system, if mantle fO_2 conditions are IW + 1–2 log units, then in the presence of graphite or diamond, fluids within the upper mantle should range from nearly pure water (fO_2 = IW + 2) to 1 : 1 CH_4 : H_2O (fO_2 = IW + 1, P = 3 GPa, T = 1,300°C) (Green et al., 1987). If we wish to understand mantle melting relationships, it is also necessary to investigate pyrolite + (C + H + O) at low oxygen fugacities.

Use of the double-capsule technique and the WC-WO$_2$ (tungsten carbide/tungsten oxide) buffer creates an environment in which the melting behaviour of Hawaiian pyrolite can be explored at fO_2 conditions very close to IW + 1 log unit (Taylor and Green, 1988, 1989). The melting relationships are summarized in Figure 7.5c. The pargasite-lherzolite assemblage is stable up to the solidus at P < 2 GPa. The melting is fluid-saturated, and at P < 2 GPa the (C + H + O) fluid at $fO_2 \sim$ IW + 1 causes depression of the solidus below the dehydration solidus for pyrolite + H_2O (implying that fH_2O in the fluid is greater than fH_2O in pargasite at P < 2 GPa). At pressures greater than 2 GPa, pargasite breaks down below the solidus, but phlogopite persists as a minor sub-solidus phase, and phlogopite + garnet lherzolite coexists with ($CH_4 + H_2O$) fluid up to the fluid-saturated solidus. Attention is drawn to the stability of pargasite to 1,175°C at 2.5 GPa in the absence of the melting reactions that are seen in the more oxidized systems. Pargasite is also stable to slightly higher pressures (3.2 GPa) in the absence of melting, possibly because of small differences in composition, such as the absence of Fe^{3+}.

At high-pressures, CH_4 is effectively an inert component in the fluid phase in relation to melting, and it raises the solidus temperature above the peridotite + H_2O solidus. Nevertheless, the overall effect of the ($CH_4 + H_2O$) fluid (fO_2 = IW + 1) is depression of the melting temperatures by about 250°C, relative to the (C + H)-free solidus. This is accompanied by solution of both $(CO_3)^{2-}$ and $(OH)^-$ in the melt phase (as also shown by the study of the Ne + Fo + Qz system, Figure 7.3b). Furthermore, the normative composition of the melt in terms of the olivine/orthopyroxene ratio (and degree of olivine saturation) closely overlies the (C+H)-free composition because of the competing effects of $(CO_3)^{2-}$ in contracting and $(OH)^-$ in expanding the olivine-phase field (Figure 7.3b). As the redox conditions are at $IW + 1$ log unit, all iron in the melt is divalent. If this highly reduced melt containing dissolved $(CO_3)^{2-}$ and $(OH)^-$ at high pressures should ascend to lower pressures, its internal oxidation state will be altered, and also it will become fluid-saturated at moderate pressures. The fluid will escape as $CH_4 + CO_2 + H_2O$ and may also precipitate graphite – the residual melt will become more oxidized ($Fe^{2+} \rightarrow Fe^{3+}$) and effectively will retain only $(OH)^-$ becaue of this degassing.

The preceding discussion illustrates the complex effects that the principal volatile species, C and H, can have on mantle melting, as well as the consequent effects on the magmas themselves arising from pressure reduction and the concomitant rapid changes in the solubilites of carbon at 1.5–2 GPa and of water at <0.5 GPa.

7.5. The Pyrolite Concept and Models of Magma Genesis
7.5.1. General Features of the Current Expression of the Pyrolite Concept
7.5.1.1. The Petrological Lithosphere and the Incipient-Melting Regime

A major difference between conventional models of the upper mantle that use the solidus for anhydrous (and carbon-free) melting (McKenzie and Bickle, 1988) and the model presented here (Figures 7.6–7.9) is that the petrological data introduce the unavoidable presence of a partially molten layer due to the presence of C+H+O volatiles. This has significant geophysical implications, in addition to its petrological significance. For example, some current geophysical and petrological models define the asthenosphere as that part of the mantle in which the mantle geotherm is inferred to be adiabatic (i.e., a convective regime). The Earth's outer layer has a geothermal gradient that exceeds the adiabatic gradient and is a thermal boundary layer, commonly equated with the lithosphere (Griffiths and Turner, Chapter 4, this volume; McKenize and Bickle, 1988). The latter authors subdivide the lithosphere into a shallow 'mechanical boundary layer' overlaying a 'thermal boundary layer' inclusive of both the less viscous (higher-temperature) part of the conductive regime and the uppermost part of the 'abiabatic' or convective regime. This thermal boundary layer is seen as inherently unsteady, with episodic advective instabilities (Griffiths and Turner, Chapter 4, this volume). However, other geophysicists define the boundary between the lithosphere and asthenosphere on the basis of rheological properties rather than the thermal properties of the upper mantle (Anderson, 1995).

In plate-tectonic theory, the concept of rigid plates correlates readily with the 'lithosphere', and their motions over and detachment from the underlying mantle correlate with the use of 'asthenosphere' for the latter region. The distribution of shear between the lithosphere and the asthenoshphere and the extent of coupling between the lithosphere and the upper part of the asthenosphere will depend on viscosity-versus-depth relationships. The ability of seismology to infer changes in seismic wavespeeds and attenuation on a regional basis, and as functions of

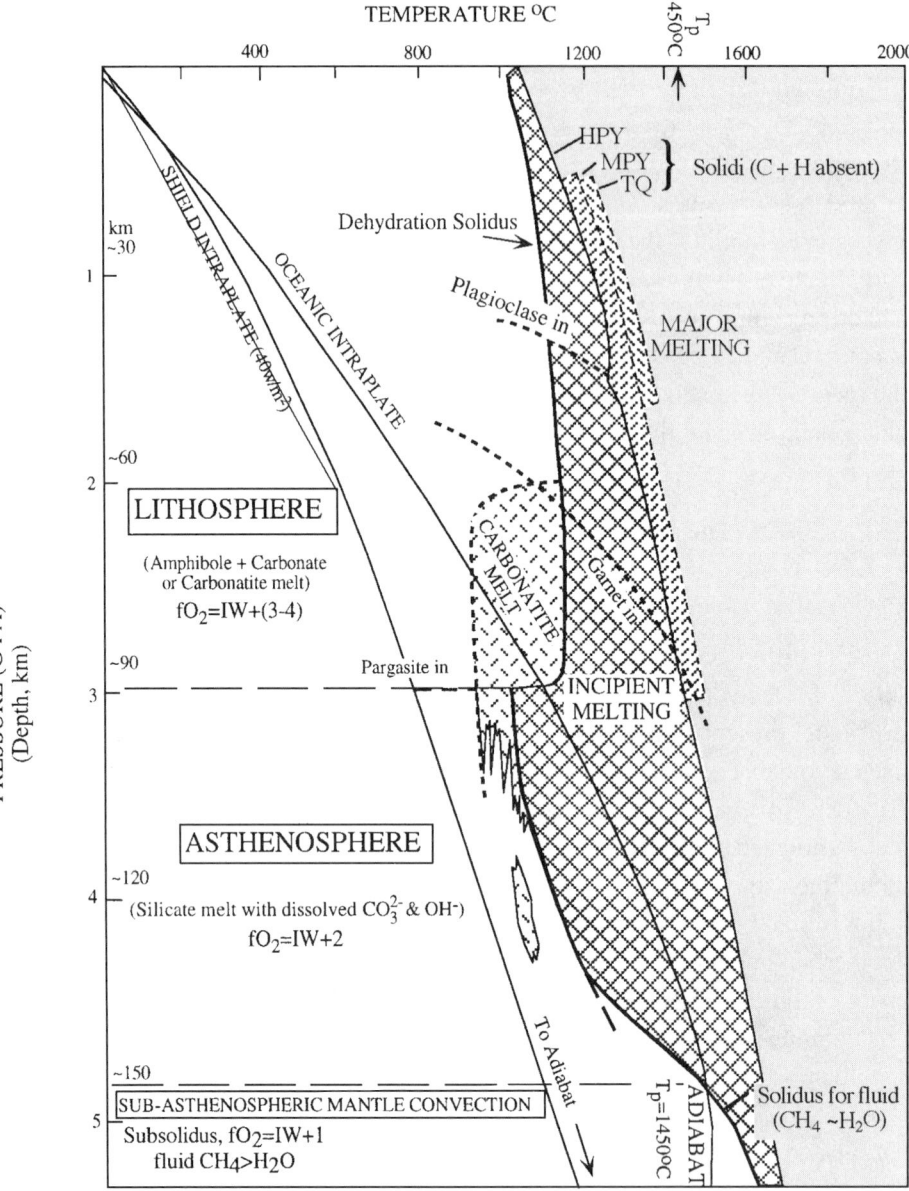

Figure 7.6. Application of the experimentally determined melting relationships for pyrolite and pyrolite $+ (C + H + O)$ to present a model for the intraplate lithosphere, asthenosphere, and sub-asthenospheric mantle. The 'oceanic intraplate' geotherm represents a geothermal gradient equally appropriate for old oceanic crust and 'young' continental crust. It is this geotherm that necessarily traverses the carbonatite stability field and incipient-melting regime for the pyrolite compositions provided that the mantle $fO_2 \geq IW + 2$ log units. Attention is drawn to the petrological base of the lithosphere, defined by the high-pressure pargasite breakdown and silicate solidus at about 95 km. The lower boundary of the asthenosphere (incipient-melting regime along the geotherm) is drawn arbitrarily at about 150 km as the intersection of a mantle adiabat of potential temperature $T_p = 1,450°C$ with the pyrolite-$(C + H + O)$ solidus for $fO_2 = IW + 1$. Below that depth, a fluid with $CH_4 \geq H_2O$ is present, but there is no silicate melt unless $fO_2 > IW + 1$ (i.e., a more oxidized region of the mantle). The carbonatite melt at more than 95 km and $T < 1,000°C$ is represented as

depth, provides the most detailed information concerning the variability of physical properties in the upper mantle. The connection between seismic wavespeeds and attenuation and effective viscosity for long-term creep is by no means immediate (e.g., Jackson and Rigden, Chapter 9, this volume; Drury and Fitz Gerald, Chapter 11, this volume). Nevertheless, it is reasonable to associate the anomalous seismic properties (low wavespeeds and high attenuation) and the low viscosity of the upper mantle beneath oceanic lithosphere with the asthenosphere.

The experimental studies of pyrolite summarized in earlier sections and in Figures 7.6 and 7.7 lead directly to the conclusion that most intraplate geothermal gradients (other than very cool 'shield' geotherms) will intersect the silicate solidus for pyrolite $+ (C + H + O)$ provided that mantle $fO_2 \geq IW + 2$ log units.

As illustrated in Figure 7.6, an oceanic intraplate geotherm crosses the silicate solidus at a depth of about 95 km and enters the incipient-melting regime. Although we have not, as yet, quantitatively determined the effects on mantle viscosity, seismic wavespeed, and seismic attenuation cause by the presence of a very small interstitial silicate melt fraction, it is reasonable to infer that passage from sub-solidus to super-solidus conditions in the mantle will have significant effects on rheological properties (cf. Drury and Fitz Gerald, Chapter 11, this volume). Stated briefly, the experimental studies of model pyrolite $+ (C + H + O)$ provide a plausible petrological argument that the transition from lithosphere to asthenosphere, defined in rheological terms (Anderson, 1995), should be seen as the transition from sub-solidus pargasite-bearing lherzolite to lherzolite containing a very small melt fraction ($\sim$1–2% melt) determined by the $(C + H)$ or $(CO_2 + H_2O)$ contents of the upper mantle (Green, 1971; Green and Liebermann, 1976).

In Figure 7.6, these arguments are illustrated by superimposing estimated geothermal gradients on the pyrolite $+ (C + H + O)$ phase diagram to illustrate the intersection of the oceanic intraplate geotherm with the silicate solidus for pyrolite. If the lithosphere has $fO_2 \geq IW + 2$ log units, then this solidus is defined by pargasite dehydration melting, and intersection with the geotherm occurs at 95 ± 5 km (Figure 7.6). This intersection provides a petrological argument for rheological change from lithosphere to asthenosphere. The base of the 'petrological lithosphere' is insensitive to temperature variations (conductive geotherm variations) of 100–150°C due to the 'back-bend' of the amphibole dehydration solidus at about 95 km ($\sim$3 GPa). Within the depth interval $\sim$95 km to $\sim$160 km, fluid is absent, and C and H are dissolved in the silicate melt fraction, which can vary in fO_2 with depth

Figure 7.6 (*cont.*). present or absent depending on local variations in fO_2. That is, if $fO_2 \geq IW + 3$, then carbonatite melt $\pm H_2O$-fluid [rather than graphite (diamond) $+ H_2O$-fluid] will be present. The 'shield intraplate' geotherm does not intersect the silicate solidus and would meet the mantle adiabat at approximately 250 km. The implication is that there is no 'petrologic asthenosphere' or region of incipient melting for such geotherms. This geotherm can intersect the carbonatite solidus only if parts of the deep Archaean mantle are oxidized ($fO_2 \geq IW + 3$) and carbonate is present.

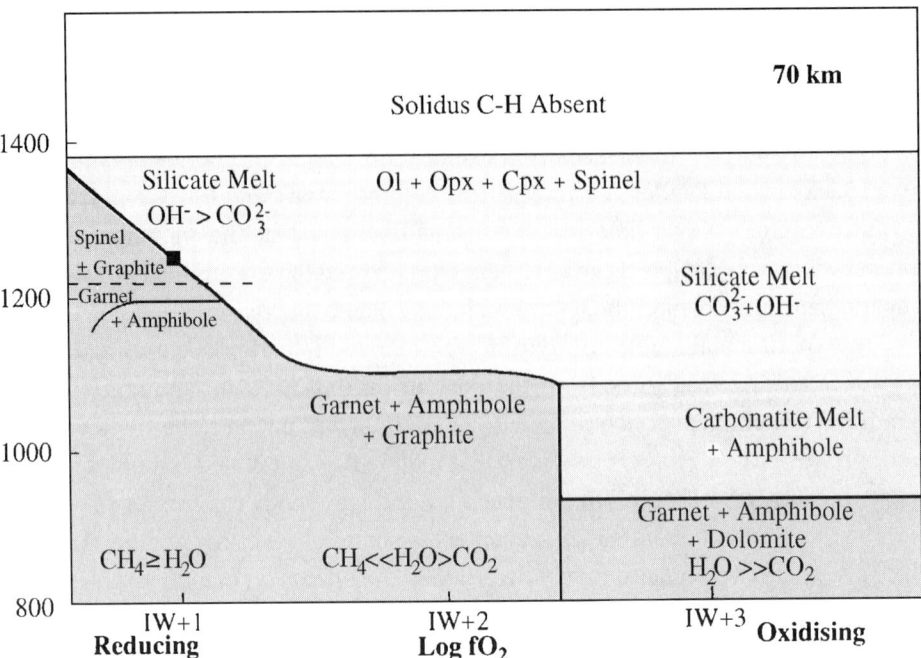

Figure 7.7a. The two parts of Figure 7.7 summarize the dramatic effects on solidus temperature and on solidus mineralogy that can be produced by small quantities of carbon and hydrogen and by variations of oxygen fugacity over 3 or 4 log units. Melting relationships are summarized at 70 km, where pargasitic amphibole plays a very significant role and where there is a field for sodic dolomitic carbonatite melt between approximately 935°C and 1,075°C. Garnet is a residual phase for melts (silicate or carbonate melts) below about 1,220°C.

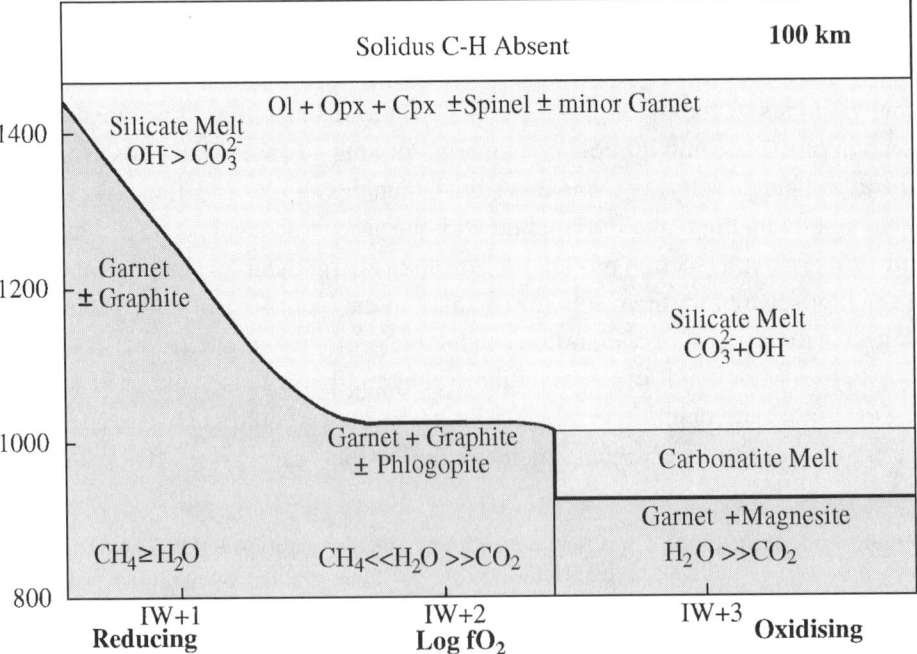

Figure 7.7b. At 100 km, pargasite is not stable, and garnet is a minor solidus phase for (C + H)-free melting, but garnet becomes more significant if the solidus is depressed by the presence of (C + H + O) fluid or by the carbon in carbonate- and H_2O-rich fluid.

because of dissolved $(CO_3)^{2-}$ and $(OH)^-$ and Fe^{2+}/Fe^{3+} variation. The concept of the 'petrological lithosphere' potentially correlates with rheological definitions of the lithosphere (Anderson, 1995). The thermal boundary layer (in the narrower sense described previously, i.e., an unstable feature in a convecting upper mantle) in the pyrolite model is the upper part of the petrological asthenosphere. It is characterized by the presence of an incipient-melt fraction and the potential for convective instabilities or overturn.

The model presented here also differs from most conventional models of magma genesis by emphasizing two very different and distinctive melting regimes. The melting regime for pyrolite $+ (C + H + O)$ at temperatures higher than the position of the $(C + H)$-free solidus is one in which the melt proportion increases rapidly with increase in temperature – *the major melting regime*. In the major melting regime, melt compositions vary in major elements and normative character along cotectics between olivine + orthopyroxene $\pm$ clinopyroxene $\pm$ chrome-spinel (Green, 1971; Jaques and Green, 1980; Falloon and Green, 1987, 1988; Falloon et al., 1988). Residual plagioclase, aluminous spinel, or garnet will have little significance in controlling either major elements or trace elements over the depth range 25–100 km. This contrasts with the melting regime between the pyrolite $+ (C + H + O)$ solidus for appropriate fO_2 conditions and the $(C + H)$-free solidus – the *incipient-melting regime*. If mantle compositions vary from enriched (HPY), through normal mantle (MPY), to depleted lherzolite (TQ), then the relevant 'solidi' for major melting are as shown in Figures 7.1, 7.6, and 7.8. The relevant solidi for the incipient-melting regime are shown in Figure 7.4 (i.e., the pargasite-lherzolite dehydration solidi).

Thus the application of the pyrolite experimental data to Earth models depends critically on assumptions about both fO_2 and the presence and abundance of $(C + H)$. This is illustrated in Figure 7.7, where two 'slices' at depths of 70 km and 100 km show plots of temperature against fO_2 and illustrate the presence or absence of melt or fluid, the sub-solidus mineralogy, the temperature of the solidus, and the nature of the melt formed at the solidus of a 'fertile' lherzolite. Attention is particularly drawn to the 70-km slice and the contrast between the oxidized regime with carbonatite and silicate melt fields at 935°C and 1,075°C, respectively, and the reduced (IW + 1) regime with absence of carbonatite, and melting only near the intersection of the amphibole dehydration reaction and the peridotite $+ (CH_4 + H_2O)$ solidus at about 1,200°C (Figures 7.5 and 7.7).

In the incipient-melting regime the melt proportion is controlled principally by the $(C + H)$ content of the source and melt compositions reflect the dissolved $(CO_3)^{2-}$ and $(OH)^-$. Contours of melting (Green, 1971; Green and Liebermann, 1976) in P–T space show very small (1–2% or less) melt proportions over large temperature and pressure ranges. Melt compositions can vary greatly in incompatible-element contents (e.g., by a factor of 10 over the 0.2% to 2% melt fraction), but their normative major-element characteristics remain very undersaturated (olivine nephelinites, olivine-rich basanites, olivine leucitites, olivine melilitites). The

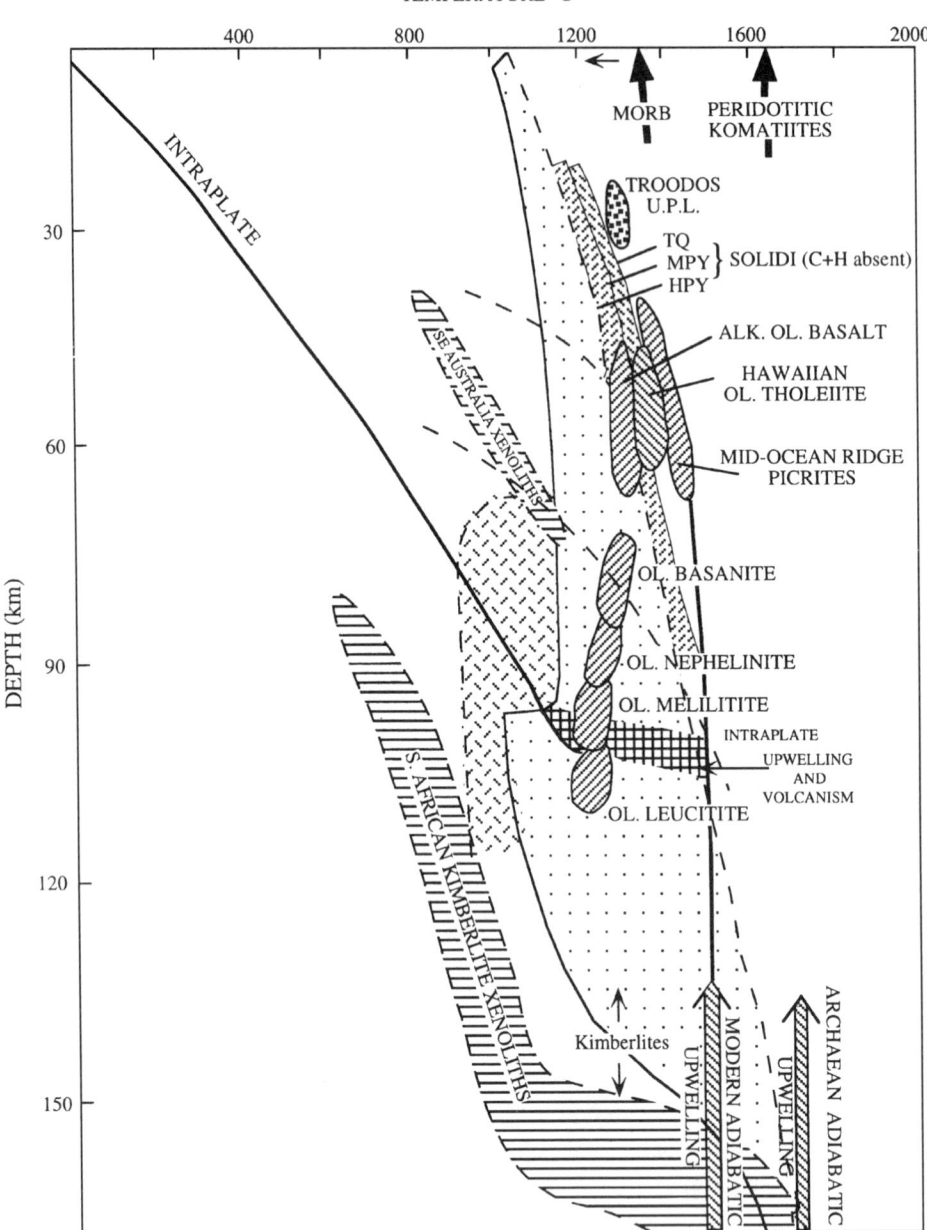

Figure 7.8. As a test of the model, the mantle model of Figure 7.6 is overlain with information derived from natural magmas regarding their conditions of origin. Intraplate magmas
containing mantle xenoliths range from alkali olivine basalts, through olivine basanites,
etc., to olivine leucitites (olivine lamproites) and kimberlites. Experimental studies cited
in the text have defined the roles of $(CO_3)^{2-}$, and $(OH)^-$ and the P–T conditions at which
these natural melts have olivine, orthopyroxene, and clinopyroxene ($\pm$garnet) as liquidus
phases. Note that all lie in the 'incipient-melting regime' and not in the 'major melting
regime'. By contrast, alkali olivine basalts represent magma segregation conditions entering
the major melting regime for Hawaiian pyrolite. Hawaiian olivine tholeiite and tholeiitic
picrite lie within the major melting regime, as do mid-ocean-ridge picrites. As noted in
the text, mid-ocean-ridge picrites have higher temperatures of magma segregation than

residual phases are olivine, orthopyroxene, and clinopyroxene, and garnet is a major residual phase throughout much of the incipient-melting regime over a large P–T field. Other minor phases (ilmenite, phlogopite, pargasite) may also be present for particular incipient-melting conditions. At the lower temperatures and high-pressures within the incipient-melting regime, both liquids and residual clinopy-roxene may acquire high Ca/Mg and Na/Ca ratios not achievable (except possibly at $P > 5$ GPa) in the major melting regime. These characteristics are reflected in the high Ca/Mg and Na/Ca ratios for melts from this P–T field. Melts also have $CaO/Al_2O_3 > 1$, and residues from extraction of incipient melts correspondingly begin to diverge from chondritic CaO/Al_2O_3 ratios to lower values.

If melt migration by flow through peridotite is very rapid even at small degrees of melting in a lithostatic stress field, then the consequences of the pyrolite melting behaviour summarized in Figure 7.6 will be melt migration upwards along the geotherm (the oceanic intraplate geotherm) and freezing (with crystallization of pargasite) at a depth of 95 ± 5 km. This zone of maximum pargasite precipita-tion in the base of the lithosphere may seal the lithosphere at 95 km to further reaction, leading to ponding of the melt fraction beneath the amphibole-rich lid. Ponded magma may be tapped by fracture propagation, leading to eruption of small-volume, highly undersaturated and explosive $(C + H)$-rich magmas in in-traplate settings (Figure 7.8). Interstitial melt may also be entrained by 'diapirs' or 'plumes' traversing the 150–90-km depth interval (Figure 7.9).

7.5.1.2. The Nature of Mantle Upwelling and Melt Retention

At present, it is possible to interpret magma genesis in terms of four end-member models due to two different upwelling and melt-retention models (Shen and Forsyth, 1995). Mantle upwelling can be modelled as either passive or dynamic (Figure 7.9), and mantle melting may involve either low melt retention (referred to as fractional melting) or high melt retention (referred to as batch melting). In passive-flow models (e.g., McKenzie and Bickle, 1988), mantle upwelling is regarded as a consequence

Figure 7.8 (*cont.*). hotspot or intraplate magmas. This figure illustrates the basis for the view that the modern mantle adiabat has a potential temperature of about 1,450°C and that all modern volcanism, whether hotspot, back-arc, or intraplate rifting, can be understood within the temperature envelope between the appropriate geotherm for the thermal boundary layer and the mantle adiabat ($T_p \approx 1,450$°C). Attention is also drawn to the inference that Archaean peridotitic komatiites imply mantle potential temperatures of at least 1,650°C. Also illustrated is the P–T field inferred for the spinel-lherzolite, pyroxenite, and garnet-pyroxenite xenoliths in the Newer Volcanics of southeastern Australia. This is considered to represent a perturbed-lithosphere geotherm during intraplate, rift-related volcanism and mantle metasomatism in southeastern Australia (O'Reilly and Griffin, 1985). The field for South African kimberlite xenoliths (Finnerty and Boyd, 1987) has two components: the field enclosing the 40-W/m² geotherm (Figure 7.6) and the high-temperature 'kink' of enigmatic origin. Here, that region and a similar kink at the lower end of the intraplate geotherm are represented as transient phenomena associated with adiabatic upwelling and magmatism linked with lithospheric plate movement and thinning.

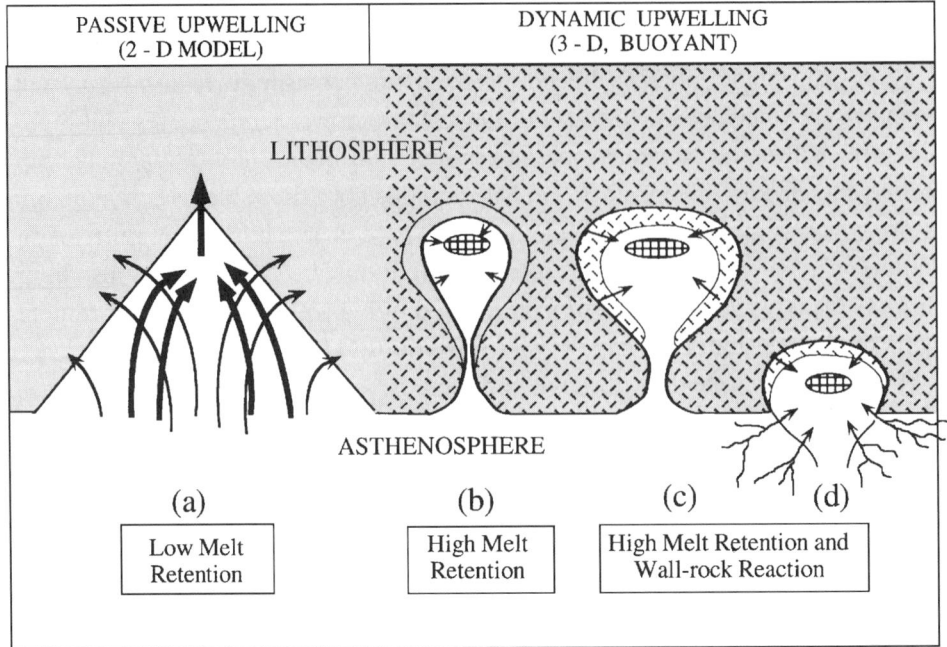

Figure 7.9. Cartoons representing differing concepts of separation of melt from mantle residue. (a) Passive upwelling beneath linear mid-ocean ridges (McKenzie and Bickle, 1988): The lithosphere is depicted as newly formed from upwelling, so that the uppermost non-crustal part of the lithosphere is most refractory, and the deepest part approaches the original asthenosphere, with little or no melt extraction. Melt extraction is deemed to be by continuous flow through porous residual peridotite, with upward flow of the melt exceeding that of the crystals. Melt focusing and pooling precede melt segregation, and the compositional characteristics of the melt are determined by the shape and residual mineralogy of the melting volume (e.g., Eggins, 1992b). (b) Dynamic melting with high melt retention (McKenzie and Bickle, 1988): The asthenosphere is depicted as responding rapidly to temperature perturbation or melt-fraction variation (e.g., by volatile flux as in Figure 7.10) by episodic rises of buoyant plumes, thermals (Griffiths and Turner, Chapter 4, this volume, Figure 4.1), or diapirs. This sketch represents an asthenospheric diapir penetrating the lithosphere, cooling and partially crystallizing against its wall rocks. Because of the concentration of deformation around diapiric margins and the existence of temperature gradients, the melt fraction becomes enriched in the diapir core until a high melt fraction and roof fracturing result in loss of the magma. The process approaches 'closed-system' batch melting, but the original asthenospheric source may become enriched in incompatible elements by fractional crystallization against the wall rocks. (c) Dynamic melting, high melt retention, and wall-rock reaction or entrainment: The model is as for part b, but the dynamics of diapirism permit wall rocks (older lithosphere) to be raised above their solidi [peridotite-$(C + H + O)$, $fO_2 \sim IW + 3$] and entrained into the buoyant diapir (Griffiths and Turner, Chapter 4, this volume). The thermal and deviatoric stress gradients drive small melt fractions to migrate from wall rock to diapiric core. The process is an open system, one in which the original asthenospheric-source geochemical signatures may be overprinted by melts highly enriched in incompatible elements, extracted from older lithosphere. (d) Dynamic melting, high melt retention, wall-rock reaction, and asthenospheric mixing: The model is as for part c, but a diapir sourced in the deeper asthenosphere may also be enriched by the incipient-melt fraction already present in the upper part of the asthenosphere (cf. Figures 7.6–7.8).

of separation of lithospheric plates. The melting regime under mid-ocean ridges can be modelled in two dimensions as a triangular shape (Figure 7.9) (Langmuir, Klein, and Plank, 1992). The composition and total volume of accumulated melt produced are independent of the form of mantle flow, being dependent only on the initial and final depths of melting (Eggins, 1992; Shen, and Forsyth, 1995). Consequently, crustal thickness is a function of the depth of initial melting (McKenzie and Bickle, 1988). In dynamic-flow models (e.g., Buck and Su, 1989) there is positive buoyancy associated with higher-temperature, depleted residual mantle or melt retention within the upwelling mantle matrix. In most conventional models of magma genesis, especially those related to MORBs, both passive upwelling and dynamic upwelling are modelled in two dimensions only (Shen, and Forsyth, 1995; Salters, 1996). However, if mantle upwelling is modelled in three dimensions, both fluid dynamical experiments (Whitehead, Dick, and Schouten, 1984) and theoretical modelling (Phipps Morgan and Forsyth, 1988; Buck and Su, 1989; Forsyth, 1992) indicate that buoyancy will cause the mantle melting regime to be more plume-like in shape. The term 'diapir' (Forsyth, 1992) is an appropriate term to describe these small (50-150-km-diameter), discrete, three-dimensional melting zones (Figure 7.9), to make a distinction from plumes, which are generally regarded as large in scale ($\gg$150 km in diameter). The concept of three-dimensional dynamic flow under mid-ocean ridges is supported by theoretical modelling (Cordery and Phipps Morgan, 1994), along-axis variations in gravity anomalies (Forsyth, 1992; Lin and Phipps Morgan, 1992), and the geochemistry of MORBs (Niu and Batiza, 1993, 1994). In dynamic-flow models, the thickness of the crust depends on the degree of melting and the volume of mantle convected through the melting region in discrete plumes, thermals, or diapirs.

In melting models with low melt retention, melting is assumed to be close to fractional, that is, melt is instantaneously separated from the mantle matrix as each increment of melt is generated (McKenzie and Bickle, 1988). The melts are then assumed to be aggregated in magma chambers or conduits before eruption. In melting models with high melt retention, the melting is assumed to occur within thermally and mechanically discrete, partially molten thermals or diapirs within a predominantly melt-free sub-axial mantle (Figure 7.9).

The evidence that high melt retention (i.e., a batch-melting process) is the more appropriate process for modelling magma genesis includes the following:

1. *Melt distribution and extraction in partially molten peridotite*. Earlier geophysical models of melt extraction relied extensively on observations from simple olivine melt systems and particularly on theoretical models of melt distribution and extraction. These types of studies suggested that melt fractions much less than 0.1% could be effectively extracted from an olivine-rich matrix (McKenzie, 1984). However, this approach is too simplistic, inasmuch as the models assumed uniform grain sizes, isotropic olivine surface energies, and the occurrence of melt

in a regular array of melt-filled tubules along three grain edges, neglecting the effects of other phases (especially pyroxenes) and assuming steady-state conditions. Faul (1997) has worked with an olivine matrix and equilibrium melt compositions and considered a range in grain sizes and anisotropic olivine surface energies. His observations of experimentally equilibrated melt distributions suggest that melt cannot be extracted over reasonable timescales for less than about 3% melt fraction (Faul, 1997). For a four-phase lherzolite, the permeability threshold may possibly be greater than 9% (Toramaru and Fuji, 1986). The existence of a significant permeability threshold in fertile mantle peridotite will enhance the buoyancy of mantle undergoing partial melting. Plumes, thermals, or diapirs within the incipient-melting regime can lead to the development of three-dimensional upwelling structures in which significant melt retention can occur (Figure 7.9) and in which relative buoyancy increases with decreasing depth during upwelling, to some limiting depth of magma escape (depth of magma segregation).

2. *The major-element chemistry of melting residues.* Because fractional melting and batch melting are end-member melting processes, the residues of partial melting should sensitively record the melting process (Johnson, Dick, and Quick, 1990). In Figure 7.10, the major-element compositions of abyssal peridotites (Dick, 1989) and of a spinel-lherzolite xenolith suite (Nickel, and Green, 1984) are compared to experimentally determined lherzolite batch-melting residues. Figure 7.10 demonstrates that there is an overlap in compositions, consistent with the inference that these peridotite suites are batch-melting residues. However, data from the abyssal peridotite suite were used by Johnson et al. (1990) in support of a model of fractional melting. Johnson et al. (1990) argued that the Ti and Zr contents in abyssal clinopyroxene could be successfully modelled by fractional melting, but not by batch melting. This conclusion is dependent on the assumptions made in the model. For example, Johnson et al. (1990) used values for the partition coefficients for Ti and Zr that were independent of clinopyroxene composition, whereas in reality the partition coefficients for Ti and Zr are sensitive to the composition of the clinopyroxene, which changes systematically during progressive partial melting, whether batch or fractional. In Figure 7.11 we compare the Ti contents of residual clinopyroxene from the batch-melting experiments of Baker and Stolper (1994) with the clinopyroxene data of Johnson et al. (1990). The experimental data illustrate the change in partitioning of Ti as clinopyroxene becomes more refractory at higher temperatures. Figure 7.11 demonstrates that batch melting can produce the range of Ti contents observed in clinopyroxene from abyssal peridotites, and the spread of data among the natural pyroxenes is consistent with a spread of source compositions, such as MORB pyrolite to Hawaiian pyrolite. We consider that the relatively large changes in the major-element compositions of both abyssal peridotites and spinel lherzolites are indicative of batch-melting

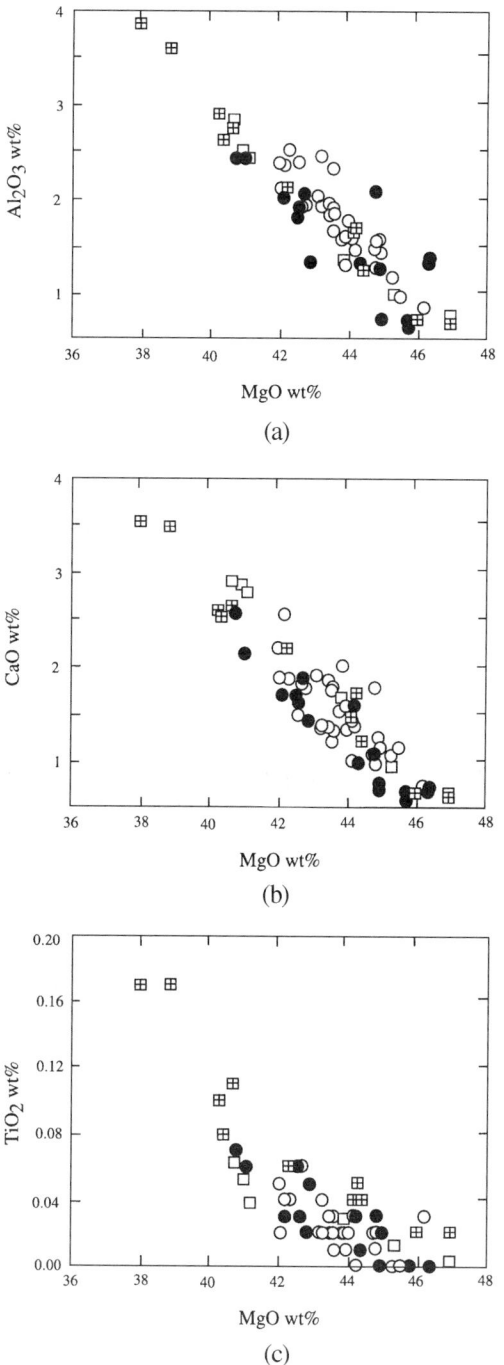

Figure 7.10. Plots of Al_2O_3, CaO, and TiO_2 against MgO for abyssal peridotites (Dick, 1989) and spinel-lherzolite xenoliths in basalt (Nickel and Green, 1984, types A1, A2, A4, C), compared with residues from batch melting (i.e., high melt retention) obtained by experimental methods: open circle, abyssal peridotites; cross in square, spinel lherzolite; filled circle, residues from MORB pyrolite (Falloon and Green, 1987, 1988); open square, residues from MM3 lherzolite (Baker and Stolper, 1994).

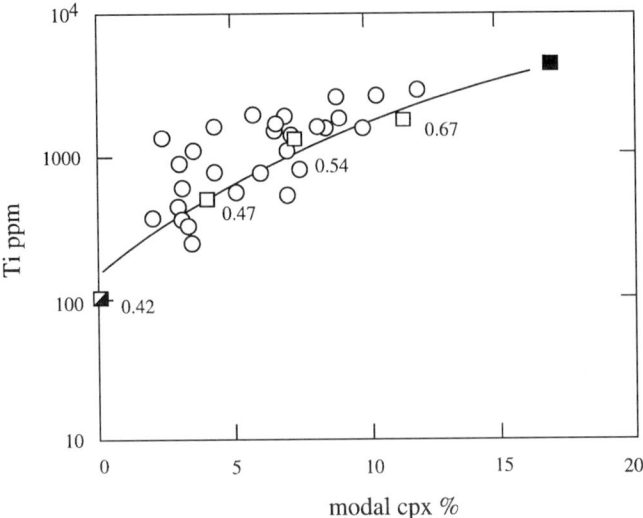

Figure 7.11. Plot of Ti content of clinopyroxene (cpx) against its modal abundance in 1-GPa batch-melting experiments of Baker and Stolper (1994). The numbers beside the data points indicated by square symbols are the TiO_2 contents (wt%) in coexisting liquids, illustrating the change in partition coefficients (Ti_{liq}/Ti_{cpx}) with changing composition of clinopyroxene (Baker and Stolper, 1994). These experimental data are compared with data for clinopyroxenes in abyssal peridotites (Johnson et al., 1990), shown as open circles. It is argued that the latter are consistent with residues from batch melting of lherzolite, particularly if the parental lherzolite had small variations in TiO_2 content. Open circle, data of Johnson et al. (1990); filled square, sub-solidus clinopyroxene; open square, clinopyroxene in lherzolite residue; half-filled square, orthopyroxene in harzburgite residue.

processes and indeed require extraction of large melt fractions (>15%) to generate the observed range. The major-element data do not support models invoking low melt retention (i.e., fractional melting). Trace-element data, particularly for incompatiable trace elements, are too readily disturbed by metasomatic events and by the mineralogical changes consequent on cooling and exsolution from magmatic temperatures to provide definitive tests for the melting and melt-extraction processes in residual mantle (Frey et al., 1978; O'Reilly and Griffin, 1988; Yaxley et al., 1991; Rudnick et al., 1993).

7.5.1.3. Consequences of High Melt Retention and Dynamic Mantle Upwelling

As a consequence of the existence of a threshold porosity for melt extraction, chemical contrasts between magmas can arise during dynamic upwelling. A 'garnet-controlled signature' could arise in a particular MORB or hotspot upwelling if a higher melt fraction and higher permeability allow melt extraction while the upwelling mantle is in the garnet-lherzolite field. With a lower volatile content, or lower temperature, another mantle upwelling with a lower melt fraction (<2%) may retain the melt while in the garnet-lherzolite stability field. Melt extraction may then occur at lower pressure, where garnet is unstable, and the upwelling approaches the major melting regime. Differences in melt fractions, and thus differences in the presence or absence of relative flow between melt and residue, can arise because of small variations in volatile content and do not necessarily require temperature differences. The important point is that the existence of a melt-fraction threshold for effective permeability and melt migration in a lithostatic melt column or upwelling can produce contrasting trace-element signatures from a single source without requiring large temperature anomalies or different geometries of the partial-melting volume. Dynamic upwelling with high melt retention, applied to the phase fields of Figure 7.7, would lead to increasing buoyancy and increasing temperature and viscosity contrasts between a diapir (dynamic upwelling) and the wall rock. There is thus competition among three processes (Figure 7.9):

1. Cooling of a diapir against the wall rock, with crystallization within the diapir and migration of smaller melt fractions from the outer cooling and deforming shell to the inner lithostatic core. The diapir is a closed system, but the core composition becomes enriched by migration of small melt fractions from the crystallizing margins. [For perfectly incompatible elements, concentration within the diapir will double, for a 20% reduction in diapir (sphere) radius.]
2. Heating of wall rock through its solidus, particularly through its dehydration solidus, and migration of melt from deforming and recrystallizing wall rocks into the diapir's core. The diapir is an open system and may sample prior enrichments and isotopic fingerprints, particularly in the lithosphere.
3. Within the asthenosphere, a diapir sourced from deeper levels may capture an incipient-melt fraction from deforming and recrystallizing wall rocks in the upper part of the asthenosphere.

The wall-rock reaction processes in the mantle were emphasized by Green and Ringwood (1967a) as an important control on the incompatible-element signatures of mantle-derived magmas. Currently, wall-rock reaction processes are being given renewed emphasis, particularly in models involving the flow of small-degree melt fractions (Navon and Stolper, 1987; Kelemen, Dick, and Quick, 1992; Nielson and Wilshire, 1993; Iwamori, 1994). In the general petrogenetic model outlined in this section, significant fractionation between incompatible elements can occur at the margins of upwelling mantle diapirs as they pass through the incipient-melting regime and lithosphere. The pressure and temperature gradients surrounding the diapir margins provide the driving force for the movement of very small melt fractions into the hot interior and the zone of melt accumulation. Although further discussion is beyond the scope of this chapter, many of the distinctive geochemical signatures of mantle-derived magmas can be explained by this mechanism.

7.5.1.4. Pyrolite Model and Mantle Heterogeneity

On the basis of mantle samples, isotope systematics in magmas and xenoliths, seismology, and our current understanding of geodynamics, we must conclude that the upper mantle is chemically and mineralogically, and possibly thermally, heterogeneous. To a first approximation (i.e., in terms of major elements and dominant mineralogy), the three compositions we have been emphasizing, Hawaiian Pyrolite, MORB pyrolite, and Tinaquillo lherzolite, are very similar – all are lherzolites with about 60% olivine and 3–4% Al_2O_3 and CaO, and the olivine is of Fo_{89-90} composition. This is taken as the mean upper-mantle composition, or 'model mantle composition', noting that second-order chemical and mineralogical heterogeneities will be required to understand magma genesis and isotopic and geochemical fingerprints. The second-order heterogeneities should have a rational explanation within the constraints of the overall model composition. Significant and large-scale compositional deviations from the mean upper-mantle composition can and do arise in the process of oceanic-crust formation and because of convergent-margin magmatic and metamorphic processes leading to island-arc and crustal accretions. These changes are in the direction of increasingly refractory (in major-element composition and mineralogy) residual lherzolite, harzburgite, and dunite, characterized by decreases in CaO and Al_2O_3, increases in the Mg# for olivine (up to Fo_{94}) and orthopyroxene, and increases in the $Cr/(Cr + Al)$ ratios for residual spinel and in $Ca/(Ca + Na)$ for any minor clinopyroxene or plagioclase. These larger-scale changes to the mean upper-mantle composition occur at relatively shallow depths in the modern Earth (i.e., are associated with melt extraction) and create a compositional contrast between the lithosphere and the mean upper-mantle composition.

The formation of the lithosphere as a layer both thermally and compositionally different from the underlying upper mantle produces two competing effects. Cooling of the oceanic lithosphere creates thermal and gravitational instabilities, leading to sinking (subduction) of lithospheric plates into the upper mantle. However, if lithospheric slabs reach temperatures similar to those of the surrounding 'mean upper mantle', then they become relatively buoyant. The density contrast and buoyancy increase with increasingly refractory character; that is, lithosphere with about 2% Al_2O_3 and CaO, and olivine of Fo_{90-91} composition, is less likely to reverse its relative buoyancy and become stably plated to the sub-continental lithosphere than is very refractory harzburgite with less than 0.5% Al_2O_3 and CaO and olivine of Fo_{93} composition. The formation of gravitationally stable and compositionally layered lithosphere thus reflects the complex interplay of temperature, composition, and large-scale compositional heterogeneities.

In addition to broad-scale heterogeneity, small-scale heterogeneity is clearly a product of the processes of magma extraction, evolution, and transport through the lithosphere. Further heterogeneity is introduced in convergent margins through serpentinization and other chemical alterations of crust and upper mantle, followed by subduction, with some proportions of these very diverse lithologies being carried into the mantle. Considered in detail, models of the upper mantle must allow for such diverse lithologies being carried into and recycled within the upper mantle.

7.5.2. Magma Genesis in the Pyrolite Model
7.5.2.1. MORB Magmatism

Conventional models (McKenzie and Bickle, 1988; Langmuir et al., 1992; Shen and Forsyth, 1995; Salters, 1996) of MORB petrogenesis assume simple two-dimensional upwelling (passive or dynamic) and very low melt retention in the melting regime (i.e., fractional melting), such that parental/primary MORB magmas are aggregates of very small melt fractions (<1 wt%) from a range of depths. A widely quoted view is that 'normal' MORB mantle (i.e., away from the influence of mantle plumes) has a potential temperature (T_p) of 1,280°C and that primary aggregated MORB has about 10 wt% MgO, producing a 7-km thickness of oceanic crust (McKenzie and Bickle, 1988). The petrological 'data' used to interpret MORB petrogenesis are in some models restricted to a small part of n-dimensional compositional space. Selected pairs of elements are modelled (e.g., Na/Fe, Lu/Hf, etc.), and from these parameters global patterns of MORB geochemistry are presented and interpreted in terms of factors such as the initial and final depths of melting, ridge depth, and amount of garnet involved in the melting process (Klein and Langmuir, 1987; Shen and Forsyth, 1995; Salters, 1996).

In the pyrolite concept, the petrological 'data' used to interpret MORB petrogenesis are the major-element phase equilibria for model pyrolite compositions and for the natural MORB magmas themselves. A useful tool is the presentation of the chemical- and phase-equilibrium data in terms of the molecular norm, allowing the relationships of liquids and residues to be plotted within the 'basalt tetrahedron' (jadeite, calcium Tschermak's molecule, quartz, olivine, and diopside) (Falloon and Green, 1987). Experimental studies of primitive MORBs and of MORB pyrolite (Green et al., 1979; Jaques and Green, 1980; Falloon and Green, 1987, 1988) conducted at The Australian National University and the University of Tasmania have established the following:

1. The most magnesian MORB glasses (9.5–10.5 wt% MgO) do not have appropriate chemical compositions (express as the CIPW molecular norm) to be in thermodynamic equilibrium with an appropriate upper-mantle peridotite residue. This conclusion is supported by the work of other researchers (Albarède, 1992; Hess, 1992; Baker and Stolper, 1994; Kushiro, 1996).
2. Even the most primitive of MORB glasses have undergone a prior history of olivine-only fractionation from parental/primary magmas richer in MgO (>10.5

wt%). Such magmas are classed as tholeiitic picrites, and arguments for picritic magmas as parental to MORBs have been presented by O'Hara (1968a,b), Green et al. (1979), Elthon and Scarfe (1984), Stolper (1980), and Falloon and Green (1987, 1988).

The experimental studies of Green et al. (1979) and Falloon and Green (1988) suggest that at least some primary N-MORB magmas must have at least 16 wt% MgO or more and represent upwelling-mantle potential temperatures (T_p) at or above 1,430°C, significantly higher than the 1,280°C suggested by McKenzie and Bickle (1988). Such high mantle potential temperatures for MORB petrogenesis are fully supported by the mineralogy and geochemistry of primitive MORB glasses. We cite two well-constrained examples:

1. Primitive parental glasses from ODP leg 148, site 896A, near the Costa Rica Rift, have the typical trace-element characteristics of depleted N-MORB. They also have MgO contents ranging up to 9.5 wt% (McNeill and Danyushevsky, 1996; A. W. McNeill, unpublished data), and are associated with 'normal' crustal thickness. The site 896A glasses, however, contain olivine phenocrysts ranging from Fo_{80} to $Fo_{91.6}$, which places an unequivocal thermodynamic constraint on the composition of more primitive MgO-rich parental magmas via the well-established relationship of Fe-Mg partitioning between olivine and coexisting basaltic melt (Roedder and Emsile, 1970; Ford et al., 1983; Langmuir et al., 1992). McNeill and Danyushevsky (1996), using the most primitive glass and the composition of the most forsterite-rich olivine, have calculated a minimum MgO content for the parental 896A picrite of 15.1 wt%. The composition of this parental picrite has been experimentally demonstrated to be in equilibrium with an upper-mantle lherzolite residue at 2 GPa and 1,450°C (T. J. Falloon, unpublished data).

2. Perfit et al. (1996) recently report the compositions for N-MORB glasses very rich in MgO (up to 10.5 wt%) from picrites recovered from the Siqueiros transform-fault zone at approximately 8°20′ N on the East Pacific Rise. As in the case of the 896A glasses, there is a range of compositions of the olivine phenocrysts in the Siqueiros glasses ($Fo_{89.5-91.5}$). On the basis of the most magnesian glass composition from the Siqueiros transform-fault zone (glass D20-20) (Perfit et al., 1996), the parental/primary magma must have a MgO content of 13 wt% to be able to crystallize an olivine of $Fo_{91.5}$ composition. The composition of this calculated parental melt is appropriate to be in equilibrium with a lherzolite residue at about 1.6–1.7 GPa at a temperature of about 1400°C, based on the experimental data of Falloon and Green (1988).

These two examples provide further evidence that picritic magmas (cf. Falloon and Green, 1988) are parental to N-MORB and that the conditions of melt segregation require mantle potential temperatures for MORB magmatism of T_p ~1,380–1,430°C (Figures 7.1 and 7.8).

The major-element compositions, the liquidus olivine compositions, and the high-pressure phase relations of the most primitive MORB provide evidence for segregation of primary magmas from lherzolite-to-harzburgite residues over the pressure range 0.8–2.0 GPa, with most deriving from deeper levels (1.7–2.0 GPa) (Falloon and Green, 1987, 1988). There is also evidence for ultra-depleted (in terms of incompatible elements) melts within the mid-ocean-ridge environment that crystallize anorthite-rich (>An_{90}) plagioclase, magnesian clinopyroxene (>Mg_{90}), and chrome-rich spinel as near-liquidus phases at low-pressure. These have been interpreted as 'second-stage melts' (Duncan and Green, 1980), formed when residual lherzolite from picritic-tholeiite melt extraction at about 2 GPa continues upwelling to yield a second melt component at shallower pressures. Sobolev and Shimizu (1993) identified such an ultra-depleted melt in glass inclusions in olivine ($Fo_{89.9}$). The major-element composition of this entrapped melt is a very close match to a 15-wt% batch melt from Tinaquillo lherzolite at 0.5 GPa and 1,250°C. This interpretation requires the source 'pyrolite' to have been an already extremely depleted composition in terms of incompatible elements. Modelling of REE fractionation (Sobolev and Shimizu, 1993) led to the interpretation of this composition as the last small melt fraction (1–2%) produced in a column-melting scenario, produced from very refractory olivine and pyroxenes and entrapped in olivine before mixing with the aggregated melt of the melt column (N-MORB). The model of Duncan and Green (1980, 1987) suggesting mixing of low-pressure second-stage melt with more voluminous N-MORB with picritic parental magmas may also explain the observations.

Studies of entrapped melt inclusions in phenocryst phases and the search for the most magnesian olivine and pyroxene phenocrysts, the most calcic plagioclases, and the earliest-formed spinel are providing a new approach to the issues of primary melt compositions, magma homogeneity, source pyrolite compositions, and magma segregation and mixing (Donaldson and Brown, 1977; Dmitriev et al., 1985; Danyushevsky, Sobolev, and Dmitriev, 1987; Sobolev et al., 1989; Sobolev and Dmitriev, 1989; Sinton et al., 1993; Kamenetsky, 1996).

Remarkable variations of incompatible-element relative abundances have been observed within melt inclusions in olivines from a single magma body (Kamenetsky and Sobolev, 1995; Shimizu, 1995; Nikogosian and Sobolev, 1996; Tsamerian and Sobolev, 1996; Kamenetsky, 1996), but such variations in incompatible elements are *not* matched by variations in major-element compositions. For example, it might be expected that the major-element composition of a melt showing strong LREE enrichment and relative HREE depletion (i.e., a garnet or clinopyroxene + garnet signature) accompanied by high K_2O content should be appropriate for a melt in equilibrium with residual garnet lherzolite (i.e., alkali picrite), indicating pressures of 3 GPa or higher. Major-element, compositions, however, appear to remain characteristic of much lower pressures, with harzburgitic or lherzolitic residues. Rather than being indicative of mixing of increments from a depth profile in column melting the data may be more consistent with two-component mixing, in which the major component has depleted (N-MORB) incompatible-element contents and a tholeiite-picrite-to-olivine-tholeiite composition. The minor component may be an incipient melt from the wall-rock environment entrained within or metamorphosed by the emplacement of the N-MORB pyrolite diapir.

Heterogeneity in MORB geochemistry (enriched or E-MORB, normal or N-MORB, and ultra-depleted MORB) may also be a consequence of preconditioning of the source by melt migration within the asthenosphere, in contrast to dynamic upwelling. Mantle heterogeneity may develop by melt migration and incompatible-element enrichment in the uppermost part of the asthenosphere or lowermost lithosphere. Plate tectonics requires relative movement between the lithospheric base and the underlying mantle, as represented by the hotspot frame of reference. This relative motion in the asthenosphere may induce melt migration by porous flow, if such flow is enhanced by deviatoric stress and deformation/recrystallization. Thus the model of Figures 7.6–7.8 leads to the expectation that the asthenosphere is a layer in which mantle chemical inhomogeneity is generated – the lower part acquires a residual signature in incompatible-element abundances and, in particular, preferentially retains those elements partitioned into garnet and clinopyroxene at $P > 3$ GPa. The upper part of the asthenosphere and base of the lithosphere acquire the complementary trace-element signature of incompatible-element enrichment. The timescales for this enrichment process may be either short (i.e., re-establishment of an enriched upper asthenosphere in new crust and lithosphere formation at spreading centres) or long (asthenosphere and base of lithosphere beneath continents).

It is emphasized that only by considering the incipient-melting regime and combining the low-degree, volatile-enriched melts of that region with the high-degree melts of the major melting regime can we see mantle melting as producing the spectrum from E-MORB through N-MORB to ultra-depleted MORB, with a garnet signature in both enrichment processes and depletion processes. This statement applies to a modern mantle in which the mantle potential temperature is 1,450°C or less. If models of *anhydrous melting* of the mantle are advocated, and a garnet signature during melt extraction is invoked either to explain U/Th disequilibrium in isotopic ratios or to explain Lu/Hf, U/Th, or REE fractionation (e.g., Salters, 1996), then this must occur with mantle potential temperatures greater than 1,470°C. This requirement for very high temperature is set by the need for the mantle adiabat to intersect the anhydrous solidus for MORB pyrolite (Figure 7.1) at $P > 2.8$ GPa if garnet is to play a part in controlling melt compositions.

7.5.2.2. Intraplate Magmatism

In Figure 7.8, a model for the intraplate setting illustrates an application of the experimental data and the additional information or assumptions required. Firstly, the upper mantle beneath the asthenosphere is assumed to be at low oxygen fugacity: $fO_2 \sim IW + 1$. The mantle adiabat is assumed to have a potential temperature of about 1,430°C, on the basis of the liquidus temperature of mid-ocean-ridge picrites (~1,330–1,350°C) and the conditions of magma segregation inferred for primitive MORB (2 GPa, 1,430–1,450°C) (Green et al., 1979; Falloon and Green, 1988). Secondly, the upper mantle is considered to have small carbon and hydrogen contents, but greater amounts than can be accommodated by solid solution in nominally anhydrous minerals such as olivine and pyroxenes. Thirdly, the oxidation state of the mantle is considered to decrease with increasing depth.

In Figure 7.8, the models of temperature distribution for intraplate settings are integrated with the experimentally determined melting relationships for lherzolite ('pyrolite'). Additional inferences about the general presence of minor carbon (as CO_2-fluid inclusions, graphite, or diamond in natural samples) and hydrogen (in hydrous minerals, primary magmas, and fluid inclusions) help to constrain the lithosphere-asthenosphere model. Attention is drawn to the above-solidus regions in Figure 7.8 and to two distinctive limits to melt ascent by porous flow within the asthenosphere or incipient-melting regime. The melt present above the silicate solidus at depths greater than 95 km (incipient-melting regime) along the 'oceanic geotherm' is olivine nephelinite to olivine melilitite, depending on $(CO_3)^{2-} : (OH)^-$ contents (Green, 1971; Frey et al., 1978). We do not yet know whether small silicate melt fractions (e.g., 1–3% melt) are 'trapped' or readily mobile because of their relative density (Faul, 1997). However, it is commonly inferred that deformation (e.g., between lithospheric plate and underlying asthenosphere) enhances melt migration. The model of Figure 7.8 predicts a region at a depth of 70–95 km in which metasomatism by reaction between 'incipient melts' and garnet or spinel lherzolite will produce variably titaniferous pargasite ± phlogopite in direct proportion to the amount of introduced melt. This reaction of $(CO_3)^{2-}$-bearing silicate melt in the lower lithosphere may leave carbonatite melt in equilibrium with garnet- or spinel-bearing pargasite lherzolite.

Within the lower part of the 'petrological lithosphere', Figures 7.6–7.8 illustrate the field for primary carbonatite liquid. Carbonatite is considered to be transitory; as noted earlier, it is formed if a $(CO_3)^{2-}$-bearing silicate melt from the asthenosphere reacts with the overlying lithosphere or if there is lithospheric thinning and asthenospheric upwelling without silicate melt extraction. Sodic dolomitic carbonatite is the residual melt as olivine nephelinite or olivine-rich basanite from the asthenosphere ($P \geq 3\,GPa, T = 1,100–1,200°C$) freezes to form pargasite ± phlogopite, provided that $fO_2 \geq IW + 2$ at $P \leq 3$ GPa. The current understanding of the high mobility of carbonatite by flow through peridotite suggests that it will move relatively rapidly. However, carbonatite also reacts rapidly with enstatite at $P \sim 2.0$ GPa, as the decarbonation reaction is encountered (Figure 7.5). The carbonatite melt is absorbed by reaction with lherzolite at depths around 60–70 km, producing a second and different zone of metasomatism in the lithosphere: sodic dolomitic carbonatite + enstatite + spinel → olivine + (diopside + jadeite) + chromite + CO_2. The metasomatized lithosphere is enriched in clinopyroxene, as illustrated by the formation of distinctive wehrlites (olivine + clinopyroxene + apatite + chromite) (Yaxley et al., 1991; Rudnick et al., 1993).

7.5.2.3. Plume Magmatism

The rationale for calculation of the Hawaiian-pyrolite composition was to produce a model source for Hawaiian magmatism, the largest and most active expression of plume or hotspot magmatism on the modern Earth. Experimental tests for self-consistency of the model are provided by the demonstration that primitive Hawaiian olivine tholeiite and tholeiitic picrite have olivine ($\geq Fo_{90}$) + orthopyroxene $\pm$ clinopyroxene as liquidus phases at some pressure and temperature conditions. The other side of this test is to show that the Hawaiian-pyrolite composition produces these same residual phases and an olivine-tholeiite or tholeiitic-picrite melt at those $P-T$ conditions. These self-consistency requirements are considered to be met to a first approximation by experimental studies (Green and Ringwood, 1967a; Green, 1971; Eggins, 1992a,b), as the ($H_2O + CO_2$) contents of the primitive tholeiite magmas lower liquidus temperatures by 50–80°C (assuming 0.5 wt% H_2O and similar CO_2 contents), but have only small effects on phase relations or phase compositions. The liquidus phase relationships for primitive Hawaiian olivine tholeiite or tholeiitic picrite indicate a harzburgitic residue (Green and Ringwood, 1967a; Falloon et al., 1988; Eggins, 1992a,b). Thus the distinctive minor-element and trace-element enrichments of Hawaiian olivine tholeiite (Mauna Loa and Kilauea Iki picrites in particular) reflect characteristics of their source 'pyrolite' (picrite melt + harzburgite residue) at the depth of magma segregation or melt pooling.

The Hawaiian-pyrolite composition was also shown to be a satisfactory source, in its major-, minor-, and trace-element inventory, for intraplate primitive magmas (i.e., those entraining mantle xenoliths and with Mg# of 70–75). These liquids range from olivine melilitite ($\sim$4% melt), through olivine nephelinite and olivine-rich basanite ($\sim$6% melt), to alkali olivine basalt ($\sim$12% melt) (Frey et al., 1978). The specific conditions of melt segregation/extraction for these magma types were determined by experimental studies on primitive melts, including demonstration of the importance of $(OH)^-$, $(CO_3)^{2-}$, and F^- (Green and Hibberson, 1970a; Green, 1971, 1973a; Brey and Green, 1975; Edgar, Green, and Hibberson, 1976; Frey et al., 1978; Foley, Taylor, and Green, 1986). The results of thse studies are summarized in Figure 7.8.

The success of the fixed-hotspot frame of reference in analyzing lithospheric plate movements over the past 100 million years has demonstrated that the cause for hotspot or island-chain volcanism lies in a region of higher viscosity below the lithosphere and asthenosphere. The concept of thermal plumes, originating either near the core-mantle boundary or at a thermal boundary layer within the transition zone, is widely accepted (Morgan, 1971; Davies, 1990; Griffiths and Campbell,

1990). However, important questions arise from petrological and geochemical observations.

The geochemistry of plume magmas strongly suggest the involvement of recycled sediment via subduction (White and Duncan, 1996). It is inferred in most plume models that the buoyancy of the plumes is thermal origin and that plumes are caused by instabilities in a thermal boundary layer, inferred to be the core–mantle boundary (Campbell, Chapter 6, this volume; Davies, Chapter 5, this volume). Temperature differences of 200°C or more between upwelling plumes and the mantle adiabat have been inferred (White and McKenzie, 1989 Campbell, Chapter 6, this volume). However, the available petrological evidence does not support such large temperature differences. We have shown that the primary MORB picrites require temperatures of 1,400–1,450°C at 1.6–2.0 GPa for their generation, and at present there is no petrological evidence that primitive Hawaiian olivine tholeiites or picrites are from magmas whose temperatures, either at the source or on eruption, were significantly higher than those of primitive MORB picrites (cf. Figure 7.8). On the contrary, the higher H_2O and CO_2 contents [which may be underestimates because of possible degassing of $(C + H + O)$ fluids] of primitive Hawaiian olivine tholeiite argue for depression of their melting temperatures by 50–80°C (at 0.5 wt% H_2O and 0.5 wt% CO_2) relative to those for equivalent $(C + H)$-free magmas. For example, a recent estimate for a parental picrite for Mauna Loa volcano has 18.5 wt% MgO in equilibrium with olivine of $Fo_{91.3}$ composition (Sobolev and Nikogosian, 1994; Garcia, Hulsebosch, and Rhodes, 1995). A comparison of this parental composition with experimentally produced partial-melt compositions from Hawaiian pyrolite (Falloon et al., 1988) suggests that it would be in equilibrium with a harzburgite residue at 2 GPa and 1,480°C. If the parental composition had had about 0.5 wt% H_2O and low CO_2 content, then that would have lowered the temperature of origin to about 1,430°C, which is no higher than the temperature of primary MORB picrite. As the liquidus depression due to volatiles is obviously significant, critical evidence for the volatile content of primitive magmas must be sought, including analysis of undegassed volatile contents of melt inclusions in primitive olivine and the search for evidence of $(C + H + O)$ fluids (and graphite) exsolved from and dissolved in glass. The postulated temperature differences of 200°C or more between 'plume' and normal mantle (MORB source) do not conform with the petrological observations. If MORBs require mantle potential temperatures of 1,380–1,430°C, then a postulated $\Delta T_p \sim 200°C$ would require Hawaiian plume temperatures of 1,580–1,630°C. Such temperatures are similar to conservative estimates for the Archaean mantle potential temperatures required for peridotitic-komatiite magmas with about 32% MgO (Green, 1975, 1981; Green et al., 1975). Mantle potential temperatures of 1,600°C or more would induce 70–80% melting of pyrolite, with dunite of Fo_{94-95} composition as residue (Figure 7.8). However, these characteristics are not those of Hawaiian magmatism.

Models other than those featuring deep-seated, thermally driven plumes for hotspot volcanism should be considered. The petrologic and geochemical requirements and constraints include the following:

1. higher volatile contents ($H_2O + CO_2$) in primitive magmas,
2. high degrees of melting (harzburgite residues) at lithospheric pressures (60–70 km) for magma segregation,
3. higher oxidation states than for MORB, at least at equivalent stages of olivine + spinel crystallization,
4. source 'pyrolite' enriched prior to magma segregation by processes occurring in the garnet-lherzolite stability field, and
5. evidence of 'older components', multi-event histories, and/or mixing of distinctive components in the magma source or in the processes of melt formation, extraction, and migration.

An alternative to the hypothesis of deep-seated, thermally driven plumes can be suggested. The subduction process has already been mentioned as producing lithospheric slabs ranging from cool depleted lherzolite to refractory harzburgite. The deep lithospheric keels of Archaean cratons also include highly depleted harzburgite (Boyd, 1989). If subducted slabs approach ambient mantle temperatures, then they should acquire neutral or positive buoyancy. A similar possibility applies to the delaminated lithospheric keels of old cratons. Both ancient lithosphere and old subducted slabs may be more oxidized than the surrounding mantle, and the presence of ($CH_4 + H_2O$) fluids at $fO_2 \approx IW + 1$ in the surrounding mantle will produce redox reactions at the slab/mantle interface. These reactions will precipitate diamond or graphite, leading to increased H_2O activity and to partial melting at the interface between the old subducted slab or lithosphere and the enclosing mantle (Figure 7.12). Opportunities exist for capture of trace-element and isotopic signatures from such slabs (peridotite + eclogite + minor crustal/sedimentary components). Because of local volatile (H_2O) contents, a partially molten boundary layer may form at the margins of the old subducted slab or old lithosphere, with sufficient compositional buoyancy to ascend as mantle plumes from below the asthenosphere and pass through it. Buoyant plumes may extend into the lithosphere to depth at least as shallow as 70 km (i.e., a hotspot). However, the primary cause of the hotspot is enhanced volatile flux, resulting in melting below the asthenosphere (Figure 7.12). No significant temperature perturbation is postulated below the relict subducted slab or old lithosphere. Regional buoyancy and uplift can be attributed to the lower density of the relict slab, to density reduction by partial melting, to vertical flow, and to the thermal effects of the upwelling plume initiated at the interface between the ambient mantle and the relict subducted slab or old lithosphere (Figure 7.12). The model provides continuity of the plume source as long as there is a redox contrast between the relict slab/old lithosphere and the ambient mantle,

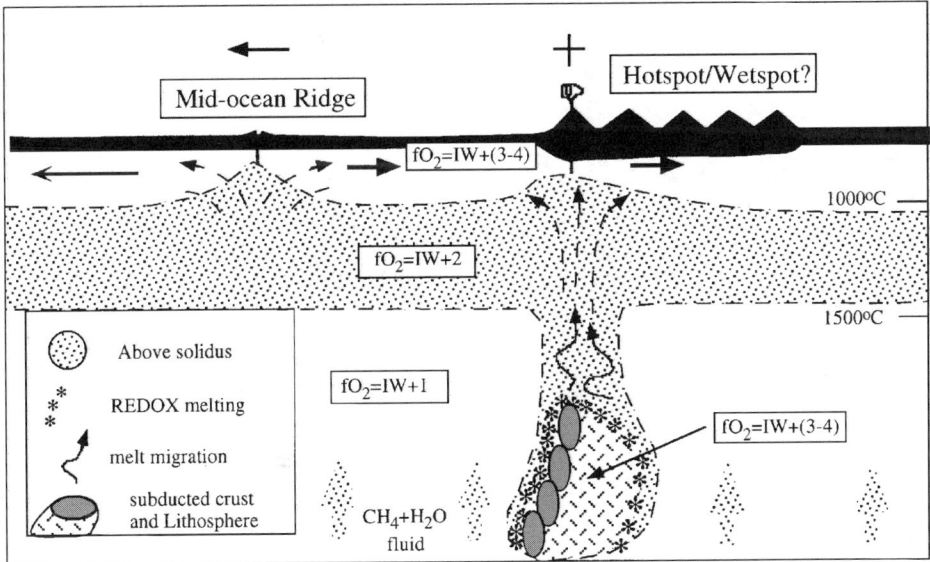

Figure 7.12. Cartoon suggesting a model for intraplate hotspot volcanism, consistent with Figure 7.8, and attributing the primary cause of the melting anomaly (hotspot) to a compositional anomaly within the mantle beneath the asthenosphere (i.e., below 150–200 km). Increased $(C + H + O)$ volatile flux is attributed to redox melting at the interface of the ambient mantle with $(CH_4 + H_2O)$ fluid ($fO_2 = IW + 1$) and old subducted slab or delaminated cratonic lithosphere ($fO_2 = IW + 3$ or 4). The geochemical anomaly (plume source) includes deep mantle fluid $(CH_4 + H_2O)$, sub-asthenospheric and asthenospheric mantle, and old subducted sources in a mixing and upwelling process. The geophysical anomaly is attributed to the relative densities of the subducted-slab/delaminated-lithosphere complex and the ambient mantle, to the melt-enhanced column above the compositional anomaly, and to the thermal anomaly of the diapirism leading to magma segregation at lithospheric depths (40–60 km).

and the concept also requires the mantle to be currently degassing $(CH_4 + H_2O)$ fluid, possibly reflecting entrapment of primordial $C + H$ (and noble gases).

7.5.2.4. *Subduction-Zone Magmatism*

Subduction of oceanic crust and lithosphere beneath island arcs or 'Andean' continental margins produces P–T conditions and composition variability such that magma genesis at convergent margins is more complex than at either mid-ocean-ridge or intraplate settings (Tatsumi and Eggins, 1995). Temperature distributions, particularly the cooling effect of a down-going slab and its persistence to at least the transition zone of the Earth's mantle, argue that the most probable source for magmatism is in the peridotite 'wedge' overlying the subducted slab (Figure 7.13a). Another possible location for melting is the subducted basaltic crust (and possible intermingled sediments) juxtaposed against higher-temperature lithosphere along the Benioff zone or at subduction-zone/transform-fault intersections.

Subducted oceanic crust is commonly inferred to have suffered variable hydration, carbonation, and oxidation. Subduction leads to progressive metamorphism

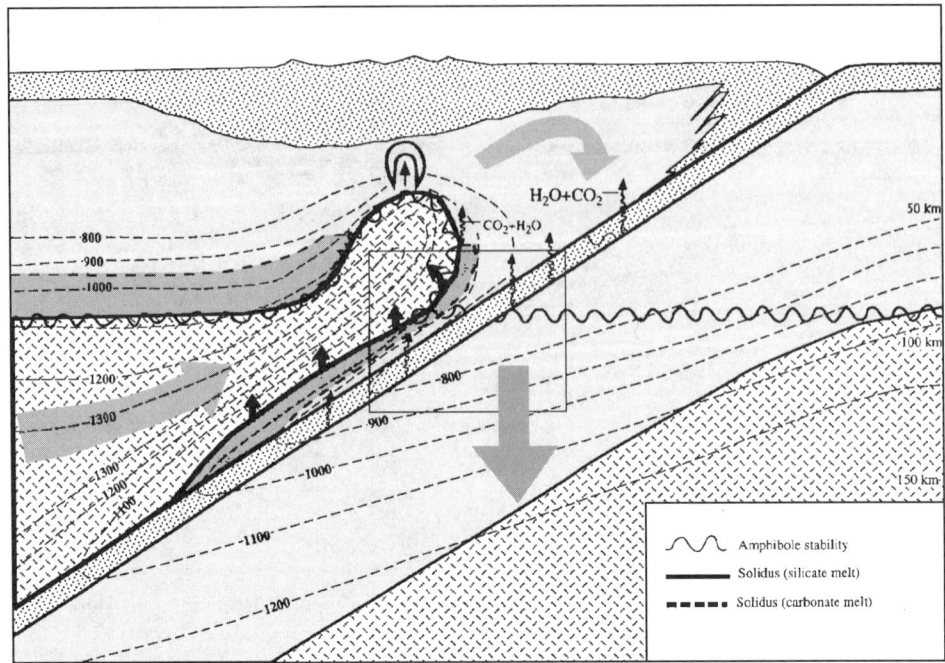

Figure 7.13a. Model of temperature distributions and lithologies at a convergent margin, leading to prediction of regions of partial melting for appropriate *P–T* conditions and fluid ($CO_2 + H_2O$) activity. The entire regime is inferred to be relatively oxidized ($fO_2 \geq IW + 3$) because of the role of the subducted slab and the release of oxidized fluids ($H_2O + CO_2 \pm SO_2$, Cl, F, etc.) from the subducted crust and uppermost mantle. Stippled area, subducted oceanic crust and crust of fore arc and magmatic arc; light shaded area, lithospheric (sub-solidus) mantle; darker shaded area, carbonatite melt in amphibole lherzolite (at depths <90–95 km) or in garnet lherzolite (at depths >90–95 km); patterned, high-temperature area, silicate melt in lherzolitic-to-harzburgitic residue. The large rectangle shows the block that is seen enlarged in Figure 7.13b.

through greenschist, blueschist, garnet-amphibolite, and eclogite assemblages. The solidi for such assemblages along the subduction trajectories can be broadly categorized as dehydration solidi for garnet amphibolite or amphibole- or mica-bearing eclogite, and such solidi lie at 700–850°C at 1.5–2.5 GPa (Figure 7.13b). For dehydration melting of garnet amphibolite or amphibole-bearing eclogite, minor carbonate is a residual phase, in equilibrium with the silicate melt. The melt component is water-rich, $(CO_3)^{2-}$-poor and rhyodacitic to dacitic in composition (Yaxley and Green, 1994). If such melts were to migrate 'up-temperature' into the overlying mantle wedge, they would react to increase the pyroxene $\pm$ garnet, phlogopite, and (at $P < 3$ GPa) pargasite contents of peridotite within the mantle wedge (Figure 7.13b).

The sampling of seafloor and submarine stages of island-arc volcanoes and the search for primitive or parental magmas in island-arc settings have led to identification of ankaramitic, picritic, and boninitic magmas as primary or parental magmas. Distinctive features, in comparison with mid-ocean-ridge magmas or intraplate

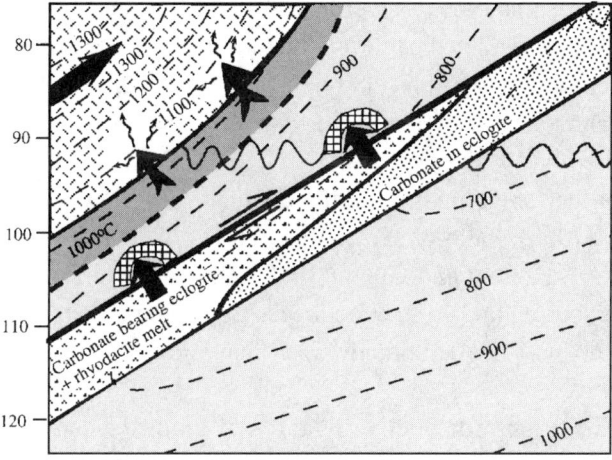

Figure 7.13b. Enlargement of the central part of Figure 7.13a to illustrate the presence of a melting region of subducted basaltic crust. At temperatures greater than 750°C, a hydrous rhyodacitic melt in equilibrium with residual carbonate-bearing eclogite may form within the subducted slab. Migration of the melt into overlying peridotite of the mantle wedge leads to reaction and crystallization within sub-solidus peridotite, enriching the latter in pyroxene, garnet, and phlogopite at depths greater than 95 km, but in pargasite and pyroxene at depths less than 95 km. By contrast, the carbonatite melts formed in the mantle wedge at $T \geq 935$°C may migrate into the overlying silicate melting regime [defined by the inverted temperature profile and the water-saturated silicate solidus at $T = 970$°C (70 km) to $T = 1,020$°C (100 km)]. If the model correctly predicts the melting regimes and their different capacities for mixing into the mantle wedge, then we would predict a carbonatite melt, highly and distinctively enriched in incompatible elements, to be a mixing component in some convergent-margin silicate melts, probably those of island-arc picrite or ankaramite type.

primary magmas, include the following:

1. Water contents, commonly in excess of 1 wt%, observed in quenched glasses, in amphibole or biotite crystallization, and in exsolved CO_2-rich or H_2O-rich fluids (vesicularity of melts).
2. Highly magnesian olivine phenocrysts (to Fo_{92}–Fo_{94}) in the most primitive magmas.
3. Variable NiO contents at similar Mg# in the most primitive olivine phenocrysts.
4. Early spinels with high Cr# (generally >70, and >85 in boninites).
5. Early spinels with high Fe^{3+}/Fe^{2+} ratios.
6. Early magnesian clinopyroxene (to Mg# 93 or 94, coexisting with ~Fo_{92} olivine) in ankaramites or ankaramitic picrites.
7. Early proto-enstatite inverted to twinned clinoenstatite (to Mg_{92}) and enstatite (to Mg_{90}) in low-calcium boninites.
8. Evidence for magma mixing of contrasted types of parental magmas (e.g., MORB-like + boninite-like) and of parental and evolved magma batches.
9. Variable minor- and trace-element signatures, leading to identification of island-arc-tholeiites, back-arc-basin basalts, mid-ocean-ridge-like basalts, or boninites

as distinctive magma types. Nevertheless, boninites may exhibit either LREE-enriched or LREE-depleted patterns, and there is consistent decoupling between (i) enrichment with large-ion lithophile elements (LILE) and (ii) depletion (or lack of enrichment) in Ti, Zr, and Nb.

The sources for the foregoing information include the following: Dallwitz, Green, and Thompson (1996), Falloon, Green, and Crawford (1987), Crawford, Falloon, and Green (1989), Falloon and Green (1989), Barsdell and Berry (1990), Eggins (1993), Sigurdsson et al. (1993), Woodhead, Eggins, and Gamble (1993), Ewart, Hergt, and Hawkins (1994), Sobolev and Chaussidon (1996), and Kamenetsky (1996).

Explanations for these distinctive features of erupted magmas in convergent-margin settings should be sought in the distinctive potential-source compositions and in the physical conditions (including variables such as fH_2O and fO_2) of convergent-margin environments. The following concepts derive from the preceding observations and are considered to be important (see also Figure 7.13):

1. Fluxing of the melting regime by transport of H_2O-rich volatiles into the mantle wedge, the source being devolatilization reactions in the down-going slab or in entrained cool peridotite at the base of the mantle wedge.
2. Magma source regions and/or magma genesis reflect more oxidizing conditions than in MORB or intraplate settings.
3. Residues from magma extraction generally are more refractory than those from MORB or intraplate settings, with low-calcium boninites, in particular, implying residues of highly refractory harzburgite, with low-Al_2O_3, low-CaO magnesian enstatite (Mg_{92}) and Cr-rich spinel Cr# > 85) (Figure 7.14).
4. Enrichments in minor elements and in incompatible elements in magmas appear complex and are unlikely to reflect a single process of migration from slab, through the cool wedge, to the higher-temperature 'anomalous mantle' beneath volcanic arcs.
5. Consideration of the melting relationships of basalt + (C + H + O) and peridotite + (C + H + O) under 'oxidizing conditions' (fO_2 > IW + 2 or 3) and in the $P-T$ regimes of convergent margins suggests the following possibilities:
   (a) Loss of H_2O-rich fluid from prograde metamorphic reactions in basaltic crust/serpentinized lithosphere, at sub-solidus temperatures in the slab. The fluids may be saline and would be effective metasomatic agents for growth of pargasitic hornblende in wedge peridotite.
   (b) Loss of hydrous rhyodacitic melt from eclogite, including carbonate-bearing eclogite. Migration of this melt into the overlying peridotite would produce pyroxene and phlogopite ($\pm$amphibole if P < 3 GPa) enrichment in peridotite (Figure 7.13b).
   (c) Formation of sodic dolomitic carbonatite melt within the peridotite wedge at

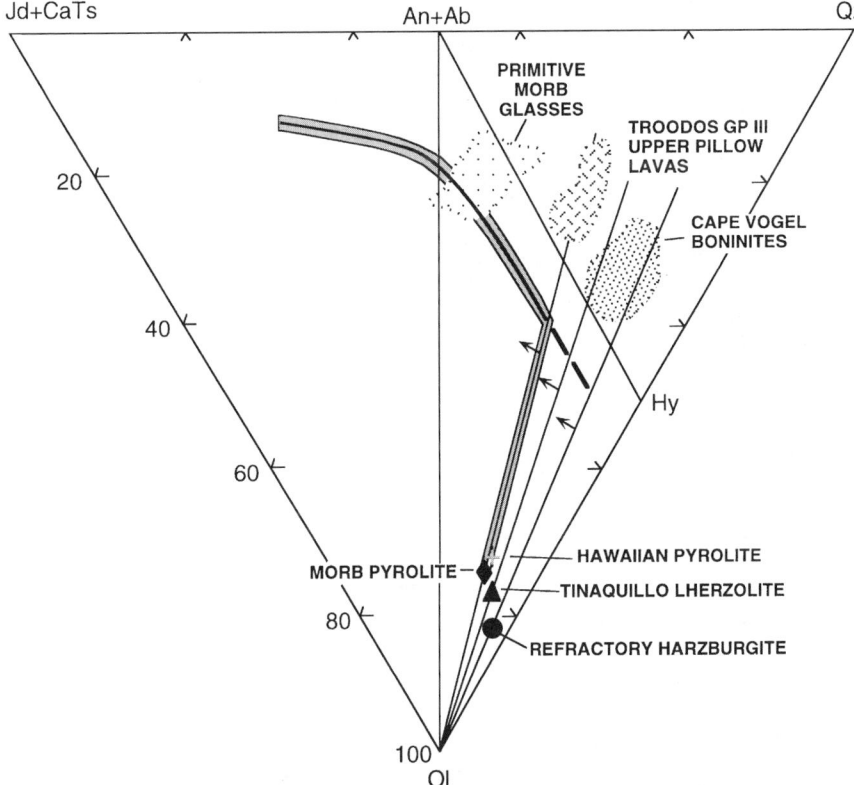

Figure 7.14. Projection onto the base of the basalt tetrahedron (Figure 7.2) of primitive or primary magma types: MORB, Troodos Group III picrites, possibly from a back-arc basin, and low-calcium boninites. The figure shows that it appears possible for MORB glasses (or olivine-rich picritic parents) to be derived from all of the indicated peridotite sources. However, not all Troodos Group III lavas can be derived from a MORB-pyrolite source, but all can be derived from a Tinaquillo-lherzolite source. Low-calcium boninites from Cape Vogel, on the other hand, require a very refractory source and cannot be derived from MORB-pyrolite or Tinaquillo-lherzolite sources in a single-stage melting process.

the $P–T$ environment between the cool, dense slab and the overlying higher-temperature peridotite (Figure 7.13). The carbonatite melt may occur with or without an additional H_2O-rich fluid phase. The carbonatite $P–T$ field is overlain by the water-saturated silicate solidus. Opportunity exists for migration of sodic dolomitic carbonatite melt, with distinctive minor- and trace-element partitioning, relative to lherzolite or harzburgite. If such carbonatite melts migrate into the silicate melting regime, they can be expected to dissolve in the silicate melt at $T > 1{,}050°C$ and $P > 2$ GPa. However, any silicate magmas segregating at these conditions would become over-saturated with $(C + H + O)$ fluids at $P \geq 1$ GPa because of the very strong pressure effect on $(CO_3)^{2-}$ solubility; fluids with $CO_2 > H_2O + CH_4$ would degas from such magmas.

(d) Given the inverted temperature profiles in the mantle wedge (as illustrated schematically in Figure 7.13) and the presence of H_2O-rich fluids, the relevant peridotite solidus is assumed to be water-saturated. Upward migration of melt into higher-temperature parts of the wedge will produce fluid-undersaturated melting, with the melt fraction determined by bulk composition, pressure, temperature, and water ($\pm CO_2$) content. Upwelling or diapirism from within the wedge will increase the melt fraction and push the residues towards harzburgite (rather than lherzolite) with high Mg# and Cr# in silicates and spinels, respectively. If upwelling or diapirism reaches depths of 30–40 km, then the decreasing solubility of $(CO_3)^{2-}$ at $P \leq 1.5$ GPa will lead to degassing of CO_2-rich fluids. The melts accordingly will become increasingly H_2O-rich as emplacement and degassing continue to lower pressures.

The requirement for source compositions more refractory than those for Hawaiian pyrolite or MORB pyrolite is well illustrated by Figure 7.14. Some of the primitive magmas of the Troodos Upper Pillow Lavas plot to the Qz-rich side of the olivine control lines passing through HPY or MPY, and thus these magmas cannot be derived from such sources – they can, however, be derived from the Tinaquillo-lherzolite composition (with or without the effect of H_2O as a fluxing agent) (cf. Figure 7.8). Similarly, the low-calcium boninites of Cape Vogel (Crawford et al., 1989) cannot be derived from Tinaquillo lherzolite, but require a more refractory harzburgite source. Melting of such refractory source rocks requires the introduction of H_2O-rich fluids. The association of variable Na_2O, K_2O, and incompatible-element contents with the refractory nature of the source harzburgite suggests the need for several melting and enrichment events.

6. Primary magmas from the mantle wedge will be constrained by peridotite melting relations, with variable $fH_2O > fCO_2$, and thus will vary from quartz-normative and boninitic (harzburgite residue, higher fH_2O, lower P) to strongly olivine-normative and picritic (harzburgite-to-lherzolite residue, lower fH_2O, higher P) or ankaramitic (lherzolite or wehrlite residue, higher fCO_2 and higher P). Because of the characteristic and essential role of water in controlling melting, primary magmas cannot reach the surface without intersecting their vapour-saturated liquidi. Primary magmas must thus be porphyritic and will commonly fractionate by separation of olivine, chromite, and clinopyroxene, and at higher degrees of crystallization, by plagioclase, orthopyroxene, amphibole, and titano-magnetite.

7. The complexity of the potential sources and processes summarized under items 5 and 6 explains the prevalence of magma mixing in convergent-margin volcanism. Mixing may occur between 'primary' magmas of different types or more commonly between evolving magmas of one type and primary magmas of a different type.

7.6. Conclusion

The pyrolite concept has proved robust in its basic premise that natural processes of mantle sampling yield both primitive, mantle-derived melts and crystalline residues of partial-melting events. Upper-mantle compositions can be calculated by combining melts and residues when the liquidus phases of olivine ($Fo_{90\pm1}$) and orthopyroxene $\pm$ spinel are known for particular primitive or parental magmas. The arbitrariness of assuming proportions of melt : olivine : orthopyroxene : spinel is removed by using the least refractory of natural high-pressure lherzolite xenoliths, or high-temperature, high-pressure lherzolite intrusions (e.g., Tinaquillo lherzolite), as the guide to matching modal mineralogy and thus the concentrations of the major oxides SiO_2, MgO, FeO, CaO, and Al_2O_3 (Figure 7.10).

The calculated compositions of Hawaiian pyrolite and MORB pyrolite (Tables 7.1A and 7.1B) and the use of a refractory lherzolite (Tinaquillo lherzolite) remain appropriate after several decades of study of their sub-solidus and melting behaviour. There is wide consensus on a mean upper-mantle composition, with Mg# of 89 ± 1, 3–4 wt% Al_2O_3 and CaO, and about 60% modal olivine (in spinel-lherzolite mineralogy), and most experimental studies have used natural or synthetic starting materials approaching this composition. These compositional parameters determine the first-order features of the upper mantle (i.e., the stability fields for plagioclase lherzolite, spinel + aluminous pyroxene lherzolite, and garnet lherzolite) and the sequence of disappearance of phases during melting, which finds expression as lherzolitic, harzburgitic, or dunitic residues. The compositional parameters are also sufficient to define the sequences of polymorphic transitions, reactions, and solid solutions that will characterize the mineralogy and phase chemistry of 'pyrolite' and residual harzburgite throughout the transition zone and the lower mantle, as reviewed by Jackson and Rigden (Chapter 9, this volume).

However, if we are to understand the melting behaviour of the upper mantle, then the concentrations of incompatible elements and the variations in Na/Ca, Ti/Al, K/Na, and Ca/Al ratios among primitive basalts show that we must consider a large range of extracted melt fractions and complementary variable degrees of depletion of any model source. Thus it becomes important to understand the effects of variable enrichment and depletion (as expressed in the minor elements Na, Ti, K, and P) on lherzolite melting behaviour and sub-solidus mineralogy.

Systematic studies of the three compositions designated as Hawaiian pyrolite (plume source, enriched intraplate source, hotspot source), MORB pyrolite (depleted modern mantle, modern asthenospheric mantle, 'normal' mantle, or N-MORB source), and Tinaquillo lherzolite (depleted lherzolite) have provided insight into the nature of the upper mantle and constraints on geophysical models.

The following major points have been demonstrated by experimental studies arising directly from the concept and testing of the 'pyrolite model' for the upper mantle:

1. The position of the boundary between plagioclase lherzolite and spinel lherzolite and the temperature of the lherzolite solidus are sensitively dependent on Na_2O content and on the normative plagioclase composition of the lherzolite composition. Models of melting, including dynamic melting, must incorporate the variation of solidus temperatures (effectively buffering liquid compositions and melt fraction) with both composition and pressure.

2. The presence of carbon and hydrogen (water) in all types of samples of the upper mantle (xenoliths, magmas, and high-pressure lherzolites) argues that an understanding of melting and phase relationships of the upper mantle must take their presence into account. Because of the sensitive relationships between oxygen activity and carbon + hydrogen speciation, the system to be explored is the peridotite-$(C + H + O)$ system.

3. Pargasitic amphibole has been shown to be stable in model mantle compositions up to the water-saturated solidus and to define the silicate solidus under water-undersaturated conditions, from low pressures to 2.8–3 GPa. The instability of pargasite at pressures greater than 3 GPa ensures that even very small quantities of water (0.1%) will cause melting at the water-saturated solidus for silicate melting at depths greater than 95 km.

4. Pargasite is stable to higher temperatures in enriched mantle (Na-, Ti-, and K-rich) than in depleted mantle, producing a paradoxical inversion of solidi between residual (Na-, Ti-, and K-poor) and more fertile compositions. These relationships demonstrate the capacity for very small amounts of water to produce geochemical inhomogeneity within the lithosphere ($P < 3$ GPa, $970°C < T < 1,150°C$).

5. Because of the dramatic effects of carbon and hydrogen in lowering the solidus temperature for pyrolite, the melting behaviour of the Earth's upper mantle is dominated by the existence of the incipient-melting regime, in which the volume of melt at a particular P and T is primarily determined by the $(C + H)$ abundance, and the composition of the melt is determined by $P–T$ conditions and the $CO_2 : H_2O$ $[(CO_3)^{2-} : (OH^-)]$ proportions. Within the incipient-melting regime, the volume of melt varies little ($\leq$1–2%) over a large temperature range, and the residual mineralogy is garnet lherzolite (i.e., liquids reflect a 'garnet signature' and its attendant REE pattern) over very large pressure and temperature intervals.

6. A second characteristic of pyrolite melting is the contrast between the incipient-melting regime and the major melting regime encountered at temperatures near and above the $(C + H)$-free solidi. In the major melting regime, the volume of melt increases rapidly with increasing temperature, garnet plays no role as a residual phase at $P < 2.8$–3 GPa (depths <90 km), and melt compositions are controlled by the movement of the olivine + orthopyroxene $\pm$ clinopyroxene cotectic as a function of pressure.

7. A petrogenetic grid has been derived from the parallel studies of the liquidus phase relationships for primary magmas and the melting behaviour of model sources (pyrolite). These studies show clearly that intraplate magmatism (particularly for the more undersaturated magmas identified with plume traces and with intraplate rifting) originates within the incipient-melting regime. In contrast, more voluminous melts of hotspots, rift magmatic provinces, and flood-basalt provinces are derived from the major melting regime at higher temperatures. The compositions of the latter suite in intraplate settings commonly indicate harzburgitic residues, higher volatile $(CO_3)^{2-} + (OH)^-$ contents, and mixture of several isotopic and geochemical components in their petrogenesis.

8. Petrological studies of picrites and experimental studies of picrites and primitive glasses argue against major differences in temperature between the most primitive magmas erupted at mid-ocean ridges (N-MORB, ultra-depleted MORB) and plume magmatism. Indeed, there is evidence from the higher $(CO_3)^{2-} + (OH)^-$ contents and incompatible-element contents that plume-derived picrites form at temperatures below or closely similar to those of MORB picrites. Similarly, for magmatism in back-arc-basins and convergent-margin environments, primitive magmas exhibit more refractory (i.e., higher-temperature) characteristics, coupled with 'melt fluxing' by $(OH)^-$, Na_2O, and K_2O, and the implied maximum mantle temperatures approach those of MORB petrogenesis.

9. It is inferred that magma genesis in the modern Earth implies and is constrained by a mantle potential temperature of about 1,430°C. In the case of plumes, there is no supportive evidence from the magmas themselves for the existence of temperatures elevated some 200°C above those of the ambient mantle ($T_p \sim 1,430$°C). However, the existence of peridotitic-komatiite magmas (melts with $\geq 30\%$ MgO) in all Archaean cratons implies mantle potential temperatures of at least 1,650°C prior to 2.5 Ga.

10. Models of the temperature distribution within the earth have been combined with estimates of mantle composition to predict the compositional characteristics of melts and residues, taking oxidation state into account. In this chapter we have summarized some of the implications for the carbonatite melting interval, the incipient-melting regime, and the major melting regime.

11. The current status of mantle petrogenesis will require a choice between models for episodic three-dimensional plume-like upwelling, with significant melt retention in the upwelling mantle material, and models of passive upwelling, with very low melt retention and efficient migration and mixing of melts over a large depth interval. The difficult questions of melt extraction, including the role of deformation, remain a barrier to a full understanding of the issues. Nevertheless, experimental studies based on the pyrolite concept, and using several bulk compositions, have provided the necessary framework for further progress.

The Ringwood concept of 'pyrolite' has proved remarkably robust and has evolved as we have gained increasing knowledge both from natural rocks and from experimental studies, which have provided a striking validation of the original approach.

References

Albarède, F. 1992. How deep do common basaltic magmas form and differentiate? *J. Geol. Geophys.* 97:10997–1009.

Anderson, D. L. 1995. Lithosphere, asthenosphere, and perisphere. *Rev. Geophys.* 33:235–49.

Baker, M. B., Hirschmann, M. M., Ghiorso, M. S., and Stolper, E. M. 1995. Compositions of low-degree partial melts of peridotite: results from experiments and thermodynamic calculations. *Nature* 375:308–11.

Baker, M. B., and Stolper, E. M. 1994. Determining the composition of high-pressure mantle melts using diamond aggregates. *Geochim. Cosmochim. Acta* 58:2811–27.

Ballhaus, C., Berry, R. F., and Green, D. H. 1990. Oxygen fugacity controls in the earth's upper mantle. *Nature* 348:437–40.

Ballhaus, C., Berry, R. F., and Green, D. H. 1991. High pressure experimental calibration of the olivine-orthopyroxene-spinel oxygen geobarometer: implications for the oxidation state of the upper mantle. *Contrib. Min. Pet.* 107:27–40.

Ballhaus, C., and Frost, B. R. 1994. The generation of oxidized CO_2-bearing basaltic melts form reduced CH_4-bearing upper mantle sources. *Geochim. Cosmochim. Acta* 58:4931–40.

Barsdell, M., and Berry, R. F. 1990. Origin and evolution of primitive island arc ankaramites from Western Epi, Vanuatu. *J. Petrol.* 31:747–77.

Birch, F. 1952. Elasticity and constitution of the earth's interior. *J. Geophys. Res.* 57:227–86.

Boyd, F. R. 1989. Compositional distinction between oceanic and cratonic lithosphere. *Earth Planet. Sci. Lett.* 96:15–26.

Brey, G., and Green, D. H. 1975. The role of CO_2 in the genesis of olivine melilitite. *Contrib. Min. Pet.* 49:93–103.

Buck, W. R., and Su, W. 1989. Focussed mantle upwelling below mid-ocean ridges due to feedback between viscosity and melting. *Geophys. Res. Lett.* 16:641–4.

Christie, D. M., Carmichael, I. S. E., and Langmuir, C. H. 1986. Oxidation state of mid-ocean ridge basalt glasses. *Earth Planet. Sci. Lett.* 79:397–411.

Cordery, M. J., and Phipps Morgan, J. 1994. Convection and melting at mid-ocean ridges. *J. Geophys. Res.* 91:9315–23.

Crawford, A. J., Falloon, T. J., and Green, D. H. 1989. Classification, petrogenesis and tectonic setting of boninites. In: *Boninites and Related Rocks,* ed. A. J. Crawford, pp. 1–49. London: Unwin Hyman.

Dallwitz, W. B., Green, D. H., and Thompson, J. E. 1966. Clinoenstatite in a volcanic rock from the Cape Vogel area, Papua. *J. Petrol.* 7:375–403.

Danyushevsky, L. V., Sobolev, A. V., and Dmitriev, L. V. 1987. Low-titanium orthopyroxene-bearing tholeiite, a new type of ocean-rift tholeiite. *Trans. (Doklady) USSR Acad. Sci.* 292:102–5.

Davies, G. F. 1990. Mantle plumes, mantle stirring and hot spot chemistry. *Earth Planet. Sci. Lett.* 99:94–109.

Dick, H. J. B. 1989. Abyssal peridotites, very slow spreading ridges and ocean ridge magmatism. In: *Magmatism in the Ocean Basins, Special Publications,* vol. 42, ed. A. D. Saunders and M. J. Norry, pp. 71–105. London: Geological Society.

Dmitriev, L. V., Sobolev, A. V., Suschevskaya, N. M., and Zapunny, S. A. 1985. Abyssal glasses petrologic mapping of the oceanic floor and Geochemical Leg 82. In: *Initial*

Reports of the Deep Sea Drilling Project, ed. H. Bougault, S. C. Cande, et al., pp. 509–18. Washington, DC: U.S. Government Printing Office.

Donaldson, C. H., and Brown, R. W. 1977. Refractory megacrysts and magnesium-rich melt inclusions within spinel in oceanic tholeiites: indicators of magma mixing and parental magma compositions. *Earth Planet. Sci. Lett.* 37:81–9.

Doukhan, N., Doukhan, J.-C., Ingrin, J., Jaoul, O., and Raterron, P. 1993. Early partial melting in pyroxenes. *Am. Min.* 78:1246–56.

Duncan, R. A., and Green, D. H. 1980. Role of multistage melting in the formation of oceanic crust. *Geology* 8:22–6.

Duncan, R. A., and Green, D. H. 1987. The genesis of refractory melts in the formation of oceanic crust. *Contrib. Min. Pet.* 96:326–42.

Edgar, A. D., Green, D. H., and Hibberson, W. O. 1976. Experimental petrology of a highly potassic magma. *J. Petrol.* 17:339–56.

Eggins, S. M. 1992a. Petrogenesis of Hawaiian tholeiites: 1. Phase equilibria constraints. *Contrib. Min. Pet.* 110:387–97.

Eggins, S. M. 1992b. Petrogenesis of Hawaiian tholeiites: 2. Aspects of dynamic melt segregation. *Contrib. Min. Pet.* 110:398–410.

Eggins, S. M. 1993. Origin and differentiation of picritic arc magmas, Ambae (Aoba), Vanuatu. *Contrib. Min. Pet.* 114:79–100.

Elthon, D., and Scarfe, C. M. 1984. High-pressure phase equilibria of a high-magnesia basalt and the genesis of primary oceanic basalts. *Am. Min.* 69:1–15.

Evans, B. W., and Trommsdorff, V. 1978. Petrogenesis of garnet lherzolite, Cima di Gagnone, Lepontine Alps. *Earth Planet. Sci. Lett.* 40:333–48.

Ewart, A., Hergt, J., and Hawkins, J. W. 1994. Major element, trace element, and isotope (Pb, Sr and Nd) geochemistry of Site 839 basalts and basaltic andesites: implications for arc volcanism. In: *Proceedings of the Ocean Drilling Program, Scientific Results*, vol. 135, ed. J. W. Hawkins, L. M. Parson, and J. F. Allan, pp. 519–31. College Station, TX: Ocean Drilling Program.

Falloon, T. J., and Green, D. H. 1987. Anhydrous partial melting of MORB pyrolite and other peridotite compositions at 10 kbar and implications for the origin of primitive MORB glasses. *Mineral. Petrol.* 37:181–219.

Falloon, T. J., and Green, D. H. 1988. Anhydrous partial melting of peridotite from 8 to 35 kbar and petrogenesis of MORB. *J. Petrol., Special Lithosphere Issue*, pp. 379–414.

Falloon, T. J., and Green, D. H. 1989. The solidus of carbonated, fertile peridotite. *Earth Planet. Sci. Lett.* 94:364–70.

Falloon, T. J., and Green, D. H. 1990. Solidus of carbonated fertile peridotite under fluid-saturated conditions. *Geology* 18:195–9.

Falloon, T. J., Green, D. H., and Crawford, A. J. 1987. Dredged igneous rocks from the northern termination of the Tofua magmatic arc, Tonga and adjacent Lau. Basin. *Aust. J. Earth Sci.* 34:487–506.

Falloon, T. J., Green, D. H., Hatton, C. J., and Harris, K. L. 1988. Anhydrous partial melting of fertile and depleted peridotite from 2 to 30 kbar and application to basalt petrogenesis. *J. Petrol.* 29:257–82.

Falloon, T. J., Green, D. H., O'Neill, H., and Ballhaus, C. G. 1996. Quest for low-degree melts. *Nature* 381:285.

Faul, U. 1997. Permeability of partially molten upper mantle rocks from experiments and percolation theory. *J. Geophys. Res.* 102:10,299–311.

Feigenson, M. D. 1986. Constraints on the origin of Hawaiian lavas. *J. Geophys. Res.* 91:9383–93.

Finnerty, A. A., and Boyd, F. R. 1987. Thermobarometry for garnet peridotites: basis for the determination of thermal and compositional structure of the upper mantle. In: *Mantle Xenoliths*, ed. P. H. Nixon, pp. 381–402. New York: Wiley.

Foley, S. F., Taylor, W. R., and Green, D. H. 1986. The role of fluorine and oxygen fugacity in the genesis of the ultrapotassic rocks. *Contrib. Min. Pet.* 94:183–92.

Ford, C. E., Russell, D. G., Craven, J. A., and Fisk, M. R. 1983. Olivine–liquid equilibria: temperature, pressure and composition dependence of the crystal/liquid cation partition coefficients for Mg, Fe^{2+}, Ca and Mn. *J. Petrol.* 24:256–65.

Forsyth, D. W. 1992. Geophysical constraints on mantle flow and melt generation beneath mid-ocean ridges. In: *Mantle Flow and Melt Generation at Mid-ocean Ridges, AGU Monographs*, vol. 71, ed. J. P. Phipps Morgan, D. K. Blackman, and J. M. Sinton, pp. 1–66. Washington, DC: American Geophysical Union.

Frey, F., and Green, D. H. 1974. The mineralogy, geochemistry and origin of lherzolite inclusions in Victorian basanites. *Geochim. Cosmochim. Acta* 38:1023–59.

Frey, F. A., Green, D. H., and Roy, S. D. 1978. Integrated models of basalt petrogenesis – a study of quartz tholeiites to olivine melilitites from southeastern Australia utilizing geochemical and experimental petrological data. *J. Petrol.* 19:463–513.

Frey, F. A., and Roden, M. F. 1987. The mantle source for the Hawaiian Islands. In: *Mantle Metasomatism*, ed. M. A. Menzies and C. J. Hawkesworth, pp. 423–463. London: Academic Press.

Garcia, M. O., Hulsebosch, T. P., and Rhodes, J. M. 1995. Olivine-rich submarine basalts from the Southwest Rift zone of Mauna Loa Volcano: implications for magmatic processes and geochemical evolution. In: *Mauna Loa Revealed: Structure Composition, History, and Hazards, AGU monographs*, vol. 92, ed. J. M. Rhodes and J. P. Lockwood, pp. 219–39. Washington, DC: American Geophysical Union.

Green, D. H. 1964a. The metamorphic aureole of the peridotite at the Lizard, Cornwall. *J. Geol.* 72:543–63.

Green, D. H. 1964b. The petrogenesis of the high-temperature peridotite intrusion in the Lizard area, Cornwall. *J. Petrol.* 5:134–88.

Green, D. H. 1964c. A re-study and re-interpretation of the geology of the Lizard peninsula, Cornwall. In: *Present Views on Some Aspects of the Geology of Cornwall and Devon*, pp. 87–114. Royal Geological Society, Cornwall.

Green, D. H. 1971. Compositions of basaltic magmas as indicators of conditions of origin: application to oceanic volcanism. *Phil. Trans. R. Soc. London*, A268:707–25.

Green, D. H. 1973a. Conditions of melting of basanite magma from garnet peridotite. *Earth Planet. Sci. Lett.* 17:456–65.

Green, D. H. 1973b. Experimental melting studies on a model upper mantle composition at high pressure under water-saturated and water-undersaturated conditions. *Earth Planet. Sci. Lett.* 19:37–53.

Green, D. H. 1975. Genesis of Archaean peridotitic magmas and constraints on Archaean geothermal gradients and tectonics. *Geology* 3:15–18.

Green, D. H. 1976. Experimental testing of 'equilibrium' partial melting of peridotite under water-saturated, high pressure conditions. *Can. Mineral.* 14:255–68.

Green, D. H. 1981. Petrogenesis of Archaean ultramafic magmas and implications for Archaean tectonics. In: *Precambrian Plate Tectonics*, ed. A. Kroner, pp. 469–89. Amsterdam: Elsevier.

Green, D. H., Falloon, T. J., and Taylor, W. R. 1987. Mantle-derived magmas – roles of variable source peridotite and variable C–H–O fluid compositions. In: *Magmatic Processes and Physicochemical Principles*, special publication 1, ed. B. O. Mysen, pp. 139–54. Philadelphia: Geochemical Society.

Green, D. H., and Hibberson, W. 1970a. Experimental duplication of conditions and precipitation of high pressure phenocrysts in a basaltic magma. *Phys. Earth Planet. Int.* 3:247–54.

Green, D. H., and Hibberson, W. 1970b. The instability of plagioclase in peridotite at high pressure. *Lithos* 3:209–21.

Green, D. H., Hibberson, W. O., and Jaques, A. L. 1979. Petrogenesis of mid-ocean ridge basalts. In: *The Earth: Its Origin, Structure and Evolution*, ed. M. W. McElhinny, pp. 265–90. London: Academic Press.

Green, D. H., and Liebermann, R. C. 1976. Phase equilibria and elastic properties of a pyrolite model for the oceanic upper mantle. *Tectonophysics* 32:61–92.

Green, D. H., Nicholls, I. A., Viljoen, R., and Viljoen, M. 1975. Experimental demonstration of the existence of peridotitic liquids in earliest Archaean magmatism. *Geology* 3:15–18.

Green, D. H., and Ringwood, A. E. 1963. Mineral assemblages in a model mantle composition. *J. Geophys. Res.* 68:937–45.

Green, D. H., and Ringwood, A. E. 1967a. The genesis of basaltic magmas. *Contrib. Min. Pet.* 15:103–90.

Green, D. H., and Ringwood, A. E. 1967b. The stability fields of aluminous pyroxene peridotite and garnet peridotite. *Earth Planet. Sci. Lett.* 3:151–60.

Green, D. H., and Ringwood, A. E. 1970. Mineralogy of peridotitic compositions under upper mantle conditions. *Phys. Earth Planet. Int.* 3:359–71.

Green, D. H., and Wallace, M. E. 1988. Mantle metasomatism by ephemeral carbonatite melts. *Nature* 336:459–62.

Griffiths, R. W., and Campbell, I. H. 1990. Stirring and structure in mantle starting plumes. *Earth Planet. Sci. Lett.* 99:67–78.

Gupta, A. K., Green, D. H., and Taylor, W. R. 1987. The liquidus surface of the system forsterite-nepheline-silica at 28 kb. *Am. J. Sci.* 287:560–5.

Hess, P. C. 1992. Phase equilibria constraints on the origin of ocean floor basalts. In: *Mantle Flow and Melt Generation at Mid-ocean Ridges, AGU Monographs*, vol. 71, ed. J. Phipps Morgan, D. K. Blackman, and J. M. Sinton, pp. 67–102. Washington, DC: American Geophysical Union.

Hirose, K., and Kushiro, I. 1993. Partial melting of dry peridotites at high pressures: determination of compositions of melts segregated from peridotites using aggregates of diamonds. *Earth Planet. Sci. Lett.* 114:477–89.

Iwamori, H. 1994. ^{238}U-^{230}Th-^{226}Ra and ^{235}U-^{231}Pa disequilibria produced by mantle melting with porous and channel flows. *Earth Planet. Sci. Lett.* 125:1–16.

Jaques, A. L., and Green, D. H. 1980. Anhydrous melting of peridotite at 0–15 kb pressure and the genesis of tholeiitic basalts. *Contrib. Min. Pet.* 73:287–310.

Johnson, K. T. M., Dick, H. J. B., and Quick, J. E. 1990. Melting in the oceanic upper mantle: an ion mocroprobe study of diopside in abyssal peridotites. *J. Geophys. Res.* 95:2661–78.

Kamenetsky, V. S. 1996. Methodology for the study of melt-inclusions in Cr-spinel and implications for parental melts of MORB from FAMOUS area. *Earth Planet. Sci. Lett.* 142:479–86.

Kamenetsky, V. S., and Sobolev, A. V. 1995. Geochemical variability for FAMOUS parental melts: a snapshot from spinel-hosted melt inclusions. *EOS, Trans. AGU* 76:270.

Kelemen, P. B., Dick, H. J. B., and Quick, J. E. 1992. Formation of harzburgite by pervasive melt/rock reaction in the upper mantle. *Nature* 358:635–41.

Klein, E. M., and Langmuir, C. H. 1987. Global correlations of ocean ridge basalt chemistry with axial depth and crustal thickness. *J. Geophys. Res.* 92:8089–115.

Kushiro, I. 1996. Partial melting of a fertile mantle peridotite at high pressures: an experimental study using aggregates of diamond. In: *Earth Processes: Reading the Isotopic Code, AGU Monographs*, vol. 95, ed. A. Basu and S. Hart, pp. 109–22. Washington, DC: American Geophysical Union.

Langmuir, C. H., Klein, E. M., and Plank, T. 1992. Petrological systematics of mid-ocean ridge basalts: constraints on melt generation beneath ocean ridges. In: *Mantle Flow and Melt Generation at Mid-ocean Ridges, AGU Monographs* vol. 71, ed. J. Phipps Morgan, D. K. Blackman and J. M. Sinton, pp. 183–280. Washington, DC: American Geophysical Union.

La Tourrette, T. Z., Kennedy, A. K., and Wassserburg, G. J. 1993. Thorium–uranium fractionation by garnet: evidence for a deep source and rapid rise of oceanic basalts. *Nature* 261:739–42.

Leeman, W. P., Budahn, J. R., Gerlach, D. C., Smith, D. R., and Powell, B. N. 1980. Origin of Hawaiian tholeiites: trace element constraints. *Am. J. Sci.* 280A:794–819.

Lin, J., and Phipps Morgan, J. 1992. The spreading rate dependence of three-dimensional mid-ocean ridge gravity structure. *Geophys. Res. Lett.* 19:13–16.

McKenzie, D. 1984. The generation and compaction of partially molten rock. *J. Petrol.* 25:713–65.

McKenzie, D., and Bickle, M. J. 1988. The volume and composition of melt generated by extension of the lithosphere. *J. Petrol.* 29:625–79.

McNeill, A. W., and Danyushevsky, L. V. 1996. Composition and crystallisation temperatures of primary melts from Hole 896A basalts: evidence from melt inclusion studies. In: *Proceedings of the Ocean Drilling Program, Scientific Results,* vol. 148, ed. J. C. Alt, H. Kinoshita, L. B. Stokking, and Michael, P. J., pp. 21–35. College Station, TX: Ocean Drilling Program.

Mengel, K., and Green, D. H. 1989. Stability of amphibole and phlogopite in metasomatized peridotite under water-saturated and water-undersaturated conditions. In: *Fourth International Kimberlite Conference,* vol. 1, ed. J. Ross, pp. 571–81. Special Publication 14. Carlton: Geological Society of Australia.

Morgan, W. J. 1971. Convection plumes in the lower mantle. *Nature* 230:42–3.

Navon, O., and Stolper, E. 1987. Geochemical consequences of melt percolation: the upper mantle as a chromatographic column. *J. Geol.* 95:285–307.

Nickel, K. G., and Green, D. H. 1984. The nature of the upper-most mantle beneath Victoria, Australia, as deduced from ultramafic xenoliths. In: *Kimberlites II: The Mantle and Crust–Mantle Relationships,* ed. J. Kornprobst, pp. 161–78. Amsterdam: Elsevier.

Nielson, J. E., and Wilshire, H. G. 1993. Magma transport and metasomatism in the mantle: a critical review of current geochemical models. *Am. Min.* 78:1117–34.

Niida, K., and Green, D. H. In press. Stability and chemical composition of pargasitic amphibole in MORB pyrolite under water-undersaturated conditions. *Contrib. Min. Pet.*

Nikogosian, I. K., and Sobolev, A. V. 1996. Ultra-depleted and ultra-enriched primary melts in Hawaii plume: evidence from melt inclusions study. *Abstracts, V. M. Goldschmidt Conference* 1:433.

Niu, Y., and Batiza, R. 1993. Chemical variation trends at fast and slow spreading ridges. *J. Geophys. Res.* 98:7887–902.

Niu, Y., and Batiza, R. 1994. Magmatic processes at the mid-Atlantic ridge ~26° S. *J. Geophys. Res.* 99:19719–40.

O'Hara, M. J. 1968a. Are any ocean floor basalts primary magma? *Nature* 220:683–6.

O'Hara, M. J. 1968b. The bearing of phase equilibria studies on the origin and evolution of basic and ultrabasic rocks. *Earth Sci. Rev.* 4:69–133.

O'Neill, H. St. C., Rubie, D. C., Canil, D., Geiger, C. A., Ross, C. R., II, Seifert, F., and Woodland, A. B. 1993. Ferric iron in the upper mantle and in transition zone assemblages: implications for relative oxygen fugacities in the mantle. *Geophys. Monogr.* 74:73–88.

O'Reilly, S. Y., and Griffin, W. L. 1985. A xenolith-derived geotherm for southeastern Australia and its geophysical implications. *Tectonophysics* 111:41–63.

O'Reilly, S. Y., and Griffin, W. L. 1988. Mantle metasomatism beneath western Victoria, Australia: 1. Metasomatic processes in Cr-diopside lherzolites. *Geochim. Cosmochim. Acta* 53:433–47.

Perfit, M. R., Fornari, D. J., Ridley, W. I., Kirk, P. D., Casey, J., Kastens, K. A., Reynolds, J. R., Edwards, M., Desonie, D., Shuster, R., and Paradis, S. 1996. Recent volcanism in the Siqueiros transform fault: picritic basalts and implictions for MORB magma genesis. *Earth Planet. Sci. Lett.* 93:2955–66.

Phipps Morgan, J., and Forsyth, D. W. 1988. Three-dimensional flow and temperature perturbations due to a transform offset: effect on oceanic crustal and upper mantle structure. *J. Geophys. Res.* 93:2955–66.

Ringwood, A. E. 1962a. A model for the upper mantle. *J. Geophys. Res.* 67:857–66.

Ringwood, A. E. 1962b. A model for the upper mantle. 2. *J. Geophys. Res.* 67:4473–7.

Ringwood, A. E. 1966. The chemical composition and origin of the Earth. In: *Advances in Earth Science*, ed. P. M. Hurley, pp. 287–356. Massachusetts Institute of Technology Press.

Roedder, P. L., and Emslie, R. F. 1970. Olivine–liquid equilibrium. *Contrib. Min. Pet.* 29:275–89.

Ross, C. S., Foster, M. D., and Myers, A. T. 1954. Origin of dunites and olivine-rich inclusions in basaltic rocks. *Am. Min.* 39:693–737.

Rudnick, R. L., McDonough, W. F., and Chappell, B. W. 1993. Carbonatite metasomatism in the northern Tanzanian upper mantle: petrographic and geochemical characteristics. *Earth Planet. Sci. Lett.* 114:463–75.

Salters, V. J. M. 1996. The generation of mid-ocean ridge basalts from the Hf and Nd isotope perspective. *Earth Planet. Sci. Lett.* 141:109–23.

Salters, V. J. M., and Hart, S.R. 1989. The Hf-paradox, and the role of garnet in the MORB source. *Nature* 342:420–2.

Schilling, J. G. 1971. Sea-floor evolution: rare earth evidence. *Phil. Trans. R. Soc. London* A268:663–706.

Shen, Y., and Forsyth, D. W. 1995. Geochemical constraints on initial and final depth of melting beneath mid-ocean ridges. *J. Geol. Geophys.* 100:2211–37.

Shimizu, N. 1995. Chromatographic chemical modifications during migration of MORB melts. *EOS, Trans. AGU* 76:266–7.

Sigurdsson, I. A., Kamenetsky, V. S., Crawford, A. J., Eggins, S. M., and Zlobin, S. K. 1993. Primitive island arc and oceanic lavas from the Hunter Ridge–Hunter Fracture Zone. Evidence from glass, olivine, and spinel composition. *Mineral. Petrol.* 47:149–69.

Sinton, C. W., Christie, D. M., Coombs, V. L., Nielsen, R. L., and Fisk, M. R. 1993. Near-primary melt inclusions in anorthite phenocrysts from the Galapagos platform. *Earth Planet. Sci. Lett.* 119:527–37.

Sobolev, A. V., and Chaussidon, M. 1996. H_2O concentrations in primary melts from suprasubduction zones amd mid-ocean ridges: implications for H_2O storage and recycling in the mantle. *Earth Planet. Sci. Lett.* 137:45–55.

Sobolev, A. V., Danyushevsky, L. V., Dmitriev, L. V., and Suschevskaya, N. M. 1989. High-alumina magnesian tholeiite as the primary basalt magma at mid-ocean ridge. *Geochem. Int.* 26:128–133.

Sobolev, A. V., and Dmitriev, L. V. 1989. Primary melts of tholeiites of oceanic rifts (TOR): evidence from studies of primitive glasses and melt inclusions in minerals. In: *Proceedings of the 28th International Geological Congress*, vol. 3, pp. 147–8. Washington, DC: IGC.

Sobolev, A. V., and Nikogosian, I. K. 1994. Petrology of long-lived mantle plume magmatism: Hawaii, Pacific, and Réunion Island, Indian Ocean. *Petrology* 2:111–44.

Sobolev, A. V., and Shimizu, N. 1993. Ultra-depleted primary melt included in an olivine from the Mid-Atlantic Ridge. *Nature* 363:151–4.

Stolper, E. 1980. A phase diagram for mid-ocean ridge basalts: preliminary results and implications for petrogenesis. *Contrib. Min. Pet.* 74:13–27.

Takahashi, E. 1986. Melting of a dry peridotite KLB-1 up to 14 GPa: implications on the origin of peridotitic upper mantle. *J. Geophys. Res.* 91:9367–82.

Tatsumi, Y., and Eggins, S. M. 1995. *Subduction Zone Magmatism*. Oxford: Blackwell Scientific.

Taylor, W. R., and Green, D. H. 1987. The petrogenetic role of methane: effect on liquidus phase relations and the solubility mechanism of reduced C – H volatiles. In: *Magmatic Processes and Physiochemical Principles*, special publication 1, ed. B. O. Mysen, pp. 121–38. Philadelphia: Geochemical Society.

Taylor, W. R., and Green, D. H. 1988. Measurement of reduced peridotite-C – O – H solidus and implications for redox melting of the mantle. *Nature* 332:239–352.

Taylor, W. R., and Green, D. H. 1989. The role of reduced C–O–H fluids in mantle partial melting. In: *Kimberlites and Related Rocks – Their Occurrence, Origin and Emplacement*, vol. 1, ed. J. Ross, pp. 592–602. Geol. Soc. Aust. Special Publication 14. Carlton: Geological Society of Australia.

Toramaru, A., and Fuji, N. 1986. Connectivity of melt phase in a partially molten peridotite. *J. Geophys. Res.* 91:9239–52.

Tsamerian, O. P., and Sobolev, A. V. 1996. Evidence for extremely low degree of melting under mid-Atlantic ridge at 14° N. *Abstracts, V. M. Goldschmidt Conference* 1:628.

Wallace, M. E., and Green, D. H. 1988. An experimental determination of primary carbonatite magma composition. *Nature* 335:343–6.

Wallace, M. E., and Green, D. H. 1991. The effect of bulk rock composition on the stability of amphibole in the upper mantle: Implications for solidus positions and mantle metasomatism. *Mineral. Petrol.* 44:1–19.

White, R. U., and McKenzie, D. P. 1989. Magmatism at rift zones: the general of volcanic continental margins and flood basalts. *J. Geophys. Res.* 94:7685–729.

White, W. M., and Duncan, R. A. 1996. Geochemistry and geochronology of the Society Islands: new evidence for deep mantle recyling. In: *Earth Processes: Reading the Isotopic Code*, ed. A. Basu and S. R. Hart, pp. 183–206. *AGU Monographs*, vol. 95. Washington, DC: American Geophysical Union.

Whitehead, J. A., Jr., Dick, H. J. B., and Schouten, H. 1984. A mechanism for magmatic accretion under spreading centres. *Nature* 312:146–8.

Woodhead, J., Eggins, S., and Gamble, J. 1993. High field strength and transition element systematics in island arc and back-arc basin basalts: evidence for multi-phase melt extraction and a depleted mantle wedge. *Earth Planet. Sci. Lett.* 114:491–504.

Wyllie, P. J. 1978. Mantle fluid compositions buffered in peridotite-CO_2-H_2O by carbonates, amphibole and phlogopite. *J. Geol.* 86:687–713.

Wyllie, P. J. 1987. Discussion of recent papers on carbonated peridotite, bearing on mantle metasomatism and magmatism. *Earth Planet. Sci. Lett.* 82:391–7.

Yaxley, G. M., Crawford, A. J., and Green, D. H. 1991. Evidence for carbonatite metasomatism in spinel peridotite xenoliths from W. Victoria, Australia. *Earth Planet. Sci. Lett.* 107:305–17.

Yaxley, G. M., and Green, D. H. 1994. Experimental demonstration of refractory carbonate-bearing eclogite and siliceous melt in the subduction regime. *Earth Planet. Sci. Lett.* 128:313–25.

Part Three

Structure and Mechanical Behaviour of the Modern Mantle

8

Seismic Structure of the Mantle: From Subduction Zone to Craton

B. L. N. KENNETT and R. D. VAN DER HILST

8.1. Introduction

Seismological techniques have provided much of the currently available information on the internal structure of the Earth, and in particular on the mantle. Early studies revealed the need for an increase in seismic velocity with depth in the Earth, and by 1915 Gutenberg was able to make a good estimate of the radius of the core. Knowledge of the Earth's internal structure was refined by iterative improvement of earthquake locations and the travel times for seismic phases through the Earth, so that in 1940 Jeffreys and Bullen were able to publish an extensive set of travel-time tables based on a model of both P-wave and S-wave velocities in the mantle. Their velocity profile was intentionally as smooth as possible, but it was not possible to avoid introducing a sharp change in velocity gradient near a depth of 400 km to account for the distinct change in the slope of travel-time curves at a distance of approximately 20° from the source, for both P and S waves. Subsequent studies have refined our conception of mantle structure to reveal the presence of discontinuities in velocity and zones of strong velocity gradients, which have been correlated with mineralogical phase changes (e.g., Jackson and Rigden, Chapter 9, this volume).

The presence of three-dimensional variations in the Earth's structure became apparent in regional differences in seismic travel times, and they became better understood once surface-wave observations demonstrated the significant differences in surface-wave dispersion between oceanic and continental regions (e.g., Knopoff, 1972). Surface-wave studies revealed the presence of a zone of decreased shear-wave velocities at depth and showed significant variations in the thickness of the overlying high-velocity zone ('lid') between different regions. The differences between the characteristics of the upper mantle in oceanic and continental regions led Dziewonski, Hales, and Lapwood (1975) to develop models with allowances for oceanic and continental character as well as an average radial model for the whole Earth.

In recent years, the quality and quantity of seismological data have improved sufficiently that it is possible to begin to resolve the three-dimensional structure within the Earth using a combination of information from the travel times for seismic phases, the free oscillations of the Earth, and long-period seismic waveforms. A consensus is developing on the largest-scale features in the aspherical structure. The ellipticity of the figure of the Earth provides a major component of spherical-harmonic order 2, and this order is also strongly represented in the velocity structure, particularly in the transition zone in the upper mantle (e.g., Masters et al., 1982). There is substantial power in the low-order heterogeneity of the Earth, and some authors (e.g., Su and Dziewonski, 1991) have suggested that the spectrum of heterogeneity is 'red' (i.e., dominated by the large-scale features). However, there is ample evidence for significant heterogeneity on smaller scales than have yet been revealed by global studies.

In particular, the upper mantle is a zone of major variability and relatively strong horizontal gradients in seismic properties. Subducted slabs are associated with large, localized contrasts in velocity. Detailed P-wave tomography based on the inversion of seismic travel times has revealed the complex patterns of subduction in many regions (e.g., van der Hilst et al., 1991). The influence of subduction is largest in the upper mantle, but in many places subducted material appears to have penetrated directly or indirectly into the lower mantle. The high seismic velocities associated with the colder subducted material seem to be the dominant mode of smaller-scale heterogeneity in the lower mantle, which otherwise appears to be characterized by relatively low gradients of heterogeneity. However, the degree of variability increases as the core–mantle boundary is approached, and the D'' layer in the 300-km zone at the base of the mantle shows considerable variability on a wide range of scales revealed by studies with many different types of probes.

8.2. The Seismic Structure of the Mantle

The variation of seismic properties within the Earth is inferred from the analysis of seismograms in a variety of ways and is dominated by a radial dependence. However, three-dimensional variations are manifest in the crust and in all parts of the mantle; currently the most effective representation of such three-dimensional structure is as a perturbation to a reference radial model.

For this radial reference model, the major sources of information come from the travel times for seismic phases and from the free oscillations of the Earth. The travel times provide constraints on the seismic wave speeds within the Earth; the frequencies of the normal modes provide additional information on the density distribution and the attenuation profile for seismic waves. A thorough discussion of seismological methods has been provided by Lay and Wallace (1995).

The zones of greatest heterogeneity lie in the uppermost part of the mantle (depths shallower than about 250 km) and near the core–mantle boundary. In such zones, horizontal gradients in velocities can approach the radial gradients, and so it is difficult to define a representation of an 'average' structure in those parts of the Earth.

Various techniques are currently being employed to build up a body of information on the three-dimensional structure of the Earth. To determine the distribution of velocities for compressional (P) waves, the dominant approach is the use of travel-time tomography, primarily based on the impressive collection of arrival-time data for different seismic phases assembled by the International Seismological Centre (ISC). Recently, additional information on travel times has become available from analyses of long-period seismograms by correlation techniques (e.g., Woodward and Masters, 1991; Bolton and Masters, 1994).

In determining the velocity distribution for shear (S) waves, much of the available information has come from techniques based on the fitting of long-period waveform segments to computed seismograms. Such information has been complemented by the use of ISC travel-time data and the long-period S times and differential times between different S phases (Woodward and Masters, 1991).

The techniques of waveform fitting used in the global context are based on summation of the normal modes of the Earth in a perturbation treatment to include the influence of three-dimensional structure. Additional information on such structure comes from analyses of the free oscillations through the local frequencies of modes and the splitting of the modal frequencies produced by interaction with heterogeneity (Masters and Ritzwoller, 1987).

8.3. The Dominant Radial Structure in the Mantle

At present, nearly all the techniques designed to assess the three-dimensional structure of the Earth are based on a representation in terms of deviations from a reference model (which normally is a spherically symmetrical model). The nature of the reference model is therefore of considerable importance.

Dziewonski et al. (1975) introduced a new style of representation for such reference models with the PEM model, which was defined in terms of a limited number of radial segments, within each of which the seismic velocities and densities were defined by polynomials (up to cubic) in radius. The advantage of such a form of representation is that the entire model is defined by a relatively small number of parameters. In consequence, the computational labour of retrieving a model from the observations is reduced.

Such parametrized models have been used extensively since 1975, notably in the PREM model of Dziewonski and Anderson (1981), which endeavoured to take account of a very wide range of information from the free oscillations of the Earth, dispersion of surface waves, travel times for the major seismic phases, and differential travel times. The PREM model allowed for the frequency dependence of seismic velocities associated with anelastic attenuation within the Earth, and it also introduced the concept of transverse isotropy in the uppermost mantle to try to reconcile the dispersion characteristics of Rayleigh and Love waves. The PREM model has been extensively used in work involving the normal modes of the Earth, and it is frequently used as a reference model in global studies.

The same style of parametrization has been adopted in the recent models *iasp91* (Kennett and Engdahl, 1991) and *sp6* (Morelli and Dziewonski, 1993) for P and S velocities derived from travel-time information. With these improved velocity models, it is possible to refine the locations of seismic events and thus obtain an updated set of empirical travel times for a wide range of seismic phases. Kennett, Engdahl, and Buland (1995) have constructed a new model, *ak135*, for the seismic velocities using such improved travel times, and in order to fit the observed behaviour they were forced to employ a more complex parametrization in the lower mantle and core.

The merits of improved reference models for seismic velocities have been demonstrated in seismic-tomography studies. Event relocation in such a model can lead to nearly as much reduction in data variance as does direct estimation of three-dimensional structure using raw ISC data, and a further improvement in fit can then be obtained by superimposing three-dimensional structure on the reference model (e.g., Inoue et al., 1990; van der Hilst et al., 1991).

8.3.1. *The Upper Mantle and Transition Zone*

Such global models for upper-mantle structure represent an average of structures from many regions, even when, as for travel-time studies, the information is dominantly from continental regions. Studies of the three-dimensional structure of the mantle indicate major contrasts, with scale lengths of the order of 3,000 km or so, and within such a zone it is possible to develop a regional representation of the major features of the velocity structure in the upper mantle. Grand and Helmberger (1984a) have used observations of long-period SH waveforms in North America to construct two characteristic models: *sna* for the shield regions, and *tna* for the tectonic regions. Grand and Helmberger (1984b) have made a comparable study for the northwest Atlantic Ocean. These three models differ markedly for depths less than 250 km. For the northern Australian region, Kennett, Gudmundsson, and Tong (1994) have produced both P and S velocity models from the same events using broadband seismograms recorded in northern Australia from the earthquake belt through Indonesia and New Guinea. This study indicated the presence of differences in structure along groups of paths extending into the transition zone, and it confirmed the findings from short-period studies for P waves summarized by Dey et al. (1993). Nolet, Grand, and Kennett (1994) provided a broad survey of heterogeneity and velocity variability in the upper mantle, including results from other areas.

Figure 8.1 shows the shear-wave velocity distributions from the North American and Australian studies, revealing the general concordance of the shield models *sna*

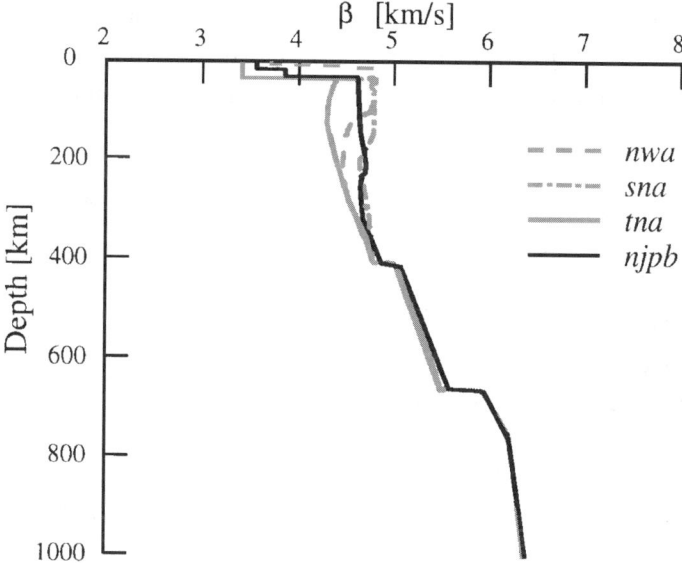

Figure 8.1. Shear-wave velocity distribution from refracted-wave studies: *sna* and *tna* models for shield and tectonic North America (Grand and Helmberger, 1984a), *nwa* for the northwest Atlantic Ocean (Grand and Helmberger 1984b), and *njpb* for the northern Australian shield (Kennett, Gudmundsson, and Tong, 1994).

and *njpb* as compared with the tectonic and oceanic models *tna* and *nwa*. The major differences between the models occur in the shallower structure. We see the major features of upper-mantle structure as revealed by refracted-wave studies displayed in these models. A zone of relatively high velocity just below the crust–mantle boundary is sustained for some depth before the shear velocity decreases (with an increase in shear-wave attenuation) and then recovers as the 410-km transition is approached. This zone of marked velocity change is represented here as a sharp discontinuity, but it may be spread over a few kilometres. The next major transition lies near a depth of 660 km, with a comparable contrast of about 4–6% in seismic velocities. Refracted-wave studies have not produced evidence for an intermediate discontinuity (Jones, Mori, and Helmberger, 1992; Cummins et al., 1992), but techniques based on the stacking of long-period seismograms (Shearer, 1991) and on ScS reverberations for SH waves (Revenaugh and Jordan, 1991) suggest the presence of a transition near 520 km. If this feature were spread over a zone 30 km in depth, or if a component of density change were involved, it would be possible to reconcile the two sets of observations.

The broadband studies in northern Australia have also revealed the presence of significant shear-wave splitting in the horizontally refracted waves used in the analysis (Tong, Gudmundsson, and Kennett, 1994). Such splitting is compatible with the presence of about 1% anisotropy in the low-velocity zone at a depth below 210 km beneath northern Australia, a zone also marked by high attenuation (Gudmundsson, Kennett, and Goody, 1994). Most other observations of shear-wave splitting have come from waves with relatively steep paths through the mantle (e.g., SKS, SKKS), and so it is difficult to localize the anisotropy that gives rise to the splitting. In Australia, there is evidence for a complex regime of anisotropy giving rise to different levels of splitting and different directions of fast polarization at different frequencies (Clitheroe and van der Hilst, 1997).

8.3.2. The Lower Mantle

As noted earlier, the upper mantle is a zone with considerable regional variability, but in the lower mantle, beneath about 750 km, there is much greater consistency between the velocities for different reference models. Indeed, the general trends in velocity gradients are very well defined, but there are some differences between models, depending on the different procedures used in their construction.

The structure of the lower mantle is illustrated in Figure 8.2 with the reference model *ak135* for which the P-wave velocity (α) and S-wave velocity (β) have been determined by analysis of travel times for a broad range of seismic phases. The distributions of density (ρ) and seismic-wave attenuation (Q_α, Q_β) have then been constructed so that the reference model will be compatible with both travel times and free oscillations (Montagner and Kennett, 1996).

From a mineralogical point of view, the bulk sound velocity V_ϕ is of major interest (e.g., Jackson and Rigden, Chapter 9, this volume), but this quantity cannot

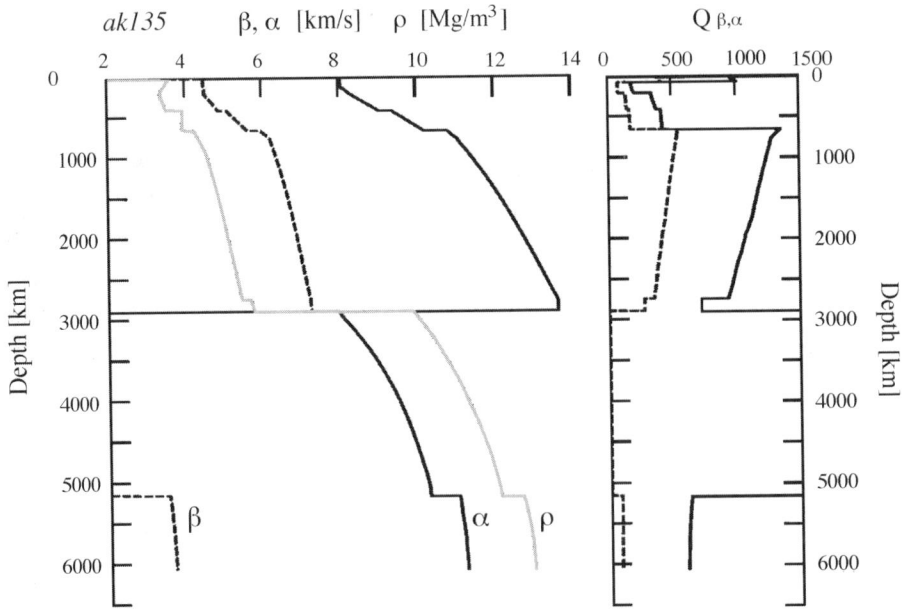

Figure 8.2. The *ak135* Earth model indicating P velocity α and S velocity β (Kennett, Engdahl, and Buland, 1995), as well as density ρ and attenuation (Q_α, Q_β) (Montagner and Kennett, 1996). The zone of density inversion in the upper mantle is an artifact of a constraint to monotonically increasing shear velocity in the upper mantle.

be extracted directly from seismic data; instead it has to be calculated by combining the P-wave velocity α and the shear-wave velocity β:

$$V_\phi = \sqrt{\alpha^2 - \frac{4}{3}\beta^2} \tag{1}$$

It is therefore important that there be consistency between the assumptions underlying the relevant P and S velocities, such as in the *sp6* and *ak135* models.

The bottom 200 km of the lower mantle (the D″ layer) is a zone of relatively strong heterogeneity, as evidenced by a wide range of studies (e.g., Lay, 1994). This zone, just above the core–mantle boundary, may well have heterogeneity comparable to that near the Earth's surface and may serve as a repository for the debris from mantle processes that potentially can be remobilized by the heat emerging from the core.

8.4. Detailed Three-Dimensional Structure in the Mantle

Nearly all of the available information on the details of the three-dimensional structure within the Earth comes from the analysis of departures of the observed properties of the seismic wavefield from the predictions made with a reference model. For a majority of studies, the significant quantity is a time shift in the arrival of a phase, measured directly for a short-period body wave, by correlation techniques for long-period body waves and surface waves, or by a waveform inversion where the record contains a number of arrivals, as, for example, the shear-wave field containing both body-wave and surface-wave components. The travel-time residuals for body waves or modified dispersion characteristics for surface waves can then be used as the basis for recovering a three-dimensional model. The essence of seismic tomography is to relate the time differential back to the structure from which it arises. Such a time residual will represent the cumulative effect of all the deviations from the reference model encountered along the propagation path. With multiple crossing paths it is possible to begin to reconstruct the spatial distributions of velocity perturbations, commonly using the assumption that the ray path is unperturbed from that in the reference model. In principle, the results should be improved by an iterative procedure, with recomputation of ray paths in the new three-dimensional model followed by inversion to determine perturbations from the new heterogeneous reference model. Such a procedure is computationally intensive, but is desirable for detailed studies of the shallower regions of the mantle, where there are large variations in seismic velocities.

The resolution of velocity variations and the quality of the resulting images of velocity structure are strongly dependent on the ray-path coverage that can be attained. A number of different procedures have been devised for handling the large systems of linearized equations that must be solved in the tomographic inversion to reconstruct velocity structure; a comprehensive discussion of current techniques has been provided by Iyer and Hirahara (1993).

The amplitudes of seismic waves have been exploited to a lesser degree. However, Romanowicz (1995) has begun to use such amplitude information to define the large-scale variations in seismic attenuation. On a smaller scale, studies of seismic amplitudes have helped to define stochastic models for the velocity variations that lie below the resolution of direct methods (e.g., Kennett and Bowman, 1990).

8.4.1. Subduction Zones and Their Environment

At depths greater than 100 km, slabs of subducted lithosphere are probably the major and best-constrained aspherical structures in the Earth's interior. They play an essential role in the cooling of the Earth's mantle by convection and the consequent recycling of oceanic lithosphere. For various reasons, the structure of subducted slabs is better known than that of plumes, the secondary mode of mantle convection (e.g., Davies, Chapter 5, this volume; Griffiths and Turner, Chapter 4, this volume). The narrow tail of a plume is more difficult to detect by seismological methods, although success in imaging a plume tail has been reported (Nataf and Van Decar, 1993). Moreover, an unambiguous indication of the presence of subducted lithosphere is the occurrence of intermediate-depth earthquakes (focal depth 70–300 km) and deep earthquakes (focal depth >300 km) beneath convergent-plate boundaries.

Thermal models of the subduction process indicate that the temperature difference between the cold core of a slab and the ambient mantle at the same depth is likely to be 500–1,000°C (e.g., Toksöz, Minnear, and Julian, 1971). This thermal anomaly changes the seismic wave speed within the subducted slab, so that its wave speed is several percent higher than that for the surrounding material at the same depth. The effect of the subduction regime is therefore to produce an environment in which seismic waves that propagate within the slab travel faster than would be expected for a global reference model. Surface-reflected phases sampling the region in the back arc above the slab will be delayed, relative to the reference, because of the combination of temperature effects and volatiles released from the slab (e.g., Green and Falloon, Chapter 7, this volume). These differences will be reflected in the arrival times of seismic waves at different stations, depending on the propagation path. The density of earthquake sources in subduction zones aids the delineation of seismic structure by allowing measurements of the deviations in arrival times from those predicted by the reference model.

The first tomographic images of subducted lithosphere were produced in the late 1970s (Hirahara, 1977; Hirahara and Mikumo, 1980) using several thousand P-wave travel times. Since then, the quantity and quality of seismic data have improved, and increased computer power now allows the simultaneous interpretation of many millions of data. In many inversions the heterogeneous structure has been assumed to be confined to the region of interest around the subduction zone. Such an assumption can be very effective for situations where most of the ray paths lie within the structure of interest. For example, tomography has yielded very detailed information about the shallow part of the subduction zone and the relationship between wave-speed anomalies and shallow seismicity and arc volcanism in back-arc regions, such as northern Honshu in Japan (Hasegawa et al., 1991). When information is used from a global distribution of stations, allowance must be made for possible structure outside the zone of interest. This can be done in part by introducing station corrections at teleseismic stations. However, with sufficient computing resources, a preferable approach is to undertake a low-resolution tomographic inversion for the velocity structure of the whole Earth with an embedded high-resolution model for the environs of the subduction zone. That approach was pioneered by Fukao et al. (1992) and taken to higher resolution by Widiyantoro and van der Hilst (1996).

In addition to the representation of the model, an important issue is the nature of the data employed and the reference model. Van der Hilst et al. (1991) have demonstrated that the depth phase pP can commonly be extracted from travel-time observations. Relocation of the earthquake sources using all available phases, with an improved reference model such as *iasp91* (Kennett and Engdahl, 1991), provides a significant improvement in data fit before three-dimensional structure is introduced, and the resulting velocity images show better definition because the path coverage is improved by the inclusion of the surface reflections pP. With well-designed regional tomographic studies, structural variations over distances of about 100 km can be detected (e.g., Spakman, 1991; van der Hilst, 1995), and such studies can provide significant information on the large-scale structure of subducted slabs and their interaction with the mantle transition zone. Global models are just beginning to approach the resolution required for the investigation of slab structure and, in particular, the influence of slabs on the lower mantle (Inoue et al., 1990; Vasco et al., 1994; van der Hilst, Widiyantoro, and Engdahl, 1996).

Slabs of subducted lithosphere, partly delineated by deep seismicity in the Wadati-Benioff zones, represent tangible trajectories of convective flow in the Earth's mantle and have therefore played a central role in attempts to resolve the controversial issue of the depth scale of mantle convection. In addition to geochemical arguments (Anderson and Bass, 1986; Ringwood and Irifune, 1988; Ringwood, 1991), the cessation of seismicity above the 660-km seismic discontinuity has been used as evidence that lower-mantle flow is separate from convection in the upper mantle and transition zone. However, a range of seismological studies, based on analyses of the patterns of travel-time residuals, have provided evidence for aseismic continuation of slabs into the lower mantle (Jordan and Lynn, 1974; Jordan, 1975, 1977; Creager and Jordan, 1984; Fischer, Creager, and Jordan, 1991). These findings have been supported by thermal models (Wortel, 1982) and used as arguments for whole-mantle convection. However, the issue of slab penetration and the style of the convective process in the mantle have remained controversial; for reviews, see Olson, Silver, and Carlson (1990), Silver, Carlson, and Olson (1988), Davies and Richards (1992), Ringwood (1991), Lay (1994), and Davies (Chapter 5, this volume).

Recently, results of numerical modelling of mantle flow (e.g., Machetel and Weber, 1991; Tackley et al., 1993, Davies, Chapter 5, this volume) and improved seismic imaging (van der Hilst et al., 1991; Fukao et al., 1992) have begun to converge to a hybrid-flow model in which neither layered convection nor whole-mantle convection is regarded as a steady-state process. The model of direct slab penetration, as advocated by Jordan and co-workers, provides an oversimplified and incomplete description of the flow pattern associated with subduction. Tomographic imaging has confirmed the conclusions reached in early studies that deep slabs are present in the lower mantle beneath Middle America (Grand, 1987, 1994; van der Hilst and Spakman, 1989) and beneath several western Pacific arcs (van der Hilst et al., 1991; Fukao et al., 1992; van der Hilst, 1995). Indeed, high-resolution global images (van der Hilst, Widiyantoro, and Engdahl, 1996) have revealed relatively narrow regions of high P-wave speed in the lower mantle beneath most major subduction zones, as can be seen in Figure 8.3, which displays a cross section through global structure at a depth of 1,300 km. However, detailed seismic imaging is providing mounting evidence that the interactions of the slabs with the transition zone between the upper mantle and the lower mantle are quite variable: Locally, some slabs penetrate into the lower mantle without much apparent obstruction, whereas elsewhere they are strongly deformed and remain stagnant (at least temporarily) in the transition zone (Zhou and Clayton, 1990; van der Hilst et al., 1991; Fukao et al., 1992).

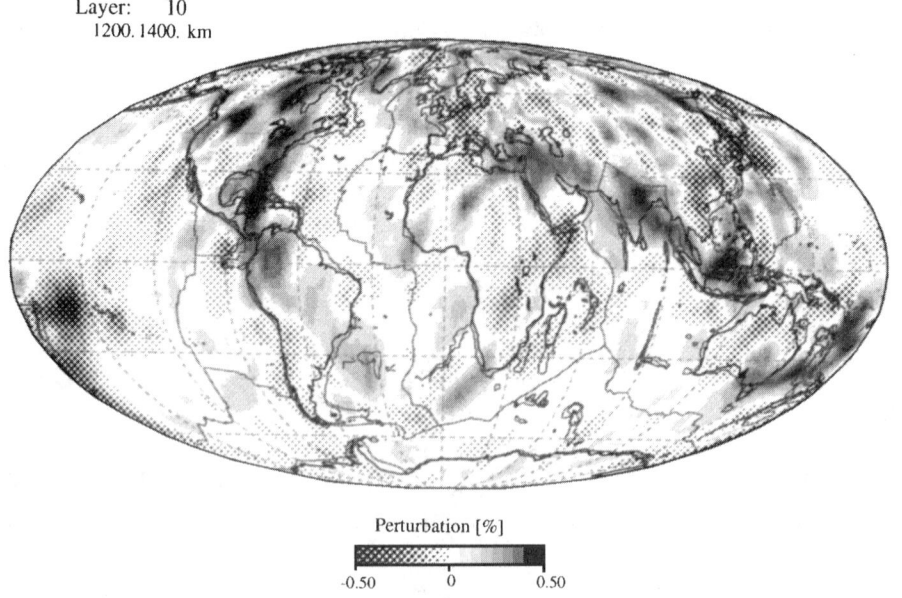

Figure 8.3. Lateral variations of P-wave speed at a depth of approximately 1,300 km determined by a tomographic inversion for the whole globe, with an approximate resolution of 3 × 3° (van der Hilst, Widiyantoro, and Engdahl, 1996).

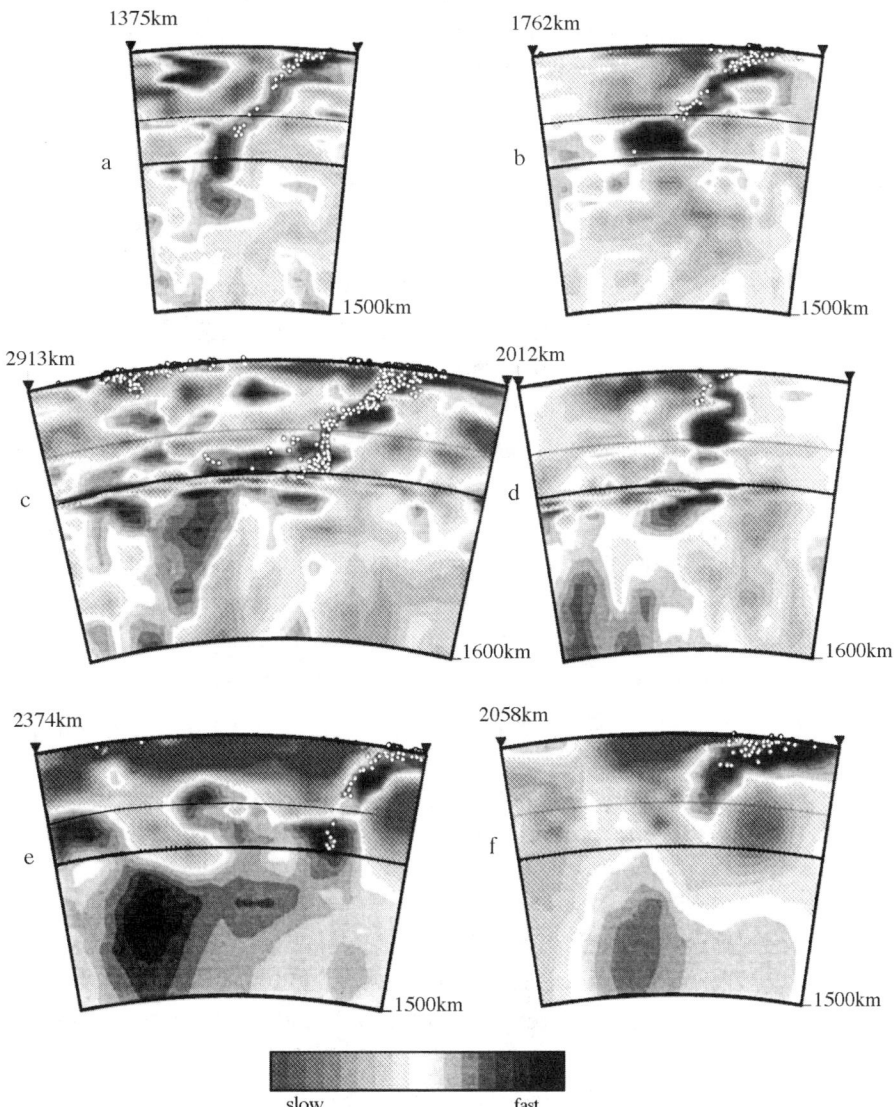

Figure 8.4. Cross sections through tomographic images of P-wave velocity structure for a number of subduction zones, illustrating the variety in styles of subduction: (a) northern Kuriles, (b) southern Kuriles, (c) northern Tonga arc, (d) Kermadec region, (e) Java, (f) Sumatra. The superimposed lines indicate the positions of the 410-km and 660-km discontinuities. Sections a and b are from the work of van der Hilst et al. (1991), c and d from van der Hilst (1995), and e and f from Widiyantoro and van der Hilst (1996).

The following examples of the complex behaviour of slabs in the transition zone beneath individual arc systems have been derived from studies of the structures of deep slabs in the western Pacific (van der Hilst et al., 1991; van der Hilst, 1995) and Indonesia (Widiyantoro and van der Hilst, 1996). Figure 8.4a depicts the trajectory of the Pacific plate subducting beneath the Sea of Okhotsk along the Kurile Trench.

The slab is represented by a steeply dipping structure of higher-than-average wave speed which is partly delineated by seismicity. The zone of high wave speeds can be detected to a depth of about 1,200 km, which suggests that the slab penetrates into the lower mantle. Figure 8.4b depicts the trajectory of the subducting Pacific plate beneath the southern part of the Kurile Trench. In contrast to the situation farther to the north, the slab appears to be deflected in the transition zone; this inference is substantiated by the recent occurrence of a deep earthquake beneath Sakhalin Island that appears to have been located in the sub-horizontal part of the slab. A comparable contrast in subduction behaviour has been detected beneath the Izu Bonin and Mariana arcs: The change from a sub-horizontal slab beneath the Izu Bonin arc to a sub-vertical slab farther to the south is consistent with the pattern of deep seismicity in that region (Okino et al., 1989; Lundgren and Giardini, 1992; van der Hilst and Seno, 1993). Lateral variations in the shape of the deep slab are not restricted to the northwest Pacific subduction systems. Figures 8.4c and 8.4d show the lateral variations in the shape of the Tonga slab (van der Hilst, 1995), and Figures 8.4e and 8.4f show the Sunda slab (Widiyantoro and van der Hilst, 1996). The images of the northern Tonga slab (Figure 8.4c) and the Java slab (Figure 8.4e) are of particular interest because they provide evidence for slab penetration into the lower mantle even though the part of the slab in the transition zone is sub-horizontal.

However, a word of caution is in order. Interpretation of the velocity images to suggest that slabs penetrate into the lower mantle ignores the possibility that the lower-mantle anomaly may be caused by thermal coupling between the transition zone and lower mantle (i.e., the cooling of lower-mantle material by thermal diffusion and the subsequent sinking of that cooled lower-mantle material). The interpretation in terms of slab penetration to a depth of 1,200 km also rejects the possibility that the 660-km discontinuity may be depressed by several hundreds of kilometres. The independent analysis of Shearer and Masters (1992) suggests depression of the depth of this discontinuity by about 50 km beneath subduction zones, and that is concordant with current mineralogical models for the associated phase changes (e.g., Jackson and Rigden, Chapter 9, this volume).

Van der Hilst and Seno (1993) noticed that the sub-horizontal slabs in the transition zone were located beneath regions characterized by recent back-arc spreading due to fast, oceanward migration of the deep-sea trenches. The Kurile slab is laid down in the transition zone beneath the South Kurile Basin, whereas the sub-horizontal slab related to subduction at the Izu Bonin Trench is located beneath the Shikoku Basin. A similar relationship applies to the sub-horizontal part of the northern Tonga slab, which is located beneath the Oligocene South Fiji Basin and the Pliocene–present Lau Basin. Van der Hilst and Seno (1993) and van der Hilst (1995) have used a combination of findings from experimental fluid dynamics (Kincaid and Olson, 1987) and numerical simulations (Christensen and Yuen, 1984; Machetel and Weber, 1991), together with petrological data pertinent to the dynamics of slabs in the transition zone (Ringwood and Irifune, 1988; Ringwood,

1991), to argue that the kinks in the subduction trajectory can be explained by the interaction between the lateral translation of the source of flow and the tendency towards vertical flow in a stratified mantle.

It is generally accepted that slabs in the transition zone encounter resistance against further, unobstructed sinking owing to a combination of increased viscosity in the lower mantle, the dynamical effects of the perovskite-forming phase changes in the subducting slab, and, possibly, a small increase in intrinsic density of material in the ambient mantle due to compositional differences. If the lateral migration of a trench is sufficiently fast relative to the sinking speed in the lower mantle, the slab can be laid down in the transition zone. This model is supported by numerical tests (Zhong and Gurnis, 1995) and fluid dynamical modelling (Griffiths, Hackney, and van der Hilst, 1995) and does not depend critically on the nature of the boundary between the upper mantle and the lower mantle. The boundary slows down radial flow but does not completely eliminate mixing between the upper mantle and the lower mantle.

Such observations can be explained by radial variations in viscosity alone (Griffiths et al., 1995; Griffiths and Turner, Chapter 4, this volume). However, a mechanism of 'megalith' formation representing the cumulation of slab material at the top of the lower mantle, as postulated by Ringwood and co-workers (Ringwood and Irifune, 1988; Ringwood, 1991; Kesson, Fitz Gerald, and Shelley, 1994), would be compatible with the observations if the lifetime of such a structure were about 10–20 million years, rather than the 100 million years suggested by Ringwood and Irifune (1988). Beneath convergent-plate margins, relatively cold material can accumulate in the transition zone and gather sufficient negative buoyancy to accomplish the phase change to a denser assemblage and subsequently be 'flushed' into the lower mantle. Such events, also referred to as 'avalanches', have been simulated in numerical models (Tackley et al., 1993; Davies, Chapter 5, this volume): When a trench is stationary in space, the accumulation of material will occur in a small region in the transition zone, thus facilitating the flushing to greater depths. In contrast, if a trench is migrating sufficiently fast, the subducted material would be laid down on the 660-km discontinuity, thus retarding the flow to greater depths.

The reshaping of Ringwood's megalith is likely to be an important phenomenon on relatively short geological timescales. Episodes of fast trench migration and concurrent back-arc spreading typically last no more than several tens of millions of years; so it is to be expected that the complex behaviour of slabs in the transition zone will have similar timescale. On much longer timescales, relative motions are less important, and the amount of mixing between the upper mantle and the lower mantle will largely depend on differences in intrinsic density. Low-resolution seismic images of lower-mantle structure probably are more representative of flux averaged over a long time period, whereas the high-resolution images of slabs from regional studies represent 'snapshots' of a complex, transient dynamical process. The observation of slab-like structural features in the lower mantle beneath the zones of subduction of old, thermally mature plates (Figure 8.3) suggests that eventually a very substantial amount of subducted material becomes entrained in lower-mantle flow.

8.4.2. Structure beneath Continental Regions

The recycling of oceanic lithosphere is very efficient. Indeed, the oldest ocean floor that currently resides at the Earth's surface was created in Jurassic times, some 200 million years ago. In contrast, the oldest parts of continents, the Archaean shields, are almost 4 billion years old. Such continental regions may be stable tectonic provinces at the present day, but the continental lithosphere and mantle can be highly heterogeneous, reflecting a long history of reworking. The imaging of continental regions by means of body-wave information is often complicated owing to the absence of sufficient natural sources. Interpretation of surface-wave data generally provides a more efficient means to study intraplate regions characterized by low levels of seismicity, albeit with somewhat lower resolution owing to the larger wavelengths of the seismic waves considered.

Global images of shear-wave structure in the upper mantle derived principally from waveform matching of long-period seismograms (particularly surface waves) supplemented by differential times between the S and SS phases (e.g., Su, Woodward, and Dziewonski, 1994) show high shear-wave speeds associated with the cratonic regions of the continents, extending to depths of 300 km or more, reminiscent of the 'tectosphere' hypothesis of Jordan (1975, 1988). The horizontal extent of such zones can be quite large, extending 3,000 km or more. However, the vertical resolving power of global studies is somewhat limited, and even in the detailed study by Grand (1994) for SH wave structure in the Western Hemisphere, vertical resolution was limited to about 75 km. As a result, it is difficult to resolve any detail in the transition between different continental regions, particularly the character of the transition at depth between the younger parts of continental assemblages and the older cratons.

8.4.2.1. Body-Wave Studies

Studies of the seismic structure beneath northern Australia have exploited the active seismicity in the seismic belt extending along the Indonesian arc, through New Guinea, and on to Vanuatu, using arrays of seismic stations. Such arrays have allowed dense sampling of the seismic wave field in a limited time and have revealed regional variations in seismic wave structure of about 1% over a horizontal distance of 1,000 km (Dey et al., 1993). These variations are associated with structure at the base of the lithosphere (which occur near 210 km of depth in this region) and within the transition zone.

The use of arrays of stations not only allows determination of radial structure but also provides constraints on the heterogeneity in seismic wave speeds. Kennett and Bowman (1990) have suggested the presence of heterogeneity with about ±1% fluctuations in P-wave speeds on scales of 200–300 km, superimposed on larger-scale variations. This analysis was based on amplitude and waveform variability for individual events recorded in a number of portable-array experiments with different array dimensions and spacings. The scale length is small enough to be difficult to resolve directly (particularly using refracted waves), and this style of heterogeneity may need to be represented in a stochastic manner. The influence of medium- and small-scale heterogeneity is likely to be particularly important for the regions where radial velocity gradients are small, as, for example, in the uppermost mantle, because horizontal velocity gradients can then dominate the behaviour of the seismic wave field (e.g., Kennett, 1993).

Grand (1994) has endeavoured to extend the analysis of S waves and their multiple reflections SS and SSS to provide coverage of three-dimensional structure from the surface down to the core–mantle boundary. The resulting images confirm the differences in the dominant structures between shields and tectonic regions, as summarized in the *sna* and *tna* models of Grand and Helmberger (1984a,b), with the largest differences in structure occurring above 250 km. There are, however, more subtle but persistent variations throughout the upper mantle, even though the degree of heterogeneity is relatively low in the transition zone.

8.4.2.2. *Surface-Wave Imagery*

In order to obtain high-resolution images of the shallower part of the Earth, we need to use seismological probes whose sensitivity is greatest in this region. Fortunately, we are able to exploit the surface-wave portion of the shear-wave field, which commonly represents the largest-amplitude portion of broadband seismograms. The fundamental-mode Love waves and Rayleigh waves are preceded by higher-mode surface waves that represent the superposition of multiply reflected S waves interacting with the free surface. The fundamental-mode energy is concentrated near the surface, but the penetration increases as the frequency is reduced because the S wavelength is longer. The higher modes provide both deeper penetration at comparable frequencies and complementary information on shallow structure, because the pattern of sampling is rather different.

The worldwide network of digital seismic stations has now developed to sufficient coverage of the Earth that over a period of years it is possible to achieve good sampling of most of the surface. Trampert and Woodhouse (1995) have used the full available data set to extract the geographical distribution of the phase velocities of surface waves at different periods. This analysis has concentrated on the dispersion of the fundamental-mode Love and Rayleigh waves at frequencies up to 0.025 Hz (40-s period) and shows features that correlate well with the shear-velocity models from earlier studies, whilst attaining higher resolution in many areas.

Much of the analysis of surface waves has been based on the assumption of direct propagation from source to receiver. However, Snieder (1988) has demonstrated that waveform inversion using scattering theory can be used to image off-path features. Laske (1995) has used the anomalies in the direction of propagation of surface-wave trains to constrain three-dimensional structure.

Although global coverage is quite good, the interstation spacing in many areas exceeds 1,000 km on the continents, and there are only a few stations in oceanic areas. In consequence, if the details of the structure in a region are required, there is a need to incorporate additional information, using, for example, deployments of additional portable broadband stations. This approach was employed by Nolet (1990) in a study of structure beneath Western Europe using a linear array (NARS) of 14 instruments extending from Malaga in southern Spain to Göteborg in Sweden. In order to exploit the full information content of the shear-wave field, Nolet introduced the two-step technique of 'partitioned waveform inversion'. In the first step, a nonlinear waveform inversion is performed on individual seismograms in order to determine the averaged shear-wave structure along the great-circle paths between source and receiver; this requires the matching of observed waveforms with comparable portions of theoretical seismograms, synthesized by mode summation. Separate analysis windows are used for the fundamental mode and the higher modes, to improve the vertical resolution of the structure (see Figure 8.6 for an example). The second step is to combine all of the different constraints on the model structure from the different seismograms in a linear tomographic inversion to construct a laterally varying velocity structure. Nolet (1990) used this approach to determine the large-scale structure in two dimensions under the linear NARS array. Kennett and Nolet (1990) have demonstrated that the method is insensitive to the presence of small-scale heterogeneity provided that analysis is restricted to frequencies below 0.15 Hz.

The partitioned-waveform-inversion approach has been been extended by Zielhuis and Nolet (1994) to determine the three-dimensional variation in shear-wave structure beneath Western Europe and has revealed a very strong contrast between the structures on either side of the Tornquist-Teisseyre zone, extending deeper than 200 km. This boundary marks the junction between Phanerozoic rocks to the southwest and Precambrian material to the northeast. The contrast of several percent in wave speed can be discerned in global velocity images, but the regional study achieves a much more effective localization.

Studies in North America and Europe can build on a good basic network of permanent seismological stations, but a rather different approach must be taken for other continents. In Australia, a sequence of temporary deployments of arrays of up to 12 portable broadband seismometers has been used to synthesize a continentwide array (van der Hilst et al., 1994). This project (SKIPPY) has been designed to exploit the high level of regional seismicity along the active plate boundaries surrounding Australia: the subduction zones to the east and north, with very high seismic activity and many deep earthquakes, and the mid-oceanic ridges to the south and west, with shallow and less frequent seismicity (Figure 8.5, top). The sequence of temporary

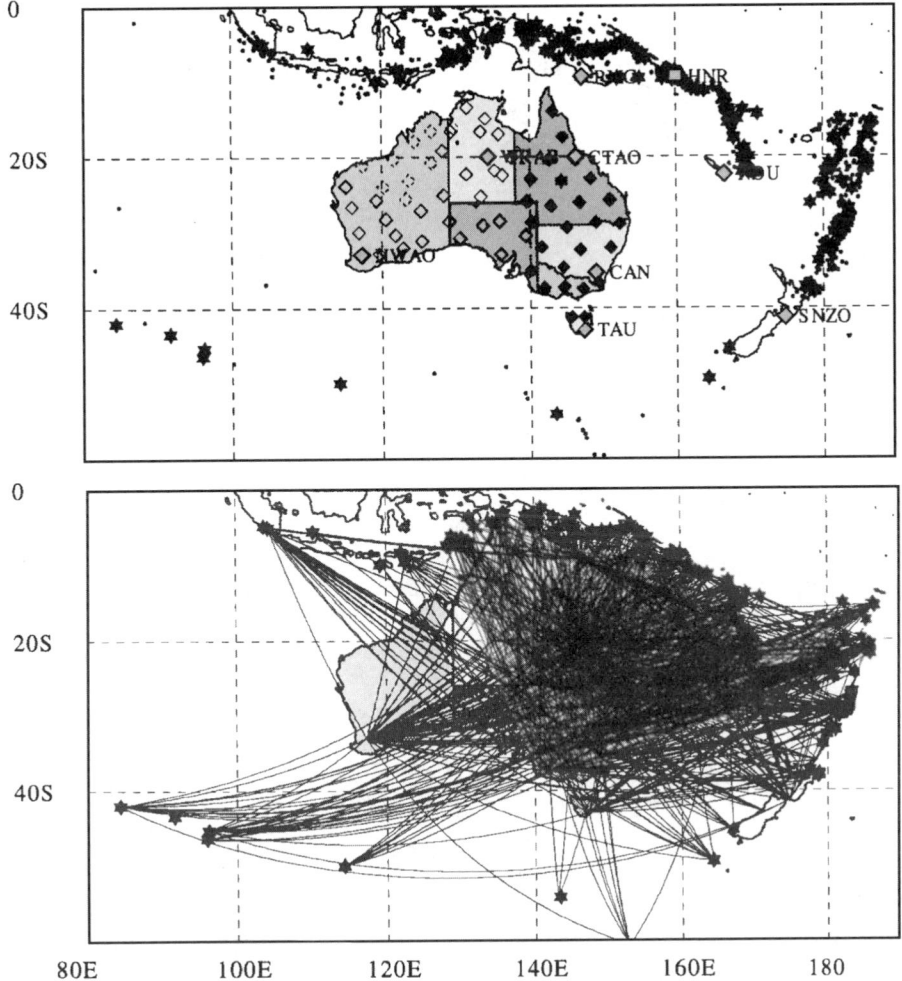

Figure 8.5. Top: Station configuration for the SKIPPY-array deployments relative to the regional seismicity. Bottom: Wave-path coverage for eastern Australia from the SK1, SK2, and BAS deployments.

deployments has led to excellent path coverage (Figure 8.5, bottom), so that it is possible to resolve structural features with horizontal scales of 200 km or more beneath the eastern part of the Australian continent and the oceanic and submerged continental regions between the continent and the convergent-plate boundaries to the north and east. This resolution is substantially better than that of the best current global models (Trampert and Woodhouse, 1995).

The structural analysis has exploited the partitioned-waveform-inversion technique, which is particularly useful for the Australasian region because of the regular occurrence of deep earthquakes beneath the surrounding convergent margins that generate strong higher-mode surface waves. The increased depth resolution that is obtained from the inclusion of higher modes is illustrated by the records of a Vanuatu earthquake at stations in Queensland (Figure 8.6, top). The observed

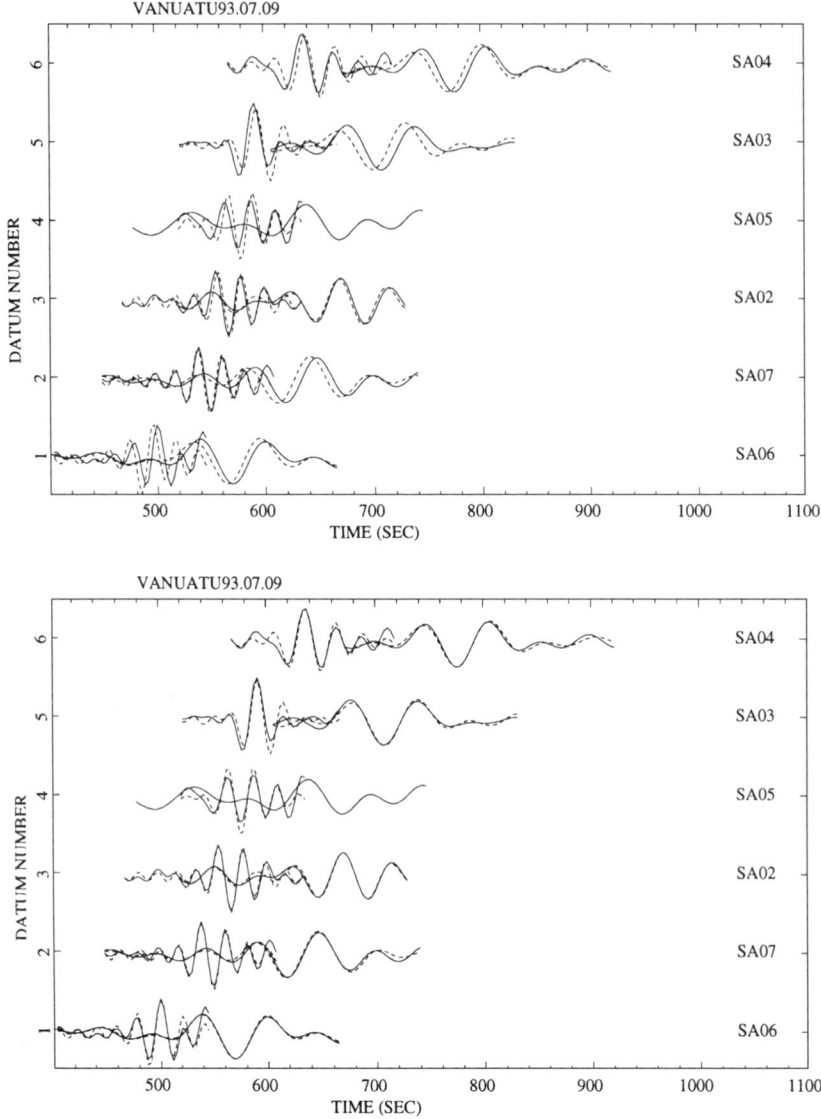

Figure 8.6. Seismogram matching for a Vanuatu event at stations in Queensland, illustrating the advantages of exploiting both fundamental and higher-mode surface waves. The observed seismograms are indicated by solid lines, and those computed from the structural models by dashed lines. Top: Initial fits. Bottom: Waveform matches after adjustment of model (Zielhuis and van der Hilst, 1996).

fundamental mode is slower than predicted, indicating lower wave speeds than in the reference model at shallow depths, but the higher modes are faster, so the wave speeds at depth need to be increased. Once the differences in wave speeds are taken into account to update the velocity model, the observations can be closely matched by theoretical records (Figure 8.6, bottom).

The Australian continent has evolved during a long geological history, with the tectonic processes that have most recently influenced the lithosphere becoming

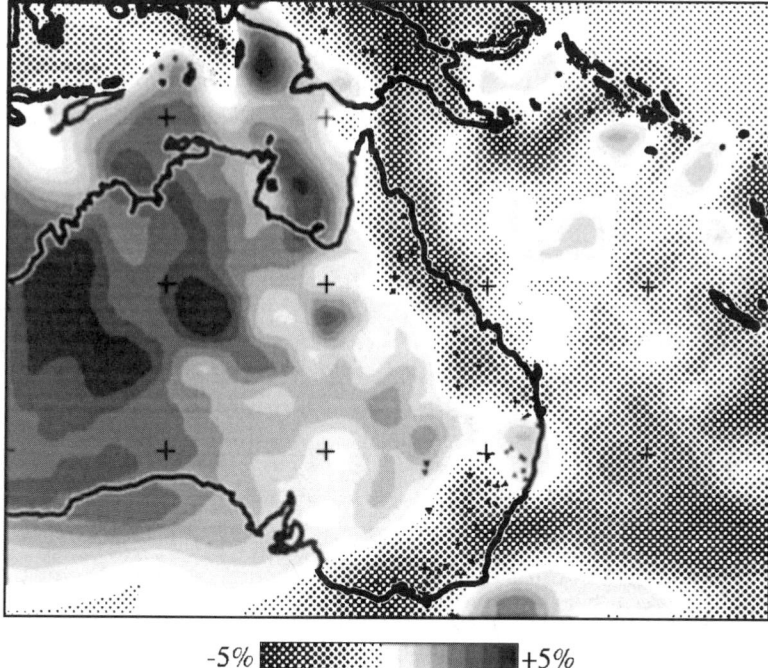

-5% ▨▨▨▨▨▨▨▨▨▨▨ +5%

Figure 8.7. Lateral variation of shear-wave speed in the upper mantle beneath eastern Australia and the surrounding region at a depth of 140 km (Zielhuis and van der Hilst, 1996).

younger from west (Archaean) to east (Phanerozoic). Proterozoic cratons form the central part of the continent. The lateral boundaries between the major tectonic divisions are not well constrained, and, in particular, their depth extent is unknown. Waveform inversion reveals large lateral variations in shear-wave speed in the upper mantle beneath the eastern Australian region (Figure 8.7). These findings have been discussed in detail by Zielhuis and van der Hilst (1996) and were obtained by analysis of the waveform data collected at a total of 25 SKIPPY stations in eastern Australia, augmented by data from the permanent observatories operated by IRIS (NWAO, CTAO, SNZO, HNR, PMG, TAU) and Geoscope (CAN, NOU).

The mantle beneath easternmost Australia, and the marginal basins between Australia and the Indo-Australia–Pacific-plate boundary, is marked by a pronounced low-wave-speed anomaly. On the continent, the lateral distribution of these low wave speeds correlates closely with the location of Neogene volcanoes, particularly strongly for seismic structure between 140 and 200 km of depth. This suggests that at least some of the low wave speeds have a thermal origin. Locally, pronounced positive wave-speed anomalies are detected to depths exceeding 200 km, as, for instance, beneath the Paleozoic basement of the New England foldbelt.

Farther to the west, beneath the Lachlan and Adelaide foldbelts, the seismic wave speeds in the upper mantle are significantly higher. The pronounced transition from the low wave speeds in the mantle beneath easternmost Australia to the moderately high wave speeds in the mantle farther to the west is relatively sharp and is well resolved by the data used. Preliminary results indicate that the regions of fast shear-wave propagation related to the central shield in the Northern Territory extend to more than 300 km in depth. However, the precise locations of such anomalies and the boundary between the Proterozoic shields of central Australia and the Phanerozoic basement to the east, the 'Tasman Line', have not been well resolved by the current data coverage (Figure 8.5b). Once additional waveform data from the later deployments covering the rest of the continent (Figure 8.5a) are included, resolution of this boundary and the deep structure of the central Australian cratons will be dramatically enhanced.

8.5. Discussion

The rapid development of techniques for estimating three-dimensional structure has changed our perception of the interior of the Earth and has led to a much stronger interaction between seismology and geodynamical studies. However, even the most impressive of the three-dimensional images can explain only part of the available information, and we must be wary in ascribing too detailed an interpretation of structure before the reliability of the images has been fully explored.

As we have noted, most three-dimensional models are derived as perturbations from a spherically symmetrical reference model. In many cases the level of heterogeneity is large enough to suggest that a full three-dimensional analysis is appropriate. The computational tools for simulation of the full seismic wave field in three-dimensional models are just beginning to be developed (e.g., Cummins et al., 1994a,b), and these will be needed to provide direct checks on the models postulated from a simplified analysis.

As the resolution of seismological structure increases, particularly on a regional scale, we will be faced with the problem of interpreting the substantial variations in seismic wave speeds in terms of both the mineralogy and the thermal characteristics of the mantle, because neither alone can provide an explanation for the classes of structures that are beginning to be revealed. Thermal influence may explain much of the large-scale heterogeneity, particularly if the temperature dependence of seismic velocities is greater at seismic frequencies than in the ultrasonic regime, where most laboratory measurements have been made (cf. Jackson and Rigden, Chapter 9, this volume). On smaller spatial scales, the temperature gradients required to match the velocity variations will be difficult to sustain against erosion through thermal conductivity, so that a compositional origin appears more attractive. Previous thermal activity can also leave residual compositional effects. Progress in the understanding of seismic heterogeneity may well depend on resolution of the detail of seismic anisotropy, which, although clearly recognized, is as yet poorly localized (e.g., Tong et al., 1994).

Resolution of many of the outstanding questions about the nature of the structure and related processes in the Earth's mantle will depend on the integration of information from many studies on a wide range of scales, such as, for example, using surface-wave analysis in combination with studies of refracted and reflected S waves to improve resolution of structure in the transition zone to the point where mineralogical questions can be answered directly (cf. Jackson and Rigden, Chapter 9, this volume).

References

Anderson, D. L., and Bass, J. 1986. Transition region of the Earth's upper mantle. *Nature* 320:321–8.

Bolton, H., and Masters, G. 1994. Scaling S to P in the mantle: Does it work? *EOS, Trans. AGU, 1994 Fall Meeting Supplement* 99:476.

Bowman, J. R., and Kennett, B. L. N. 1990. An investigation of the upper mantle beneath NW Australia using a hybrid seismograph array. *Geophys. J. Int.* 101:411–24.

Christensen, U. R., and Yuen, D. A. 1984. The interaction of subducting lithospheric slab with a chemical or phase boundary. *J. Geophys. Res.* 89:389–402.

Clitheroe, G., and van der Hilst, R. D. In press. Complex anisotropy in the Australian lithosphere from shear wave splitting in broad-band SKS records. *Geophys. Res. Lett.*

Creager, K. C., and Jordan, T. H. 1984. Slab penetration into the lower mantle. *J. Geophys. Res.* 89:3031–49.

Cummins, P. R., Geller, R. J., Hatori, T., and Takeuchi, N. 1994a. DSM complete synthetic seismograms: P-SV, spherically symmetric case. *Geophys. Res. Lett.* 21:1663–6.

Cummins, P. R., Geller, R. J., and Takeuchi, N. 1994b. DSM complete synthetic seismograms: SH, spherically symmetric case. *Geophys. Res. Lett.* 21:533–6.

Cummins, P. R., Kennett, B. L. N., Bowman, J. R., and Bostock, M. G. 1992. The 520 km discontinuity? *Bull. Seismol. Soc. Am.* 82:323–36.

Davies, G. F., and Richards, M. A. 1992. Mantle convection. *J. Geol.* 100:151–206.

Dey, S. C., Kennett, B. L. N., Bowman, J. R., and Goody, A. 1993. Variations in the upper mantle structure under northern Australia. *Geophys. J. Int.* 114:304–10.

Dziewonski, A. M., and Anderson, D. L. 1981. Preliminary reference Earth model. *Phys. Earth. Planet. Int.* 25:297–358.

Dziewonski, A. M., Hales, A. L., and Lapwood, E. R. 1975. Parametrically simple Earth models consistent with geophysical data. *Phys. Earth. Planet. Int.* 10:12–48.

Fischer, K. M., Creager, K. C., and Jordan, T. H. 1991. Mapping the Tonga slab. *J. Geophys. Res.* 96:14403–27.

Fukao, Y., Obayashi, M., Inoue, H., and Nenbai, M. 1992. Subduction zones stagnant in the lower mantle. *J. Geophys. Res.* 97:4809–22.

Grand, S. P. 1987. Tomographic inversion for shear velocity beneath the North American plate. *J. Geophys. Res.* 92:14065–90.

Grand, S. P. 1994. Mantle shear structures beneath the Americas and surrounding oceans. *J. Geophys. Res.* 99:11591–621.

Grand, S. P., and Helmberger, D. V. 1984a. Upper mantle shear structure of North America. *Geophys. J. Royal. Astron. Soc.* 76:399–438.

Grand, S. P., and Helmberger, D. V. 1984b. Upper mantle shear structure beneath the northwest Atlantic Ocean. *J. Geophys. Res.* 89:11465–75.

Griffiths, R. W., Hackney, R. I., and van der Hilst, R. D. 1995. A laboratory investigation of trench migration and the fate of subducted slabs. *Earth Planet. Sci. Lett.* 133:1–17.

Gudmundsson, O., Kennett, B. L. N., and Goody, A. 1994. Broadband observations of upper mantle seismic phases in northern Australia and the attenuation structure in the upper mantle. *Phys. Earth Planet. Int.* 84:207–26.

Hasegawa, A., Zhao, D., Hori, S., Yamamoto, A., and Horiuchi, S. 1991. Deep structure of the northeastern Japan arc and its relationship to seismic and volcanic activity. *Nature* 352:683–9.

Hirahara, K. 1977. A large-scale three-dimensional seismic structure under the Japan islands and the Sea of Japan. *J. Phys. Earth* 28:221–41.

Hirahara, K., and Mikumo, T. 1980. Three-dimensional seismic structure of subducting lithospheric plates under the Japan islands. *Phys. Earth Planet. Int.* 21:109–11.

Inoue, H., Fukao, Y., Tanabe, K., and Ogata, Y. 1990. Whole mantle P-wave mantle tomography. *Phys. Earth. Planet Int.* 59:294–328.

Iyer, H., and Hirahara, K. 1993. *Seismic Tomography*. London: Chapman & Hall.

Jeffreys, H., and Bullen, K. E. 1940. *Seismological Tables*. London: British Association Seismological Committee.

Jones, L. E., Mori, J., and Helmberger, D. V. 1992. Short-period constraints on the proposed transition zone discontinuity. *J. Geophys. Res.* 97:8765–74.

Jordan, T. H. 1975. The continental tectosphere. *Rev. Geophys.* 13:1–12.

Jordan, T. H. 1977. Lithospheric slab penetration into the lower mantle beneath the Sea of Okhotsk. *J. Geophys. Res.* 43:473–96.

Jordan, T. H. 1988. Structure and formation of continental tectosphere. *J. Petrol. Special Lithosphere Issue*, pp. 11–37.

Jordan, T. H., and Lynn, W. S. 1974. A velocity anomaly in the lower mantle. *J. Geophys. Res.* 79:2679–85.

Kennett, B. L. N. 1991. Seismic velocity gradients in the upper mantle. *Geophys. Res. Lett.* 18:1115–18.

Kennett, B. L. N. 1993. Seismic structure and heterogeneity in the upper mantle. In: *Relating Geophysical Structures and Processes: the Jeffreys Volume*, pp. 53–66. AGU Monograph 76, IUGG vol. 16. Washington, DC: American Geophysical Union.

Kennett, B. L. N., and Bowman, J. R. 1990. The velocity structure and heterogeneity of the upper mantle. *Phys. Earth. Planet. Int.* 59:134–44.

Kennett, B. L. N., and Engdahl, E. R. 1991. Traveltimes for global earthquake location and phase identification. *Geophys. J. Int.* 105:429–65.

Kennett, B. L. N., Engdahl, E. R., and Buland, R. 1995. Constraints on the velocity structure in the Earth from travel times. *Geophys. J. Int.* 122:108–24.

Kennett, B. L. N., Gudmundsson, O., and Tong, C. 1994. The upper-mantle S and P velocity structure beneath northern Australia from broad-band observations. *Phys. Earth Planet. Int.* 86:85–98.

Kennett, B. L. N., and Nolet, G. 1990. The interaction of the S-wavefield with upper mantle heterogeneity. *Geophys. J. Int.* 101:751–62.

Kesson, S. E., Fitz Gerald, J. D., and Shelley, J. M. G. 1994. Mineral chemistry and density of subducted basaltic crust at lower-mantle pressures. *Nature* 372:767–9.

Kincaid, C., and Olson, P. 1987. An experimental study of subduction and slab migration. *J. Geophys. Res.* 92:13832–40.

Knopoff, L. 1972. Observation and inversion of surface wave dispersion. *Tectonophysics* 13:497–520.

Laske, G. 1995. Global observations of off-great-circle propagation of long-period surface waves. *Geophys J. Int.* 123:245–59.

Lay, T. 1994. The fate of descending slabs. *Annu. Rev. Earth Planet. Sci.* 22:33–61.

Lay, T., and Wallace, T. 1995. *Modern Global Seismology*. Orlando: Academic, Press.

Lundgren, P. R., and Giardini, D. 1992. Seismicity, shear-failure and modes of deformation in deep subduction zones. *Phys. Earth Planet. Int.* 74:63–74.

Machetel, P., and Weber, P. 1991. Intermittent layered convection in a model mantle with an endothermic phase change at 670 km. *Nature* 350:55–7.

Masters, G., Jordan, T. H., Silver, P. G., and Gilbert, F. 1982. Aspherical earth structure from fundamental spheroidal mode data. *Nature* 298:609–13.

Masters, G., and Ritzwoller, M. 1987. Low frequency seismology and three-dimensional structure – observational aspects. In: *Mathematical Geophysics*, ed. N. J. Vlaar, G. Nolet, M. J. R. Wortel, and S. A. P. L. Cloetingh, pp. 1–30. Dordrecht: D. Riedel.

Montagner, J.-P., and Kennett, B. L. N. 1996. How to reconcile body-wave and normal mode reference Earth models? *Geophys. J. Int.* 125:229–48.

Morelli, A., and Dziewonski, A. M. 1993. Body-wave traveltimes and a spherically symmetric *P*- and *S*-wave velocity model. *Geophys. J. Int.* 112:178–94.

Nataf, H. C., and Van Decar, J. 1993. Seismological detection of mantle plume? *Nature* 364:115–120.

Nolet, G. 1990. Partitioned waveform inversion and two-dimensional structure under the network of autonomously recording seismographs. *J. Geophys. Res.* 95:8499–512.

Nolet, G., Grand, S. P., and Kennett, B. L. N. 1994. Seismic heterogeneity in the upper mantle. *J. Geophys. Res.* 99:23753–66.

Okino, K., Ando, M., Kaneshima, S., and Hirahara, K. 1989. A horizontally lying slab. *Geophys. Res. Lett.* 16:1059–63.

Olson, P., Silver, P. G., and Carlson, W. W. 1990. The large-scale structure of convection in the Earth's mantle. *Nature* 344:209–15.

Revenaugh, J. S., and Jordan, T. H. 1991. Mantle layering from ScS reverberations. 2. The transition zone. *J. Geophys. Res.* 96:19763–811.

Ringwood, A. E. 1991. Phase transformations and their bearing on the constitution and dynamics of the mantle. *Geochim. Cosmochim. Acta* 55:2083–110.

Ringwood, A. E., and Irifune, T. 1988. Nature of the 650-km seismic discontinuity: implications for mantle dynamics and differentiation. *Nature* 331:131–6.

Romanowicz, B. 1995. A global tomographic model of shear attenuation in the upper mantle. *J. Geophys. Res.* 100:12375–94.

Shearer, P. M. 1991. Constraints on upper mantle discontinuities from observations of long-period reflected and converted waves. *J. Geophys. Res.* 96:18147–82.

Shearer, P. M., and Masters, T. G. 1992. Global mapping of topography on the 660-km discontinuity. *Nature* 355:791–6.

Silver, P. G., Carlson, W. W., and Olson, P. 1988. Deep slabs, geochemical heterogeneity, and the large-scale structure of mantle convection: investigations of an enduring paradox. *Annu. Rev. Earth Planet. Sci.* 16:477–541.

Sneider, R. 1988. Large-scale waveform inversion of surface waves for lateral heterogeneity: 2. Application to surface waves in Europe and the Mediterranean. *J. Geophys. Res.* 93:12067–80.

Spakman, W. 1991. Delay-time tomography of the upper mantle below Europe, the Mediterranean, and Asia minor. *Geophys. J. Int.* 107:309–32.

Su, W.-J., and Dziewonski, A. M. 1991. Predominance of long-wavelength heterogeneity in the mantle. *Nature* 352:121–6.

Su, W.-J., Woodward, R. L., and Dziewonski, A. M. 1994. Degree 12 model of shear velocity heterogeneity in the mantle. *J. Geophys. Res.* 99:6945–81.

Tackley, P. J., Stevenson, D. J., Glatzmaier, G. A., and Schubert, G. 1993. Effects of an endothermic phase transition at 670 km depth on spherical mantle convection. *Nature* 361:699–704.

Toksöz, M. N., Minnear, J. W., and Julian, B. R. 1971. Temperature field and geophysical effects of a downgoing slab. *J. Geophys. Res.* 76:1113–18.

Tong, C., Gudmundsson, O., and Kennett, B. L. N. 1994. Shear wave splitting in refracted waves returned from the upper mantle transition zone beneath northern Australia. *J. Geophys. Res.* 99:15783–97.

Trampert, J., and Woodhouse, J. H. 1995. Global phase velocity maps of Love and Rayleigh waves between 40 and 150 seconds. *Geophys. J. Int.* 122:675–90.

van der Hilst, R. D. 1995. Complex morphology of subducted lithosphere in the mantle beneath the Tonga trench. *Nature* 374:154–7.

van der Hilst, R. D., Engdahl, E. R., Spakman, W., and Nolet, G. 1991. Tomographic imaging of subducted lithosphere below northwest Pacific island arcs. *Nature* 353:37–43.

van der Hilst, R. D., Kennett, B. L. N., Christie, D. R., and Grant, J. 1994. Project SKIPPY explores the lithosphere and mantle beneath Australia. *EOS, Trans. AGU* 75:177, 180–1.

van der Hilst, R. D., and Seno, T. 1993. Effects of relative plate motion on the deep structure and penetration depth of slabs below the Izu-Bonin and Mariana Island arcs. *Earth Planet. Sci. Lett.* 120:375–407.

van der Hilst, R. D., and Spakman, W. 1989. Importance of the reference model in linearized tomography and images of subduction below the Caribbean Plate. *Geophys. Res. Lett.* 16:1093–6.

van der Hilst, R. D., Widiyantoro, S., and Engdahl, E. R. 1996. Global slab structure from high resolution tomographic imaging. *EOS, Trans. AGU* 77:137.

Vasco, D. W., Johnson, L. R., Pulliam, R. J., and Earle, P. S. 1994. Robust inversion of IASP91 travel time residuals for mantle P and S velocity structure. *J. Geophys. Res.* 99:11727–3755.

Widiyantoro, S., and van der Hilst, R. 1996. Structure and evolution of lithospheric slab beneath the Sunda arc, Indonesia. *Science* 271:1566–70.

Woodward, R. L., and Masters, G. 1991. Global upper mantle structure from long-period differential travel times. *J. Geophys. Res.* 96:6351–78.

Wortel, R. 1982. Seismicity and rheology of subducting slabs. *Nature* 296:553–6.

Zhong, S., and Gurnis, M. 1995. Mantle convection with plates and mobile, faulted plate margins. *Science* 267:838–43.

Zhou, H.-W., and Clayton, R. W. 1990. P and S wave travel-time inversions for subducting slab under the island arcs of the northwest Pacific. *J. Geophys. Res.* 95:6829–54.

Zielhuis, A., and Nolet, G. 1994. Shear wave velocity variations in the upper mantle below Central Europe. *Geophys. J. Int.* 117:695–715.

Zielhuis, A., and van der Hilst, R. 1996. Upper mantle shear velocity beneath eastern Australia from inversion of waveforms from Skippy portable arrays. *Geophys. J. Int.* 127:1–16.

9

Composition and Temperature of the Earth's Mantle: Seismological Models Interpreted through Experimental Studies of Earth Materials

IAN JACKSON and SALLY M. RIGDEN

9.1. Introduction

In an extraordinarily influential paper, Birch (1952) first demonstrated the utility of finite-strain equations-of-state in modelling the elastic properties of the Earth's deep interior. He used the seismologically well constrained radial variation of the compressional- and shear-wave speeds, V_P and V_S, and hence of the 'seismic parameter'

$$\phi = V_P^2 - \left(\tfrac{4}{3}\right) V_S^2 \tag{1}$$

and the approximation of constant gravitational acceleration g in evaluating the left-hand side of the equation

$$1 - g^{-1}\partial\phi/\partial r = (\partial K_S/\partial P)_S + (\tau\alpha\phi/g)\{1 + (\partial K_S/\partial T)_P/\alpha K_S\} \tag{2}$$

This equation relates the radial variation of ϕ to material properties (specifically, the adiabatic bulk modulus or incompressibility K_S, its P and T derivatives, and the thermal expansivity α) for a self-compressed homogeneous layer subject to a superadiabatic temperature gradient τ. Birch argued that the value of $1 - g^{-1}\partial\phi/\partial r$ thus obtained for the lower mantle, which decreases with increasing depth from about 4 at low pressure towards 3 in the lowermost mantle, is consistent with values of $\partial K/\partial P$ derived from laboratory isothermal-compression data (notably Bridgman's measurements on the very compressible alkali metals). Birch therefore concluded that the lower mantle is 'reasonably homogeneous' and that departures from adiabaticity embodied in the second term on the right-hand side of equation (2) are small.

Assuming both homogeneity and adiabaticity, he proceeded to use finite-strain equations-of-state in demonstrating that the density ρ_0 and seismic parameter ϕ_0 for the (hot) adiabatically decompressed lower mantle are about $4.0\,\mathrm{g}\cdot\mathrm{cm}^{-3}$ and $51\,\mathrm{km}^2\cdot\mathrm{s}^{-2}$, respectively. These values are distinctly higher than those for the common silicate minerals of the crust and upper mantle. Birch noted, however, that the oxides MgO, TiO_2, and Al_2O_3 display appropriate combinations of high density

405

and incompressibility. Accordingly, Birch concluded that 'in the "transitional layer" at depths between about 200 and 900 km, the rate of rise of velocity is too great for a homogeneous layer, and indicates a gradual change of composition, or phase, or both'. More specifically, he boldly speculated that 'beginning at about 200 to 300 km depth, there is a gradual shift towards high-pressure modifications of the ferromagnesian silicates, probably close-packed oxides, with the transition complete at about 800 to 900 km ... such new phases are required to account for the high elasticity of the deeper part of the mantle'.

Since then, the reality of pressure-induced phase transformations has been thoroughly demonstrated, especially in the laboratories of Ringwood, Akimoto, and Liu, firstly in germanate and other structural analogues and later in the silicate minerals of the upper mantle. Phase transformations *must* therefore be responsible for much of the seismic structure of the transition zone and lower mantle, but the question concerning the relative importance of changes in phase and in chemical composition, first posed by Birch, remains to be resolved unequivocally.

This review consists of two main parts. In the first part of the chapter, the background required for analysis of the elasticity, composition, and temperature of the mantle is assembled by summarizing the phenomenal progress made during the past four decades in the complementary fields of seismology, experimental petrology, and mineral/rock physics. In the second part of this chapter, the background material of the first part is applied to the interpretation of selected aspects of the seismic structure of the upper mantle, transition zone, and lower mantle. In the final sections of the chapter, an attempt is made to summarize our current understanding of the elasticity, composition, and temperature of the mantle and to identify some of the contemporary research frontiers.

9.2. Improved Resolution of the Seismic Structure of the Mantle
9.2.1. Generalities

The past 30 years have seen substantial refinement in our understanding of the (average) radial structure of the Earth's mantle, as well as progress towards robust resolution of the superimposed lateral variability of seismic wave speeds (e.g., Nolet, Grand, and Kennett, 1994; Kennett and van der Hilst, Chapter 8, this volume). These studies have provided a firm basis for subdivision of the mantle into the following layers, each with its own distinctive characteristics:

Upper mantle (depth $z < 410$ km), with a high degree of lateral variability in wave speeds (exceeding $\pm 4\%$) and relatively strong attenuation

Transition zone ($410 < z < 660$ km), bounded by major global compressional- and shear-wave velocity discontinuities, with a further minor discontinuity at 520 km and generally high velocity–depth gradients

Lower mantle (660 < z < 2,600 km), with smooth velocity–depth gradients closely
   consistent with the notion of adiabatic compression of material that is homogeneous in chemical composition and phase (the uppermost part, at depths of
   660–750 km, is anomalous, with very strong velocity–depth gradients)

D″ layer in the lowermost mantle (2,600 < z < 2,900 km), with anomalous velocities and velocity gradients, a high degree of lateral variability, and strong
   attenuation

These advances have been made possible through major developments in both
seismic instrumentation (arrays, broadband seismometers, digital recording) and
data analysis (synthetic seismograms, stacking, inversion techniques).

9.2.2. Radial Structure of the Transition Zone

Much effort has been devoted to improved resolution of the radial structure of the
transition zone. Refraction surveys conducted with arrays of densely spaced stations
first demonstrated unequivocally the existence of major, essentially discontinuous
increases in seismic wave speeds near 410 and 660 km (Niazi and Anderson, 1965;
Johnson, 1967). Modelling of the waveforms (mainly of refracted body waves
of relatively long period) has provided additional constraints on velocity–depth
structure (e.g., Wiggins and Helmberger, 1973; Grand and Helmberger, 1984a;
Bowman and Kennett, 1990).

The interaction of seismic waves with zones of strong velocity–depth gradient in the mantle results also in a rich variety of reflected and converted phases.
These are extremely variably observed on individual seismograms, but they emerge
very clearly from systematic analyses of large, long-period datasets. By combining
(stacking) many thousands of seismograms to produce travel-time–distance images, Shearer (1991) showed that reflected and converted phases associated with
the major mantle discontinuities at depths near 410 and 660 km can be consistently
observed. An additional discontinuity near a depth of 520 km was also resolved,
although the observability of this feature was later questioned (Bock, 1994). Subsequent modelling of a reflectivity profile derived from the stacked SS precursors
resulting from underside reflections yielded estimated shear-wave impedance (density × wave speed) contrasts of 6.7%, 2.9%, and 9.9% across discontinuities at
average depths of 420, 519, and 663 km and constraints on the high gradient in
the depth interval 660–750 km (Shearer, 1996). Similarly, Revenaugh and Jordan
(1991) identified reflections from the 410-km and 660-km discontinuities through
analysis of long-period ScS reverberations and found evidence for additional minor
discontinuities near 520, 710, and 900 km. These studies also revealed variability
(of order 20–30 km peak-to-peak) in the apparent depths of the two major discontinuities. However, the relatively large corrections required for heterogeneity in the

overlying upper mantle render the absolute depths much less certain. Accordingly, the interesting suggestions that the topography, of amplitude less than about 30 km, on the two major discontinuities might be negatively correlated (Revenaugh and Jordan, 1991) and that the regional depression of the 660-km discontinuity might be associated with subduction (e.g., Shearer and Masters, 1992) must be regarded as more tentative.

In order to be observable through interaction with a seismic wave of wavelength λ, it is necessary that a 'discontinuous' increase in wave speed in the transition zone occur within a depth interval less than $\lambda/4$ (Richards, 1972). The long-period studies described earlier ($\lambda = 120$–180 km) were therefore unable to distinguish between a genuine 'first-order' discontinuity and a 30-km-thick interval with a steep velocity gradient. For this reason, short-period waves reflected and converted at the discontinuities hold the key to the crucial issue of their sharpness. Observation of 0.5–1-Hz PKPPKP (P′P′) precursors interpreted as underside reflections (e.g., Engdahl and Flinn, 1969; Whitcomb and Anderson, 1969; Nakanishi, 1988) and of P/SV converted phases (e.g., Vinnik, 1977; Petersen et al., 1993; Yamazaki and Hirahara, 1994) is prima facie evidence that the discontinuities, especially that at 660 km, are sharp within 5 km. However, these inferences need to be carefully evaluated, because P′P′ precursors are observed only rarely in individual short-period records (e.g., Davis, Kind, and Sacks, 1989). Significantly, the stacking of many such records for an individual earthquake can result in considerable enhancement of these weak and variably observed signals (Vidale and Benz, 1992; Castle and Creager, 1997). Possible reasons for the variability in observation of P′P′ precursors include defocusing caused by the topography on the discontinuities (Davis et al., 1989) and the influence of temperature on the width of the two-phase (olivine + wadsleyite) loop in the $(Mg, Fe)_2SiO_4$ phase diagram (Helffrich and Bina, 1994).

9.2.3. Lateral Heterogeneity

Early observations of lateral variability of seismic wave speeds in the upper mantle, correlated with surface tectonics, came from studies of the regional variation of traveltime residuals (e.g., Cleary and Hales, 1966; Doyle and Hales, 1967) and from regional refraction surveys (e.g., Wiggins and Helmberger, 1973). Analyses of surface-wave dispersion along great-circle paths led to a 'tectonic regionalization' of the upper mantle that emphasized the differences among continental, oceanic, and tectonic paths (Toksöz and Anderson, 1966; Dziewonski, 1971). This early evidence for lateral variability of seismic wave speeds in the upper mantle that is correlated with tectonic provinces has been overwhelmingly confirmed in more recent global and regional tomographic studies (e.g., Nataf, Nakanishi, and Anderson, 1986; Grand, 1994; Su, Woodward, and Dziewonski, 1994; Zielhuis and Nolet, 1994; Zielhuis and van der Hilst, 1996). In particular, there is a consistent association in the most recent global models between the mid-ocean-ridge system and low wave speeds and between continental shields and high upper-mantle wave speeds. The high velocities attributable to old continental roots appear to continue downward to at least 250–300 km, in accordance with Jordan's (1975) notion of the continental 'tectosphere' (e.g., Nolet et al., 1994).

The degree of wave-speed heterogeneity is, in general, a decreasing function of depth within the mantle. For the recent global model of Su et al. (1994), in which the lateral heterogeneity in shear-wave speed is modelled to spherical-harmonic degree 12 (resolving length ~1,700 km), the peak-to-peak amplitude of the variability decreases from ±3–5% above 200-km depth to no more than ±1% throughout most of the lower mantle. The heterogeneity rises again to ±2% in the D″ layer of the lowermost mantle. Broadly similar findings have been reported at higher spatial resolution (~300 km) in the traveltime tomographic study of the Western Hemisphere by Grand (1994), with maximum V_{SH} heterogeneity of about ±4% at depths shallower than 175 km. Tomographic inversion of the waveforms of regional surface waves has revealed a similar degree of lateral heterogeneity beneath Europe and Australia (Zielhuis and Nolet, 1994; Zielhuis and van der Hilst, 1996).

Spectra of lateral velocity variations throughout the mantle, rather than deterministic tomographic models, were obtained by Gudmundsson, Davies, and Clayton (1990) from the scatter evident in a large global set of compressional-wave traveltimes. In this study, and in a subsequent application of the same method to shear waves (Davies, Gudmundsson, and Clayton, 1992), it was again concluded that the heterogeneity is strongly concentrated in the upper mantle at depths shallower than 400 km and, importantly, that much of it is contributed by small-scale (<300 km) features.

Comparisons of tomographic inversions for δV_P and δV_S have revealed significant correlations between the heterogeneities in compressional-wave and shear-wave speeds, both in space and in amplitude. The amplitude ratio $(\delta V_S/V_S)/(\delta V_P/V_P) \doteq \partial \ln V_S/\partial \ln V_P$ of about 2 (Hales and Doyle, 1967; Woodhouse and Dziewonski, 1989; Li, Giardini, and Woodhouse, 1991; Davies et al., 1992; Robertson and Woodhouse, 1995), most tightly constrained for the upper mantle, is substantially larger than would be expected on the basis of high-frequency laboratory data for anomalies of thermal origin. However, it is likely that $(\partial \ln V_S/\partial \ln V_P)_P$ is substantially greater at seismic frequencies as a consequence of viscoelastic relaxation (see Section 9.5).

Finally, it must be stressed that although much of the long-wavelength heterogeneity appears to be consistently resolved in a variety of tomographic inversions, there is compelling evidence for both strong smaller-scale seismic heterogeneity and pronounced anisotropy, especially in the upper mantle. Although there is clear observational evidence for elastic anisotropy in the upper mantle, it is often ignored, especially in regional tomographic studies, where both the radial reference model and the inferred three-dimensional perturbations are usually treated as isotropic. Presumably some of the wave-speed variability caused by anisotropy will be mapped into lateral variability of isotropic elastic properties that typically is attributed to compositional or thermal variability (see Section 9.5). These effects will be particularly important in analyses of structure at relatively small scales, where the lattice preferred orientation of anisotropic minerals (notably olivine) is more likely to be maintained across length scales comparable to those of the tomographic model.

9.2.4. Attenuation

Attenuation of seismic waves in the mantle is dominated by strain-energy losses in shear. Observations of the amplitudes of multiple ScS and sScS waves at near-vertical incidence provide tight constraints on the radially averaged shear-mode attenuation Q^{-1} and also indicate that most of this loss occurs in the upper mantle (e.g., Kovach and Anderson, 1964). Additional constraints on the variation of Q^{-1} with depth come mainly from surface-wave dispersion and free-oscillation measurements (e.g., Anderson, Ben-Menahem, and Archambeau, 1965; Anderson and Hart, 1978; Widmer, Masters, and Gilbert, 1991; Romanowicz, 1995; Durek and Ekström, 1996). Recent models for the radial variation of Q^{-1} feature a low-loss 'lid' in the uppermost mantle (Q^{-1} usually <0.002), overlying a zone of very high average attenuation ($Q^{-1} \sim 0.015$) at depths of 80–220 km in the upper mantle. Q^{-1} is markedly lower at greater depths, averaging 0.006 between 200 and 700 km depth, and 0.003 for most of the lower mantle, although somewhat higher values are found for the D'' layer.

Although there is strong evidence for regional variability of upper-mantle attenuation, especially from studies of the amplitudes and waveforms of surface waves, the amplitudes will also be strongly influenced by wave-speed heterogeneity. For this reason, formal tomographic inversion for attenuation has only recently been attempted (e.g., Romanowicz, 1995). The resulting model, comparable to a degree-6 spherical-harmonic expansion, and therefore of relatively low spatial resolution (resolving length $\sim$3,000 km), reveals maximum lateral variability of $\pm50\%$ at depths of 200–300 km. The regions of higher- and lower-than-average attenuation in the upper mantle are correlated with oceans and continents, respectively, in accord with the lateral variability in wave speeds. The heterogeneity in Q^{-1} decreases with increasing depth, but is still significant in the transition zone, where features tentatively attributed to the major Hawaiian and northeastern African plumes are imaged.

9.3. Mineralogy of the Deep Mantle
9.3.1. Composition and Temperature of the Upper Mantle

Basaltic magmas and the xenoliths that they entrain provide abundant direct samples of the mantle to depths of about 200 km. Their compositions demonstrate that the upper mantle is overwhelmingly peridotitic, with about 60% (by volume) olivine, the balance being mainly pyroxenes and lesser amounts of spinel or garnet (e.g., Ringwood, 1975; Basaltic Volcanism Studies Project, 1981; Green and Falloon, Chapter 7, this volume). The complementary relationship between the basaltic magmas produced by partial melting in the upper mantle and the refractory, olivine-rich residua provided the basis for the construction and testing of the 'pyrolite' compositional model (Ringwood, 1962a; Green and Ringwood, 1967; Green, Hibberson, and Jaques, 1979; Green and Falloon, Chapter 7, this volume). Average upper-mantle pyrolite is dominated by the five oxides SiO_2, MgO, FeO, Al_2O_3, and CaO (e.g., Green and Falloon, Chapter 7, this volume), which together represent about 99% (by weight) of the bulk composition (Figure 9.1a). Interpretation of

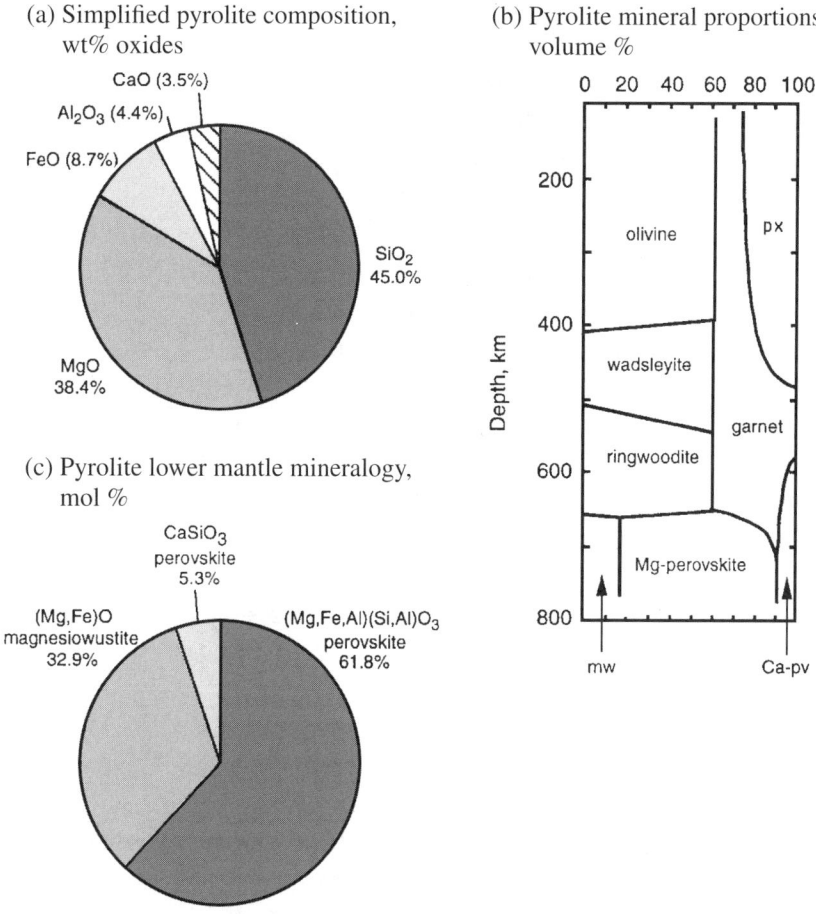

Figure 9.1. The pyrolite model for the composition of the Earth's mantle. (a) Chemical composition (expressed as wt%) of the five most abundant oxide components, which together account for 99% of the bulk (Green et al., 1979; Green and Falloon, Chapter 7, this volume). (b) Pressure-induced mineralogical changes in pyrolite across the pressure range of the upper mantle, transition zone, and uppermost lower mantle. Abbreviations: px, pyroxenes; mw, magnesiowüstite; Ca-pv, $CaSiO_3$ perovskite. The phase labelled 'garnet' in the transition zone is a complex solid solution involving partial octahedral co-ordination of silicon. (Adapted from Irifune, 1993, 1994.) (c) The mineralogy (mol%) of the pyrolite model for the lower mantle.

mid-ocean-ridge basalts (MORBs) as primary melts formed directly by the partial melting of the mantle source suggests a potential temperature T_0 near 1,300°C (1,570 K) at the zero-pressure 'foot' of the average upper-mantle adiabat (e.g., Green and Ringwood, 1967; Herzberg, 1992; White and McKenzie, 1995). Alternatively, if MORBs are derived from picritic precursors through olivine fractionation (e.g., O'Hara, 1968; Green et al., 1979; Green and Falloon, Chapter 7, this volume), a much higher potential temperature near 1,450°C (1,720 K) is required.

9.3.2. High-Pressure Phase Equilibria

Demonstration of the existence of pressure-induced phase transformations in silicate minerals has been closely linked to the progressive development of experimental apparatus capable of the required extremes of pressure and temperature. During the 1950s, Ringwood used analogue compounds, especially germanates, to provide insights into the high-pressure behaviour of their silicate relatives. For example, he showed how the mutual solid solubilities between a silicate olivine and a germanate spinel could be used to provide a robust estimate of the relative stabilities of the olivine and spinel forms (polymorphs) of the silicate. The free-energy difference thus calculated could be combined with an estimate of the difference in molar volume to predict the transformation pressure in the silicate end member (e.g., Ringwood, 1956; Ringwood, 1962b). By 1970, opposed-anvil apparatus had been used to demonstrate directly the olivine $\rightarrow \beta$-phase (wadsleyite) $\rightarrow$ spinel (ringwoodite) phase transformations in $(Mg, Fe)_2SiO_4$ and the progressive incorporation of pyroxene into complex garnet solid solutions (Akimoto and Fujisawa, 1966; Ringwood and Major, 1966, 1970; Ringwood, 1967). [The $(Mg, Fe)SiO_3$ garnet, subsequently discovered in naturally shocked meteorites, is known as majorite.] Liu (1974, 1976; Liu and Ringwood, 1975) exploited the then newly developed laser-heated diamond-anvil apparatus to reveal the stability, at still higher pressures, of $CaSiO_3$ and $MgSiO_3$ perovskites.

Detailed equilibria involving these and other high-pressure phases have subsequently been determined, mainly in multiple-anvil high-pressure apparatus (e.g., Akaogi and Akimoto, 1979; Irifune and Ringwood, 1987; Ito and Takahashi, 1989; Katsura and Ito, 1989; Gasparik, 1990; Irifune, 1994). These studies have provided determinations of P–T phase boundaries and/or the compositions and relative proportions of coexisting phases. The changes in mineralogy observed for material of upper-mantle 'pyrolite' composition, as pressure is increased progressively to values representative of the uppermost part of the lower mantle, are illustrated in Figure 9.1b. The major phase transformations in the $(Mg, Fe)_2SiO_4$ component occur over relatively narrow pressure intervals and therefore have the potential to provide explanations for the two major seismic discontinuities in the transition zone (see Section 9.6). The major transformations in the $(MSi, Al_2)O_3$ component

($M = $ Mg, Fe, Ca) yield comparable or larger overall increases in density and elastic wave speeds, but such changes will be spread out over many tens of kilometers of depth (Figure 9.1b), presumably contributing to the generally steep velocity–depth gradients of the transition zone. For a lower-mantle composition similar to that of the upper mantle, the (Mg, Fe)SiO_3 perovskite and (Mg, Fe)O magnesiowüstite phases together account for about 93% of the bulk by mass or volume, or 95 mol% (Figure 9.1c), given the experimental evidence that all of the Al_2O_3 budget is accommodated in dilute solid solution (~6 mol%) in the (Mg, Fe)SiO_3 perovskite phase (Irifune, 1994). The balance (7 wt%) is cubic $CaSiO_3$ perovskite. Only very limited mutual solid solubility has been reported between the $MgSiO_3$ and $CaSiO_3$ perovskites (Irifune et al., 1989; Kesson, Fitz Gerald, and Shelley, 1994).

Similar progress, as reviewed by Irifune (1993), has been made in determining the *P–T* phase equilibria for the basaltic and harzburgitic compositions represented in chemically differentiated oceanic lithosphere. During subduction, the basaltic layer is expected to transform progressively from eclogite to garnetite and finally to perovskitite. On account of its enrichment relative to pyrolite in FeO and Al_2O_3, this layer is generally more dense than pyrolite. However, the stability field for garnet is wider for this more aluminous compositon, and as a consequence the usual density relationship is reversed in the depth interval 650–800 km, where the basaltic layer is expected to be about 3% less dense than the surrounding mantle. For refractory harzburgite, densities generally are about 1% less than for pyrolite, with the exception of the region immediately above the 660-km discontinuity, where the narrower stability field for garnet in this *less* aluminous composition results in 'early' transformation to the perovskite-bearing assemblage. It follows from these considerations of the relative densities and thicknesses of the three layers of a differentiated slab that for the depth interval 650–800 km, a thermally equilibrated slab will be, on average, about 0.8% (or 0.03 g · cm^{-3}) less dense than the surrounding mantle.

9.3.3. Calorimetry

High-temperature enthalpies of solution and specific heats have now been measured for all of the major high-pressure phases of the Earth's mantle. This calorimetric information has been combined with experimentally determined equilibria to provide a robust, internally consistent thermodynamic framework for the computation of phase diagrams (e.g., Akaogi, Ito, and Navrotsky, 1989; Fei, Mao, and Mysen, 1991). Of particular importance in discussions of the influence of isochemical phase transformations on the dynamics of the Earth's interior (e.g., Christensen and Yuen, 1984; Davies, Chapter 5, this volume) is the temperature sensitivity of the transition pressure. For a univariant, first-order transformation, the competing phases coexist in thermodynamic equilibrium only along a unique curve in P–T space whose gradient is given by the Clapeyron-Clausius equation

$$dP/dT = \Delta S/\Delta V \qquad (3)$$

where ΔS and ΔV are respectively the changes in entropy and volume across the phase transition. ΔV is invariably negative for pressure-induced phase transformations, but ΔS is more variable, because the lattice vibrational and configurational contributions to the entropy vary widely with the details of crystal structure and substitutional disorder.

9.3.4. Transformational and Chemical Buoyancies of Subducting Slabs

Phase-equilibrium and calorimetric studies indicate that the Clapeyron slopes for the major olivine $\rightarrow$ wadsleyite and ringwoodite $\rightarrow$ perovskite + magnesiowüstite transformations are respectively positive and negative; the preferred values are $+3$ and -2 MPa $\cdot$ K^{-1} (Bina and Helffrich, 1994) (Figure 9.2). The resulting deflections

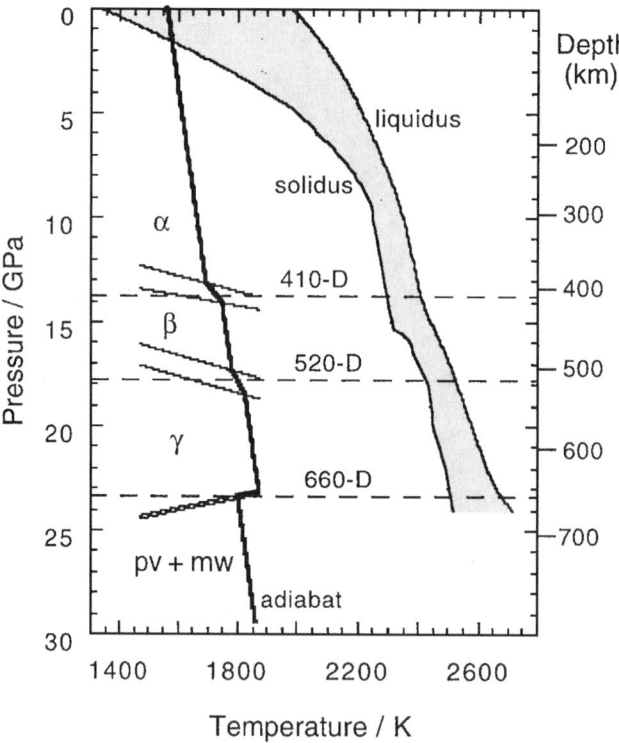

Figure 9.2. An upper-mantle adiabat with potential temperature 1,550–1,600 K, which is consistent with basalt eruption temperatures and with the depths to the major seismic discontinuities (labelled 410-D, etc.) interpreted in terms of phase transformations in a pyrolite model mantle. The procedure employed in construction of the adiabatic temperature profile is essentially that of Ito and Katsura (1989). The shaded region is the melting interval for anhydrous peridotite as determined by Green et al. (1979) and Herzberg and Zhang (1996).

of the equilibrium phase boundaries[1] in thermally buoyant convecting material serve, respectively, to augment and detract from the thermal buoyancy, and therefore represent 'destabilizing' and 'stabilizing' influences on the flow of which the slab is part. The effective Clapeyron slopes for the gradual pyroxene–garnet and garnet–perovskite transitions are less well established. However, calorimetric data suggest respectively negative and positive Clapeyron slopes for the pyroxene–garnet and garnet–perovskite transformations in $MgSiO_3$ (Yusa, Akaogi, and Ito, 1993), meaning that the configurational entropy associated with Mg–Si disorder on the octahedral sites in majorite garnet outweighs the unusually high vibrational entropy of the perovskite phase (Navrotsky, 1989). It is therefore likely that the destabilizing and stabilizing influences of the olivine $\rightarrow$ wadsleyite and ringwoodite $\rightarrow$ perovskite + magnesiowüstite transformations, respectively, will be substantially offset by opposing influences of the transformations in the $(MSi, Al_2)O_3$ component (Ringwood, 1993; Weidner and Wang, 1995).

Under these circumstances, it is probable that chemical-buoyancy effects associated with differentiation of subducting slabs (Anderson, 1979; Irifune and Ringwood, 1993; Kesson et al., 1994), rather than transformational buoyancy, will be the most important perturbation to the thermal buoyancy of slabs being subducted through the transition zone. In the following comparison of the dynamical effects of chemical differentation of a slab with the effects of temperature-sensitive phase-transformation pressures, it is assumed that the slab retains its mechanical integrity (i.e., that gravitational segregation of the basaltic and harzburgitic components does not occur) (e.g., Davies, Chapter 5, this volume; Griffiths and Turner, Chapter 4, this volume). The average density deficit of 0.03 g · cm^{-3} for a thermally equilibrated subducting slab relative to the ambient mantle for the depth interval 650–800 km, as estimated earlier, will contribute the same buoyancy as would a downward deflection of about 15 km ($\sim$0.5 GPa) of the 660-km discontinuity (density contrast $\sim$0.3 g · cm^{-3}). A depression of similar magnitude caused by deflection of an equilibrium phase boundary within the cooler interior of the slab would require a Clapeyron slope of about −1 to −2 MPa · K^{-1} for an average temperature contrast of 500–250 K.

For a hot plume ($\Delta T_0 = 200$–300 K) of recycled, differentiated lithosphere (Davies, Chapter 5, this volume; Campbell, Chapter 6, this volume) ascending through the uppermost part of the lower mantle, the chemical buoyancy would likewise be positive, and therefore destabilizing, and equivalent in effect to a Clapeyron slope of about +1 MPa · K^{-1}. Somewhat larger effects would be expected if the plume source were enriched in the basaltic component, as suggested by Campbell (Chapter 6, this volume). In view of the likelihood of substantial cancellation (discussed earlier) between the transformational buoyancies for the M_2SiO_4 and $(MSi, Al_2)O_3$ components, these chemical buoyancies might significantly influence

---

[1] The possibility that olivine might persist metastably to greater depths in the cool interior of a subducting slab is reviewed by Drury and Fitz Gerald (Chapter 11, this volume).

the behaviour of plates and plumes undergoing thermal convection (e.g., Davies, Chapter 5, this volume).

9.3.5. Transition-Zone Adiabats

Computation of adiabats through the transition zone suggests temperatures near 1,450°C and 1,600°C at the depths of the major 410- and 660-km discontinuities for a potential temperature T_0 of 1,300°C (1,600 K), consistent with the pressures required for the olivine $\rightarrow$ wadsleyite and ringwoodite $\rightarrow$ perovskite + magnesiowüstite tranformations (Ito and Katsura, 1989; Akaogi et al., 1989; Ita and Stixrude, 1992). The relationship between this inferred geotherm and the solidus temperature determined for the pressure range of the upper mantle and transition zone is also shown in Figure 9.2. The additional increase in temperature along this adiabat as it traverses the lower mantle is about 550 K (see Section 9.7 and Figure 9.8c).

The higher potential temperature ($\Delta T_0 = 150$ K) required for the derivation of MORBs from a picritic precursor would change the depths inferred for the olivine $\rightarrow$ wadsleyite and ringwoodite $\rightarrow$ perovskite + magnesiowüstite transformations from the phase equilibria discussed earlier by +13 and −9 km, respectively (Figure 9.2). Given all of the uncertainties, especially in pressure calibration at high temperature in multiple-anvil apparatus, these perturbations are insufficient to seriously compromise the general consistency between the P–T phase equilibria and the seismologically determined depths to the major discontinuities.

9.3.6. Further Pressure-induced Phase Transformations in the Deep Mantle?

The very wide range of structural distortions away from the cubic parent structure in perovskite-structured compounds raises the possibility of pressure- and/or temperature-induced changes in the symmetry of the (Mg, Fe)SiO_3 and $CaSiO_3$ perovskite phases. Twinning, suggestive of transformations of $MgSiO_3$ perovskite from structures of higher (tetragonal and cubic) symmetry to orthorhombic during recovery from synthesis at 26 GPa and 1,600°C, was reported by Wang et al. (1990). This interpretation, however, was undermined by subsequent x-ray-diffraction measurements (Funamori and Yagi, 1993) that demonstrated a wide P–T stability field for the orthorhombic phase, inclusive of the synthesis conditions of the earlier study. Recent theoretical calculations suggest that in fact the orthorhombic phase of $MgSiO_3$ should be thermodynamically preferred over the alternative tetragonal and cubic polymorphs over the entire pressure range of the lower mantle (Stixrude and Cohen, 1993; Wentzcovitch, Ross, and Price, 1995). In summary, although transformations to phases of higher symmetry might be possible at very high temperatures (Meade, Mao, and Hu, 1995), it seems most probable that the orthorhombic (Mg, Fe)SiO$_3$ perovskite phase remains stable relative to perovskites of higher symmetry along plausible geotherms throughout the lower mantle.

The inference from high-pressure x-ray-diffraction data (Liu and Ringwood, 1975; Mao et al., 1989; Wang, Weidner, and Guyot, 1996) and *ab initio* calculations (e.g., Sherman, 1993; Wentzcovitch et al., 1995) that CaSiO$_3$ adopts the ideal cubic perovskite crystal structure for the entire pressure range of the lower mantle, even at 300 K, has recently been questioned (Stixrude et al., 1996). These latter authors conclude from an elaborate 'first-principles' analysis that the cubic phase is dynamically unstable at low temperatures and high pressure relative to a tetragonal distorted perovskite structure, based on a 5° rotation of the SiO$_6$ octahedra, sufficiently small to have been overlooked in high-pressure x-ray-diffraction data. Stixrude et al. (1996) predict that this phase will transform to the cubic structure at a temperature near 2,200°C for a pressure of 80 GPa.

Another possibility worthy of serious consideration is the disproportionation of perovskites to marginally denser assemblages of simple binary oxides. The familiar rocksalt-structured polymorphs of CaO and FeO are replaced at pressures of 60–70 GPa by CsCl- and NiAs-structured polymorphs with density increases of about 11% and 4%, respectively (Jeanloz et al., 1979; Fei and Mao, 1994). Evidence has also been obtained at pressures as low as 50 GPa for a silica polymorph with the CaCl$_2$ structure, which is about 1% denser than stishovite, although its thermodynamic stability under hydrostatic stress remains to be established (Tsuchida and Yagi, 1989; Kingma et al., 1995). Finally, the alumina end member in solid solution in (Mg, Fe)SiO$_3$ perovskite is some 2.5% less dense than corundum (Irifune, Koizumi, and Ando, 1996). It is therefore possible to envisage an assemblage of simple oxides at least comparable in density to the perovskite-dominated assemblage of the uppermost part of the lower mantle (Figure 9.1). However, the higher vibrational and configurational entropies of the Mg-perovskite phase are likely to result in a $T \Delta S$ term in the free energy that would outweigh any slight $P \Delta V$ advantage for the oxide mixture.

9.4. Elasticity and Equations-of-State for Mantle Minerals and Rocks
9.4.1. Experimental Approaches

In parallel with the improvements in seismological resolution of the Earth's structure and the revolution in our understanding of the relevant phase equilibria (reviewed in Sections 9.2 and 9.3) have come phenomenal advances in our knowledge of the physical properties of mantle minerals. The information required for interpretation of seismological models includes P–V–T equation-of-state data (constraining compressibility and thermal expansivity) and elastic wave speeds and attenuation under the wide range of P–T conditions prevailing in the mantle. The development and application of an array of different experimental approaches for measurement of thermoelastic properties (Figure 9.3, Table 9.1) has

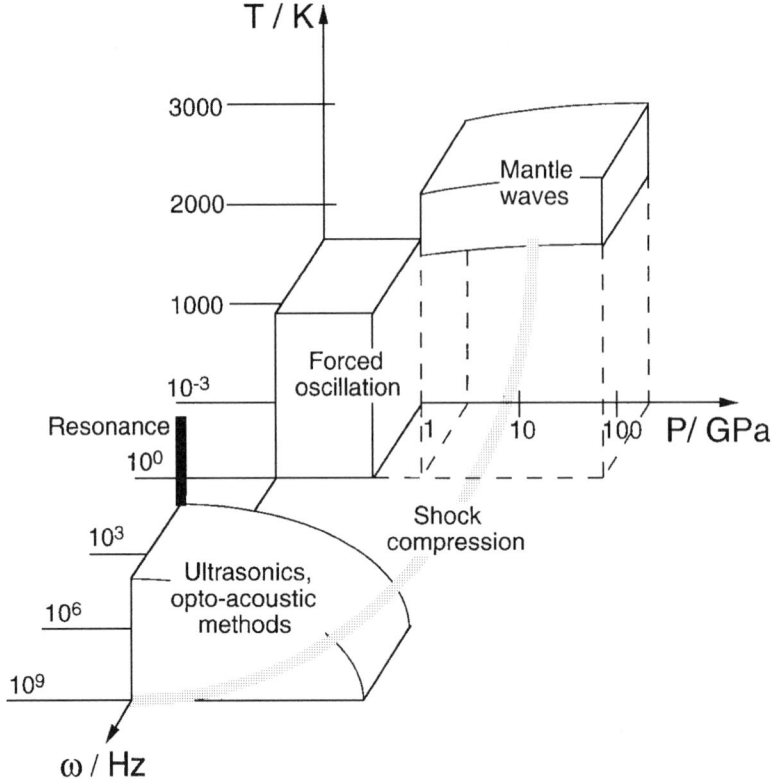

Figure 9.3. A comparison of the capabilities in P–T–ω space of various experimental techniques (see Table 9.1) used to probe the elastic/anelastic behaviour of mantle materials. Development and application of techniques with the potential to shorten the extrapolation to the P–T–ω regime of teleseismic wave propagation is a major frontier in mineral/rock physics.

been complemented by dramatic progress towards robust *ab initio* calculation of the electron densities, structures, and properties of oxide and even complex silicate minerals (e.g., Cohen, 1987; Wolf and Bukowinski, 1987; Hemley, Jackson, and Gordon, 1987; Cohen et al., 1989; Sherman, 1993; Lacks and Gordon, 1995; Sherman and Jansen, 1995; Wentzcovitch et al., 1995; Stixrude et al., 1996).

The result is a fairly comprehensive understanding of the incompressibility, thermal expansivity, and elastic moduli for the important mantle minerals [olivine, pyroxenes, garnets (inclusive of majorite), wadsleyite, ringwoodite, ilmenite, stishovite, and the silicate perovskites] at or near ambient pressure and temperature and at megahertz-to-gigahertz frequencies. By comparison, our knowledge of the pressure and temperature dependences of wave speeds and elastic moduli is more rudimentary. However, $\partial K/\partial P$ and $\partial K/\partial T$ (equivalently, $\partial\alpha/\partial P$) are being increasingly well constrained by synchrotron-based P–V–T studies (Table 9.1). Ultrasonic wave-propagation measurements of $\partial V_i/\partial P$ ($i =$ P, S) at frequencies of 10–100 MHz have been performed for the key phases of the transition zone, and similar determinations of $\partial V_i/\partial T$ are under way. Resonance measurements, also at

Table 9.1. *Experimental investigations of thermoelastic properties of mantle materials:*
key methodologies and outcomes (1960–96)

Methodology	Outcome	References
1960s		
Shock compression of mantle materials	Pressure-induced structural changes: dense, incompressible behaviour like close-packed oxides	McQueen, Marsh, and Fritz (1967), Trunin et al. (1965), Ahrens, Anderson, and Ringwood (1969), Jeanloz and Ahrens (1980)
Ultrasonic studies of rocks and cm-size crystals of mantle minerals	$V_{P,S}(P)$ to 1 GPa for rocks; $c_{ij}(P)$ to 1 GPa and $c_{ij}(T)$ to a few 100 K for upper-mantle minerals	Birch (1960), Simmons (1964), Graham and Barsch (1969), Kumazawa and Anderson (1969)
1970s		
Ultrasonic experiments on synthetic polycrystals	$V_{P0,S0}$ for structural analogues for high-pressure phases	Liebermann (1975)
X-ray diffraction in diamond-anvil apparatus	$V(P) -> $ average incompressibility $K_T = K_{T0} + K'_{T0}P_{max}/2$ (implied K_{T0}–K'_{T0} trade-off)	Mao et al. (1969), Yagi, Mao, and Bell (1982)
1980s		
Brillouin scattering from $\sim$100-μm microcrystals	c_{ij0} for high-pressure phases	Sawamoto et al. (1984), Weidner et al. (1984), Yeganeh-Haeri (1994)
Resonance measurements	$c_{ij}(T)$ to $T \gg \Theta$ for upper-mantle minerals and oxides	Ohno (1976), Sumino et al. (1976), Anderson and Goto (1989), Isaak, Anderson, and Goto (1989), Isaak (1992)
Dilatometric measurement of thermal expansion	α to $\sim$1,000 K on most high-pressure phases; pioneering x-ray measurements on $MgSiO_3$ perovskite	Suzuki (1975), Suzuki, Ohtani, and Kumazawa (1979), Knittle, Jeanloz, and Smith (1986)
1990s		
Ultrasonic interferometry on $\sim$3-mm polycrystals	$\partial V_{P,S}/\partial P$ for transition-zone phases; $\partial V_{P,S}/\partial T$ in progress	Niesler and Jackson (1989), Gwanmesia et. al. (1990), Rigden, Gwanmesia, and Liebermann (1994)
Opto-acoustic techniques (BS, ISS) and ultrasonic interferometry at 10-MHz to GHz frequencies, coupled with diamond- and multiple-anvil apparatus	$c_{ij}(P)$ and $V_{P,S}(P)$ to $\sim$10 GPa	Zaug et al. (1993), Duffy et al. (1995), Li et al. 1996b; Spetzler et al. (1996), Knoche, Webb, and Rubie (in press)

(cont.)

Table 9.1. *(cont.)*

Methodology	Outcome	References
Torsional forced-oscillation/ creep methods	G and Q^{-1} in high-T, low-ω regime	Berckhemer et al. (1982), Jackson, Paterson, and Fitz Gerald (1992), Getting et al. (1997)
Synchrotron-based x-ray diffraction coupled with diamond-anvil and multiple-anvil apparatus	$V(P, T)$ and hence α and $(\partial K_T / \partial T)_P$ for high-pressure phases	Fei et al. (1992), Mao et al. (1991), Wang et al. (1994), Funamori et al. (1996)
IR/Raman spectroscopy coupled with diamond-anvil apparatus	Mode Grüneisen parameters γ_i and hence γ_{th}	Williams, Jeanloz, and McMillan (1987), Hemley et al. (1989), Chopelas and Boehler (1992), Chopelas (1996), Gillet, Guyot, and Wang (1996)

Note: V, P, and T are respectively volume, pressure, and temperature; α, Θ, c_{ij}, K, G, Q^{-1}, V_P, V_S, and γ_{th} are respectively thermal expansivity, Debye temperature, single-crystal elastic-stiffness moduli, bulk modulus, shear modulus, attenuation (in shear), compressional- and shear-wave speeds, and the thermodynamic Grüneisen parameter; K' is the pressure derivative of K, and the subscript 'zero' refers to ambient laboratory conditions; BS, Brillouin scattering; ISS, implusive stimulated scattering.

megahertz frequencies, have provided crucial insight into the temperature dependence of single-crystal elastic moduli c_{ij} and of derived aggregate properties such as K and G at temperatures well beyond the Debye temperature Θ, conventionally regarded as the threshold for high-temperature behaviour of the crystal lattice. The combination of high-frequency opto-acoustic and ultrasonic interferometric techniques with diamond-anvil and multiple-anvil apparatus is dramatically shortening the inevitable extrapolation in pressure (Figure 9.3). Low-frequency forced-oscillation and creep methods are providing direct experimental access to seismic frequencies (Figure 9.3) and are beginning to highlight the importance of the dispersion and attenuation associated with viscoelastic relaxation.

9.4.2. Equations-of-State

9.4.2.1. Isothermal Compression

A vast body of experimental data now demonstrates that the 300 K isothermal compression of most materials is adequately described by finite-strain equations-of-state, based on the Taylor expansion of the Helmholtz free energy

$$\Psi = a_1\varepsilon^2 + a_2\varepsilon^3 + a_3\varepsilon^4 + \cdots \tag{4}$$

in powers of the Eulerian strain

$$\varepsilon = [1 - (\rho/\rho_0)^{2/3}]/2 = [1 - (V/V_0)^{-2/3}]/2 \tag{5}$$

truncated at fourth or even third order (e.g., Birch, 1952; Mao and Bell, 1979; Jeanloz and Knittle, 1986; Knittle and Jeanloz, 1987). V and ρ are respectively the molar volume and density; the subscript zero identifies their zero-pressure values. Differentiation of equation (4) with respect to volume V leads to the fourth-order expression for pressure:

$$P = -d\Psi/dV = -(1 - 2\varepsilon)^{5/2}\{C_1\varepsilon + C_2\varepsilon^2/2 + C_3\varepsilon^3/6\} \tag{6}$$

(e.g., Davies and Dziewonski, 1975). The corresponding expressions for the in-compressibility or bulk modulus

$$K = (1 - 2\varepsilon)^{5/2}\{C_1 + (C_2 - 7C_1)\varepsilon + (C_3 - 9C_2)\varepsilon^2/2 - 11C_3\varepsilon^3/6\}/3 \tag{7}$$

and its first and second pressure derivatives K' and K'' follow by further differentiation. The term in ε^3 is incomplete and therefore is sometimes omitted from the fourth-order expression for K (e.g., Davies and Dziewonski, 1975; Bina and Helffrich, 1992), with the result that K is no longer identically equal to $-V(\partial P/\partial V)$. This internal inconsistency is avoided here by retention of the final term in equation (7) (Ita and Stixrude, 1992; Jackson, in press). The relationships between the coefficients C_i and the bulk modulus and its pressure derivatives evaluated at zero pressure are as follows:

$$C_1 = 3K_0$$
$$C_2 = 9K_0(4 - K_0')$$
$$C_3 = 27K_0\left[K_0K_0'' - K_0'(7 - K_0') + 143/9\right] \tag{8}$$

9.4.2.2. Thermal Pressure

The effect of temperature can be incorporated into the equation-of-state in a variety of ways. Of these, the most widely used are the 'pointwise' addition of the thermal pressure required to maintain constant volume and the 'global' approach in which the zero-pressure volume V_0 and the finite-strain coefficients C_i are regarded as temperature-dependent (e.g., Jackson and Rigden, 1996).

Explicit incorporation of the simple but empirically successful Debye model for the thermal energy E_{th} yields a parametrically economical Mie-Grüneisen description of the thermal pressure:

$$\Delta P_{th}(V, T) = P(V, T) - P(V, 300\,\text{K})$$
$$= \gamma(V)\{E_{th}[\Theta(V), T] - E_{th}[\Theta(V), 300\,\text{K}]\}/V \tag{9}$$

$P(V, 300\,\text{K})$ is the pressure on the principal (300 K) isotherm that often is most appropriately modelled by the third-order Eulerian expression [equation (6), with $C_3 = 0$]. Θ is the Debye temperature, with a volume dependence described by

$$\gamma = -(\partial \ln \Theta/\partial \ln V)_T \tag{10}$$

The volume dependence of the Grüneisen parameter γ is given by

$$(\partial \ln \gamma / \partial \ln V)_T = q = \text{constant} \tag{11}$$

which yields, upon integration,

$$\gamma(V) = \gamma_0 (V/V_0)^q \tag{12}$$

with $\gamma_0 = \gamma(V_0)$.

One of the major attractions of equation (9) as a P–V–T equation-of-state is that it provides a complete thermodynamic framework, which facilitates *internally consistent* conversion between the isothermal conditions of much laboratory experimentation and the isentropic conditions prevailing along mantle adiabats (Stixrude and Bukowinski, 1990; Bina and Helffrich, 1992; Jackson and Rigden, 1996).

9.4.2.3. Hot Finite-Strain Isentropes

The finite-strain equation-of-state [equation (6)] with suitably adjusted coefficients has also been widely used to describe adiabatic compression (e.g., Davies and Dziewonski, 1975; Duffy and Anderson, 1989; Rigden et al., 1991), and recent modelling suggests a close consistency between these two approaches (Jackson, in press). Equations (5) and (7) yield for the seismic parameter

$$\phi = K_S/\rho = (1 - 2\varepsilon)\{C_1(1 - 7\varepsilon) + C_2(\varepsilon - 9\varepsilon^2/2) + C_3(\varepsilon^2/2 - 11\varepsilon^3/6)\}/3\rho_0 \tag{13}$$

The terms within the { } brackets in equation (13) are grouped in such a way as to emphasize the fact that the coefficients of the powers of ε are not all independent, an important consideration in fitting seismological data (Jackson, in press). Equations (6) and (13) provide an internally consistent method for modelling (at fourth order) the compression of the Earth's interior. The more familiar third-order expressions are retrieved by setting $C_3 = 0$.

9.5. Variability of Seismic Wave Speeds and Attenuation in the Upper Mantle
9.5.1. Heterogeneity: Thermal versus Compositional

The upper mantle is distinguished from the transition zone and most of the lower mantle (all but the D'' layer) by a high degree of lateral variability of seismic wave speeds, by marked elastic anisotropy, and by strong attenuation, especially in shear (Kennett and van der Hilst, Chapter 8, this volume). It was observed in Section 9.2 that the neglect of anisotropy, especially in regional tomographic studies, might introduce bias into the inferred lateral variability of isotropic seismic wave speeds. This presents a significant problem for interpretation, because the anisotropic variation of wave speeds in olivine-rich ultramafic xenoliths (typically 7–12%) (e.g., Babuška, 1984) is substantially greater than the likely compositional variability (discussed later). The focus of the following discussion is the relative importance of the compositional and thermal contributions to the wave-speed variability ($\delta V_S > \pm 4\%$) on the largest spatial scales.

Much of the compositional variation within the suite of ultramafic xenoliths brought to the Earth's surface by basaltic magmas is the result of varying degrees of prior fractionation (e.g., Green and Falloon, Chapter 7, this volume). Jordan (1979) showed that the extraction of basaltic magma by 20% partial melting of a fertile garnet lherzolite changes the calculated densities and compressional- and shear-wave speeds of the residuum by -1.6%, $+0.1\%$, and $+0.5\%$, respectively. He further argued that the lower intrinsic density of such barren peridotite might stabilize the relatively cool continental tectosphere against thermal advection. The insensitivity of V_P to the compositional variation associated with basalt extraction is interpreted as the result of the near-cancellation between the effects of elimination of the high-velocity garnet and low-velocity clinopyroxene phases and the effects of the more magnesian character and hence higher wave speeds of the residual olivine and orthopyroxene (Leven, Jackson, and Ringwood, 1981). Calculated densities and compressional-wave speeds for a suite of garnet-lherzolite xenoliths from the Massif Central in France similarly span very limited ranges, $\pm 0.9\%$ and $\pm 0.4\%$, respectively (Sobolev et al., 1996). Spinel lherzolites from the same region are slightly more variable ($\pm 0.8\%$) in calculated compressional-wave speed.

Also represented in the xenolith suite, *but at a much lower overall abundance*, are compositions interpreted as products of melting, fractional crystallization, and metasomatism (e.g., eclogites, pyroxenites, and amphibole- or phlogopite-bearing hydrous lherzolites). Inclusion of these lithologies has the effect of broadening (to perhaps -3% to $+1\%$) the allowed compositional variability of seismic wave speeds in the upper mantle (Leven et al., 1981; O'Reilly, Jackson, and Bezant, 1990). Under these circumstances, the contribution of major-element compositional heterogeneity to the observed wave-speed variability of the upper mantle must be relatively minor. Volatiles could, however, play an important role, especially in explaining low wave speeds in the mantle wedge above subduction zones, by drastically lowering the solidus and thereby allowing incipient melting at temperatures well below the anhydrous solidus (Green and Falloon, Chapter 7, this volume).

The temperature sensitivity of wave speeds in upper-mantle minerals and rocks is constrained mainly by ultrasonic studies at megahertz frequencies far removed from those of seismic wave propagation (see Section 9.4). For the dominant upper-mantle mineral, olivine, the value of $(\partial \ln V_S / \partial T)_P$ is -8×10^{-5} K^{-1} at megahertz frequencies (Isaak, 1992). It follows that lateral variation of $\pm 4\%$ in V_S attributed to a thermal origin is mapped into anomalies of amplitude ± 500 K. Thermal anomalies of this magnitude, except in the immediate vicinity of subducting slabs, are implausible on a variety of grounds, including gross incompatibility with the observed heat flow and gravity (e.g., Sobolev et al., 1996).

9.5.2. Solid-State Viscoelastic Relaxation

Along with the variability of wave speeds, the generally strong attenuation ($Q^{-1} \sim$ 0.015) of seismic shear waves in the upper mantle, and its lateral variability ($\pm 50\%$), must also be explained. Goetze (1977) argued that solid-state viscoelastic relaxation might play an important role. He reasoned, mainly by analogy with the better-understood mechanical behaviour of metals, that both the temperature sensitivity of shear-wave speed and the associated strain-energy dissipation are likely to be greater at seismic frequencies than at ultrasonic frequencies (Figure 9.4).

During the past two decades, new experimental methods have begun to provide the necessary laboratory information on geological materials at high temperatures and seismic frequencies (Berckhemer et al., 1982; Gueguen et al., 1989; Getting et al., 1991, 1997; Jackson, Paterson, and Fitz Gerald, 1992; Jackson and Paterson, 1993). Thus, a consensus is emerging concerning the strong attenuation ($Q^{-1} \sim 0.01$) in upper-mantle materials at high sub-solidus temperatures (Figure 9.5a). The impressive consistency among such findings for ultramafic rocks measured in different laboratories and for olivine single crystals is suggestive of dominantly intragranular (dislocation-related?) rather than grain-boundary relaxation (Jackson et al., 1992).

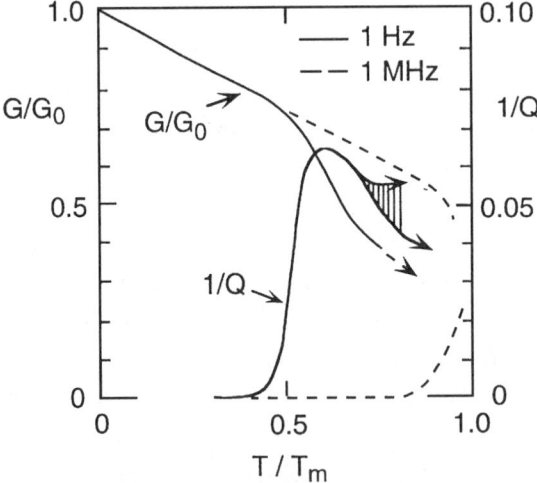

Figure 9.4. Schematic illustration of the temperature dependence of the shear modulus G and attenuation $1/Q$ expected for viscoelastic crystalline material at high sub-solidus temperatures. As a consequence of thermally activated viscoelastic relaxation, the properties measured at ultrasonic frequencies (~ 1 MHz) and at seismic frequencies (~ 1 Hz) are expected to diverge as the melting temperature (T_m) is approached. (Adapted from Goetze, 1977.)

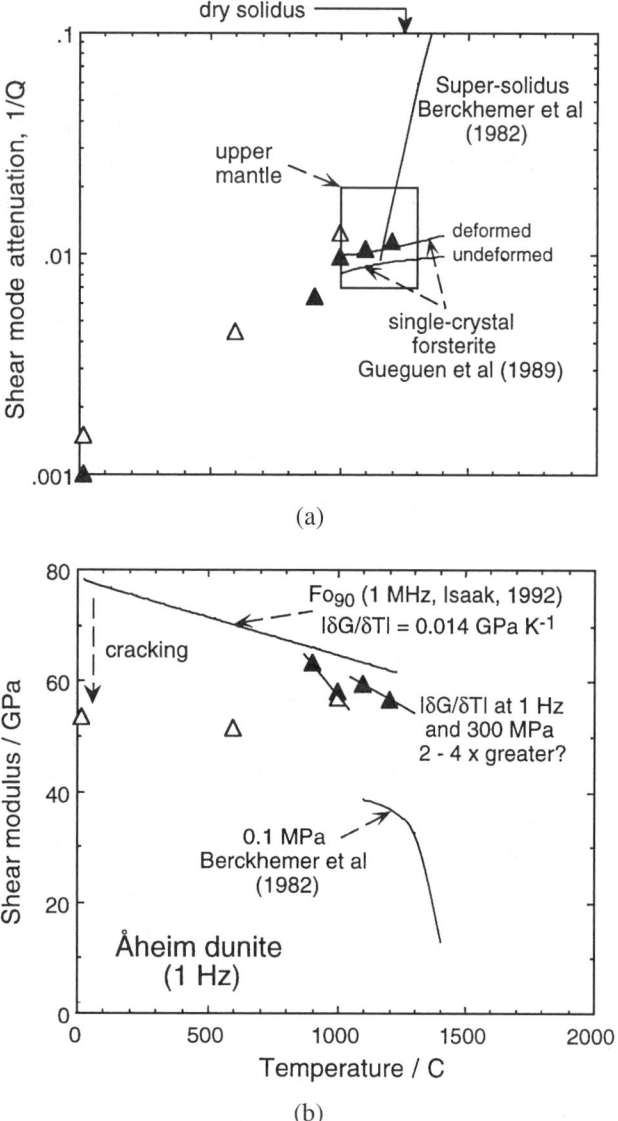

Figure 9.5. Shear modulus G and attenuation $1/Q$ for Åheim dunite measured (except where otherwise indicated) at 1 Hz with torsional forced oscillation techniques. Open and solid triangles denote results for multiple specimens cored respectively normal and parallel to a pronounced layer-silicate foliation and then fired at 1,200°C for 24 hours under a suitably controlled atmosphere in order to effect complete dehydration (Jackson et al., 1992, and unpublished data). G and $1/Q$ were measured at low strains ($<3 \times 10^{-6}$) on these anhydrous specimens at 300 MPa and the indicated temperatures; the solid symbols plotted at 900°C and 1,000°C represent results obtained in a subsequent experiment on the specimen recovered after testing at 1,100°C. (a) The shear mode attenuation measured on anhydrous Åheim dunite at high sub-solidus temperatures is comparable to that reported for single-crystal forsterite and values determined seismologically for the upper mantle. (b) The shear moduli measured at 300 MPa for anhydrous Åheim dunite are much higher than those measured at ambient pressure by Berckhemer et al. (1982), reflecting the role of confining pressure in partially, but not completely, suppressing thermal microcracking.

The attenuation and dispersion (frequency dependence) of the shear modulus G and wave speed V_S are inextricably and quantitatively coupled, as illustrated in Figure 9.4, through the well-established theory of linear viscoelasticity. The substantial shear-mode attenuation measured for these ultramafic materials at high sub-solidus conditions is therefore inevitably associated with pronounced dispersion of the shear modulus and wave speed. However, definitive experimental determination of the magnitude of the shear-modulus dispersion through comparison of measurements of $G(T)$ at ultrasonic (megahertz) frequencies and at seismic frequencies (<1 Hz) has been complicated by the difficulty in separating the effects of viscoelastic relaxation and thermal cracking in relatively coarse grained natural rocks (Jackson et al., 1992). However, the unpublished findings from our laboratory that have been least compromised by thermal cracking suggest a temperature sensitivity at high sub-solidus conditions and 1 Hz that is two to four times greater than that at the megahertz frequencies of ultrasonic experimentation (Figure 9.5b). Pending the conduct of definitive laboratory measurements, Karato (1993) has employed the linear theory of viscoelasticity along with seismologically measured attenuation to estimate the expected dispersion and thus $|\partial \ln V_S / \partial T|$ at seismic frequencies. His conclusion, consistent with the observations mentioned earlier, is that the seismic-frequency derivative is likely to be at least twice as great as that measured at the megahertz frequencies of ultrasonic experiments. Torsional forced-oscillation experiments in progress on fine-grained synthetic olivine aggregates (Tan, Jackson, and Fitz Gerald, 1997) are beginning to provide definitive measurements of both $\partial G / \partial T$ and Q^{-1} at seismic frequencies and high sub-solidus temperatures. Systematic manipulation of the microstructure (especially grain size) of these synthetic materials may also allow the positive identification of the microscopic defects responsible for the macroscopic viscoelastic behaviour. Such a mechanistic understanding is required to underpin the seismological application of these experimental findings.

The fact that viscoelastic relaxation is thermally activated means that it is likely to scale approximately with homologous temperature T/T_m, where T_m is the relevant melting temperature. Comparison of geotherms with the experimentally determined solidus for the pyrolite model mantle suggests that the dry solidus is most closely approached in the upper mantle[2] (Figure 9.2). Viscoelastic relaxation and the associated dispersion and attenuation of seismic waves will accordingly be most pronounced in that region. With increasing depth into the transition zone and lower mantle, the geotherm falls progressively further below the solidus, and both reduced attenuation and the gradual recovery of the modulus deficit associated with intense viscoelastic relaxation at shallower depths are to be expected. The recovery of this modulus deficit is expected to contribute to the steepening of average velocity–depth gradients, especially for V_S through the transition zone (see Section 9.6).

---

[2] If temperatures exceed those of the pyrolite + (C, H, O) solidus, a zone of incipient melting will be encountered (Green and Falloon, Chapter 7, this volume).

9.6. Discontinuities, Velocity Gradients, and the Composition of the Transition Zone

Over the past decade there have been many attempts to develop and test compositional models for the transition zone by comparison of calculated wave velocities with those from seismological models (Weidner, 1986; Anderson and Bass, 1986; Duffy and Anderson, 1989; Gwanmesia et al., 1990; Rigden et al., 1991; Ita and Stixrude, 1992; Agee, 1993; Irifune, 1993). These studies have incorporated the progressively tighter constraints on the thermoelastic properties of the relevant high-pressure phases that have been emerging from laboratory studies (see Sections 9.3 and 9.4).

9.6.1. Modelling Strategy

Here we illustrate the state of the art by modelling the variation of bulk sound speed

$$V_\phi = \phi^{1/2} = \left[V_P^2 - (4/3)V_S^2\right]^{1/2} = (K_S/\rho)^{1/2} \qquad (14)$$

with depth through the transition zone for a model mantle of pyrolite composition (Figure 9.1). This approach, previously employed by Ita and Stixrude (1992), provides an attractive alternative to the modelling of the compressional- and shear-wave speeds, which are more directly inferred from seismological studies, for the following reasons: Because V_ϕ is a function only of density ρ and the adiabatic bulk modulus K_S, it can be calculated with some confidence from appropriate P–V–T equations-of-state, into which the constraints on $\partial K/\partial T$ beginning to come from synchrotron-based x-ray-diffraction studies are readily incorporated. By comparison, our knowledge of the shear modulus is much more rudimentary. $\partial G/\partial T$ is yet to be measured, even at the megahertz-to-gigahertz frequencies of laboratory wave-propagation studies, for most of the minerals of the transition zone and lower mantle. Furthermore, the potentially significant dispersion between these high frequencies and the millihertz-to-hertz frequencies of seismic waves is yet to be quantified (see Section 9.5).

The preferred equation-of-state for modelling the density and compressibility of mantle minerals combines the Birch-Murnaghan finite-strain description of isothermal compression at 300 K with the Mie-Grüneisen-Debye approximation to the superimposed thermal pressure (see Section 9.4). The values of the required parameters V_0, K_0, K_0', Θ_0, γ_0, and q for each of the major minerals of the transition zone are presented in Table 9.2. Those for $(Mg, Fe)_2SiO_4$ wadsleyite are derived from a recent analysis of all the relevant acoustic and static compression data (Jackson and Rigden, 1996). A similar approach has been taken with ringwoodite – the preferred set of equation-of-state parameters being obtained by fitting the 300 K compression data of Hazen (1993) and Meng et al. (1994) along with the thermal expansion measured by Suzuki, Ohtani, and Kumazawa (1979). For

CaSiO$_3$ perovskite, V_0, K_0, and K'_0 are constrained by compression studies (Mao et al., 1989; Wang and Weidner, 1994) (see Section 9.7). Other thermoelastic parameters are assumed to be the same as for MgSiO$_3$ perovskite. For the remaining important phases, the equation-of-state parameters are those estimated by Ita and Stixrude (1992).

At each of a series of representative depths between 400 km and 770 km, the proportions and compositions of coexisting phases appropriate for the pyrolite bulk composition were identified (Irifune and Ringwood, 1987; Irifune, 1993, 1994) (Figure 9.1b). Physical properties, including density, bulk modulus, and entropy, were then calculated at common P–T values for each phase from the equation-of-state parameters in Table 9.2. The bulk modulus of the aggregate was calculated as the Hill average of the Voigt and Reuss bounds, and the temperature required to maintain constant total entropy was obtained by interpolation. In this way, the variation of bulk sound speed with depth for the pyrolite composition was calculated along an adiabat with a potential temperature of 1,550 K consistent with the petrological evidence (see Section 9.3 and Figure 9.2).

Representative *seismological* profiles for bulk sound speed and density through the transition zone were calculated by Ita and Stixrude (1992) as averages of existing density profiles and by combining a large number of independent P-wave and S-wave profiles. In order to maximize internal consistency, we have chosen, instead, for this study, to focus on radial profiles that were determined in the same region with similar techniques for both P and S waves. Four such pairs of body-wave studies are available: models GCA and TNA, representing profiles under tectonically active North America (Grand and Helmberger, 1984b; Walck, 1984); S25 and SNA, representing shield models of North America (Grand and Helmberger, 1984b; Le Fevre and Helmberger, 1989); ATLP and NWA, sampling regions in the northwestern Atlantic Ocean (Grand and Helmberger, 1984a; Zhao and Helmberger, 1993); NJPB, a model for both P and S waves from northern Australia (Kennett, Gudmundsson, and Tong, 1994). The bulk sound velocity as a function of depth was calculated for each of these pairs of models.

9.6.2. Calculated versus 'Observed' Bulk Sound Speeds

The bulk sound velocity calculated for pyrolite along the 1,550 K adiabat is compared with these seismological models in Figure 9.6. Although the calculated sound speed is systematically about 2% higher than that determined seismologically, the magnitudes of the major discontinuities and the velocity–depth gradients are well approximated by the calculated properties. The average gradient between 400 km and 660 km varies between 0.0025 and 0.0035 km $\cdot$ s^{-1} $\cdot$ km^{-1} for the seismological models, whereas the pyrolite gradient is 0.003 km $\cdot$ s^{-1} $\cdot$ km^{-1}. For comparison, the gradient for bulk sound velocity in the lower mantle is about 0.0015 km $\cdot$ s^{-1} $\cdot$ km^{-1}.

Table 9.2. *Equation-of-state parameters for the major mantle minerals*

Equation-of-state parameters	Olivine	Wadsleyite	Ringwoodite	Diopside	Pyrope	Almandine	Grossular	Majorite	CaSiO$_3$ perovskite	(Mg$_{1-x}$Fe$_x$)SiO$_3$ perovskite	(Mg$_{1-x}$Fe$_x$)SiO$_3$ perovskite	(Mg$_{1-x}$Fe$_x$)O
V_0 cm^3	44.08	40.96	39.92	66.11	113.19	115.23	125.30	114.15	27.46	24.46 + 1.03x	24.434 + 1.21x	11.247 + 1.00x
K_{S0} GPa	128.0	176.0	185.0	114.0	173.0	177.0	168.0	160.0	281.0	266.5[d]	264.0	162.5 + 11.5x
$(\partial K_S/\partial P)_T$	5.0	4.8	5.0	4.5	4.9	4.9[b]	4.9[b]	4.9[b]	4.0	3.82[d]	4.0	4.13
Θ_0 K	898[a]	969[a]	989[a]	941	981	981[b]	981[b]	981[b]	1,000[c]	1,017	1,000	673
γ_0	1.14	1.39	1.39	1.06	1.24	1.24[b]	1.24[b]	1.24[b]	1.33[c]	1.96	1.33	1.41
q	1.0	1.0	1.0	1.0	1.0	1.0	1.0	1.0	1.0	2.5	1.0	1.3
Source	(1)	(2)	(1)	(1)	(1)	(1)	(1)	(1)(3)	(4)	(1)	(2)	(2)

[a]Weighted average of values for Mg and Fe end members.
[b]Same value as for pyrope.
[c]Same value as for MgSiO$_3$ perovskite.
[d]Consistent with $K_{T0} = 263$ GPa, $K'_{T0} = 3.9$.

Sources: Data from (1) Ita and Stixrude (1992), (2) Jackson and Rigden (1996), (3) Pacalo (1993), and (4) Wang and Weidner (1994).

This demonstration that phase transformations in a model mantle of pyrolite composition provide an adequate explanation for the structure (major discontinuities and gradients) of the transition zone accords with the earlier findings of Ita and Stixrude (1992). Refinement of the relevant phase equilibria since that time indicates that the pyroxene-to-garnet transformation is complete by about 480 km, with the last 18% of pyroxene transforming at depths between 400 and 480 km, slightly deeper than was previously thought (Irifune, 1993). Also, it now appears that the wadsleyite–ringwoodite phase transformation occurs at depths between 510 and 540 km. The garnet–perovskite phase transformation begins with the gradual exsolution of about 7% $CaSiO_3$ perovskite between 580 and 720 km. The residual garnet transforms progressively to (Mg, Fe, Al)(Si, Al)O_3 perovskite at depths of 660–730 km (Irifune, 1994). These recent observations suggest completion of the phase changes at somewhat greater depths than those assumed by Ita and Stixrude (1992). However, *their important conclusion that these transformations combine to produce steep gradients for bulk sound speed throughout the transition zone, comparable to those observed seismologically, remains unaffected.* The 520-km discontinuity, often interpreted in terms of the wadsleyite–ringwoodite transformation, is conspicuous by its virtual absence from Figure 9.6. Given the small contrast (~1%) in wave speeds across this discontinuity and the substantial width of the binary loop (Figure 9.2), it is probable that it is observable only in studies at relatively

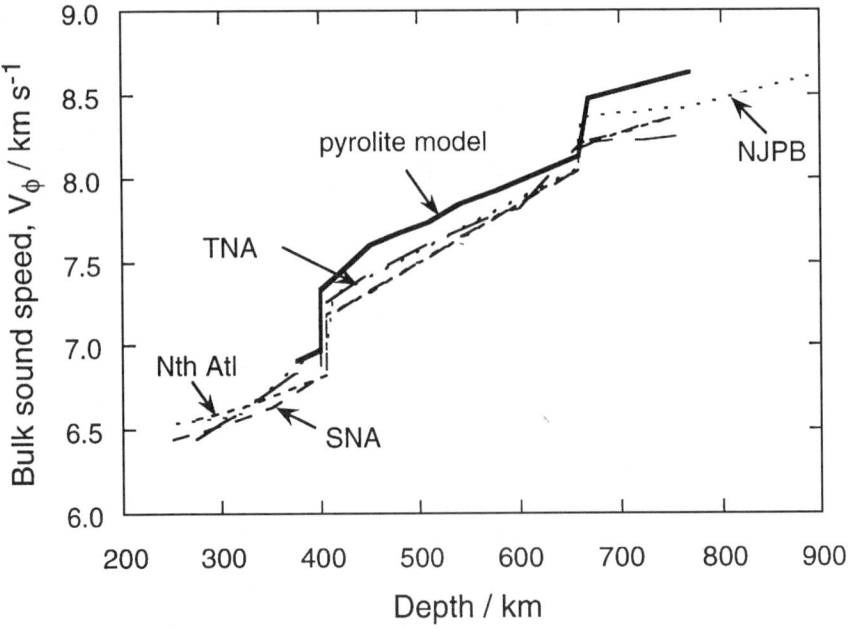

Figure 9.6. Bulk sound speed V_ϕ versus depth through the transition zone from several representative seismological profiles compared with calculations for the pyrolite model. 'TNA', 'SNA', and 'Nth Atl' denote the bulk sound speeds derived respectively from models for tectonically active and shield regions of North America and for the northwestern Atlantic Ocean. 'NJPB' is a model for northern Australia.

long periods and near-normal incidence, where the reflectivity is determined by the contrast in impedance (wave speed × density, ∼3%), rather than wave speed alone (Rigden et al., 1991; Shearer, 1996).

9.6.3. Uncertainties in the Extrapolation of Bulk Sound Speeds

The bulk sound speeds calculated for pyrolite thus provide a satisfactory match to the amplitudes of the discontinuities and the gradients in the transition zone. However, the fact that the calculated velocities are consistently about 2% too high (Figure 9.6) suggests that there may be a systematic error in the extrapolation of thermoelastic properties to transition-zone $P-T$ conditions. Opto-acoustic and ultrasonic measurements of single-crystal elastic moduli within the $P-T$ environment of diamond-anvil and multiple-anvil apparatus are beginning to provide the data needed to address this question. Measurements on single-crystal olivine to pressures of 12 and 16 GPa by Zaug et al. (1993) and Duffy et al. (1995), respectively, suggest lower average values for the pressure derivatives K' and G' than are obtained by finite-strain extrapolation through derivatives measured at lower pressures, although some of the details remain to be resolved (Chen, Li, and Liebermann, 1996). Similarly, ultrasonic data for polycrystalline specimens of the olivine and wadsleyite polymorphs of Mg_2SiO_4, measured to pressures greater than 12 GPa (Li, Gwanmesia, and Liebermann, 1996a), have yielded average values for K' that are lower by about 0.6 than the values for K_0' derived from previous measurements over more limited ranges of pressure (Table 9.2). Such behaviour might be expected if the dominant microscopic compression mechanisms by which the changing volume of the unit cell is accommodated were to change with increasing pressure. An extreme example is provided by orthopyroxene, in which K' decreases by almost 50% (from about 11 to 6) over the pressure interval 0–3 GPa (Webb and Jackson, 1993; Angel and Hugh-Jones, 1994).

The effect of a reduction in K_0' by 0.6 is a reduction of 8 GPa (5%) in the bulk modulus calculated for olivine at 14 GPa and 1,450°C; a similar change would result from increasing $|\partial K_S / \partial T|$ by 0.006 GPa · K^{-1}. This simple analysis suggests that the systematic differences between the calculated and seismologically determined bulk sound speeds (Figure 9.6) probably can be largely accounted for by systematic errors in extrapolation; any remaining discrepancy will be well within the residual uncertainties in the thermoelastic properties.

9.6.4. Compressional- and Shear-Wave Speeds

The seismologically observed contrasts in compressional- and shear-wave speeds across the 410-km discontinuity, interpreted in terms of the olivine–wadsleyite transformation, provide a potentially very stringent test for the composition of the transition zone (e.g., Weidner, 1986; Duffy and Anderson, 1989). Using values for K'_0 and G'_0 derived from ultrasonic interferometric measurements at pressures below 3 GPa, Gwanmesia et al. (1990) demonstrated that unexceptional values of the as-yet-unmeasured temperature derivatives allow the velocity contrasts across the 410-km discontinuity to be reconciled with those expected for the olivine–wadsleyite transformation in a pyrolite model mantle. However, it has recently been argued by Duffy et al. (1995) that the lower compressional- and shear-wave speeds for olivine, implied by their Brillouin scattering measurements to 16 GPa, suggest that the contrast at 410 km depth between the wave speeds for the olivine and high-pressure wadsleyite phases is too large to be compatible with a pyrolite composition. If the lower derivatives measured on olivine by Duffy et al. (1995) over an extended range of pressures result from changes in the compression mechanism, as suggested earlier, then similar effects are to be expected in other mantle minerals, including wadsleyite. The lower average values for K' for both olivine and wadsleyite recently measured over an extended pressure range by Li et al. (1996a) undermine the argument of Duffy et al. (1995). Pending the outcome of measurements in progress on the temperature dependence of the elastic moduli for wadsleyite (S. M. Rigden, personal communication; Suzuki, Sakai, and Katsura, 1996), any conclusion that the amplitudes of the 410-km seismic discontinuities are inconsistent with the $(Mg, Fe)_2SiO_4$ content of pyrolite is without foundation.

The relatively steep seismologically determined gradients for V_P and especially V_S through the transition zone are substantially greater than those expected in an isochemical $(Mg, Fe)_2SiO_4$ model mantle subject to the known phase transformations (Duffy and Anderson, 1989; Rigden et al., 1991). Inclusion of the gradual pyroxene–garnet and garnet–perovskite phase transformations affecting the $[(Mg, Fe, Ca)Si, Al_2]O_3$ component of pyrolite has the potential to explain the V_P and V_S gradients as well (Weidner, 1986), although significantly shallower gradients have been calculated in some analyses (Agee, 1993; Irifune, 1993). Given the tendency for attenuation (Q^{-1}) to decrease with increasing depth into the mantle (see Sections 9.2 and 9.5), it is almost inevitable that the average gradient for shear-wave speed and, to a lesser extent, compressional-wave speed in the transition zone will be increased by the progressive recovery of the modulus deficit associated with pronounced viscoelastic relaxation in the upper mantle (Rigden et al., 1991). For example, the recovery of 60% of a 5% velocity deficit, as suggested by the $Q^{-1}(z)$ models discussed in Section 9.2, averaged over the 250-km depth interval between the major discontinuities, would contribute $0.0006 \text{ km} \cdot \text{s}^{-1} \cdot \text{km}^{-1}$, or about 20%, to the observed average gradient $\delta V_S / \delta z$.

9.6.5. Sharpness of the Discontinuities

Finally, we ask whether or not the observed sharpness of the major discontinuities (i.e., a substantial contrast in wave speed across a depth interval no greater than a quarter of a wavelength for a short-period compressional wave) (see Section 9.2) is consistent with their attribution to phase transformations in an isochemical mantle. The olivine $\rightarrow$ wadsleyite and ringwoodite $\rightarrow$ perovskite + magnesiowüstite transformations, presumed responsible for the 410-km and 660-km discontinuities, respectively, have two-phase fields with widths of 0.7 GPa and less than 0.15 GPa, corresponding to depth ranges of 20 and 4 km, respectively (Ito and Takahashi, 1989; Katsura and Ito, 1989) (Figure 9.2). However, application of the 'lever rule' that governs the proportions of coexisting phases indicates that these proportions change in a sigmoidal rather than linear manner (D. J. Weidner, personal communication, 1994; Helffrich and Bina, 1994; Stixrude, 1995), with a rate of change twice the average that prevails in the central part of the binary loop. This concentration of much of the change of phase within a pressure interval substantially narrower than the binary loop itself might contribute significantly to the short-period seismological observability, especially for the olivine–wadsleyite transformation. The temperature sensitivity of the width of the olivine + wadsleyite two-phase field (Figure 9.2) has been identified as a possible cause of the apparent variability in the reflectivity of the 410-km discontinuity (Helffrich and Bina, 1994).

In summary, it appears that the increasingly well resolved radial seismic structure of the transition zone continues to be compatible with the simplest possible petrological model, namely, that of pressure-induced phase transformations within an isochemical mantle. Although the temperature dependence of seismic wave speeds in transition-zone minerals remains to be determined, the depths, velocity contrasts, and sharpness of the major 410-km and 660-km discontinuities are explained, within the residual uncertainties, by the transformation of olivine to wadsleyite and the disproportionation of ringwoodite to yield $(Mg, Fe)SiO_3$ perovskite plus magnesiowüstite. The generally steep velocity–depth gradients for the transition zone are attributed to the combination of the pyroxene–garnet, wadsleyite–ringwoodite, and garnet–perovskite transformations. The gradient in shear-wave speed, in particular, probably contains a significant component arising from the recovery of the modulus deficit associated with intense viscoelastic relaxation in the upper mantle.

9.7. Composition and Temperature of the Lower Mantle
9.7.1. Uniformity and Adiabaticity

In marked contrast to the situation in the overlying transition zone, seismic wave speeds vary smoothly with depth in the lower mantle. Moreover, the parameter $1 - g^{-1}\partial\phi/\partial r$ takes values consistent with adiabatic compression of material that is homogeneous in composition and phase (see Section 9.1). Under these circumstances, the variation of density with depth can be calculated from models of seismic wave speed versus depth by integration of the Adams-Williamson relationship

$$d\rho/\rho = -g(r)\, dr/\phi(r) \tag{15}$$

with ϕ given by equation (1). Such density–depth models have generally required only very minor modifications in order to satisfy the additional constraints derived since the 1960s from observations of the Earth's free oscillations (e.g., Dziewonski and Anderson, 1981; see, however, Montagner and Kennett, 1996). It must be concluded either that conditions of homogeneity and adiabaticity are indeed closely approached or that the free-oscillation constraints are reasonably permissive.

Recognition that the Rayleigh number for the Earth's mantle far exceeds the critical value for the onset of thermal convection (e.g., Griffiths and Turner, Chapter 4, this volume) further strengthens the case for temperature gradients no steeper than marginally superadiabatic. However, the episodic transfer of large volumes of relatively cold upper-mantle material into the lower mantle, as suggested by some recent numerical models of mantle convection (Machetel and Weber, 1991; Honda et al., 1992; Peltier and Solheim, 1992; Tackley et al., 1993; Weinstein, 1993; Davies, Chapter 5, this volume), could result in transiently perturbed temperature profiles, with significantly *subadiabatic* gradients for large regions of the lower mantle that could persist for periods of several hundred million years.

Ignoring this latter possible complication, it will be assumed here that the lower mantle is homogeneous in composition and phase and that its temperature distribution is adiabatic, with a potential temperature T_0. The phase-equilibrium studies reviewed in Section 9.3 indicate that for a wide range of plausible chemical compositions the mineralogy of the lower mantle will be dominated by $(Mg, Fe)SiO_3$ perovskite and $(Mg, Fe)O$ magnesiowüstite (e.g., O'Neill and Jeanloz, 1990; Irifune, 1993). The next most important components (Figure 9.1) are Al_2O_3, accommodated in solid solution in magnesian silicate perovskite, and $CaSiO_3$, which forms a separate cubic perovskite phase. These two minor components will initially be neglected in the following analysis, although enough is now known about the thermoelastic properties of $CaSiO_3$ perovskite and about the effects of the coupled substitution $Al_2/MgSi$ on density and elastic properties to allow later assessment of the adequacy of this approximation.

Two different approaches to analysis of the elasticity of the lower mantle have been widely used. In the first, pioneered by Birch, the fitting of an equation-of-state to a seismological model for the lower mantle provides for its adiabatic decompression (i.e., extrapolation of its properties along the adiabat to zero pressure). These properties (density, incompressibility, etc.) of the lower-mantle material at the foot of the adiabat ($T = T_0$, $P = 0$) can then be compared with those expected for appropriate compositions and mineralogies from laboratory data (see Section 9.4). Alternatively, if enough is known about the pressure dependence of the physical properties of the candidate phases, their properties can be extrapolated to lower-mantle $P–T$ conditions for direct comparison with seismological models. Consistency between the outcomes from these two alternative strategies has recently been demonstrated (Jackson, in press). The alternative approaches and their implications for the composition and temperature of the lower mantle will be briefly reviewed next.

9.7.2. *Adiabatic Decompression of the Lower Mantle*

By comparison with the radial variations of the elastic wave speeds V_P and V_S, which are tightly constrained by high-resolution traveltime studies, the radial variation of density is relatively poorly constrained by the Earth's mass and moment of inertia and the frequencies of certain modes of free oscillation. This poses a difficulty in fitting finite-strain equations-of-state to seismological models for the lower mantle, because the density and its zero-pressure value ρ_0 are required for calculation of the strain. Some analyses of the elasticity of the lower mantle have been based solely on the use of equation (6) to fit the radial covariation of pressure and density prescribed by a suitable gross Earth model. However, there are two shortcomings of this approach. Firstly, there are wide ranges of covariance for the equation-of-state parameters ρ_0, K_0, and K_0' about the global minimum, especially for the more flexible fourth-order fit (Bukowinski and Wolf, 1990; Jackson, in press). Secondly, because the density–depth model is essentially that obtained by integration of equation (15), and pressure, similarly, is an integral quantity given by

$$P(r) = -\int \rho(r)g(r)\,dr \tag{16}$$

neither quantity reflects very directly the seismological information of highest resolution concerning the elasticity of the lower mantle, namely, $V_P(r)$ and $V_S(r)$.

The parallel fitting of equations (6) and (13) to the variations of pressure and seismic parameter with density, as prescribed by a one-dimensional (radial) Earth model, can overcome these two difficulties. As the trial value for ρ_0 is varied, the different combinations of K_0, K_0', and K_0'' that best fit $P(\varepsilon)$ and $\phi(\varepsilon)$ are identified. These domains of covariance intersect at relatively large angles, defining quite precisely the values for ρ_0 and K_0 that will allow $P(\varepsilon)$ and $\phi(\varepsilon)$ to be simultaneously well fitted (Jackson, in press). The compromise thus made, which involves a small $P(\varepsilon)$ misfit penalty, appropriately weights the most robust seismological information (V_P, V_S, and thus ϕ) more heavily. The properties (ρ_0, K_0, K_0', and $K_0 K_0''$) of the adiabatically decompressed PREM lower mantle (Dziewonski and Anderson, 1981) thus obtained from third- and fourth-order fits [labelled III(P, ϕ) and IV(P, ϕ), respectively] are broadly consistent with the results of previous analyses (Table 9.3), but are to be preferred on the grounds of internal consistency and more appropriate weighting of the most robust seismological data. The properties of the decompressed *ak135* lower mantle (Kennett, Engdahl, and Buland, 1995; Montagner and Kennett, 1996) are less sensitive than those of the PREM model to the choice between third- and fourth-order finite-strain equations-of-state (Table 9.3, Figure 9.7).

This procedure, in which the V_P and V_S data are combined to best constrain the compression of the lower mantle, suffers from the disadvantage that it does not provide estimates for the zero-pressure shear-mode properties G_0, G_0', and G_0''. Finite-strain expansions of the compressional [$M_P = \rho V_P^2 = K + (4/3)G$] and

Table 9.3. *Properties of the adiabatically decompressed lower mantle*

Analysis	Fitted	ρ_0 (g·cm^{-3})	K_0 (GPa)	K_0'	$K_0 K_0''$	G_0 (GPa)	G_0'	$K_0 G_0''$
PREM								
Jackson	III(P, ϕ)	3.985	212.5	3.89(1)	(−3.8)	134.1	1.56	(−2.4)
(in press)	IV(P, ϕ)	3.984	206.0	4.2(1)	−6(1)	130.0	1.75	−4.0
Jeanloz and	IIIP	4.003	222.5	3.76	(−3.7)	—	—	—
Knittle	III(M_P, G)	—	216.5	3.80	(−3.1)	135.4	1.52	(−2.1)
(1986)	IVP	3.994	214.5	4.05	−5.2	—	—	—
	IV(M_P, G)	—	204.7	4.38	−7.5	130.1	1.77	−4.0
ak135								
Jackson	III(P, ϕ)	4.007	213.5	4.1(2)	(−4)			
(in press)	IV(P, ϕ)	4.002	212.0	4.1(1)	−3.7			

Note: III and IV denote third- and fourth-order finite-strain fits; the label P indicates that pressure only was fitted; (P, ϕ) and (M_P, G) indicate simultaneous fits to pressure and seismic parameter, and to compressional and shear moduli, respectively.

shear moduli have been widely used in modelling the elastic properties of the mantle (e.g., Davies and Dziewonski, 1975; Jeanloz and Knittle, 1986, and Table 9.3; Duffy and Anderson, 1989) although the convergence of these series has not been clearly demonstrated. A satisfactory alternative is provided by the observation that the relationship

$$G = AK + BP \tag{17}$$

with $A = 0.631(1)$ and $B = -0.899(6)$, describes very accurately the variation of the shear modulus in the PREM lower mantle (Stacey, 1995). It follows from equation (17) that the shear modulus and its first and second pressure derivatives, evaluated at zero pressure, are given by

$$G_0 = AK_0$$
$$G_0' = AK_0' + B \tag{18}$$
$$G_0'' = AK_0''$$

Values for G_0, G_0', and $K_0 G_0''$, thus associated with those for K_0, K_0', and K_0'' derived from the III(P, ϕ) and IV(P, ϕ) fits to PREM, are included in Table 9.3.

9.7.3. Decompressed Lower Mantle Interpreted as a Perovskite and Magnesiowüstite Mixture

The properties thus established for the adiabatically decompressed lower mantle can next be compared with those expected at zero pressure and the potential temperature T_0 for binary mixtures of the dominant $(Mg_x, Fe_{1-x})SiO_3$ perovskite and $(Mg_y, Fe_{1-y})O$ magnesiowüstite phases. With the bulk composition specified by the mole fractions X_{Pv} of perovskite and $X_{Mg} = [MgO]/\{[MgO] + [FeO]\}$, and the partitioning of Fe and Mg between the two phases represented by the distribution coefficient

$$k = (Fe/Mg)_{Mw}/(Fe/Mg)_{Pv} \qquad (19)$$

the compositions x and y of the respective phases are readily determined. Following Kesson and Fitz Gerald (1991), k is set equal to 4, although the density and bulk modulus for a pyrolite composition (calculated as described later) vary negligibly (by only 0.01% and 0.1%, respectively) for variation of k from 2 to 6. Once the phase compositions are established for chosen X_{Pv} and X_{Mg}, the molecular weights and STP molar volumes V_0 are calculated from molar averages of the respective end-member properties (Table 9.2). All other thermoelastic properties are assumed to be independent of composition. This is well known to be a good approximation for the bulk modulus K (e.g., Anderson, 1976) and is also a reasonable approximation for other properties such as α and $(\partial K/\partial T)_P$ (Isaak, 1995).

The Mie-Grüneisen-Debye equation-of-state [equation (9)], with the total pressure $P(V, T)$ set equal to zero, is then used with the parameters from Table 9.2 to calculate the relative volume $V(T_0)/V_0$ and bulk modulus $K(T_0)$ at the chosen potential temperature T_0 and zero pressure for each phase (Jackson, in press). The density and bulk modulus for the binary mixture are then calculated – the latter as the Hill average of the Voigt and Reuss bounds, which differ by about 4%. Use of the Hill average, rather than the more closely spaced Hashin-Shtrikman bounds, avoids the need to include the shear modulus at this stage of the modelling. This is desirable because of the absence of experimental constraints on $\partial G/\partial T$ for the perovskite phase and the possible complicating effects of dispersion associated with viscoelastic relaxation.

The densities and bulk moduli calculated in this way for various choices of the compositional parameters X_{Pv} and X_{Mg} and potential temperature T_0 are compared in Figure 9.7 with the corresponding properties for the decompressed lower mantle derived from the intersections of the covariance domains for the $III(P, \phi)$ and $IV(P, \phi)$ finite-strain fits described earlier. The properties for the simplified pyrolite composition (large diamond symbol plotted in Figure 9.7) are calculated at a potential temperature T_0 of 1,600 K, which is consistent with the eruption temperatures for mid-ocean-ridge basalts modelled as primary magmas and with the depths of the major seismic discontinuities interpreted as phase transformations in an isochemical mantle (see Section 9.3). It is evident that the calculated high-temperature density and bulk modulus for pyrolite provide a reasonable match to the corresponding properties for the decompressed lower mantle, although the bulk modulus given by the $III(P, \phi)$ decompressions is overestimated by about 2%. Somewhat *less silicic* compositions with X_{Mg} and T_0 unchanged would actually provide a better match, but given the residual uncertainties inherent in this comparison, are not required by the data.

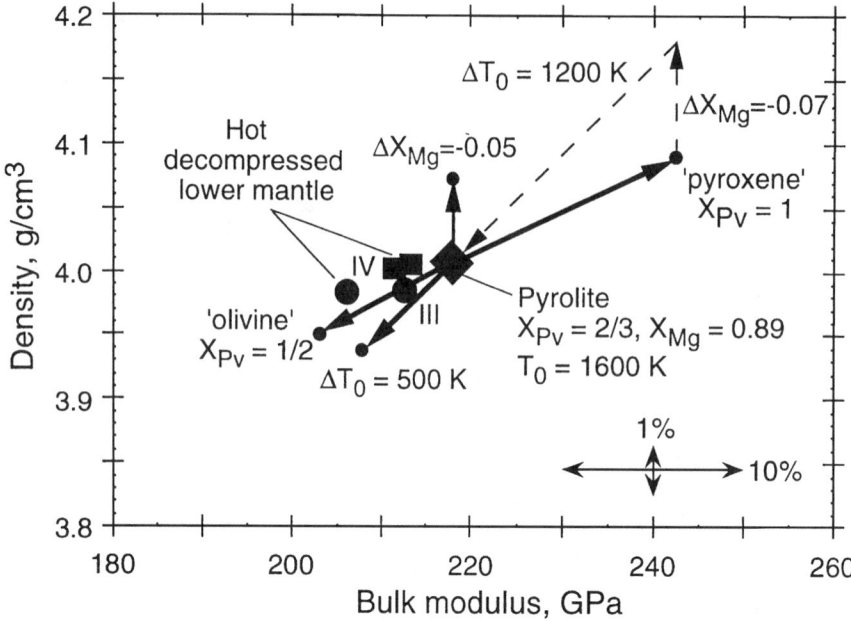

Figure 9.7. Comparison of the density and bulk modulus for the adiabatically decompressed lower mantle with those for (Mg, Fe)SiO_3 perovskite + (Mg, Fe)O magnesiowüstite mixtures for various potential temperatures (T_0) and bulk compositions (defined by the parameters X_{Mg} and X_{Pv}). The large solid circles labelled III and IV represent the results of the simultaneous fits to $\phi(\varepsilon)$ and $P(\varepsilon)$ for the PREM lower mantle, as described in the text, at third- and fourth-order, respectively. The solid squares represent the corresponding results for decompression of the *ak135* lower mantle. The large diamond symbol represents the calculated properties for the perovskite-magnesiowüstite mixture for the upper-mantle composition ($X_{Pv} = 2/3$, $X_{Mg} = 0.89$; Figure 9.1) and a potential temperature of 1,600 K. The effects of increasing FeO content ($\Delta X_{Mg} = -0.05$) and of increasing potential temperature ($\Delta T_0 = 500$ K), at constant silica stoichiometry, are shown by the 'north'-trending and 'southwest'-trending vectors drawn from the diamond. The effect of changing silica stoichiometry is illustrated by the points labelled $X_{Pv} = 1$ and $X_{Pv} = 1/2$ for pyroxene and olivine stoichiometries, respectively. The triangle outlined by the broken lines illustrates the composition–temperature trade-off discussed in the text.

In principle, there is a wide trade-off possible between values of the compositional parameters and the potential temperature (e.g., Jackson, 1983). The nature of this trade-off is evident from the relative orientations and lengths of the ΔX_{Pv}, ΔX_{Mg}, and ΔT_0 vectors in Figure 9.7. Thus, it is possible to choose combinations of ΔX_{Pv}, ΔX_{Mg}, and ΔT_0 such that the vector sum of the resulting perturbations in $K(T_0)$–$\rho(T_0)$ space will be zero. All such combinations of composition and temperature will match the density and bulk modulus of the decompressed lower mantle equally well. For example, compositions enriched, relative to pyrolite, in silica and FeO are allowed – provided that the potential temperature is increased appropriately. A combination of silica enrichment to pyroxene stoichiometry ($\Delta X_{Pv} = 1/3$), substantial iron enrichment ($\Delta X_{Mg} = -0.07$), and a much higher potential temperature ($\Delta T_0 = +1,200$ K) will reproduce the density and bulk modulus calculated for the pyrolite ($T_0 = 1,600$ K) model. Simple scaling will allow identification of other models with intermediate compositions and potential temperatures and the same high-temperature values for ρ and K_S.

Reconciliation of the high potential temperatures associated with models substantially more silicic than pyrolite with the relatively well constrained adiabat for the upper mantle and transition zone (Figure 9.2) presents a major difficulty, as follows: A large temperature increment (e.g., 1,200 K for a lower mantle of pyroxene stoichiometry) would have to be supported conductively across a pair of thermal boundary layers, presumably several hundred kilometers in total thickness, bounding separately convecting regions above and below. Given that the discontinuities and velocity gradients for the transition zone are adequately explained by phase transformations in a pyrolite model mantle (see Section 9.6), any compositional/thermal boundary would have to be located deeper (i.e., within the lower mantle). In this scenario, the regions above and below the compositional/thermal boundary layers would exhibit comparable values for density and bulk modulus, albeit along adiabats substantially offset in temperature. Within the boundary layers it would be necessary for the thermal and compositional gradients to be very closely matched if the boundary were to be seismologically unobservable, as suggested by Jeanloz (1991). Moreover, given that at constant composition and phase, viscosity decreases approximately fourfold for each 100 K increase in temperature, it would be most fortuitous if a conductively supported temperature increment of even a few hundred degrees could be reconciled with the evidence for a modest *increase* in viscosity (about 30-fold) across the depth of the mantle (e.g., Lambeck and Johnston, Chapter 10, this volume). The lack of seismological and rheological evidence for the thermal boundary layers implied by layered convection is just one of many difficulties faced by layered models of mantle convection (Davies and Richards, 1992; Davies, Chapter 5, this volume).

Finally, it should be noted that the composition–temperature trade-off described earlier is resolvable, in principle, through incorporation into the analysis of another constraint – in the form of the shear modulus. However, so much less is known about the pressure and temperature dependence of the shear modulus for $(Mg, Fe)SiO_3$ perovskite that its inclusion is deferred until later in this section.

9.7.4. Projection of Laboratory Data to Mantle P–T Conditions

There will inevitably be some sensitivity of the properties of the adiabatically decompressed lower mantle to any departure from the assumptions of homogeneity and adiabaticity, as well as to the choice of the equation-of-state (Figure 9.6, Table 9.3). Included in the latter category is the extrapolatory bias discussed by Bukowinski and Wolf (1990). It is therefore of interest to examine the alternative approach in which the properties of the perovskite-magnesiowüstite mixture

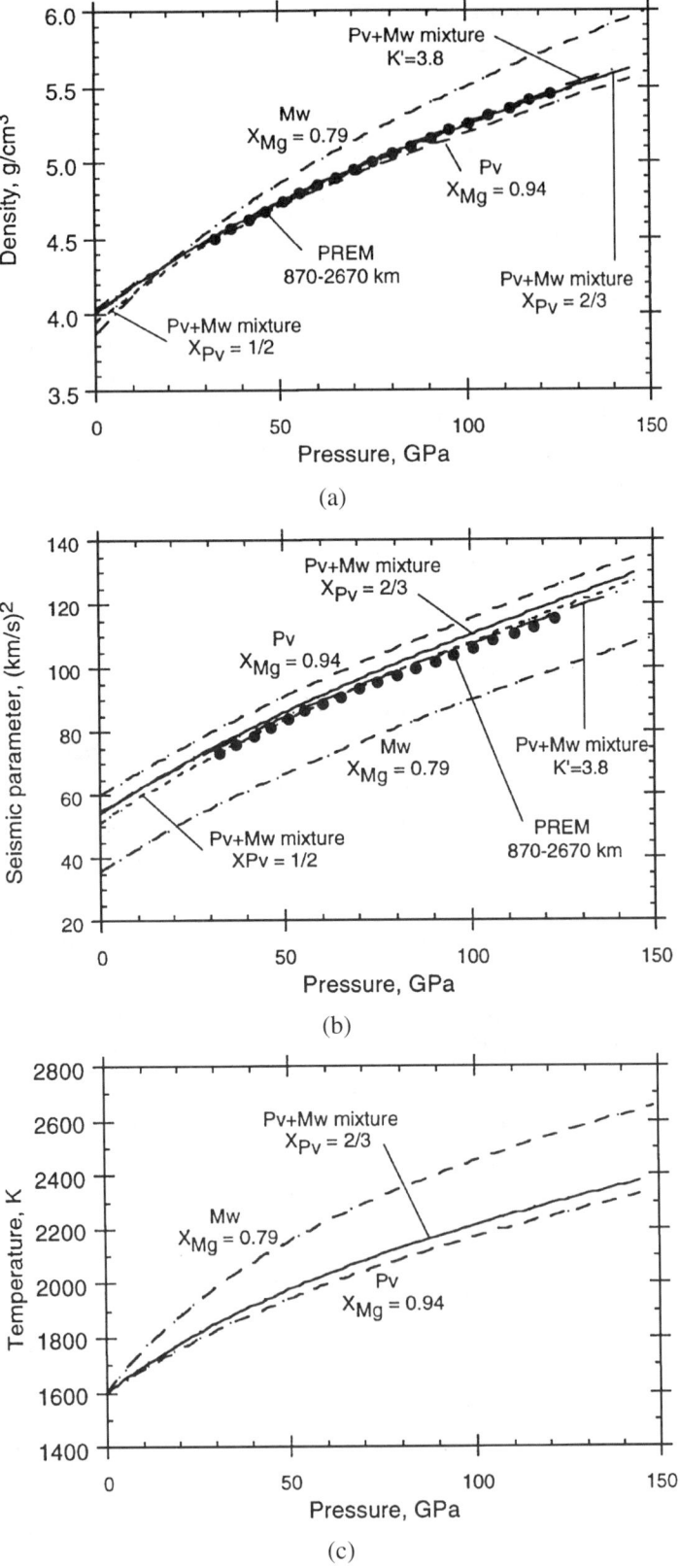

(a)

(b)

(c)

are extrapolated to lower-mantle $P-T$ conditions for direct comparison with the seismological model.

The Mie-Grüneisen-Debye equation-of-state [equation (9)] used earlier to compute the physical properties of perovskite + magnesiowüstite mixtures at zero pressure and high temperature can also be used with the parameters of Table 9.2 to calculate high-temperature isentropes, as outlined in Section 9.6. For example, the density, seismic parameter, and temperature calculated in this way for the pyrolite composition ($X_{Pv} = 2/3$, $X_{Mg} = 0.89$) along the $T_0 = 1,600$ K isentrope are compared with the values for PREM lower mantle in Figure 9.8. Also plotted are the variations of ρ and ϕ with pressure along the respective $T_0 = 1,600$ K isentropes for each of the coexisting perovskite ($X_{Mg} = 0.94$) and magnesiowüstite ($X_{Mg} = 0.79$) phases. It is evident from Figure 9.8a that this composition–temperature model provides a reasonable match to the density of the lower mantle. However, it is the density gradient, given by

$$(\partial \rho / \partial P)_S = \rho/K_S = \phi^{-1} \tag{20}$$

rather than the density itself, that is most tightly constrained by seismological observations [equation (14)]. Close inspection of Figure 9.8a reveals that the calculated density gradient given by the slope of the solid line is systematically too low. This misfit of the seismological data is much more clearly seen in Figure 9.8b, which shows that the calculated seismic parameter for the mixture of simplified pyrolite composition is consistently about 4% higher than that for PREM throughout the lower mantle. As in the analysis of the properties of the decompressed lower mantle (Figure 9.7), the Mie-Grüneisen-Debye equation-of-state with the preferred values of the thermoelastic parameters for (Mg, Fe)SiO$_3$ perovskite (Table 9.2) tends to match the density but somewhat overestimate the incompressibility K_S and hence ϕ.

The sensitivities of the calculated density and seismic parameter to reasonable variations in the thermoelastic parameters for the perovskite phase (second column from the right-hand side, Table 9.2) have recently been explored (Jackson, in press). Reduction of K' for the perovskite phase from 4 to 3.8 substantially reduces the discrepancy between the calculated properties and those for PREM, especially for the seismic parameter (Figure 9.8b). Reasonable variations of other parameters

Figure 9.8. Projections of the properties of perovskite-magnesiowüstite mixtures (Table 9.2, last two columns) to lower-mantle $P-T$ conditions for comparison with the PREM seismological model (solid circles). (a) Computed variation of density with pressure along adiabats (potential temperature 1,600 K) for (Mg, Fe) silicate perovskite and magnesiowüstite with compositions and relative proportions appropriate for pyrolite (solid line). (b) As for part a, but for seismic parameter $\phi = K_S/\rho = V_P^2 - (4/3)V_S^2$. The effects of reducing K' for perovskite from 4.0 to 3.8 and of changing the silica stoichiometry from $X_{Pv} = 2/3$ to $X_{Pv} = 1/2$ are illustrated by the long-dash line and short-dash line, respectively. (c) Increases in temperature along the 1,600 K adiabats for each of the individual phases considered separately and for the ($X_{Pv} = 2/3$) mixture.

such as γ_0 and q will have less influence on the calculated value of ϕ because of compensating changes in density and bulk modulus. For example, the combination $q = 2$ and $\gamma_0 = 1.41$, which fits the $P-V-T$ data for $MgSiO_3$ perovskite as well as the preferred values of Table 9.2, yields increases of 50% and 11% respectively in the average values of $|(\partial K_S/\partial T)_P|$ and α for the temperature interval 300–1,600 K at atmospheric pressure. The calculated value for ϕ along the 1,600 K adiabat for the perovskite + magnesiowüstite mixture of pyrolite composition is accordingly reduced by an amount that decreases from about 3% at zero pressure to 1% at 140 GPa.

The perovskite phase alone, with $X_{Mg} = 0.94$, also matches the average density of the lower mantle reasonably well (within about 1%), but overestimates by about 10% its incompressibility and seismic parameter (Figure 9.8a). Admixture of a substantial proportion of the much more compressible magnesiowüstite phase is clearly required if the variation of the well-constrained seismic parameter is to be matched. With $K'_{Pv} = 4$ (Table 9.2), there exists a prima facie case for a somewhat less silicic composition for the lower mantle (broken curve labelled $X_{Pv} = 1/2$ in Figure 9.8b); as noted earlier, however, a satisfactory match is obtained with the upper-mantle composition and a lower value of K'_{Pv} near 3.8. This finding is consistent with the fact that K' values near 3.8 are obtained from third-order finite-strain decompressions of the PREM lower mantle (Table 9.3). In order to reconcile these inferences, it is required that $K'_{SS} = (\partial K_S/\partial P)_S$ not increase significantly with increasing temperature at zero pressure. It has recently been demonstrated that the Mie-Grüneisen-Debye description of the thermal pressure [equation (9)], as applied to wadsleyite and $MgSiO_3$ perovskite, yields near-zero values of $(\partial K'_{SS}/\partial T)_P$ (Jackson and Rigden, 1996). Under these circumstances, thermal corrections to K'_{SS} for seismological application are negligible.

It is thus apparent that consistent findings have emerged from analyses (1) in which the lower mantle has been considered adiabatically decompressed, for comparison with laboratory thermoelastic data, and (2) in which the laboratory data have been projected to the $P-T$ conditions of the lower mantle. This gratifying result is in part a consequence of the fact that Eulerian isentropes faithfully describe the $V(P)_S$ calculated from an Eulerian principal isotherm through the Mie-Grüneisen equation-of-state (see Section 9.4). A totally satisfactory description of the variations of ϕ and ρ throughout the lower mantle can be obtained with the upper-mantle composition, within the framework specified by the preferred equation-of-state [equation (9)], provided that K'_{Pv} is significantly less than 4. Otherwise, a somewhat *less* silicic composition approaching olivine stoichiometry ($X_{Pv} = 0.5$) would be required (Figure 9.8b). Perovskite alone, with the unexceptional thermoelastic properties that have emerged from recent $P-V-T$ studies of the Mg end member, is far too incompressible to match the seismic parameter of the lower mantle, unless the lower mantle is much hotter ($\Delta T_0 \sim 1,200$ K; Figure 9.7) than the upper mantle.

9.7.5. Comparison with Previous Studies

The findings from this analysis are broadly consistent with those from previous investigations, in which relatively low values for α and normal values for $\partial K/\partial T$ for the perovskite phase were employed (e.g., Jackson, 1983; Bukowinski and Wolf, 1990; Wang et al., 1994; Zhao and Anderson, 1994; Stacey, 1996). In marked contrast, analyses based on very high average expansivities ($>4 \times 10^{-5}\,\mathrm{K}^{-1}$) and abnormally large $|\partial K/\partial T|$ values ($\sim 0.05\,\mathrm{GPa \cdot K}^{-1}$) for perovskite (Jeanloz and Knittle, 1989; Stixrude et al., 1992) have consistently favoured strong silica enrichment of the lower mantle (see also Zhao and Anderson, 1994). Representative of such studies is the work of Stixrude et al. (1992), whose approach has been followed here, with the important difference that their values for the parameters Θ, γ_0, and q were constrained by the P–V–T data for $(Mg_{0.9}Fe_{0.1})SiO_3$ perovskite (Knittle, Jeanloz, and Smith, 1986; Mao et al., 1991) rather than the data that have subsequently become available for the $MgSiO_3$ end member. The need to reconcile the unusually strongly temperature-dependent thermal expansivity reported by Knittle et al. (1986; see discussion by Hill and Jackson, 1990) with the much lower expansivities measured at high pressure (Mao et al., 1991) led Stixrude et al. (1992) and other analysts (e.g., Bina, 1995) to use exotic values for some of the thermoelastic parameters (Table 9.2, third column from right-hand side). The combination of higher-than-usual values for both γ_0 (1.96) and q (2.5) ensures high and very strongly temperature-dependent expansivity ($2.3 \times 10^{-5}\,\mathrm{K}^{-1}$ at 300 K to $7.0 \times 10^{-5}\,\mathrm{K}^{-1}$ at 2,000 K!) and unusually strong temperature sensitivity of the bulk modulus. With these equation-of-state parameters, $(\partial K_T/\partial T)_P$ increases in magnitude from $-0.043\,\mathrm{GPa \cdot K}^{-1}$ at 300 K to $-0.077\,\mathrm{GPa \cdot K}^{-1}$ at 2,000 K. Between 500 K and 2,000 K, $(\partial P/\partial T)_V = \alpha K_T$ increases by more than 30%, from $8.3 \times 10^{-3}\,\mathrm{GPa \cdot K}^{-1}$ to $11.1 \times 10^{-3}\,\mathrm{GPa \cdot K}^{-1}$, in marked contrast to the common observation that αK_T is almost independent of temperature for $T \gtrsim \Theta/2$ (Anderson, 1984).

The use of this physically implausible combination of values for γ_0 and q inevitably results in very different conclusions concerning the composition and temperature of the lower mantle. The density and seismic parameter calculated for the model ($X_{Pv} = 1$, $X_{Mg} = 0.88$, $T_0 = 1,660$ K) preferred by Stixrude et al. (1992) do provide a good match to the PREM model (Figure 9.9a,b). The effect of the larger, but more strongly volume-dependent, value of γ is a significant increase (from about 550 K to 750 K) in the temperature rise along the lower-mantle adiabat (Figures 9.8c and 9.9c).

9.7.6. Influence of the Neglected CaSiO₃ and Al₂O₃ Components

Static compression studies of the unquenchable cubic $CaSiO_3$ perovskite phase have yielded a robust estimate for its zero-pressure density and have improved the

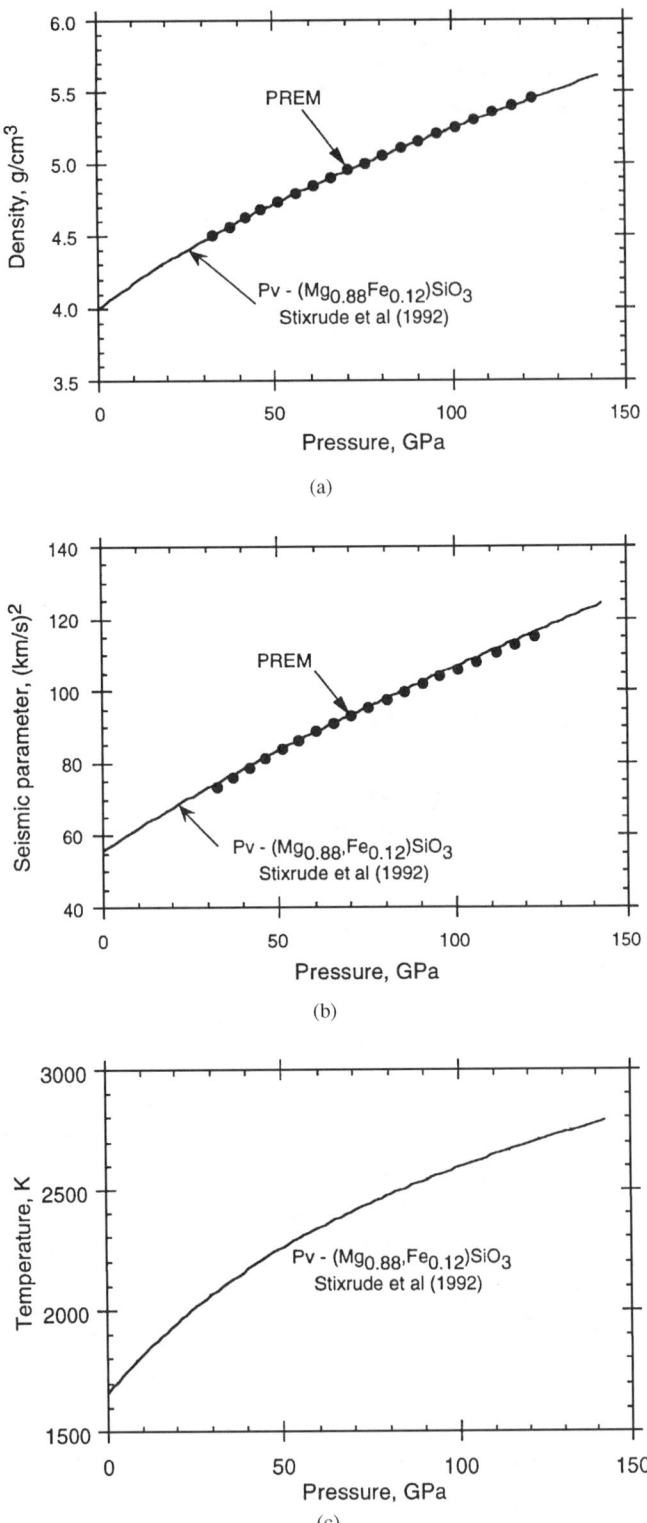

Figure 9.9. As for Figure 9.8 with the equation-of-state parameters of Stixrude et al. (1992) for $(Mg_{0.88}, Fe_{0.12})SiO_3$ perovskite (third column from right-hand side, Table 9.2).

constraints on the bulk modulus and its pressure derivative. In a diamond-anvil study to 130 GPa at 300 K, Mao et al. (1989) obtained $K_0 = 281$ GPa with $K'_0 = 4$ (Table 9.2). Tighter constraints on K_0, from a multiple-anvil $P-V-T$ study to 13 GPa by Wang et al. (1996), indicate a much lower value (232 GPa), which can be reconciled with the Mao et al. (1989) data through a higher value (4.8) for K'_0. These alternative (K_0, K'_0) combinations, and others derived from recent *ab initio* calculations [$K_0 = 300$ GPa, $K'_0 = 4$, (Sherman, 1993), $K_0 = 254$ GPa, $K'_0 = 4.4$ (Wentzcovitch et al., 1995)], define essentially indistinguishable average incompressibilities over the pressure range of the lower mantle. Comparison of the Wang et al. (1996) values for ρ_0 and K_0 with those for $MgSiO_3$ perovskite indicates that the density and seismic parameter for $CaSiO_3$ perovskite are respectively 3% greater and 14% less than those for the $MgSiO_3$ phase. Given that $CaSiO_3$ makes up 7%, by mass or volume, of the pyrolite composition, its inclusion in the foregoing analysis would be expected to raise the overall density by about 0.2% and reduce the zero-pressure seismic parameter by about 1%. The higher value of K'_0 deduced by Wang et al. (1996) for $CaSiO_3$ perovskite implies that the perturbation to the seismic parameter should decrease substantially with increasing pressure.

For the three-dimensional, corner-connected octahedral framework structure of Mg perovskite, substitution of Al_2 for MgSi results in systematic expansion of the lattice (e.g., Irifune, Koizumi, and Ando, 1996), although the site occupancies of the Al atoms remain to be determined. For the 6 mol% dissolved Al_2O_3 expected in $(Mg, Fe)SiO_3$ perovskite for a pyrolite bulk composition, the density will be reduced by 0.3%. If the product KV of bulk modulus and molar volume were constant for these solid solutions, as was reasonably assumed by Ita and Stixrude (1992), K would be 0.4% lower than that for $(Mg, Fe)SiO_3$, with ϕ essentially unchanged.

It is therefore concluded that the effect of incorporation of both the $CaSiO_3$ and Al_2O_3 components of pyrolite, often neglected in analyses of lower-mantle elasticity, would be a negligible change (about -0.1%) in density and a change of less than 1% in ϕ. This analysis is consistent with the conclusions from other studies concerning the invisibility of the $CaSiO_3$ component in the lower mantle (Mao et al., 1989; Wang and Weidner, 1994). The effect of neglecting the $CaSiO_3$ and Al_2O_3 components in analyses of lower-mantle elasticity and density clearly will be much smaller than the residual uncertainties in key thermoelastic parameters [especially K' and q or $(\partial K/\partial T)_P$] for the dominant $(Mg, Fe)SiO_3$ phase.

9.7.7. The Shear Modulus of a Perovskite-rich Lower Mantle

Because little is yet known about the pressure and temperature dependences of the shear modulus for $(Mg, Fe)SiO_3$ perovskite, interpretation of the shear-mode properties of the lower mantle is necessarily less secure. The state of the art is illustrated in Figure 9.10, in which the bulk and shear moduli for the adiabatically

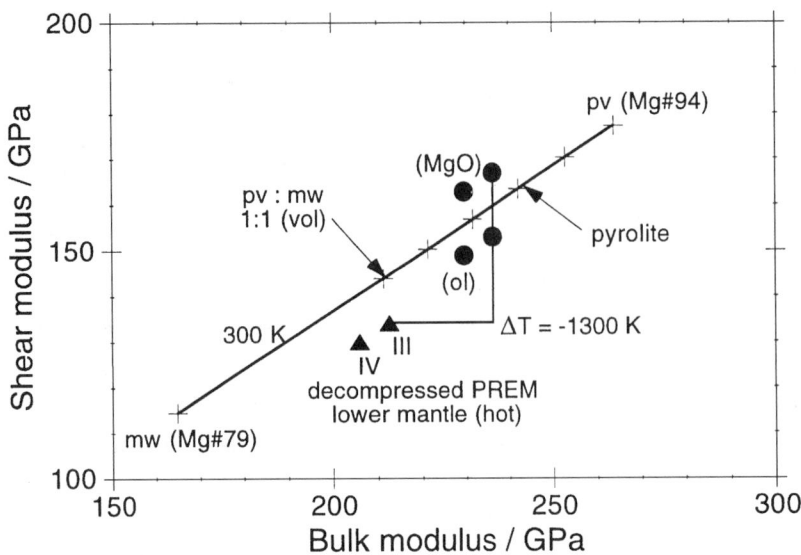

Figure 9.10. Comparison of the bulk and shear moduli for the adiabatically decompressed PREM lower mantle (solid triangles labelled III and IV for third- and fourth-order fits; Table 9.3) with the corresponding properties for perovskite + magnesiowüstite mixtures. The solid circles represent the moduli for the decompressed lower mantle adjusted to 300 K for comparison with the properties for perovskite + magnesiowüstite mixtures. The value for $\partial K_s/\partial T$ is prescribed by the parameters of Table 9.2, whereas the impact of a range of assumed values for $\partial G/\partial T$, from -0.014 GPa·K^{-1} (olivine) to -0.024 GPa·K^{-1} (MgO), is illustrated.

decompressed lower mantle (at the unknown potential temperature T_0; Table 9.3) are compared with those calculated for perovskite + magnesiowüstite mixtures at STP. The interesting question is whether or not, and for which compositions, the properties of the hot decompressed lower mantle can be reconciled with those of the two-phase mixture at STP through appropriate combinations of T_0 and the temperature sensitivities of the elastic moduli. For the purpose of this calculation, compositions of the coexisting perovskite and magnesiowüstite phases appropriate for the pyrolite bulk composition are chosen. The elastic moduli K and G for magnesiowüstite are assumed to vary linearly with composition between the values appropriate for stoichiometric end members (Jackson, in press). In the absence of experimental data, the elastic moduli for perovskite are assumed to be independent of composition.

Both K and G for the hot decompressed lower mantle are substantially smaller than the values for relatively perovskite-rich mixtures at 300 K, indicating the need for relatively large thermal corrections. The horizontal and vertical lines in Figure 9.10 indicate the magnitudes of the upward corrections to K and G, respectively, that would be required for a downward adjustment to 300 K from a potential temperature of 1,600 K. The correction to K is now constrained to some degree by thermoelastic data (Table 9.2). The adjustment to G, on the other hand, is based on the high-frequency (anharmonic) values of $|\partial G/\partial T|$ for MgO and olivine, as plausible upper and lower bounds. It is concluded that average values for $\partial G/\partial T$ of -0.020 to -0.025 GPa $\cdot$ K^{-1}, near the upper end of the plausible range, are required in order to reconcile the experimentally determined shear moduli for perovskite and magnesiowüstite with the values for a lower mantle of pyrolite composition ($\sim$20 vol% magnesiowüstite) at a potential temperature of 1,600 K. For more silicic compositions, higher temperatures T_0 and/or larger values of $|\partial G/\partial T|$ are clearly required. For example, Zhao and Anderson (1994) showed that $\partial G/\partial T = -0.035$ GPa $\cdot$ K^{-1} admits a wide range of perovskite-rich compositions. Wang and Weidner (1996) have calculated the temperature-averaged values for $(\partial K_S/\partial T)_P$ and $(\partial G/\partial T)_P$ needed to satisfy the radial (PREM) and lateral variations $[(\partial \ln V_S/\partial \ln V_P)_P = 1.8\text{--}2.0]$ of seismic wave speeds for compositional models of varying silica content. Two distinct trends emerge from the analyses of the radial and lateral variabilities, and these intersect at a composition near pyrolite with $0.015 < |(\partial K_S/\partial T)| < 0.020$ and $0.020 < |(\partial G/\partial T)_P| < 0.035$ GPa $\cdot$ K^{-1}, in general agreement with the analysis presented here.

9.8. Summary

The major conclusion from this review is that our current knowledge of the seismic structure of the mantle is consistent with the hypothesis that the mantle is grossly uniform in chemical composition throughout. That this conclusion has not changed since the publication of Birch's classic paper in 1952 is not the result of neglect of this field of scientific inquiry. On the contrary, this attractively simple hypothesis has survived four decades of progressively more rigorous testing. During this period there have been remarkable advances in the seismological delineation of the radial structure of the mantle, especially in the transition zone, and in the characterization of the superimposed lateral variability. In parallel with these developments there has been a revolution in our understanding of the chemical and physical behaviour of mantle minerals. An attempt has been made in the first part of this review to highlight the most important of these advances, which together allow a much more informed assessment of the elasticity, composition, and temperature of the Earth's mantle.

The combination of strong lateral variability of wave speeds and intense attenuation in the upper mantle, along with the findings emerging from low-frequency forced-oscillation experiments, suggest an important role for sub-solidus viscoelastic relaxation. Consequently, the temperature sensitivity of the shear-wave speed at seismic frequencies is almost certainly greater, by at least a factor of 2, than that at the megahertz frequencies of ultrasonic measurements. It is suggested that thermal rather than compositional anomalies must be responsible for most of the lateral wave-speed variability, although the neglect of anisotropy, especially in regional tomographic studies, precludes a definitive interpretation. Recovery of the deficit in shear-wave speed associated with intense viscoelastic relaxation in the upper mantle is likely to contribute significantly to the relatively steep velocity–depth gradients of the transition zone.

Given the inevitability that phase transformations contribute much, if not all, of the radial structure of the transition zone, the major focus of this review has been the testing of the simplest possible model, namely, that of grossly uniform chemical composition throughout. Although alternative, largely ad hoc, compositional models have, for the most part, not been considered here, an effort has been made to demonstrate the sensitivity of the various analyses to the remaining uncertainties, especially in the thermoelastic properties of the high-pressure mineral phases.

The salient features of the seismological models that appear to be satisfactorily explained by known phase transformations in a model mantle of pyrolite composition are as follows: Firstly, the depths to the two major seismic discontinuities near 410 and 660 km are readily interpreted in terms of the olivine ↔ wadsleyite and ringwoodite ↔ perovskite + magnesiowüstite equilibria and a transition-zone adiabat with a potential temperature T_0 of 1,550–1,600 K (or even significantly higher T_0), in accord with other petrological constraints such as basalt eruption temperatures. Tentative indications of anti-correlated topography, on the order of 30 km in amplitude, on these discontinuities are consistent with experimentally determined Clapeyron slopes dP/dT for these transitions.

Secondly, evidence, in the form of short-period reflectivity, that much of the impedance contrast across these discontinuities, especially that at 660 km, must occur within a quarter of a wavelength of a short-period compressional wave (i.e., ∼5 km) is readily reconciled with the experimentally determined spinel-disproportionation equilibria. For the 410-km discontinuity, the olivine + wadsleyite field is about 20 km wide at ambient transition-zone temperature, meaning that both the sigmoidal (rather than linear) variation of phase proportions with pressure across the two-phase field and the tendency for its width to decrease with increasing temperature may need to be invoked to explain the more scattered observations of its short-period reflectivity. The contrasts in wave speeds across the major seismological discontinuities are also broadly consistent with the expected changes consequent upon the olivine → wadsleyite and ringwoodite → perovskite + magnesiowüstite transformations. Measurements in progress to determine the temperature dependences of the bulk and shear moduli for the main transition-zone phases and other studies that are beginning to eliminate the need for lengthy extrapolation in pressure will allow a more robust analysis.

Thirdly, the character of the more ephemeral 520-km discontinuity, seen only in long-period observations at near-normal incidence, is consistent with expectations that the wadsleyite–ringwoodite transformation should occur over the depth interval 510–540 km. Moreover, the generally steep velocity–depth gradients throughout the transition zone are attributed mainly to the major pyroxene–garnet and garnet-perovskite transformations occurring at depths of about 350–480 km and 580–720 km, respectively.

Fourthly, radial models for the lower mantle, excluding the uppermost 100–200 km and the lowermost 300 km, are relatively featureless – consistent with the notion of chemical and mineralogical uniformity, and in accord with the absence of major pressure-induced phase transformations beyond 30 GPa. Calculated values for the seismic parameter ϕ and the bulk sound speed $V_\phi = \phi^{1/2}$ for the pyrolite model, with the preferred set of thermoelastic parameters for $(Mg, Fe)SiO_3$ perovskite, are systematically higher than those observed seismologically. These findings reflect the extremely high calculated incompressibility of the perovskite phase throughout the $P–T$ domain of the lower mantle. At face value, this suggests that a lower-mantle composition substantially *less silicic* than pyrolite is required. However, the residual uncertainties in $\partial K/\partial P$ and q or $\partial K/\partial T$ for perovskite are more than sufficient to accommodate the pyrolite composition. Greater difficulties, in the form of additional thermal boundary layers supporting large temperature increments for which there is no evidence in the radial-wave-speed profiles or the viscosity structure, are faced by the compositional models more silicic than pyrolite that have been widely advocated. $\partial G/\partial T$, yet to be measured for silicate perovskite, will have to be relatively large in magnitude to accommodate pyrolite, let alone more silicic compositions.

Finally, possible changes in the symmetry of the perovskite phases, especially $CaSiO_3$, and the possible disproportionation, deep in the lower mantle, of the perovskites to yield mixtures of recently discovered dense polymorphs of their constituent binary oxides, cannot be firmly excluded. However, the changes in density and elastic properties associated with such transformations are expected to be very small, possibly beyond seismological resolution.

9.9. Future Prospects

9.9.1. Temperature Dependence of the Elasticity of Transition-Zone Minerals

Further testing of the remarkably resilient hypothesis of gross chemical uniformity of the mantle will require, above all else, additional thermoelastic data for mantle minerals. Polycrystalline specimens of the main transition-zone minerals have already been measured by ultrasonic interferometry for the pressure dependence of their elastic wave speeds, although there remain significant uncertainties in extrapolation (as discussed later). Similar methods should provide determinations of $\partial V_{P,S}/\partial T$, at relatively low temperatures, in the near future. Such data, extrapolated to transition-zone $P–T$ conditions, will allow more accurate modelling of transition zone discontinuities and velocity gradients.

9.9.2. Perovskite Elasticity

Much more needs to be learned about perovskite elasticity. Although the single-crystal elastic moduli have been determined by Brillouin spectroscopy at ambient conditions, the pressure and temperature dependences of K are thus far constrained only by $V(P, T)$ measurements. For more accurate measurements of $\partial K/\partial P$ and especially $\partial K/\partial T$, and for insight into the pressure and temperature dependences of G, acoustic-wave-propagation experiments will be required. Given the difficulty of preparing the relatively large polycrystals of the silicate perovskites needed for megahertz-frequency ultrasonic interferometry, the pursuit of ultrasonic studies on structural analogues and their solid solutions with the silicate end members should prove rewarding. Better experimental constraints on mixed derivatives like $\partial^2 K/\partial T\partial P$ are also needed. These may come from ultrasonic interferometric measurements on single crystals, or polycrystals of very high acoustic quality, under the well-controlled $P\text{-}T$ conditions accessible in gas-medium apparatus.

9.9.3. Shortening the Extrapolation in Pressure, Temperature, and Frequency

It will also be very important to continue to develop and apply new experimental methods with the potential to shorten significantly the currently lengthy extrapolation between the laboratory and the $P\text{-}T\text{-}\omega$ domain of seismic wave propagation in the mantle. High-frequency ultrasonic interferometry, combined with the transition-zone $P\text{-}T$ capacity of multiple-anvil high-pressure apparatus, has the potential to shorten the extrapolation in pressure and temperature. Even wider ranges of pressure are potentially accessible to opto-acoustic methods in diamond-anvil apparatus. Growing evidence of the importance of the dispersion and attenuation of seismic waves through viscoelastic relaxation in the solid state, as reviewed earlier, also places a premium on experimentation at seismic frequencies, now becoming accessible through the application of forced-oscillation methods.

9.9.4. Incorporation of Anisotropy into Seismic Tomography and Improved Resolution of Transition-Zone Structure

The effects of azimuthal wave-speed variability and shear-wave birefringence associated with the strong anisotropy of olivine-rich rocks in the uppermost 200 km of the Earth's mantle are larger in magnitude than the common compositional perturbations. It is therefore vital that methods be developed to include anisotropy, especially in regional-scale tomographic studies of the upper mantle.

The trade-off between the velocity contrasts across the major discontinuities and the generally steep velocity–depth gradients of the transition zone has the potential to become the limiting factor in elucidation of the composition of the transition zone. More systematic studies of the frequency dependence of the reflectivity of the discontinuities than have thus far been attempted might provide more detailed information concerning the rates of change of wave speeds with depth in the immediate neighbourhood of the discontinuities. Also very important will be any progress that can be made towards determining a density distribution for the lower mantle that will be genuinely independent of the assumptions of adiabaticity and homogeneity (i.e., the Adams-Williamson approximation).

9.9.5. Phase Equilibria, Crystal Chemistry, and Calorimetry

It remains to be robustly established that the phase transformations between 10 and 30 GPa that replace the familiar upper-mantle peridotitic assemblage with a mixture of silicate perovskites and magnesiowüstite define the stable assemblage for the entire mantle. There is the possibility of P–T-induced symmetry changes in the orthorhombic (Mg, Fe, Al)(Si, Al)O_3 and cubic $CaSiO_3$ perovskites, as well as the possible thermodynamic stability at deep-mantle pressures of a very dense mixture of simple oxides. The crystallographic basis for the Al_2/MgSi substitution in (Mg, Fe)SiO_3 perovskite needs to be clarified. In addition, more phase-equilibrium and calorimetric work is needed to characterize the pyroxene–garnet and garnet-perovskite equilibria in complex multicomponent systems. Such information would provide much more realistic estimates for the impact of transformational buoyancy on mantle convection. Finally, there remains some debate, centred on different models for the generation of mid-ocean-ridge basalts, concerning the potential temperature appropriate for the oceanic upper mantle. It is most desirable that this ambiguity be resolved through further studies of the relevant equilibria and of the melt-extraction process.

References

Agee, C. B. 1993. Petrology of the mantle transition zone. *Annu. Rev. Earth Planet. Sci.* 21:19–42.

Ahrens, T. J., Anderson, D. L., and Ringwood, A. E. 1969. Equations-of-state and crystal structures of high-pressure phases of shocked silicates and oxides. *Rev. Geophys.* 7:667–707.

Akaogi, M., and Akimoto, S. 1979. High-pressure phase equilibria in a garnet lherzolite, with special reference to Mg^{2+}-Fe^{2+} partitioning among constituent minerals. *Phys. Earth Planet. Int.* 19:31–51.

Akaogi, M., Ito, E., and Navrotsky, A. 1989. Olivine-modified spinel-spinel transitions in the system Mg_2SiO_4-Fe_2SiO_4: calorimetric measurements, thermochemical calculation, and geophysical application. *J. Geophys. Res.* 94:15671–85.

Akimoto, S., and Fujisawa, H. 1966. Olivine-spinel transition in system Mg_2SiO_4-Fe_2SiO_4 at 800°C. *Earth Planet. Sci. Lett.* 1:237–40.

Anderson, D. L. 1976. The 650 km mantle discontinuity. *Geophys. Res. Lett.* 3:347–9.

Anderson, D. L. 1979. The upper mantle transition region: eclogite? *Geophys. Res. Lett.* 6:433–6.

Anderson, D. L., and Bass, J. D. 1986. Transition region of the Earth's upper mantle. *Nature* 320:321–8.

Anderson, D. L., Ben-Menahem, A., and Archambeau, C. B. 1965. Attenuation of seismic energy in the upper mantle. *J. Geophys. Res.* 70:1441–8.

Anderson, D. L., and Hart, R. S. 1978. Attenuation models of the Earth. *Phys. Earth Planet. Int.* 16:289–306.

Anderson, O. L. 1984. A universal thermal equation-of-state. *J. Geodyn.* 1:185–214.

Anderson, O. L., and Goto, T. 1989. Measurement of elastic constants of mantle-related minerals at temperatures up to 1800 K. *Phys. Earth Planet. Int.* 55:241–53.

Angel, R. J., and Hugh-Jones, D. A. 1994. Equations of state and thermodynamic properties of enstatite pyroxenes. *J. Geophys. Res.* 99:19777–83.

Babuška, V. 1984. P wave velocity anistropy in crystalline rocks. Geophys. *J. Royal Astron. Soc.* 76:113–19.

Basaltic Volcanism Studies Project. 1981. *Basaltic Volcanism on the Terrestrial Planets.* New York: Pergamon.

Berckhemer, H., Kampfmann, W., Aulbach, E., and Schmeling, H. 1982. Shear modulus and Q of forsterite and dunite near partial melting from forced oscillation experiments. *Phys. Earth Planet. Int.* 29:30–41.

Bina, C. R. 1995. Confidence limits for silicate perovskite equations of state. *Phys. Chem. Min.* 22:375–82.

Bina, C. R., and Helffrich, G. R. 1992. Calculation of elastic properties from thermodynamic equation-of-state principles. *Annu. Rev. Earth Planet. Sci.* 20:527–52.

Bina, C. R., and Helffrich, G. R. 1994. Phase transition Clapeyron slopes and transition zone seismic discontinuity topography. *J. Geophys. Res.* 99:15,853–60.

Birch, F. 1952. Elasticity and constitution of the earth's interior. *J. Geophys. Res.* 57:227–86.

Birch, F. 1960. The velocity of compressional waves in rocks to 10 kilobars. I. *J. Geophys. Res.* 65:1083–102.

Bock, G. 1994. Synthetic seismogram images of upper mantle structure: no evidence for a 520-km discontinuity. *J. Geophys. Res.* 99:15,843–51.

Bowman, J. R., and Kennett, B. L. N. 1990. An investigation of the upper mantle beneath northwestern Australia using a hybrid seismograph array. *Geophys. J. Int.* 101: 411–24.

Bukowinski, M. S. T., and Wolf, G. H. 1990. Thermodynamically consistent decompression: implications for lower mantle composition. *J. Geophys. Res.* 95:12583–93.

Castle, J. C., and Creager, K. C. 1997. Seismic evidence against a mantle chemical discontinuity near 660 km depth beneath Izu-Bonin. *Geophys. Res. Lett.* 24:241–4.

Chen, G., Li, B., and Liebermann, R. C. 1996. Selected elastic moduli of single-crystal olivines from ultrasonic experiments to mantle pressures. *Science* 272:979–80.

Chopelas, A. 1996. Thermal expansivity of lower mantle phases MgO and $MgSiO_3$ perovskite at high pressure derived from vibrational spectroscopy. *Phys. Earth Planet. Int.* 98:3–16.

Chopelas, A., and Boehler, R. 1992. Raman spectroscopy of high pressure $MgSiO_3$ phases synthesized in a CO_2 laser heated diamond anvil cell: perovskite and clinopyroxene. In: *High Pressure Research: Application to Earth and Planetary Science*, ed. Y. Syono and M. H. Manghnani, pp. 101–8. Tokyo: Terra Scientific.

Christensen, U. R., and Yuen, D. A. 1984. The interaction of a subducting lithospheric slab with a chemical or phase boundary. *J. Geophys. Res.* 89:4389–402.

Cleary, J. R., and Hales, A. L. 1966. An analysis of the travel times of P waves to North American stations in the distance range 32–100°. *Bull. Seismol. Soc. Am.* 56:467–89.

Cohen, R. E. 1987. Elasticity and equation-of-state of $MgSiO_3$ perovskite. *Geophys. Res. Lett.* 14:1053–6.

Cohen, R. E., Boyer, L. L., Mehl, M. J., Pickett, W. E., and Krakauer, H. 1989. Electronic structure and total energy calculations for oxide perovskites and superconductors. In: *Perovskite: A Structure of Great Interest to Geophysics and Materials Science*, ed. A. Navrotsky, and D. J. Weidner, pp. 55–66. Washington, DC: American Geophysical Union.

Davies, G. F., and Dziewonski, A. M. 1975. Homogeneity and constitution of the Earth's lower mantle and outer core. *Phys. Earth Planet. Int.* 10:336–43.

Davies, G. F., and Richards, M. A. 1992. Mantle convection. *J. Geol.* 100:151–206.

Davies, J. H., Gudmundsson, O., and Clayton, R. W. 1992. Spectra of mantle shear wave velocity structure. *Geophys. J. Int.* 108:865–82.

Davis, J. P., Kind, R., and Sacks, I. S. 1989. Precursors to P′P′ re-examined using broad-band data. *Geophys. J. Int.* 99:595–604.

Doyle, H. A., and Hales, A. L. 1967. An analysis of the travel times of S waves to North American stations, in the distance range 28° to 82°. *Bull. Seismol. Soc. Am.* 57:761–71.

Duffy, T. S., and Anderson, D. L. 1989. Seismic velocities in mantle minerals and the mineralogy of the upper mantle. *J. Geophys. Res.* 94:1895–912.

Duffy, T. S., Zha, C., Downs, R. T., Mao, H. K., and Hemley, R. J. 1995. Elasticity of forsterite to 16 GPa and the composition of the upper mantle. *Nature* 378:170–3.

Durek, J. J., and Ekström, G. 1996. A radial model of anelasticity consistent with long-period surface-wave attenuation. *Bull. Seismol. Soc. Am.* 86:144–58.

Dziewonski, A. M. 1971. On regional differences in dispersion of mantle Rayleigh waves. *Geophys. J. Royal Astron. Soc.* 22:289–325.

Dziewonski, A. M., and Anderson, D. L. 1981. Preliminary reference Earth model. *Phys. Earth Planet. Int.* 25:297–357.

Engdahl, E. R., and Flinn, E. A. 1969. Seismic waves reflected from discontinuities within Earth's upper mantle. *Science* 163:177–9.

Fei, Y., and Mao, H. K. 1994. In situ determination of the NiAs phase of FeO at high pressure and temperature. *Science* 266:1678–80.

Fei, Y., Mao, H. K., and Mysen, B. O. 1991. Experimental determination of element partitioning and calculation of phase relations in the MgO-FeO-SiO_2 system at high pressure and high temperature. *J. Geophys. Res.* 96:2157–69.

Fei, Y., Mao, H. K., Shu, J., Parthasarathy, G., Bassett, W. A., and Ko, J. 1992. Simultaneous high-P, high-T X-ray diffraction study of β-$(Mg, Fe)_2SiO_4$ to 26 GPa and 900 K. *J. Geophys. Res.* 97:4489–95.

Funamori, N., and Yagi, T. 1993. High pressure and high temperature in situ X-ray observation of $MgSiO_3$ perovskite under lower mantle conditions. *Geophys. Res. Lett.* 20:387–90.

Funamori, N., Yagi, T., Utsumi, W., Kondo, T., Uchida, T., and Funamori, M. 1996. Thermoelastic properties of $MgSiO_3$ perovskite determined by in situ X-ray observations up to 30 GPa and 2000 K. *J. Geophys. Res.* 101:8257–69.

Gasparik, T. 1990. Phase relations in the transition zone. *J. Geophys. Res.* 95:15751–69.

Getting, I. C., Dutton, S. J., Burnley, P. C., Karato, S., and Spetzler, H. A. 1997. Shear attenuation and dispersion in MgO. *Phys. Earth Planet. Int.* 99:249–57.

Getting, I. C., Paffenholz, J., and Spetzler, H. A. 1990. Measuring attenuation in geological materials at seismic frequencies and amplitudes. In: *The Brittle–Ductile Transition in Rocks*, ed. A. G. Duba, W. B. Durham, J. W. Handin, and H. F. Wang, pp. 239–43. Washington, DC: American Geophysical Union.

Getting, I. C., Spetzler, H. A., Karato, S., and Hanson, D. R. 1991. Shear attenuation in olivine. *EOS, Trans. AGU* 72:451.

Gillet, P., Guyot, F., and Wang, Y. 1996. Microscopic anharmonicity and equation-of-state of $MgSiO_3$ perovskite. *Geophys. Res. Lett.* 23:3043–6.

Goetze, C. 1977. A brief summary of our present day understanding of the effect of volatiles and partial melt on the mechanical properties of the upper mantle. In: *High-Pressure Research: Applications in Geophysics*, ed. M. H. Manghnani and S. Akimoto, pp. 3–23. New York: Academic Press.

Graham, E. K., and Barsch, G. R. 1969. Elastic constants of single-crystal forsterite as a function of temperature and pressure. *J. Geophys. Res.* 74:5949–60.

Grand, S. P. 1994. Mantle shear structure beneath the Americas and surrounding oceans. *J. Geophys. Res.* 99:11591–621.

Grand, S. P., and Helmberger, D. V. 1984a. Upper mantle shear structure beneath the northwest Atlantic Ocean. *J. Geophys. Res.* 89:11465–75.

Grand, S. P., and Helmberger, D. V. 1984b. Upper mantle shear structure of North America. *Geophys. J. Royal Astron. Soc.* 76:399–438.

Green, D. H., Hibberson, W. O., and Jaques, A. L. 1979. Petrogenesis of mid-ocean ridge basalts. In: *The Earth: Its Origin, Structure and Evolution*, ed. M. W. McElhinny, pp. 265–99. London: Academic Press.

Green, D. H., and Ringwood, A. E. 1967. Genesis of basaltic magmas. *Contrib. Min. Pet.* 15:103–90.

Gudmundsson, O., Davies, J. H., and Clayton, R. W. 1990. Stochastic analysis of global traveltime data: mantle heterogeneity and random errors in the ISC data. *Geophys. J. Int.* 102:25–43.

Gueguen, Y., Darot, M., Mazot, P., and Woirgard, J. 1989. Q^{-1} of forsterite single cyrstals. *Phys. Earth Planet. Int.* 55:254–8.

Gwanmesia, G. D., Rigden, S. M., Jackson, I., and Liebermann, R. C. 1990. Pressure dependence of elastic wave velocity for β-Mg_2SiO_4 and the composition of the Earth's mantle. *Science* 250:794–7.

Hales, A. L., and Doyle, H. A. 1967. P and S travel time anomalies and their interpretation. *Geophys. J. Royal Astron. Soc.* 13:403–15.

Hazen, R. M. 1993. Comparative compressibilities of silicate spinels: anomalous behaviour of $(Mg, Fe)_2SiO_4$. *Science* 259:206–9.

Helffrich, G., and Bina, C. R. 1994. Frequency dependence of the visibility and depths of mantle seismic discontinuities. *Geophys. Res. Lett.* 21:2613–16.

Hemley, R. J., Cohen, R. E., Yeganeh-Haeri, A., Mao, H. K., Weidner, D. J., and Ito, E. 1989. Raman spectroscopy and lattice dynamics of $MgSiO_3$ perovskite at high pressure. In: *Perovskite: A Structure of Great Interest to Geophysics and Materials Science*, ed. A. Navrotsky and D. J. Weidner, pp. 35–44. Washington, DC: American Geophysical Union.

Hemley, R. J., Jackson, M. D., and Gordon, R. G. 1987. Theoretical study of the structure, lattice dynamics and equations of state of perovskite-type $MgSiO_3$ and $CaSiO_3$. *Phys. Chem. Min.* 14:2–12.

Herzberg, C. 1992. Depth and degree of melting of komatiites. *J. Geophys. Res.* 97:4521–40.

Herzberg, C., and Zhang, J. 1996. Melting experiments on anhydrous peridotite KLB-1: compositions of magmas in the upper mantle and transition zone. *J. Geophys. Res.* 101:8271–95.

Hill, R. J., and Jackson, I. 1990. The thermal expansion of $ScAlO_3$ – a silicate pervoskite analogue. *Phys. Chem. Min.* 17:89–96.

Honda, S., Balachandar, S., Yuen, D. A., and Reuteler, D. 1992. Three-dimensional mantle dynamics with an endothermic phase transition. *Geophys. Res. Lett.* 20:221–4.

Irifune, T. 1993. Phase transformations in the Earth's mantle and subducting slabs; implications for their compositions, seismic velocity and density structures and dynamics. *Island Arc* 2:55–71.

Irifune, T. 1994. Absence of an aluminous phase in the upper part of the Earth's lower mantle. *Nature* 370:121–33.

Irifune, T., Koizumi, T., and Ando, J. 1996. An experimental study of the garnet–perovskite transformation in the system $MgSiO_3$-$Mg_3Al_2Si_3O_{12}$. *Phys. Earth Planet. Int.* 96:147–58.

Irifune, T., and Ringwood, A. E. 1987. Phase transformations in primitive MORB and pyrolite compositions to 25 GPa and some geophysical implications. In: *High Pressure Research in Geophysics*, ed. M. H. Manghnani and Y. Syono, pp. 231–42. Tokyo: Terra Scientific.

Irifune, T., and Ringwood, A. E. 1993. Phase transformations in subducted oceanic crust and buoyancy relationships at depths of 600–800 km in the mantle. *Earth Planet. Sci. Lett.* 117:101–10.

Irifune, T., Susaki, J., Yagi, T., and Sawamoto, H. 1989. Phase transformations in diopside $CaMgSi_2O_6$ at pressures up to 25 GPa. *Geophys. Res. Lett.* 16:187–90.

Isaak, D. G. 1992. High-temperature elasticity of iron-bearing olivines. *J. Geophys. Res.* 97:1871–85.

Isaak, D. G. 1995. Effects of the Mg/(Mg + Fe) ratio on thermoelastic parameters of mantle minerals. In: *Abstracts, IUGG XXI General Assembly*, p. A354.

Isaak, D. G., Anderson, O. L., and Goto, T. 1989. Measured elastic moduli of single-crystal MgO up to 1800 K. *Phys. Chem. Min.* 16:704–13.

Ita, J., and Stixrude, L. 1992. Petrology, elasticity and composition of the mantle transition zone. *J. Geophys. Res.* 97:6849–66.

Ito, E., and Katsura, T. 1989. A temperature profile of the mantle transition zone. *Geophys. Res. Lett.* 16:425–8.

Ito, E., and Takahashi, E. 1989. Post-spinel transformations in the system Mg_2SiO_4-Fe_2SiO_4 and some geophysical implications. *J. Geophys. Res.* 94:10637–46.

Jackson, I. 1983. Some geophysical constraints on the chemical composition of the Earth's lower mantle. *Earth Planet. Sci. Lett.* 62:91–103.

Jackson, I. In press. Elasticity, composition and temperature of the Earth's lower mantle: a reappraisal. *Geophys. J. Int.*

Jackson, I., and Paterson, M. S. 1993. A high-pressure, high-temperature apparatus for studies of seismic wave dispersion and attenuation. *PAGEOPH* 141:445–66.

Jackson, I., Paterson, M. S., and Fitz Gerald, J. D. 1992. Seismic wave dispersion and attenuation in Åheim dunite: an experimental study. *Geophys. J. Int.* 108:517–34.

Jackson, I., and Rigden, S. M. 1996. Analysis of *P-V-T* data: constraints on the thermoelastic properties of high-pressure minerals. *Phys. Earth Planet. Int.* 96:85–112.

Jeanloz, R. 1991. Effects of phase transitions and possible compositional changes on the seismological structure near 650 km depth. *Geophys. Res. Lett.* 18:1743–6.

Jeanloz, R., and Ahrens, T. J. 1980. Equations of state of FeO and CaO. *Geophys. J. Royal Astron. Soc.* 62:505–28.

Jeanloz, R., Ahrens, T. J., Mao, H. K., and Bell, P. M. 1979. B1–B2 transition in calcium oxide from shock-wave and diamond-cell experiments. *Science* 206:829–30.

Jeanloz, R., and Knittle, E. 1986. Reduction of mantle and core properties to a standard state by adiabatic decompression. In: *Chemistry and Physics of Terrestrial Planets*, ed. S. K. Saxena, pp. 275–309. Berlin: Springer-Verlag.

Jeanloz, R., and Knittle, E. 1989. Density and composition of the lower mantle. *Phil. Trans. R. Soc. London* A328:377–89.

Johnson, L. R. 1967. Array measurements of P velocities in the upper mantle. *J. Geophys. Res.* 72:6309–25.

Jordan, T. H. 1975. The continental tectosphere. *Rev. Geophys.* 13:1–12.

Jordan, T. H. 1979. Mineralogies, densities and seismic velocities of garnet lherzolites and their geophysical implications. In: *The Mantle Sample: Inclusions in Kimberlites and Other Volcanics*, ed. F. R. Boyd and H. O. A. Meyer, pp. 1–14. Washington, DC: American Geophysical Union.

Karato, S. 1993. Importance of anelasticity in the interpretation of seismic tomography. *Geophys. Res. Lett.* 20:1623–6.

Katsura, T., and Ito, E. 1989. The system Mg_2SiO_4-Fe_2SiO_4 at high pressure and temperatures. Precise determination of stabilities of olivine, modified spinel and spinel. *J. Geophys. Res.* 94:15663–70.

Kennett, B. L. N., Engdahl, E. R., and Buland, R. 1995. Constraints on seismic velocities in the Earth from traveltimes. *Geophys. J. Int.* 122:108–24.

Kennett, B. L. N., Gudmundsson, O., and Tong, C. 1994. The upper mantle S and P velocity structure beneath northern Australia from broad-band observations. *Phys. Earth Planet. Int.* 86:85–98.

Kesson, S. E., and Fitz Gerald, J. D. 1991. Partitioning of MgO, FeO, NiO, MnO and Cr_2O_3 between magnesian silicate perovskite and magnesiowüstite: implications for the origin of inclusions in diamond and the composition of the lower mantle. *Earth Planet. Sci. Lett.* 111:229–40.

Kesson, S. E., Fitz Gerald, J. D., and Shelley, J. M. G. 1994. Mineral chemistry and density of subducted basaltic crust at lower-mantle pressures. *Nature* 372:22–9.

Kingma, K. J., Cohen, R. E., Hemley, R. J., and Mao, H. K. 1995. Transformation of stishovite to a denser phase at lower-mantle pressures. *Nature* 374:243–5.

Knittle, E., and Jeanloz, R. 1987. Synthesis and equation-of-state of (Mg, Fe)SiO$_3$ perovskite to over 100 GPa. *Science* 235:669–70.

Knittle, E., Jeanloz, R., and Smith, G. L. 1986. Thermal expansion of silicate perovskite and stratification of the Earth's mantle. *Nature* 319:214–16.

Knoche, R., Webb, S. L., and Rubie, D. C. In press. Measurements of acoustic wave velocities at *P–T* conditions of the Earth's mantle. In: *Proceedings of the U.S.-Japan Conference on High-Pressure Research*. Washington, DC: American Geophysical Union.

Kovach, R. L., and Anderson, D. L. 1964. Attenuation of shear waves in the upper and lower mantle. *Bull. Seismol. Soc. Am.* 54:1855–64.

Kumazawa, M., and Anderson, O. L. 1969. Elastic moduli, pressure derivatives and temperature derivatives of single-crystal olivine and single-crystal forsterite. *J. Geophys. Res.* 74:5961–72.

Lacks, D. J., and Gordon, R. G. 1995. Calculations of pressure-induced phase transitions in mantle minerals. *Phys. Chem. Min.* 22:145–50.

Le Fevre, L. V., and Helmberger, D. V. 1989. Upper mantle P velocity structure of the Canadian shield. *J. Geophys. Res.* 94:17749–65.

Leven, J. H., Jackson, I., and Ringwood, A. E. 1981. Upper mantle seismic anisotropy and lithospheric decoupling. *Nature* 289:234–9.

Li, B., Gwanmesia, G. D., and Liebermann, R. C. 1996a. Sound velocities of olivine and beta polymorphs of Mg$_2$SiO$_4$ at Earth's transition zone pressures. *Geophys. Res. Lett.* 23:2259–62.

Li, B., Jackson, I., Gasparik, T., and Liebermann, R. C. 1996b. Elastic wave velocity measurement in multi-anvil apparatus to 10 GPa using ultrasonic interferometry. *Phys. Earth Planet. Int.* 98:79–91.

Li, X., Giardini, D., and Woodhouse, J. H. 1991. The relative amplitudes of mantle heterogeneity in P velocity, S velocity and density from free-oscillation data. *Geophys. J. Int.* 105:649–57.

Liebermann, R. C. 1975. Elasticity of olivine (α), beta (β) and spinel (γ) polymorphs of germanates and silicates. *Geophys. J. Royal Astron. Soc.* 42:899–929.

Liu, L. 1974. Silicate perovskite from phase transformations of pyrope-garnet at high pressure and temperatures. *Geophys. Res. Lett.* 1:277–80.

Liu, L. 1976. The high-pressure phases of MgSiO$_3$. *Earth Planet. Sci. Lett.* 31:200–8.

Liu, L., and Ringwood, A. E. 1975. Synthesis of a perovskite-type polymorph of CaSiO$_3$. *Earth Planet Sci. Lett.* 28:209–11.

McQueen, R. G., Marsh, S. P., and Fritz, J. N. 1967. Hugoniot equation of state of twelve rocks. *J. Geophys. Res.* 72:4999–5036.

Machetel, P., and Weber, P. 1991. Intermittent layered convection in a model with an endothermic phase change at 670 km. *Nature* 350:1836–9.

Mao, H. K., and Bell, P. M. 1979. Equations of state of MgO and ε-Fe under static pressure conditions. *J. Geophys. Res.* 84:4533–6.

Mao, H. K., Chen, L. C., Hemley, R. J., Jephcoat, A. P., and Wu, Y. 1989. Stability and equation of state of CaSiO$_3$-perovskite to 134 GPa. *J. Geophys. Res.* 94:17889–94.

Mao, H. K., Hemley, R. J., Fei, Y., Shu, J. F., Chen, L. C., Jephcoat, A. P., Wu, Y., and Bassett, W. A. 1991. Effect of pressure, temperature and composition on the lattice parameters and density of (Mg, Fe)SiO$_3$ perovskite to 30 GPa. *J. Geophys. Res.* 96:8069–79.

Mao, H. K., Takahashi, T., Bassett, W. A., Weaver, J. S., and Akimoto, S. 1969. Effect of pressure and temperature on the molar volumes of wüstite and of three (Fe, Mg)$_2$SiO$_4$ spinel solid solutions. *J. Geophys. Res.* 74:1061–9.

Meade, C., Mao, H. K., and Hu, J. 1995. High-temperature phase transition and dissociation of (Mg, Fe)SiO$_3$ perovskite at lower mantle pressure. *Science* 268:1743–5.

Meng, Y., Fei, Y., Weidner, D. J., Gwanmesia, G. D., and Hu, J. 1994. Hydrostatic compression of γ-Mg$_2$SiO$_4$ to mantle pressures and 700 K. Thermal equation of state and related thermoelastic properties. *Phys. Chem. Min.* 21:407–12.

Montagner, J.-P., and Kennett, B. L. N. 1996. How to reconcile body-wave and normal-mode reference Earth models? *Geophys. J. Int.* 125:229–48.

Nakanishi, I. 1988. Reflections of P'P' from upper mantle discontinuities beneath the Mid-Atlantic Ridge. *Geophys. J.* 93:335–46.

Nataf, H.-C., Nakanishi, I., and Anderson, D. L. 1986. Measurements of mantle wave velocities and inversion for lateral heterogeneities and anisotropy. 3. Inversion. *J. Geophys. Res.* 91:7261–307.

Navrotsky, A. 1989. Thermochemistry of perovskites. In: *Perovskite: A Structure of Great Interest to Geophysics and Materials Science*, ed. A. Navrotsky and D. J. Weidner, pp. 67–80. Washington, DC: American Geophysical Union.

Niazi, M., and Anderson, D. L. 1965. Upper mantle structure of western North America from apparent velocities of P waves. *J. Geophys. Res.* 70:4633–40.

Niesler, H., and Jackson, I. 1989. Pressure derivatives of elastic wave velocities from ultrasonic interferometric measurements on jacketed polycrystals. *J. Acoust. Soc. Am.* 86:1573–85.

Nolet, G., Grand, S. P., and Kennett, B. L. N. 1994. Seismic heterogeneity in the upper mantle. *J. Geophys. Res.* 99:23753–66.

O'Hara, M. J. 1968. Are ocean-floor basalts primary magmas? *Nature* 220:683–6.

Ohno, I. 1976. Free vibration of a rectangular parallelepiped crystal and its application to determination of the elastic constants of orthorhombic crystals. *J. Phys. Earth* 24:355–79.

O'Neill, B., and Jeanloz, R. 1990. Experimental petrology of the lower mantle: a natural peridotite taken to 54 GPa. *Geophys. Res. Lett.* 17:1477–80.

O'Reilly, S. Y., Jackson, I., and Bezant, C. 1990. Equilibration temperatures and elastic wave velocities for upper mantle rocks from eastern Australia: implications for the interpretation of seismological models. *Tectonophysics* 185:67–82.

Pacalo, R. E. G. 1993. Elasticity and structure of high pressure majorite garnets and super hydrous phase B. Ph.D. thesis, State University of New York at Stony Brook.

Peltier, W. R., and Solheim, L. P. 1992. Mantle phase transitions and layered chaotic convection. *Geophys. Res. Lett.* 19:321–4.

Petersen, N., Vinnik, L., Kosarev, G., Kind, R., Oreshin, S., and Stammler, K. 1993. Sharpness of the mantle discontinuities. *Geophys. Res. Lett.* 20:859–62.

Revenaugh, J., and Jordan, T. H. 1991. Mantle layering from ScS reverberations. 2. The transition zone. *J. Geophys. Res.* 96:19763–80.

Richards, P. G. 1972. Seismic waves reflected from velocity gradient anomalies within the earth's upper mantle. *Z. Geophys.* 38:517–27.

Rigden, S. M., Gwanmesia, G. D., Fitz Gerald, J. D., Jackson, I., and Liebermann, R. C. 1991. Spinel elasticity and seismic structure of the transition zone of the mantle. *Nature* 354:143–5.

Rigden, S. M., Gwanmesia, G. D., and Liebermann, R. C. 1994. Elastic wave velocities of a pyrope-majorite garnet to 3 GPa. *Phys. Earth Planet. Int.* 86:35–44.

Ringwood, A. E. 1956. The olivine–spinel transition in the Earth's mantle. *Nature* 178:1303–4.

Ringwood, A. E. 1962a. A model for the upper mantle. *J. Geophys. Res.* 67:857–67.

Ringwood, A. E. 1962b. Prediction and confirmation of the olivine–spinel transition in Ni_2SiO_4. *Geochem. Cosmochim. Acta* 26:457–69.

Ringwood, A. E. 1967. The pyroxene–garnet transformation in the Earth's mantle. *Earth Planet. Sci. Lett.* 2:255–63.

Ringwood, A. E. 1975. *Composition and Petrology of the Earth's Mantle*. New York: McGraw-Hill.

Ringwood, A. E. 1993. Role of the transition zone and 660 km discontinuity in mantle dynamics. *Phys. Earth Planet. Int.* 86:5–24.

Ringwood, A. E., and Major, A. 1966. Synthesis of Mg_2SiO_4-Fe_2SiO_4 spinel solid solutions. *Earth Planet. Sci. Lett.* 1:241–5.

Ringwood, A. E., and Major, A. 1970. The system Mg_2SiO_4-Fe_2SiO_4 at high pressure and temperatures. *Phys. Earth Planet. Int.* 3:89–108.

Robertson, G. S., and Woodhouse, J. H. 1995. Evidence for proportionality of P and S heterogeneity in the lower mantle. *Geophys. J. Int.* 123:85–116.

Romanowicz, B. 1995. A global tomographic model of shear attenuation in the upper mantle. *J. Geophys. Res.* 100:12375–94.

Sawamoto, H., Weidner, D. J., Sasaki, S., and Kumazawa, M. 1984. Single-crystal elastic properties of the modified spinel (beta) phase of Mg_2SiO_4. *Science* 224:749–51.

Shearer, P. M. 1991. Constraints on upper mantle discontinuities from observations of long-period reflected and converted phases. *J. Geophys. Res.* 96:18147–82.

Shearer, P. M. 1996. Transition zone velocity gradients and the 520-km discontinuity. *J. Geophys. Res.* 101:3053–66.

Shearer, P. M., and Masters, T. G. 1992. Global mapping of topography on the 660-km discontinuity. *Nature* 355:791–6.

Sherman, D. M. 1993. Equation of state, elastic properties, and stability of $CaSiO_3$ perovskite: first principles (periodic Hartree-Fock) results. *J. Geophys. Res.* 98:19795–805.

Sherman, D. M., and Jansen, H. J. F. 1995. First-principles prediction of the high-pressure phase transition and electronic structure of FeO: implications for the chemistry of the lower mantle and core. *Geophys. Res. Lett.* 22:1001–4.

Simmons, G. 1964. Velocity of shear waves in rocks to 10 kilobars. *J. Geophys. Res.* 69:1123–30.

Sobolev, S. V., Zeyen, H., Stoll, G., Werling, F., Altherr, R., and Fuchs, K. 1996. Upper mantle temperatures from teleseismic tomography of French Massif Central including effects of composition and mineral reactions. *Earth Planet. Sci. Lett.* 139:147–63.

Spetzler, H., Shen, A., Chen, G., Hermannsdorfer, G., Schulze, H., and Weigel, R. 1996. Ultrasonic measurements in a diamond-anvil cell. *Phys. Earth Planet. Int.* 98:93–9.

Stacey, F. D. 1995. Theory of thermal and elastic properties of the lower mantle and core. *Phys. Earth Planet. Int.* 89:219–45.

Stacey, F. D. 1996. Thermoelasticity of (Mg, Fe)SiO_3 perovskite and a comparison with the lower mantle. *Phys. Earth Planet. Int.* 98:65–79

Stixrude, L. 1995. Mantle composition and structure of mantle discontinuities. In: *Abstracts, IUGG XXI General Assembly*, p. B383.

Stixrude, L., and Bukowinski, M. S. T. 1990. Fundamental thermodynamic relations and silicate melting with implications for the constitution of D″. *J. Geophys. Res.* 95:19311–25.

Stixrude, L., and Cohen, R. E. 1993. Stability of orthohombic $MgSiO_3$-perovskite in the Earth's lower mantle. *Nature* 364:613–16.

Stixrude, L., Cohen, R. E., Yu, R., and Krakauer, H. 1996. Prediction of phase transition in $CaSiO_3$ perovskite and implications for lower mantle structure. *Am. Mineral.* 81:1293–6.

Stixrude, L., Hemley, R. J., Fei, Y., and Mao, H. K. 1992. Thermoelasticity of silicate perovskite and magnesiowüstite and stratification of the Earth's mantle. *Science* 257:1099–101.

Su, W., Woodward, R. L., and Dziewonski, A. M. 1994. Degree 12 model of shear velocity heterogeneity in the mantle. *J. Geophys. Res.* 99:6945–80.

Sumino, Y., Ohno, I., Goto, T., and Kumazawa, M. 1976. Measurement of elastic constants and internal friction on single-crystal MgO by rectangular parallelepiped resonance. *J. Phys. Earth* 24:263–73.

Suzuki, I. 1975. Thermal expansion of periclase and olivine and their anharmonic properties. *J. Phys. Earth* 23:145–9.

Suzuki, I., Ohtani, E., and Kumazawa, M. 1979. Thermal expansion of γ-Mg_2SiO_4. *J. Phys. Earth* 27:53–61.

Suzuki, I., Sakai, M., and Katsura, T. 1996. Elastic moduli of modified spinel with composition of natural olivine measured by means of RST (abstract). *EOS, Trans. AGU* 77(22):W134.

Tackley, P. J., Stevenson, D. J., Glatzmaier, G. A., and Schubert, G. 1993. Effects of an endothermic phase transition at 670 km depth in a spherical model of convection in the Earth's mantle. *Nature* 361:699–704.

Tan, B. H., Jackson, I., and Fitz Gerald, J. D. 1997. Shear wave dispersion and attenuation in fine-grained synthetic olivine aggregates: preliminary results. *Geophys. Res. Lett.* 24:1055–8.

Toksöz, M. N., and Anderson, D. L. 1966. Phase velocities of long-period surface waves and structure of the upper mantle. *J. Geophys. Res.* 71:1649–58.

Trunin, R. F., Gon'shakova, V. I., Simakov, G. V., and Galdin, N. E. 1965. A study of rocks under the high pressures and temperatures created by shock compression. *Izv. Earth Physics* 9:1–12.

Tsuchida, Y., and Yagi, T. 1989. A new, post-stishovite high-pressure polymorph of silica. *Nature* 340:217–20.

Vidale, J. E., and Benz, H. M. 1992. Upper-mantle seismic discontinuities and the thermal structure of subduction zones. *Nature* 356:678–83.

Vinnik, L. P. 1977. Detection of waves converted from P to SV in the mantle. *Phys. Earth Planet. Int.* 15:39–45.

Walck, M. C. 1984. The P wave upper mantle structure beneath an active spreading centre: the Gulf of California. *Geophys. J. Royal Astron. Soc.* 76:697–723.

Wang, Y., Guyot, F., Yeganeh-Haeri, A., and Liebermann, R. C. 1990. Twinning in $MgSiO_3$ perovskite. *Science* 248:468–71.

Wang, Y., and Weidner, D. J. 1994. Thermoelasticity of $CaSiO_3$ perovskite and implications for the lower mantle. *Geophys. Res. Lett.* 2:895–8.

Wang, Y., and Weidner, D. J. 1996. $(\partial\mu/\partial T)_P$ of the lower mantle. *PAGEOPH* 146:533–49.

Wang, Y., Weidner, D. J., and Guyot, F. 1996. Thermal equation of state of $CaSiO_3$ perovskite. *J. Geophys. Res.* 101:661–72.

Wang, Y., Weidner, D. J., Liebermann, R. C., and Zhao, Y. 1994. P-V-T equation-of-state of $(Mg, Fe)SiO_3$ perovskite: constraints on composition of the lower mantle. *Phys. Earth Planet. Int.* 83:13–40.

Webb, S. L., and Jackson, I. 1993. The pressure dependence of the elastic moduli of single-crystal orthopyroxene $(Mg_{0.8}Fe_{0.2})SiO_3$. *Eur. J. Mineral.* 5:1111–19.

Weidner, D. J. 1986. Mantle model based on measured physical properties of minerals. In: *Chemistry and Physics of Terrestrial Planets*, ed. S. K. Saxena, pp. 251–74. Berlin: Springer-Verlag.

Weidner, D. J., Sawamoto, H., Sasaki, S., and Kumazawa, M. 1984. Single-crystal elastic properties of the spinel phase of Mg_2SiO_4. *J. Geophys Res.* 89:7852–60.

Weidner, D. J., and Wang, Y. 1995. Chemical and Clapeyron-induced buoyancy at the 660 km discontinuity. In: *Abstracts, IUGG XXI General Assembly*, p. B383.

Weinstein, S. A. 1993. Catastrophic overturn of the Earth's mantle driven by multiple phase changes and internal heat generation. *Geophys. Res. Lett.* 20:101–4.

Wentzcovitch, R. M., Ross, N. L., and Price, G. D. 1995. Ab initio study of $MgSiO_3$ and $CaSiO_3$ perovskites at lower-mantle pressures. *Phys. Earth Planet. Int.* 90:101–12.

Whitcomb, J. H., and Anderson, D. L. 1969. Reflection of P′P′ seismic waves from discontinuities in the mantle. *J. Geophys. Res.* 75:5713–28.

White, R. S., and McKenzie, D. 1995. Mantle plumes and flood basalts. *J. Geophys. Res.* 100:17543–85.

Widmer, R., Masters, G., and Gilbert, F. 1991. Spherically symmetric attenuation within the earth from normal mode data. *Geophys. J. Int.* 104:541–53.

Wiggins, R. A., and Helmberger, D. V. 1973. Upper mantle structure of the western United States. *J. Geophys. Res.* 78:1870–80.

Williams, Q., Jeanloz, R., and McMillan, P. 1987. Vibrational spectrum of $MgSiO_3$ perovskite: zero-pressure Raman and mid-infrared spectra to 27 GPa. *J. Geophys. Res.* 92:8116–28.

Wolf, G. H., and Bukowinski, M. S. T. 1987. Theoretical study of the structural properties and equations of state of $MgSiO_3$ and $CaSiO_3$ perovskites: implications for lower mantle composition. In: *High Pressure Research in Mineral Physics*, ed. M. H. Manghnani and Y. Syono, pp. 313–31. Tokyo: Terra Scientific.

Woodhouse, J. H., and Dziewonski, A. M. 1989. Seismic modelling of the Earth's large-scale three-dimensional structure. *Phil. Trans. R. Soc. London* 328:291–308.

Yagi, T., Mao, H. K., and Bell, P. M. 1982. Hydrostatic compression of perovskite-type $MgSiO_3$. In: *Advances in Physical Geochemistry*, vol. 2, ed. S. K. Saxena, pp. 317–25. Berlin: Springer-Verlag.

Yamazaki, A., and Hirahara, K. 1994. The thickness of upper mantle discontinuities, as inferred from short-period J-array data. *Geophys. Res. Lett.* 21:1811–14.

Yeganeh-Haeri, A. 1994. Synthesis and re-investigation of the elastic properties of single-crystal magnesium silicate perovskite. *Phys. Earth Planet. Int.* 87:111–21.

Yusa, H., Akaogi, M., and Ito, E. 1993. Calorimetric study of $MgSiO_3$ garnet and pyroxene: heat capacities, transition enthalpies, and equilibrium phase relations in $MgSiO_3$ at high pressures and temperatures. *J. Geophys. Res.* 98:6453–60.

Zaug, J. M., Abramson, E. H., Brown, J. M., and Slutsky, L. T. 1993. Sound velocities in olivine at earth mantle pressures. *Science* 260:1487–9.

Zhao, L. S., and Helmberger, D. V. 1993. Upper mantle compressional velocity structure beneath the northwest Atlantic Ocean. *J. Geophys. Res.* 98:14185–96.

Zhao, Y., and Anderson, D. L. 1994. Mineral physics constraints on the chemical composition of the Earth's lower mantle. *Phys. Earth Planet. Int.* 85:273–92.

Zielhuis, A., and Nolet, G. 1994. Shear-wave velocity variations in the upper mantle beneath central Europe. *Geophys. J. Int.* 117:695–715.

Zielhuis, A., and van der Hilst, R. D. 1996. Mantle structure beneath the eastern Australian region from partitioned waveform inversion. *Geophys. J. Int.* 127:1–16.

10

The Viscosity of the Mantle: Evidence from Analyses of Glacial-Rebound Phenomena

KURT LAMBECK and PAUL JOHNSTON

10.1. Introduction

Our knowledge of the bulk composition of the Earth is well constrained by observations that include cosmic elemental abundances and laboratory analyses of rock and mineral samples originating from the mantle (e.g., Ringwood, 1975; O'Neill and Palme, Chapter 1, this volume). Also well constrained are the bulk elastic properties of the Earth through the analysis of seismic wave propagation (e.g., Kennett and van der Hilst, Chapter 8, this volume) and analyses of tides and the planet's rotation. But less satisfactory is our understanding of the time-dependent or viscous response. Various geophysical observations indicate that stress and strain in the planet as a whole are not in phase, as seen in observations of the Earth's tides, and that the mantle creeps when subjected to stress, as demonstrated by crustal rebound after removal of ice loads. But we do not have a complete description of the solid Earth's departures from elasticity. Laboratory experimentation on terrestrial materials indicates that the nonelastic response of the mantle is dependent on the defect nature of the solid, such as dislocation density and dislocation mobility, which in turn are functions of the ambient temperature, pressure, and nonhydrostatic stress. Thus, at seismic frequencies, the response to an applied oscillatory stress is out of phase because of the finite diffusion time of the atoms around the dislocation, whereas for longer-term motions associated with tectonic stresses the response is described in terms of a solid-state viscosity. This viscosity is an important property entering into quantitative evaluations of the thermal history and the convective regime of the mantle. Yet it remains a relatively poorly known quantity, and the major unresolved issues include the nature of its depth dependence and lateral variability, particularly within the upper mantle. There are numerous reasons for this state of affairs, not least being the difficulty of interpreting the geophysical data regarding attenuation or relaxation and the lack of an adequate basis for extrapolating the behaviour observed in laboratory experiments (e.g., Jackson and Rigden, Chapter 9, this volume) to the solid Earth as a whole.

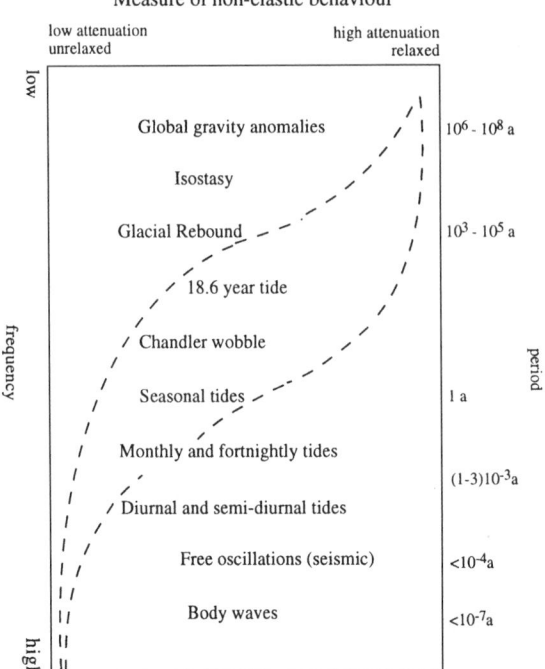

Figure 10.1. Schematic spectrum of Earth-deforming processes that can be observed by a range of techniques. The rheology function describing the nonelastic behaviour over this frequency range is largely unknown, and two possible examples of how attenuation or stress relaxation may vary with frequency are illustrated by the dashed lines.

A range of geophysical observations provides some information on the departures of the Earth's response from elastic behaviour. Figure 10.1 shows the spectrum of possible processes, ranging from periods of seconds to 10^8 years. At the very high frequencies (i.e., periods of 10^{-4} years or shorter), the response is essentially elastic, whereas at the low-frequency end of the spectrum (periods of 10^7 years or longer), the response is essentially that of a fluid. The nature of the response function between these two limits is, however, essentially unknown, and its determination remains a major objective of geophysical research.

The geophysical observations of interest here concern the response or deformation $\mathcal{D}$ to an applied force $\mathcal{F}$. The deformation may take various forms: a surface displacement measured with seismic sensors; a change in the gravity field, as in the case of earth-tide measurements; a change in the planet's rotation, as in the case of the Chandler wobble or changes in the length-of-day on decadal timescales; a change in sea level, as in the case of the isostatic adjustments to changes in the Earth's ice sheets. If the corresponding forces are known, then it becomes possible to estimate a response function $\mathcal{R}$ for the Earth on the time and length scales that are characteristic of the particular phenomenon under consideration. The resulting response function is, however, likely to be complex, because the different geophysical problems have different timescales, cause different stress levels, or preferentially stress different parts of the Earth. Thus the resultant response function derived from analysis of any single process involves effective parameters that describe the response characteristic of the particular applied force.

The geophysical challenges are twofold: to infer the appropriate response functions for the Earth for the different processes and to interpret these functions in terms of the mantle rheology. The first is complicated by the fact that the force responsible for the observed deformation may not be well known, and aspects of it may have to be inferred from the observations themselves. The second challenge is complicated by the fact that the estimated response functions are global responses, reflecting not only the solid Earth but also the atmosphere and oceans. Thus, an estimate of the tidal response of the Earth derived from analyses of the perturbations in satellite orbits contains a solid-Earth component and ocean and atmosphere components, and the last two have to be evaluated independently in order to assess the response function for the solid Earth. Likewise, the response function for the Earth inferred from analyses of the Chandler wobble is contaminated by the ocean and atmosphere responses.

The geophysical problems may therefore be considered as falling into one of two categories: (1) where the force $\mathcal{F}$ is known, the deformation $\mathcal{D}$ is observed, and the response function $\mathcal{R}$ is inferred; (2) where $\mathcal{R}$ is assumed known, and $\mathcal{F}$ is inferred from the observed $\mathcal{D}$. An example of the first class of problems occurs in tide analyses, where the gravitational force giving rise to the tide in the solid Earth is well known and the Earth's response can be measured as changes in the tilt of its surface, in gravity, or in its rotation. Also representative of the first category is the response of the crust to volcanic or sediment loads on timescales of 10^6 years. Here geological data can provide a description of the loading history, and the observed quantity may be the deflection of the surface or the gravity anomaly produced by the redistribution of the surface load and the deformation. Included amongst the second class of problems is the study of the decade-scale changes in the Earth's rotation. Here the cause of the irregular rotation is essentially unknown, and the response function is usually assumed to be known. Another example is provided by mantle convection, where a quantification of the forces driving the mantle motion is sought, and the viscosity is usually assumed to be known. In reality, many of these analyses involve a combination of the two classes of problems, where the observations are used to infer aspects of both $\mathcal{F}$ and $\mathcal{R}$. The glacial-rebound problem is one example of this, where the history of the ice load is only partly known from independent evidence or arguments, and aspects of the load have to be inferred from the same observations that constrain the Earth's response function $\mathcal{R}$. Thus, when searching for the rheology of the solid Earth, unique solutions are rarely possible, and one often learns as much about the forcing functions or about the fluid parts of the Earth as about the mantle's viscosity structure. In some instances a separation of the various aspects of a problem is possible, as is the case for the response of the Earth to the waxing and waning of the ice sheets on timescales of 10^3–10^5 years. But here also one learns much about surface processes in the course of understanding the mantle viscosity.

In this chapter we focus on this latter problem: the inference of mantle viscosity from the change in the Earth's shape and gravity field as ice sheets grow and decay. The resulting redistribution of the surface load of ice and meltwater has several consequences that are of geophysical significance on both global and regional scales. On a global scale, the Earth deforms under the new ice–water load distribution, and the inertia tensor and gravity field for the planet, comprising the solid Earth, ice, and oceans, become time-dependent. This has two immediate consequences: The Earth's rotation is modified, and the orbits of near-Earth satellites are perturbed from what they would be in a time-constant gravity field. The former includes changes in the angular velocity about the rotation axis (a change in the length-of-day) and in the direction of this axis relative to the Earth itself (polar motion). Over intervals for which some observational record exists, both changes are of a secular nature: a positive acceleration of the Earth or a decrease in the length-of-day, and a drift in the position of the rotation axis relative to the crust along the 70° west longitude meridian. The length-of-day record extends more than 2,000 years back in time, but other physical processes modify this quantity as well, and the challenge is to isolate these from the deglaciation-induced changes. The polar-motion record extends about 100 years back in time, and the interpretation of this record is ambiguous (Lambeck, 1980, 1988).

The time-dependent gravity field introduces perturbations in the orbits of near-Earth satellites, with the most readily observed one being a linear change in the rate of precession of the orbit about the Earth's rotation axis. The observational record of less than 20 years is short when compared to the characteristic loading time-constant or the characteristic viscous relaxation-time, and other forces can perturb a satellite orbit in addition to the deglaciation-induced changes in the gravity field. Unique interpretations are again out of the question.

More locally, the Earth's surface is deformed in response to the changing ice–water distribution. The primary deformation is in the vertical direction, but horizontal displacements may also be important. Only recently has it become possible to measure these deformations using the precise geodetic-positioning methods provided by satellite and radio-astronomy technology, and it will be some years before useful records become available. More important is the deformation of the crust relative to sea level, because long geological records of such movements are abundant, even though the observations themselves usually are not very precise. Sea levels change relative to the Earth's centre of mass because of the changing ocean volumes as ice sheets grow and decay, but the crust is also displaced radially as a result of the variations in surface loads. An observation of relative sea level is therefore a combination of these two contributions: the changing ocean volume, and the displacement of the surface of the solid Earth. These sea-level observations will form the primary data source discussed in this chapter.

The glacial-rebound problem is not a new one. T. F. Jamieson, in 1865, suggested that the elevated shorelines in northern Europe are indicative of the Earth's adjustment to the removal of the late Pleistocene ice, and N. S. Shaler, in 1874, reporting on the changing levels along the coast of Maine, New England, made a similar suggestion (Andrews, 1974). Attempts to devise quantitative models for the Earth's rebound to the melting of the most recent glacial cycle, between about 20,000 and 10,000 years ago, were made by Haskell, Niskanen, Vening Meinesz, and Gutenberg in the 1930s and 1940s, and their conclusion, that the average viscosity of the mantle is about $1–3 \times 10^{21}$ Pa $\cdot$ s (e.g., Cathles, 1975), is still valid today.

Over the past two decades this topic has enjoyed a renaissance, for several reasons. Foremost was the development of the plate-tectonics paradigm: A dynamical description of mantle convection requires a knowledge of mantle viscosity (e.g., Davies, Chapter 5, this volume). Glacial rebound is the most appropriate phenomenon available for estimating this property at timescales approaching those characteristic of mantle convection. Secondly, a wealth of geomorphological and geological evidence of raised and submerged shorelines and other indicators of ancient sea levels has become available, constrained in time by radiocarbon ages. Such data, not only from the formerly glaciated areas but also from areas much farther away, have been critical in constraining the rebound models. Thirdly, there have been major improvements in the conception of the glacial-rebound model deriving from the work of McConnell (1968), O'Connell (1971), Walcott (1970, 1972), Peltier (1974), Cathles (1975), Farrell and Clark (1976), Peltier and Andrews (1976), and Clark, Farrell, and Peltier (1978), and together, their findings have led to the modern formulation of the glacio-isostatic problem.

10.2. Surface Loading of a Spherical Earth
10.2.1. Deformation of Spherical-Earth Models by Surface Loads
10.2.1.1. Governing Equations

The basic equations governing the deformation of a model Earth describe the inertia (or motion), the link between density and gravitational potential, and continuity conditions (e.g., Fung, 1965; Malvern, 1969). Large-scale changes in glacial loads occur on timescales of thousands of years, and so the effects of rotation and of the acceleration term in the inertia equation, which operate on much shorter scales, can be neglected. The initial state of the body is assumed to be one of hydrostatic equilibrium, and the equations are expressed in terms of incremental changes in deformation and stress. Second-order terms containing the products or squares of the increments are ignored. With these assumptions, the incremental forms of the three equations in a Lagrangian reference frame are (e.g., Wolf, 1991)

$$\nabla \cdot \tau^{(\delta)} + \nabla\left(\rho^{(0)}\mathbf{u} \cdot \nabla\psi^{(0)}\right) + \rho^{(\Delta)}\nabla\psi^{(0)} + \rho^{(0)}\nabla\psi^{(\Delta)} = 0 \tag{1}$$

$$\nabla^2\psi^{(\Delta)} = -4\pi G\rho^{(\Delta)} \tag{2}$$

$$\rho^{(\Delta)} = -\nabla \cdot \left(\rho^{(0)}\mathbf{u}\right) \tag{3}$$

where **u** is the displacement, τ is the Cauchy stress tensor, ρ is density, ψ is the gravitational potential, and G is the gravitational constant. Equation (1) describes the balance of forces from spatial gradients in stress (first term) and perturbations of gravitational forces due to changes in density (third term) and in the gravitational potential (fourth term). Before deformation, pressure increases with depth because of the weight of overlying material. As the Earth is deformed, the initial stress is advected, as indicated by the second term in equation (1). Equation (2) describes the change in gravitational potential caused by changes in density, and equation (3) is a statement of the conservation of mass. The superscripts zero, lowercase delta, and capital delta in equations (1)–(3) refer to the initial state, to material increments, and to local increments, respectively. For any field variable f, the material increment $f^{(\delta)}$ is the change from the initial state for a particle, and the local increment $f^{(\Delta)}$ is the change in the field variable at a location. If the original location of the particle in the Lagrangian reference frame is **X**, the relation between the material increment $f^{(\delta)}(\mathbf{X})$ and the local increment $f^{(\Delta)}(\mathbf{X})$ is

$$f^{(\delta)}(\mathbf{X}) = f^{(\Delta)}(\mathbf{X}) + \mathbf{u}(\mathbf{X}) \cdot \nabla f^{(0)}(\mathbf{X}) \tag{4}$$

where the second term on the right-hand side describes the advection of the initial field.

10.2.1.2. Constitutive Law

In addition to these fundamental physical equations, a constitutive relation between the incremental stress field ($\tau^{(\delta)}$) and strain field (ϵ) is required. For high-frequency deformations, an elastic relation is mostly satisfactory, and the constitutive equation is

$$\tau_{ij}^{(\delta)} = \left(\kappa - \frac{2}{3}\mu \right) \epsilon_{kk}\delta_{ij} + 2\mu\epsilon_{ij} \tag{5}$$

where μ is the elastic shear modulus, κ is the bulk modulus, and δ_{ij} is the Kronecker-delta tensor ($= 1$ if $i = j$, 0 otherwise). The repeated indices follow the summation convention (i.e., $\epsilon_{kk} = \epsilon_{11} + \epsilon_{22} + \epsilon_{33}$).

Glacial rebound occurs on an intermediate timescale, where the Earth exhibits aspects of both solid and fluid behaviour, and one of the simplest representative constitutive laws is that of a Maxwell solid, for which

$$\dot{\tau}_{ij}^{(\delta)} + \frac{\mu}{\eta}\left(\tau_{ij}^{(\delta)} - \frac{1}{3}\tau_{kk}^{(\delta)}\delta_{ij} \right) = \left(\kappa - \frac{2}{3}\mu \right)\dot{\epsilon}_{kk}\delta_{ij} + 2\mu\dot{\epsilon}_{ij} \tag{6}$$

where η is the kinematic viscosity, and a dot over a variable indicates time differentiation. The Maxwell body is one example of a general class of linear viscoelastic materials in which at high frequencies the material responds as if elastic, and at low frequencies the material flows as a viscous fluid. This choice is largely dictated by mathematical expediency, rather than by physical reality, but it nevertheless appears to work remarkably well over the frequency range and stress-magnitude range of the glacial-rebound problem. The consequence of this choice is that the viscosity values are effective parameters that describe the viscoelastic behaviour of the Earth under glacial-loading conditions, but cannot be readily extrapolated to the substantially different frequency–stress conditions of other phenomena. Other, more complex forms for the linear viscoelastic body have sometimes been used in describing glacial rebound (e.g., Peltier, Drummond, and Tushingham, 1986; Rümpker and Wolf, 1996), but this does not appear to have offered any major improvements in the matching of model predictions to the observational evidence.

10.2.1.3. Boundary Conditions

The solution of the equations must be subjected to a number of boundary conditions, including the requirement of regularity at the centre of the Earth to ensure that physical parameters remain finite there. At the outer surface, the boundary conditions require that the radial stress τ_{rr} be balanced by the normal surface load, that the tangential stress elements $\tau_{r\theta}$ and $\tau_{r\lambda}$ be zero, and that a condition for the continuity of a modified gravitational potential gradient be satisfied (e.g., Longman, 1962). If $L^{(\delta)}$ is the surface-load increment (mass per unit area) at the surface $r = R$, then these boundary conditions are

$$\tau_{rr}^{(\delta)}(R, \theta, \lambda) = -g^{(0)}(R)L^{(\delta)}(R, \theta, \lambda) \tag{7}$$

$$\tau_{r\theta}^{(\delta)} = \tau_{r\lambda}^{(\delta)} = 0 \tag{8}$$

$$\hat{\mathbf{r}} \cdot \left[\nabla \psi^{(\Delta)} - 4\pi G \rho^{(0)} \mathbf{u} \right]_{-}^{+} = -4\pi G L^{(\delta)}(R, \theta, \lambda) \tag{9}$$

where θ is the colatitude and λ is the longitude of the point of application, $\hat{\mathbf{r}}$ is a unit radial vector, and $g^{(0)}$ is the initial gravity $(= -\partial \psi^{(0)} / \partial r)$. The symbol $[f]_{-}^{+}$ signifies $\lim_{h \to 0} f(R+h) - f(R-h)$. Boundary conditions also apply at interior surfaces across which the material properties are discontinuous. These include the core–mantle boundary and chemical- or phase-transformation boundaries in the crust and mantle, as well as boundaries of discontinuity that may be introduced by the discretization of the radial parameters of density and rheology.

10.2.1.4. Formulation in Spherical Co-ordinates

For Earth models in which the initial state is one of spherical symmetry, the natural formulation for the equations of motion is in terms of spherical co-ordinates, in which the loads and deformations are defined in a spectral domain using spherical-harmonic functions. If a linear rheology is also assumed, then a surface-load harmonic of a given degree will lead to deformation, stress, and gravity changes in the Earth model that will be described in the same spherical-harmonic form, with the constants of proportionality defined by a set of radially varying parameters known collectively as Love numbers.

Any function on a unit sphere, such as a surface load L, can be expanded into surface spherical harmonics Y_{nm} as

$$L(\theta, \lambda) = \sum_{n=0}^{\infty} \sum_{m=-n}^{n} L_{nm} Y_{nm}(\theta, \lambda) = \sum_{n=0}^{\infty} L_n Y_n(\theta, \lambda) \tag{10}$$

with

$$Y_{nm}(\theta, \lambda) =$$
$$\left(\frac{(2n+1)(2-\delta_0^m)(n-|m|)!}{(n+|m|)!}\right)^{1/2} P_{nm}(\cos\theta) \binom{\cos m\lambda}{\sin|m|\lambda} \quad \begin{array}{l} (n \geq m \geq 0) \\ (0 > m \geq -n) \end{array}$$
$$\tag{11}$$

The P_{nm} are the associated Legendre polynomials of degree n and order m. (The functions as defined here are fully normalized, such that the mean square amplitude for each over the surface of the sphere is unity.) The displacements, stresses, and change in the gravitational potential can likewise be expanded in terms of the Y_{nm}. For example, for the surface load $L(\theta, \lambda)$ from equation (10), whose gravitational potential $\psi_1(r, \theta, \lambda)$ at $r > R$ is

$$\psi_1(r, \theta, \lambda) = \sum_n \psi_{1,n}(r) \left(\frac{R}{r}\right)^n Y_n(\theta, \lambda)$$

$$= 4\pi RG \sum_n \frac{1}{2n+1} \left(\frac{R}{r}\right)^n L_n Y_n(\theta, \lambda) \tag{12}$$

the new potential after deformation is defined as

$$\psi = \psi_1(r) + \psi_2(r) \tag{13}$$

with

$$\psi_2^{(\Delta)}(r) = \sum_n k_n(r) \psi_{1,n}(r) \tag{14}$$

Also, the displacement vector $\mathbf{u}(r, \theta, \lambda)$ for $r < R$ is defined in terms of Love numbers and spherical-harmonic functions as

$$\mathbf{u}(r, \theta, \lambda) = \sum_n \frac{\psi_{1,n}(r)}{g^{(0)}(r)} \left(\frac{r}{R}\right)^n [h_n(r) Y_n(\theta, \lambda)\hat{\mathbf{r}} + rl_n(r)\nabla Y_n(\theta, \lambda)] \tag{15}$$

where ∇Y_n is the gradient of the spherical-harmonic function, and h, l, and k are the Love numbers appropriate for this particular surface-loading problem (Love, 1911; Jeffreys, 1959). The first two are the displacement Love numbers, and they define the radial and tangential deformations within the planet; the third is the potential Love number.

The equations for the entire deformation of a spherical elastic model reduce to six coupled first-order partial-differential equations (Alterman, Jarosch, and Pekeris, 1959):

$$\frac{\partial \mathbf{y}_n(r, t)}{\partial r} = A_n(r, t)\mathbf{y}_n(r, t) \qquad (n = 0, 1, \ldots, \infty) \tag{16}$$

where

$$\mathbf{y}_n = (u_n, v_n, \tau_{rn}, \tau_{\theta n}, \psi_n, Q_n)^T \tag{17}$$

is a vector of coefficients describing the radial and tangential displacements (u_n, v_n), the radial and tangential stresses (τ_{rn}, $\tau_{\theta n}$), the gravitational potential ψ_n, and a function of the potential gradient Q_n. The matrix $A_n(r)$ is a function of the elastic moduli $\mu(r)$ and $\kappa(r)$, gravity $g(r)$, and density $\rho(r)$ and spherical-harmonic degree n. The (u_n, v_n, ψ_n) relate to the Love numbers according to

$$\begin{pmatrix} h_n(r, t) \\ l_n(r, t) \\ 1 + k_n(r, t) \end{pmatrix} = \frac{2n+1}{4\pi R G L_n(t)} \begin{pmatrix} g^{(0)}(r) u_n(r, t) \\ g^{(0)}(r) v_n(r, t) \\ \psi_n(r, t) \end{pmatrix} \tag{18}$$

10.2.1.5. Elastic Solutions

Solutions to the elastic equations for the spherical geometry are well known, both for the free oscillations of the Earth, in which the acceleration terms in the inertia equation are important (e.g., Alterman et al., 1959), and for surface-loading problems. For the latter, the equations have been discussed by Longman (1962) and Farrell (1972). The appropriate boundary conditions at the surface have been discussed by Longman (1962) and follow from equations (7)–(9). The special boundary conditions at the core–mantle interface have been discussed by Dahlen and Fels (1978), and those where the discontinuity moves with respect to the material, such as at phase-transformation boundaries, have been discussed by Dehant and Wahr (1991) and Johnston, Lambeck, and Wolf (1996). Equation (16) can be solved analytically for models consisting of uniform incompressible shells (Yuen, Sabadini, and Boschi, 1982) or numerically for more complicated elastic models.

10.2.1.6. Viscoelastic Solutions: Laplace-Transform Methods

The elastic solutions have been widely used for modelling the response of the Earth to short-duration loads, such as ocean tides or atmospheric-pressure fluctuations, but for the glacial-rebound problem the time dependence of the response to the changing surface loads must be included. One of the primary reasons for adopting the linear-viscoelastic-response model, equation (6), to describe this behaviour is that it permits the correspondence principle to be used. This states that a viscoelastic problem can be transformed to the equivalent elastic problem by taking the Laplace transform of the elastic formulation for the fundamental physical equations, the constitutive equations, and the boundary conditions (Bland, 1960; Peltier, 1974).

The Laplace transform of a function $f(t)$ is defined by

$$\mathcal{L}\{f(t)\} = \tilde{f}(s) = \int_0^\infty e^{-st} f(t) \, dt \tag{19}$$

It has the useful property that time derivatives in the time domain are transformed to polynomials of the Laplace-transform variable s in the Laplace domain; that is,

$$\mathcal{L}\left\{\frac{df}{dt}\right\} = s\tilde{f}(s) - f^{(0)} \tag{20}$$

The first step in the application of the correspondence principle to the viscoelastic problem is to take the Laplace transform of the equations (1)–(3) and (6)–(9). The correspondence principle states that the viscoelastic formulation in the Laplace-transform domain, is identical to that of the equivalent elastic problem in the time domain, and the solution of the former follows from the inverse Laplace transform of the equivalent elastic solution. Because there are no time derivatives in equations (1)–(3), after Laplace transformation three analogous equations are obtained in which each parameter $f(X, t)$ is replaced by $\tilde{f}(X, s)$. Also, the transform of the constitutive equation (6) is

$$\tilde{\tau}_{ij}^{(\delta)}(s) = \left(\kappa - \frac{2\tilde{\mu}(s)}{3}\right)\tilde{\epsilon}_{ij}(s)\delta_{ij} + 2\tilde{\mu}(s)\tilde{\epsilon}_{ij}(s) \tag{21}$$

with the Laplace-transformed shear modulus

$$\tilde{\mu}(s) = \frac{\mu s}{s + \mu/\eta} \tag{22}$$

Equation (21) is formally identical to the elastic constitutive equation (5). Thus the first-order partial-differential equations (16) in the Laplace-transform domain are

$$\frac{\partial \tilde{\mathbf{y}}_n(r, s)}{\partial r} = A_n(r, s)\tilde{\mathbf{y}}_n(r, s) \qquad (n = 0, 1, \dots, \infty) \tag{23}$$

with

$$\tilde{\mathbf{y}}_n = (\tilde{u}_n, \tilde{v}_n, \tilde{\tau}_{rn}, \tilde{\tau}_{\theta n}, \tilde{\psi}_n, \tilde{Q}_n)^T \tag{24}$$

The boundary conditions are also Laplace-transformed, and those at the outer surface are (Longman, 1962; Wu and Peltier, 1982)

$$\tilde{\tau}_{rn}(R) = -g^{(0)}\tilde{L}_n, \qquad \tilde{\tau}_{\theta n}(R) = 0, \qquad \tilde{Q}_n(R^-) + \frac{n+1}{R}\tilde{\psi}_n(R) = 4\pi G\tilde{L}_n \tag{25}$$

where $\tilde{L}(s)$ is the Laplace-transformed load. Likewise, the definitions of the Laplace-transformed Love numbers are

$$\begin{pmatrix} \tilde{h}_n(r, s) \\ \tilde{l}_n(r, s) \\ 1 + \tilde{k}_n(r, s) \end{pmatrix} = \frac{2n+1}{4\pi R G\tilde{L}_n(s)} \begin{pmatrix} g^{(0)}(r)\tilde{u}_n(r, s) \\ g^{(0)}(r)\tilde{v}_n(r, s) \\ \tilde{\psi}_{1,n}(r, s) \end{pmatrix} \tag{26}$$

10.2.1.7. Inversion of the Laplace Transform

An issue of some importance in applying this methodology to the glacial-rebound problem is the inversion of the solutions for the Laplace-transformed parameters back into the time domain. For Earth models consisting of a number of uniform incompressible layers, analytic solutions exist, and the solution vector takes the normal-mode form (Wu, 1978; Peltier, 1985)

$$\tilde{\mathbf{y}}_n(r, s) = \tilde{L}_n(s) \left(\mathbf{y}_n^E(r) + \sum_{j=1}^{M} \frac{\mathbf{y}_n^j(r)}{s - s_n^j} \right) \tag{27}$$

where s_n^j are the negative inverse relaxation times. Such a form can also be assumed for a more realistic compressible-Earth model with continuously varying properties (Schapery, 1962; Peltier, 1974), and although the resulting solutions are only approximate, they have been shown to give reliable results by comparison with the analytic results for the layered incompressible case (Mitrovica and Peltier, 1992). Other general methods for obtaining the inverse transform have been discussed by Davies and Martin (1979) and Fang and Hager (1995).

10.2.2. The Sea-Level Equation
10.2.2.1. Description of the Load

To obtain realistic solutions of the equations, a high-resolution description of the coupled ocean–ice surface load is required. The ice load is defined here in terms of ice columns at discrete time intervals within which the change in the load is assumed to be linear. The model starts at a reference epoch that is far enough in the past for any loading history prior to that time to have no significant influence on the predicted geological or geophysical observables for the past 20,000 years or so. The description of the meltwater load requires that the ocean surface remain an equipotential surface at all times and that the ocean–ice mass be conserved.

The total change in the ice–ocean load is

$$L^{(\delta)}(\theta, \lambda, t) = \rho_I \Delta I(\theta, \lambda, t) + \rho_W \Delta W(\theta, \lambda, t) \tag{28}$$

where ΔI is the change in the effective ice height I, ΔW is the change in water depth W, and ρ_I and ρ_W are the densities of ice and water, respectively. (If the ice is grounded below sea level, then I is an effective ice thickness that includes a correction for the displacement of water by part of the ice load.) The change in the ice load is the difference in the ice height between a time t and the initial time t_0, and the change in the water load is the change in sea level, defined over the oceans only. This latter spatial constraint is introduced through the ocean function $\mathcal{O}(\theta, \lambda, t)$, defined to be zero on land and unity on the oceans. Because the ocean geometry changes with changing water depths, this function is variable in time. If $\zeta(\theta, \lambda, t)$ is the sea-level change, defined with respect to the water level at the reference time t_0, then the change in water depth at time t with respect to the initial condition at time t_0 is

$$\Delta W(\theta, \lambda, t) = \int_{t_0}^{t} \mathcal{O}(\theta, \lambda, t') \frac{\partial \zeta(\theta, \lambda, t')}{\partial t'} \, dt' \tag{29}$$

To take advantage of the Love-number formulation, the surface loads are expanded in series of surface harmonics as well as functions of time. Thus the change in the water depth ΔW is expanded into spherical harmonics as

$$\Delta W = \sum_{n=0}^{\infty} \sum_{m=-n}^{n} \Delta W_{nm} Y_{nm} \tag{30}$$

and, likewise, the change in ice thickness ΔI at any time is expanded as

$$\Delta I = \sum_{n=0}^{\infty} \sum_{m=-n}^{n} \Delta I_{nm} Y_{nm} \tag{31}$$

where the coefficients ΔW_{nm} and ΔI_{nm} are functions of time. Then the total surface load is

$$\Delta L = \sum_n \sum_m (\rho_W \Delta W_{nm} + \rho_I \Delta I_{nm}) Y_{nm}$$

$$= \sum_n \sum_m \Delta L_{nm} Y_{nm} = \sum_n \Delta L_n Y_n \tag{32}$$

10.2.2.2. Components of Sea-Level Change

Records of past sea levels take the form of the heights or depths of older shoreline features, of age t, measured relative to the present (at time t_p) ocean surface. The measured quantity is therefore the change in sea level relative to the crust, which itself moves vertically because of the response of the Earth to the change in the ice–water load. This relative change in sea level is defined as

$$\Delta\zeta(\theta, \lambda, t) = \zeta(\theta, \lambda, t) - \zeta(\theta, \lambda, t_p) \tag{33}$$

The sea-level change (33) comprises several components. The first is the eustatic component $\Delta\zeta^e(t)$, defined as the mean change in relative sea level over the entirety of the oceans, and because of the need to conserve mass, it provides a measure of the change in the volume of continent-based ice. The zero-degree harmonic of a function defines its average value, and

$$\Delta\zeta^e(t) = \int_{t_0}^{t} \frac{1}{\mathcal{O}_{00}(t')} \frac{dW_{00}(t')}{dt'} dt' \approx -\frac{\rho_I}{\rho_W} \frac{\Delta I_{00}(t)}{\mathcal{O}_{00}} \tag{34}$$

The additional components allow for the sea-level change due to the deformation of the crust under the load, proportional to $h_n \Delta L_n$, and due to the change in the geoid as surface mass and mass in the interior of the planet are displaced, proportional to $(1 + k_n) \Delta L_n$. The combined deformational effect is

$$\Delta\zeta^d(\theta, \lambda, t) = \frac{\psi^{(\Delta)}(R)}{g^{(0)}(R)} - u_r(R)$$

$$= \frac{4\pi G R}{g^{(0)}(R)} \sum_{n=1}^{\infty} \left(\frac{1 + k_n(R, t) - h_n(R, t)}{2n + 1} * \Delta L_n(t) \right) Y_n(\theta, \lambda) \tag{35}$$

$$= \sum_{n=1}^{\infty} \sum_{m=-n}^{n} \Delta\zeta_{nm}^d(t) Y_{nm}(\theta, \lambda) \tag{36}$$

where the operator $*$ represents multiplication for an elastic Earth and convolution for a viscoelastic-Earth model.

Because equation (35) does not conserve the total mass of ice and water, a small time-dependent but spatially uniform factor $d\zeta_0(t)$ must be added to $\Delta\zeta^d$, where

$$d\zeta_0(t) \approx -\frac{1}{\mathcal{O}_{00}} \sum_{n=1}^{\infty} \sum_{m=-n}^{n} \Delta\zeta_{nm}^d(t)\mathcal{O}_{nm} \qquad (37)$$

The expressions for the components $\Delta\zeta^e$, $\Delta\zeta^d$, and $d\zeta_0$ were first derived for a step-loading history by Farrell and Clark (1976) and were extended to the case of linear loading phases by Nakada and Lambeck (1987). Explicit expressions allowing for changes in the ocean geometry have been given by Johnston (1993), and the best method currently available for calculation of sea-level change has been described by Mitrovica and Peltier (1991). The total relative sea-level change is given by

$$\Delta\zeta(\theta, \lambda, t) = \Delta\zeta^e(t) + \Delta\zeta^d(\theta, \lambda, t) + d\zeta_0(t) \qquad (38)$$

where the load consists of the ice and water contributions. Alternatively, the change can be schematically written as

$$\Delta\zeta(\theta, \lambda, t) = \Delta\zeta^e(t) + \Delta\zeta^I(\theta, \lambda, t) + \Delta\zeta^W(\theta, \lambda, t) \qquad (39)$$

The first term is the eustatic-sea-level function, which is a function of time only, and the other two terms, the glacio- and hydro-isostatic components, represent the perturbations resulting from the ice-load and water-load redistributions, including the changes in gravitational attraction. Note that the water-load component depends on the sea-level change through equation (29) and is calculated by iteration from an initial approximate value.

10.2.3. Changes in the Long-Wavelength Component of the Gravitational Potential

The modification of the gravitational potential of a yielding-Earth model produced by a redistribution of surface mass is given by equation (13). The conventional notation for the static part of the Earth's potential $U(r, \theta, \lambda)$ is

$$U(r, \theta, \lambda) = -\frac{Gm_E}{r}\left[1 + \sum_{n=2}^{\infty}\left(\frac{R}{r}\right)^n \sum_{m=-n}^{n} \bar{C}_{nm}Y_{nm}(\theta, \lambda)\right] \qquad (r \geq R) \quad (40)$$

where the $\bar{C}_{nm}$ are the fully normalized Stokes coefficients representing integrals of functions of mass distribution within the Earth (e.g., Lambeck, 1988). The major term is $\bar{C}_{20}$, the second-degree zonal Stokes coefficient, which relates to the second-degree moment-of-inertia elements I_{ii} by

$$\bar{C}_{20} = \frac{-1}{\sqrt{5}m_E R^2}\left[I_{33} - \frac{1}{2}(I_{11} + I_{22})\right] \qquad (41)$$

and gives a measure of the Earth's equatorial flattening (its value is about -500×10^{-6}, whereas the other coefficients are on the order of 10^{-6} and smaller); m_E is the mass of the Earth, and R is the mean equatorial radius of the Earth. For the glacially loaded planet, the total gravitational potential is the sum of equations (13) and (40), and the combined potential can be written in the generalized form in which the Stokes coefficients have a time dependence according to

$$\dot{C}_{nm} = \frac{4\pi R^2}{m_E} \frac{d}{dt} \left(\frac{1 + k_n(R, t)}{2n + 1} * \Delta L_{nm}(t) \right) \tag{42}$$

This is the quantity that can be observed from repeated measurements of the Earth's gravitational potential (see Section 10.4).

10.3. Sea-Level Changes of Glacio-hydro-isostatic Origin
10.3.1. *Spatial Variability of Sea-Level Change*

The primary observed response to the melting of the last great ice sheets of latest Pleistocene time is the change in sea level, evinced by old shorelines located above or below the present position of the sea, depending on which terms in equation (39) are most important. In the formerly glaciated areas (near-field sites), the dominant contribution to the relative sea-level change is the crustal response to the changing ice load, and this exceeds the eustatic change. This is well illustrated by the elevated palaeo-shorelines observed in the Gulf of Bothnia in Scandinavia (Donner, 1995) and in Hudson Bay in Canada (Fulton, 1989). In the former locality, for example, rates of relative sea-level fall in excess of 20 mm/yr have been inferred for some localities once the areas became ice-free, and because of the viscous nature of the mantle, the rebound and sea-level change have continued long after the last ice was removed (Figure 10.2a). When the ice sheet is small, such as that over Britain, or for locations near the margins of the larger ice sheets, as in southern Norway or northern Denmark, the crustal rebound is much reduced, with a magnitude comparable to that of the eustatic contribution, but of opposite sign. Thus, sea levels have remained within a few tens of meters of the present level since late glacial time (Figure 10.2b). Farther away, beyond the areas of former glaciation (intermediate-field sites), the response of the crust to the removal of the ice is one of subsidence, caused by the flow of the underlying mantle material towards the rebounding, formerly loaded

Figure 10.2. Predicted and observed sea levels at several locations at increasing distances from the centre of ice loading. The eustatic (e.s.l.) and isostatic (ice and water) contributions are shown separately on the left-hand side. Note the different time and sea-level scales used. The total predicted sea levels (continuous curve) are compared with the observations on the right-hand side. (a) The Angerman River, northern Sweden, where the glacio-isostatic signal (ice) dominates the eustatic change. (b) The Firth of Forth, Scotland, where the isostatic rebound and eustatic change are more comparable in magnitude. (c) Dungeness, southern England, where the eustatic change dominates over the glacio- and hydro-isostatic contributions. (d) Gulf of Carpentaria, Australia, where the hydro-isostatic contribution led to the small-amplitude, mid-Holocene highstand. All predictions are for a standard three-layer model defined by equation (45).

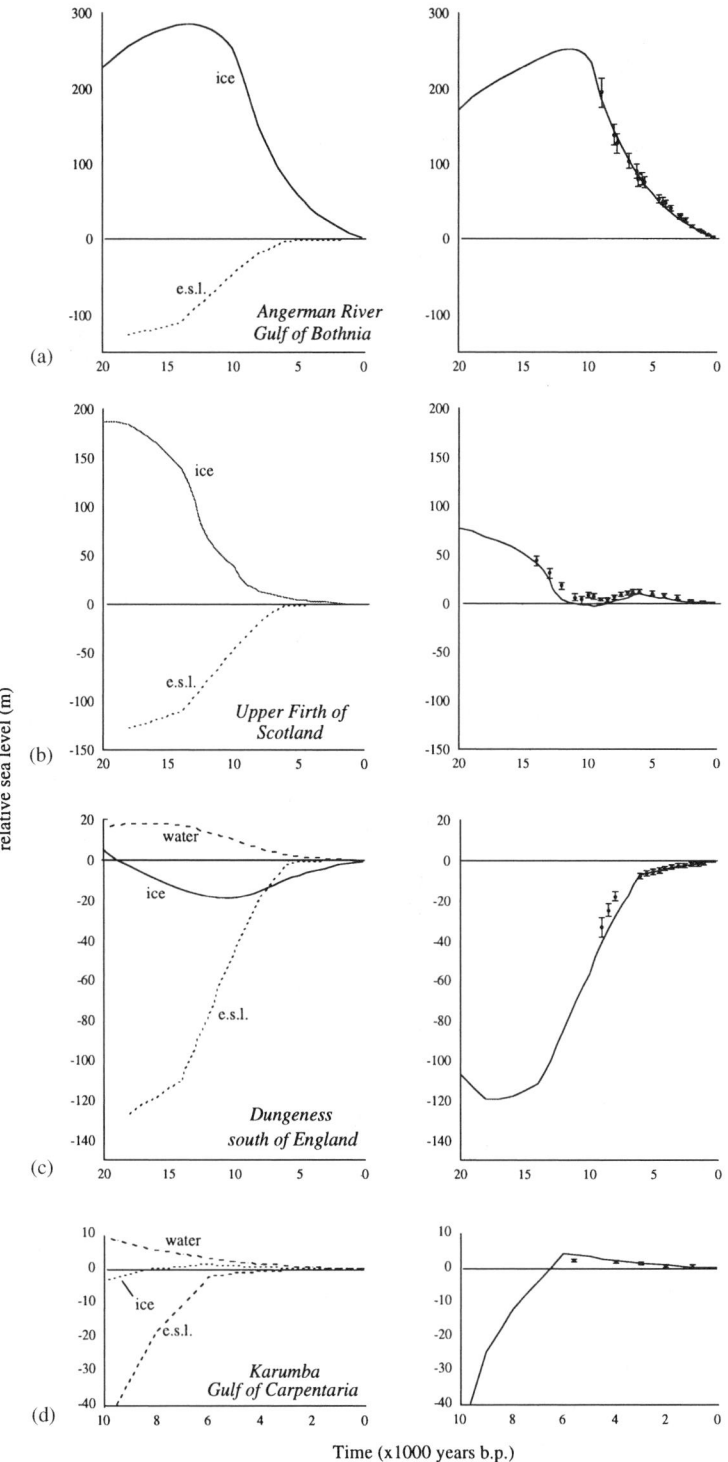

relative sea level (m)

Time (x1000 years b.p.)

areas. At such sites the sea level appears to have been rising up to the present time, even though all melting may have ceased much earlier. This is the case, for example, for southern England and for the Atlantic margin of France (Figure 10.2c). For these locations, the hydro-isostatic contributions become important when compared with the glacio-isostatic effect.

For continental-margin sites far from the former ice limits (far-field sites) the dominant perturbation to the eustatic-sea-level function is the hydro-isostatic term, although residual glacio-isostatic contributions remain (as discussed later). Here the rising water loads the oceanic lithosphere, inducing flow of the underlying mantle towards the adjacent continent or the former areas of glaciation. Thus the ocean floor subsides under the load, dragging down with it the coastal zone. But only after the eustatic rise has become very small or zero will the hydro-isostatic contribution become dominant and the sea levels appear to fall (Figure 10.2d). This typically occurred after about 6,000 B.P.

Sites on small islands far from the ice margins will move with the seafloor in response to the water loading, and the hydro-isostatic term vanishes. Now the residual perturbation is the small glacio-isostatic term, which does not vanish even for sites far from the former ice margin. When an ice sheet grows, the crust beneath it subsides, but beyond the ice limits a broad zone of crustal uplift develops because the volume of the solid Earth must, to a first approximation, be conserved. Where such a crustal-uplift zone occurs in an oceanic area, water is displaced, and a sea-level rise will be recorded throughout the far-field. When the ice sheets melt, the uplifted "peripheral bulge" subsides, the water-holding capacity of the ocean increases, and, in the absence of all other factors, sea level falls globally. Thus, at small oceanic islands away from the bulge the relative sea level will appear to have fallen during late Holocene time, even when no further melting occurred (Mitrovica and Peltier, 1991).

The schematic sea-level curves illustrated in Figure 10.2 indicate that the sea-level response to the changing ice loads is spatially quite complex, as was recognized by Walcott (1972) and Clark et al. (1978). Figure 10.3 illustrates some of the spatial complexity in sea-level response for three different locations at 6,000 B.P. The first is for northwestern Europe (Figure 10.3a), where the primary contributions are the vanishing loads over Great Britain and Fennoscandia and the increasing water load in the Atlantic and in the gradually flooding North Sea. The centres of rebound over Scotland and Fennoscandia are clearly defined, and a narrow zone of subsidence between the two developed as a consequence of both the ice and water loads. At this epoch, when the nominal eustatic sea level was zero, the actual changes over the region ranged from -25 m to more than 30 m. Clearly, the concept of a regional sea-level curve for this region, as sometimes proposed (Mörner, 1980), has little merit.

The second example is from the eastern Mediterranean region of Greece and the Aegean Sea (Figure 10.3b). This region lay some 2,500 km from the northern European ice sheets, sufficiently close for the glacio-isostatic terms to be important,

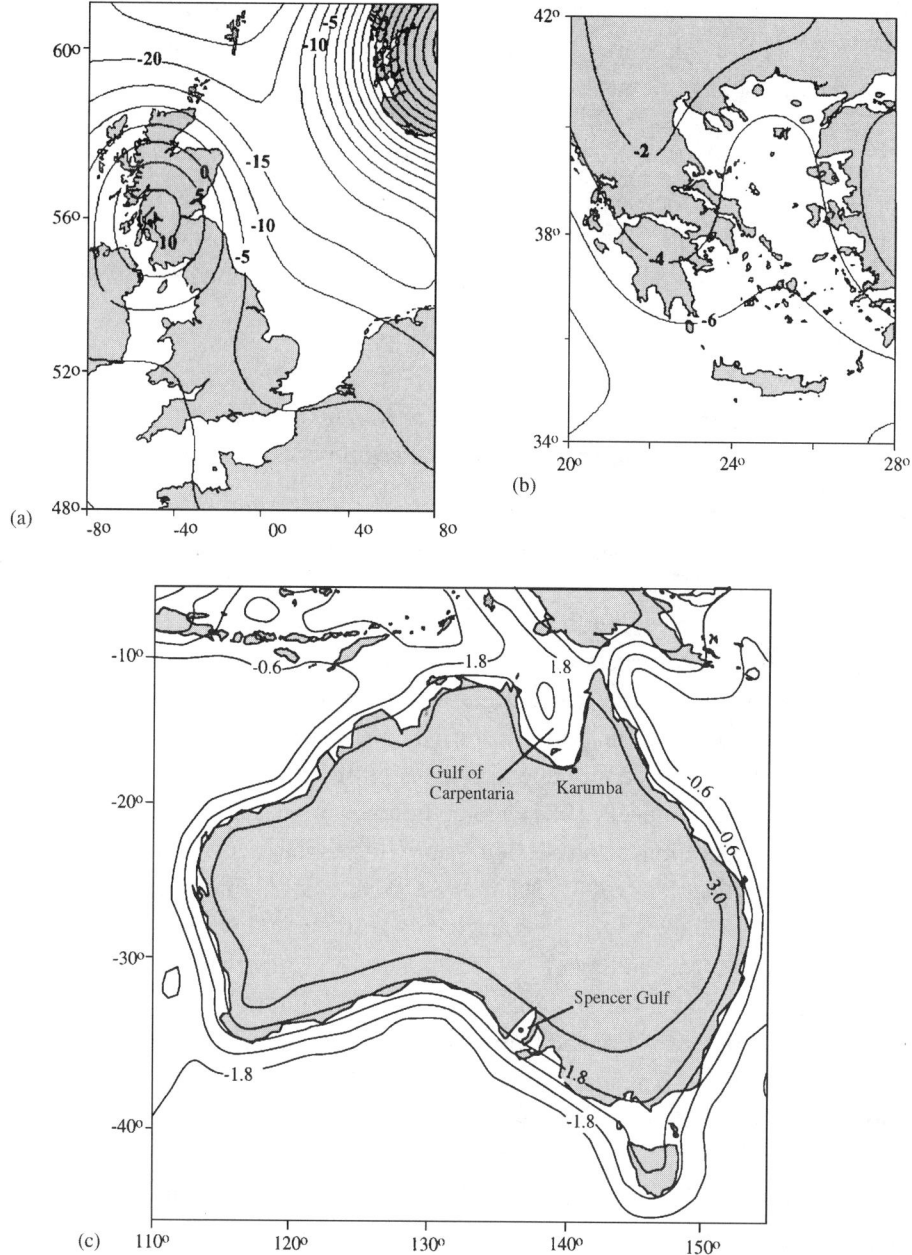

Figure 10.3. Predicted patterns of spatial variability in relative sea level for three regions at 6,000 B.P. (a) North Sea and Britain. (b) Greece and the Aegean. (c) Australia. The contours represent the positions of mean sea level in meters, relative to the present-day mean sea level. All predictions are for the standard model of equation (45).

such that even when melting had ceased, the sea continued to encroach upon the land. The water load is also important here, producing the sea-level contours that broadly follow the coastal geometry of the Aegean, Ionic, and Adriatic seas. The spatial variability predicted for this region is large and within the range of observational detection using archaeological and geological evidence for the past 6,000 years or so (Flemming, 1978). That means that inferences of tectonic rates for vertical motion from such observations need to be corrected for the isostatic factors (Lambeck, 1995a). The third example (Figure 10.3c) is for the Australian region, where the primary departure from eustasy is manifested by a small-amplitude highstand of mid-Holocene age, formed when the bulk of deglaciation had ceased (Lambeck and Nakada, 1990). This is primarily the hydro-isostatic effect. Major spatial variability is predicted, and observed, for example, along the narrow gulfs of South Australia, as discussed later.

10.3.2. Requirements for Solution of the Glacial-Rebound Problem and the Separation of Parameters

Prediction of sea-level change through time requires (1) a formulation to take into consideration all physical aspects of the Earth that affect the amplitudes and rates of deformation in response to surface loading, (2) a set of parameters to describe the elasticity and viscosity of the Earth, (3) models of the ice sheets through time, (4) a model of the ocean geometry though time, which requires that the sea-level change itself be known, and (5) a high-resolution mathematical solution for the sea-level equation, as outlined in Section 10.2. The first requirement includes the introduction of realistic depth-dependence for elastic moduli and density (Peltier, 1974; Nakada and Lambeck, 1987) and compressibility (Wu and Yuen, 1991), the effects of any migration of phase boundaries in response to the surface loading (Fjeldskaar and Cathles, 1984; Dehant and Wahr, 1991; Johnston et al., 1997), and the effects of lateral variations in mantle response (Gasperini and Sabadini, 1989; Kaufmann, Wu, and Wolf, 1997). Generally, neither the second nor third requirement is satisfied, and parameters describing both the rheology and aspects of the ice sheet must be estimated from comparisons between the model predictions and the observations. Thus, a further requirement is a set of observations of sea-level changes relative to the land, changes that are free from tectonic influences of other than glacio-isostatic origin.

Our knowledge of the geographical distribution of the ancient ice sheets is limited, and generally it is not possible to obtain realistic inferences for the mantle viscosity without considering the uncertainties inherent in the loading models. The geographical limits of the principal ice sheets over the northern continents are generally well known for the Last Glacial Maximum and for the retreat of the ice margin, largely because of the geomorphological features left behind by the retreating ice. Over the shallow seas the limits of grounded ice sheets and retreat are generally less well known. Also unknown is the height of the ice, except in areas where mountaintops extended out above the ice and retain evidence in the form of trimlines, as in Scotland (e.g., Ballantyne and Harris, 1994). What may be better known are the profiles across ice sheets, derived from considerations of the rheology of ice and the shear-stress conditions at the base of an ice sheet and from observations of existing glaciers (e.g., Paterson, 1971). Models of the ice sheets have been developed through such considerations, although results based on similar observational constraints can vary significantly depending on the assumptions made about the shear stresses that can be supported at the rock–ice interface. [See the models by Boulton et al. (1977, 1985) for the British ice sheet, where the maximum and minimum reconstructions differ in terms of ice volume by a factor greater than 5.] A simple procedure to overcome some of these limitations is to introduce a height scale factor as an unknown in the ice models for the time of maximum glaciation (as discussed later) and to allow the form of the ice-height profiles to follow the predictions based on the glaciological theories and on the observed spatial limits of the ice sheet.

Provided that an adequate observational data base exists, some separation of the various unknown or partially known parameters can be achieved because of the spatial and temporal variability of the sea-level change since the time of the Last Glacial Maximum, about 18,000 years ago. Sea levels far from the former ice sheets, for example, provide a constraint on the total amount of meltwater added into the oceans. Generally, levels were about 120–130 m lower than today (Chappell, 1974; Nakada and Lambeck, 1988), with some of the range in observed values reflecting the uncertainty in the data and some reflecting the spatial variability in sea levels due to the glacio-hydro-isostatic adjustments of the crust. This variability is illustrated in Figure 10.4a for a profile across the eastern Australian margin and shelf for two different Earth models. The model for a 50-km-thick lithosphere, for example, predicts the 18,000-year-old shoreline at a coastal site to occur at a depth of about 115 m, compared with about 135 m at sites 200–300 km offshore. Thus the inference of total ice volume on the basis of the last-glacial-maximum shorelines requires an isostatic correction irrespective of whether the evidence comes from far-field continental-margin sites or from small mid-ocean islands.

The rise in far-field sea level during the melting phase provides a good estimate of the integrated rate of melting of the ice sheets, because the predictions for any one site are relatively insensitive to the choice of Earth-model parameters (Nakada and Lambeck, 1989). The predictions do vary, however, from site to site, particularly across continental margins, because of the hydro-isostatic sea-level term, as illustrated in Figure 10.4b: the sea-level rise for the offshore sites occurred later than that for the coastal site by as much as 1,000 years during the past 10,000 years. Here again, the inference of eustatic sea level from such observations requires the application of isostatic corrections. This model prediction is also important for understanding the development of coral reefs during the period of rapid sea-level rise: offshore reef-corals are consistently younger than nearshore reef-corals at the same depth below present sea level.

Observations of sea-level change from the far field before about 6,000 B.P. constrain mainly the eustatic-sea-level function. The timing of the highstand at about

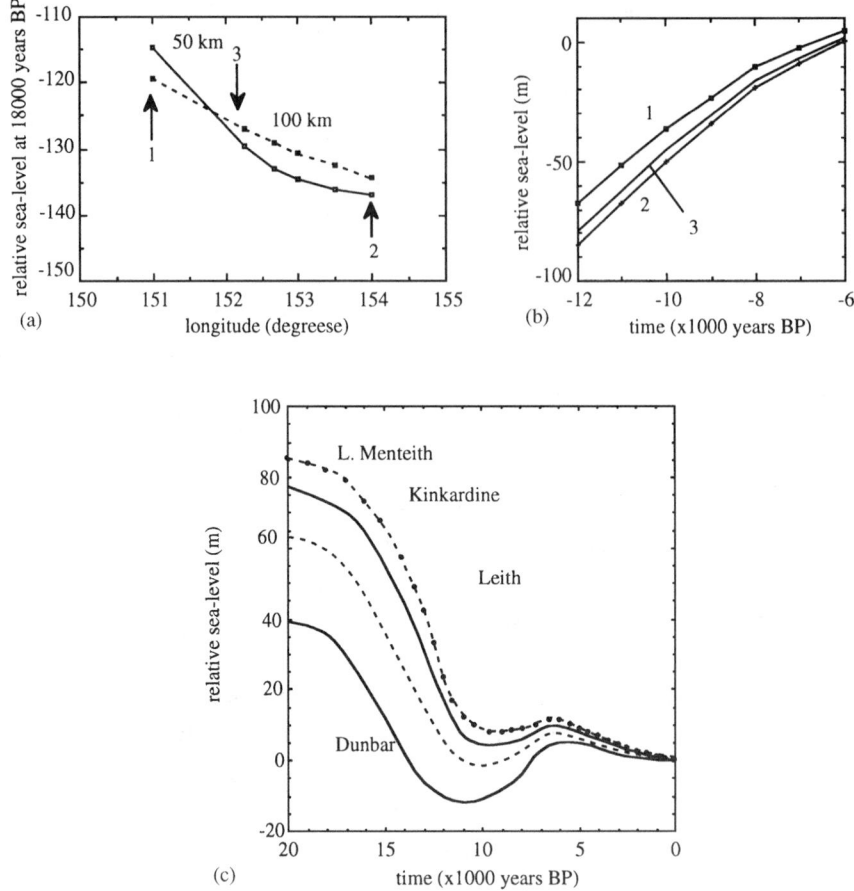

Figure 10.4. (a) Predictions of relative sea level at the time of the Last Glacial Maximum at 18,000 B.P. for sites along a section across the Australian margin (in Queensland at 22° south). The arrow numbered 1 marks the coastline, and that numbered 2 marks the edge of the continental shelf. The predictions are for standard mantle-viscosity models defined by equation (45), with a lithospheric thickness of 50 km for the continuous curve and 100 km for the broken curve. (b) Predicted sea levels at three sites from the section shown in part a for the period 12,000–6,000 B.P. (c) Predicted sea levels at sites along the Firth of Forth, from sites near the entrance to the Firth (Dunbar) to the upper Forth Valley (Lake Menteith). All predictions are for the standard model of equation (45).

6,000 B.P. (Figure 10.2d) constrains the end of the major phase of deglaciation, whereas its spatial variability determines largely the response function for the solid Earth. The amplitude of this highstand is a function of three terms: any change in ocean volume (the eustatic sea level) over the past 6,000 years, a small and spatially quite uniform contribution from the response to the distant glacial unloading, and a more important contribution from the water-load term. This latter is spatially quite variable when coastline geometries are complex, as around the Gulf of Carpentaria in Australia, or when sites lie effectively inland, as for Spencer Gulf in South Australia (Figure 10.3c). The spatial differences among these highstands are therefore functions primarily of the Earth's response to the water loading, and observations of such variability provide a measure of the mantle's viscosity, with the actual amplitudes providing measures of viscosity and eustasy (the ice-load terms being quite predictable away from the ice sheets). Separation of some of the parameters again becomes possible (Nakada and Lambeck, 1989).

The form of the late-glacial fall in sea level observed at sites near the former centres of glaciation establishes constraints on mantle viscosity. Observational evidence can be found only once the areas are ice-free, and the available records are, unfortunately, relatively short, going back at most 9,000 years for the Gulf of Bothnia in Scandinavia and about 6,000–7,000 years for the Hudson Bay area in Canada. The eustatic contribution to these curves is based on the inferences made from the far-field records, and the gravitational attraction of the sea by the ice will be small or zero, because the ice was mostly gone by the time evidence for sea-level change could be recorded at such sites. The other parameter that helps to determine the sea-level curve for sites of former glaciation is the ice height at the time of maximum glaciation. But if the shape of the height profiles for the ice is assumed known, and the single unknown is the ice height, then this parameter can be separated from the mantle viscosity: the latter determines largely the shape of the sea-level curve, whereas both the viscosity and the ice height determine the magnitude of the total change.

Observations from near the margins of the ice sheets constrain all aspects of the rebound model. The period of near-constant sea level, at about 10,000 B.P. in Figure 10.2b, for example, determines the time for which the eustatic and isostatic contributions are of the same magnitude but of opposite sign. If the former is known from the far-field analyses, then constraints are placed on the rheology. These constraints become particularly valuable if the spatial variability of the height–time relation for these stillstands can be inferred from the observational evidence, as is the case for some regions of Scotland and Norway. In the former, for example, a rock platform (the Main Rock Platform) generally a few metres above present sea level (Gray, 1978; Dawson, 1988) is believed to have been formed or reshaped at the time of the 10,000-B.P. stillstand (Stone et al., 1996), and observations of the gradient of this feature provide important constraints on the mantle rheology if the ice limits are known. The small highstand observed in these curves for about 6,000 B.P. (Figure 10.2b) provides a measure of the time at which the melting of the ice sheets ceased. If this is everywhere the same, then the melting ceased abruptly; but if the timing was spatially variable, having been earlier towards the centre of the rebound, as appears to have been the case in Scotland, that indicates that some melting continued after the peak was established. The amplitude of this mid-Holocene highstand and the subsequent fall in sea level provide a measure of the mantle viscosity.

Figure 10.4c illustrates the predicted sea levels for late-glacial and post-glacial times for sites along the Firth of Forth in Scotland. These illustrate well the spatial variability of some of the key features defining the sea-level response. For example, the minimum sea level occurs deeper and earlier the farther the site (e.g., Dunbar) is from the centre of the ice load (near Lake Menteith). Evidence for this feature is mainly hidden by marine sediments deposited during the subsequent sea-level rise, and few reliable observations exist. Better defined is the shoreline formed during the highstand at about 6,000 B.P., and it is smaller and later at the eastern sites (Dunbar, Leith) than at the western sites (Lake Menteith, Kincardine). The relative sea-level maximum at about 6,000 B.P. led to the formation of the Main Postglacial shoreline found widely in eastern Scotland.

The predicted sea-level curves for these former ice-margin sites depend greatly on the assumed ice models for the region, and inferences about mantle viscosity could be partly ice-model-dependent. Hence, in the first step of any analysis, observations are used only for regions where reasonable constraints exist for the ice sheet. Then, if the mantle parameters have been estimated, it becomes possible to make estimates that will improve the ice models for less well constrained areas, and through an iterative procedure it becomes possible to separate the various parameters (Lambeck, 1993a,b).

Further separation of mantle parameters can be achieved because of the broad wavelength spectrum of the surface load. The largest ice loads have dimensions of up to 3,000–4,000 km, and these can be expected to stress the deeper mantle, whereas the small ice loads, such as that formerly over the British Isles, stress only the shallower mantle. However, the water load, of the dimensions of the oceans, will stress the entire mantle, and for the British Isles, for example, it contributes some 10–20% of the total observed sea-level change (Lambeck et al., 1996), with the actual amount exhibiting some dependence on lower-mantle viscosity. This is illustrated in Figure 10.5 in the form of differences in predicted sea levels for two Earth models that differ only in their lower-mantle viscosity, 10^{22} Pa · s as compared with 10^{21} Pa · s. For the time of the glacial maximum, the differences attain several tens of metres and are primarily the consequence of the Fennoscandian ice load. For 6,000 B.P. the differences are of the order ± 4 m for the region, larger than the observational accuracy of much of the data at this time. Thus the sea-level data from most sites will generally contain some information regarding the lower-mantle viscosity, as well as upper-mantle structure.

Further constraints on the lower-mantle viscosity follow from observations of the global response to glacial unloading. It can be shown (as discussed later) that the glacially induced time dependence of the $\bar{C}_{20}$ potential coefficient from equation (42) is relatively insensitive to the lithospheric thickness and upper-mantle viscosity and that the main dependence is on both the lower-mantle viscosity and the very recent melting history of the ice sheets. The palaeo-sea-level data are not very sensitive to this latter part, but modern tide-gauge records are, so that a judicious combination of observations can lead to a further separation of parameters.

These examples of the spatial and temporal variability of the Earth's response to glacial unloading illustrate how it is possible to infer aspects of the mantle's

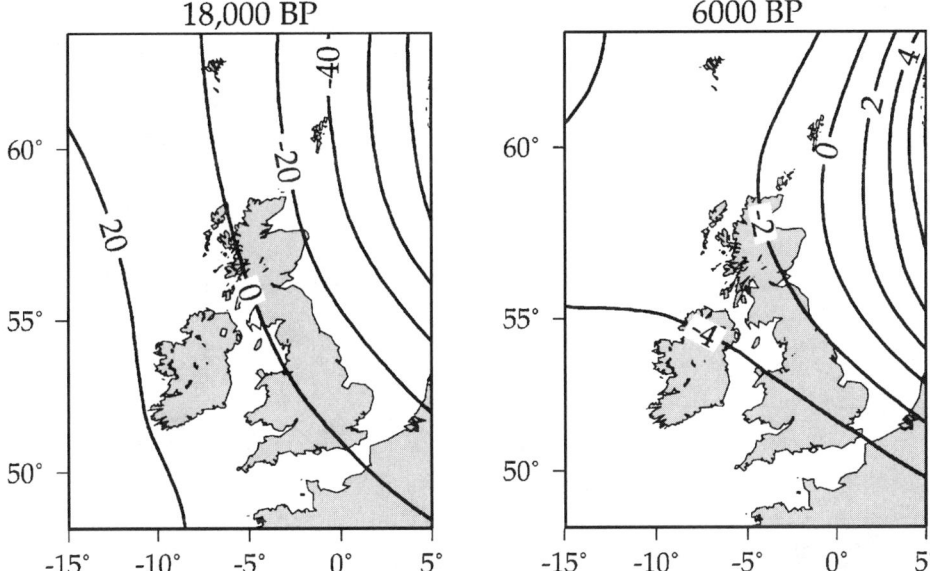

Figure 10.5. Spatial variability in the difference in sea level predicted for two three-layer Earth models, each with a lithospheric thickness of 65 km and an upper-mantle viscosity of 4×10^{20} Pa · s. Their lower-mantle viscosities η_{lm} are 10^{22} and 10^{21} Pa · s, respectively. The predicted differences (high-η_{lm} model minus low-η_{lm} model) are contoured in metres for two epochs, at 18,000 and 6,000 B.P. The observational accuracy for 6,000 B.P. typically is better than 1 m.

viscosity structure. The analysis, as sketched earlier, assumes that lateral variation in viscosity is not important – an assumption that, in light of lateral variation in other properties of the Earth, such as seismic shear-wave velocities, attenuation, and surface heat flow, may not be tenable.

10.3.3. An Example of Analysis of Sea-Level Change

Evidence for sea-level changes that are attributable to the glacial rebound of the crust and to eustatic changes is most prevalent for coastal areas near the centres of former glaciation over northern Europe and North America. One area where a detailed analysis of such evidence has been made encompasses the British Isles and the North Sea. Figure 10.6 illustrates some of the evidence collated for this region by Lambeck (1993b). Near central Scotland, the interplay of the local glacio-isostatic rebound and eustatic change is clearly evident, and the crustal subsidence due to mantle flow primarily towards Scandinavia produces the characteristic curves for southern England. Another important data base is provided by the spatial variation in the position of shorelines, such as the Main Rock Platforms or the Main Postglacial Shoreline introduced earlier. In eastern Scotland, for example, the shoreline features that have been identified all slope upwards from east to west (Sissons, 1983) and point to the centre of rebound being over central or western Scotland from at

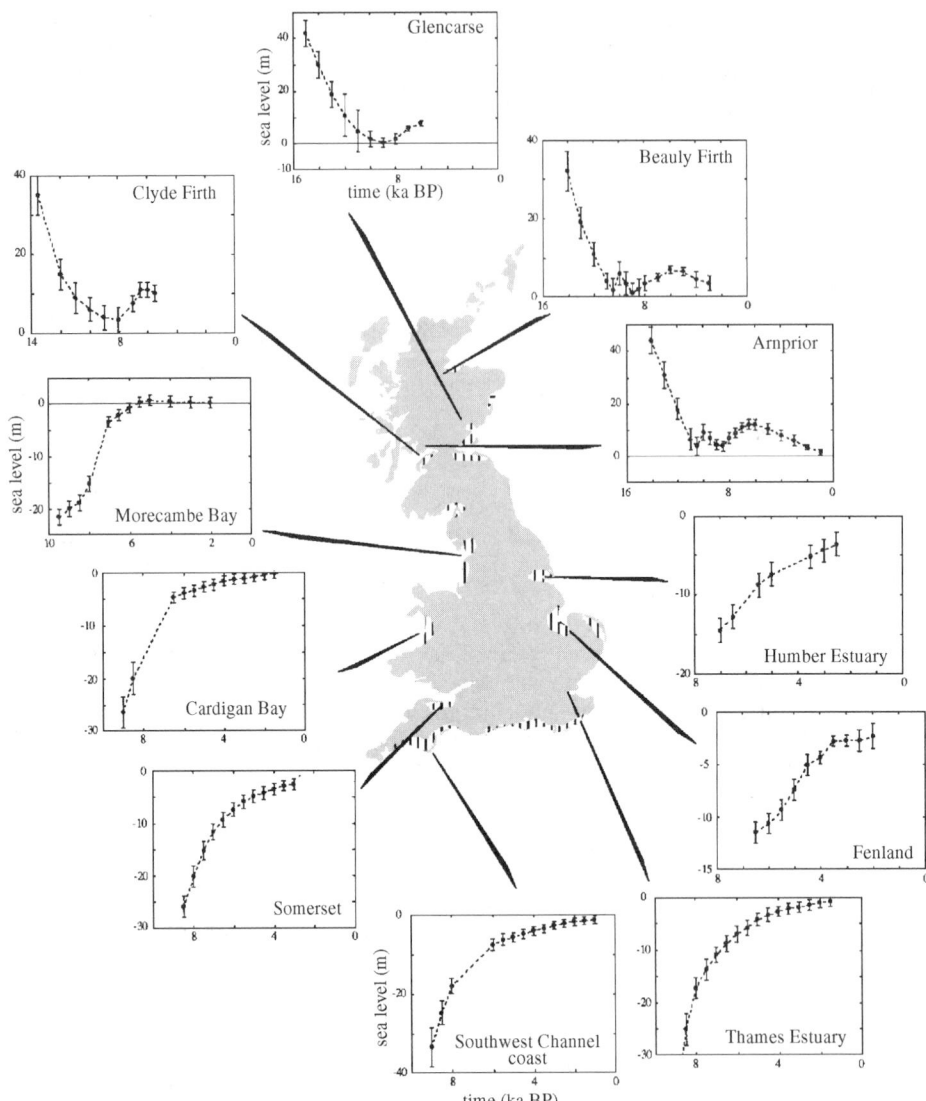

Figure 10.6. Observed relative sea-level changes for some of the sites in England, Scotland, and Wales used in the analysis for Earth-model parameters. Note the use of different scales for height and time.

least about 14,000 B.P. onwards. These observations can help to resolve one important question about the former ice sheet: Did it, during the Last Glacial Maximum, extend over the North Sea, as postulated by Denton and Hughes (1981), such that the British and Scandinavian ice sheets formed a single entity, with a thick ice ridge over the North Sea, having a maximum ice thickness of more than 1,500 m? Rebound models based on these maximum reconstructions predict Lateglacial shoreline gradients over eastern Scotland that are much smaller than those actually observed, or even of opposite sign for the oldest features, and therefore such ice models are incompatible with the observational evidence, unless this ice bridge vanished much earlier than 18,000 B.P. These conclusions remain valid for a large range of plausible Earth models, and the maximum-ice-sheet reconstructions in which the ice extended over much of the continental shelves around the British Isles at the time of the Last Glacial Maximum are inconsistent with the qualitative patterns of observed sea-level change. Geomorphological evidence, from which it should be possible to establish the ice limits over these shelves, is often inconclusive because of inadequate age constraints on glacial or periglacial features, and it is not always obvious whether the features date from the last peak in glaciation or from an earlier one. In developing new models for the ice sheet, the gross constraints from the preliminary modelling have been imposed, and the geological and geomorphological indicators have then been used to place the ice limit at a particular location. For example, the Earth-model-independent conclusion of no major ice over the North Sea is imposed on the ice model, and the limits of the ice sheet at the time of maximum glaciation are then placed at the Wee Bankie Moraines, offshore from eastern Scotland, for which independent, but not wholly conclusive, evidence suggests that they were formed during the Last Glacial Maximum. Further details of this ice model have been given by Lambeck (1993b, 1995c).

The other requirement to solve the sea-level equation is the construction of models for the major ice sheets. The model for Fennoscandia is based on the ice-retreat isochrons proposed by Andersen (1981), but without the extension of ice over the North Sea, and with a reduction in ice over the Russian platform. This model has been constrained by recent observations of Holocene shorelines, or an absence thereof, by rebound modelling for the area of the Barents and Kara seas (Lambeck, 1996a), and by a preliminary inversion of Scandinavian sea-level data for the ice height. The maximum ice thickness, centred over the Gulf of Bothnia, was about 2,800 m. The ice model for North America and Greenland is derived from the model of Peltier and Andrews (1976). [This model is preferred to some of the subsequent models derived by Peltier and colleagues (e.g., Tushingham and Peltier, 1991) because it is less dependent of assumptions about the Earth's rheology.] The Antarctic ice model has been discussed previously (Nakada and Lambeck, 1988), and its importance here is mainly to ensure that the total ice volume and the total rate of addition of meltwater into the oceans are consistent with the adopted eustatic-sea-level curve.

The equation relating the model unknowns to the observations is

$$\Delta\zeta_0(\theta, \lambda, t) + \varepsilon_0(\theta, \lambda, t) = \Delta\zeta^e(t) + \delta\zeta^e(t) + \beta\Delta\zeta^{\mathrm{Br}}(\theta, \lambda, t) + \Delta\zeta^{\mathrm{ff}}(\theta, \lambda, t)$$

$$= \Delta\zeta_{\mathrm{predicted}} \tag{43}$$

where

$\Delta\zeta_0$ = observed sea level, reduced to mean sea level, at location (θ, λ)
      and time t, with a standard deviation of σ
ε_0 = assumed observational error

$\Delta\zeta^e$ = eustatic-sea-level function for the combined ice sheets

$\delta\zeta^e$ = correction term to $\Delta\zeta^e$

β = scale parameter for the British ice sheet

$\Delta\zeta^{Br}$ = predicted deformational contributions to sea-level change from the ice load and water load of the British ice sheet for specified Earth-model parameters

$\Delta\zeta^{ff}$ = predicted deformational contributions from the more distant (far-field) ice sheets of Fennoscandia, Laurentia, Barents, and Antarctica

Equation (43) is solved for the rheological parameters (the lithospheric thickness H_1 and the effective viscosities η_i of the i mantle layers), the ice-height scale parameter, and the corrective function $\delta\zeta^e$ to the eustatic-sea-level curve. This last term is also subject to the condition that the corrected eustatic function must be consistent with estimates derived from other regions. An iterative procedure has been adopted, as discussed by Lambeck (1993b) and Lambeck et al. (1996), in which for each Earth model k, defined within a large parameter space K, the corresponding β and $\delta\zeta^e$ are estimated according to the requirement that the variance function

$$\Sigma_k = \frac{1}{M} \sum_{m=1}^{M} \left[\frac{\Delta\zeta_o^m - \Delta\zeta_{\text{predicted}}^m}{\sigma^m} \right]^2 \tag{44}$$

be a local minimum. M, the total number of observations, each of standard deviation σ^m, is 424 in this case. A systematic search is then conducted through the Earth-model space $k = 1, \ldots, K$ to find the overall minimum variance. The expected value for Σ_k is unity. Figure 10.7 illustrates some typical results for a three-layer model defined by a lithosphere of thickness H_1 and a two-layered mantle, with an upper-mantle (above 670 km) viscosity η_{um} and a lower-mantle viscosity η_{lm}. The results are illustrated here in η_{um}–η_{lm} space for $H_1 = 65$ km, but the actual search is conducted through a range of H_1 values. These results correspond to a

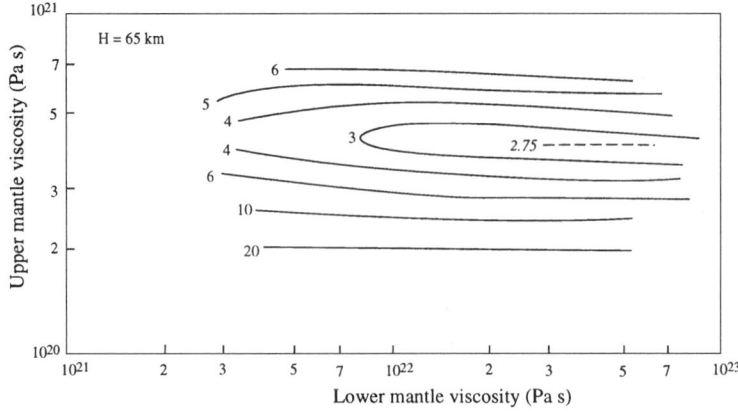

Figure 10.7. Minimum-variance function Σ_k for the three-layer models with a lithospheric thickness of 65 km, with the eustatic correction term $\delta\zeta^e$ in equation (43) set to zero.

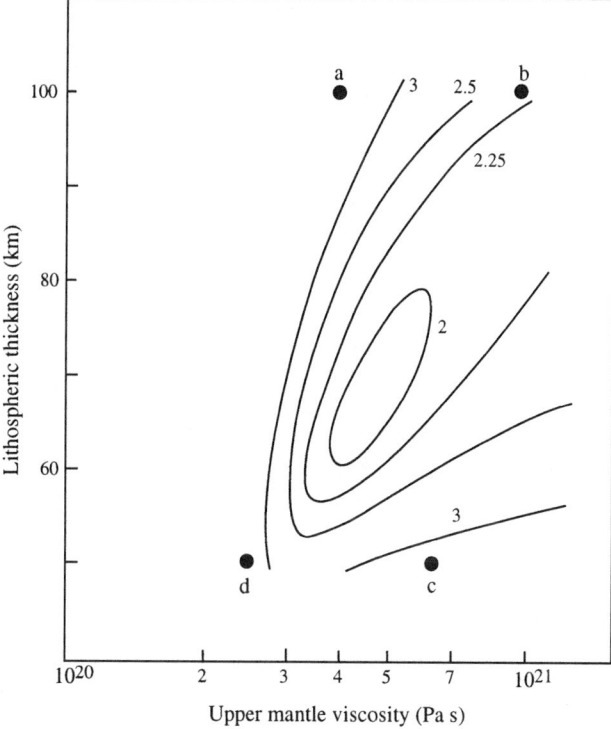

Figure 10.8. Minimum-variance function Σ_k for the three-layer models with $\eta_{lm} = 10^{22}$ Pa · s. The solutions include the eustatic correction term $\delta\zeta^e$ in equation (43). The region outside the space defined by the points marked a–d yields solutions for the eustatic sea-level correction that are inconsistent with analyses of the sea-level evidence from far-field locations.

solution in which the correction $\delta\zeta^e$ is not included. The minimum variance occurs when $\eta_{lm} \gtrsim 10^{22}$ Pa · s, as is also the case for other values for H_l in the range $30 < H_l < 120$ km. Although the resolving power for η_{lm} is low, the results are consistent with those from similar analyses for other parts of the world that point to an average lower-mantle viscosity significantly higher than the average value for the mantle above a depth of 670 km (Lambeck and Nakada, 1990; Lambeck, 1993c). Figure 10.8 shows further results for models with $\eta_{lm} = 10^{22}$ Pa · s for the full solution of equation (43) (including the corrective term $\delta\zeta^e$). Solutions beyond the domain of parameter space, enclosed by the points marked a–d, require eustatic corrections $\delta\zeta^e$ that are inconsistent with evidence from other regions, and the optimum three-layer solution is defined by

$$H_l \simeq 65\text{–}70 \text{ km}$$

$$\eta_{um} \simeq 4\text{–}5 \times 10^{20} \text{ Pa} \cdot \text{s} \tag{45}$$

$$\eta_{lm} \gtrsim 10^{22} \text{ Pa} \cdot \text{s}$$

with $\Sigma_k = 1.90$. Earth models in which the upper-mantle viscosity is stratified give a somewhat smaller overall minimum variance, and the solution that best satisfies the observational data is defined by (Lambeck et al., 1996)

$$55 < H_1 < 60 \text{ km}$$
$$(2 < \eta_2 < 4) \times 10^{20} \text{ Pa} \cdot \text{s for } (H_1 < H \leq 200) \text{ km}$$
$$(4 < \eta_3 < 6) \times 10^{20} \text{ Pa} \cdot \text{s for } (200 < H \leq 400) \text{ km} \qquad (46)$$
$$\eta_4 \sim 2 \times 10^{21} \text{ Pa} \cdot \text{s for } (400 < H \leq 670) \text{ km}$$
$$\eta_{lm} \gtrsim 10^{22} \text{ Pa} \cdot \text{s for } (670 < H < H_{CMB}) \text{ km}$$

with $\Sigma_k = 1.35$. The solution (46) corresponds to a five-layer model in which viscosity contrasts are permitted at depths of 200, 400, and 670 km. The current inversion is constrained by the assumptions that $\eta_2 \leq \eta_3 \leq \eta_4 \leq \eta_{lm}$ and that no other minima develop with a smaller overall variance outside of the limits $3 \times 10^{19} \leq (\eta_2, \eta_3, \eta_4) \leq 5 \times 10^{21} \text{ Pa} \cdot \text{s}, 10^{21} \leq \eta_{lm} \leq 10^{23} \text{ Pa} \cdot \text{s}, \text{and } 30 \leq H_1 \leq 120 \text{ km}$. For nearly all predictive purposes regarding sea-level change or shoreline evolution, there is little difference between the two solutions (45) and (46), and both provide very satisfactory agreement between prediction and observation at most of the sites. These mantle parameters are effective parameters in that they describe the response of the planet to surface loading on timescales of the order of 10^4 years, with stress differences of the order of 10 MPa, based on the assumption of a Maxwell rheology. Models with a viscosity inversion, other than at the base of the lithosphere, have not been examined. Also, the models used here are based on the assumption that the phase boundaries at depths of 400 and 670 km (e.g., Jackson and Rigden, Chapter 9, this volume) migrate with the displacement fields at these depths. But models in which the phase boundary remains an isobaric surface (Johnston et al., 1997) produce essentially the same results, except that the value for the β parameter corresponding to the least-variance solution is reduced by about 10%. Although there is a suggestion in solution (46) of a viscosity gradient from the base of the effective lithosphere to a depth of 400 km, the evidence is not compelling. In particular, a low-viscosity channel immediately beneath the base of the lithosphere is not required by the observations. The solutions do appear to favour models in which the viscosity of the transition zone is greater than that of the uppermost mantle by a factor of perhaps 3 or 4, but it remains to be determined whether or not similar analysis for longer-wavelength loads will require the same trend. The further increase in viscosity across the 670-km boundary does appear to be a robust feature of the analysis. However, the resolving power for that region of the mantle is poor, and no attempt has been made to determine whether or not there is any structure in the viscosity below this boundary. But, as suggested by findings such as those illustrated in Figure 10.5, the larger ice sheets do stress the lower mantle and may provide further constraints on the mantle structure.

An important aspect of these findings is that they demonstrate the trade-offs among the upper-mantle parameters. If in the three-layer models, for example, the lithospheric thickness is fixed at a relatively large value of, say, 100 km, the estimate for the upper-mantle viscosity is increased to about 10^{21} Pa · s, but if a thinner lithosphere is imposed, the estimate for the viscosity is reduced to about 2×10^{20} Pa · s. Because an observational record from the formerly glaciated regions exists only for the post-glacial stage, models with large H_1 (small initial deflection) and high η_{um} (slow relaxation) produce rebound curves for the post-glacial stage similar to (but not identical with) those for models with small H_1 (large initial deflection) and low η_{um} (rapid relaxation) (Lambeck et al., 1996). But either estimate represents only a local minimum-variance solution, and the results illustrate well the need to conduct the search throughout the entire parameter space defined by H_1 and the η_i. The trade-offs among the parameters become more pronounced when models with greater depth dependence of viscosity in the upper mantle are considered – to the extent that if a very low value for H_1 is imposed on the model, the solution will favour a low-viscosity channel immediately below the lithosphere. [This trade-off is also seen to a minor degree between the different H_1 and viscosity values found for the three- and five-layer model results, (45) and (46).] Using an inverse-theory approach, Mitrovica (1996) also demonstrated the trade-off between upper- and lower-mantle viscosities.

These trade-offs may explain, in part, some of the differences in the published estimates for the mantle viscosity. In the work by Fjeldskaar (1994), for example, the lithosphere is assumed to be relatively thin (<50 km), and this leads to models with low viscosity in the upper most mantle. Applying equation (43) to a four-layer model with a 30-km-thick lithosphere (layer 1), a low-viscosity channel down to a depth of 150 km (layer 2), and the remaining upper mantle with a uniform viscosity (layer 3), we find viscosities of about 5×10^{19} Pa · s for the second layer and 2×10^{21} Pa · s for layer 3. But the variance factor is about 50% greater than that for the optimum three-layer solution discussed earlier. If, on the other hand, a thick lithosphere is imposed on the model, of say 100–120 km, as in the recent models by Peltier and colleagues (Tushingham and Peltier, 1991; Mitrovica and Peltier, 1993), then the solution leads to an overestimation of the upper-mantle viscosity and to models in which the contrast between upper- and lower-mantle values is less than that found here. But again, such solutions represent only local minima within the larger Earth-model parameter domain created when the lithospheric thickness is properly treated as an unknown (Mitrovica, 1996).

10.4. Changes in the Earth's Gravitational Potential

Information on the mantle's viscosity is also contained in the time dependence of some of the very long wavelength components of the Earth's gravitational potential caused by the changing mass distribution on and within the Earth. This is conveniently expressed through the time dependence of the Stokes coefficients $\bar{C}_{nm}$, equation (42). The most direct measure of such dependence for some of the zonal coefficients is through analysis of the perturbations in the motions of near-Earth satellites resulting from the changing gravitational potential, as first noted by Yoder et al. (1983) and Rubincam (1984).

The motion of a near-Earth satellite is determined primarily by the gravitational force, whose potential, defined in an Earth-fixed reference frame with one axis parallel to the rotation axis, is given by equation (40). For a body in which the density distribution is a function of radial distance only, all the Stokes coefficients vanish, and in the absence of other forces, the satellite motion is Keplerian. Viewed from space, the orbit's orientation, shape, and size remain fixed, and the only time-dependent behaviour is the periodic motion of the satellite along its trajectory and the rotation of the Earth beneath the orbit. The gravitational attraction of any departures from this spherical symmetry, however, will perturb this idealized motion. Both the orientation and shape of the orbit undergo cyclic and secular changes, and it can be shown that each Stokes coefficient induces a characteristic perturbation into the satellite motion (Kaula, 1966; Lambeck, 1988). The precise tracking of the satellite then provides a measure of the spectrum of these perturbations, from which the Stokes coefficients and the gravitational potential can be inferred. The majority of the perturbations are small and periodic, but secular changes in the orbit's orientation occur in response to the terms of even degree and zero order, of which the $n = 2$ term is most important. These are the zonal ($m = 0$) Stokes coefficients of even degree. For these terms, the orbit, *inter alia*, precesses in space about the Earth's rotation axis such that the intersection of the plane of the orbit with the equatorial plane moves at a linear rate along the equator. Introducing the usual Keplerian elements (Figure 10.9), the principal observational consequence of these even $\bar{C}_{n0}$ terms is a secular variation in the angle Ω, the longitude of the ascending node, given by (e.g., Lambeck, 1988)

$$\dot{\Omega} = \sum_{\substack{n=2 \\ n \text{ even}}}^{\infty} \bar{N}_n(a, e, I)\bar{C}_{n0}, \tag{47}$$

where the $\bar{N}_n$ are functions of a (the semi-major axis), e (the eccentricity), and I (the inclination of the orbit). Separation of these zonal Stokes coefficients can be achieved through observations of the secular rates for satellites with orbits of different shapes and inclinations.

If the Stokes coefficients are time-dependent because of the redistribution of mass on and within the Earth, then the longitude of the ascending node will exhibit an acceleration

$$\ddot{\Omega} = \sum_{\substack{n=2 \\ n \text{ even}}}^{\infty} \bar{N}_n(a, e, I)\dot{\bar{C}}_{n0} \tag{48}$$

(The orbital elements a, e, and I relate to the energy and angular momentum of the orbital motion and do not undergo secular change in a conservative system.) Observations of the acceleration of the node for a number of satellites with different orbital parameters can therefore permit, in principle, estimation of the time

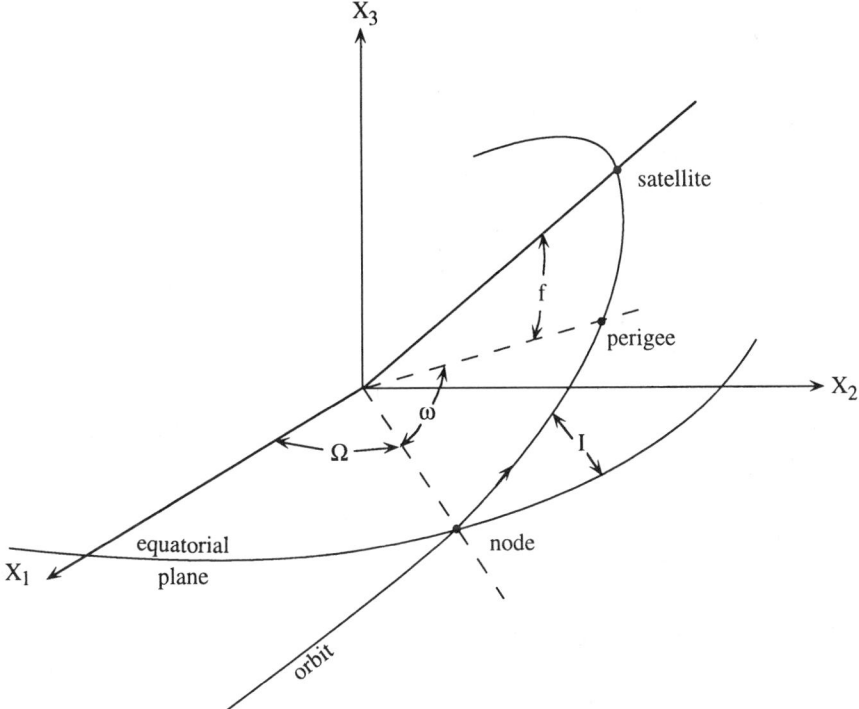

Figure 10.9. Orbital geometry for a near-Earth satellite. The $X_1 X_2$ axes lie in the Earth's equatorial plane, and the X_3 axis corresponds to the Earth's rotation axis. The orbital plane intersects the equatorial plane along the line of nodes. The orientation of the orbital plane in space is defined by the inclination I of this plane to the equatorial plane and by the angle between the line of nodes and the X_1 axis, the longitude of the ascending node Ω. The orientation of the orbit within this plane is defined by the argument of perigee, ω, the angle between the node and perigee. The shape of the orbit is defined by the eccentricity, e, and the semi-major axis, a. The position of the satellite in its orbit is specified by the angle f. These six elements a, e, I, ω, Ω, and f define the orbit and satellite position in the inertial frame X_i. For Keplerian motion, all elements, with the exception of f, are constant.

dependence of the even-degree zonal coefficients in the expansion of the Earth's gravitational potential.

Several processes contribute to the time dependence of the gravitational potential. Foremost are the tidal deformations of the Earth. Though primarily diurnal and subdiurnal, long-period zonal tides also occur, with periods up to about 19 years, so that the $\bar{C}_{20}$ coefficient is time-dependent, and the longitude of the node can be expected to exhibit long-period oscillations from this source. Seasonal redistributions of mass within the hydrosphere–atmosphere also contribute to periodic changes in the zonal potential coefficients (Gegout and Cazenave, 1993). But the contribution of interest here is the change [equation (42)] associated with the deglaciation and concomitant redistribution of mass within and on the Earth.

To understand the glacially induced rate of change of the second-degree zonal coefficient ($\dot{\bar{C}}_{20}$), the effects of the ice sheets and the effects of the associated sea-level changes are examined separately. Because of the distribution of the continents, and in particular because Antarctica occupies the South Pole, a uniform lowering of sea level removes more mass from the equatorial regions than from polar regions and, for a rigid-Earth model, leads to an increase in $\bar{C}_{20}$ ($\bar{C}_{20}$ is a negative quantity, as the transport away from the equator decreases the planet's flattening). Also, the formation of polar ice caps on this rigid-earth model moves mass to high latitudes and further increase $\bar{C}_{20}$. For a yielding-earth model, after the formation of ice caps, mantle material flows away from beneath the ice-loaded polar regions to lower latitudes, and the $\bar{C}_{20}$ coefficient now decreases and acts in opposition to the initial contribution from the shift in the surface load. Mantle flow resulting from the unloading of the ocean lithosphere by the decreased water load further decreases $\bar{C}_{20}$. When the ice is removed, $\bar{C}_{20}$ initially decreases because of the surface transport of mass from high latitudes to middle and low latitudes and then increases again in response to the mantle flow. The magnitude of the response to the formation of the ice sheets is about 10 times the response to the change in sea level, because the ice distribution is much more strongly concentrated near the poles, as compared with the almost uniform distribution of oceans (Mitrovica and Peltier, 1993). Therefore, calculations that neglect the change in sea level (e.g., Rubincam, 1984) are not committing errors that will significantly affect their conclusions concerning mantle viscosity.

Although the acceleration of a node of a satellite ($\ddot{\Omega}$) is influenced by all of the even-degree $\dot{\bar{C}}_{n0}$ terms in equation (48), an effective value for $\dot{\bar{C}}_{20}$ can be obtained from $\ddot{\Omega}$ by dividing it by $\bar{N}_2$, in effect neglecting all terms of degree greater than 2 in equation (48), assuming them to be much smaller than $\bar{N}_2 \dot{\bar{C}}_{20}$. However, because the coefficients $\bar{N}_n$ can be computed with great precision, it is more logical to predict all of the $\dot{\bar{C}}_{n0}$ terms for a particular post-glacial-rebound model, to compute the total post-glacial-rebound effect on the satellite orbits, and to compare this directly with the observed acceleration of the node. The orbital parameters, the observed values for $\dot{\Omega}$ and $\ddot{\Omega}$, and the first four values for $\bar{N}_n$ for the LAGEOS and Starlette satellites are shown in Table 10.1. Because the second-degree term dominates in the expression for $\dot{\Omega}$, and $\bar{C}_{20}$ is negative, the sign of the angular velocity of the node is opposite that for $\bar{N}_2$, whose sign is governed by the inclination of the orbit.

As for observations of the present-day rate of change in sea level, the satellite observations of $\ddot{\Omega}$ correspond to the present epoch, long after the main deglaciation event has been completed, and a potentially important contribution to this acceleration arises from any present-day changes in ice volume (e.g., Peltier, 1988). As discussed earlier, the geological evidence for sea-level change points to there having been a continuing but small reduction in ice volume after the cessation of the main phase of deglaciation, although whether or not that has continued up to the present day has not been determined. Direct evidence for deglaciation of the present-day remnant ice sheets of Greenland and Antarctica is equivocal, but other evidence points to a reduction in mountain glaciers over this century at a rate that is sufficient to raise the sea level by about 0.5 ± 0.3 mm/yr (Meier, 1984), and this

Table 10.1. *Orbital parameters for LAGEOS I and Starlette*

Property	LAGEOS	Starlette
Semi-major axis (km)	12,270	7,340
Eccentricity	0.004	0.021
Inclination (degrees)	109.9	49.8
$\dot{\Omega}$ (rad $\cdot$ s^{-1})	6.929×10^{-8}	-7.965×10^{-7}
$\ddot{\Omega}$ (rad $\cdot$ s^{-2})	$-5.3 \pm 0.6 \times 10^{-23}$	$6.8 \pm 0.7 \times 10^{-22}$
$\bar{N}_2(\text{s}^{-1})$	-1.425×10^{-4}	1.639×10^{-3}
$\bar{N}_4(\text{s}^{-1})$	-7.083×10^{-5}	8.680×10^{-5}
$\bar{N}_6(\text{s}^{-1})$	-1.823×10^{-5}	-1.459×10^{-3}
$\bar{N}_8(\text{s}^{-1})$	-1.547×10^{-6}	-4.482×10^{-4}

can be expected to make a contribution to $\dot{C}_{20}$ and $\ddot{\Omega}$. [See Lambeck (1980) for the analogous problem of the contribution of this change in mountain glaciation to the change in length-of-day.] Analysis of tide-gauge records suggests that globally the sea level may have been rising at a rate of about 1.0–1.5 mm/yr for the past century and that the rate may have been as high as 2.0–2.5 mm/yr for the past few decades (e.g., Gornitz, Lebedeff, and Hansen, 1982; Barnett, 1984; Emery and Aubrey, 1991; Nakiboglu and Lambeck, 1991). Thus, in an analysis of the observed $\ddot{\Omega}$ for mantle viscosity, the potential contribution from present-day glacial melting must be considered as well.

The procedure employed for estimating Earth-model parameters that are consistent with the observations of $\ddot{\Omega}$ is similar to that used for analysis of data on past sea-level changes. The ice model, in the first instance, is assumed known and includes the ice-growth stage of the glacial cycle. The Earth models k (see Section 10.3.3) are defined by a set of parameters (in the space K) describing the viscous properties for the mantle, and for each Earth model the accelerations $\ddot{\Omega}$ are predicted for the two satellites. The model-parameter space that is consistent with the observed values for $\ddot{\Omega}$ is then identified. To allow for contributions from any present-day melting of the continent-based ice sheets, the ice model includes recent melting in which the source of the meltwater can be specified to be Greenland, Antarctica, mountain glaciers, or any combination of these potential source regions. Satisfactory solutions are those that give the same values for the earth- and ice-model parameters as estimated from each of the two satellite orbits.

Tests have shown that the assumption of incompressibility of the mantle is not important for the predicted accelerations, and this assumption is made here for computational convenience. The ice model corresponds to the global model used in the preceding section, with the incorporation of a small amount of recent or ongoing melting of selected parts of the ice sheets. Initial tests of the ice model without recent melting indicated that the predicted $\ddot{\Omega}$ is not sensitive to the choice of lithospheric thickness, in accordance with earlier analyses by Peltier (1985). This arises because the principal contributions to $\ddot{\Omega}$ come from the low-degree terms in the gravitational potential with spatial wavelengths that are much greater than the thickness of the lithosphere. Thus, a value of 80 km is adopted, consistent with the value found from sea-level analyses for different parts of the world. The Earth parameter space explored is similar to that used for the sea-level analysis in Section 10.3.3. For the same reason that the dependence of $\ddot{\Omega}$ on lithospheric thickness is weak, the dependence on upper-mantle viscosity is also weak (Figure 10.10a). Hence, only three-layer mantle models are considered, with the mid-mantle boundary at 670 km depth.

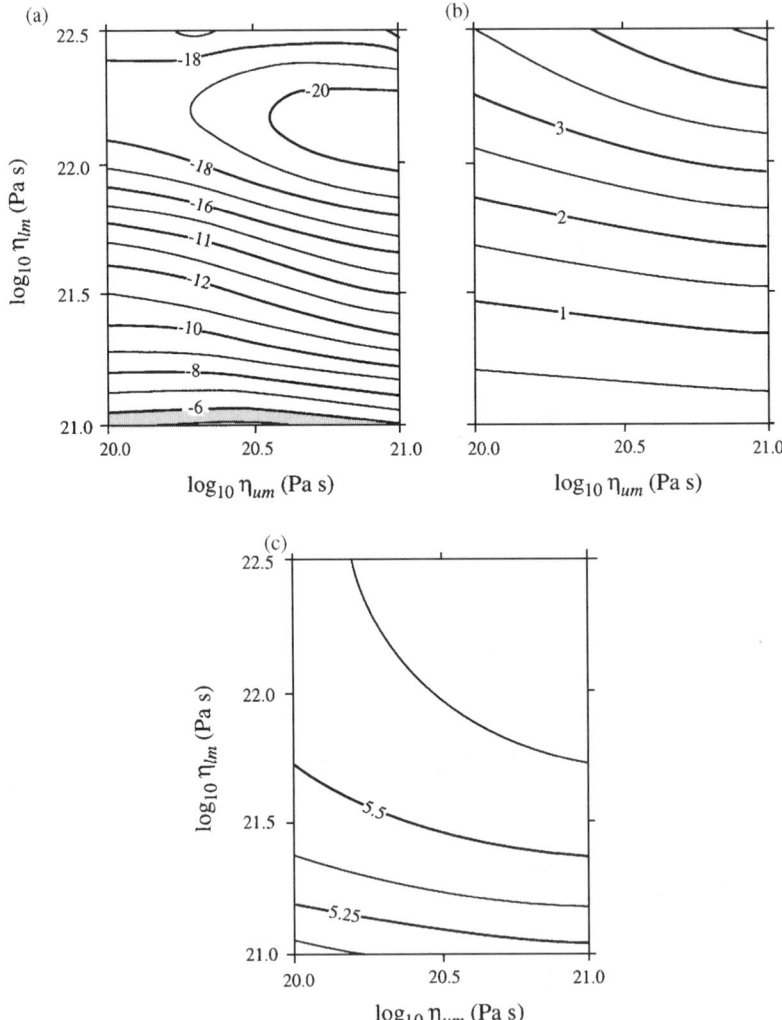

Figure 10.10. Predictions of the acceleration (units of 10^{-23} s^{-2}) of the node $\ddot{\Omega}$ for the LAGEOS orbit as a function of mantle viscosity due to (a) the late Pleistocene and early Holocene glacial cycles, with no deglaciation after 6,000 B.P., (b) uniform melting of continent-based ice for the past 6,000 years in Antarctica and Greenland, in the ratio 2 : 1, such that the eustatic sea-level rise has been 0.5 mm/yr for that period, (c) uniform melting of polar ice at a rate of 0.5 mm/yr for the past 50 years, with contributions from both hemispheres, in the same ratio as in part b.

Figure 10.10a illustrates the predicted $\ddot{\Omega}$ for the LAGEOS satellite as a function of upper-mantle viscosity (the horizontal axis) and lower-mantle viscosity (the vertical axis), and the near-horizontal nature of the contours of constant $\ddot{\Omega}$ indicate the weak dependence on upper-mantle viscosity. Of note is that in the absence of any glacial melting after 6,000 B.P., the observed value for $\ddot{\Omega}$ (Table 10.1) implies a viscosity close to the classical value inferred from Fennoscandian uplift observations, by Haskell (1937) and others, of about 10^{21} Pa · s (the shaded area in Figure 10.10a). This agrees with inferences by, for example, Rubincam (1984) and Yuen et al. (1982), although slightly higher estimates have been obtained by Peltier (1986) and Yuen and Sabadini (1984). The assumption that the 670-km seismic discontinuity is also a viscosity boundary is based partly on the observation that this boundary provides a partial barrier to subducting slabs in the mantle, as shown by seismic tomographic images (e.g., Kennett and van der Hilst, Chapter 8, this volume), and also on the possibility that changes in mineralogy across the 670-km boundary may produce changes in viscosity (e.g., Drury and Fitz Gerald, Chapter 11, this volume). If the viscosity boundary is assumed to be deeper, and the viscosity of the mantle above it is fixed at 10^{21} Pa · s, a larger viscosity contrast is inferred (Yuen and Sabadini, 1984; Peltier and Jiang, 1996). However, unique solutions for the lower-mantle structure are not possible from these data alone, and the subsequent analysis is restricted to the limited class of Earth models already described, with particular emphasis on the effects of recent glacial history on the predicted values for $\ddot{\Omega}$.

Figures 10.10b and 10.10c illustrate the contribution to $\ddot{\Omega}$ for two different models of the late Holocene melting of continent-based ice from both Greenland and Antarctica. In the first model (Figure 10.10b), ice-sheet decay occurs linearly over the past 6,000 years, so as to contribute 0.5 mm/yr to the eustatic sea-level rise, with the Greenland and Antarctic contributions arbitrarily assumed to be in the ratio 1 : 2. In the second model (Figure 10.10c), the melting occurs only for the past 50 years, at a rate sufficient to raise eustatic sea level by 0.5 mm/yr for this interval, again with the Greenland and Antarctic contributions in the ratio 1 : 2. Comparison of the three results illustrates the opposing effects of the essentially instantaneous response to melting (Figure 10.10c) and the subsequent mantle relaxation, with the magnitudes for $\ddot{\Omega}$ being less in model 1 (Figure 10.10b) than in model 2 (Figure 10.10c), despite the present-day rate of melting being the same in both cases.

The magnitude of the change $\ddot{\Omega}$ for the recent melting (Figure 10.10c), when compared with the response to the earlier melting (Figure 10.10a), shows the importance of the former and suggests that observations of $\ddot{\Omega}$ could be used as indicators of recent change in the continent-based ice volumes, provided that the mantle rheology can be determined independently from the palaeo-sea-level record and that the models of past melting, including any change to the ice sheets since the end of the principal deglaciation phase, are sufficiently well known. Otherwise, the inversion of the satellite data for the present-day sea-level rise remains inherently nonunique.

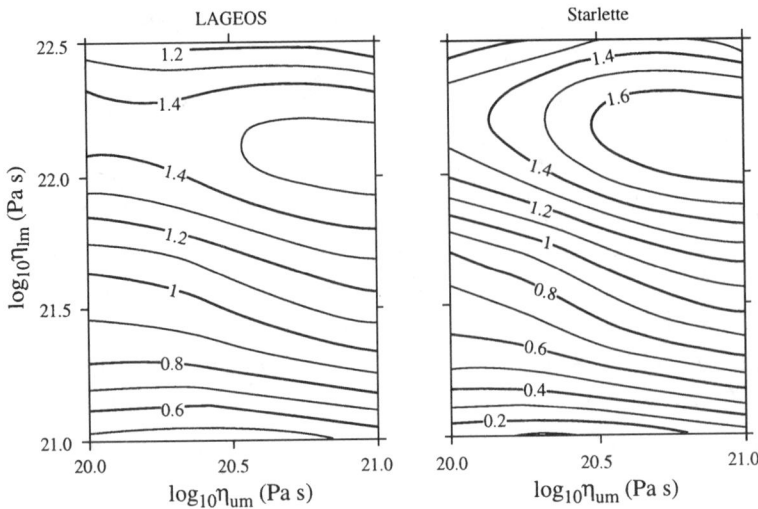

Figure 10.11. Predicted rates of present-day sea-level change (mm/yr) as functions of upper- and lower-mantle viscosities inferred from the observed accelerations of the nodes of the satellites LAGEOS and Starlette. (These predictions include the 0.5 mm/yr from the late Holocene melting.) For the preferred upper-mantle viscosity of 4–5×10^{20} Pa · s and a lower-mantle viscosity of 10^{22} Pa · s, both datasets indicate a consistent recent sea-level rise of about 1.5 mm/yr.

Whether or not the results from the present analysis of the palaeo-records of sea-level change are adequate for estimating the present-day rate of sea-level change perhaps can best be determined by performing the analysis. Figure 10.11(left) illustrates the total rate of present-day melting as a function of the Earth model used, and expressed in terms of equivalent rate of sea-level rise, required to explain the difference between the observed value for the LAGEOS satellite and the predicted value, where the latter is based on the two contributions illustrated in Figures 10.10a and 10.10b. If the viscosity model favoured by the palaeo-sea-level data is adopted ($\eta_{lm} \simeq 10^{22}$, $\eta_{um} \simeq 4$–5×10^{20} Pa · s), then the present-day sea-level rise is about 1.5 mm/yr, a rate that would be representative of the past two decades, for which satellite data are available, and also is consistent with estimates from tide-gauge records. The predicted sea-level rise for this earth-model inferred from the Starlette satellite is consistent with the LAGEOS estimate (Figure 10.11, right). If a lower-mantle viscosity of 10^{21} Pa · s were adopted, then the two estimates for the present-day rise would be inconsistent – a total of about 0.5 mm/yr for LAGEOS and 0.2 mm/yr for Starlette. Both estimates are significantly less than the global rise in sea level inferred from tide-gauge data, particularly from the records for the past two or three decades.

10.5. Conclusions

The melting of the last great ice sheets of the Late Pleistocene left a number of traces in the geological record that can provide insight into the Earth's response to the changing surface loads of ice and water. Of these, the positions of past shorelines above or below their present-day levels are most important. Inversions of such observations for mantle parameters are not, however, independent of assumptions made about the melting models, and further progress in understanding the waxing and waning of the ice sheets is essential if improved and unambiguous solutions for mantle viscosity are to be found. However, because of the very considerable temporal and spatial variation in the global pattern of sea-level change, some separation of parameters can be achieved for the better-constrained ice sheets, such as those over northwestern Europe. Here, the evidence concerning sea-level change along continental margins far from the former ice margins points to a viscosity for the mantle that exhibits some considerable depth dependence.

An important aspect of the solutions reported here is that considerable trade-off can occur between parameters that describe the mantle response – for example, between upper-mantle viscosity and lithospheric thickness. Thus, unless the search for appropriate parameters is conducted through a large model-parameter space, the resulting solutions may not be the most appropriate, even though they may lead to a plausible description of some of the associated deglaciation phenomena. Figure 10.12 illustrates the preferred three-layer model, based on (1) the foregoing analysis of the glacial rebound in Great Britain, where the findings constrain mostly the upper-mantle structure, (2) a preliminary analysis of the glacial rebound in Fennoscandia (Lambeck, Johnston, and Nakada, 1990; Lambeck, 1993c), and (3) an analysis of sea-level changes in the Australasian region (Lambeck and Nakada, 1990). The most robust feature of this solution is that the average lower mantle has a higher viscosity than that of the upper mantle by a factor of perhaps 20 to 30, and there is some preliminary evidence that the viscosity of the upper mantle may be laterally variable (Nakada and Lambeck, 1991). This latter inference is based on analysis of rebound parameters suggesting that the viscosity of the upper mantle exhibits a spatial variation that is consistent with the global-scale variation in seismic wave attenuation and shear-wave velocity. But the findings at this stage are more suggestive of further directions for research, rather than articles of faith.

Some radial structure within the upper-mantle viscosity is suggested by analysis of the observational evidence from the British Isles, with the possibility that the transition zone between about 400 and 670 km may have a viscosity intermediate between that of the uppermost mantle and that of the region below 670 km (the five-layer model illustrated in Figure 10.12). Further analyses of intermediate-size ice sheets, such as that over Scandinavia, will be required to determine whether or not this is a necessary feature of the mantle in northwestern Europe as a whole.

One of the suggestions drawn from the palaeo-sea-level observations is that melting of the ice sheets may not have ceased at 6,000 B.P. and that eustatic sea level may have continued to rise, by 2–3 m over the past 6,000 years. Analyses of recent rebound or sea-level changes should therefore include the possibility that ice and ocean volumes have not been constant, as is indeed indicated by the measured

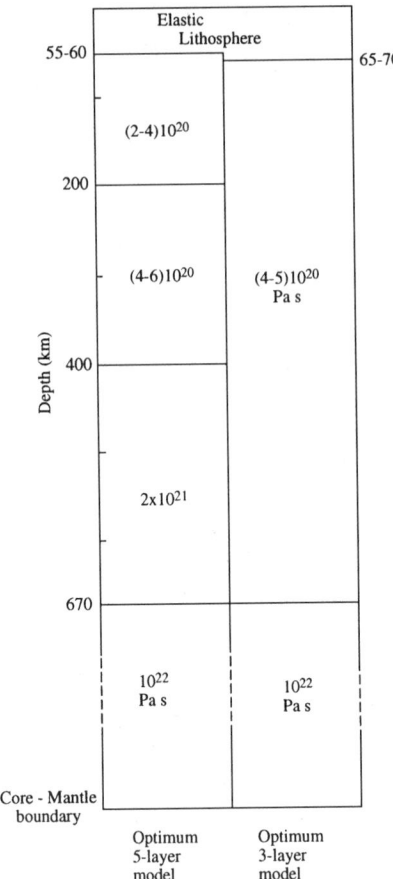

Figure 10.12. Parametrized viscosity structure for the mantle. The optimum five-layer model on the left is based mainly on the British Isles analysis. The three-layer model on the right is consistent with these data as well as with the Scandinavian evidence. The different lithospheric thicknesses for the two models reflect the trade-offs that occur between this parameter and the upper-mantle viscosity.

rates of present-day sea-level rise of 1–2 mm/yr. The recently inferred change in the Earth's dynamical flattening, the $\bar{C}_{20}$ coefficient, is particularly sensitive to the recent melting history of the ice sheets, and the analysis of this coefficient for mantle viscosity should include, as unknown, any very recent changes in ocean and ice volumes. When this is done, the findings regarding lower-mantle viscosity appear to be consistent with both the palaeo-sea-level evidence and with the analyses of tide-gauge records showing a global rise of about 1.5 mm/yr.

Although a consistent picture of the mantle's viscosity is beginning to emerge, much remains to be done before any findings can be said to be definitive. Improved ice models, independent of assumptions about the Earth's rheology, are necessary, particularly for the North American ice sheet (e.g., Fulton, 1989) (for which our models are less well constrained than for the case of northwestern Europe) and for Antarctica (where major uncertainties remain about the limits of the ice sheet during the Last Glacial Maximum). Improved glacial-rebound observations are also desirable, particularly further back in time, so as to constrain the early part of the rebound history, which is much more sensitive to Earth-model parameters and which offers a better separation of these parameters than does the essentially post-glacial record currently available. Improved geodetic data, including observations of the acceleration of the orbital nodes of satellites and the present-day rates of crustal rebound, will also lead to improved solutions. But despite the expressed reservations about the current state of knowledge, the picture that has emerged is sufficiently consistent to be able to use the outcomes from the models in different spheres of Earth-science research, whether it be in using the models to infer from sea-level records the ice-sheet limitations and thicknesses for poorly known ice domes, such as that over the Barents-Kara region (Lambeck, 1995b), or to estimate rates of tectonic uplift of coastal environments, as in the Aegean Sea (Lambeck, 1995a), or to predict the evolution of shorelines in areas of pre-historical significance (Lambeck, 1996b).

References

Alterman, Z., Jarosch, H., and Pekeris, C. L. 1959. Oscillations of the Earth. *Proc. R. Soc. London* A252:80–95.

Andersen, B. G. 1981. Late Weichselian ice sheets in Eurasia and Greenland. In: *The Last Great Ice Sheets*, ed. G. H. Denton and T. J. Hughes, pp. 1–65. New York: Wiley.

Andrews, J. T. (ed.). 1974. *Glacial Isostasy*. Stroudsburg, PA: Dowden, Hutchinson and Ross.

Ballantyne, C. K., and Harris, C. 1994. *The Periglaciation of Great Britain*. Cambridge University Press.

Barnett, J. P. 1984. The estimation of "global" sea level change – problem of uniqueness. *J. Geophys. Res.* 89:7980–8.

Bland, D. R. 1960. *The Theory of Linear Viscoelasticity*. Oxford: Pergamon Press.

Boulton, G. S., Jones, A. S., Clayton, K. M., and Kenning, M. J. 1977. A British ice-sheet model and patterns of glacial erosions and deposition in Britain. In: *British Quaternary Studies: Recent Advances*, ed. R. W. Shotton, pp. 231–46. Oxford: Clarendon Press.

Boulton, G. S., Smith, G. D., Jones, A. S., and Newsome, J. 1985. Glacial geology and glaciology of the last mid-latitude ice sheets. *J. Geol. Soc. London* 142:447–74.

Cathles, L. M. 1975. *The Viscosity of the Earth's Mantle*. Princeton University Press.

Chappell, J. 1974. Geology of coral terraces, Huon Peninsula, New Guinea: a study of Quaternary tectonic movements and sea-level changes. *Geol. Soc. Am. Bull.* 95:553–70.

Clark, J. A., Farrell, W. E., and Peltier, W. R. 1978. Global changes in postglacial sea level: a numerical calculation. *Quaternary Res.* 9:265–98.

Dahlen, F. A., and Fels, S. B. 1978. A physical explanation of the static core paradox. *Geophys. J. Royal Astron. Soc.* 55:317–32.

Davies, B., and Martin, B. 1979. Numerical inversion of the Laplace transform: a survey and comparison of methods. *J. Computational Phys.* 33:1–32.

Dawson, A. G. 1988. The main rock platform (main late glacial shoreline) in Ardnamurchan and Moidart, western Scotland. *Scott. J. Geol.* 24:163–74.

Dehant, V., and Wahr, J. M. 1991. The response of a compressible, non-homogenous Earth to internal loading: theory. *J. Geomag. Geoelectr.* 43:157–78.

Denton, G. H., and Hughes, T. J. (eds.). 1981. *The Last Great Ice Sheets*. New York: Wiley.

Donner, J. 1995. *The Quaternary History of Scandinavia*. Cambridge University Press.

Emery, K. O., and Aubrey, D. G. 1991. *Sea Levels, Land Levels and Tide Gauges*. Berlin: Springer-Verlag.

Fang M., and Hager, B. H. 1995. The singularity mystery associated with a radially continuous Maxwell viscoelastic structure. *Geophys. J. Int.* 123:849–65.

Farrell, W. E. 1972. Deformation of the Earth by surface loads. *Rev. Geophys. Space Phys.* 10:761–97.

Farrell, W. E., and Clark, J. A. 1976. On postglacial sea level. *Geophys. J.* 46:79–116.

Fjeldskaar, W. 1994. Viscosity and thickness of the asthenosphere detected from the Fennoscandian uplift. *Earth Planet. Sci. Lett.* 126:399–410.

Fjeldskaar, W., and Cathles, L. M. 1984. Measurement requirements for glacial uplift detection of non-adiabatic density gradients in the mantle. *J. Geophys. Res.* 89:10115–24.

Flemming, N. C. 1978. Holocene eustatic changes and coastal tectonics in the northeast Mediterranean: implications for models of crustal consumption. *Phil. Trans. R. Soc. London* A289:405–58.

Fulton, R. J. 1989. Quaternary geology of the Canadian Shield. In: *Quaternary Geology of Canada and Greenland*, ed. R. J. Fulton, pp. 175–317. Geological survey of Canada.

Fung, Y. C. 1965. *Foundations of Solid Mechanics*. Englewood Cliffs, NJ: Prentice-Hall.

Gasperini, P., and Sabadini, R. 1989. Lateral heterogeneities in mantle viscosity and post-glacial rebound. *Geophys. J.* 98:413–28.

Gegout, P., and Cazenave, A. 1993. Temporal variations of the Earth gravity field for 1985–1989 derived from LAGEOS. *Geophys. J. Int.* 114:347–59.

Gornitz, V., Lebedeff, S., and Hansen, J. 1982. Global sea level trend in the past century. *Science* 215:1611–14.

Gray, J. M. 1978. Low-level shore platforms in the south-west Scottish Highlands: altitude, age and correlation. *Trans. Inst. Brit. Geogr.* 3:151–63.

Haskell, N. A. 1937. The viscosity of the asthenosphere. *Am. J. Sci.* 233:22–8.

Jeffreys, H. 1959. *The Earth*. 4th ed. Cambridge University Press.

Johnston, P. 1993. The effect of spatially non-uniform water loads on the prediction of sea-level change. *Geophys. J. Int.* 114:615–34.

Johnston, P., Lambeck, K., and Wolf, D. 1997. Material versus isobaric internal boundaries in the Earth and their influence on glacial rebound. *Geophys. J. Int.* 129:252–68.

Kaufmann, G., Wu, P., and Wolf, D. 1997. Some effects of lateral heterogeneities in the upper mantle on postglacial land uplift close to continental margins. *Geophys. J. Int.* 128:175–87.

Kaula, W. M. 1966. *Theory of Satellite Geodesy*. Waltham, MA: Blaisdell.

Lambeck, K. 1980. *The Earth's Variable Rotation: Geophysical Causes and Consequences*. Cambridge University Press.

Lambeck, K. 1988. *Geophysical Geodesy: The Slow Deformations of the Earth*. Oxford University Press.

Lambeck, K. 1993a. Glacial Rebound of the British Isles. I: Preliminary model results. *Geophys. J. Int.* 115:941–59.

Lambeck, K. 1993b. Glacial Rebound of the British Isles. II: A high resolution, high-precision model. *Geophys. J. Int.* 115:960–90.

Lambeck, K. 1993c. Glacial rebound and sea-level change: an example of a relationship between mantle and surface processes. *Tectonophysics* 223:15–37.

Lambeck, K. 1995a. Late Pleistocene and Holocene sea-level change in Greece and southwestern Turkey: a separation of eustatic, isostatic and tectonic contributions. *Geophys. J. Int.* 122:1022–44.

Lambeck, K., 1995b. Constraints on the Late Weichselian ice sheet over the Barents Sea from observations of raised shorelines. *Quaternary Sci. Rev.* 14:1–16.

Lambeck, K. 1995c. Late Devensian and Holocene shorelines of the British Isles and

North Sea from models of glacio-hydro-isostatic rebound. *J. Geol. Soc. London* 152:437–48.

Lambeck, K. 1996a. Limits on the areal extent and thickness of the Barents Sea Ice Sheet in Late Weischelian time. *Palaeogeogr. Palaeoclimatol. Palaeoecol. (Global and Planetary Change Section)* 12:41–51.

Lambeck, K. 1996b. Sea-level change and shoreline evolution in Aegean Greece since Upper Palaeolithic time. *Antiquity* 70:588–611.

Lambeck, K., Johnston, P., and Nakada, M. 1990. Holocene glacial rebound and sea-level change in NW Europe. *Geophys. J. Int.* 103:451–68.

Lambeck, K., Johnston, P., Smither, C., and Nakada, M. 1996. Glacial rebound of the British Isles. III. Constraints on mantle viscosity. *Geophys. J. Int.* 125:340–54.

Lambeck, K., and Nakada, M. 1990. Late Pleistocene and Holocene Sea-Level Change along the Australian Coast. *Palaeogeogr. Palaeoclimatol. Palaeoecol. (Global and Planetary Change Section)* 89:143–76.

Longman, I. M. 1962. A Green's function for determining the deformation of the Earth under surface-mass loads. 1. Theory. *J. Geophys. Res.* 67:845–50.

Love, A. E. H. 1911. *Some Problems of Geodynamics.* Cambridge University Press.

McConnell, R. 1968. Viscosity of the mantle from relaxation time spectra of isostatic adjustment. *J. Geophys. Res.* 73:7089–105.

Malvern, L. E. 1969. *Introduction to the Mechanics of a Continuous Medium.* Englewood Cliffs, NJ: Prentice-Hall.

Meier, M. F. 1984. Contributions of small glaciers to global sea level. *Science* 226:1418–21.

Mitrovica, J. X. 1996. Haskell [1935] revisited. *J. Geophys. Res.* 101:555–69.

Mitrovica, J. X., and Peltier, W. R. 1991. On postglacial geoid subsidence over the equatorial oceans. *J. Geophys. Res.* 96:20053–71.

Mitrovica, J. X., and Peltier, W. R. 1992. A comparison of methods for the inversion of viscoelastic relaxation spectra. *Geophys. J. Int.* 108:410–14.

Mitrovica, J. X., and Peltier, W. R. 1993. The inference of mantle viscosity from an inversion of the Fennoscandian relaxation spectrum. *Geophys. J. Int.* 114:45–62.

Mörner, N.-A. 1980. The northeast European sea-level laboratory and regional Holocene eustasy. *Palaeogeogr. Palaeoclimatol. Palaeoecol.* 29:281–300.

Nakada, M., and Lambeck, K. 1987. Glacial rebound and relative sealevel variations: a new appraisal. *Geophys. J. Royal Astron. Soc.* 90:171–224.

Nakada, M., and Lambeck, K. 1988. The melting history of the Late Pleistocene Antarctic ice sheet. *Nature* 333:36–40.

Nakada, M., and Lambeck, K. 1989. Late Pleistocene and Holocene sea-level change in the Australian region and mantle rheology. *Geophys. J. Int.* 96:497–517.

Nakada, M., and Lambeck, K. 1991. Late Pleistocene and Holocene sea-level change; evidence for lateral mantle viscosity structure? In: *Glacial Isostasy, Sea Level and Mantle Rheology.* ed. R. Sabadini, K. Lambeck, and E. Boschi, pp. 79–94. Dordrecht: Kluwer Academic.

Nakiboglu, S. M., and Lambeck, K. 1991. Secular sea-level change. In: *Glacial Isotasy, Sea Level and Mantle Rheology.* ed. R. Sabadini, K. Lambeck, and E. Boschi, pp. 237–58. Dordrecht: Kluwer Academic.

O'Connell, R. J. 1971. Pleistocene glaciation and the viscosity of the lower mantle. *Geophys. J.* 23:299–327.

Paterson, W. S. B. 1971. *The Physics of Glaciers.* Oxford: Pergamon.

Peltier, W. R. 1974. The impulse response of a Maxwell Earth. *Rev. Geophys.* 12:649–69.

Peltier, W. R. 1985. The LAGEOS constraint on deep mantle viscosity: results from a new normal mode method for the inversion of viscoelastic relaxation spectra. *J. Geophys. Res.* 90:9411–21.

Peltier, W. R. 1986. Deglaciation-induced vertical motion of the North American continent and transient lower mantle rheology. *J. Geophys. Res.* 91:9099–123.

Peltier, W. R. 1988. Global sea level and Earth rotation. *Science* 240:895–901.

Peltier, W. R., and Andrews, J. T. 1976. Glacial-isostatic adjustment. 1. The forward problem. *Geophys J. Royal Astron. Soc.* 46:605–46.

Peltier, W. R., Drummond, R. A., and Tushingham, A. M. 1986. Postglacial rebound and transient lower mantle rheology. *Geophys. J. Royal Astron. Soc.* 87:79–116.

Peltier, W. R., and Jiang, X. 1996. Mantle viscosity from the simultaneous inversion of multiple data sets pertaining to postglacial rebound. *Geophys. Res. Lett.* 23:503–6.

Ringwood, A. E. 1975. *Composition and Petrology of the Earth's Mantle*. New York: McGraw-Hill.

Rubincam, D. P. 1984. Postglacial rebound observed by LAGEOS and the effective viscosity of the lower mantle. *J. Geophys. Res.* 89:1077–87.

Rümpker, G., and Wolf, D. 1996. Viscoelastic relaxation of a Burgers half-space – implications for the interpretation of the Fennoscandian uplift. *Geophys. J. Int.* 124:541–55.

Schapery, R. A. 1962. Approximate methods of transform inversion for viscoelastic stress analysis. In: *Proceedings of the 4th U.S. National Congress of Applied Mechanics*, pp. 1075–85. New York: ASME.

Sissons, J. B. 1983. Shorelines and isostasy in Scotland. In: *Shorelines and Isostasy*, ed. D. E. Smith and A. G. Dawson, pp. 209–25. London: Academic Press.

Stone, J., Lambeck, K., Fifield, L. K., Evans, J. M., and Cresswell, R. G. 1996. A Lateglacial age for the Main Rock Platform, Western Scotland. *Geology* 24:707–10.

Tushingham, A. M., and Peltier, W. R. 1991. ICE-3G: a new global model of Late Pleistocene deglaciation based upon geophysical predictions of postglacial sea-level change. *J. Geophys. Res.* 96:4497–523.

Walcott, R. I. 1970. Flexural rigidity, thickness, and viscosity of the lithosphere. *J. Geophys. Res.* 75:3941–54.

Walcott, R. I. 1972. Past sea levels, eustasy and deformation of the Earth. *Quaternary Res.* 2:1–14.

Wolf, D. 1991. Viscoelastodynamics of a stratified, compressible planet: incremental field equations and short- and long-time asymptotes. *Geophys. J. Int.* 104:401–17.

Wu, J., and Yuen, D. 1991. Post-glacial relaxation of a viscously stratified, compressible mantle. *Geophys. J. Int.* 104:331–49.

Wu, P. 1978. The response of a Maxwell earth to applied surface mass loads: glacial isostatic adjustment. Master's thesis, University of Toronto.

Wu, P., and Peltier, W. R. 1982. Viscous gravitational relaxation. *Geophys. J. Royal Astron. Soc.* 70:435–85.

Yoder, C. F., Williams, J. G., Dickey, J. O., Schultz, B. E., Eanes, R. J., and Tapley, B. D. 1983. Secular variations of Earth's gravitational J_2 coefficient from LAGEOS and non-tidal acceleration of Earth rotation. *Nature* 303:757–62.

Yuen, D. A., and Sabadini, R. 1984. Secular rotational motions and the mechanical structure of a dynamical viscoelastic Earth. *Phys. Earth Planet. Int.* 36:391–412.

Yuen, D. A., Sabadini, R., and Boschi, E. 1982. Viscosity of the lower mantle as inferred from rotational data. *J. Geophys. Res.* 87:10745–62.

11

Mantle Rheology: Insights from Laboratory Studies of Deformation and Phase Transition

MARTYN R. DRURY and JOHN D. FITZ GERALD

The deformation of Earth materials is not fundamentally different from that of man-made polycrystalline, polyphase solids, such as metals, alloys, and ceramics. We already have extensive bodies of knowledge concerning fracture and flow in such materials in general, and that information should be reviewed briefly before we proceed to discuss the rheology of rocks. The discussion of mechanical behaviour at elevated temperature and pressure inside the Earth will be divided into two parts: processes in the upper mantle, and processes in the deep mantle (involving the transition zone and the lower mantle). We shall follow the terminology of Ringwood (1975) for the subdivisions of the Earth's mantle (Jackson and Rigden, Chapter 9, this volume). Later sections will review phase transformations in the Earth and the possible mechanical consequences, and a final section will briefly address future developments.

11.1. General Introduction to Deformation of Crystalline Solids

A solid material subjected to stress that exceeds its elastic limit will become permanently deformed or strained, and the accumulation of the increments of strain is time-dependent. The term 'plastic' denotes the nonelastic part of the strain that is not recovered following the removal of the applied stress. In general, to analyze or predict the elastic and plastic responses of any stressed solid, one must know not only the state of stress but also the environmental conditions and the nature of the solid.

If a material deforms in a brittle manner, the permanent strain field is discontinuous. Fractures, cracks, and faults are common, easily recognized expressions of brittle failure at a macroscopic scale. Usually, discontinuities are also present at finer scales as microfractures and microfaults.

If the strain field is truly continuous, then the material will show ductile, not brittle, behaviour: The material will maintain its coherence, yet be changed in external size and shape. In crystalline materials, the agents of ductile deformation are defects – local disruptions to the perfect unit-cell repeat that characterizes every grain of any crystalline substance.

11.1.1. Macroscopic Description of Deformation: Environmental Variables

The primary independent variable in deformation can be considered to be either stress σ or strain ε. Generally, strain is considered to depend upon stress. Following Frost and Ashby (1982, chap. 1) and Poirier (1985), ductile deformation can be formalized via a constitutive equation in which the rate of strain $\dot{\varepsilon}$ is expressed in terms of other parameters:

$$\dot{\varepsilon} = f(\sigma, T, P, S_i, M_j) \tag{1}$$

where T is temperature, P is pressure, S_i are the variables that describe the microstructural state of the material (grain size, densities of defects, etc.), and M_j are the material properties (lattice parameters, bond strengths, activation parameters, etc.).

In a macroscopic description, the total $\dot{\varepsilon}$ generally will have several components, each resulting from deformation due to a distinct process and described by its own constitutive equation. Poirier (1985, pp. 79–81) discussed the possible interdependence of deformation processes. However, it is common that, given a set of state parameters, one deformation mechanism will contribute most of the $\dot{\varepsilon}$.

Frost and Ashby (1982), Poirier (1985), and Evans and Kohlstedt (1995) have provided thorough reviews of the many different mechanisms of deformation, each characterized by a particular rate equation. Here we shall briefly consider the possibilities currently thought to be most important in the deformation of materials deep in the Earth's interior.

Power-Law Creep.[1] This is a rheology with the constitutive relation

$$\dot{\varepsilon} = A\sigma^n d^{-m} \exp[-(E^* + PV^*)/RT] \tag{2}$$

where A, n, d, m, E^*, and V^* are respectively a constant, the stress exponent, grain size, grain-size exponent, activation energy, and activation volume for the process that controls creep. $Q = E^* + PV^*$ is the activation enthalpy. For power-law flow, $\dot{\varepsilon}$ is usually grain-size-independent ($m = 0$), but it is very sensitive to σ (n is commonly 2 or greater). Table 2 of Evans and Kohlstedt (1995) details relationships for many different mechanisms of this type. Power-law creep generally dominates at intermediate stress levels and temperatures that are medium to high (>0.4 times the melting temperature T_m).

---

[1] Strictly speaking, 'creep' in the metallurgical sense is the flow of a material at relatively high temperatures in response to constant stress. However, its common usage in the materials literature has grown to encompass many types of ductile deformations.

Linear Creep. This type of rheology can also be expressed using equation (2), but with a stress exponent $n = 1$ and the grain-size exponent $m = 2$ or 3. With such grain-size sensitivity but low stress sensitivity, this type of creep is the most efficient mechanism at low stresses and small grain sizes.

Low-Temperature Plasticity and Power-Law Breakdown. These types of behaviour dominate at low temperatures and high stresses. The references already cited should be consulted for details of their mechanisms and constitutive relations.

11.1.2. Microscopic Description of Deformation: Defects and Mechanisms

Defects enable plastic deformation: For a crystal to deform ductilely on the macroscopic scale, microscopic defects must be present, and these defects must be changed in position or abundance to produce the deformation. Crystalline regions between defects remain perfect during deformation, and thus macroscopic coherence is retained. Creation and migration of defects are thermally activated processes with kinetics that are exponentially temperature-dependent, and this results in strong dependence of the macroscopic strain rate on temperature, as in equation (2).

Defects fall into three broad categories, and these categories are central to understanding the range of mechanisms by which plastic deformation occurs. In more macroscopic terms, these mechanisms provide a firm basis for rationalizing the different forms of constitutive equations introduced earlier.

Point Defects. Vacancies in the crystal lattice, atoms interstitial to the lattice, and substitutional or impurity atoms can all be important in deformation. Diffusion-controlled creep is deformation involving diffusive motions of point defects, and it has a linear-creep constitutive relation. If temperature is high, it is diffusion through the crystal lattice that dominates, and the grain-size exponent $m = 2$ (Nabarro-Herring creep). At lower temperatures, grain-boundary diffusion is relatively more important, and $m = 3$ (Coble creep). In compounds, there must be co-operative diffusion of defects involving all of the atomic species in the crystal is such a way as to maintain stoichiometry and charge neutrality (e.g., Evans and Kohlstedt, 1995). The strain rate is effectively controlled by the slowest-diffusing species.

Line Defects or Dislocations. Each dislocation has an associated crystallographic slip vector, the Burgers vector **b**, which indicates the relative atomic motion (slip) developed within the crystal as this dislocation passes. The surface that traces the motion of a dislocation is known as its glide surface, and such surfaces are commonly planar. The combination of **b** and glide plane {e.g., [001](100)} denotes the dislocation slip system. If the dislocation line is parallel to **b**, it is a screw dislocation; if the line is perpendicular to **b**, it is an edge dislocation; otherwise the dislocation has 'mixed character', with edge and screw components. The simplest

motion of a dislocation is slip, which can be envisaged as the breaking of inter-
atomic bonds ahead of the dislocation line, advance of the dislocation together with
an increment of slip displacement corresponding to **b**, and re-formation of perfect
crystal structure behind; in this way, motion is conservative of mass. Edge dislocations (or the edge components of mixed dislocations) also move by climb, where
the dislocation moves 'out of' its glide plane; this is nonconservative of mass and
requires diffusion of atoms to or from the defect. Deformation of materials arising
from dislocations generally involves both glide and climb components. Whereas it
is straightforward to envisage atomic configurations associated with dislocations in
monatomic crystals, the configurations must be much more complex in compounds,
particularly in minerals characterized by large unit cells containing many atoms.
Dislocations enable power-law creep [equation (2)], so this type of deformation
is often referred to as dislocation creep. Note, though, that dislocations are also
essential for ductile deformation of crystals at stresses above the upper limit for
power-law creep. In some situations, the rate of dislocation creep will be limited
by diffusion (i.e., by climb activity), but strain can still be dominated by dislocation glide. A particular class of creep at low stresses involves dislocation motion
dominated by climb; this mechanism, which exhibits a linear dependence of $\dot{\varepsilon}$ on
σ, is known as Harper-Dorn creep (Poirier, 1985, chap. 4).

Planar Defects. Several types of planar defects can be involved in deformation
(e.g., stacking faults, twins, and grain boundaries). Grain boundaries are the most
important type, because they mark discontinuities in atomic arrangements between
adjacent crystals or phases.[2] Grain boundaries interact with both point defects
and dislocations, so they are connected with both diffusion-controlled creep and
dislocation creep. Grain boundaries are also structures that allow grains to move
relative to each other in an aggregate – the phenomenon of grain-boundary sliding.
Even in diffusion-controlled creep, much of the strain is from grain sliding, but
because a change of grain shape (by diffusion) is necessary for grain translation,
it is still diffusion that limits the *rate* of macroscopic strain. The structure of grain
boundaries obviously is also involved in the dynamics of sliding, and if grains
ultimately lose physical contact, then the material could even deform in a brittle
rather than a ductile manner.

Many reviews provide excellent opportunities for further reading: Poirier (1985),
Chapter 2 of Frost and Ashby (1982), Hull and Bacon (1984) and Hobbs, Means,
and Williams (1976) are particularly recommended as dealing with the general phenomena of defects and their involvement in plastic deformation of solid materials.

The presence of defects raises a material's total energy, so that minimization
of system energy effectively drives the removal of defects from materials. This

---

[2] The term 'grain boundary', when used in a formal sense, denotes a boundary between two crystals of the same
substance, whereas the interface between two different substances is known as an interphase boundary. In this
chapter, 'grain boundary' is used as a general term encompassing both true grain and interphase boundaries.

tendency for elimination of defects constantly competes with all processes that generate defects. At any T above zero Kelvin, a certain density of point defects will exist in dynamic equilibrium between spontaneous creation and removal. High densities of dislocations are produced by dislocation multiplication and are characteristic of crystal regions where high degrees of plastic strain have developed. The defect densities can be reduced through recovery, also resulting in polygonization, the reconfiguration of dislocations into low-energy, organized, three-dimensional dislocation networks that constitute walls which divide crystals into subgrains (Poirier, 1985, chap. 6). For polycrystals, grain growth is a process that reduces total energy by decreasing the grain-boundary area per unit volume. However, many highly deformed materials will recrystallize at $T \geq T_m/2$ to aggregates of grains that are distinctly smaller and strain-free. Recrystallization can be either static (no external stress) or dynamic (external stress present), and it reduces the stored strain energy at the cost of raising the total grain surface energy. For dynamic recrystallization, the higher the external stress, the smaller the recrystallized grain size (and the smaller the subgrain size). Consequently, in a system actively deforming at elevated T, there is dynamic competition between grain growth and grain-size reduction due to recrystallization, and also dynamic competition between dislocation multiplication and recovery and polygonization within each grain. Defect structures are sensitive (though not unique) indicators of the deformation history; hence their study microscopically is a fundamental tool contributing to our understanding of natural and laboratory deformations. The reviews cited earlier should be consulted for more details.

11.1.3. Deformation-Mechanism Maps

Frost and Ashby (1982) thoroughly analyzed the powerful graphical tool of the deformation-mechanism map. A map of parameter space, generally restricted, for simplicity, to two of the variables (Figure 11.1), is divided into fields in which individual deformation mechanisms dominate. Commonly, temperature and stress are selected as the map variables. Strain-rate contours are generally superimposed, the spacings of these contours indicating the sensitivity of strain to each of the map variables. Inspection of Figure 11.1 will reveal how efficiently much of the mechanistic information in the preceding few paragraphs can be presented. Some care is needed in reading deformation-mechanism maps: Although each field indicates that a particular mechanism is dominant, this is not to the exclusion of others, and so the contributions to strain from competing mechanisms remain significant, particularly close to field boundaries.

The numerous Frost and Ashby (1982) maps are all normalized parameter spaces; map co-ordinates are T/T_m (called the homologous temperature) and σ/μ, where μ is the shear modulus. This normalization creates dimensionless parameters with limited ranges of values and is a scaling mechanism by which the behaviours of different materials can be compared. Frost and Ashby (1982) and Poirier (1985,

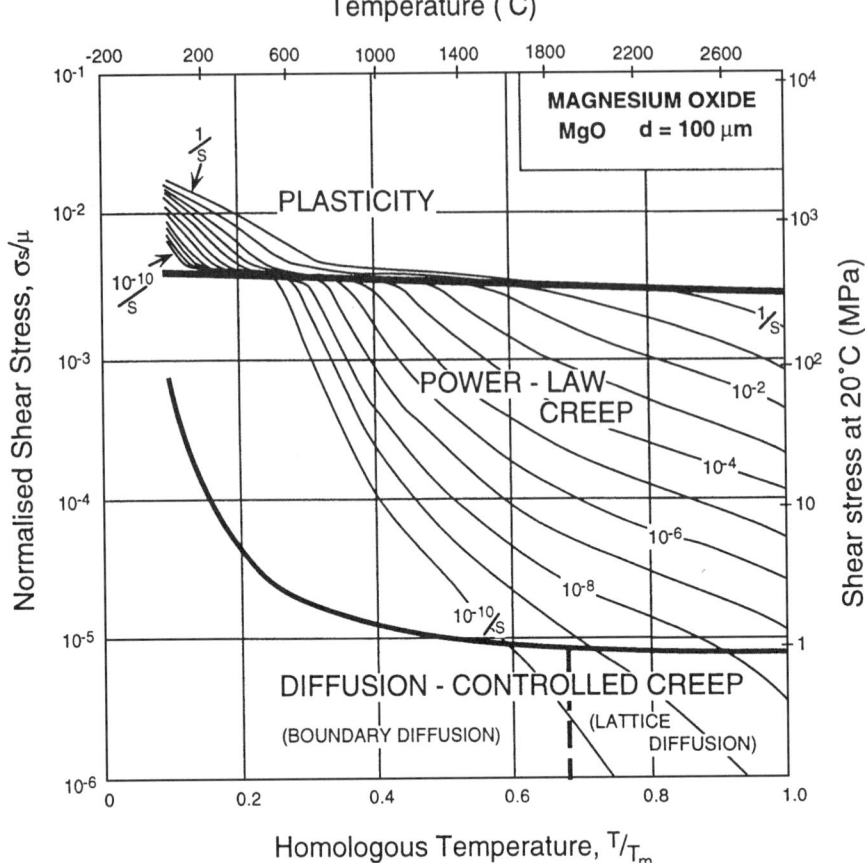

Figure 11.1. A deformation-mechanism map for MgO with grain size of 100 μm. Fields are labelled according to the dominant mechanism of deformation. Contours of strain rate cross these fields. (Adapted from Frost and Ashby, 1982.)

chap. 9) have discussed scaling using different parameter groups and have illustrated how this is one way of relating similar materials in so-called isomechanical groups. It follows that scaling can be used to predict some mechanical properties that have not been or cannot be measured. The importance of this for Earth materials is discussed later. In principle, dimensionless parameters can be used to reduce constitutive equations to simpler forms.

11.1.4. *Transformations and Rheology*

If a solid aggregate is transformed in some way, its mechanical behaviour will generally be affected, both during the transformation and subsequently. The classes of transformations most relevant to materials in the Earth's mantle are phase transitions (one mineral changes to a different polymorph with the same stoichiometry) and isochemical reactions (in which several mineral phases are involved as reactants or products). A transformation will have potential mechanical effects if it changes

1. molar volume (change in aggregate density)
2. molar volume (generation of high internal stresses)
3. grain size
4. resistance to plastic deformation (conversion of any phase to a different crystal structure; this extends to mixed phases in the case of reactions)
5. degree of crystallographic preferred orientation
6. microstructure (e.g., change in defect density, generation of twins or stacking faults)

The field of endeavour dealing with interrelationships between materials undergoing transformations and their rheologies is generally known as 'transformation plasticity', and it has developed largely through research into metallic and ceramic materials. It is timely that attention is being paid to similar aspects of Earth materials. As a definition of transformation plasticity, an adaptation of that of Paterson (1983) is recommended, namely, it is 'the degree to which plastic deformation occurs more readily in a polycrystalline material while it is actively undergoing a ... transformation'. Of the items listed earlier, only internal stress (item 2) can change the rheology of a material *while transformation is occurring*. For the other five items, a change in deformation is possible only *after transformation has occurred*, so changes in these cannot result in true transformation plasticity. For reviews of the phenomenon in Earth sciences, see Poirier (1985, chap. 8), Meike (1993), and McLaren and Meike (1996).

Some of the effects that transformations can have on rheology have been introduced (items 1–6). It should be appreciated that stresses can also influence the progress of transformations. This is exemplified by martensitic phase transitions, where one crystal structure transforms to another by the action of diffusionless, atomic-scale shear, possibly accompanied by minor 'shuffling' to relieve unfavourable atomic arrangements. The effect of an appropriately oriented *external* shear stress can enhance or retard martensitic transformations. As another example, for phase transitions or reactions characterized by large volume changes, high external pressures (mean stresses) will tend to favour the denser phase or assemblage of phases. Therefore, in assessing the response to external stress (i.e., the rheology) of a transforming material, it must be borne in mind that the same stress could feed back and affect the progress of the transformation.

Greenwood and Johnson (1965) have produced one of the most carefully considered treatments of materials actively undergoing phase transitions through experiment and analysis of polymorphic changes with temperature in metallic polycrystals. By changing the temperature through a phase transition and then back again, they found that such materials showed irreversible strains (even though the volume changes were completely reversible) due to plastic deformation, and these strains could be increased linearly by small external stresses. The important characteristics of the phase transitions were the associated volume changes and the differences amongst the yield strengths of different polymorphs. The volume change at each transition generated internal stresses that resulted in plastic deformation of the weaker phase, even though the external stress was below the supposed yield stress for that phase. The levels of strain developed per cycle were low (less than or equal to the volume change, which, for these transitions, is of order 1%), so that continued cycling would be needed to develop high strains. This result is independent of the nature of the transition (i.e., martensitic versus reconstructive in character), and the macroscopic analysis did not need to take into account details of plastic deformation (such as dislocation-glide systems).

The Greenwood and Johnson model applies to transformational plasticity *sensu stricto* and can be modified to allow for other effects, such as dislocation recovery (Paterson, 1983). It is included here only as one particularly well developed example of plastic deformation affected by phase transitions, which is potentially applicable to any transformation involving volume change. It does not cover situations where a difference in rheology is observed not during transformation but afterwards (e.g., items 1 and 3–6). One case in point, relevant to Section 11.5.3 on transformational faulting, involves a major reduction in aggregate grain size accompanying phase transition, leading to a marked softening associated with subsequent deformation through grain-size-sensitive flow.

11.2. Deformation of Earth Materials

The brittle and ductile behaviours for some of the simplest mineralogical systems in the Earth's crust and upper mantle have been extensively investigated experimentally, and their rheological laws have been evaluated. By assuming that monomineralic aggregates (e.g., of quartz or olivine) represent rocks of the Earth, then assuming appropriate compositions and T and P as functions of depth, the mechanical behaviour of the Earth can be modelled via the constitutive equations introduced earlier and experimentally calibrated. Analysis of the validity of the assumptions involved is beyond the scope of this chapter. Though many complications exist, such as lateral variations in the Earth associated with the tectonic setting, some major general conclusions can be reached.

A very simple model has been selected, from the work of Hopper and Buck (1993). To represent the rheology of the Earth, a constant strain rate is assumed (e.g., $10^{-14}\,s^{-1}$ in Figure 11.2), and the stress required to sustain that rate (termed 'strength') is plotted as a function of depth. In the upper crust, deformation is dominantly brittle, with a strength that is rate-, temperature-, and material-independent to a first approximation. Lockner (1995) has reviewed some possibilities, including the commonly used empirical relation of Byerlee whereby strength at depth is limited by a coefficient of friction of 0.6. Lockner has also introduced additional complications, such as the mechanical and chemical effects of fluids and the velocity dependence of friction. For a discussion of the complexities involved in the

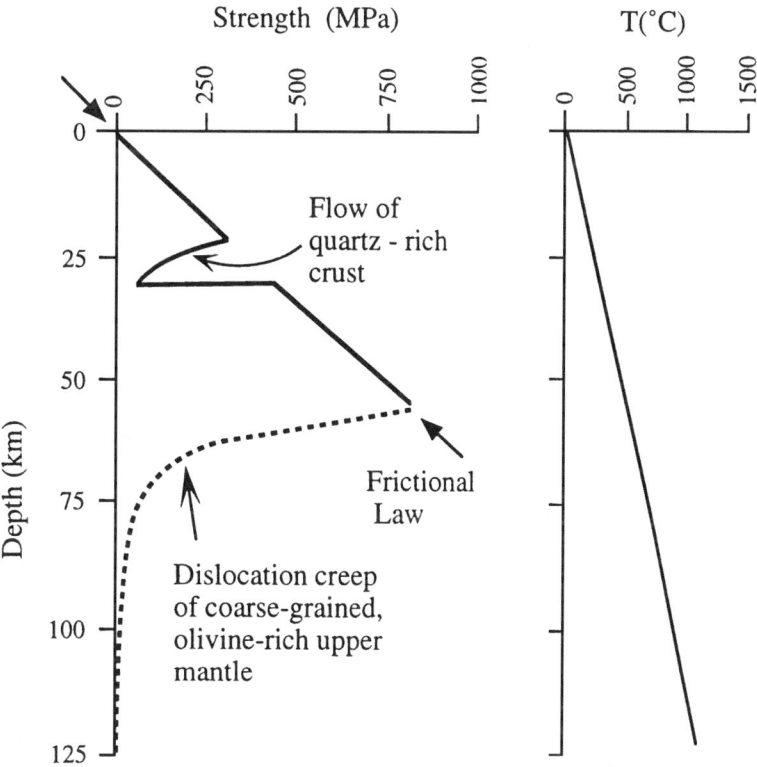

Figure 11.2. A very simple model for the strength of the Earth at depth. Strength here is the level of stress required to maintain a strain rate of 10^{-14} s^{-1}. The original source should be consulted for the rheological parameters used in the model. (Adapted from Hopper and Buck, 1993.)

brittle–ductile transition, see Rutter (1986) and Evans et al. (1990). Deep below the Mohorovicic discontinuity (the Moho) in olivine-rich upper mantle, deformation undoubtedly is ductile and is due to dislocation or diffusion-controlled flow, as summarized by Karato and Wu (1993). The shallow upper mantle is a complex region where the distinction between lithosphere and asthenosphere becomes important (Davies, Chapter 5, this volume) and aspects of mechanical behaviour at depth are sensitive to the tectonic setting (Ranalli and Murphy, 1986; Molnar, 1988).

Deformation-mechanism mapping is especially useful for representing the effects of stress, temperature, and grain size on flow in the Earth (Ranalli, 1982; Poirier, 1985). Karato and Wu (1993) used such maps to argue the case for a transition in deformation mechanism at depth from dislocation creep at large grain size to diffusion creep[3] at small grain size. Again using maps, Rutter and Brodie (1988) and Karato and Wu (1993) pointed to the possibility of diffusional flow at higher stresses even at the top of the upper mantle during continental rifting (Hopper and Buck, 1993).

---

[3] On a matter of terminology, note that it is common in the Earth-science literature to find the term 'diffusion creep' in place of the more general 'diffusion-controlled creep'. Consistent with this usage, we shall employ only the shorter form throughout the remainder of this chaper.

In Newtonian materials such as simple fluids, $\dot{\varepsilon}$ varies linearly with σ, and the proportionality constant $(\sigma/\dot{\varepsilon})$ is known as the viscosity (η). For models of the deeper Earth, viscosity is commonly selected as the parameter that expresses the linear dependence of mechanical response on stress (Lambeck and Johnston, Chapter 10, this volume; Karato, 1981). Materials deforming ductilely by linear creep (e.g., diffusion or Harper-Dorn creep) are adequately represented by a viscosity, because the associated stress exponent is unity. However, for materials deforming by other mechanisms, flow can be represented only by an effective viscosity that is stress-dependent.

Prediction of mechanical properties at depth in the Earth can involve considerable extrapolation from limited sets of mechanical data measured under very unrealistic conditions in the laboratory: particularly the $\dot{\varepsilon}-T$ combination (Paterson, 1987), but also P. It is therefore important that the systematics of parameter variation be well understood. In this context, the approach of parameter scaling becomes critical. Sammis et al. (1977) estimated viscosity throughout the mantle using a scaling approach for the activation parameters E^* and V^* [equation (2)] and by assuming that creep activation follows the same $P-T$ trends as does atomic volume. Karato (1981), Poirier (1985, chap. 5), and others have refined this approach, but with different scaling to allow further predictions of viscosity down to the base of the mantle. On the basis of all these findings, it appears that viscosity in the lower mantle could be remarkably constant over a very large depth range, and this is often assumed in studies of the Earth's deformation processes over different time and length scales (Lambeck and Johnston, Chapter 10, this volume) (see Section 11.4.5).

Borch and Green (1987) investigated another scaling relationship that is potentially useful for predicting flow stresses in the upper mantle. It relies on the observation that mechanical properties, like other thermally activated material properties, often scale with homologous temperature. A nontrivial difference, though, is that Borch and Green applied the scaling to chemically complex, polyphase materials, where T_m is controlled by eutectic and other phenomena such that melting occurs over a range (solidus T, not equal to liquidus T) that is significantly below T_m for any of the individual phases in isolation. Yet it is likely to be one phase (e.g., olivine in pyrolite composition) that controls the aggregate strength, and it is difficult to rationalize why that phase's mechanical behaviour should scale with the aggregate T_m! Nevertheless, Borch and Green (1987) claimed success in predicting viscosity for a chemically complex upper mantle using an empirical relation between strength and homologous temperature.

Superplasticity is a phenomenon first observed in metals whereby an aggregate can be exceptionally elongated (strains on the order of 1,000%) without necking of the sample or significant straining of individual grains; see the reviews by Edington, Melton, and Cutler (1976) and Poirier (1985, chap. 7). For the phenomenon of structural superplasticity in the laboratory, a very small grain size is critically important, as is the action of grain-boundary sliding. Similar behaviour has been inferred for mylonitic shear zones (Boullier and Gueguen, 1975). Of course, at the very low $\dot{\varepsilon}$ predicted for the Earth's interior, it could be that the natural grain sizes compatible with the phenomenon will be much larger than those (1–10 μm) established in laboratory tests. As noted by Poirier (1985), it is regrettable that the term 'superplasticity' has been used in different ways in the Earth-science literature. Some authors use 'superplasticity' to refer to a deformation mechanism of diffusion-controlled grain-boundary sliding, whereas others use the term to imply large strains without failure.

11.3. The Upper Mantle

A volume fraction of 20–30% of one mineral is sufficient to control the rheology of a rock provided that this phase is significantly weaker than all other phases (Carter, 1976), but this does depend on the degree of intermixing between grains of different phases (Handy, 1994). Consequently, because olivine is both the volumetrically dominant and the weakest mineral in the upper mantle, it should control the rheology in most cases. Information on deformation mechanisms and the rheology of the upper mantle can be obtained from deformation experiments and from studies of mantle rocks that outcrop at the surface (Nicolas and Poirier, 1976). Because deformation at mantle conditions is overwhelmingly plastic in nature, the following section relating to the upper mantle will ignore brittle deformation. However, brittle processes and seismicity can be locally and transiently important in the uppermost mantle (e.g., Molnar, 1988). Supporting evidence is provided by the occurrence of pseudotachylyte, a type of glassy rock inferred to be produced by melting during seismic shear events (Sibson, 1975), which occurs rarely in rocks from the upper mantle, such as the Balmuccia Peridotite (Obata and Karato, 1995).

11.3.1. Deformation Experiments on Olivine

Various aspects of olivine deformation have previously been reviewed by Goetze (1978), Tsenn and Carter (1987), Karato (1989a), and Kohlstedt, Evans, and Mackwell (1995). Creep rates for single crystals are strongly dependent on crystal orientation (Durham and Goetze, 1977; Bai, Mackwell, and Kohlstedt, 1991), and several creep mechanisms can occur (Bai and Kohlstedt, 1992a; Jin, Bai, and Kohlstedt, 1994a). At high temperatures, slip in the [c] direction, [001], is harder (i.e., requires greater σ) than slip in the [a] direction, [100]. The creep rate for the weak [100](010) slip system depends on the chemical activity of oxygen and silica (or orthopyroxene), whereas slip on the hard [001](010) slip system is insensitive to these parameters.

Recrystallized single crystals (Karato, Toriumi, and Fuji, 1982) and dried natural rocks (Chopra and Paterson, 1984) follow flow laws similar to those for [c] slip (Bai et al., 1991), suggesting that creep in these polycrystals is controlled by the

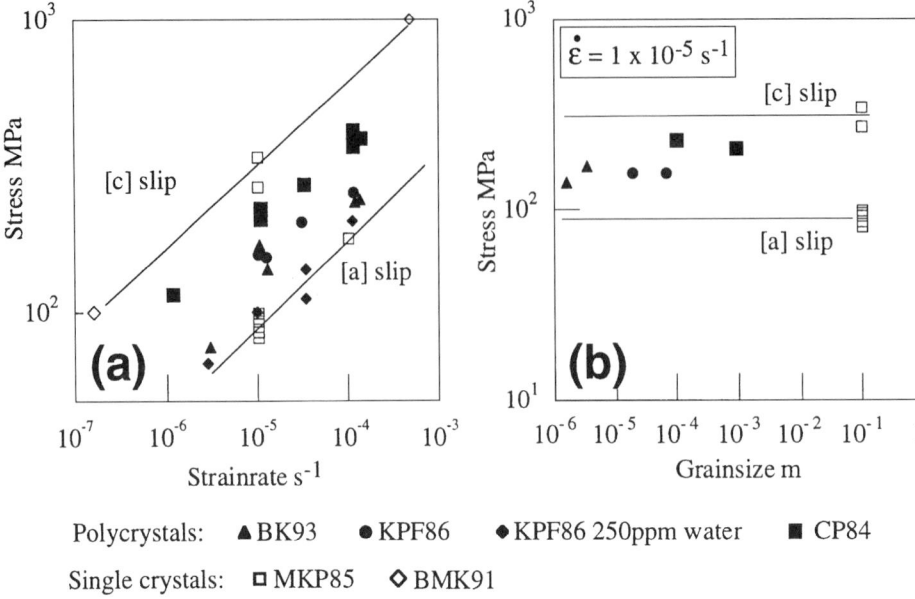

Figure 11.3. Selected experimental data for olivine at 1,300°C and 300 MPa, with oxygen fugacity controlled by the iron-wüstite buffer. BK93, data from Beeman and Kohlstedt (1993), extrapolated using an activation energy $Q = 385$ kJ/mol and an fO_2 exponent $p = 0.33$. KPF86, Karato et al. (1986). CP84, Chopra and Paterson (1984). MKP85, Mackwell et al. (1985). BMK91, Bai et al. (1991). (a) Polycrystals have a range of strength that falls between the strong and weak single-crystal levels. Note that there is also a correlation between strength and drying time. CP84 samples were dried for 60 hours at 1,200°C, and KPF86 for 40 hours; KPF86 with water at 250 ppm was dried for 20 hours. (b) Stress versus grain size for the data at strain rate 10^{-5} s^{-1}: Polycrystals are slightly weaker at smaller grain size.

strongest slip system (Figure 11.3a). The creep strength of synthetic polycrystals is comparable to that of the weaker [a] slip system (Beeman and Kohlstedt, 1993) (Figure 11.3a). This implies that dislocation creep in fine-grained polycrystals is controlled by some combination of [a] slip and other accommodating mechanisms, such as grain-boundary sliding (Fitz Gerald and Chopra, 1982; Hirth and Kohlstedt, 1995a). The 'diamond-grain structures' (Figure 11.4) found in synthetic olivine indicate significant grain-boundary sliding (Karato, Paterson, and Fitz Gerald, 1986; Drury and Humphreys, 1988). Figure 11.3b shows that creep strength for polycrystals appears to be slightly grain-size-sensitive.

A transition to low-temperature plasticity and power-law breakdown in olivine occurs between 200 and 600 MPa (Tsenn and Carter, 1987). Power-law-breakdown creep should occur only in the shallow lithosphere in parallel with brittle processes (Kirby, 1980; Kohlstedt et al., 1995).

At very low stress, a transition can occur to Harper-Dorn creep (Ruano, Wadsworth, and Sherby, 1988). The mechanism involves dislocation climb and is grain-size-insensitive. The transition to Harper-Dorn creep occurs if the dislocation

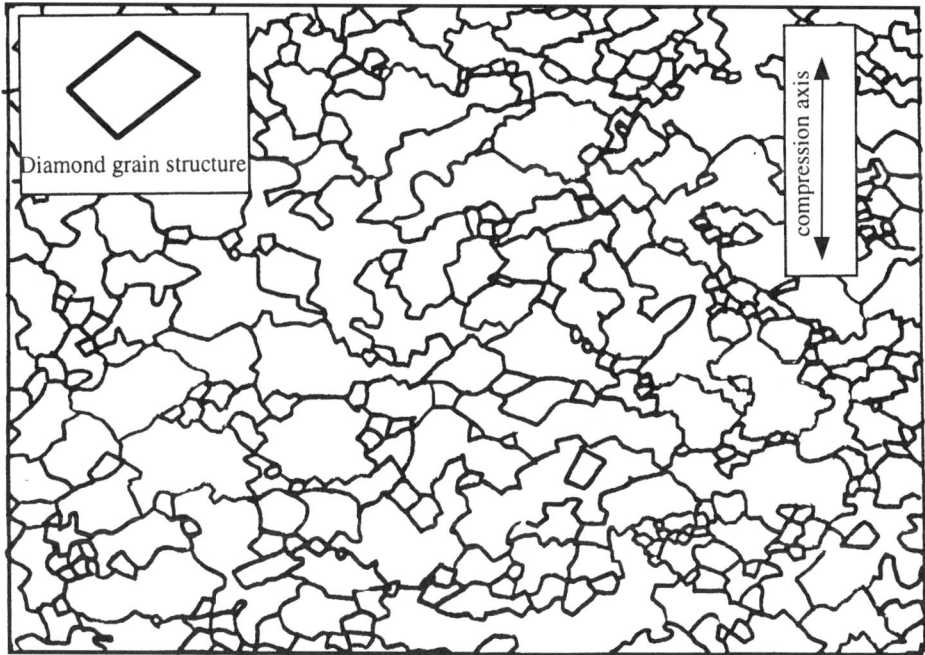

Figure 11.4. Diamond-shaped grain boundaries in an olivine polycrystal deformed to 19% strain at 1,250–1,300°C. Many of these grain boundaries are aligned at an angle of 30–50° to the compression axis. Such alignments are indicative of coupling between sliding and migration of grain boundaries.

density becomes stress-independent, which is expected at stresses below about 1 MPa for olivine (Langdon, Degghan, and Sammis, 1982).

A transition to diffusion creep occurs at small grain size and low stress in olivine rocks (Cooper and Kohlstedt, 1984; Chopra, 1986; Karato et al., 1986; Hirth and Kohlstedt, 1995b). The creep rate increases with increasing oxygen fugacity (Hirth and Kohlstedt, 1995b). Hirth and Kohlstedt (1995b) have suggested that diffusion creep is controlled by diffusion of Si via grain boundaries, although the role of grain-boundary sliding has not been assessed.

11.3.2. Influence of Pressure on Olivine Deformation

The influence of pressure on creep is usually expressed through an activation volume V^* [equation (2)]. The activation volume is the most difficult parameter to measure experimentally because of the difficulty of making accurate stress measurements in very high pressure ($P > 1$ GPa) deformation apparatus. Table 11.1 lists V^* for dislocation and diffusion processes. Recent studies indicate similar values of V^* for both dislocation creep and recovery (Bussod et al., 1993a; Karato, Rubie, and Yan, 1993; Kohlstedt et al., 1995).

V^* for diffusion creep can be controlled by a wide range of processes (Gordon, 1985). Karato and Wu (1993) have suggested that a V^* of 6 cm^3/mol (dislocation

Table 11.1. *Data on pressure sensitivity of dislocation and diffusion processes in olivine polycrystals*

Process	Conditions	V^* (cm^3/mol)	Source
Dislocation creep	0.7–2.2 GPa	28	Borch and Green (1989)
Dislocation creep	6–12 GPa	5–10	Bussod et al. (1993a)
Recovery	[a] dislocations (7–10 GPa)	6	Karato et al. (1993)
Recovery	all dislocations (0–2 GPa)	14–19	Karato and Ogawa (1982)
Diffusion			
(Mg) lattice	up to 10 GPa	1–3.5	Chakraborty and Farver (1994)
(Mg) grain boundary	up to 10 GPa	not more than 1	Farver et al. (1994)

recovery) could also apply to diffusion creep controlled by lattice diffusion. V^* theoretically scales with activation enthalpy (Karato and Wu, 1993), so that V^* for diffusion creep should be 60–80% of V^* for dislocation creep. Experimental data are needed to test these hypotheses as a basis for extrapolation of flow laws to depth.

11.3.3. Influence of Water on Olivine Deformation

The presence of trace amounts of water, either as a free fluid or dissolved in mantle minerals, has a drastic softening effect (Carter and Ave Lallement, 1970; Blacic, 1972; Chopra and Paterson, 1981; Karato et al., 1986). Work on olivine single crystals (Mackwell, Kohlstedt, and Paterson, 1985) suggests a dependence of creep rate on water fugacity, and therefore a sensitivity of deformation to the composition and pressure of any fluids present. Experiments on water solubility in olivine have shown that the water-defect concentration is proportional to water fugacity (Bai and Kohlstedt, 1992b, 1993). Incorporation of water as a point defect into the olivine structure could increase the concentration of extrinsic diffusing species controlling deformation and thus result in weakening. Consistent with that idea, the introduction of water is generally found to lower the activation enthalpy for both dislocation creep and diffusion creep (Chopra and Paterson, 1981, Rutter and Brodie, 1988).

Results on wet natural polycrystals have shown variations of creep strength that correlate with the initial grain size (Chopra and Paterson, 1981; Chopra, 1986). This was explained by Chopra and Paterson (1981) as an effect of water on grain boundaries. Studies on single crystals have clearly indicated that water has some effect on the motions of dislocations (Mackwell et al., 1985). Other effects of water that could have some significance for creep behaviour include (1) enhancements of dynamic recrystallization and grain growth rates (Chopra and Paterson, 1984; Karato, 1989a) and (2) segregation of hydrogen along dislocation cores, which could lead to the climb dissociation of [c] dislocations (Drury, 1991). We conclude that the mechanisms of 'water weakening' are still poorly understood, and there is great uncertainty about the relative roles of grain-boundary processes and intragranular processes, and even about the qualitative flow laws as functions of water content (McLaren, 1991, pp. 336–41; Hirth and Kohlstedt, 1996).

11.3.4. Influence of Melts on Olivine Deformation

During the past decade there has been increasing interest in the potential effects of melt in lowering the resistance to solid-state flow, as recently reviewed by Kohlstedt and Zimmerman (1996). Creep in the presence of melt is important in the mantle beneath oceanic spreading ridges and above mantle plumes. The seismic low-velocity zone could be associated with incipient melting (e.g., Anderson and Sammis, 1970; Ringwood, 1975; Green and Falloon, Chapter 7, this volume) or with solid-state viscoelastic relaxation (e.g., Jackson and Rigden, Chapter 9, this volume). If melts are present, the low-velocity zone might, in principle, also be a zone of low viscosity.

In the case of dislocation creep, melt has only a modest influence on strength if the melt fraction is less than 5–6% (Beeman and Kohlstedt, 1993). The strength can be reduced by an order of magnitude at higher melt contents (Jin, Green, and Zhou, 1994b; Hirth and Kohlstedt, 1995a). Hirth and Kohlstedt (1995a) have suggested that deformation at high melt contents occurs by grain-boundary sliding accommodated by [a] dislocation creep. Melt-enhanced recrystallization (Bussod, 1990; Jin et al., 1994b) or a transition to diffusion creep could also produce weakening. The creep strength for super-solidus lherzolite reported by Jin et al. (1994b) is similar to that predicted from the diffusion-creep data of Hirth and Kohlstedt (1995b). In experiments with wet olivine, melt exerts a modest influence on creep strength (Bussod, 1990), but produces an activation enthalpy (200 kJ/mol) significantly lower than that for nominally melt-free, wet olivine (Table 11.2).

For diffusion creep, early studies found that 1–3% melt decreased the viscosity by a factor of 2–5 (Cooper and Kohlstedt, 1984). This was explained by the distribution of melt in 'tubes' along the grain edges – so as to minimize surface free energy. For diffusion creep, deformation is controlled by melt-free olivine grain boundaries, as for melt-free rocks. Large weakening effects occur at higher melt contents (Hirth and Kohlstedt, 1995b). For diffusion creep where melt is present as layers 50–100 nm thick along the grain boundaries, viscosity is reduced by a factor of 20–50.

When both water and melt are present, water partitions preferentially into the melt phase. Introduction of melt could therefore reduce the water concentration in olivine crystals and produce hardening (Karato, 1986; Kohlstedt and Chopra, 1994; Hirth and Kohlstedt, 1995b). In contrast, adding wet melt to a dry polycrystal could induce intragranular water weakening without melt-induced weakening (Beeman and Kohlstedt, 1993). One problem in the separation of these effects is that small volumes of melt were probably present in nominally melt-free samples, such as the water-added samples of Karato et al. (1986).

Table 11.2. *Flow laws for olivine polycrystals*[a]

Creep	A_t (MPa^{-n} · m^m · s^{-1})	n	m	p^b	Q_t (kJ/mol)	Reference fO$_2$ buffer
Power-law dislocation creep						
Dry [c] slip[c]	2.884×10^4	3.6	0	0	535	Fe-FeO
Dry [a] slip[c]	2.48	3.5	$\geq 0^c$	0.33	385	Fe-FeO
Wet[d]	9.55×10^3	3.35	0	[d,e]	444	Fe-FeO
Diffusion creep						
Dry[f]	2.46×10^{-10}	1.0	3	0.15	347	Ni-NiO
Wet[g]	1.5×10^{-12}	1.0	3	[e]	250	Fe-FeO
Melt present[f]	3.0×10^{-9}	1.0	3	0.15	347	Ni-NiO

[a] $\dot{\varepsilon}_{(s^{-1})} = A_t (\sigma_{(\text{MPa})})^n (d_{(\text{m})})^{-m} \exp[-Q_t \text{ (kJ/mol)}/RT_{(\text{K})}]$.

[b] The strain rate at a different oxygen buffer can be calculated from $\dot{\varepsilon} = \dot{\varepsilon}_{\text{ref}} \times (fO_2/fO_2 \text{ ref})^p$; $\dot{\varepsilon}_{\text{ref}}$ and fO_2 ref are the pre-exponential constant and the oxygen fugacity at the reference buffer.

[c] The flow law for [c] slip is from Chopra and Paterson (1984). The flow law for [a] slip is based on the work of Karato et al. (1986), Hitchings et al. (1989), Beeman and Kohlstedt (1993), and Hirth and Kohlstedt (1995b). Note that this flow law applies only to small grain sizes close to the transition to diffusion creep.

[d] The flow law for Anita Bay dunite of Chopra and Paterson (1981) is preferred. This flow law should apply to a situation where the olivine is water-saturated and in equilibrium with a free fluid at pressures of around 300 MPa.

[e] The effect of oxygen fugacity is not known; the flow law was obtained mainly at the iron-wüsite buffer. The activation enthalpy is from Chopra (1986).

[f] Based on Hirth and Kohlstedt (1995a). For the cases of high melt contents, the activation enthalpies have not been measured, so it is possible that creep is controlled by diffusion in the melt with an activation enthalpy of 160–200 kJ/mol (Bussod, 1990).

[g] Flow law from Karato et al. (1986) and Chopra (1986).

11.3.5. Olivine Microstructures in Experimental Studies
11.3.5.1. Grain Size

The grain size of olivine polycrystals influences rheology, controlling the deformation mechanisms and the diffusion-creep strength (Karato et al., 1986). During deformation, the initial grain size is not usually stable, because of dynamic recrystallization or grain growth (Drury and Urai, 1990).

During diffusion creep, grain growth can occur (Karato et al., 1986). Grain growth in single-phase olivine is fast, and water increases the growth rate (Karato, 1989b). When extrapolated to upper-mantle conditions, these rate laws predict that small grain size would not persist in a single-phase olivine rock. Fast grain growth will restrict the operation of diffusion-creep processes (Karato, 1989b; Handy, 1989). However, upper-mantle peridotites contain 40–50% of phases other than olivine. In such polyphase materials, grain growth is inhibited, and a stable grain size develops that depends on the grain size and volume fraction of the secondary phases (Olgaard and Evans, 1988).

For dislocation creep, recrystallization during deformation produces a stress-dependent grain size (Twiss, 1977). The stable grain size is maintained by the parallel operation of grain growth and grain-size-reducing processes (Derby and Ashby, 1987). Some studies (de Bresser, Spiers, and Reis, 1994) have indicated that grain size can also be temperature-dependent, but for olivine, the recrystallized grain size seems to be insensitive to temperature and water content (van der Wal et al., 1993).

Data on the effect of melt content on recrystallized grain size are conflicting. Van der Wal et al. (1993) reported that the grain size showed no correlation with melt content (for low melt fractions), but Hirth and Kohlstedt (1995b) found grain sizes to be smaller by a factor of 4 in olivine aggregates containing 3–5% basalt.

11.3.5.2. Melt Distributions

Small volume fractions of melt can influence those physical properties of rocks that are sensitive to the grain-scale distribution of melt. Established theory suggests that the equilibrium distribution of melt in polycrystals is determined by differences between the surface energies of grain boundaries and melt–crystal interfaces (Waff and Bulau, 1982). These energies control the dihedral angle (θ), the angle subtended at the junction between two crystal–melt interfaces and a grain or interphase boundary. Equilibrium theory predicts that when $\theta = 0°$, melt films will occur. If $\theta > 0°$, then grain boundaries will be melt-free, and melt will be restricted to a connected network of tubes along grain edges. Previous studies had found dihedral angles of 20–50° in upper-mantle rocks (Kohlstedt, 1992). However, recent studies have shown that wide melt layers, up to 2–5 μm thick, occur along flat grain boundaries that appear to be low-index crystal planes (e.g., Bussod and Christie 1991; Waff and Faul, 1992; Jin et al., 1994b; Hirth and Kohlstedt, 1995b). This has been explained by anisotropy of the energies of crystal–melt interfaces (Waff and Faul, 1992) and by stress-induced wetting (Jin et al., 1994b). Kohlstedt and Zimmerman (1996) have discussed how melt geometry can be influenced by strain or by stress during deformation.

The established theory does not include the effects of adsorption of melt at crystal surfaces. Thin adsorbed melt films could be stable even with a dihedral angle $\theta > 0°$, because of the surface forces that can act to stabilize thin melt films (Hess, 1994). Until recently it was thought that melt films were never present along olivine grain boundaries. However, Jin et al. (1994b) have shown that during deformation, melt layers can spread along grain boundaries, and in a synthetic olivine-orthopyroxene rock, Drury and Fitz Gerald (1996) have found thin nanoscale melt films coexisting with melt bodies characterized by a finite dihedral angle. These considerations imply that direct high-resolution studies of grain boundaries are necessary to determine if melt films are stable and that the dihedral angle alone is an inadequate criterion. To fully describe the distribution of melt in a polycrystal, with rheological consequences in mind, it is important to distinguish between nanoscale melt films and bulkier melt bodies such as tubes and wide layers (Figure 11.5).

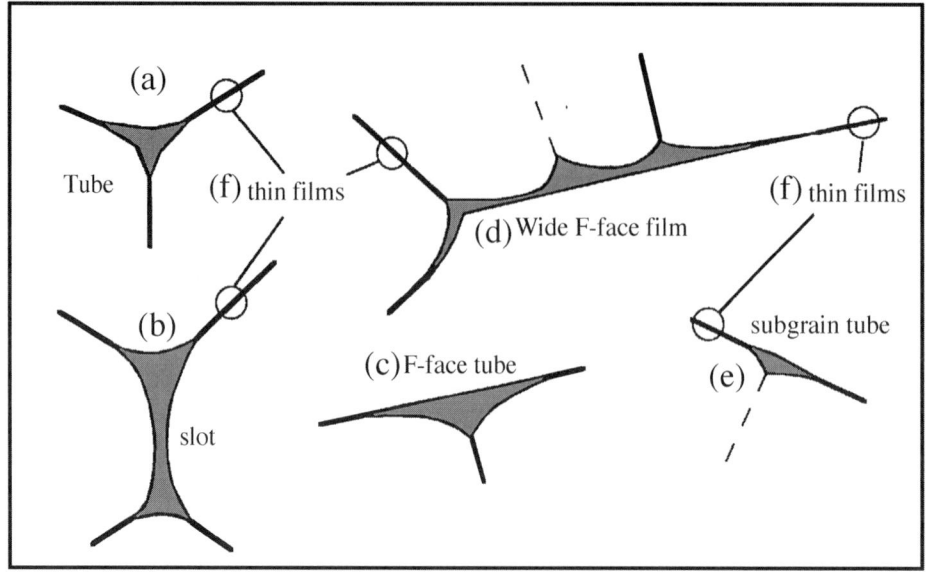

Figure 11.5. Schematic diagram showing types of melt bodies found in upper-mantle poly-crystals. (a) Grain-edge tube with dihedral angles between 20° and 50°. (b) Melt 'slot' formed by coalescence of two grain-edge tubes during deformation. (c) Tube formed along flat (F-face) olivine grain boundary. (d) Wide F-face layer formed by coalescence of F-face tubes along boundary of large grain. (e) Melt tube formed along intersection between subgrain boundary and grain boundary. (f) Adsorbed nanoscale melt films along grain boundaries.

11.3.6. Crystallographic Preferred Orientation

The development of a crystallographic preferred orientation (CPO) is influenced by several factors, including the active slip systems and the kinematics, dynamics, and extent of deformation (Nicolas and Poirier, 1976; Schmid and Casey, 1986; Karato, 1987; Wenk and Christie, 1991). In the dislocation-creep regime, strong preferred orientations develop, whereas with diffusion creep there may be some preferred orientation, but it is usually weak (Karato, 1988; Schmid, Panozzo, and Bauer, 1987). Experiments have shown that in uniaxial compression, the [b] axis of olivine tends to become aligned with the compression direction, the shortest axis of the strain ellipsoid (Ave Lallement and Carter, 1970; Nicolas, Boudier, and Boullier, 1973). For simple shear, Zhang and Karato (1995) have shown that at high strains the easy-slip direction [a] becomes aligned sub-parallel to the shear direction, and the easy-slip plane becomes aligned sub-parallel to the shear plane.

Similar CPOs are found in naturally deformed olivine rocks, with a strong [a] maximum aligned at 0–30° to the inferred shear direction. Most CPOs in natural peridotites are consistent with easy [a] slip, apart from those in some rocks deformed at lower temperatures (700–900°C), which indicate easy [c] slip (Gueguen and Nicolas, 1980; Vissers et al., 1995).

The development of a strong CPO can affect the rheology. In simple shear, the slip-plane alignment could result in a rheology controlled by the weak [a] slip system (Gueguen and Nicolas, 1980). In contrast, during compression, crystals become aligned in an orientation unfavourable for slip, and the development of the CPO should produce significant hardening (P. N. Chopra, personal communication, 1991).

A rock with a strong CPO will have anisotropic seismic wavespeeds (Mainprice, Vauchez, and Montagner, 1993). Strong seismic anisotropy can be useful as an indication of dislocation creep (Karato, 1988) and the flow dynamics (Zhang and Karato, 1995). Indeed, Karato (1992) has suggested that the Lehmann discontinuity, variably observed at depths near 200 km, may be explainable by a transition from anisotropy caused by dislocation creep to elastic isotropy associated with diffusion creep at greater depth.

11.3.7. Flow Laws for Olivine Polycrystals

A synthesis of flow laws for olivine polycrystals is given in Table 11.2. Four flow regimes are possible (Figure 11.6), and flow laws can be derived from experimental data apart from Harper-Dorn creep, which has not been verified experimentally in

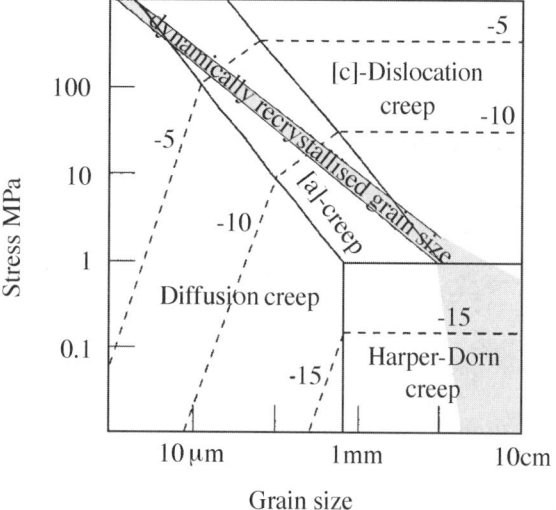

Figure 11.6. Deformation-mechanism map for olivine to illustrate general aspects that will be featured in subsequent figures representing deformation of polycrystals under specific conditions. Strain-rate contours (dashed) for 10^{-5}, 10^{-10}, and 10^{-15} s^{-1} are shown. In the [a] dislocation-creep field, deformation occurs by a combination of [a] dislocation creep and grain-boundary sliding. The shaded band shows the size of dynamically recrystallized olivine grains in the dislocation-creep field. It is shown here that this stress–grain-size relationship also holds through the hypothetical Harper-Dorn creep field; however, the grain-size scaling law there could be different from that in dislocation creep through the existence of some dependence on the stress exponent n, as suggested by Derby and Ashby (1987).

olivine. The flow laws in Table 11.2 apply to rocks in which the oxygen fugacity (fO_2) is controlled by oxide/silicate buffers. The dependence of creep rate on fO_2 is usually expressed as

$$\dot{\varepsilon} = A_c \sigma^n d^{-m} (fO_2)^p \exp(-Q_c/RT) \tag{3}$$

In polycrystal experiments and in the upper mantle, fO_2 and temperature do not vary independently. They are related by buffering reactions, where fO_2 is related to temperature by

$$fO_2 = C \exp(-Q_b/RT) \tag{4}$$

So the flow law at buffered fO_2 can be rewritten as

$$\dot{\varepsilon} = A_t \sigma^n d^{-m} \exp(-Q_t/RT) \tag{5}$$

The activation enthalpy ($Q_t = Q_c + pQ_b$) and the pre-exponential constant ($A_t = A_c C^p$) derived from a set of buffered deformation experiments are thus dependent on the oxygen-fugacity buffer. For example, Q_b for the iron-wüstite buffer is 521 kJ/mol, and p is 0.33, contributing about 170 kJ/mol to the apparent activation enthalpy of [a] dislocation creep (e.g., Bai and Kohlstedt, 1992a). Table 11.2 lists the constants for flow laws obtained with either the Fe-FeO buffer or the Ni-NiO buffer. The strain rate for a different oxygen buffer can be calculated from $\dot{\varepsilon} = \dot{\varepsilon}_{ref}(fO_2/fO_{2\ ref})^p$. In single crystals, room-pressure measurements allow fO_2 and temperature to be varied independently, so that Q_c can be measured directly (Bai et al., 1991) for [a] and [c] dislocation creep.

11.3.8. Deformation Mechanisms and Rheology of the Upper Mantle

It is important to determine whether the upper mantle has wet or dry rheology. Chopra and Paterson (1984) showed that only trace amounts of water (200–300 ppm) are required to produce water weakening. Estimates of the bulk water content within the Earth include 5–35 ppm for oceanic lithosphere, 100–200 ppm for the MORB source, 200–500 ppm for the OIB source (Bell and Rossman, 1992), and less than 200 ppm at the base of the upper mantle (Wood, 1995). If water weakening depends on the bulk concentration of water, then wet rheology should apply to most of the upper mantle (Tsenn and Carter, 1987).

The magnitude of the water-weakening effect, however, does not depend directly on the bulk water content, but on the thermodynamic activity of water (Mackwell et al., 1985; Karato et al., 1986; Kohlstedt et al., 1995). Very little is known concerning the variation of water fugacity in the upper mantle. A free aqueous fluid phase is not expected to occur throughout most of the upper mantle, because of the stability of hydrous solid phases such as amphibole and clinohumite (Liu, 1993).

Given the wide range of bulk water contents and variations in hydrous-phase assemblages, it is likely that the most hydrous regions, such as above subduction zones and in hydrated lithospheric shear zones, are water-weakened (Karato and Wu, 1993). Kohlstedt et al. (1995) have suggested that continental lithosphere should have a wet rheology, and oceanic lithosphere a dry rheology. However, it could be that water weakening need not be invoked for flow in the upper mantle (Karato and Wu, 1993; Ranalli, 1995); the hardest flow law (for [c] slip) in Table 11.1, for dry olivine, predicts a viscosity of $3 \times 10^{20\pm1}$ Pa · s at 1,300°C, at a strain rate of 1×10^{-15} s^{-1}, compatible with geophysical estimates of upper-mantle viscosity (Lambeck and Johnston, Chapter 10, this volume).

The question whether deformation in the upper mantle occurs by dislocation creep or diffusion creep can be addressed via deformation-mechanism maps (Karato et al., 1986; Ranalli, 1995). However, as noted by Frost and Ashby (1982), one must be careful not to attribute too much precision in the deformation-mechanism maps. Uncertainties in flow-law constants translate into compounded uncertainties, of at least one order of magnitude, in stress and grain size and in the positions of mechanism field boundaries on the deformation-mechanism maps. The flow laws in Table 11.2, extrapolated to different oxygen-fugacity buffers, have been used to construct deformation-mechanism maps for several upper-mantle environments. The factors used to extrapolate from an oxygen buffer to another buffer are listed in the legend for each map. The transition between dislocation creep and diffusion creep is taken as the condition where the diffusion-creep strain rate is equal to the [a] dislocation-creep strain rate (Hirth and Kohlstedt, 1995b). The transition between [a] dislocation creep and [c] dislocation creep is less well constrained, as shown in Figure 11.3a (see fig. 2c of Hirth and Kohlstedt, 1995b). The majority of strain in the dislocation-creep field results from [a] slip, but other deformation modes must accommodate incompatibilities of strain between grains. This accommodation strain occurs by [c] slip in the [c] dislocation-creep field, and by diffusion creep and grain-boundary sliding in the [a] dislocation-creep field. With decreasing grain size, transition from [c] dislocation creep to [a] dislocation creep will occur gradually as the diffusion-creep strain rate becomes significant compared to the [c] slip strain rate. On the deformation-mechanism maps of Figure 11.7, the transition between the fields of [a] and [c] dislocation creep was selected on the basis of the condition that the diffusion-creep strain rate equal 2% of the [c] slip strain rate. The strain-rate contours in the [a] dislocation-creep regime are schematic and not based on a formal multimechanism flow law. The relative importance of [a] slip, [c] slip, grain-boundary sliding, and diffusion creep in this regime, and how these processes combine to accommodate deformation, is unknown.

Figure 11.7a shows a deformation-mechanism map constructed for the sub-lithospheric mantle beneath an oceanic spreading ridge for peridotites with 5–8% melt. Studies of oceanic peridotites from ophiolites (Ceuleneer and Rabinowinc, 1992) have revealed dynamically recrystallized grain sizes in the range of 1–10 mm, suggesting stress levels of around 3–0.3 MPa. For those stress and grain-size ranges, [a] dislocation creep is predicted to dominate, in agreement with the CPOs for ophiolites (Nicolas, 1989).

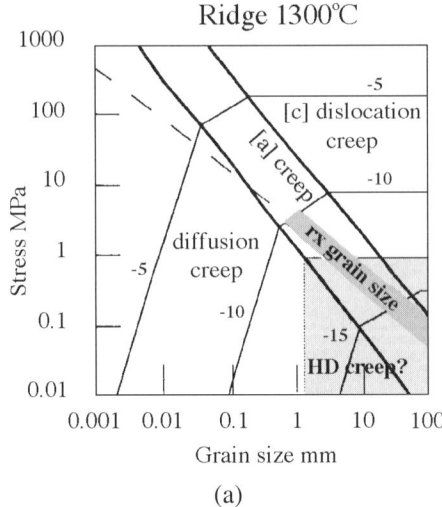

(a)

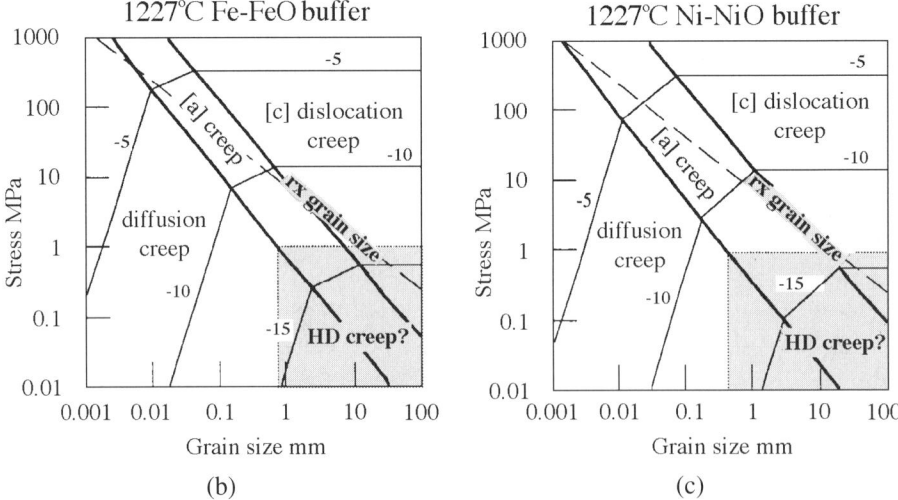

(b)   (c)

Figure 11.7a–c. Deformation-mechanism maps for dry olivine at high temperatures (>1,000°C). (a) 1,300°C, 1,000 MPa, and oxygen fugacity at one log unit below the quartz-fayalite-magnetite buffer. The stress–recrystallized-grain-size relationship 'rx grain size' is based on the data of Hirth and Kohlstedt (1995b). For [a] dislocation creep, $\dot{\varepsilon} = 7.234\dot{\varepsilon}_{(Fe-FeO)}$; for diffusion creep, $\dot{\varepsilon} = 0.58\dot{\varepsilon}_{(Ni-NiO)}$; see Table 11.2. (b) Pressure = 300 MPa; oxygen fugacity at Fe-FeO buffer. For [a] dislocation creep, $\dot{\varepsilon} = \dot{\varepsilon}_{(Fe-FeO)}$; for diffusion creep, $\dot{\varepsilon} = 0.227\dot{\varepsilon}_{(Ni-NiO)}$; see Table 11.2. (c) Pressure = 300 MPa; oxygen fugacity at Ni-NiO buffer. For [a] dislocation creep, $\dot{\varepsilon} = 26.15\dot{\varepsilon}_{(Fe-FeO)}$; for diffusion creep, $\dot{\varepsilon} = \dot{\varepsilon}_{(Ni-NiO)}$; see Table 11.2. For parts b and c, data for "rx grain size" of van der Wal et al. (1993) have been used.

The deformation-mechanism maps in Figures 11.7b and 11.7c represent two oxygen fugacities at a temperature of 1,227°C that might exist towards the base of the conductive lithosphere. Dislocation creep should occur, with dynamically recrystallized grain sizes (1–20 mm) produced at geologic strain rates (10^{-12}–10^{-15} s^{-1}) at both the reducing conditions (Fe-FeO buffer) and the relatively oxidizing conditions

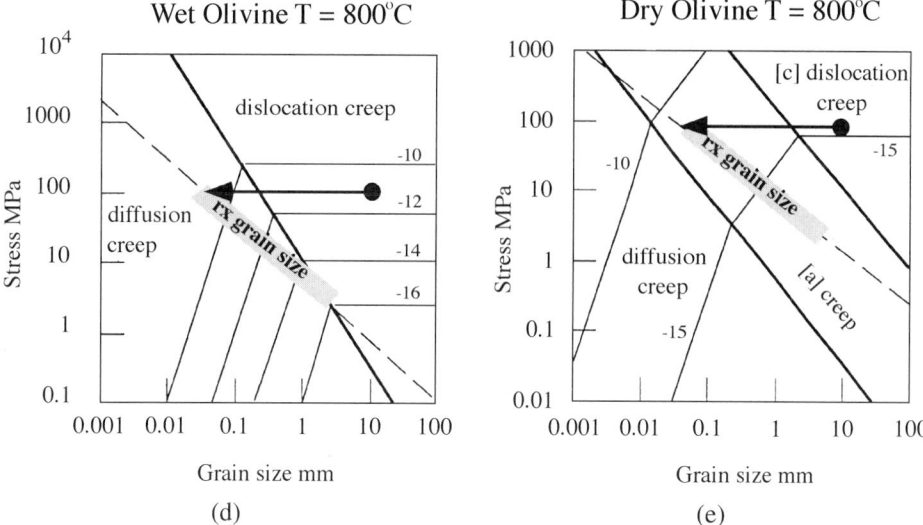

(d)   (e)

Figure 11.7d–e. Deformation-mechanism maps for olivine at low temperatures (<900°C). (d) Wet olivine, based on the flow laws of Karato et al. (1986). (e) Dry olivine at a pressure of 1,000 MPa, and oxygen fugacity at the quartz-fayalite-magnetite buffer. For [a] dislocation creep, $\dot{\varepsilon} = 23.44\dot{\varepsilon}_{(Fe-FeO)}$; for diffusion creep, $\dot{\varepsilon} = 0.789\dot{\varepsilon}_{(Ni-NiO)}$; relationship "rx grain size" from van der Wal et al. (1993). The arrows show the effects of grain-size reduction produced by dynamic recrystallization. In wet olivine, recrystallization could lead to a transition to diffusion creep. In dry olivine, recrystallization results in a transition from [c] dislocation creep to [a] dislocation creep.

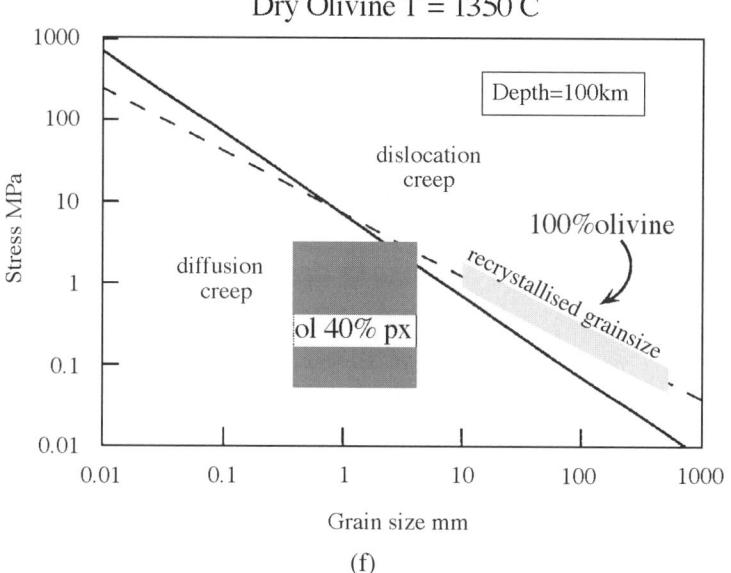

(f)

Figure 11.7f. Deformation-mechanism map for dry olivine at high temperature based on flow laws of Karato and Wu (1993). The grain size in a single-phase olivine rock is set by dynamic recrystallization, and the dominant deformation mechanism is dislocation creep. In contrast, for a common polyphase upper-mantle rock, the 40% pyroxene will restrict grain growth, so the grain size might be stabilized around 0.5–5 mm. In that case, diffusion creep would become the dominant mechanism.

(Ni-NiO buffer). Although some change is indicated in the creep accommodation (viz., amount of [c] slip) related to oxygen fugacity, it is not possible for diffusion creep to operate if dynamic recrystallization controls grain size. Diffusion creep could occur only at stresses below 1.0 MPa in rocks with a grain size of 0.5 mm (or stresses below 0.1 MPa with a 3-mm grain size) stabilized by secondary phases.

At lower temperatures, below about 900°C, the dominant mechanisms could be different (Figures 11.7d and 11.7e). Rutter and Brodie (1988) showed that at a low temperature the diffusion-creep regime in wet olivine expands, such that the grain-size-versus-stress scaling law for recrystallized grains falls partly within the diffusion-creep regime (Figure 11.7d). A transition from dislocation creep to diffusion creep could be induced by a grain-size reduction due to recrystallization and would be expected to be accompanied by a large increase in strain rate (bold arrow, Figure 11.7d). Although the viability of this mechanism change has been seriously challenged (e.g., Poirier, 1985, pp. 188–90), if such a transition could occur, it would be one way of developing localized shear zones in the lithosphere (Rutter and Brodie, 1988; Handy, 1989; Drury et al., 1991; Karato and Wu, 1993). Figure 11.7e indicates that recrystallization in dry olivine could also lead to a transition from [c]- to [a]-controlled dislocation creep, possibly associated with a significant strain-rate increase and strain localization.

A deformation-mechanism map for the sub-lithospheric upper mantle is shown in Figure 11.7f. At stress levels of 0.1–1 MPa in the convecting mantle, the recrystallized grain size should range from 10 to 300 mm. Dislocation creep is the dominant mechanism at such coarse grain sizes. Secondary phases, however, can limit the recrystallized grain size to 1–10 mm. With these smaller grain sizes, diffusion creep could dominate.

If the activation volume for diffusion creep is less than that for dislocation creep, a transition to diffusion creep would be favoured by increasing depth in the upper mantle (Karato and Wu, 1993). A similar effect could occur if oxygen fugacity were to decrease with depth (Ballhaus, 1995), because diffusion creep would become favoured over [a] dislocation creep. There could also be a transition from Si- to O-controlled diffusion creep at depths between 300 and 500 km (Ranalli, 1995), which could result in a change in deformation mechanism in the deep upper mantle.

Predictions of the dominant deformation mechanisms from Figure 11.7 are broadly in agreement with the features of naturally deformed peridotites (Nicolas, 1989; Drury et al., 1991). Deformation that occurs at temperatures less than about 950°C is concentrated in narrow, fine-grain shear zones, whereas higher-temperature deformation involves dislocation creep in kilometer-wide deformation zones. The development of fine-grained (50–5-μm) mantle shear zones deforming by diffusion creep at $T < 900$–950°C could result in a drastic lithospheric weakening (Vissers et al., 1995). The rheology of some natural shear zones (Figure 11.8) could be controlled by diffusion creep in the orthopyroxene or polyphase bands, which tend to

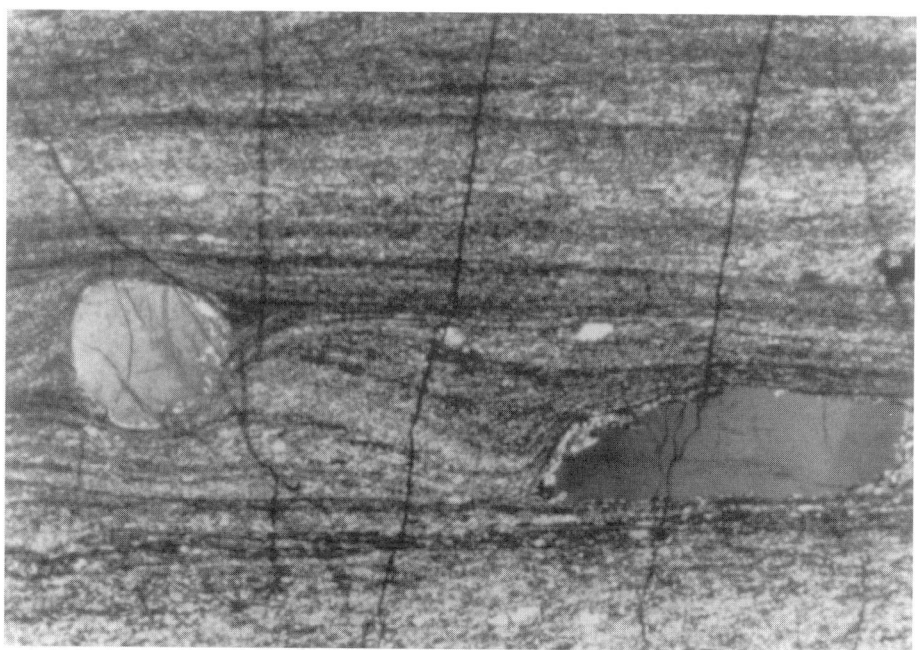

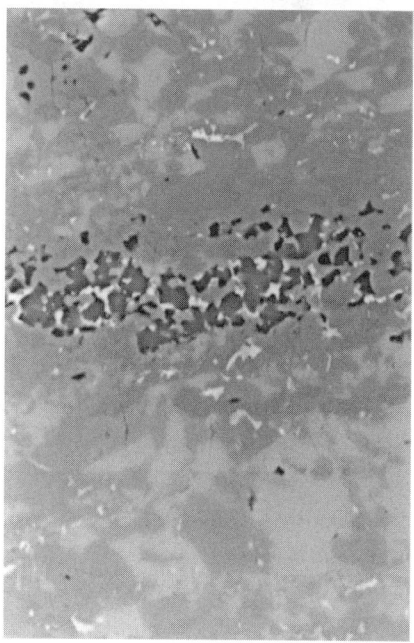

Figure 11.8. Microstructure in naturally deformed, upper-mantle peridotite from a litho-spheric shear zone. This rock was deformed at $T-P$ conditions estimated to have been 700–900°C and 500–800 MPa. During deformation, large clinopyroxene (cpx), orthopy-roxene (opx), and spinel grains break down into a fine-grain mixture of olivine (ol) plus plagioclase (pl). The polarized-light micrograph (top) shows is a banded microstructure with relatively coarse ol + opx + cpx and finer-grained (dark) bands of ol + pl; the width of the field is 3.5 mm. Also shown is an SEM image of the compositional banding (using backscattered electrons). The central band containing pl (dark grey phase) is finer-grained than the adjacent ol-dominated bands; the width of the field here is 110 μm.

have small grain sizes compared with olivine bands (Boullier and Gueguen, 1975; Vissers et al., 1995).

11.4. The Deep Mantle

The deformation behaviour of the deep mantle should be controlled by the minerals wadsleyite [β-(Mg, Fe)$_2$SiO$_4$], ringwoodite [γ-(Mg, Fe)$_2$SiO$_4$ spinel], and majorite (garnet) in the transition zone and by the minerals (Mg, Fe)SiO$_3$ perovskite and magnesiowüstite in the lower mantle. For the sequence of phase changes expected in the mantle, see Jackson and Rigden (Chapter 9, this volume). Constraints on the creep properties of deep-mantle materials can be determined in several ways. Minerals with common crystal structure and bonding can be described as isomechanical groups whose normalized mechanical properties are the same for each group member (Frost and Ashby, 1982; Ranalli, 1995). If the properties of an isomechanical group can be determined from experiments on analogue phases, stable at lower pressures than their deep-mantle counterparts, then the creep properties of the latter can be predicted. Using this approach, Karato (1989c) ranked the normalized creep strengths in the order garnet > spinel $\geq$ olivine $\geq$ perovskite $\geq$ MgO. In principle, the creep properties of deep-mantle phases can be also predicted by modelling diffusion kinetics (e.g., Vocadlo et al., 1995), combined with theoretical flow laws for diffusion creep and diffusion-controlled dislocation creep. It is important to note that our knowledge about the rheology of deep-mantle phases is largely qualitative and preliminary compared with our more detailed and quantitative knowledge of the deformation of the dominant upper-mantle mineral, olivine.

Relative creep strengths can also be studied using deep-mantle assemblages synthesized under nonhydrostatic stress (Poirier et al., 1986). Direct measurements of the plastic yield strength of deep-mantle phases are possible using hardness testing at room temperature and pressure (Karato, Fujino, and Ito, 1990), at high pressure in the diamond-anvil cell (Meade and Jeanloz, 1990), and at high pressure and temperature using multiple-anvil high-pressure apparatus and a synchrotron x-ray source (Weidner, Wang, and Vaughan, 1994). Extrapolation of room P–T data to mantle conditions is possible in principle, but difficult in practice (e.g., Karato et al., 1995a).

11.4.1. Direct Information on Yield Strength and Relative Creep Resistance

Meade and Jeanloz (1990) found that at room temperature and 25 GPa pressure, olivine and perovskite had similar shear strengths (5–6 GPa), with γ-(Mg, Fe)$_2$SiO$_4$ significantly stronger (8 GPa). MgO at 25 GPa pressure has the lowest shear strength (2–3 GPa) of any of the deep-mantle phases (Meade and Jeanloz, 1988). At lower pressures, olivine becomes much weaker than perovskite (Meade and Jeanloz, 1990; Karato et al., 1990). This is all consistent with Karato's (1989c) ranking from analogue phases.

Samples synthesized at high pressure in multi-anvil or diamond-anvil apparatus are commonly deformed. Information on the strength of different phases can be obtained by observing intragranular strain and dislocation substructures. For example, if the deviatoric stress level falls between the yield strengths for two phases, then the weaker phase will yield and develop a high density of dislocations, and the stronger phase will remain dislocation-free. These kinds of observations indicate that β- and γ-(Mg, Fe)$_2$SiO$_4$ are stronger than α-(Mg, Fe)$_2$SiO$_4$ (Boland and Liebermann, 1983; Rubie and Champness, 1987; Remsberg et al., 1988), that garnets are stronger than silicate spinels (Madon and Poirier, 1980; Ingrin and Poirier, 1995), and that MgSiO$_3$ perovskite is stronger than magnesiowüstite (Poirier et al., 1986).

11.4.2. Rheological Data for Deep-Mantle Minerals and Their Analogues
11.4.2.1. Spinels

Creep-strength data for crystals with the spinel structure are limited and insufficient to determine whether or not the spinels form an isomechanical group. Karato (1989b) showed that the normalized strengths for two spinels bracketed the strength of α-(Mg, Fe)$_2$SiO$_4$ olivine. Preliminary reports indicate that the strength of γ-Mg$_2$GeO$_4$ is twice that of α-Mg$_2$GeO$_4$ deformed at the same temperature and strain rate by dislocation creep (Tingle, Green, and Borch, 1991; Tingle, Green, and Scholz, 1993). α-Mg$_2$GeO$_4$ has creep strength similar to that of α-(Mg, Fe)$_2$SiO$_4$ (Vaughan and Coe, 1981; Tingle et al., 1991).

Bussod, Katsura, and Sharp (1993b) reported that the effective viscosity for mantle β-(Mg, Fe)$_2$SiO$_4$ was about 5 times greater than that for olivine at temperatures and pressures appropriate to the transition zone. The β phase, deformed by dislocation creep accompanied by dynamic recrystallization, develops a strong CPO, which means that the upper part of the transition zone could be seismically anisotropic (Sharp, Bussod, and Katsura, 1994). Mg-Ni inter-diffusion has been measured in α-, β-, and γ-(Mg, Ni)$_2$SiO$_4$ (Farber, Williams, and Ryerson, 1994); it is three orders of magnitude faster in the β and γ phases than in α-(Mg, Ni)$_2$SiO$_4$. On the basis of those data, Farber et al. (1994) suggested that diffusion creep could be relatively fast in the transition zone compared with the upper mantle. This suggestion may not be correct, however, because the creep rates for silicates are commonly limited by oxygen or silicon diffusion.

11.4.2.2. Garnets

Karato et al. (1995a) have conducted a comprehensive study of the creep of anhydrous garnets and have found that garnets constitute an isomechanical group that can be described by a single normalized flow law and that garnets have the highest normalized creep strength amongst the mantle phases.

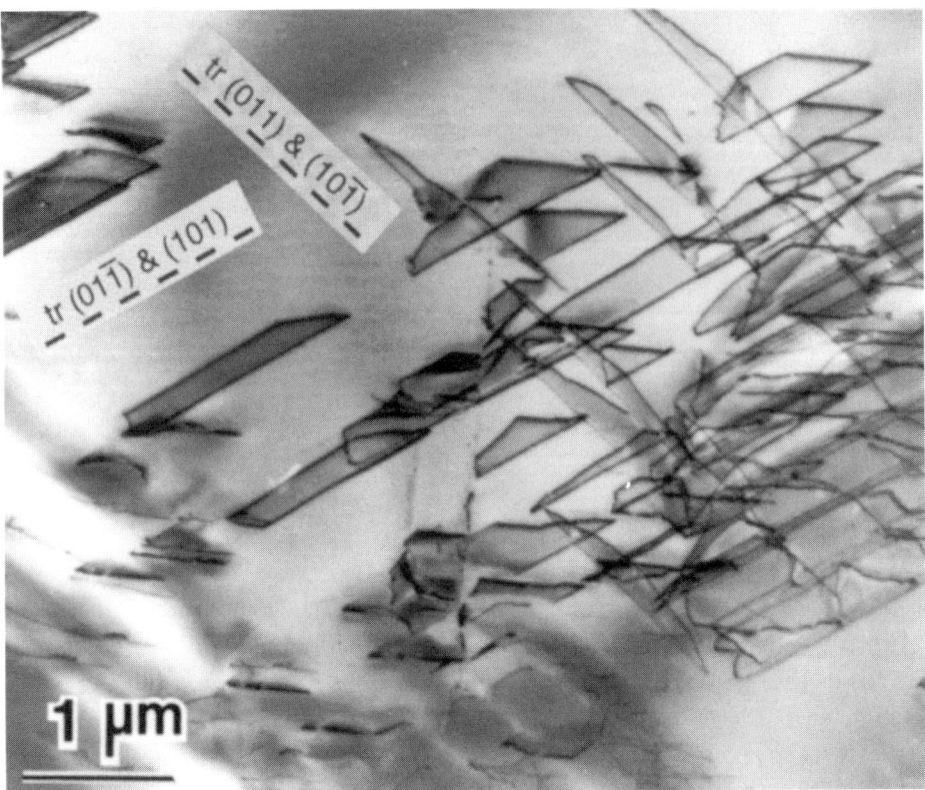

Figure 11.9. Dislocations in $BaTiO_3$ crept 1% at 1,400°C. Dislocation segments in this thin-foil specimen all lie in {110} planes of the crystal; the orientations of the traces of $(01\bar{1})$, (101), (011), and $(10\bar{1})$ are indicated. These dislocations belong to the population of defects associated with macroscopic {110} slip. (Brightfield transmission electron-microscope image from Doukhan and Doukhan, 1986, with permission.)

11.4.2.3. Perovskite

There have been extensive studies on perovskite single crystals (Doukhan and Doukhan, 1986; Poirier, Beauchesne, and Guyot, 1989). In addition to power-law dislocation creep, some perovskites deform by Harper-Dorn creep (Poirier et al., 1983; Beauchesne and Poirier, 1990). Dislocations (Figure 11.9) can slip on (100) or {110} glide planes (Poirier et al., 1989). In $SrTiO_3$, {110} slip is more difficult. If perovskites define an isomechanical group, they should be describable by a single normalized flow law of the form $n \ln(\sigma/\mu) - \ln \dot{\varepsilon} = c_1 + c_2(T_m/T)$, where c_1 and c_2 are constants. Rewritten, $\ln(\sigma/\mu) - (1/n)\ln \dot{\varepsilon}$ should be linearly related to (T_m/T) for an isomechanical group. Figure 11.10 shows that some perovskites can be described, to a first approximation, as composing an isomechanical group for {110} slip (Wang et al., 1993b), but not for (100) slip, where different perovskites have significantly different normalized strengths (Poirier et al., 1989; Wang et al., 1993b). Even the data for {110} slip show considerable scatter (Figure 11.10a), amounting to variations in viscosity by a factor of 8 at constant strain rate, and by

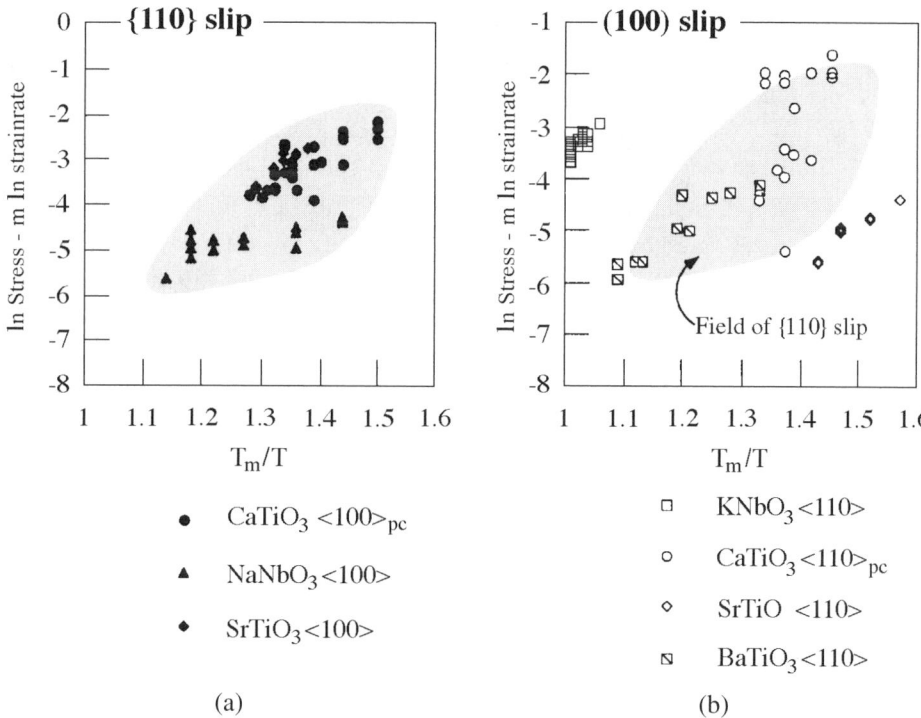

(a)   (b)

Figure 11.10. Plot of normalized creep data for perovskite crystals. The vertical axis shows the natural log of the stress normalized by the shear modulus minus m times the natural log of the strain rate, where $m = 1/n$, and n is the stress exponent in the creep law. The horizontal axis shows the reciprocal of the homologous temperature. Crystals compressed along the $\langle 100 \rangle$ direction deform by slip on $\{110\}$ planes. Crystals compressed along $\langle 110 \rangle$ deform by (100) slip; $\langle 100 \rangle_{pc}$ denotes the pseudo-cubic $\langle 100 \rangle$ directions in tetragonal $CaTiO_3$. (a) Creep data for crystals deformed along $\langle 100 \rangle$ (with slip on $\{110\}$ planes) appear to define an isomechanical group. (b) Creep data for crystals deformed along $\langle 110 \rangle$ [slip on (100)]. In this case, distinct differences occur between perovskites of different compositions. Note that (100) slip in $KNbO_3$ (open squares) is harder than $\{110\}$ slip in other perovskites (shaded field), so creep in $KNbO_3$ could be controlled by (100) rather than $\{110\}$ slip. Also note that the strength of (100) slip in $BaTiO_3$ (open squares with diagonals) is similar to the strength of $\{110\}$ in other perovskites, which implies either that $BaTiO_3$ is plastically isotropic or that $\{110\}$ slip in $BaTiO_3$ is stronger than in the perovskites plotted in part a.

a factor of 400 at constant stress. Furthermore, $\{110\}$ slip in $BaTiO_3$ and $KNbO_3$ may not fit with the 'isomechanical group' of perovskites shown in Figure 11.10a (see Figure 11.10b).

Further work clearly will be required to assess whether the $\{110\}$ slip seen in perovskites really reflects a single isomechanical group or a set of subgroups. Data from isomechanical groups of analogues can be used for a first approximation to predict the viscosity of the lower mantle, provided that (1) dislocation creep of $MgSiO_3$ perovskite is controlled by $\{110\}$ slip and not by (100) slip and (2) better constraints on normalized flow laws for $\{110\}$ are obtained.

Diffusion creep occurs in fine-grained titanate perovskites (Carry and Mocellin, 1986; Karato and Li, 1992). No systematics can be readily identified from these preliminary published reports, though different creep mechanisms appear to operate in $BaTiO_3$ and $CaTiO_3$, judging from differences in the stress exponents and activation enthalpies. Karato and Li (1992) found a significant enhancement of creep rate in $CaTiO_3$ at a flow stress of about 5 MPa resulting from transition, at $1,240°C$, from the orthorhombic to the tetragonal form, which is now thought to be cubic (Vogt and Schmahl, 1993). Karato and Li also calculated that diffusion creep would be favoured over dislocation creep for $(Mg, Fe)SiO_3$ perovskite in the lower mantle, particularly at the small grain sizes likely to result from transformation of ringwoodite to perovskite plus magnesiowüstite. The polyphase nature of the transformation product will stabilize a smaller grain size, thus favouring diffusion creep. By deforming fine-grained $CaTiO_3$ in shear, Karato et al. (1995b) experimentally determined a stress exponent near unity and a grain-size exponent of 2 and thus inferred that the lower mantle could be deforming by diffusion creep, a conclusion supported by the very weak CPO developed in the same aggregates.

Theoretical studies of diffusion have provided estimates of the temperature (E^*) and pressure (V^*) dependences of activation for oxygen diffusion in $MgSiO_3$. Gautason and Muehlenbachs (1993) used an empirical relation based on a link between anion porosity (the space in a unit cell not occupied by anions) and diffusivity to predict an activation enthalpy $E^* = 300-350$ kJ/mol. Wall and Price (1989) obtained estimates of the activation parameters $E^* = 452$ kJ/mol and $V^* = 1.6$ cm^3/mol from calculations of defect formation and migration enthalpies.

11.4.2.4. Magnesiowüstite

The rheology of MgO and $(Mg, Fe)O$ can be studied directly, and deformation-mechanism maps for MgO (Figure 11.1) have been presented by Frost and Ashby (1982). Single crystals exhibit power-law creep, with n of 3–5 at high stress. The activation enthalpy for power-law creep appears to be similar to that for oxygen self-diffusion (250–460 kJ/mol) (Frost and Ashby, 1982; Cannon and Langdon, 1988). The activation enthalpy assumes different values for intrinsic and extrinsic oxygen diffusion. Karato (1981) estimated V^* at 10 cm^3/mol for oxygen diffusion. In MgO polycrystals deformed by power-law creep, Langdon (1975) showed that significant grain-boundary sliding had occurred, and thus the activity of that process was not restricted to the diffusion-creep regime.

Diffusion creep in $(Mg, Fe)O$ has been reviewed by Gordon (1985). The addition of impurities such as FeO to MgO results in a dramatic increase in creep rates, about two orders of magnitude for an addition of 5.3 cation% Fe (Tremper et al., 1974). The rate of diffusion creep increases with oxygen fugacity. Microstructural scaling laws have been established for MgO for subgrains (Huther and Reppich, 1973) and dislocation density (Bilde-Sorensen, 1972). These relations might be used to infer values for these microstructural parameters in lower-mantle $(Mg, Fe)O$ from estimates of the flow stress.

Although not shown in Figure 11.1, Harper-Dorn creep is seen at low flow stresses (Ruano et al., 1992).

11.4.3. Role of Water-related Defects

Ringwood (1975) noted that the transition-zone phases are likely to have high solubilities of OH in comparison with upper-mantle phases. More recently, Smyth (1987) predicted substantial solubility of OH in wadsleyite. Experimental studies have confirmed these predictions, showing, for wadsleyite, concentrations up to 3.3 wt% H_2O (5.1×10^5 $H/10^6$ Si) (Young et al., 1993; Inoue, Yurimoto, and Kudoh, 1995), for ringwoodite, 4.5×10^5 $H/10^6$ Si (2.9 wt% H_2O) (Kohlstedt et al., 1996), and for $MgSiO_3$ perovskite, 700 $H/10^6$ Si (60 ppm H_2O by weight) (Meade, Reffner, and Ito, 1993).

Considering the large influence of just 200–300 ppm H_2O on the creep of olivine (Chopra and Paterson, 1984), these solubility data suggest that the rheologies of all deep-mantle phases could be strongly influenced by water (Young et al., 1993; Meade et al., 1993). The magnitude of any strength effect is extremely difficult to assess, considering the uncertainties about the weakening mechanism involving water (i.e., H or OH or H_2O), even for the extensively studied phases quartz and olivine (McLaren, 1991). A large content of OH does not necessarily equate to weakness, as illustrated by the high creep strength of amphiboles compared with pyroxenes (H. W. Green II, personal communication, 1995). Much more analysis of the defect chemistry of water in deep-mantle minerals and the interactions of water-related defects with creep will be required to address this important question.

Water could also influence the rheology of garnets. The high creep strength of garnets is related to the complex crystal structure and the difficulty of glide (Karato et al., 1995a). The Burgers vectors are long, and dissociation is common. Smith (1985) has shown that the extent of dissociation is strongly influenced by the presence of water. Water solubility in garnets is relatively high (Bell and Rossman, 1992), and thus 'water' weakening could also occur in this mineral.

11.4.4. Role of Lithologic Heterogeneity and Polyphase Assemblages

Discussions about which minerals control mantle rheology usually are based on assemblages calculated from average or bulk compositions. There is no reason that rock compositions on the scale of millimeters to centimeters should be the same as the larger-scale averages, and it is important that small-scale compositional variations be considered when modelling mantle rheology. A useful extreme to consider is material of pyrolite composition on the scale of hundreds of metres, consisting of thin interlayers of peridotite (lherzolite to dunite) and basaltic compositional bands (Allègre and Turcotte, 1986). In the upper-mantle and transition zone, the rheology would be controlled by $(Mg, Fe)_2SiO_4$ for both the uniform and small-scale layered model. In a lower mantle with small-scale layering, the abundance of (Mg, Fe)O could range locally from 30% (peridotitic compositions) to 50% (dunitic compositions). Because (Mg, Fe)O is weaker than perovskite, large-scale rheology could be controlled by the layers of dunitic-to-harzburgitic composition.

In general, the grain size for a two-phase solid is controlled by the grain size and volume fraction of the minor phase (Olgaard and Evans, 1988). Lower-mantle regions containing (Mg, Fe)O are therefore expected to have a smaller grain size than any perovskite-only regions. Madon et al. (1989) showed that coarsening of (Mg, Fe)O + perovskite mixtures slows considerably with increasing pressure. Stabilization of small grain sizes in the lower mantle could favour diffusion creep, resulting in reduced viscosity. Futhermore, theoretical studies of diffusion creep in mixtures of two phases that share chemical components (Wheeler, 1992) indicated that diffusion creep could be relatively fast in a two-phase material if there was a difference in the effective diffusivities of the component oxides. This is because strain (of a perovskite + magnesiowüstite aggregate, for example) can occur by a combination of fast-diffusing components (e.g., MgO) moving along the interphase boundaries, with slow-diffusing components (e.g., SiO_2) moving across interphase boundaries. Diffusion of SiO_2 across a perovskite–magnesiowüstite boundary is equivalent to interphase-boundary migration. Strain is then accommodated not only by diffusion along interphase boundaries but also by the changes in volume associated with interphase-boundary migration (Wheeler, 1992).

11.4.5. Deformation Mechanisms and Rheology in the Deep Mantle

Figure 11.11 shows viscosity profiles for the mantle that have been calculated from theoretical creep laws (Ranalli, 1987, 1991) and from studies on analogue phases (Karato et al., 1995a). The profiles in Figure 11.11 are for dislocation creep at a constant strain rate of 10^{-15} s^{-1} and dry conditions. The viscosity profiles suggest that a pyrolite transition zone would be slightly more viscous than the upper mantle. Parts of the transition zone could have high viscosity if the rocks are locally garnet-rich (Figure 11.11b), although only 10–20% spinel or pyroxene in a garnetite could be sufficient to control the rheology. Where subduction zones extend into the transition zone, the transformation kinetics will control the microstructure and phase assemblage (see Section 11.5). Rheology in this part of the transition zone could be very different from 'normally' convecting mantle at the same depth.

In the lower mantle the average geotherm will depend on the mode of convection; see Jackson and Rigden (Chapter 9, this volume) for a discussion of the trade-off between the inferred composition and the temperature of the lower mantle. Viscosity profiles for layered convection (labelled as 'hot' in Figure 11.11) show a low-viscosity zone in the uppermost lower mantle, associated with a thermal boundary layer that would separate the convecting regions. Viscosity profiles for whole-mantle convection will show a steady increase in viscosity with depth. These 'cold'

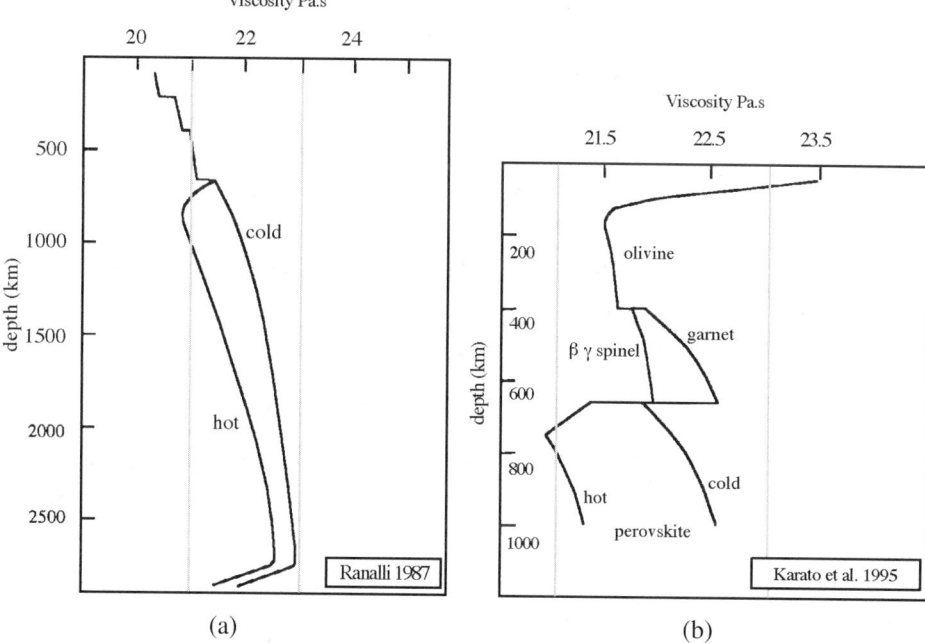

Figure 11.11. Effective viscosity–depth profiles for the Earth's mantle. (a) Data from Ranalli (1987), showing viscosity at a constant strain rate of 10^{-15} s^{-1} for dislocation creep. (b) Data from Karato et al. (1995a), based on studies of deep-mantle analogues. In the lower mantle, the 'cold' viscosity profile is for an adiabatic thermal gradient throughout the mantle, and the 'hot' viscosity profile is for the case of layered convection with a thermal boundary layer at the base of the transition zone.

profiles in Figure 11.11, especially that from Ranalli (1987), are quite similar to the depth variation of viscosity interpreted from glacial-rebound studies (Lambeck and Johnston, Chapter 10, this volume).

The changes in viscosity across the seismic discontinuities at 410 and 670 km are predicted to be small: one order of magnitude across the transition zone in Figure 11.11a. In the Karato et al. (1995a) model there is a small increase in viscosity at 410 km and a small decrease at 670 km. Below 670 km the viscosity is predicted to increase with depth, so that below 1,000 km the average lower-mantle viscosity should be one to two orders of magnitude greater than that of the upper mantle (Figure 11.11a).

These viscosity profiles do not include diffusion-creep processes or possible weakening effects due to water or additional crystalline phases. Karato and Li (1992) have suggested that diffusion creep will dominate in the lower mantle. If this is so, the viscosity would be lower than that shown in Figure 11.11b. If MgO were locally to control rheology in the lower mantle, an even lower viscosity would result. As an illustration, for an average mantle temperature of $0.5T_{\mathrm{m}}$ (Zerr and Boehler, 1994), with $\sigma/\mu = 10^{-5}$, MgO should have a viscosity of about 10^{18} Pa · s (Frost and Ashby, 1982)!

As mentioned earlier, viscosity extrapolations by Karato (1981) and Sammis et al. (1977) were based on a scaling approach to estimating thermodynamic parameters and yielded viscosity profiles quite different from those in Figure 11.11. Viscosity variations would have been larger had effective viscosity been calculated at constant stress (Karato, 1981; Ranalli, 1991).

Given the uncertainties in deformation mechanisms and the uncertain role of weakening due to water and additional phases, predictions of viscosity for the deep mantle based on measured rheologies of mantle phases and isomechanical groups remain subject to large uncertainties.

11.5. Transformations in the Mantle

Most discussions in the literature relating to transformations have long concentrated either on the effects on density (particularly in the mantle transition zone, where phase changes and reactions have been identified as functions of depth) or on the effects on elastic properties (e.g., Jackson and Rigden, Chapter 9, this volume). The potential consequences of density changes, in terms of the dynamics of the Earth's mantle (Davies, Chapter 5, this volume), have been matters of active debate for at least 30 years. The topic was extensively reviewed by Ringwood (1975), who presented a comprehensive analysis of the range of mineral phases found by direct experimentation to be stable under mantle conditions and of the transformations between them. Amongst many others, Ringwood continued after that early review to experiment and to develop theories about mantle dynamics; for a relatively recent summary, see Ringwood (1991).

There are formidable complexities in laboratory investigations of the mechanisms and kinetics describing transformations of mantle phases, as well as in the dependent geophysical interpretations. Nevertheless, so important is the topic that efforts have persisted and patterns have emerged, albeit with many qualifications. The following discussion of phase transformations for the mineral compositions $(Mg, Fe)_2SiO_4$ and $(Mg, Fe)SiO_3$ illustrates many of the experimental and theoretical difficulties that have been faced, many of which have been overcome.

11.5.1. Phase Transitions in $(Mg, Fe)_2SiO_4$ in the Mantle

Ringwood and Major (1966) experimentally transformed the α phase (olivine) of $(Mg, Fe)_2SiO_4$ into its cubic spinel polymorph, now known as either γ phase or ringwoodite (mineral name, Mg end member). In doing so, they noted the existence of a spinel-related phase, later named the β phase and given the mineral name wadsleyite. From crystal-structure determinations for these phases (e.g., Morimoto et al., 1970), it was shown that the β and γ phases are quite closely related, and both are significantly different from the α phase. Figure 11.12 shows an illustration of the similarity between β and γ structures. For more detailed contrasts between

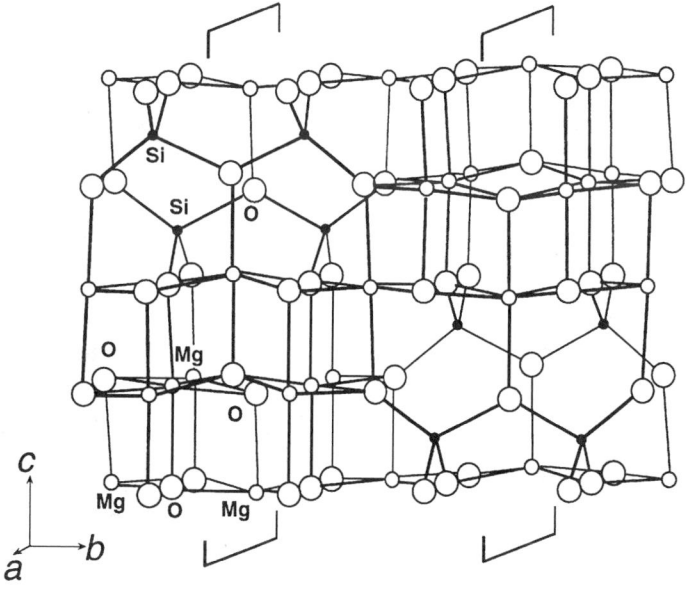

Orthorhombic crystal structure of Mg-wadsleyite β-Mg$_2$SiO$_4$

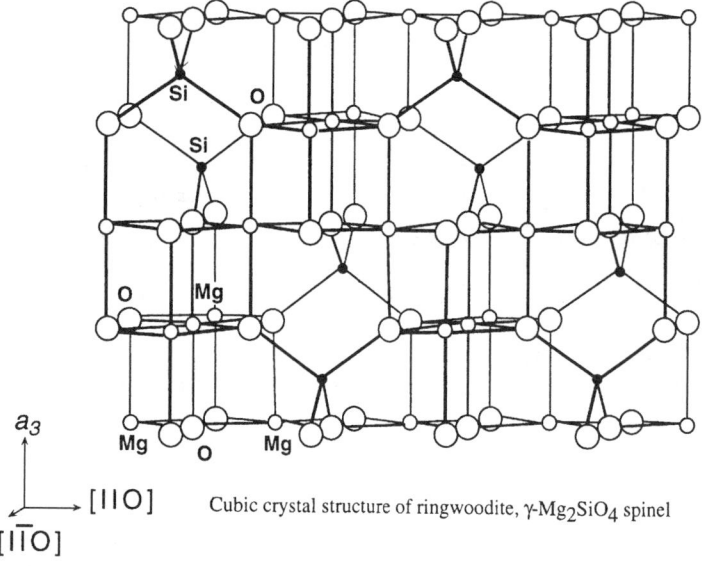

[110]   Cubic crystal structure of ringwoodite, γ-Mg$_2$SiO$_4$ spinel

[1̄10]

Figure 11.12. Schematic crystal structures for orthorhombic Mg wadsleyite (β phase) and cubic ringwoodite (γ phase). A single unit cell of the β phase is drawn above a comparable block of γ phase that has a complex relationship to its unit-cell edges (see the crystallographic axes beside both drawings). If the central half of the β-phase unit cell (viz., that part within the indicator markers above and below the drawing) were translated by $\frac{1}{2}[a \pm c]$ relative to the outer parts and the bonds were reformed, the resulting structure would closely resemble that of the block of γ phase illustrated below. On that basis, the $\beta \rightarrow \gamma$ transition in (Mg, Fe)$_2$SiO$_4$ is thought likely to involve a martensitic type of shear. (Adapted from Morimoto et al., 1970.)

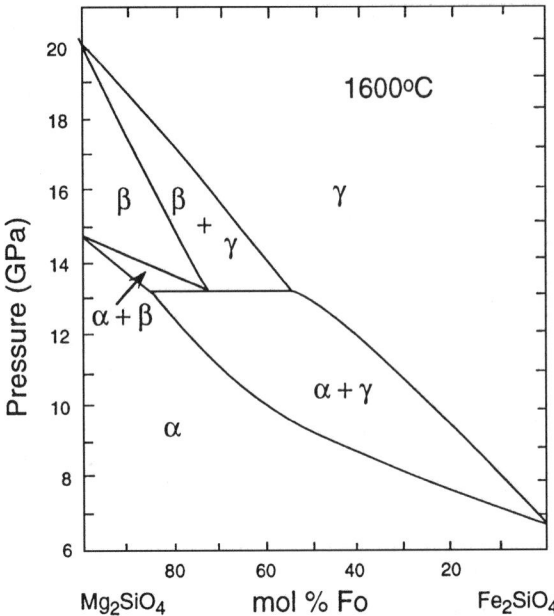

Figure 11.13. Calculated boundaries for α–β–γ transitions in $(Mg, Fe)_2SiO_4$ at $1,600°C$ based on calorimetry, but closely consistent with phase equilibria determined directly by Katsura and Ito (1989). (Adapted from Akaogi et al., 1989.)

these two and the α-phase structure, see Sung and Burns (1978) and Madon and Poirier (1983). All three phases have been shown to be thermodynamically stable in the Mg-Fe system, with stability fields that are now experimentally well established (Figure 11.13).

The close relations between the β and γ phases have led many to conclude that transition between these two phases can be accomplished by a diffusionless martensitic transformation (e.g., Price, 1983). In contrast, transition from α to either β or γ must involve major reconstruction of the polyhedral groupings and cation arrangements, though Poirier (1981) identified a complex 'synchro-shear' process for the α–γ transition in which even the close-packed oxygen lattice must be reconfigured between hcp and ccp. Madon and Poirier (1983) pointed out the potential complexities of this transition.

The mechanisms of the transitions among the three polymorphs have been studied in numerous transmission-electron-microscopy investigations of experimental products and natural equivalents preserved in meteorites. Initially, experimental investigations used analogue materials such as Mg_2GeO_4 (Vaughan and Coe, 1981) because the phase transitions there were more easily accessible to the apparatus available at that time. Vaughan and Coe demonstrated that the γ phase is intrinsically somewhat stronger (dislocation glide requires higher stresses) than its α counterpart, as predicted from the crystal structures. However, the significant reduction in grain size that accompanied the $\alpha \rightarrow \gamma$ transition implies that the aggregate creep was subsequently diffusionally controlled. It should be emphasized here that aspects of some analogue systems are inappropriate for comparison with (Mg, Fe)$_2$SiO$_4$ – for example, there is no stable β phase of Mg$_2$GeO$_4$. Co$_2$SiO$_4$ is a better analogue in this regard (Remsberg et al., 1988; Remsberg and Liebermann, 1991). Even the end member Mg$_2$SiO$_4$ undergoes a sequence of transitions, with increasing pressure, which is different from that for compositions with appreciable Fe content [Mg/(Mg + Fe) < 0.8 for $T = 1,600°$C, Figure 11.13].

The fields of phase stability in Figure 11.13 are moved to lower P (and lower Fe content) by decrease in T. Akaogi, Ito, and Navrotsky (1989) analyzed the variation through thermodynamic evaluations and represented the effect in P–T space (Figure 11.14) for a mantle of pyrolite composition, with an olivine composition Mg$_{1.78}$Fe$_{0.22}$SiO$_4$. In Figure 11.14, the dP/dT (i.e., Clapeyron slope) for each of the $\alpha \leftrightarrow \beta \leftrightarrow \gamma$ transitions is clearly positive. In addition, note the appearance, and widening, of the field $\alpha + \gamma$ at low T, so that the sequence of transitions for material in cold, subducted slabs is quite different from that developed in other mantle regions. Important consequences of this will be discussed later.

Green and Burnley (1989) and Green et al. (1992) reviewed experimental evidence for the stress dependence of the mechanisms for the $\alpha \to \gamma$ and $\alpha \to \beta$ transitions, including an evaluation of the relevance of several analogue systems to the mantle composition. The $\alpha \to \gamma$ transition occurs through incoherent nucleation and growth at conditions of hydrostatic pressure or low differential stress, whereas extremely high shear stress (>450 MPa for Mg$_2$GeO$_4$, similar for Mg$_2$SiO$_4$) (Fujino and Irifune, 1992) induces a diffusionless martensitic transition mechanism, of the type predicted by Poirier (1981). The magnitude of this shear stress must preclude the martensitic mechanism from operating within the mantle transition zone. Green and co-workers also concluded that the $\alpha \to \beta$ transition occurs only through incoherent nucleation and growth. It has been proposed that the $\alpha \to \beta$ transition occurs via some intermediate phase, either ordered (Madon and Poirier, 1983) or disordered (Guyot, Gwanmesia, and Liebermann, 1991). The $\beta \leftrightarrow \gamma$ transitions, consistent with their relatively minor atomic reconfigurations (Figure 11.12), produce many examples of crystallographically oriented intergrowths and adjoining grains (e.g., Brearley, Rubie, and Ito, 1992), implying that a martensitic mechanism has been involved, at least in the nucleation stage, possibly followed by more normal growth stages (Price, 1983). Brearley et al. (1992) identified a fraction ($\sim$10%) of metastable γ phase produced when α-Mg$_2$SiO$_4$ was transformed to the stable β phase in the appropriate, near-hydrostatic conditions. In (Mg, Fe)$_2$SiO$_4$, Brearley and Rubie (1994) and Rubie and Brearley (1994) found that only the $\gamma \to \beta$ transition occurred martensitically, whereas $\beta \to \gamma$ involved 'grain-boundary nucleation and interface-controlled growth'.

In summary, the nature of the phase-transition mechanisms likely to occur in (Mg, Fe)$_2$SiO$_4$ within the mantle above 650 km depth are becoming better understood. Our knowledge is now approaching the point where the kinetics of

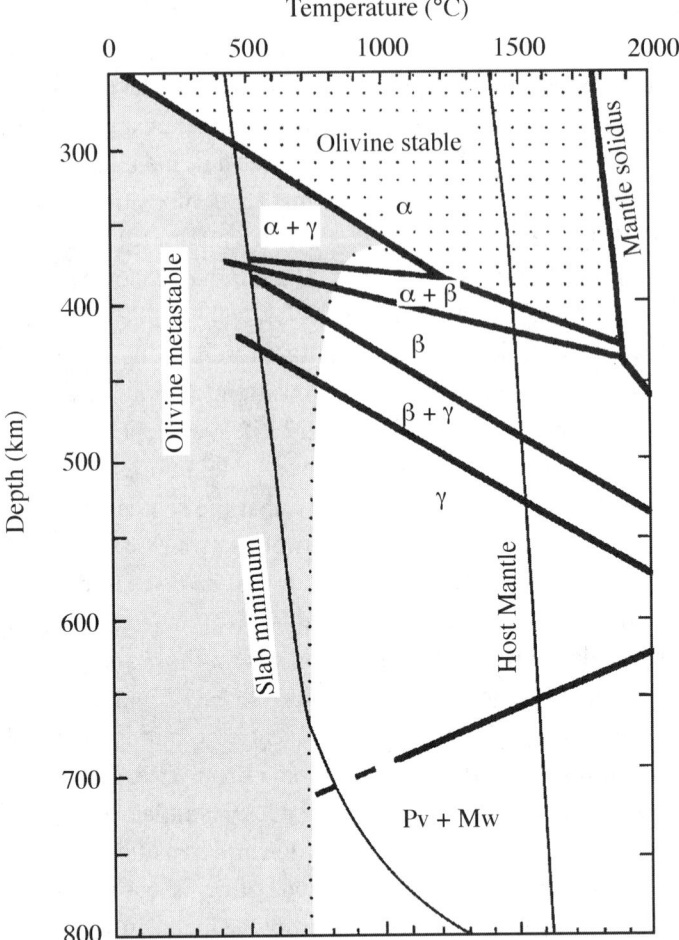

Figure 11.14. Phase relations inferred as functions of temperature and depth within the Earth for the compostion $Mg_{1.78}Fe_{0.22}SiO_4$, based on calculations by Akaogi et al. (1989). Kirby et al. (1991) proposed that the temperature at depth will fall between that for some mantle geotherm (their 'host mantle') and that for the coldest parts of subducting lithospheric slabs (their 'slab minimum'). They also proposed that olivine exists not only in the field of its equilibrium stability ('olivine stable' field, lightly stippled region) but also in a field of metastability ('olivine metastable', grey region) at low temperatures, with the boundary of this field being somewhat sensitive to pressure. Slow rates of transition from the olivine structure into the β and γ phases would account for the postulated metastability. The transformation, with a negative dP/dT, of $Mg_{1.78}Fe_{0.22}SiO_4$ at depth into perovskite plus magnesiowüstite is also indicated. (Adapted from Kirby et al., 1991.)

these transitions can be reasonably predicted and applied to the dynamics of sub-ducted pyrolite/harzburgite slabs. By reference to the equilibrium phase diagram (Figure 11.14) it was pointed out that $(Mg, Fe)_2SiO_4$ in cool subducting slabs would undergo transitions at lesser depths and initially in a sequence different (because of the $\alpha + \gamma$ field) than in the warmer surrounding mantle. The situation becomes more complicated if rates of phase transition and of slab heating are considered. As

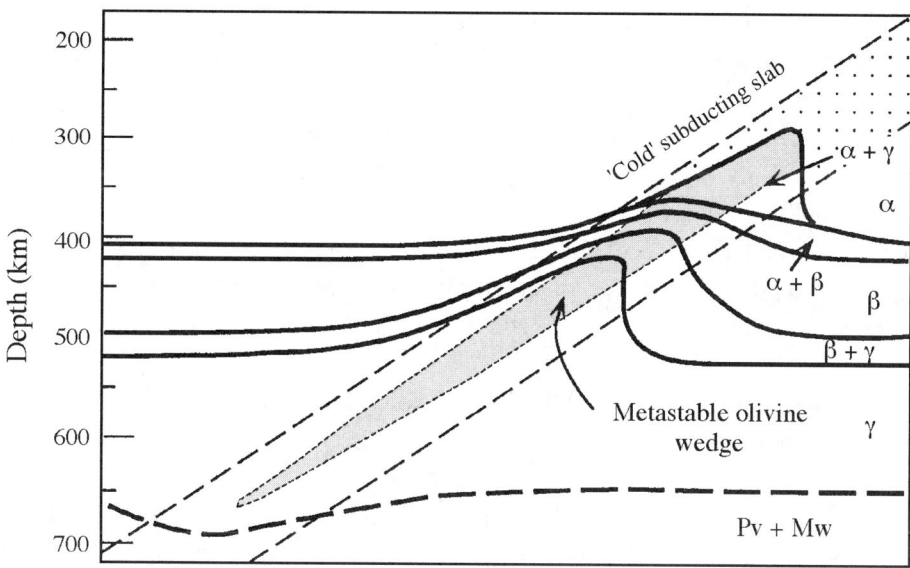

Figure 11.15. $(Mg, Fe)_2SiO_4$ phases in subducting lithosphere corresponding to the phase relations, temperature profiles, and metastability fields of Figure 11.14. In the colder slab regions, the equilibrium phase-transition boundaries (solid lines) deflect upwards, reflecting positive dP/dT for the corresponding phase transitions. Coexistence of $\alpha + \gamma$ as the equilibrium phase assemblage is restricted to shallow regions mainly within the cold slab. More importantly, because of slow transition rates at temperatures below about 700°C, a field of olivine metastability is postulated to extend in a wedge shape to depth in the interior of the slab. There have been some suggestions that the metastability field might not extend as deep as the approximately 650 km indicated here. The γ phase transforms into perovskite plus magnesiowüstite, as indicated (negative dP/dT, equilibrium boundary) by the thick dashed line deflected downwards. (Adapted from Kirby et al., 1991, and Green and Houston, 1995.)

pointed out by Sung and Burns (1976), although the equilibrium phase boundaries should be deflected to shallower depths inside a subducting slab, the sluggish kinetics (at $T < 700°C$) of transition from α-$(Mg, Fe)_2SiO_4$ should result in a metastable field of olivine, as shaded in Figure 11.14. Effectively, the dynamic phase boundary is deflected to great depth! Using a simple rate model allowing only grain-boundary nucleation, Sung and Burns (1976) calculated that metastability could extend more than 200 km below the approximately 400-km depth of equilibrium transition, in the form of a wedge of relict peridotite. Kirby, Durham, and Stern (1991) superimposed the Sung and Burns analysis on the phase equilibria from Akaogi et al. (1989) to produce Figure 11.15, showing the deflected equilibrium phase boundaries and a metastable wedge. Rubie and Ross (1994) proposed a rate model for grain-boundary nucleation that extended the model of Sung and Burns (1976) by incorporating the latent heat released at the time of olivine transition. The Rubie-Ross calculation, based on selected subduction rates and grain sizes, indicates that the depth limit for olivine metastability is 550 km; at this depth the transition proceeds to completion because of fast nucleation and local temperature increases. Liu and

Yund (1995) cautioned that consideration of grain-boundary (i.e., heterogeneous) nucleation alone might be unjustified, and they developed an analysis including homogeneous (i.e., within-grain) nucleation showing that if this mechanism is controlling, the metastability limit could be shallower than 500 km.

One possibility now being addressed is that the mantle phases coexisting with olivine could have some influence over the transitions. Sharp and Rubie (1995) have identified a catalytic effect of clinoenstatite on the $(Mg, Fe)_2SiO_4$ transitions into the γ phase and possibly the β phase and have speculated on consequent implications for reaction kinetics – such a catalytic effect could reduce the metastability depth even further.

Does weakening accompany these transitions? Rubie and Brearley (1994) and Brearley and Rubie (1994) speculated that the martensitic character of the $\gamma \rightarrow \beta$ transition would weaken $(Mg, Fe)_2SiO_4$ at the time of phase transition. Sharp et al. (1994) identified the CPO developed by the α–β transition under external stress and pointed to likely effects on aggregate seismic properties, but rheological effects could also follow. However, the effect most widely discussed with reference to rheology is the change in grain size associated with transitions, as summarized in the later section on transformational faulting.

11.5.2. Transformations of M(Si, Al)O₃ in the Mantle

Within the perovskite family, phase transitions are known in $CaTiO_3$ (e.g., Karato and Li, 1992) and have been reported in $(Mg, Fe)SiO_3$ (Wang, Guyot, and Liebermann, 1992). Information is limited, particularly for the effects of stress on the fields of phase stability and for the controls of transformation on rheology. In addition, there have been subsequent studies on the equilibrium phases in the $(Mg, Fe)SiO_3$ system that have challenged the earlier conclusions (e.g., Jackson and Rigden, Chapter 9, this volume).

Even less is known about the mechanical aspects of phase transitions in the garnet family at high pressure, though Wang, Gasparik, and Liebermann (1993) identified microstructural effects that could have developed by cubic–tetragonal transition in $MgSiO_3$ majorite garnet. Perhaps more importance should be attached to the large volume change [16.9% for the $MgSiO_3$ end member (Yusa et al., 1993) likely to generate large internal stresses] associated with the garnet-to-perovskite transition at the base of the mantle transition zone. Again, however, the mechanical effects of this ultra-high-pressure transition have not been investigated.

The pyroxene family features classical transformations that have long fascinated mineralogists (Buseck, Nord, and Veblen, 1980). Angel and Hugh-Jones (1994) reviewed the volume changes and phase relations among the ortho, high-clino, and low-clino polymorphs of enstatite. The transition from ortho to low-clino involves an atomic shear mechanism, with complex and irrational displacements involving dislocations, with partial Burgers vectors, as summarized in McLaren (1991). Coe and Kirby (1975) analyzed the possible effects of external shear stresses on transition conditions, with the monoclinic form being promoted by shear. However, the mechanical effects of pressure-induced transitions in the mantle involving the high-clino polymorph of enstatite have not been studied, and such studies will be difficult, because this is a non-quenchable phase. The transition from the high-clino to the ortho polymorph occurs within the upper-mantle, so this effect of transformation on rheology will occur outside the mantle transition zone.

Hogrefe et al. (1994) investigated the kinetics of transformation of $MgSiO_3$ at pressures of 16–21 GPa and reported the following observations and interpretations: At 1,000°C, the high-clino polymorph of enstatite can undergo a nonequilibrium transition directly to $MgSiO_3$ ilmenite. Even at 1,600°C, the dissociation of pyroxene into a stable assemblage of the β phase plus stishovite was found to be slow relative to the transition of coexisting olivine to the β phase. If this behaviour is representative of materials in the mantle, it follows that a wedge-shaped region of metastable pyroxene might be even more extensive than the proposed metastable olivine wedge, with implications for subduction dynamics.

Because of the crystallographic complexities of some of the lower-pressure members of the $M(Si, Al)O_3$ family and the relatively high pressures of many of the transformations relevant to the deeper mantle, it is likely that detailed knowledge about transformations and their geophysical effects will remain limited, compared with that for the M_2SiO_4 minerals for the foreseeable future. However, in view of the mineralogic constitutions of the materials in the deep Earth, we must continue to investigate these phases and transformations and reactions involving phases of other stoichiometries.

11.5.3. Transformational Faulting

The change in volume due to formation of denser phases in the mantle transition zone has long been suggested as the source for deep earthquakes (below 350 km) which cannot arise from brittle failure and frictional sliding. However, the energy-radiation patterns for the deep earthquakes are dominated by shear, and simple 'implosion' from phase transitions cannot account for this. Kirby et al. (1991) reviewed earlier work that pointed to the stress state in the metastable olivine wedge of a subducting slab being one of down-dip compression, because of the outer and warmer parts of the slab having been transformed to spinel.

Experimental work by H. W. Green and colleagues has explained how shear could be initiated within slabs, leading to the deep earthquakes. Green and Houston (1995) have carefully reviewed this literature, so a summary will suffice here. Investigations of olivine analogues and, recently, silicate-olivine systems have all shown a characteristic high-pressure-faulting phenomenon leading to shear failure at extremely high confining pressures (14-GPa faulting has been demonstrated) (Green et al., 1992). The phenomenon is initiated in olivine aggregates by formation of microscopic lens-shaped domains of denser phases, either β or γ, perpendicular to the maximum compressive stress. These lenses are termed 'anticracks' (Green and Burnley, 1989), because there is a 'reverse' sense (relative to tensional cracks) firstly of stresses and displacements in the vicinity of the lens structures and secondly of the orientations of the lenses. Importantly, the grain size of the denser phase(s) inside anticrack lenses is extremely small ($\ll 1\ \mu$m in germanate materials), most likely because of highly efficient nucleation at the phase boundary. Green and Burnley (1989) suggested that flow in the fine-grained material is grain-size-sensitive, so that it is very weak, even superplastic, and the anticracks can grow and link into a through-going shear zone that becomes a transformation fault (Figure 11.16). Each fault zone is lined with the denser phase, again with extremely fine grain size and low strength. It has been shown experimentally that acoustic energy will be emitted once the fault structures develop, providing an analogy for radiation of earthquake energy from the mantle transition zone.

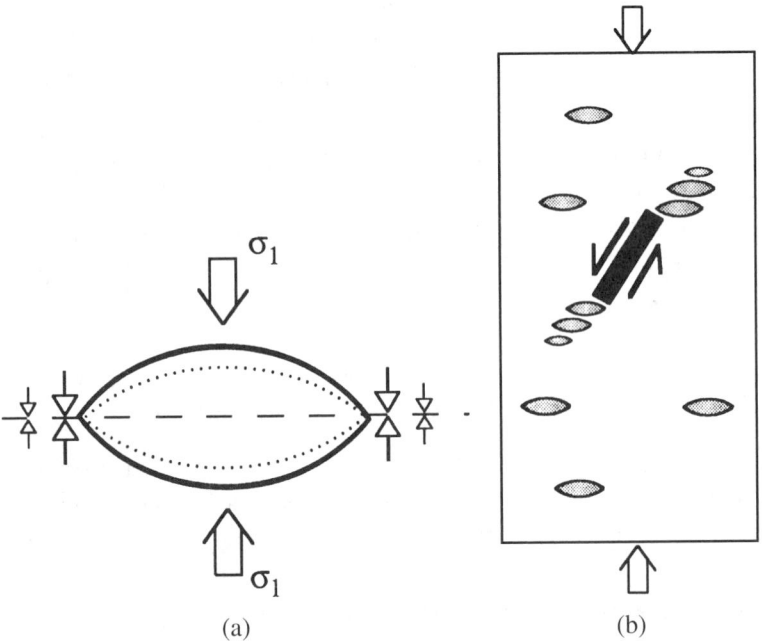

(a)   (b)

Figure 11.16. A model of an anticrack and a schematic for transformation faulting observed in experiments. (a) Under the influence of the maximum principal applied stress σ_1, the α-phase material inside a lens-shaped region (solid line) undergoes a phase transition into a denser phase (either β or γ phase), simultaneously collapsing into a thinner lens whose shape and size are indicated by the dotted line. Residual compressive stress is concentrated at the lens tips in the surrounding untransformed material. Notice that the equatorial plane of the anticrack (dashed line) is normal to σ_1, whereas conventional cracks would be expected to align parallel to σ_1. Lens aspect ratios and relative volumes have been exaggerated for clarity. (b) Lenses of transformed material occur throughout the stressed aggregate. In one location, adjacent en-echelon lenses have linked to form a through-going fault. This fault consists of transformed material in an untransformed matrix. (Adapted from Green and Houston, 1995.)

Green and Houston (1995) have argued the importance of thermal runaway for rapid phase transitions and catastrophic faulting. In this regard, it appears essential that the transition, in addition to involving a large volume change, be strongly exothermic (i.e., involve a large negative entropy change). On that basis, Brearley and Rubie (1994) suggested that transformational faulting should be more favourable for the $\alpha \rightarrow \gamma$ transition than for the $\alpha \rightarrow \beta$ transition, but would not be expected for the $\beta \rightarrow \gamma$ transition, in $(Mg, Fe)_2SiO_4$. The calculations of Rubie and Ross (1994) regarding nucleation at grain boundaries during transition from olivine have demonstrated the marked effects of the release of latent heat energy and point to the likelihood of fast production of fine-grained material at the olivine metastability limit, ideal for transformational faulting. From unpublished work by Green and Zhou on $CdTiO_3$, Green and Houston (1995) have cited data supporting the necessity for exothermic transitions: Transformation faulting was not detected for the densifying structural transition ilmenite $\rightarrow$ perovskite (which is endothermic), but was demonstrated to accompany the reverse transition.

It is hoped that this brief review, which has concentrated on the phase transitions in $(Mg, Fe)_2SiO_4$, will have instilled some appreciation of the role of phase transformations in the mechanical behaviour of the Earth's mantle. Full appreciation of this clearly complex topic is not important initially; more valuable is some basic understanding of how the numerous transformations in the Earth's mantle could be coupled with rheological processes and properties. As yet, there has been no clear demonstration that transformational plasticity is a feature of the mantle, but because many transitions and reactions involving large volume changes occur in the depth range 300–1,000 km, this remains a definite possibility. However, until the phenomenon is confirmed in conditions appropriate to the deep Earth, interpretations invoking transformational plasticity, such as the ductility of subducted plates at the 660-km discontinuity proposed by Ringwood (1991) (Figure 11.14) should be treated as extremely speculative.

11.6. Future Developments in Research

There has been rapid progress in the study of deformation mechanisms and the rheology of mantle rocks and minerals during the past 25 years. The pace of such advances is likely to increase because of the proliferation of rock-deformation laboratories around the world and because of new developments in experimental methods, microstructural characterization, and numerical modelling.

Technological developments in deformation apparatus (Paterson, 1970, 1990; Green and Borch, 1989) and improvements in experimental methodology (Zeuch and Green, 1984; Karato et al., 1986; Hitchings, Paterson, and Bitmead, 1989; Bai et al., 1991; Beeman and Kohlstedt, 1993) have facilitated the collection of accurate data under conditions appropriate to the crust and upper-mantle. Experimental studies often have been limited to relatively low strains, and this poses problems in extrapolation to high-strain processes in the Earth (White, 1979). Recent developments in experimental techniques for shear (Zhang and Karato, 1995; Zimmerman et al., 1995) and torsional deformation (Olgaard and Paterson, 1996) will allow studies of high-strain deformation phenomena. It may turn out that flow laws for high strains will be significantly different from low-strain flow laws if strain-softening (or even hardening) processes are important.

Experimental deformation studies are currently being extended to higher pressures to enable investigations of the rheology of deep-mantle phases and their analogues (Meade and Jeanloz, 1990; Sotin and Poirier, 1990; Weidner, Wang, and Vaughan, 1994). This is necessary to keep up with the advances that have occurred in experimental petrology, where well-developed technologies have allowed easier attainment of pressures and temperatures equivalent to those in the deep mantle through multi-anvil and diamond-cell apparatus. Bussod et al. (1993a) have described how quantitative, large-strain, high-temperature deformation is now possible in multi-anvil apparatus at conditions appropriate to the transition zone. In this approach, the flow stress is estimated from measurements of deformation-induced microstructures, such as dynamically recrystallized grain size and dislocation density, that scale with stress (Wang et al., 1988; Rubie et al., 1993). This approach has complications related to the possible effects of melt content (Hirth and Kohlstedt, 1995a). Also, complications follow from the work of van der Wal (1993), who found that recrystallized grain size was related to the flow stress for steady-state deformation, but did not remain 'in equilibrium' with the flow stress during relaxation tests where the strain rate and stress decreased with time. Despite these complications, the new findings represent an important first step toward direct determination of the rheology of deep-mantle phases. Recent work (Karato and Rubie, 1996) has led to further progress in very high pressure deformation studies. The development of load cells for stress measurement in multi-anvil apparatus remains as an important challenge for experimentalists (D. C. Rubie, personal communication).

Exciting discoveries have been reported in the past five years from studies of xenoliths that sample the deep upper mantle and transition zone (Sautter, Haggerty, and Field, 1991) and mineral inclusions to sample even the lower mantle (Kesson and Fitz Gerald, 1991). It has recently been proposed that some large (800-m-diameter) orogenic peridotite bodies could have been derived from the transition zone (Dobrzhinetskaya, Green, and Wang, 1996). The structures and microstructures of these rocks could preserve some record of deformation and metamorphic processes at great depth in the Earth (Doukhan, Doukhan, and Sautter, 1994), although in many cases deformation during transport to the Earth's surface will have obliterated any microstructures representative of in situ conditions. Dobrzhinetskaya et al. (1996) have suggested that the CPO for olivine in the Alpe Arami peridotite was inherited from pre-existing β-$(Mg, Fe)_2SiO_4$. That would imply that dislocation creep is important in the mantle transition zone.

Figure 11.17. Scanning-electron-microscopic image of polished synthetic forsterite formed using forward-scattered electrons. This sample was deformed at 950°C and a confining pressure of 600 MPa in the diffusion-creep regime. Contrast between grains and subgrains arises from differences in orientations, which can be measured using electron backscattering patterns. Pores within grains and along grain boundaries are visible as black features. Most grains are subgrain-free and have no orientation gradients. Width of image, 25 μm. (Courtesy of R. McDonnell and C. J. Spiers.)

There have been continuing developments in the techniques used for microstructural characterization of deformed materials, such as improvements in high-resolution transmission electron microscopy for studying dislocations, grain boundaries, and phase intergrowths at a unit-cell scale. Improved scanning-electron-microscopic techniques permit easier and faster determination of CPOs, and orientation-contrast imaging has greatly enhanced the utility of backscattered-electron imaging for characterizing granular microstructures (Figure 11.17) in fine-grained polycrystals (Lloyd et al., 1991). Developments in digital imaging and analysis methods in both the light-microscopic (Panozzo Heilbronner, 1993; Panozzo Heilbronner and Pauli, 1993) and electron-microscopic arenas will lead to increased application of quantitative analysis for microstructures of deformed materials.

A potentially powerful method of simulating the plastic deformation of crystalline materials has been developed during the past decade. The behaviour of small numbers of dislocations is now well understood (e.g., Hirth and Lothe, 1976), though the collective behaviour of large numbers of dislocations is poorly understood. This method of 'dislocation dynamics' (Amadeo and Ghoniem, 1990) is an extension of molecular dynamics, but instead of simulating atomic configurations and interactions, large numbers of interacting dislocations are modelled. In dislocation-dynamical simulations, the interaction rules and mobilities of individual dislocations must be specified. The mobility terms can be constrained on the basis of experiments on dislocation glide, climb, and recovery (e.g., Goetze, and Kolstedt, 1973; Goetze, 1978; Karato et al., 1993) or by *ab initio* calculations of diffusion (e.g., Wall and Price, 1989). Goetze (1978) pointed out that it is possible ('with some patience') to conduct experiments on dislocation mobility where the dislocation velocities are the same as in the Earth; extrapolation to slow strain rates is then not needed.

The newly developing experimental and theoretical approaches to studying the deformation mechanisms and rheology of the deep Earth can also be applied to the rocks of the upper mantle and crust. In this respect the crust and upper mantle could serve as a useful 'testing ground' in the development of these methods. The microphysical (Ranalli, 1991, 1995) or materials-based approach to Earth rheology always requires extrapolation from laboratory to geological timescales (Paterson, 1987) over a huge range of pressures. As noted by Paterson (1987), extrapolated flow laws normally provide an upper-bound for viscosity or strength, because there is always the possibility of an unpredicted transition to a different deformation mechanism in the Earth that usually will correspond to lower viscosity.

Extrapolation of a constitutive flow law should be undertaken only if there is firm theoretical justification for the flow law and good agreement between experimental data and the theoretical microphysical model (e.g., Spiers and Schutjens, 1990; Ranalli, 1995). Microstructural studies of the experimentally deformed materials are needed to verify the range of processes being considered in the theory. In this respect, finding agreement between a theoretical law and experimental creep data is not enough.

In the case of upper-mantle and crustal rocks, it is feasible to check extrapolations of experimental flow laws by studies of naturally deformed rocks that provide information about deformation mechanisms dominating in particular geological environments. Only after good agreement is established between experimental predictions and the constraints from natural rocks at shallow levels is it reasonable to extend the predictions to deeper levels in the Earth. The large difference between laboratory and Earth timescales remains as a formidable problem; however, this should not be used as justification for ignoring constraints that arise from microphysical calculations of Earth rheology – geophysical models of Earth dynamics must be consistent with the laws of solid-state physics. As an example, it could be valid to ignore the pressure-sensitivity of solid-state flow when modelling lithospheric processes, but it is physically meaningless to use a viscosity that is not pressure sensitive when modelling mantle convection. Of course, using simplified rheological laws conveniently allows the influences of different variables to be considered separately; however, great caution is required in applying such laws directly to the Earth.

Future progress in the microphysical approach to Earth rheology can be expected from the application of new experimental designs and theoretical simulations of solid-state diffusion and flow. If the past 20 years can be used as a guide, we can expect that numerical simulations will gain increasing importance in this research field. Experimental work and microstructural studies will retain their central and essential roles, however, mainly because flow processes in Earth materials are complex, subtle, and often surprising.

11.7. Conclusion

A remarkable catalogue of information about the materials of the Earth's mantle has now been gathered from laboratory studies. However, most of those studies have concentrated on materials composed of only a single mineral phase, some of which are analogues, and some of the data are open to question because of disagreements between different studies. Consequently, there are limitations in extrapolating these results to polyphase materials of the real Earth, but some imaginative manipulations of the available data have undeniably led to exciting inferences about processes that might be operating at depth. Undoubtedly, direct laboratory studies of materials and their physical properties, coupled with modelling and theoretical developments arising from such data, will continue as one of the key frontiers for advancing our understanding, both geochemical and geophysical, of the dynamics in the Earth's mantle.

References

Akaogi, M., Ito, E., and Navrotsky, A. 1989. Olivine-modified spinel-spinel transitions in the system Mg_2SiO_4-Fe_2SiO_4: calorimetric measurements, thermochemical calculation, and geophysical application. *J. Geophys. Res.* 94:15671–85.

Allègre, C. J., and Turcotte, D. L. 1986. Implications of a two-component marble-cake mantle. *Nature* 323:123–7.

Amadeo, R. J., and Ghoniem, N. M. 1990. Dislocation dynamics. I. A proposed methodology for deformation micromechanics. *Phys. Rev.* B41:6958–67.

Anderson, D. L., and Sammis, C. G. 1970. Partial melting in the upper mantle. *Phys. Earth Planet. Int.* 3:228–52.

Angel, R. J., and Hugh-Jones, D. A. 1994. Equations of state and thermodynamic properties of enstatite pyroxenes. *J. Geophys. Res.* 99:19777–83.

Ave Lallement, H. G., and Carter, N. L. 1970. Syntectonic recrystallization of olivine and modes of flow in the upper mantle. *Geol. Soc. Am. Bull.* 81:2203–20.

Ave Lallement, H. G., Mercier, J. C. C., Carter, N. L., and Ross, J. V. 1980. Rheology of the upper mantle: inferences from peridotite xenoliths. *Tectonophysics* 70:85–113.

Bai, Q., and Kohlstedt, D. L. 1992a. High temperature creep of olivine single crystals. III. Mechanical results for unbuffered samples and creep mechanisms. *Philos. Mag.* 82:65–74.

Bai, Q., and Kohlstedt, D. L. 1992b. Substantial hydrogen solubility in olivine and implications for water storage in the mantle. *Nature* 357:672–4.

Bai, Q., and Kohlstedt, D. L. 1993. Effects of chemical environment on the solubility and incorporation mechanism for hydrogen in olivine. *Phys. Chem. Min.* 19:460–71.

Bai, Q., Mackwell, S. J., and Kohlstedt, D. L. 1991. High temperature creep of olivine single crystals. 1. Mechanical results for buffered samples. *J. Geophys. Res.* 96:2441–63.

Ballhaus, C. 1995. Is the upper mantle metal saturated? *Earth Planet. Sci. Lett.* 132:75–86.

Beauchesne, S., and Poirier, J. P. 1989. Creep of barium titanite perovskite: a contribution to a systematic approach to the viscosity of the lower mantle. *Phys. Earth Planet. Int.* 55:187–199.

Beauchesne, S., and Poirier, J. P. 1990. In search of systematics for the viscosity of perovskites: creep of potassium tantalate and niobate. *Phys. Earth Planet. Int.* 61:182–98.

Beeman, M. L., and Kohlstedt, D. L. 1993. Deformation of fine-grained aggregates of olivine plus melt at high temperatures and pressures. *J. Geophys. Res.* 98:6443–52.

Bell, D. R., and Rossman, G. R. 1992. Water in Earth's mantle: the role of nominally anhydrous minerals. *Science* 255:1391–7.

Bilde-Sorensen, J. B. 1972. Dislocation structures in creep-deformed polycrystalline MgO. *J. Am. Ceram. Soc.* 55:606–10.

Blacic, J. D. 1972. Effect of water on the experimental deformation of olivine. In: *Flow and Fracture of Rocks*, ed. H. C. Heard, I. Y. Borg, N. L. Carter, and C. B. Raleigh, pp. 109–15. AGU Monograph 16. Washington, DC: American Geophysical Union.

Boland, J. N., and Liebermann, R. C. 1983. Mechanism of the olivine to spinel phase transformation in Ni_2SiO_4. *Geophys. Res. Lett.* 10:87–90.

Borch, R. S., and Green, H. W. 1987. Dependence of creep in olivine on homologous temperature and its implications for flow in the mantle. *Nature* 330:345–8.

Borch, R. S., and Green, H. W. 1989. Deformation of peridotite at high pressure in a new molten salt cell: comparison of traditional and homologous temperature treatments. *Phys. Earth Planet. Int.* 55:269–76.

Boullier, A. M., and Gueguen, Y. 1975. SP-mylonites: origin of some mylonites by superplastic flow. *Contrib. Min. Pet.* 50:93–104.

Brearley, A. J., and Rubie, D. C. 1994. Transformation mechanisms of San Carlos olivine to $(Mg, Fe)_2SiO_4$ β-phase under subduction zone conditions. *Phys. Earth Planet. Int.* 86:45–67.

Brearley, A. J., Rubie, D. C., and Ito, E. 1992. Mechanisms of the transformations between the α, β and γ polymorphs of Mg_2SiO_4 at 15 GPa. *Phys. Chem. Min.* 18:343–58.

Buseck, P. R., Nord, G. L., Jr., and Veblen, D. R. 1980. Subsolidus phenomena in pyroxenes. In: *Reviews in Mineralogy*, vol. 7, ed. C. T. Prewitt, pp. 117–211. Washington, DC: Mineralogical Society of America.

Bussod, G. 1990. The experimental deformation of spinel lherzolite at subsolidus and hypersolidus conditions. Ph.D. thesis, University of California, Los Angeles.

Bussod, G., and Christie, J. C. 1991. Textural development and melt topology in spinel lherzolite experimentally deformed at hypersolidus conditions. In: *Orogenic Lherzolites and Mantle Processes*, ed. M. A. Menzies, C. Dupuy, and A. Nicolas, pp. 17–39. Oxford University Press.

Bussod, G. Y., Katsura, T., Rubie, D. C., and Bussod, G. 1993a. The large volume multi-anvil press as a high *P-T* deformation apparatus. *Pure Appl. Geophys.* 141:579–99.

Bussod, G. Y., Katsura, T., and Sharp, T. G. 1993b. Experimental deformation of α and β $Mg_{1.8}Fe_{0.2}SiO_4$ at transition zone temperatures and pressures. *EOS, Trans. AGU* 74:597.

Cannon, W. R., and Langdon, T. G. 1988. Creep of ceramics. Part 2. An examination of flow mechanisms. *J. Mat. Sci.* 23:1–20.

Carry, C., and Mocellin, A. 1986. Superplastic creep of fine grained $BaTiO_3$ in a reducing environment. *Comm. Am. Ceram. Soc.* 69:C215–16.

Carter, N. L. 1976. Steady-state flow of rocks. *Rev. Geophys. Space Phys.* 14:301–60.

Carter, N. L., and Ave Lallement, H. G. 1970. High-temperature flow of dunite and peridotite. *Geol. Soc. Am. Bull.* 81:2181–202.

Ceuleneer, G., and Rabinowinc, M. 1992. Mantle flow and melt migration beneath oceanic ridges: models derived from observations in ophiolites. In: *Mantle Flow and Melt Generation at Mid-Ocean Ridges*, ed. J. Phipps Morgan, D. K. Blackman, and J. M. Sinton, pp. 123–54. AGU Monograph 71. Washington, DC: American Geophysical Union.

Chakraborty, S., and Farver, J. R. 1994. Mg tracer diffusion in synthetic forsterite and San Carlos olivine as a function of P, T and fO_2. *Phys. Chem. Min.* 21:489–500.

Chopra, P. N. 1986. The plasticity of some fine-grained aggregates of olivine at high pressure and temperature. In: *Mineral and Rock deformation: Laboratory Studies*, ed. B. E. Hobbs, and H. C. Heard, pp. 25–34. Washington, DC: American Geophysical Union.

Chopra, P. N., and Paterson, M. S. 1981. The experimental deformation of dunite. *Tectonophysics* 78:453–73.

Chopra, P. N., and Paterson, M. S. 1984. The role of water in the deformation of dunite. *J. Geophys. Res.* 89:7861–76.

Coe, R. S., and Kirby, S. H. 1975. The orthoenstatite to clinoenstatite transformation by shearing and reversion by annealing: mechanism and potential applications. *Contrib. Min. Pet.* 52:29–55.

Cooper, R. F., and Kohlstedt, D. L. 1984. Solution-precipitation enhanced diffusional creep of partially molten olivine-basalt aggregates during hot-pressing. *Tectonophysics* 107:207–33.

de Bresser, J. H. P., Spiers, C. J., and Reis, J. P. J. 1994. Temperature dependence of the recrystallised grain size vs. flow stress relation for the Mg alloy Magnox Al80. *EOS, Trans. AGU* 75:586–7.

Derby, B., and Ashby, M. F. 1987. On dynamic recrystallisation. *Scripta Metallurgica* 21:879–84.

Dobrzhinetskaya, L., Green, H. W., II, and Wang, S. 1996. Alpe Arami: a periodite massif from depths of more than 300 kilometers. *Science* 271:1841–5.

Doukhan, N., and Doukhan, J. C. 1986. Dislocations in perovskites BaTiO$_3$ and CaTiO$_3$. *Phys. Chem. Min.* 13:403–10.

Doukhan, N., Doukhan, J. C., and Sautter, V. 1994. Ultradeep (greater than 300 kilometers) ultramafic xenoliths: TEM preliminary results. *Phys. Earth Planet. Int.* 82:195–207.

Drury, M. R. 1991. Hydration-induced climb dissociation of dislocations in naturally deformed mantle olivine. *Phys. Chem. Min.* 18:106–16.

Drury, M. R., and Fitz Gerald, J. D. 1996. Grain boundary melt films in an experimentally deformed olivine-orthopyroxene rock: implications for melt distribution in upper mantle rocks. *Geophys. Res. Lett.* 23:701–4.

Drury, M. R., and Humphreys, F. J. 1988. Microstructural shear criteria associated with grain boundary sliding during ductile deformation. *J. Struct. Geol.* 10:83–9.

Drury, M. R., and Urai, J. L. 1990. Deformation-related recrystallization processes. *Tectonophysics* 172:235–53.

Drury, M. R., Vissers, R. L. M., Hoogerduijn Stating, E. H., and van der Wal, D. 1991. Shear localisation in upper mantle peridotites. *Pure Appl. Geophys.* 137:439–60.

Durham, W. B., and Goetze, C. 1977. Plastic flow of oriented single crystals of olivine. 1. Mechanical data. *J. Geophys. Res.* 82:5737–53.

Edington, J. W., Melton, K. N., and Cutler, C. P. 1976. Superplasticity. *Progress in Materials Science* 21:61–171.

Evans, B., Fredrich, J. T., and Wong, T.-F. 1990. The brittle–ductile transition in rocks: recent experimental and theoretical progress. In: *The Brittle–Ductile Transition in Rocks*, ed. A. G. Duba, W. B. Durham, J. W. Handin, and H. F. Wang, pp. 1–20. AGU Monograph 56. Washington, DC: American Geophysical Union.

Evans, B., and Kohlstedt, D. L. 1995. Rheology of rocks. In: *Rock Physics and Phase Relations – A Handbook of Physical Constants*, ed. T. J. Ahrens, pp. 148–65. Washington, DC: American Geophysical Union.

Farber, D. L., Williams, Q., and Ryerson, F. J. 1994. Diffusion in Mg_2SiO_4 polymorphs and chemical heterogeneity in the mantle transition zone. *Nature* 371:693–5.

Farver, J. R., Yund, R. A., and Rubie, D. C. 1994. Magnesium grain boundary diffusion in forsterite aggregates at 1000°–1300°C and 0.1 MPa to 10 GPa. *J. Geophys. Res.* 99:19809–19.

Fitz Gerald, J. D., and Chopra, P. N. 1982. Deformation mechanisms in dunite – the results of high temperature testing. In: *Sixth International Conference on the Strength of Metals and Alloys*, vol. 3, ed. R. C. Gifkins, pp. 735–40. Oxford: Pergamon.

Fitz Gerald, J. D., and Chopra, P. N. 1984. Distribution of strain in some geological materials experimentally deformed at high pressure and temperature. In: *Deformation of Ceramic Materials*, vol. 2, ed. R. E. Tressler and R. C. Bradt, pp. 321–8. New York: Plenum.

Frost, H. J., and Ashby, M. F. 1982. *Deformation-Mechanism Maps. The Plasticity and Creep of Metals and Ceramics*. Oxford: Pergamon.

Fujino, K., and Irifune, T. 1992. TEM studies on the olivine to modified spinel transformation in Mg_2SiO_4. In: *High-Pressure Research: Application to Earth and Planetary Sciences*, ed. Y. Syono and M. H. Manghnani, pp. 237–43. Tokyo: Terra Scientific.

Fujino, K., Nakazaki, H., Momoi, H., Karato, S., and Kohlstedt, D. L. 1992. TEM observation of dissociated dislocations with $b = [010]$ in naturally deformed olivine. *Phys. Earth Planet. Int.* 78:131–7.

Gautason, B., and Muehlenbachs, K. 1993. Oxygen diffusion in perovskite: implications for electrical conductivity in the lower mantle. *Science* 260:518–21.

Goetze, C. 1978. The mechanisms of creep in olivine. *Phil. Trans. R. Soc. London* A288:99–119.

Goetze, C., and Kohlstedt, D. L. 1973: Laboratory study of dislocation climb and diffusion in olivine. *J. Geophys. Res.* 78:5961–71.

Gordon, R. S. 1985. Diffusional creep phenomena in polycrystalline oxides. In: *Point Defects in Minerals*, ed. R. N. Schock, pp. 132–40. AGU Monograph 31. Washington, DC: American Geophysical Union.

Green, H. W., and Borch, R. S. 1989. A new molten salt cell for precision stress measurement at high pressure. *Eur. J. Min.* 1:213–19.

Green, H. W., and Burnley, P. C. 1989. A new self-organizing mechanism for deep-focus earthquakes. *Nature* 341:733–6.

Green, H. W., and Houston, H. 1995. The mechanics of deep earthquakes. *Annu. Rev. Earth Planet. Sci.* 23:169–213.

Green, H. W., Young, T. E., Walker, D., and Scholz, C. H. 1992. The effect of nonhydrostatic stress on the α–β and α–γ olivine phase transformations. In: *High-Pressure Research: Application to Earth and Planetary Sciences*, ed. Y. Syono, and M. H. Manghnani, pp. 229–35. Tokyo: Terra Scientific.

Greenwood, G. W., and Johnson, R. H. 1965. The deformation of metals under small stresses during phase transformations. *Proc. R. Soc. London* 283A:403–22.

Gueguen, Y., and Nicolas, A. 1980. Deformation of mantle rocks. *Annu. Rev. Earth Planet. Sci.* 8:119–44.

Guyot, F., Gwanmesia, D., and Liebermann, R. C. 1991. An olivine to beta phase transformation mechanism in Mg_2SiO_4. *Geophys. Res. Lett.* 18:89–92.

Handy, M. R. 1989. Deformation regimes and the rheological evolution of fault zones in the lithosphere: the effects of pressure, temperature, grain size and time. *Tectonophysics* 163:119–52.

Handy, M. R. 1994. Flow laws for rocks containing two non-linear viscous phases: a phenomenological approach. *J. Struct. Geol.* 16:287–301.

Hess, P. C. 1994. Thermodynamics of thin fluid films. *J. Geophys. Res.* 99:7219–29.

Hirth, G., and Kohlstedt, D. L. 1995a. Experimental constraints on the dynamics of the partially molten upper mantle; deformation in the diffusion creep regime. *J. Geophys. Res.* 100:1981–2001.

Hirth, G., and Kohlstedt, D. L. 1995b. Experimental constraints on the dynamics of the partially molten upper-mantle. 2. Deformation in the dislocation creep regime. *J. Geophys. Res.* 100:15441–9.

Hirth, G., and Kohlstedt, D. L. 1996. Water in the oceanic upper mantle: implications for rheology, melt extraction and the evolution of the lithosphere. *Earth Planet. Sci. Lett.* 144:93–108.

Hirth, J. P., and Lothe, J. 1976. *Theory of Dislocations.* New York: McGraw-Hill.

Hitchings, R. S., Paterson, M. S., and Bitmead, J. 1989. Effects of iron and magnetite additions in olivine-pyroxene rheology. *Phys. Earth Planet. Int.* 55:277–91.

Hobbs, B. E., Means, W. D., and Williams, P. F. 1976. An Outline of Structural Geology. New York: Wiley.

Hogrefe, A., Rubie, D. C., Sharp, T. G., and Seifert, F. 1994. Metastability of enstatite in deep subducting lithosphere. *Nature* 372:351–3.

Hopper, J. R., and Buck, W. R. 1993. The initiation of rifting at constant tectonic force: role of diffusion creep. *J. Geophys. Res.* 98:16213–21.

Hull, D., and Bacon, D. J. 1984. *Introduction to Dislocations.* 3rd ed. Oxford: Pergamon.

Huther, W., and Reppich, B. 1973. Dislocation structure during creep of MgO single crystals. *Philos. Mag.* 28:363–71.

Ingrin, I., and Poirier, J. P. 1995. TEM observations of several spinel-garnet assemblies: toward the rheology of the transition zone. *Terra Nova* 7:509–15.

Inoue, T., Yurimoto, H., and Kudoh, Y. 1995. Hydrous modified spinel $Mg_{1.75}SiH_{0.5}O_4$: a new water reservoir in the mantle transition zone. *Geophys. Res. Lett.* 22:117–20.

Jin, Z. M., Bai, Q., and Kohlstedt, D. L. 1994a. High temperature creep of olivine crystals from four localities. *Phys. Earth Planet. Int.* 82:55–64.

Jin, Z. M., Green, H. W., and Zhou, Y. 1994b. Melt topology in partially molten mantle peridotite during ductile deformation. *Nature* 372:164–7.

Karato, S. I. 1981. Rheology of the lower mantle. *Phys. Earth Planet. Int.* 24:1–14.

Karato, S.-I. 1986. Does partial melting reduce the creep strength of the upper-mantle? *Nature* 319:309–10.

Karato, S. I. 1987. Seismic anisotropy due to lattice preferred orientation of minerals: kinematic or dynamic? In: *High-Pressure Research in Mineral Physics*, ed. M. H. Manghnani and Y. Syono, pp. 455–71. Washington, DC: American Geophysical Union.

Karato, S. I. 1988. The role of recrystallization in the preferred orientation of olivine. *Phys. Earth Planet. Int.* 51:107–22.

Karato, S. I. 1989a. Defects and plastic deformation in olivine. In: *Rheology of Solids and of the Earth*, ed. S. I. Karato and M. Toriumi, pp. 176–208. Oxford University Press.

Karato, S. I. 1989b. Grain growth kinetics in olivine aggregates. *Tectonophysics* 168:255–73.

Karato, S. I. 1989c. Plasticity–crystal structure systematics in dense oxides and its implications for the creep strength of the Earth's deep interior: a preliminary result. *Phys. Earth Planet. Int.* 55:234–40.

Karato, S. I. 1992. On the Lehmann discontinuity. *Geophys. Res. Lett.* 22:2255–8.

Karato, S. I., Fujino, K., and Ito, E. 1990. Plasticity of $MgSiO_3$ perovskite: the results of microhardness tests on single crystals. *Geophys. Res. Lett.* 17:13–16.

Karato, S. I., and Li, P. 1992. Diffusion creep in perovskite: implications for the rheology of the lower mantle. *Science* 255:1238–40.

Karato, S. I., and Ogawa, M. 1982. High-pressure recovery in olivine: implications for creep mechanisms and creep activation volume. *Phys. Earth Planet. Int.* 28:102–17.

Karato, S.-I., and Rubie, D. C. 1996. Towards an experimental study of rheology of the deep mantle. *EOS, Trans. AGU (Suppl.)* 77:715.

Karato, S. I., Rubie, D. C., and Yan, H. 1993. Dislocation recovery in olivine under deep mantle conditions: implications for creep and diffusion. *J. Geophys. Res.* 98:9761–8.

Karato, S. I., Paterson, M. S., and Fitz Gerald, J. D. 1986. Rheology of synthetic olivine aggregates: influence of grain size and water. *J. Geophys. Res.* 91:8151–76.

Karato, S. I., Toriumi, M., and Fuji, T. 1982. Dynamic recrystallization and high temperature rheology of olivine. In: *High-Pressure Research in Geophysics*, vol. 12, ed. S. Akimoto and M. H. Manghnani, pp. 171–89. Tokyo: Terra Scientific.

Karato, S. I., Wang, Z., and Fujino, K. 1994. High temperature creep of YAG single crystals. *J. Mat. Sci.* 29:6458–62.

Karato, S. I., Wang, Z., Liu, M., and Fujino, K. 1995a. Plastic deformation of garnets: systematics and implications for the rheology of the mantle transition zone. *Earth Planet. Sci. Lett.* 130:13–30.

Karato, S. I., and Wu, P. 1993. Rheology of the upper-mantle: a synthesis. *Science* 260:771–8.

Karato, S. I., and Zhang, S. 1995. Dynamic recrystallization in experimentally sheared olivine aggregates. *EOS, Trans. AGU (Suppl.)* 76:579.

Karato, S. I., Zhang, S., and Wenk, H. R. 1995b. Superplasticity in earth's lower mantle: evidence from seismic anisotropy and rock physics. *Science* 270:458–61.

Katsura, T., and Ito, E. 1989. The system Mg_2SiO_4-Fe_2SiO_4 at high pressure and temperatures. Precise determination of stabilities of olivine, modified spinel and spinel. *J. Geophys. Res.* 94:15663–70.

Kesson, S. E., and Fitz Gerald, J. D. 1991. Partitioning of MgO, FeO, NiO, MnO and Cr_2O_3 between magnesian silicate perovskite and magnesiowüstite: implications for the origin of inclusions in diamond and the composition of the lower mantle. *Earth Planet. Sci. Lett.* 111:229–40.

Kirby, S. H. 1980. Tectonic stresses in the lithosphere: constraints provided by the experimental deformation of rocks. *J. Geophys. Res.* 85:6353–63.

Kirby, S. H., Durham, W. B., and Stern, L. A. 1991. Mantle phase changes and deep-earthquake faulting in subducting lithosphere. *Science* 252:216–25.

Kohlstedt, D. L. 1992. Structure, rheology, and permeability of partially molten rocks at low melt fractions. In: *Mantle Flow and Melt Generation at Mid-Ocean Ridges*, ed. J. Phipps Morgan, D. K. Blackman, and J. M. Sinton, pp. 103–22. AGU Monograph 71. Washington, DC: American Geophysical Union.

Kohlstedt, D. L., and Chopra, P. N. 1994. Influence of basaltic melt on the creep of polycrystalline olivine under hydrous conditions. In: *Magmatic Systems*, ed. M. P. Ryan, pp. 37–53. San Diego: Academic Press.

Kohlstedt, D. L., Evans, B., and Mackwell, S. J. 1995. Strength of the lithosphere: constraints imposed by laboratory experiments. *J. Geophys. Res.* 100:17587–602.

Kohlstedt, D. L., Keppler, H., and Rubie, D. C. 1996. Solubility of water in the α, β and γ phases of $(Mg, Fe)_2SiO_4$. *Contrib. Mineral. Petrol.* 123:345–57.

Kohlstedt, D. L., and Zimmerman, M. E. 1996. Rheology of partially molten mantle rocks. *Annu. Rev. Earth Planet. Sci.* 24:41–62.

Langdon, T. G. 1975. Grain-boundary sliding during creep of MgO. *J. Am. Ceram. Soc.* 58:92.

Langdon, T. G., Degghan, A., and Sammis, C. G. 1982. Deformation of olivine, and the application to lunar and planetary interiors. In: *Sixth International Conference on the Strength of Metals and Alloys*, vol. 3, ed. R. C. Gifkins, pp. 757–62. Oxford: Pergamon.

Liu, L. G. 1993. Effects of H_2O on the phase behaviour of the forsterite-enstatite system at high pressures and temperatures revisited. *Phys. Earth Planet. Int.* 76:209–18.

Liu, M., and Yund, R. A. 1995. The elastic strain energy associated with the olivine–spinel transformation and its implications. *Phys. Earth Planet. Int.* 89:177–97.

Lloyd, G. E., Schmidt, N. H., Mainprice, D., and Prior, D. J. 1991. Crystallographic textures. *Min. Mag.* 55:331–45.

Lockner, D. A. 1995. Rock failure. In: *Rock Physics and Phase Relations – A Handbook of Physical Constants*, ed. T. J. Ahrens, pp. 127–47. Washington, DC: American Geophysical Union.

Mackwell, S. J., Kohlstedt, D. L., and Paterson, M. S. 1985. The role of water in the deformation of olivine single crystals. *J. Geophys. Res.* 90:11319–33.

McLaren, A. C. 1991. *Transmission Electron Microscopy of Minerals and Rocks.* Cambridge University Press.

McLaren, A. C., and Etheridge, M. A. 1976. A transmission electron microscope study of naturally deformed orthopyroxene. *Contrib. Min. Pet.* 57:163–77.

McLaren, A. C., and Meike, A. 1996. Transformation plasticity in single and two-component polycrystals in which only one component transforms. *Phys. Chem. Min.* 23:439–51.

Madon, M., Guyot, F., Peyronneau, J., and Poirier, J. P. 1989. Electron microscopy of high pressure phases synthesized from natural olivine in diamond anvil cell. *Phys. Chem. Min.* 16:320–30.

Madon, M., and Poirier, J. P. 1980. Dislocations in spinel and garnet high-pressure polymorphs of olivine and pyroxene; implications for mantle rheology. *Science* 207:66–8.

Madon, M., and Poirier, J. P. 1983. Transmission electron microscope observation of α, β and γ (Mg, Fe)$_2$SiO$_4$ in shocked meteorites: planar defects and polymorphic transitions. *Phys. Earth Planet. Int.* 33:31–44.

Mainprice, D., Vauchez, A., and Montagner, J. P. 1993. Preface. *Phys. Earth. Planet. Int.* 78:R7–11.

Meade, C., and Jeanloz, R. 1988. Yield strength of MgO to 40 GPa. *J. Geophys. Res.* 93:3261–9.

Meade, C., and Jeanloz, R. 1990. The strength of mantle silicates at high pressures and room temperature: implications for the viscosity of the mantle. *Nature* 348:533–5.

Meade, C., and Jeanloz, R. 1991. Deep-focus earthquakes and recycling of water into the earth's mantle. *Science* 252:68–72.

Meade, C., Reffner, J. A., and Ito, E. 1993. Synchrotron infrared absorbance measurements of hydrogen in MgSiO$_3$ perovskite. *Science* 264:1558–60.

Meike, A. 1993. A critical review of investigations into transformation plasticity. In: *Defects and Processes in the Solid State: Geoscience Applications – The McLaren Volume*, vol. 14, ed. J. N. Boland and J. D. Fitz Gerald, pp. 5–25. Amsterdam: Elsevier Science.

Molnar, P. 1988. Continental tectonics in the aftermath of plate tectonics. *Nature* 335:131–7.

Morimoto, N., Akimoto, S., Koto, K., and Tokonami, M. 1970. Crystal structures of high pressure modifications of Mn$_2$GeO$_4$ and Co$_2$SiO$_4$. *Phys. Earth Planet. Int.* 3:161–5.

Nicolas, A. 1989. *Structures of Ophiolites and Dynamics of Oceanic Lithsophere.* Dordrecht: Kluwer.

Nicolas, A., Boudier, F., and Bouiller, A. M. 1973. Mechanisms of flow in naturally and experimentally deformed peridotites. *Am. J. Sci.* 273:853–76.

Nicolas, A., and Poirier, J. P. 1976. *Crystalline Plasticity and Solid State Flow in Metamorphic Rocks.* New York: Wiley.

Obata, M., and Karato, S. 1995. Ultramafic pseudotachylite from the Balmuccia Peridotite, Ivrea Verbano zone, northern Italy. *Tectonophysics* 242:313–28.

Olgaard, D. H., and Evans, B. 1988. Grain growth in synthetic marbles with added mica and water. *Contrib. Min. Pet.* 100:246–60.

Olgaard, D. L., and Paterson, M. S. 1996. Shear deformation to large strains in high temperature torsion experiments. *EOS, Trans. AGU (Suppl.)* 77:710.

Panozzo Heilbronner, R. 1993. Controlling the spatial distribution of deformation in experimentally deformed and dehdrated gypsum. In: *Defects and Processes in the Solid State: Geoscience Applications – The McLaren Volume*, ed. J. N. Boland and J. D. Fitz Gerald, pp. 169–94. Amsterdam: Elsevier Science.

Panozzo Heilbronner, R., and Pauli, C. 1993. Integrated spatial and orientation analysis of quartz c-axes by computer-aided microscopy. *J. Struct. Geol.* 15:369–82.

Paterson, M. S. 1970. A high temperature, high pressure apparatus for rock deformation. *Int. J. Rock Mech. Min. Sci.* 7:517–26.

Paterson, M. S. 1983. Creep in transforming polycrystalline materials. *Mechanics of Materials* 2:103–9.

Paterson, M. S. 1987. Problems in the extrapolation of laboratory rheological data. *Tectonophysics* 133:33–43.

Paterson, M. S. 1990. Rock deformation experimentation. In: *The Brittle–Ductile Transition in Rocks, The Heard Volume*, ed. A. G. Duba, W. B. Durham, J. W. Handin, and H. F. Wang, pp. 187–94. AGU Monograph 56. Washington, DC: American Geophysical Union.

Poirier, J. P. 1981. Martensitic olivine–spinel transformation and plasticity of the mantle transition zone. In: *Anelasticity in the Earth*, ed. F. D. Stacey, M. S. Paterson, and A. Nicholas, pp. 113–17. Geodynamics Series vol. 4. Washington, DC: American Geophysical Union.

Poirier, J. P. 1985. *Creep of Crystals. High-Temperature Deformation Processes in Metals, Ceramics and Minerals*. Cambridge University Press.

Poirier, J. P., Beauchesne, S., and Guyot, F. 1989. Deformation mechanisms of crystals with perovskite structure. In: *Perovskite: A Structure of Great Interest to Geophysics and Materials Science*, ed. A. Navrotsky and D. J. Weidner, pp. 119–24. AGU Monograph 45. Washington, DC: American Geophysical Union.

Poirier, J. P., Peyronneau, J., Gesland, J. Y., and Brebec, G. 1983. Viscosity and conductivity of the lower mantle: an experimental study on a $MgSiO_3$ analogue, $KZnF_3$. *Phys. Earth Planet. Int.* 32:273–87.

Poirier, J. P., Peyronneau, J., Madon, M., Guyot, F., and Revcoleshi, A. 1986. Eutectoid phase transformation of olivine and spinel into perovskite and rock salt structures. *Nature* 321:603–5.

Price, G. D. 1983. The nature and significance of stacking faults in wadsleyite, natural β-$(Mg, Fe)_2SiO_4$ from the Peace River meteorite. *Phys. Earth Planet. Int.* 33:137–47.

Ranalli, G. 1982. Deformation maps in grain-size–stress space as a tool to investigate mantle rheology. *Phys. Earth Planet. Int.* 29:42–50.

Ranalli, G. 1987. *Rheology of the Earth. Deformation and Flow Processes in Geophysics and Geodynamics*. Boston: Allen & Unwin.

Ranalli, G. 1991. The microphysical approach to mantle rheology. In: *Glacial Isostacy, Sea Level and Mantle Rheology*, ed. R. Sabadini, pp. 343–78. Dordrecht: Kluwer.

Ranalli, G. 1995. *Rheology of the Earth*. 2nd ed. London: Chapman & Hall.

Ranalli, G., and Murphy, D. C. 1986. Rheological stratification of the lithosphere. *Tectonophysics* 132:281–95.

Remsberg, A. R., Boland, J. N., Gasparik, T., and Liebermann, R. C. 1988. Mechanism of the olivine–spinel transformation in Co_2SiO_4. *Phys. Chem. Min.* 15:498–506.

Remsberg, A. R., and Liebermann, R. C. 1991. A study of the polymorphic transformations in Co_2SiO_4. *Phys. Chem. Min.* 18:161–70.

Ringwood, A. E. 1975. *Composition and Petrology of the Earth's Mantle*. New York: McGraw-Hill.

Ringwood, A. E. 1991. Phase transformations and their bearing on the constitution and dynamics of the mantle. *Geochim. Cosmochim. Acta* 55:2083–110.

Ringwood, A. E., and Major, A. 1966. Synthesis of Mg_2SiO_4-Fe_2SiO_4 spinel solid solutions. *Earth Planet. Sci. Lett.* 1:241–5.

Ruano, O. A., Wadsworth, J., and Sherby, O. D. 1988. Harper-Dorn creep in pure metals. *Acta Met.* 36:1117–28.

Ruano, O. A., Wolfenstine, J., Wadsworth, J., and Sherby, O. D. 1992. Harper-Dorn creep in single-crystalline magnesium oxide. *J. Am. Ceram. Soc.* 75:1737–41.

Rubie, D. C., and Brearley, A. J. 1994. Phase transitions between β and γ $(Mg, Fe)_2SiO_4$ in the Earth's mantle: mechanisms and rheological implications. *Science* 264:1445–8.

Wood, B. J., Bryndzia, L. T., and Johnson, K. E. 1990. Mantle oxidation state and its relationship to tectonic enviroment and fluid speciation. *Science* 248:337–45.

Wright, K., Price, G. D., and Poirier, J. P. 1992. High-temperature creep of the perovskites $CaTiO_3$ and $NaNbO_3$. *Phys. Earth Planet. Int.* 74:9–22.

Young, T. E., Green, H. W., Hofmeister, A. M., and Walker, D. 1993. Infrared spectroscopic investigation of hydroxyl in β-$(Mg, Fe)_2SiO_4$ and coexisting olivine: implications for mantle evolution and dynamics. *Phys. Chem. Min.* 19:409–22.

Yusa, H., Akaogi, M., and Ito, E. 1993. Calorimetric study of $MgSiO_3$ garnet and pyroxene: heat capacities, transition enthalpies, and equilibrium phase relations in $MgSiO_3$ at high pressures and temperatures. *J. Geophys. Res.* 98:6453–60.

Zerr, A., and Boehler, R. 1994. Constraints on the melting temperature of the lower mantle from high-pressure experiments on MgO and magnesiowüstite. *Nature* 371:506–8.

Zeuch, D. H., and Green, H. W. 1984. Experimental deformation of a synthetic dunite at high temperature and pressure. I. Mechanical behaviour, optical microstructure and deformation mechanism. *Tectonophysics* 110:233–62.

Zhang, S., and Karato, S. 1995. Lattice preferred orientation of olivine aggregates deformed in simple shear. *Nature* 375:774–7.

Zimmerman, M. E., Kohlstedt, D. L., and Karato, S.-I. 1995. Shear localization in deformed olivine-basalt aggregates. *EOS, Trans. AGU (Suppl.)* 76:559–60.

Index